Foundation Design
and Construction

Foundation Design and Construction

Sixth Edition

M J TOMLINSON
CEng, FICE, FIStructE

with contributions by
R BOORMAN BSc, M.Eng, MICE,
FIStructE

 LONGMAN

Addison Wesley Longman Limited
Edinburgh Gate, Harlow
Essex CM20 2JE
and Associated Companies throughout the world

First published by Pitman Publishing Limited 1963
Second edition 1969
Third edition 1975, reprinted 1976, 1977
Fourth edition 1980
First paperback edition 1976, reprinted 1976, 1977, 1978
Second paperback edition 1980, reprinted 1981, 1983, 1984
Fifth edition published by Longman Scientific & Technical 1986,
reprinted 1987, 1988, 1989, 1990, 1991, 1992, 1994
Sixth edition 1995
Reprinted 1996 and 1998

British Library Cataloguing in Publication Data
A catalogue entry for this title is available from the British Library

ISBN 0-582-22697-X

Library of Congress Cataloging-in-Publication Data
A catalog entry for this title is available from the Library of Congress

Produced by Addison Wesley Longman Singapore Pte Ltd
Printed in Singapore

Contents

Preface to the First Edition

The author's aim has been to provide a manual of foundation design and construction methods for the practising engineer. The book is not intended to be a textbook on soil mechanics, but it does include examples of the applications of this science to foundation engineering. The principles of the science are stated only briefly; the reader should refer to the relevant textbooks for explanations of its theory. It is hoped that the limitations and pitfalls of soil mechanics have been clearly set out – undue reliance on soil mechanics can be dangerous if foundation designs are based on inadequate data or on the use of wrong investigational techniques.

Professor Peck has listed three attributes necessary to the practice of subsurface engineering; these are a knowledge of precedents, familiarity with soil mechanics, and a working knowledge of geology. He believes the first of these to be by far the most important. Regarding soil mechanics, he states:

> The everyday procedures now used to calculate bearing capacity, settlement, or factor of safety of a slope, are nothing more than the use of the framework of soil mechanics to organize experience. If the techniques of soil testing and the theories had not led to results in accord with experience and field observations, they would not have been adopted for practical, widespread use. Indeed, the procedures are valid and justified only to the extent that they have been verified by experience. In this sense, the ordinary procedures of soil mechanics are merely devices for interpolating among the specific experiences of many engineers in order to solve our own problems, or which we recognize to fall within the limits of previous experience.

The author has included information on ordinary foundations, including the economic design of house foundations, as a help to architects and builders in the use of the present-day techniques of investigation and construction. The application of soil mechanics science to the carrying capacity of pile foundations of all types is a comparatively new development, but is coming to be recognized as having advantages over older methods using dynamic formulae, and this subject is fully treated. It is hoped that the information given on large-diameter bored-pile foundations will be helpful in aiding the design of foundations of tall multi-storey buildings. Experience in recent years has shown the economies which these high-capacity piles can give over the more conventional types where heavy foundation loads are to be carried.

The background information on soil mechanics has been mainly drawn from Terzaghi and Peck's *Soil Mechanics in Engineering Practice* (John Wiley). For examples of constructional problems the author has drawn freely on his experiences with George Wimpey & Co. Ltd, and he is indebted to Dr L. J. Murdock, DSc, MICE, manager of their Central Laboratory, for permission to publish this information, together with illustrations and photographs. General information on current design practice in Great Britain has been obtained from the Institution of Civil Engineers Code of Practice No. 4 (1954): *Foundations*, with the kind permission of the Institution.

The author gratefully acknowledges the help and criticism of his colleagues in the preparation of the book, and in particular of A. D. Rae, BSc, for checking the manuscript and proofs. Thanks are especially due to Professor H. O. Ireland, of the University of Illinois, for critical reading of the manuscript and advice on its application to American engineering practice. The illustrations are the work of Mrs W. Alder and Mrs P. Payne.

AMERSHAM   M. J. T.
[1963]

Preface to the Sixth Edition

In this edition all chapters have been revised to bring them up-to-date with recent developments in foundation design and construction techniques. There has been some re-arrangement of the text. The section on pier foundations in Chapter 6 has been brought forward to a new section of the text on deep foundations in Chapter 2. The remainder of the material in Chapter 6 has been merged into a new chapter on bridge foundations dealing with land and over-water bridges. To make space for the new material much of the previous text concerning site investigation techniques has been deleted. It was felt that these have become a routine operation and are well-known to engineers.

At the time of preparing this edition a draft of Eurocode 7: Foundations, was in circulation. However the completed work had not been approved by the EC member states for publication in draft form. Guidance has been given in this edition on the application of the present draft to foundation design, together with worked examples of the use of limit state design methods as they appear in the draft Code compared with design by permissible stress methods currently in use.

The Author is grateful to Mr Roy Whitlow, formerly of the University of the West of England, for his helpful suggestions for some of the new material in this edition; to Mr Peter Winfield of Grove Projects for assistance on some of the structural aspects of foundation design; and to Mr Roger Boorman of Wimpey Construction Limited for revising his contributions on computer-aided design, for assistance with worked examples, and for providing new examples of the application of computer techniques. There has been a major advance in these methods since the publication of the fifth edition of this book, particularly in their availabilty to the practising engineer. Developments in hardware have produced powerful, fast and compact machines available at reasonable cost. Hence most of the applications referred to in this edition, including finite element studies, can now be implemented on standard desk-top or portable computers. Additionally a wide range of software covering such applications is now readily available for purchase from specialist vendors.

Finally, I am glad to have this opportunity of acknowledging the help and encouragement of my wife for her work in typing, checking proofs and the many other tasks she has undertaken in the preparation of this and the previous five editions over the past thirty years.

Grateful acknowledgement is made to the following firms and individuals who have given permission to reproduce illustrations, photographs and other information:

American Society of Civil Engineers – Figs 2.14, 2.28, 2.54, 2.55, 5.23(a), 6.31, 7.11, 7.14, 7.24, 7.44, 7.46, 7.50, 9.45, 10.12, 11.33.
Bachy Ltd. – Figs 8.29, 10.30.
Baumaschine und Bautechnik – Fig. 6.55
Dr M.D. Bolton – Figs 6.13, 6.14, 6.15
Boston Society of Civil Engineers – Table 2.6
British Steel – Tables 8.4 and 10.1
BSP International Foundations Ltd – Figs 8.3, 8.5, 8.6, 8.8
Building Research Establishment – Figs 3.1, 3.3, 3.4, 3.5, 3.8, 3.9, 13.1
Butterworth-Heinemann Ltd – Figs 2.17, 2.50, 2.51, 10.31, Table 13.1
Burlington Engineers Ltd – Figs 8.2, 8.4, 8.7
Cementation Piling and Foundations Ltd – Figs 8.32, 11.40
Construction Industry Research and Information Association – Figs 5.25, 5.27, 5.29, 5.36, 5.43, 7.12, 9.17, 9.19, 9.20, 11.7, 11.9
W. K. Cross Esq – Figs 6.39, 6.40
Columbia University Press – Fig. 10.6
Crown Copyright, Controller of HM Stationery Office – Figs 3.1, 3.3, 3.4, 3.5, 3.6

C.S.I.R.O, Australia – Fig. 3.7

Danish Geotechnical Institute – Figs 2.8, 2.9, 2.10, 2.11, 2.12, 7.37

D'Appolonia Consulting Engineers Inc – Fig. 9.42

Edmund Nuttall Ltd – Figs 6.53, 6.54, 9.34

Engineering News Record – Figs 6.42, 9.10

Ernst & Sohn – Figs. 7.5, 11.22

Fugro-McClelland Limited – Fig. 7.7

Dr D. W. Hight – Fig. 7.5

Honshu-Shikoku Bridge Authority – Fig. 6.57

Indian Roads Congress – Figs 6.32, 6.35

Institution of Structural Engineers – Figs 4.36, 5.12, 6.8, 6.10, 12.1, 12.4, 12.25, 12.26

Japan Society ISSMFE – Fig. 9.42

Jim Mackintosh Photography – Figs 5.7, 12.20

David Lee Photography – Figs 6.44, 6.45

Longman Group Ltd – Figs 6.53, 6.54

Metropolitan Expressway Public Corporation, Yokohama – Fig. 6.52

Nanyang Technological Institute – Figs 10.13, 10.24, 10.25

National Research Council of Canada – Figs 2.37, 7.16, 7.19, 7.20

Director of Technical Services, Northumberland County Council – Figs 6.19, 6.20, 6.21

Norwegian Geotechnical Institute – Fig. 2.31

Offshore Technology Conference – Figs 7.47, 7.48, 7.53

Professor Osterberg – Fig. 7.16

Ove Arup and Partners – Figs 4.38, 12.18

Ove Arup Computing Systems – Fig. 9.11

Pentech Press – Figs 2.21, 2.36, 3.1, 3.2, 3.5

J. F. S. Pryke Esq – Figs 12.12, 12.24

Ronald White – Fig. 1.15

S. Serota Esq – Fig. 10.31

Roger Bullivant Ltd – Fig. 8.30

Professor I.M. Smith – Figs 2.50, 2.51

Soil Mechanics Limited – Fig. 8.25

Speed-Shore (UK) Ltd – Fig. 9.18

Dr S. M. Springman – Figs 6.13, 6.14, 6.15

Steen Consultants – Fig. 4.40

Swedish Geotechnical Institute – Fig. 1.4

Thomas Telford Publications (and Institution of Civil Engineers) – Figs 1.5, 1.7, 2.22, 2.26, 2.27, 2.33, 2.43, 2.44, 2.53, 3.2, 4.1, 4.35, 4.38, 5.8, 5.9, 5.44, 5.45, 6.18, 6.24, 6.28, 6.46, 6.49, 6.51, 6.56, 6.58, 7.41, 9.5, 9.39, 9.47, 9.48, 10.22, 10.23, 10.32, 10.33, 10.34, 11.14, 11.42, 12.21, 12.22, 12.23, Table 10.2

Trafalgar House Technology – Fig. 6.28

Transport Research Laboratory – Figs 6.13, 6.14, 6.15

Tudor Engineering Company – Fig. 6.31

Turner Visual Group Limited – Fig. 9.7

University of Illinois – Table 2.10

U. S. Army, Waterways Experiment Station – Figs 7.45, 11.11

U. S. National Committee, ISSMFE – Fig. 2.34

Westpile Limited – Fig. 8.28

John Wiley & Sons Inc – Figs 2.13, 9.36, 11.12

John Wiley and Sons Limited – Figs 2.50, 2.51, 5.23(b)

Wimpey Group Services Limited – Figs 1.15, 5.7, 5.16, 5.17, 8.11, 8.20, 9.4, 9.6, 9.7, 10.10, 11.21, 11.22, 12.20

Copies of British Standards 5930, 8004 and other Codes of Practice referred to in the text may be purchased from the British Standards Institution at BSI Sales, Linford Wood, Milton Keynes, MK14, 6LE

Figs 2.17, 2.34, 4.37, 4.38, 7.15, 8.31 and 11.43 are reproduced with permission from A. A. Balkema, PO Box 1675, NL-3000 BR, Rotterdam, The Netherlands.

1 Site Investigations and Soil Mechanics

1.0 General requirements

A site investigation in one form or another is always required for any engineering or building structure. The investigation may range in scope from a simple examination of the surface soils with or without a few shallow trial pits, to a detailed study of the soil and ground water conditions to a considerable depth below the surface by means of boreholes and *in-situ* and laboratory tests on the materials encountered. The extent of the work depends on the importance and foundation arrangement of the structure, the complexity of the soil conditions, and the information which may be available on the behaviour of existing foundations on similar soils.

The draft of Eurocode 7, *Geotechnics*[1.1] places structures and earthworks into three 'geotechnical categories'. Light structures such as buildings with column loads up to 250 kN or walls loaded to 100 kN/m, low retaining walls, and single or two-storey houses are placed in geotechnical category 1. Provided that the ground conditions and design requirements are known from previous experience and the ground is not sloping to any significant degree, the qualitative investigations in this category can be limited to verifying the design assumptions at the latest during supervision of construction of the works. Verification is deemed to consist of visual inspection of the site, sometimes with inspection of shallow trial pits, or sampling from auger borings.

Category 2 structures are those for which quantitative geotechnical studies are required with the involvement of qualified engineers with relevant experience.

Conventional substructures such as shallow spread footings, rafts, and piles are included in this category, as well as retaining walls, bridge piers and abutments, excavations and excavation supports, and embankments. A two-stage investigation is required for this category. The first stage includes boreholes, *in-situ* tests, and laboratory tests. A detailed study of the ground-water conditions is required.

Structures in category 3 include buildings with exceptional loads, multi-storey basements, dams, large bridges and tunnels, heavy machinery foundations, and offshore platforms. Structures on expansive or collapsing soils, or in areas of exceptionally high seismic activity are included.

The investigations required for category 3 are those deemed to be sufficient for category 2, together with any necessary additional specialized studies. If test procedures of a specialized or unusual nature are required, the procedures and interpretations should be documented with reference to the tests.

Thorough investigations are necessary for engineering structures founded in deep excavations. As well as providing information for foundation design, they provide essential information on the soil and ground-water conditions to contractors tendering for the work. Thus, money is saved by obtaining realistic and competitive tenders based on adequate foreknowledge of the ground conditions. A reputable contractor will not gamble on excavation work if its cost amounts to a substantial proportion of the whole project; a correspondingly large sum will be added to the tender to cover for the unknown conditions. Hence the saying, 'You pay for the borings whether you have them or not.'

It follows that contractors tendering for excavation work must be supplied with all the details of the site investigation. It is not unknown for the engineer to withhold certain details such as ground-water level observations under the mistaken idea that this might save money in claims by the contractor if water levels are subsequently found to be different from those encountered in the borings. This is a fallacy; the contractor will either allow in his tender for the unknown risks involved or will take a gamble. If the gamble fails and the water conditions turn out to be worse than assumed, then a claim will be made.

An engineer undertaking a site investigation may engage local labour for trial pit excavation or hand auger boring, or may employ a contractor for boring and soil sampling. If laboratory testing is required the boring contractor can send the samples to his own or to an independent testing laboratory. The engineer then undertakes the soil mechanics analysis for foundation design. Alternatively a specialist organization offering comprehensive facilities for boring, sampling, field and laboratory testing, and soil mechanics analysis may undertake the whole investigation. A single organization has an advantage of providing the essential continuity and close relationship between field, laboratory, and office work. It also permits the boring and testing programme to be readily modified in the light of information made available as the work proceeds. Additional samples can be obtained, as necessary, from soil layers shown by laboratory testing to be particularly significant. *In-situ* testing can be substituted for laboratory testing if desired. In any case, the engineer responsible for the day-to-day direction of the field and laboratory work should keep the objective of the investigation closely in mind and should make a continuous appraisal of the data in the same way as is done at the stage of preparing the report. In this way vital information is not overlooked; the significance of such features as weak soil layers, deep weathering of rock formations, and sub-artesian water pressure can be studied in such greater detail, as may be required while the field-work is still in progress.

Whatever procedure the engineer adopts for carrying out his investigation work it is essential that the individuals or organizations undertaking the work should be conscientious and completely reliable. The engineer has an important responsibility to his employers in selecting a competent organization and in satisfying himself by checks in the field and on laboratory or office work that the work has been undertaken with accuracy and thoroughness.

1.1 Information required from a site investigation

For geotechnical categories 2 and 3 the following information should be obtained in the course of a site investigation for foundation engineering purposes:

(a) The general topography of the site as it affects foundation design and construction, e.g. surface configuration, adjacent property, the presence of watercourses, ponds, hedges, trees, rock outcrops, etc., and the available access for construction vehicles and plant.

(b) The location of buried services such as electric power and telephone cables, water mains, and sewers.

(c) The general geology of the area with particular reference to the main geological formations underlying the site and the possibility of subsidence from mineral extraction or other causes.

(d) The previous history and use of the site including information on any defects or failures of existing or former buildings attributable to foundation conditions, and the possibility of contamination of the site by toxic waste materials.

(e) Any special features such as the possibility of earthquakes or climatic factors such as flooding, seasonal swelling and shrinkage, permafrost, or soil erosion.

(f) The availability and quality of local constructional materials such as concrete aggregates, building and road stone, and water for constructional purposes.

(g) For maritime or river structures information on normal spring and neap tide ranges, extreme high and low tidal ranges and river levels, seasonal river levels and discharges, velocity of tidal and river currents, and other hydrographic and meteorological data.

(h) A detailed record of the soil and rock strata and ground-water conditions within the zones affected by foundation bearing pressures and construction operations, or of any deeper strata affecting the site conditions in any way.

(j) Results of laboratory tests on soil and rock samples appropriate to the particular foundation design or constructional problems.

(k) Results of chemical analyses on soil, fill materials, and ground water to determine possible deleterious effects on foundation structures.

(l) Results of chemical and bacteriological analyses on contaminated soils, fill materials, and gas emissions to determine health hazard risks.

Items (a)–(g) above can be obtained from a general reconnaissance of the site (the 'walk-over' survey), and from a study of geological memoirs and maps and other published records. A close inspection given by walking over the site area will often show significant indications of subsurface features. For example, concealed swallow holes (sink holes) in chalk or limestone formations are often revealed by random depressions and marked irregularity in the ground surface; soil creep is indicated by wrinkling of the surface on a hillside slope, or leaning trees; abandoned mine workings are shown by old shafts or heaps of mineral waste; glacial deposits may be indicated by mounds or hummocks (drumlins) in a generally flat topography; and river or lake deposits by flat low-lying areas in valleys. The surface indications of ground water are the presence of springs or wells, and marshy ground with reeds (indicating the presence of a high water table with poor drainage and the possibility of peat). Professional geological advice should be sought in the case of large projects covering extensive areas.

On extensive sites, aerial photography is a valuable aid in site investigations. Photographs can be taken from model aircraft or balloons. Skilled interpretations of aerial photographs can reveal much of the geomorphology and topography of a site. Geological mapping from aerial photographs as practised by specialist firms is a well-established science.

Old maps as well as up-to-date publications should be studied, since these may show the previous use of the site and are particularly valuable when investigating backfilled areas. Museums or libraries in the locality often provide much information in the form of maps, memoirs, and pictures or photographs of a site in past times. Local authorities should be consulted for details of buried services, and in Britain the Geological Survey and British Coal for information on coal-mine workings. Some parts of Britain were worked for coal long before records of workings were kept, but it is sometimes possible to obtain information on these from museums and libraries. If particular information on the history of a site has an important bearing on foundation design, for example the location of buried pits or quarries, every endeavour should be made to cross-check sources of information especially if they are based on memory or hearsay. People's memories are notoriously unreliable on these matters.

Items (h), (j), and (k) of the list are obtained from boreholes or other methods of subsurface exploration, together with field and laboratory testing of soils or rocks. It is important to describe the type and consistency of soils in the standard manner laid down in standard codes of practice. In Britain the standard descriptions and classifications of soils are set out in the British Standard Code of Practice: *Site Investigations*, BS5930.

Rocks should be similarly classified in accordance with the standard procedure of codes of practice; BS 5930 requires rocks to be described in the following sequence:

> Colour
> Grain size (the grain size of the mineral or rock fragments comprising the rock)
> Texture (e.g. crystalline, amorphous, etc.)
> Structure (a description of discontinuities, e.g. laminated, foliated, etc.)
> State of weathering
> ROCK NAME
> Strength (based on the uniaxial compression test)
> Other characteristics and properties

Of the above descriptions, the four properties of particular relevance to foundation engineering are the structure,

state of weathering, discontinuity spacing, and uniaxial compression strength.

The discontinuity spacing is defined in two ways:

(1) The rock quality designation (RQD) which is the percentage of rock recovered as sound lengths which are 100 mm or more in length.
(2) The fracture index which is the number of natural fractures present over an arbitrary length (usually 1 m).

The structure is of significance from the aspect of ease of excavation by mechanical plant, and also the frequency and type of discontinuity affects the compressibility of the rock mass. The state of weathering, discontinuity spacing, and uniaxial compression strength can be correlated with the deformation characteristics of the rock mass and also with the skin friction and end-bearing of piles.

In stating the description of rock strength in borehole records the classification adopted in BS 5930 should be followed. Thus:

Classification	Uniaxial compressive strength (MN/m^2)
Very weak	Under 1.25
Weak	1.25–5
Moderately weak	5–12.5
Moderately strong	12.5–50
Strong	50–100
Very strong	100–200
Extremely strong	Over 200

1.2 Site investigations of foundation failures

From time to time it is necessary to make investigations of failures or defects in existing structures. The approach is somewhat different from that of normal site investigation work, and usually takes the form of trial pits dug at various points to expose the soil at foundation level and the foundation structure, together with deep trial pits or borings to investigate the full depth of the soil affected by bearing pressures. A careful note is taken of all visible cracking and movements in the superstructure since the pattern of cracking is indicative of the mode of foundation movement, e.g. by sagging or hogging. It is often necessary to make long-continued observations of changes in level and of movement of cracks by means of tell-tales. Glass or paper tell-tales stuck on the cracks by cement pats are of little use and are easily lost or damaged. The tell-tales should consist of devices specially designed for the purpose or non-corrodible metal plugs cemented into holes drilled in the wall on each side of the crack and so arranged that both vertical and horizontal movements can be measured by micrometer gauges. Similarly, points for taking levels should be well secured against removal or displacement. They should consist preferably of steel bolts or pins set in the foundations and surrounded by a vertical pipe with a cover at ground level. The levels should be referred to a well-established datum point at some distance from the affected structure; ground movements which may have caused foundation failure should not cause similar movement of the levelling datum. The Building Research Establishment in Britain has developed a number of devices such as tiltmeters and borehole extensometers for monitoring the movements of structures and foundations.

A careful study should be made of adjacent structures to ascertain whether failure is of general occurrence, as in mining subsidence, or whether it is due to localized conditions. The past history of the site should be investigated with particular reference to the former existence of trees, hedgerows, farm buildings, or waste dumps. The proximity of any growing trees should be noted, and information should be sought on the seasonal occurrence of cracking, for example if cracks tend to open or close in winter or summer, or are worse in dry years or wet years. Any industrial plant in which forging hammers or presses cause ground vibrations should be noted, and inquiries should be made about any construction operations such as deep trenches, tunnels, blasting, or piling which may have been carried out in the locality.

1.3 Borehole layout

Whenever possible boreholes should be sunk close to the proposed foundations. This is important where the bearing stratum is irregular in depth. For the same reason the boreholes should be accurately located in position and level in relation to the proposed structures. Where the layout of the structures has not been decided at the time of making the investigation a suitable pattern of boreholes is an evenly spaced grid of holes. For extensive areas it is possible to adopt a grid of boreholes with some form of *in-situ* probes, such as dynamic or static cone penetration tests, at a closer spacing within the borehole grid. Eurocode 7 recommends that the exploration points forming the grid should normally be at a mutual spacing of 20–40 m. Trial pits for small foundations, such as strip foundations for houses, should not be located on or close to the intended foundation position because of the weakening of the ground caused by these relatively large and deep trial excavations.

The required number of boreholes which need to be sunk on any particular location is a difficult problem which is closely bound up with the relative costs of the investigation and the project for which it is undertaken. Obviously the more boreholes that are sunk the more is

known of the soil conditions and greater economy can be achieved in foundation design, and the risks of meeting unforeseen and difficult soil conditions which would greatly increase the costs of the foundation work become progressively less. However, an economic limit is reached when the cost of borings outweighs any savings in foundation costs and merely adds to the overall cost of the project. For all but the smallest structures, at least two and preferably three boreholes should be sunk, so that the true dip of the strata can be established. Even so, false assumptions may still be made about stratification.

The depth to which boreholes should be sunk is governed by the depth of soil affected by foundation bearing pressures. The vertical stress on the soil at a depth of one and a half times the width of the loaded area is still one-fifth of the applied vertical stress at foundation level, and the shear stress at this depth is still appreciable. Thus, borings in soil should always be taken to a depth of at least one to three times the width of the loaded area. In the case of narrow and widely spaced strip or pad foundations the borings are comparatively shallow (Fig. 1.1(*a*)), but for large raft foundations the borings will have to be deep (Fig. 1.1(*b*)) unless rock is present within the prescribed depth. Where strip or pad footings are closely spaced so that there is overlapping of the zones of pressure the whole loaded area becomes in effect a raft foundation with correspondingly deep borings (Fig. 1.1(*c*)). In the case of piled foundations the ground should be explored below pile-point level to cover the zones of soil affected by loading transmitted through the piles. Eurocode 7 recommends a depth of five shaft diameters below the expected toe level or at least 5 m below this level. It is usual to assume that a large piled area in uniform soil behaves as a

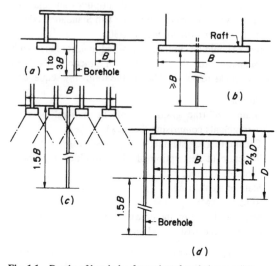

Fig. 1.1 Depths of boreholes for various foundation conditions.

raft foundation with the equivalent raft at a depth of two-thirds of the length of the piles (Fig. 1.1(*d*)).

The 'rule-of-thumb' for a borehole depth of one and a half times the foundation width should be used with caution. Deep fill material could be present on some sites and geological conditions at depth could involve a risk of foundation instability.

Where foundations are taken down to rock, either in the form of strip or pad foundations or by piling, it is necessary to prove that rock is in fact present at the assumed depths. Where the rock is shallow this can be done by direct examination of exposures in trial pits or trenches; but when borings have to be sunk to locate and prove bedrock it is important to ensure that boulders or layers of cemented soils are not mistaken for bedrock. This necessitates percussion boring or rotary diamond core drilling to a depth of at least 3 m in bedrock in areas where boulders are known to occur. On sites where it is known from geological evidence that boulders are not present a somewhat shallower penetration into rock can be accepted. In some areas boulders larger than 3 m have been found, and it is advisable to core the rock to a depth of 6 m for important structures. Mistakes in the location of bedrock in boreholes have in many cases led to costly changes in the design of structures and even to failures.

Direct exposure of rock in trial pits or trenches is preferable to boring, wherever economically possible, since widely spaced core drillings do not always give a true indication of shattering, faulting, or other structural weakness in the rock. Where rock lies at some depth below ground level, it can be examined in large-diameter boreholes drilled by equipment described in Section 8.14. Because of the cost, this form of deep exploration is employed only for important structures.

Eurocode 7 recommends that where the possibility of base uplift in excavations is being investigated the pore-water pressures should be recorded over a depth below ground-water level equal to or greater than the excavation depth. Even greater depths may be required where the upper soil layers have a low density. When boreholes are sunk in water-bearing ground which will be subsequently excavated, it is important to ensure that they are backfilled with concrete or well-rammed puddled clay. If this is not done the boreholes may be a source of considerable inflow of water into the excavations. In a report on an investigation for a deep basement structure in the Glasgow area the author gave a warning about the possibility of upheaval of clay at the bottom of the excavation, due to artesian pressure in the underlying water-bearing rock. After completing the basement the contractor was asked whether he had had any trouble with this artesian water. The answer was that 'the only trouble we had with water was up through your borehole'. In another case, large bored piles

with enlarged bases were designed to be founded within an impervious clay layer which was underlain by sand containing water under artesian pressure. The risks of somewhat greater settlement due to founding in the compressible clay were accepted to avoid the difficulty of constructing the piles in the underlying, less compressible sand. However, considerable difficulty was experienced in excavating the base of one of the piles because of water flowing up from the sand strata through an unsealed exploratory borehole.

1.4 Exploration in soils

1.4.1 Investigation methods

Methods of determining the stratification and engineering characteristics of subsurface soils are as follows:

> Trial pits
> Hand auger borings
> Mechanical auger borings
> Light cable percussion borings
> Rotary open hole drilling
> Wash borings
> Wash probings
> Dynamic cone penetration tests
> Static cone penetration tests
> Vane shear tests
> Pressuremeter tests
> Dilatometer tests
> Plate bearing tests

Detailed descriptions of the above methods as used in British practice are given in BS 5930: *Site Investigations*. Brief comments on the applicability of these methods to different soil and site conditions are given in the following Sections 1.4.2–1.4.5.

1.4.2 Trial pits and borings

Trial pits are generally used for geotechnical category 1 investigations. They are useful for examining the quality of weathered rocks for shallow foundations. Trial pits extended to trenches provide the most reliable means of assessing the state of deposition and characteristics of filled ground (see Section 1.10).

Hand and mechanical auger borings are also suitable for category 1 investigations in soils which remain stable in an unlined hole. When carefully done augering causes the least soil disturbance of any boring method.

Light cable percussion borings are generally used in British practice. The simple and robust equipment is well suited to the widely varying soil conditions in Britain, including the very stiff or dense stony glacial soils, and weathered rocks of soil-like consistency.

Large-diameter undisturbed samples (up to 250 mm) can be recovered for special testing.

Rotary open hole drilling is generally used in USA, Middle East, and Far Eastern countries. The rotary drills are usually tractor or skid-mounted and are capable of rock drilling as well as drilling in soils. Hole diameters are usually smaller than percussion-drilled holes, and sample sizes are usually limited to 50 mm diameter.

Bentonite slurry or water is used as the drilling fluid, but special foams have recently been developed to assist in obtaining good undisturbed samples.

Wash borings are small-diameter (about 65 mm) holes drilled by water flush aided by chiselling. Sampling is by 50 mm internal diameter standard penetration test equipment (see below) or 50–75 mm open-drive tubes.

Wash probings are used in over-water soil investigations. They consist of a small-diameter pipe jetted down and are used to locate rock head or a strong layer overlain by loose or soft soils, for example in investigations for dredging. There is no positive identification of the soils and sampling is usually impracticable.

1.4.3 Soil sampling

There are two main types of soil sample which can be recovered from boreholes or trial pits:

(a) *Disturbed samples*, as their name implies, are samples taken from the boring tools: examples are auger parings, the contents of the split-spoon sampler in the standard penetration test (see Section 1.4.5), sludges from the shell or wash-water return, or hand samples dug from trial pits.
(b) *Undisturbed samples*, obtained by driving a thin-walled tube into the soil, represent as closely as is practicable the true *in-situ* structure and water content of the soil. It is important not to overdrive the sampler as this compresses the contents. It should be recognized that no sample taken by driving a tube into the soil can be truly undisturbed.

Disturbance and the consequent changes in soil properties can be minimized by careful attention to maintaining a water balance in the borehole. That is, the head of water in the borehole must be maintained, while sampling, at a level corresponding to the piezometric pressure of the pore water in the soil at the level of sampling. This may involve extending the borehole casing above ground level or using bentonite slurry instead of water to balance high piezometric pressures. The care in sampling procedure and the elaborateness of the equipment depends on the class of work which is being undertaken, and the importance of accurate results on the design of the works.

BS 5930 recommends five quality classes for soil sampling following a system developed in Germany. The classification system, the soil properties which can be determined reliably from each class, and the appropriate sampling methods are shown in Table 1.1.

In cohesive soils sensitive to disturbance, quality classes 1 and 2 require a good design of sampler such as a piston or thin-walled sampler which is jacked or pulled down into the soil and not driven down by blows of a hammer. Class 1 and 2 sampling in soils insensitive to disturbance employs open-drive tube samplers which are hammered into the soils by blows of a sliding hammer or careful hand-cut samples taken from trial pits. There is a great difference in cost between piston and open sampling, but the engineer should recognize the value of good quality if this can result in economies in design; for example, good-quality sampling means higher indicated shear strengths, with higher bearing pressures and consequently reduced foundation costs. In certain projects good sampling may mean the difference between a certain

Table 1.1   Quality classification for soil sampling

Quality class (as BS 5930)	Soil properties that can be determined reliably (BS 5930)	Sampling method
1	Classification, moisture content, density, strength, deformation, and consolidation characteristics	*Soils sensitive to disturbance:* thin wall piston sampler *Soils insensitive to disturbance:* thick or thin wall open sampler *Soils containing discontinuities (fabric) affecting strength, deformation, and consolidation:* large-diameter thin or thick wall sampler (piston or open)
2	Classification, moisture content, density	Thin- or thick-wall open sampler
3	Classification and moisture content	Disturbed sample of cohesive soils taken from clay cutter or auger in dry borehole
4	Classification	Disturbed sample of cohesive soils taken from clay cutter or auger in boreholes where water is present
5	None, sequence of strata only	Disturbed samples of non-cohesive soils taken from shell in cable percussion boring or recovered as debris flushed from rotary drilling or wash boring

construction operation being judged possible or impossible, for example the placing of an embankment on very soft soil for a bridge approach. If shear strength as indicated by poor-quality sampling is low, then the engineer may decide it is impossible to use an embanked approach and will have to employ an expensive piled viaduct. On the other hand in 'insensitive' clays such as stiff glacial till the sampling procedure has not much effect on shear strength and thick-wall open samplers may give quite adequate information. Also, elaborate samplers such as the fixed piston types may be incapable of operation in clays containing appreciable amounts of large gravel.

The presence of discontinuities in the form of pockets or layers of sand and silt, laminations, fissures, and root holes in cohesive soils is of significance to their permeability, which in turn affects their rate of consolidation under foundation loading, and the stability of slopes of foundation excavations. The use of large-diameter sample tubes may be justified to assess the significance of such discontinuities or 'fabric' to the particular foundation problem.[1.2]

The engineer should study the foundation problem and decide what degree of elaborateness in sampling is economically justifiable, and he should keep in mind that *in-situ* tests such as the vane or cone tests may give more reliable information than laboratory tests on undisturbed samples. If *in-situ* tests are adopted, elaborateness in undisturbed sampling is unnecessary and the 'simple' class is sufficient to give a check on identification of soil types. A good practice, recommended by Rowe[1.2] is to adopt continuous sampling in the first boreholes drilled on a site. An open-drive sampler with an internal split sleeve is used to enable the samples to be split longitudinally for examination of the soil fabric. The critical soil layers can be identified and the appropriate class of sampling or *in-situ* testing adopted.

BS 5930 gives details and dimensions of five types of soil samplers for use in boreholes. These are:

Thin-walled samplers
General-purpose 100 mm diameter open-tube sampler
Split-barrel standard penetration test sampler
Thin-walled stationary piston sampler
Continuous sampler

Thin-walled samplers which are pushed rather than hammered into the soil cause the minimum of moisture content changes and disturbance to the fabric of the soil. Sample diameters are generally 75–100 mm, but tubes up to 250 mm can be provided for special purposes. The thin-wall sampler is suitable for use in very soft to soft clays and silts.

One type of thin-walled sampler, not described in BS 5930, is the *Laval sampler* developed in Canada for sampling soft clays.[1.3] It has been shown to provide samples of a quality equal to those obtained by conventional hand-cut block sampling. The tube is hydraulically pushed into a mud-supported borehole to recover samples 200 mm in diameter and 300 mm long. The tube is overcored before withdrawal.

The general-purpose 100 mm diameter open-tube sampler was developed in UK as a suitable device for sampling the very stiff to hard clays, gravelly glacial till, and weak weathered rocks such as chalk and marl. In this respect the detachable cutting shoe is advantageous. It can be discarded or reconditioned enabling many reuses of the equipment. However, the relatively thick-walled tube and cutting shoe do cause some disturbance of the fabric of the soil and moisture content changes within the sample. The equipment is suitable for geotechnical category 2 investigations.

Chandler *et al.*[1.4] have described the detailed examination of samples of stiff to hard London Clay recovered by the general-purpose tube with internal and external diameters of 105 and 118 mm respectively, and cutting shoe internal and external diameters of 103 and 117 mm respectively. Moisture contents across the diameter of the samples were compared with those from adjacent block samples. They showed a reduction in moisture content around the sheared periphery of the tube samples. This was followed by migration of pore water from the centre to the periphery and consequent increase of effective stress within the sample. The measured stress increase was 15–45 kN/m^2 in soft to firm clay from shallow depths, to 250–300 kN/m^2 in the very stiff unweathered clay. It was evident that these pore pressure increases would have resulted in an over-optimistic measurement of undrained shear strengths and an excessive value of the coefficient of each pressure at rest, K_0. Equalization of pore pressures required a period of 3 weeks or more, suggesting that it is advantageous to test stiff clay samples as quickly as possible after sampling in order to minimize the detrimental effects of pore-pressure changes.

The split-barrel standard penetration test sampler is used to make the *in-situ* soil test described in Section 1.4.5. The tube has an internal diameter of 35 mm and recovers a disturbed sample suitable for classes 3 and 4 in Table 1.1. Some indication of layering or laminations can be seen when the sampler is taken apart.

Thin-walled stationary piston samplers are suitable for quality class 1 in Table 1.1 and for geotechnical category 3 investigations. Diameters range from 75 to 100 mm with special types up to 250 mm. They recover good samples of very soft to soft clays and silts, and sandy soils can sometimes be recovered. Special thin-wall piston samples are used in stiff clays.

The Delft continuous sampler is an example of this type. It is made in 29 and 66 mm diameters with a penetration generally up to 18 m, but samples up to 30 m can be recovered in favourable soil conditions. It is designed to be pushed into the ground using the 200 kN thrust of the standard cone penetration test sounding machine (see below).

The samples from the 66 mm tubes are retained in plastic liners which can be split longitudinally to examine the stratification and fabric of the soil.

1.4.4 Spacing of soil samples

It is frequently specified that soil samples should be taken at intervals of 1.5 m and at each change of strata in boreholes. While this spacing may be adequate if a large number of boreholes is to be drilled, there can be a serious deficiency in quantitative soil information if the size of the area under investigation warrants only a few boreholes. The lack of information is particularly noticeable where structures with shallow foundations are proposed. Thus it is quite usual for the first sample to be taken just below the topsoil, say from 0.2 to 0.7 m. The next, at the 1.5 m spacing, is from 1.7 to 2.2 m. Exactly the same depths are adopted for all the boreholes on the site. It is normal to place foundations in clay at a depth of 0.9 or 1.0 m. Thus there is no information on soil shear strength and compressibility at and for a distance of 0.8 m below foundation level, probably within a zone where there are quite large variations in soil characteristics due to the effects of surface desiccation (Fig. 1.2). Where only a few boreholes are to be sunk it is a good practice to adopt continuous sampling for the first few metres below ground level or to stagger the sampling depths where the 1.5 m spacing is adopted.

1.4.5 *In-situ* testing of soils

Tests to determine the shear strength or density of soils *in situ* are a valuable means of investigation since these characteristics can be obtained directly without the disturbing effects of boring or sampling. They are particularly advantageous in soft sensitive clays and silts or loose sands. They must not be used as a substitute for borings but only as a supplementary method of investigation. One cannot be sure of identifying the types of soil they encounter and the tests give no information on ground-water conditions.

The *vane shear test apparatus* was developed to measure the shear strength of very soft and sensitive clays, but in Scandinavian countries the vane test is also regarded as a reliable means of determining the shear strength of stiff-fissured clays. The standard equipment and test procedure are described in BS 1377 (Test 18). The vane test is performed by rotating a four-bladed vane, 101.6 mm long × 50.8 mm wide overall, in the soil below the bottom of a borehole or by pushing down and rotating the vane rod independently of boring. Thus the test is performed in soil unaffected by boring disturbance. However, it has been observed that the undrained shear strength of a clay as measured by the vane test can differ quite appreciably from the actual field strength as measured from the behaviour of full-scale earthworks. Bjerrum[1.5] concluded that the difference is caused by the anisotropy of the soil and the difference in the rate of loading between a rapidly executed field vane and the slow application of loading from foundations and earthworks. Bjerrum's correction factors to vane test results correlated with the plasticity index of clays are shown in Fig. 1.3. These factors should

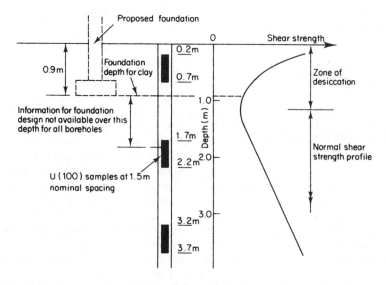

Fig. 1.2 Lack of information on shear strength at foundation level due to adoption of uniform sampling depths in all boreholes.

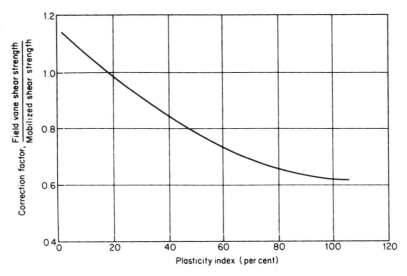

Fig. 1.3 Correction factors for undrained shear strength of normally consolidated clays obtained by field vane tests (after Bjerrum[1.5]).

be applied to vane test results to obtain the equivalent undrained shear strengths for foundation bearing capacity calculations using the methods described in Section 2.3.

From the results of this test or subsequent laboratory tests the clays are classified in accordance with BS 5930 as follows:

Term	Undrained shear strength (kN/m^2)
Very soft	Less than 20
Soft	20–40
Soft to firm	40–50
Firm	40–75
Firm to stiff	75–100
Stiff	75–150
Very stiff or hard	Greater than 150

The standard penetration test (SPT) (BS 1377 (Test 19)) is made in boreholes by means of the standard 50.8 mm outside diameter split-spoon sampler (sometimes known as the Raymond sampler). It is a very useful means of determining the approximate *in-situ* density of cohesionless soils and, when modified by a cone end, the relative strength or deformability of rocks. The sampler is driven to penetration of 450 mm by repeated blows of a 63.5 kg monkey falling through 760 mm, actuated by an automatic trip device. Only the number of blows for the last 300 mm of driving is recorded as the standard penetration number (*N*-value). It is standard practice to count the number of blows for every 75 mm of penetration in the full 450 mm of driving. By this means the depth of any disturbed soil in the bottom of the borehole can be assessed and the level at which any obstructions to driving, such as cobbles, large gravel, or cemented layers,

are met can be noted. Normally not more than 50 blows (including the number of blows required to seat the sampler below the disturbed zone) are made in the test. If the full 300 mm penetration below the initial seating drive is not achieved, i.e. when 50 blows have been made before full penetration is achieved, then both the depth at the start of the test and the depth at which it is concluded must be given in the borehole record, suitable symbols being used to denote whether the test was concluded within or below the initial seating drive. After withdrawal from the borehole the tube is taken apart for examination of the contents.

In gravelly soil and rocks the open-ended sampler is replaced by a cone end. Investigations have shown a general similarity in *N*-values for the two types in soils of the same density.

Although the applications of the test are wholly empirical, very extensive experience of their use has enabled a considerable knowledge of the behaviour of foundations in sands and gravels to be accumulated. Relationships have been established between *N*-values and such characteristics as density and angle of shearing resistance as described in Section 2.3.2.

BS 5930 gives the following relationship between the SPT *N*-values and the relative density of a sand:

N-value (blows/300 mm of penetration)	Relative density
Below 4	Very loose
4–10	Loose
10–30	Medium–dense
30–50	Dense
Over 50	Very dense

The use of the SPT for calculating allowable bearing pressures of spread foundations is shown in Section 2.3.2 and for piled foundations in Section 7.4. Stroud[1.6] has established relationships between the N-value, undrained shear strength, modulus of volume compressibility, and plasticity index of clays as shown in Fig. 1.4. However, the adoption of the SPT for determining the shear strength and deformability of clay soils is not recommended in preference to the direct method of making laboratory tests on undisturbed samples. This is because the relationships which have been established between the SPT and the strength and deformability of clays are wholly empirical, taking no account of such factors as time effects, anisotropy, and the fabric of the soil. When laboratory tests are made, these factors can be taken into account in the test procedure which can be selected in a manner appropriate to the soil characteristic and the type, rate, and duration of the load which will be applied to the soil.

Various corrections are necessary to the standard penetration test values before using them to calculate allowable bearing pressures and settlements. These adjustments take account of looseness and fineness of the soil, the effects of overburden pressure, and the position of the water table. The procedure for making such adjustments is described in Sections 2.3.2 and 2.6.5.

Dynamic probing employs various forms of rod with or without cone or other specially enlarged ends which are driven down into the soil by blows of a drop hammer. The number of blows for a given distance of penetration is recorded. The Borros penetrometer is used in Britain and other European countries. It employs a 63 kg hammer impacting a 50.5 mm cone at a rate of 20 blows per

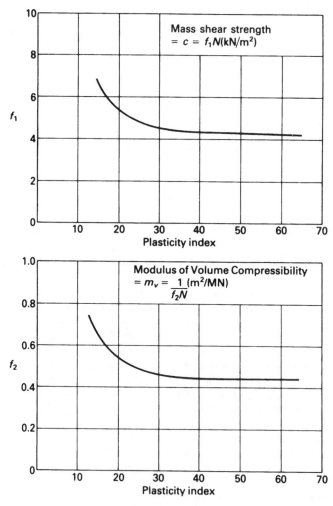

Fig. 1.4   Relationship between mass shear strength, modulus of volume compressibility, plasticity index, and SPT N-values (after Stroud[1.6])

minute. The number of blows required for a penetration of 100 mm is denoted as n. The torque on the cone is measured to provide an additional means of interpreting the data. There is very little information published in Britain on correlation between n-values and the SPT N-value or q_c values from static cone penetration tests. Cearns and McKenzie[1.7] have published relationships between n and the SPT N for sands, gravels, and chalk as shown in Fig. 1.5. Dynamic probing is a useful means of supplementing conventional boring and *in-situ* penetration tests, and is particularly advantageous in delineating areas of weak soils overlying stronger strata and for locating cavities in weak rock formations.

The static (Dutch) cone penetration test (CPT) is used widely in European countries and to a lesser extent in Britain and North America for investigations in cohesionless soils. Three types of penetrometer are in general use (Fig. 1.6). In all three types the cone end has a base area of 1000 mm^2 and an apex angle of 60°. The mantle cone shown in Fig. 1.6(*a*) was developed by the Delft Soil Mechanics Laboratory. Separate determinations of cone resistance and skin friction on the sleeve tubes, and the combined resistance of cone and tubes are obtained over

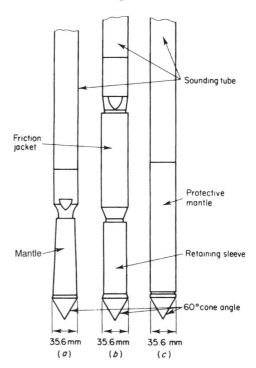

Fig. 1.6   Types of static cone penetrometer. (*a*) Mantle cone, (*b*) Friction jacket cone, (*c*) Electrical cone.

stages of 200 mm (sometimes lesser stages are adopted down to 100 mm). In the case of the friction jacket (or Begemann) cone shown in Fig. 1.6(*b*) the skin friction is measured over a short cylindrical jacket mounted above the cone which can be jacked down independently of the cone and the sleeve tubes above it.

The electrical cone (Fig. 1.6(*c*)) was developed in The Netherlands by Fugro NV. With this equipment both cone and sleeve tubes are jacked down continuously and together. The thrust on the cone end and on a 120 mm length of cylindrical sleeve are measured separately by electrical load cells installed at the lower end of the penetrometer. A full description of the CPT equipment, the method of operation, and the application of the test results to foundation design has been given by Meigh.[1.8]

It has been found that the cone resistance as measured by the three types does not differ significantly, but there are, of course, differences in the measured skin friction. Empirical methods have been developed whereby the type of soil can be identified by the separate and combined end and frictional resistances. The author does not rely on such measurements, but prefers to use the cone resistance only to obtain factors from which shear strength and deformability of soils can be estimated (see Sections 2.3.2 and 2.6.5) with the soil identification being obtained from adjacent conventional boreholes.

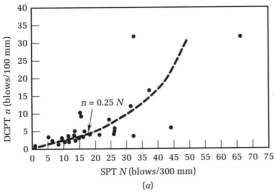

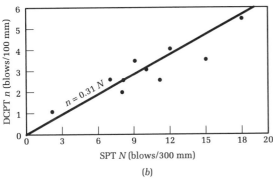

Fig. 1.5   Relationship between dynamic CPT n and STP N. (*a*) In sands and gravels. (*b*) In chalk. (after Cearns and McKenzie).

The static cone test is a valuable method of recording variations in the *in-situ* density of loose sandy soils or laminated sands and clays in conditions where the *in-situ* density is disturbed by boring operations, thus making the SPT unreliable in evaluation. Only limited penetration can be achieved by the cone in coarse gravelly soils. An empirical relationship between CPT and SPT results and the particle size distribution of soils is shown in Fig. 1.7.

The Delft Soil Mechanics Laboratory have continued to develop the CPT for obtaining a number of soil characteristics by *in-situ* testing. These include the following:

(a) The piezocone (to obtain pore-water pressures simultaneously with cone resistance);
(b) The nuclear backscattering probe (to measure *in-situ* soil density);
(c) The seismocone (to measure seismic velocity);
(d) The chemoprobe (see Section 1.10).

Pressuremeters are not described in the current edition of BS 5930. Mair and Wood[1.10] have given a detailed description of the five types listed in Table 1.2 together with information on their method of installation, the test procedures, the interpretation of the test results, and their applications to foundation design. A model specification for pressuremeter testing has been prepared by Clarke and Smith.[1.11]

Essentially, the pressuremeter consists of a cylindrical rubber membrane expanded against the sides of a borehole which is either predrilled in the case of the Menard-type pressuremeters or formed by the equipment in the case of the Camkometer and push-in types. The expansion of the membrane is measured directly by feeler gauges, or indirectly by measuring the volume of water or oil

Table 1.2 Types of pressuremeter

Type	Installation method	Measurement system	Diameter (mm)
Menard pressuremeter (type GB)	Lowered into preformed hole at base of borehole	Membrane expanded by water pressure. Volume measured at surface	32, 44, 58, and 74
Oyo elastmeter (type 100)	Lowered into preformed hole at base of borehole	Membrane expanded by water pressure. Expansion measured by displacement transducer	70
Self-boring pressuremeter (Camkometer)	Drilled into soil by integral unit	Membrane expanded by gas. Expansion measured by three strain gauged feeler arms	82
Cambridge *in-situ* high-pressure dilatometer	Lowered into preformed hole at base of borehole	Membrane expanded by oil pressure. Expansion measured by six strain gauged feeler arms	74
Building Research Establishment push-in pressuremeter	Pushed into soil at base of borehole, or into under-size pre-cored hole	Membrane expanded by oil pressure. Volume measured at surface	78

required for the increased diameter. The Camkometer is preferred for use in soft clays, silts, and sands because of the difficulty in maintaining the stability of an open preformed hole in these soils, except perhaps by using bentonite or foam as the circulating fluid. The Menard and Oyo types are suitable for firm to stiff clays and weak

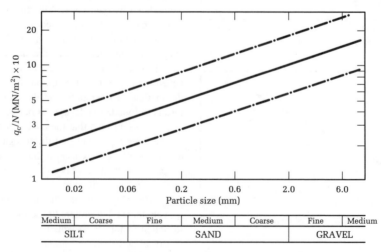

Fig. 1.7 Relationship between q_c/N and grain size. Values of N are not corrected for overburden pressure (after Burland and Burbidge[1.9]).

weathered rocks, and the push-in types for soft to firm clays and silts. Pressuremeters cannot be used in gravels.

The Menard pressuremeters, when correctly operated, produce a pressure–volume curve of the type shown in Fig. 1.8(*b*) and the Camkometer and push-in types a pressure–cavity strain curve as shown in Fig. 1.8(*a*).

The pressure–volume or strain curves require calibration and correction. This is usually done by the test operator. Mair and Wood strongly recommend that the

engineer commissioning the tests should ask for the raw data and calibration data to ensure that the corrections have been made properly before commencement of interpretation. The shear modulus (*G*) of the soil is best obtained from the slope of the unload–reload cycle after the expansion has reached the plastic stage. The elastic modulus is derived from the expression

$$E = 2G(1 + \nu), \qquad (1.1)$$

where ν is the Poisson's ratio of the soil which varies as to whether undrained or drained conditions are operating in the foundation design; *E* is sometimes referred to as the deformation modulus because the soil does not behave elastically at any stage of the pressuremeter test. Hence *G* or *E* should be referred to as the pressuremeter modulus and the strain amplitude should be defined, e.g. the initial tangent modulus G_i or a secant modulus G_s at a shear strain of 50 per cent of the peak shear stress or at a defined percentage strain.

The undrained shear strength of clays and the drained strength of sands can be obtained from the pressuremeter tests using the methods described by Mair and Wood. Only the self-boring pressuremeter can be used to obtain the ϕ'-value of sands. Mair and Wood[1.10] point out that ϕ'-values obtained in this way can be higher than those obtained by empirical correlations with the SPT or CPT, but are in agreement with values obtained from triaxial tests. Undrained strengths derived from the pressuremeter are usually very much higher than those obtained from triaxial compression tests. This is partly due to sample disturbance in the case of triaxial testing, but mainly due to the different method of test. Hence it is essential to know the particular test method when using undrained shear strength values to obtain the ultimate bearing of clays by the methods given in Section 2.3.2 or the skin friction and base resistance of piles (Sections 7.7–7.9).

The self-boring pressuremeter provides a good method of obtaining the coefficient of horizontal earth pressure at rest (K_0) in terms of effective stresses. This is possible only in clays.

The flat-type dilatometer (Marchetti dilatometer) is a 95 mm wide spade-shaped probe with an expandable metal-faced pressure cell 60 mm in diameter on one face of the probe. The device is pushed or hammered into the soil either directly from the surface or from the bottom of a borehole. Readings to determine the gas pressure required for initial movement of the cell, and for 10 mm movement of the cell into the soil, are taken at 200 mm intervals of depth. The device has some similarities with the CPT equipment rather than the pressuremeter, but it causes less disturbance of the soil than the standard cone.

The cell pressure readings are interpreted empirically[1.13] to provide the predominant grain size, the K_0

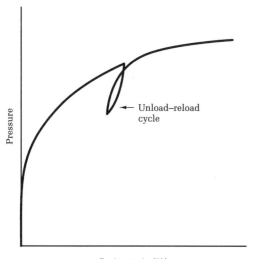

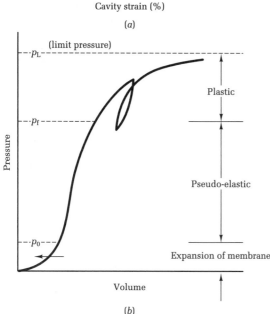

Fig. 1.8 Types of corrected curves for pressuremeter tests. (*a*) Self-boring pressuremeter (Camkometer). (*b*) Menard pressuremeter.

value, and the dilatometer modulus (related to the deformation modules) of the soil.

Plate bearing tests are made by excavating a pit to the predetermined foundation level or other suitable depth below ground level, and then applying a static load to a plate set at the bottom of the pit. The load is applied in successive increments until failure of the ground in shear is attained or, more usually, until the bearing pressure on the plate reaches some multiple, say two or three, of the bearing pressure proposed for the full-scale foundations. The magnitude and rate of settlement under each increment of load is measured. After the maximum load is reached the pressure on the plate is reduced in successive decrements and the recovery of the plate is recorded at each stage of unloading. This procedure is known as the *maintained load test* and is used to obtain the deformation characteristics of the ground. Alternatively, the load can be applied at a continuous and controlled rate to give a penetration of the plate of 2.5 mm/min. This is known as the *constant rate of penetration test* and is applicable to soils where the failure of the ground in undrained shear is required, as defined by gross settlement of the plate; or where there is no clear indication of failure with increasing load, the ultimate bearing capacity is defined by the load causing a settlement of 15 per cent of the plate diameter.

Although such tests appear to answer all the requirements of foundation design, the method is subject to serious limitations and in certain cases the information given by the tests can be wildly misleading. In the first place it is essential to have the bearing plate of a size which will take account of the effects of fissures or other discontinuities in the soil or rock. A 300 mm plate is the minimum size which should be used which is suitable for obtaining the undrained shear strength of stiff fissured clays. If deformation characteristics are required from these soils, a 750 mm plate should be provided in conjunction with the maintained load procedure. It is essential to make the plate tests in soil or rock of the same characteristics, as will be stressed by the full-scale foundation. Misleading information will be given if, for example, the tests are made in the stiff crust of weathered clay overlying a soft clay as illustrated in Fig. 1.9. A 1000 mm plate is generally the economic limit, since a 1000 mm plate loaded say to 800 kN/m^2 will require some 63 t of kentledge, which is expensive to hire including the costs of transport and handling. The cost of a single plate bearing test with a 300–600 mm plate with 50 t of kentledge is about three times the cost of a 12 m deep borehole (in soft ground) complete with *in-situ* and laboratory testing. A single plate bearing test on a site is, in any case, far from sufficient since the ground is generally variable in its characteristics both in depth and laterally. At least three

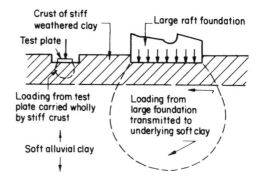

Fig. 1.9

tests, and preferably more, are required to obtain representative results.

Economies in plate bearing tests on rock can be made by jacking against cable or rod anchorages grouted into drill-holes in the rock, instead of using kentledge. Single anchors have been used successfully. The anchor cable, which is not bonded to the rock over its upper part, is passed through a hole drilled in the centre of the test plate. A test of this type can be made at the bottom of a borehole.

The level of the water table has an important effect on the bearing capacity and settlement of sands. Thus a plate bearing test made some distance above the water table will indicate much more favourable results than will be given by the large full-scale foundation which transmits stresses to the ground below the water table. The plate bearing test gives no information whereby the magnitude and rate of long-term consolidation settlement in clays may be calculated.

In spite of these drawbacks, the plate bearing test cannot be ruled out as a means of site investigation, since in certain circumstances it can give information which cannot be readily obtained by other means. For example, the bearing capacity and deformation characteristics of certain types of rocks such as broken shales or variably weathered materials cannot be assessed from *in-situ* pressuremeter tests or laboratory tests due to difficulties in sampling. In these ground conditions it is necessary to sink a number of trial pits down to or below foundation level. The pits are carefully examined and three or four are selected in which the ground appears to be more heavily weathered than the average, and plate bearing tests are made in these selected pits. Alternatively it may be desirable to select the pits in the weakest and strongest ground so that a range of deformation moduli can be obtained to assess likely differential settlement. The largest practicable size of plate should be used and it is advisable to dig or probe below plate level on completion

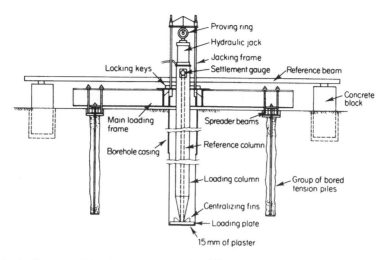

Fig. 1.10 Rig for plate loading test made in a borehole (after Marsland[1.12]).

of the test to find out if there are any voids or hard masses of material present which might affect the results.

Plate bearing tests made in fill materials consisting of brick or stone rubble are of doubtful value because of the large particle size and wide variation in density of such material. However, meaningful results can be obtained from tests made in filled ground consisting of sands, gravels, colliery waste, or boiler ash.

The procedure for making plate bearing tests is fully described in BS 5930. A typical arrangement for a test using a set-up developed by the Building Research Establishment[1.12] is shown in Fig. 1.10. This rig is suitable either for a test in an open pit or at the bottom of a large-diameter borehole drilled using the equipment described in Section 8.14.1.

1.5 Exploration in rocks

1.5.1 General requirements

Investigations into rock formations for foundation engineering purposes are concerned first with the allowable bearing pressures for spread foundations or working loads on piles, and second with the conditions which are likely to be met if excavations have to taken into the rock strata for deep foundations. The engineer must therefore have information on the depth of any weathering of the rock, the presence of any shattered zones or faults susceptible to movement, the possibility of the occurrence of deep drift-filled clefts, buried glacial valleys, swallow holes, or concealed cavities, and the quantity of water likely to be pumped from excavations. Much of this information can be obtained in a general way by advice from a geologist

from his knowledge of local conditions and the study of published maps and memoirs. Indeed the advice of a geologist in connection with the siting of any important project on rock formations is very necessary. An essential part of exploration of rock masses to aid the interpretation of borehole data and field and laboratory tests is a detailed study of the spacing, thickness, and orientation of joints in the rock mass, together with a study of the composition and consistency of any weathered rock or other material infilling the joints. If detailed observations of joint characteristics are made at locations of pressuremeter or plate bearing tests, then the test results will be applicable to other parts of the site where similar jointing conditions exist in the rock formation.

Weathering or other disturbances of the surface of a rock stratum are likely to necessitate varying foundation levels on a site, and it is often difficult if not impossible to assess from the results of boreholes a definite foundation level for a structure. For example, some types of rock such as marl or limestone soften as a result of seepage of water down fissures forming zones of weakened rock of soft clayey consistency surrounding strong unweathered material. If a few boreholes only are put down in these conditions they may encounter only unweathered rock, giving a false picture of the true site conditions. Conversely, boreholes striking only softened weathered rock might suggest the presence of a deep stratum of soft clay overlying hard rock, whereas such soft material might only exist in comparatively narrow fissures. Glacial action can have caused deep and irregular disturbance of the surface of bedrock, for example the breaking-up and bodily movement of shales, or the tilting of large blocks of massively bedded rock.

The surface of friable, and therefore erodible, rocks may be intersected by narrow drift-filled valleys or clefts which again may be undiscovered by borings or trial pits, although a geologist would anticipate their occurrence. These conditions may require major redesign or relocation of foundations when the actual bedrock surface is revealed at the construction stage of the project.

There are three methods in general use for subsurface exploration in rocks. These are

(1) Test pits
(2) Drilled shafts
(3) Rotary core drilling

1.5.2 Test pits

Test pits are the most satisfactory means of assessing foundation conditions in rock, since the exposed bedrock surface can be closely inspected. The dip of the strata can be measured and it is often possible to assess the extent of weathering in layers or fissures. The strength of the rock and its ease of excavation can be determined by trial with a pick or compressed-air tools. If necessary, blocks or cylinders of the rock can be cut for laboratory tests. However, test pits are only economical when bedrock lies fairly close (say within 3 m) of the ground surface. They should be used instead of boreholes when rock level is shallower than 2 m below ground level, but for depths between 2 and 3 m a few pits can be dug to supplement the evidence given by boreholes.

1.5.3 Drilled shafts

Where rock lies deeper than about 3 m, hand-excavated test pits become very costly. When it is essential to assess the character of rock lying at some depth below the surface in connection with the design of important structures, the exploration can be conducted from shafts drilled by a rotary mechanical auger (see Section 8.14.1). The shafts should have a minimum diameter of 750 mm (preferably 1 m), and should be supported throughout by a steel tube liner which can be raised from the bottom of the shaft as required to permit examination of the rock surface or to make tests such as plate loading tests with the load applied in a horizontal direction. Full safety precautions should be observed when descending the shafts for geological examination or *in-situ* tests. These are described in BS 5573: 1978, *Safety Precautions in the Construction of Large-diameter Boreholes for Piling and Other Purposes.*

1.5.4 Rotary core drilling and rock testing

Rotary core drilling is regarded as the most satisfactory method of assessing the character of rock formations which lie at depth below the ground surface. Specimens of rock in the form of cylindrical cores are recovered from the drill-holes by means of a core barrel.

Core barrels are made in various types and sizes depending on the depth of hole, type of rock, and size of specimen required. The barrels are also made in various lengths, 1.5 and 3.0 m being common lengths for site investigation work. Core diameters and the corresponding borehole diameters for core barrels conforming to BS 4019, Part 1, are listed in Table 1.3. BS 5930 lists a number of other types and diameters including core barrels designed for wire-line equipment used in offshore drilling. The Mazier barrel is designed to achieve good recovery in weak friable rocks and hard clays. It consists of a triple-tube barrel, the inner tube being stationary and fitted with a spring-loaded cutting shoe. This pushes the inner tube in advance of the cutting bit, so keeping the core from contact with the drilling fluid. For most purposes core diameters of N-size (54.5 mm) or H-size (76 mm) for the weaker rocks are suitable. For very weak friable rocks or heavily fissured strong rocks it may be desirable to adopt coring in the P to Z designations (92–165 mm diameter).

It is false economy to use a small-diameter core barrel for the sake of cheaper drilling rates if this results in a poor core recovery, or even no recovery at all in weak or shattered rocks. It is most important to recover the *weakest* materials when assessments of safe bearing pressures are being made on rock cores. This requires, as nearly as possible, 100 per cent core recovery. Failure to obtain specimens of the weak rocks in formations consisting of alternating beds of strong and weak material will prevent any reasonable assessment of the bearing capacity of such formations. A case occurred on a building site where a strong red sandstone was overlain by weakly cemented and friable weathered sandstone. For the sake of economy, small-diameter core drilling had been used in the site investigation, with the result that the weathered sandstone did not yield a core but was returned to the surface as a

Table 1.3   British Standard core barrel sizes

Reference on borehole record	Core barrel design	Nominal core diameter (mm)	Nominal diameter of hole (mm)
B	BWF, BWG, or BWN	42.0	60.0
N	NFW, NWG, or NWM	54.5	76.0
H	HWF or HWG	76.0	99.0
P	PWF	92.0	121.0
S	SWF	112.5	146.0
U	UWF	140.0	175.0
Z	ZWF	165.0	200.0

sand suspension in the wash water. The weathered rock was thus identified as a sand and a correspondingly low bearing pressure was adopted for the foundation design. When, at the construction stage, the foundations were excavated, compressed-air tools were required to break up the friable rock. A much higher bearing pressure could have been used, at least 1000 kN/m^2, but at that stage it was too late to change the design.

The point load test[1.13] is a quick and cheap method of obtaining an indirect measurement of the uniaxial compression strength of a core specimen. It is particularly useful in closely jointed rocks lacking in cores of sufficient length for making uniaxial compression tests in the laboratory. The point load test equipment is easily portable and suitable for use in the field. The tests are made in axial and diametrical directions on the cores or block samples. The failure load to break the specimen is designated as the point load strength (I_s) which is then corrected to the value of I_s which would have been measured by a diametral test on a 50 mm diameter core using a standard method of correction[1.14] to obtain $I_{s(50)}$. Some values collected by the author of the ratio of the uniaxial compression strength of weak rocks to the corresponding axial point load strengths are shown in Table 1.4. The value of $q_c/I_{s(50)}$ of 24 is frequently quoted for sandstone.

Since rotary core drilling is a fairly expensive procedure it is important to give care and attention to preserving the cores recovered from the drill-holes. The value of the investigation is lost if the cores are mislaid or become mixed up in the boxes. The moisture content of weak rocks and weathered rocks should be maintained at its true *in-situ* value if laboratory tests are required to be made. This can be done by coating selected cores with wax or wrapping them in aluminium foil or plastic sheeting.

A frequent practice in Britain is to adopt cable percussion boring in the soil overburden then to continue with this method through very weak to weak weathered rock until the material becomes too strong to advance the hole. Only at this stage is core drilling commenced. This is a *bad practice* because the important engineering charac-

teristics of the weathered rock cannot be assessed properly from chippings brought to the surface by percussion drilling. Attempts to hammer standard U 100 or SPT samplers result in poor recovery of shattered material.

Cable percussion boring should be stopped as soon as it is evident that a rock formation has been reached. Core drilling with bentonite, foam, or air flush with a barrel of sufficient diameter should then be adopted. The equipment should be capable of recovering good cores of a very weak weathered rock. Laboratory tests on the rock cores can be supplemented by pressuremeter or dilatometer testing in separate boreholes drilled specially for this purpose using open-hole (non-coring) techniques.

1.6 Ground water

1.6.1 Water level observations

Reliable information on ground-water levels within the depth proposed for excavations and pile borings, and within the zone of influence of foundation pressures, is vital to many aspects of foundation design and construction. Regrettably, observations of ground-water conditions are all too often neglected in site investigation work. This is because in cable percussion drilling the operation of the tools causes wide variations in water level and it is often necessary to add water during drilling. Frequently insufficient time is given for the first 'strike' of ground-water in a borehole or trial pit to reach an equilibrium level. In wash boring or rotary core drilling no observations of any value are possible while drilling is in progress. Merely to observe water levels on conclusion of drilling is inadequate, since these levels may take days to recover to equilibrium conditions. Ground-water levels should be monitored over as long a period as possible by measurements in one or more standpipes or piezometers installed in boreholes. A simple standpipe (Fig. 1.11(*a*)) consisting of a PVC tube with a slotted end and surrounded by a granular filter (see Section 11.3.4) or plastic filter fabric is satisfactory for granular soils or permeable rocks. In silts or clays more sensitive equipment is required. The hydraulic piezometer (Fig. 1.11(*b*)) consists of a porous element connected by twin small-bore plastic tubing to a remote reading station where pressures are measured by a mercury manometer or a Bourdon gauge. Where ground water under pressure is met in aquifers confined by impermeable strata above and below, it is necessary to install standpipes or piezometers within each separate aquifer, sealed from the adjacent strata by expandable plugs or a bentonite–cement grout. Details of standpipes, piezometers, and their method of installation are given in BS 5930.

Table 1.4 Relationships between uniaxial compression strength (q_c) and point load strength ($I_{s(50)}$) of some weak rocks

Rock description	Average q_c (MN/m^2)	$q_c/I_{s(50)}$
Magnesian limestone	37	25
Upper Chalk (Humberside)	3–8	18
Carbonate siltstone/sandstone (UAE)	2	12
Mudstone/siltstone (Coal Measures)	11	23
Tuffaceous rhyolite (Korea)	20–70	10
Tuffaceous andesite (Korea)	40–140	10

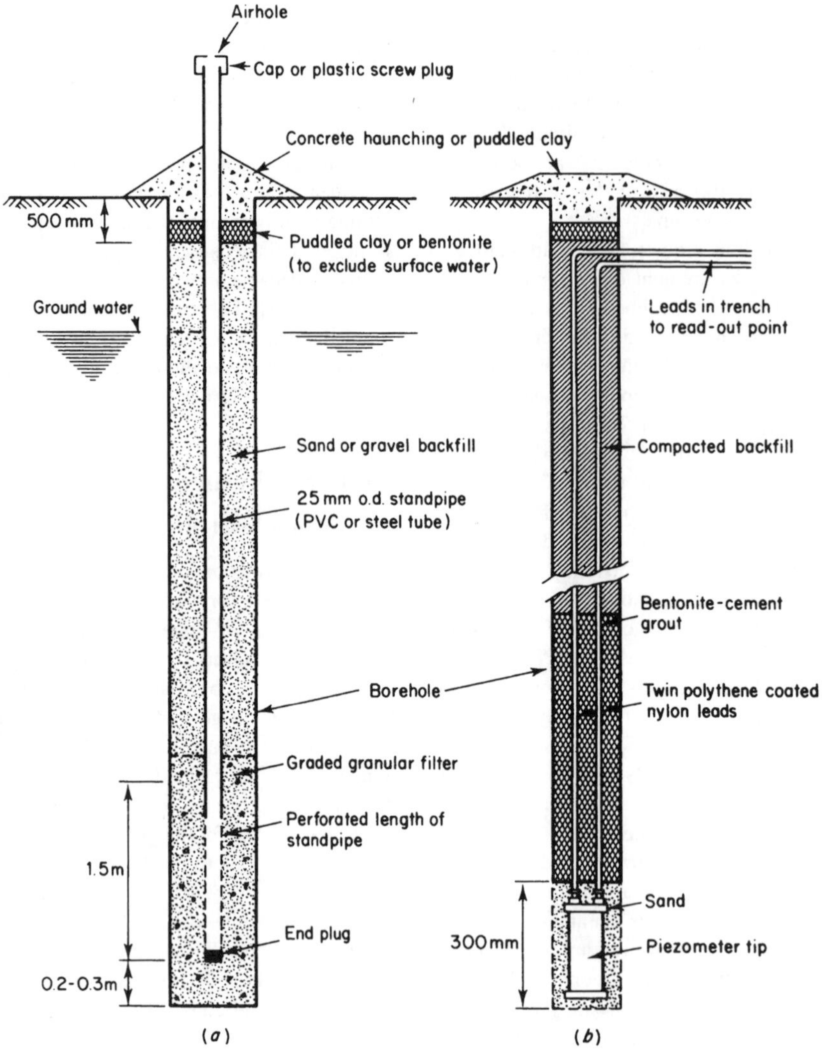

Fig. 1.11　Types of piezometer for ground-water level observations. (*a*) Standpipe. (*b*) Hydraulic piezometer.

1.6.2 Sampling ground water

Because ground-water samples are required for chemical analyses to determine the risks of aggressive action on buried foundation structures (see Chapter 13) or ground-water lowering installations, it is essential that they are not diluted by water used to assist drilling. The ground water should be sampled immediately it is struck in a pit or borehole. If samples are required during progress of drilling, the borehole should be pumped or baled dry and water allowed to seep in before samples are taken. Ground water in confined aquifers should preferably be sampled from standpipes or piezometers sealed into these strata. Site testing of pH values may be required immediately

after extraction of the sample where a measure of the acidity of the ground water is required in connection with aggressive action on concrete or steel (see Section 13.2). This is because oxidation of the water may occur on exposure to air in a sample jar, leading to a marked increase in acidity. Water samples for bacterial analysis in connection with corrosion studies should be taken in special sterilized containers.

1.6.3 Field permeability tests

Field tests to determine *in situ* the permeability of soils and rocks are used in connection with ground-water lowering schemes when it is necessary to estimate the number and diameter of pumping wells and the size of the pumps, for a given size of excavation (see Section 11.2.2). These tests are also used to determine the *rate* of consolidation of soil strata under foundation loading more accurately than is possible with conventional laboratory oedometer tests (see Section 2.6.6) and in investigations for dams and impounding reservoirs when information is required on the rate of seepage beneath the dam or into the surrounding ground. These tests can be carried out quite economically in site investigation boreholes where relatively simple tests can give reliable indications of field permeability. BS 5930 describes two types of test. In the *variable head* test a piezometer tube is filled with water and then the water in the tube is allowed to fall until it reaches equilibrium with the water level in the surrounding ground. The falling head of water in the tube is measured at intervals of time from the commencement of the test. Alternatively, the water can be pumped or baled from the piezometer tube and the rising head is recorded until equilibrium is reached. In the *constant head* test, water is allowed to flow into the piezometer tube at such a steady rate that the head of water in the tube attains a constant value, when the rate of flow required to obtain these conditions is recorded.

Pumping tests in which the ground water is pumped from a well with observations of the draw-down in the surrounding water table in an array of standpipes or piezometers are time-consuming and costly. They are usually performed only in connection with trials for large-scale ground-water lowering installations (see Section 11.2.2).

Details of all the above tests and methods of recording and interpretation are given in BS 5930.

1.7 Borehole records

The first stage in the preparation of borehole records is the site log or 'journal' which gives a record of the soil or rock strata as determined by visual examination in the field. All relevant data on ground-water levels, reduced levels of strata changes, depths of samples, and records of any *in-situ* tests are shown on these site records which provide preliminary information for the designer and enable the laboratory testing programme to be drawn up. The second stage is the final record which is included in the engineering report on the site investigation. In this record the descriptions of the soils and rocks are amended where necessary in the light of information given by laboratory testing and examination of the samples and detailed assessment by a geologist. In both stages of borehole record presentation it is important to adopt a consistent method of describing soil and rock types. The standard

method of description and recording in borehole records and daily logs is given in BS 5930.

Where rotary core drilling in rocks has been undertaken, an essential addition to the standard practice of showing percentage total core recovery is a statement of the *rock quality designation* (RQD) and fracture index for each run of the core barrel (see Section 1.1).

1.8 Investigations for foundations of works over water

The equipment and techniques for sinking boreholes through water are essentially the same as for land borings. For boring close to land or existing structures or in shallow water, it is economical to use staging to carry the rig. This need only be of light construction such as tubular steel scaffolding since the drilling equipment is of no great weight. Where the borehole is within about 50 m of the land the platform can be connected to the shore by a catwalk. For greater distances it may be more economical to erect the platform as an island, either lowering it as a unit from a crane barge or in the form of a raft which can be sunk by admitting water into buoyancy chambers. For boring in water deeper than about 10 m or at some distance from the shore it is convenient to mount the drilling rig on a platform cantilevered over the side of a barge or pontoon. Dumb craft of this type are suitable for working in sheltered waters or where a harbour is reasonably close to hand to which the craft can be towed at times of storms. However, when boring in open unsheltered waters away from harbours it is advisable to use a powered craft which, at the onset of a gale, can cast off its moorings and ride out the storm at anchor or seek shelter at the nearest available place. With increased availability of the smaller types of jack-up platforms, greater use is being made of this equipment for site investigations in shallow or moderate water depths.

Drilling from a floating craft in water depths up to about 30 m requires some slight modifications in technique and there are complications in making some of the *in-situ* tests such as vane tests and static cone tests.

For investigations in deep water, say for the foundations of petroleum drilling and production platforms, special techniques are necessary since equipment such as the standard penetration test sampler or cone penetrometer cannot be used within a drill 'string' which may be 200 m or more in length below platform level on the drill-ship. Moorings are not used to hold the ship in position. This is achieved by an assembly of thruster-type propellers controlled by signals from transponders installed on the sea bed around the borehole position. For investigations of this type, boreholes are usually drilled entirely by water or mud-flush rotary methods with sampling by core barrels lowered to the drill head by wire rope, latched into position, and lifted to the surface again by wire rope. Soil samples recovered by these 'wireline' core barrels are not usually of good quality and reliance is placed more on *in-situ* testing to obtain shear strength and deformation values. Various devices have been developed for latching into core barrels for cone pressuremeter or vane testing. There is, as yet, no standardization of equipment or test methods, and information is rather scattered in the proceedings of the numerous conferences on offshore technology held in Europe and North America.

1.9 Geophysical methods of site investigation

It is possible to determine stratification of soils and rocks by geophysical methods which measure changes in certain physical characteristics of these materials, for example the magnetism, density, electrical resistivity, elasticity, or a combination of these properties. However, such methods are of limited value in foundation engineering since they only record changes in stratification where the layers have appreciably different geophysical properties, and the only useful information they give is the level of the interfaces between the various strata. Vital information on groundwater conditions is usually lacking. Geophysical methods in their present state of development do not give direct quantitative data on shear strength, compressibility, or particle-size distribution, but measurements of seismic velocity can be helpful in assessing the effect of weathering and discontinuities on the compressibility of rock masses (see Section 2.7). At best geophysical surveying is a means of filling in data on strata changes between widely spaced boreholes. On large sites geophysical methods can show economies due to the rapidity with which extensive areas can be covered. Before embarking on such surveys, the engineer should consider whether other methods such as wash boring or probing, or static and dynamic cone penetration tests would not be preferable, since these methods are also rapid and have the advantage of giving quantitative data on the *in-situ* density or the shear strength of soils. Generally, geophysical methods are best suited to deep investigations in rock strata, for example for dams or tunnels where the stratification of rocks at depth is required, and for investigations in soils containing many cobbles or boulders where probings or cone tests are impracticable. British Standard 5930 describes the suitability for engineering purposes of the following techniques:

(a) Electrical resistivity
(b) Gravimetric
(c) Magnetic

(d) Seismic refraction and reflection
(e) Side scan sonars
(f) Borehole logging

Ground radar is a recently developed method not described in BS 5930. It is used for the detection of buried objects at shallow depths in a manner similar to magnetometer surveying.

1.10 Investigations of filled and contaminated ground

Investigations of filled ground are required to deal with three different aspects. These are:

(1) The engineering characteristics such as compressibility, permeability, and the presence of buried obstructions.
(2) The effect of organic and mineral substances on the durabilty of foundation structures and services.
(3) The toxicity of contaminants in the fill, or in ground water seeping into natural soil underlying the fill, which could cause health risks to personnel undertaking the investigations, subsequently to construction operatives and the general public, and eventually to the occupiers of the completed works.

Foundation design and construction methods for items (1) and (2) are dealt with in Section 3.6 and Chapter 13 respectively. Considerations of toxic contaminants are beyond the scope of this book except for descriptions of methods for isolating buildings and their occupants from these substances. Categories of potential toxicity and the corresponding health risks are described by Solomon and Powrie.[1.15]

Investigations of filled ground are best undertaken by visual examination and sampling in test pits. Intensive investigations are needed where toxic contaminants are suspected. The British Standard Draft Code of Practice[1.16] recommends at least 25 sampling points for a site of 1 ha ($10\,000\,m^2$). Guidance of investigations, and statutory controls for building on contaminated ground is given in a number of British Government and European Community directives listed in ref. 1.15.

Careful attention must be paid to precautions against health risks to personnel engaged in the investigations. These may require special protective clothing and the possible need for self-contained breathing apparatus.

Plate bearing tests or other forms of *in-situ* testing described in Section 1.4.5 are of limited value in filled ground. This is because of the wide-ranging variability in the composition and density of material placed by uncontrolled tipping. Obstructions in the fill usually prevent the use of probe-type devices. Investigations of backfilled opencast mineral workings are particularly difficult because of the depth to which test pits are needed for an adequate exploration of the extent of any voids or arching in the rock debris.

The contamination of natural soils underlying the fill due to seepage was mentioned above. The chemoprobe, described by Olie *et al.*,[1.17] provides a means of making investigations at close spacings of soluble contaminants in the ground water. The pH value, redox potential, and the electrical and hydraulic conductivity are measured directly. Sampling in the form of slugs discharged from the probe is possible, but qualitative identification of contaminants can only be made on the samples.

1.11 Laboratory tests on soils

The physical characteristics of soils can be measured by means of laboratory tests on samples extracted from boreholes or trial pits. The results of shear strength tests can be used to calculate the ultimate bearing capacity of soils or the stability of slopes in foundation excavations and embankments. Laboratory tests also provide data from which soils can be classified and predictions made of their behaviour under foundation loading. From the laboratory test information, methods of treating soils can be devised to overcome difficulties in excavations, especially in dealing with ground-water problems. It is important to keep in mind that natural soil deposits are variable in composition and state of consolidation; therefore it is necessary to use considerable judgement based on common sense and practical experience in assessing test results and knowing where reliance can be placed on the data and when they should be discarded. It is dangerous to put blind faith in laboratory tests, especially when they are few in number. The test data should be studied in conjunction with the borehole records and other site observations, and any estimations of bearing pressures or other engineering design data obtained from them should be checked as far as possible with known conditions and past experience.

Laboratory tests should be as simple as possible. Tests using elaborate equipment are time-consuming and therefore costly, and are liable to serious error unless carefully and conscientiously carried out by highly experienced technicians. Such methods may be quite unjustified if the samples are few in number, or if the cost is high in relation to the cost of the project. Elaborate and costly tests are justified only if the increased accuracy of the data will give worthwhile savings in design or will eliminate the risk of a costly failure, as in the case of geotechnical category 3 investigations.

An important point in favour of carrying out a reasonable amount of laboratory testing is that an increasing amount of valuable data is built up over the years relating test results to foundation behaviour, for example stability and settlement, enabling engineers to use laboratory tests with greater confidence. At the very least the test results give a check on field descriptions of boreholes based on visual examination and handling of soil samples, and are a useful corrective to 'wishful thinking' by engineers in their first impression of the strength of a soil as it appears in the borehole or trial pit.

The soil mechanics tests made in accordance with BS 1377 which concern the foundation engineer are as follows:

(a)　Visual examination
(b)　Natural moisture content
(c)　Liquid and plastic limits
(d)　Particle-size distribution
(e)　Unconfined compression
(f)　Triaxial compression
(g)　Shear box
(h)　Vane
(i)　Consolidation
(j)　Permeability
(k)　Chemical analyses

Visual tests carried out in the laboratory are for noting the colour, texture, and consistency of the disturbed and undisturbed samples received from the site. This should be undertaken as a routine check on the field engineer or boring foreman's descriptions.

Natural moisture content test results are compared and related to the liquid and plastic limits of the corresponding soil types in order to arrange the programme for shear strength tests and to ensure that tests on the softer soils (as suggested by the higher moisture content) are not omitted.

Liquid and plastic limit tests are made on cohesive soils for classification purposes and for predicting their engineering properties. The plasticity chart (Fig. 1.12) can be used to predict the compressibility of clays and silts. To use this chart it is necessary to know whether the soil is of organic or inorganic origin. The usual procedure is to make liquid and plastic limit tests on a few selected samples of each main soil type found in the boreholes. By comparing the results and plotting the data on the plasticity chart the various soil types can be classified in a rough order of compressibility and samples selected accordingly for consolidation tests if these are required.

The particle-size distribution test is a form of classification test for which sieve analysis or a combination of sieve analysis and sedimentation or hydrometer analysis is used to obtain grading curves which can be plotted on the chart shown in Fig. 1.13. The grading curves are of no direct value in assessing allowable bearing pressure, and gen-

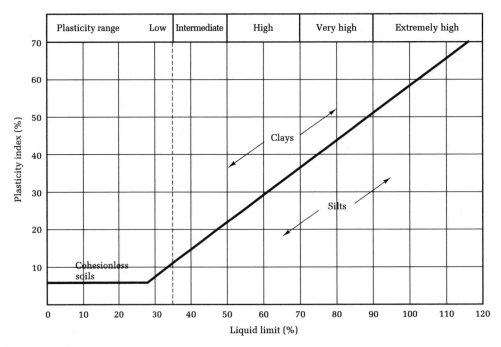

Fig. 1.12   Plasticity chart.

erally this type of test need not be made in connection with any foundation investigation in clays or in the case of sands and gravels where the excavation is above the water table. The particle-size distribution test is, however, of particular value in the investigation of problems of excavation in permeable soils below the water table, when the results can be used to ascertain which of several geotechnical processes are feasible for ground-water lowering or grouting treatment, as described in Section 11.3.1.

The *shear strength* of soil can be used directly to calculate the ultimate bearing capacity of a foundation, as described in Chapters 2, 4, and 7, and to calculate earth pressure on sheeted excavations as described in Section 9.6.2.

The unconfined compression test is the simplest form of shear strength test. It cannot be made on cohesionless soils or on clays and silts which are too soft to stand in the machine without collapsing before the load is applied. In the case of fissured or brittle soils the results are lower than the true *in-situ* strength of these soils.

The triaxial compression test is a more adaptable form of shear strength test which can be applied to a wider range of soil types than the unconfined compression test, and the conditions of tests and observations made can be varied to suit a wide range of engineering problems. It is used to determine the cohesion (*c*) and the angle of shearing resistance (ϕ) of a soil as defined by the Coulomb–Mohr equation, for three conditions.

Undrained shear (total stresses)

$$s_u = c_u.$$

Drained shear strength of sands and normally consolidated clays (effective stresses)

$$s = \sigma_n' \tan \phi'.$$

Drained shear strength of over-consolidated clays

$$s = c_u' + \sigma_n' \tan \phi'.$$

Drained residual (large strain) of clays

$$s_r = c_r' + \sigma_n' \phi_r'.$$

The three main types of triaxial test are:

(1) Undrained
(2) Consolidated-undrained
(3) Drained

In the *undrained test* the specimen is not allowed to drain during the application of the all-round pressure or during the application of the deviator stress, and therefore the pore pressure is not allowed to dissipate at any stage of the test. In the case of a saturated cohesive soil, this test procedure reproduces the conditions which occur when the soil

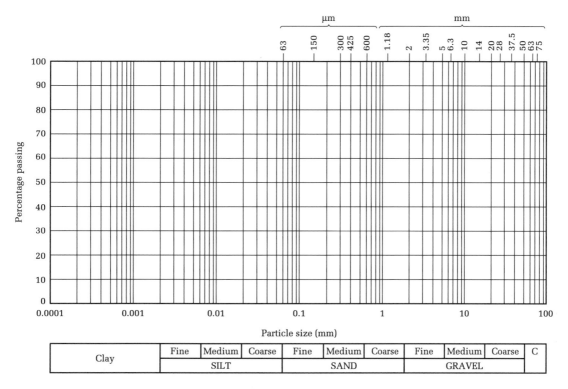

Fig. 1.13   Grading chart.

beneath the full-scale foundation is subjected to load or when earth is removed from an open or sheeted excavation. In these conditions the pore pressures in the soil beneath the loaded foundation or in the soil behind the face of an excavation have no time to dissipate during the time of application of stress. The analyses to determine the ultimate bearing capacity of the foundation soil or the initial stability of excavations are carried out in terms of *total stresses*.

The test procedure for the *consolidated-undrained test* is to allow the specimen to drain while applying the all-round pressure; thus the specimen is allowed to consolidate fully during this stage of the test. Drainage is not allowed during the application of the deviator stress.

In the case of the *drained test*, drainage of pore water from the specimen is allowed both during the stage of consolidation under all-round pressure and during the application of the deviator stress. The time allowed for consolidation under all-round pressure and for application of the deviator stress must be slow enough to ensure that no build-up of pore pressure occurs at any stage of the test. The procedure for consolidated-undrained and drained tests corresponds to the conditions when the soil below foundation level is sufficiently permeable to allow dissipa-

tion of excess pore-water pressure during the period of application of foundation loading, or when pore-water pressure changes can occur due to external influences at any time during the life of a structure. The long-term stability of excavated slopes is also investigated by means of consolidated-undrained or drained tests. These long-term stability problems are analysed in terms of the *effective stress* (see Section 9.3). The reader is referred to a book by Bishop and Henkel[1.18] for a detailed and authoritative account of the triaxial test and an extensive bibliography of the application of the various types of test to the solution of engineering problems. Elaboration of triaxial testing techniques such as the insertion of probes or other devices into the test specimen to measure pore pressures and deformations at small strains can be justified for category 3 investigations and to obtain small strain values of Young's modulus for use in finite element analysis (see Section 2.8.2).

Triaxial tests are limited to clays, silts, peats, and weak rocks. The angle of shearing resistance of sands and gravels is best determined empirically by means of *in-situ* tests (see Section 1.4.5).

The vane shear test is more applicable to field conditions than to the laboratory. However, the laboratory vane test has a useful application where satisfactory undisturbed samples of very soft clays and silts have been obtained by the procedures described in Section 1.4.3 and where it is impossible to prepare specimens, because of their softness, from the tubes for shear strength tests using the unconfined or triaxial apparatus.

The shear box test can be used to determine the shearing resistance of soils, but it is not used in preference to the triaxial test because of difficulties in controlling drainage conditions, and the fact that the failure plane is predetermined by the apparatus. However, the reversing shear box provides a useful means of obtaining the residual or long-term shear strength used in calculating the stability of earth slopes where failure may take place on an ancient slip surface.

The ring shear test[1.19] is also used to obtain the large strain parameters c'_r and ϕ'_r.

Consolidation test results are used to calculate the magnitude and rate of consolidation of the soil beneath foundations. The test is more accurately described as a one-dimensional consolidation test, because the sample is enclosed in a metal ring and the load is applied in one direction only. The apparatus used is known as the oedometer, or sometimes as the consolidometer. From the results the coefficient of consolidation (c_v) is obtained which, with proper instrumentation, enables the *rate* of settlement of the full-scale structure to be calculated. The load–settlement data obtained from the full cycle of loading and unloading are used to draw a pressure–voids ratio curve from which the coefficient of volume compressibility (m_v) is derived. This is used to calculate the *magnitude* of consolidation settlement under any given loading.

Consolidation tests are restricted to clays and silts since the theories on which settlement calculations are based are limited to fine-grained soils of these types. The *rate* of settlement calculated from the coefficient of consolidation as obtained by oedometer tests on conventional 75 mm specimens may be grossly in error. This is because the 'fabric' of the soil, i.e. the presence of fissures, laminations, root-holes, etc. may not be properly represented in a 75 mm specimen.[1.2] Where soils exhibit a type of fabric which will influence their permeability, and hence their rate of consolidation, the consolidation tests should be made on 200 or 250 mm diameter specimens. Alternatively, the rate of consolidation can be deduced from observations of the rate of settlement of full-scale structures on similar soil types. The use of consolidation test data to calculate the magnitude and rate of settlement of foundations is described in Section 2.6.6. The settlement of structures founded on sands is usually estimated from field test data as described in Section 2.6.5.

Permeability tests can be made in the laboratory on undisturbed samples of clays and silts, or on sands or gravels which are compacted in cylindrical moulds to the same density as that in which they exist in their natural state (as determined from *in-situ* tests). However, the results of laboratory tests on a few samples from a vertical borehole are of rather doubtful value in assessing the representative permeability of the soil for calculating the quantity of water to be pumped from a foundation excavation or the rate of settlement of large foundations. It is preferable to determine the permeability of the soil on a given site by means of tests in boreholes or field pumping tests as described in Section 1.6.3.

Chemical analyses of soils and ground water are required to assess the possibility of deterioration of buried steel and concrete foundation structures. In the case of steel structures such as permanent sheet piling or steel bearing piles it is usually sufficient to determine the pH value and chloride content of the soil and ground water. For concrete structures the sulphate content and pH value are normally required. Although the pH value, which is a measure of the degree of acidity or alkalinity of the soil or ground water, cannot be used directly to determine the nature or amount of acid or alkaline material present, it is a useful index in considering if further information is required to decide on the precautions to be taken in protecting buried concrete structures. For example, a low pH value indicates acid conditions, which might result from naturally occurring matter in the soil or which might be due to industrial wastes dumped on the site. In the latter case, detailed chemical analyses would be needed to determine the nature of the substances present, to assess the health risks to construction operatives and in the long term to the occupants of the site, and to assess their potential aggressiveness towards concrete. A full discussion on the subject of chemical attack on foundation structures, including the procedure for sampling, is given in Chapter 13.

1.12 Laboratory tests on rock specimens obtained by rotary core drilling

For foundation engineering purposes, laboratory tests on rock cores obtained by rotary drilling are usually limited to point load tests (Section 1.5.4) and uniaxial compression tests. The results are used for purposes of strength classification and for empirical correlations to obtain the compressibility of a rock mass (Section 2.7). Failure in shear of a rock mass under spread foundation loading is unlikely to occur, hence triaxial testing to obtain shear strength parameters is not normally required.

Triaxial and consolidation tests can be made on specimens of completely or highly weathered rocks of soil-like consistency recovered in open-drive samplers. The compression tests should not be made on unconfined specimens, since these are likely to exhibit brittle failure, with consequent underestimate of the bearing capacity value. Triaxial compression tests should be made using high lateral pressures comparable with the vertical deviator stresses, so enabling Mohr's circles of stress to be drawn from which the cohesion and angle of shearing resistance can be derived.

Where laboratory compression tests on rock cores are required the procedure should be as follows:

(a) Examine all rock cores and the results of point load tests made in the field.
(b) Select cores representative of the weakest and strongest rocks for test.
(c) Examine the drilling records to ensure that there is no weaker rock from which there has been no core recovery (if the drilling technique has not been satisfactory the weakest rocks may have crumbled away). If this is the case selected holes should be redrilled with a more careful technique or a larger core barrel to ensure recovery of the weakest rocks. If this cannot be done the idea of laboratory testing should be abandoned.
(d) Make uniaxial compression tests on the rock cores.
(e) Make further point load tests as necessary to establish the relationship between point load and uniaxial compression strength.
(f) On the basis of the test results classify the rock in terms of strength (see Section 1.1).

The references describing the procedure for compression tests on rock specimens and for a number of other tests of relevance to foundation engineering are listed in BS 5930.

1.13 The foundation engineering report

The engineering report on a foundation investigation is a consideration of all available data from boreholes, trial pits, site observations, historical records, and laboratory tests. Eurocode 7 requires the preparation of a geotechnical report and its submission to the owner/client to be part of the foundation design process. Most reports follow a fairly stereotyped pattern under the following headings.

1.13.1 Introduction

This should tell the reader for whom the investigation was undertaken, the reason for the investigation, how (briefly) the work was carried out, and the time of year the job was done. It should state the terms of reference, for example whether the investigation was merely to obtain a limited amount of factual data for assessment by the design engineer, or whether a full investigation was required with boring, laboratory tests, and an analysis of the results to consider possible methods of foundation design and construction and to calculate the allowable bearing pressures. If the scope of the investigation has been limited, on the grounds of cost or for other reasons, to such an extent that the engineer regards it as inadequate, he should state the reasons for this limitation in the introduction or at another appropriate point in the report. If, because of such limitations, the soil conditions are subsequently found to be different from those inferred, and the cost of the work is thereby increased, the engineer cannot be held to blame. However, he will be regarded as negligent if he bases his conclusions on inadequate data without qualifying his report.

It is sometimes the practice for a site investigation report to be divided into two parts: (1) a factual report, and (2) an interpretative report. Part 1 includes all the factual data from the desk study, field investigation, and laboratory testing. Part 2 comprises the interpretation of the data by an engineering geologist or by the engineer or both of these. Because much of Part 2 is an expression of opinion, it is not unusual for this document to be withheld from contractors tendering for the construction work. Opinions on such matters as excavation difficulties, quantities of ground water to be pumped, and the installation of piles often have to be made necessarily on limited data, and when difficult conditions are encountered at the construction stage these give good grounds for a claim by the contractor.

An experienced contractor has the capability to make his own judgement on these matters from the factual information in a sufficiently detailed Part 1 report, or can obtain the necessary specialist advice. Therefore while the author can appreciate the arguments of those who advocate supplying both factual and interpretative reports to tendering contractors, he believes that as a general rule it is in the best interests of the employing authority to supply only the former. An exception to this might be a foundation design specifically adopted to accommodate certain inferred ground conditions when it would be reasonable to inform tenderers of the reasons for the particular design.

1.13.2 General description of the site

This part of the report should describe the general configuration and surface features of the site, noting the presence of any trees, hedges, old buildings, cellars, quarries, mine shafts, marshy ground, ponds, watercourses, filled areas, roads, and tracks. Any useful information derived from historical records on previous usage of the site should be described, and other observations should cover such factors as flooding, sea or wind erosion, subsidence, earthquakes, or slope instability; any nearby buildings showing signs of settlement cracking should be noted.

1.13.3 General geology of the area

Notes should be given on the geology of the site, comparing published information on maps, memoirs, etc., with conditions found in the boreholes. Attention should be drawn to any known faults, quarries, springs, swallow holes, mines or shafts, or other features which will have a bearing on the foundation works.

1.13.4 Description of soil conditions found in boreholes (and trial pits)

This is a general description of the soil conditions with reference to the configuration of the ground and variations in level of the various strata and the ground-water table. A detailed description is not required. The written matter should not be a mere catalogue of the borehole records, for the reader of the report can get a much clearer picture of these by studying the records himself. If a number of boreholes have been sunk on a site it is a good plan to draw one or more sections through the site to show the variations and level of particular strata which may be of significance in the engineering problem (Fig. 1.14). A single drawing is better than pages of written matter. However, there can be risks in including a diagram of this type in a site investigation report. It is not practicable to

include detailed information on the profile such as the presence locally of obstructions or boulders, the variations in ground-water levels in successive soil strata and the extent and variability of weathering in rock formations. Costly mistakes can then be made if the lazy design engineer, or a contractor's estimator pressed for time, bases engineering judgement only on the diagram without a proper study of the borehole records.

1.13.5 Laboratory test results

A long description of the test results should not be given. The descriptive matter should be limited to a brief mention of the various types of tests which were made and attention drawn to any results which are unusual or of particular significance. For details of the results the reader should be referred to a table of results with charts and diagrams of such tests as particle size analysis, triaxial compression (Mohr's circles of stress), and consolidation tests (pressure–voids ratio curves).

The test procedure should be described only in the case of non-standard tests specially devised for the investigation as required by Eurocode 7.

1.13.6 Discussion of results of investigation in relation to foundation design and construction

This is the heart of the report and the writer should endeavour to discuss the problem clearly and concisely without 'ifs' and 'buts'. For readability this section of the

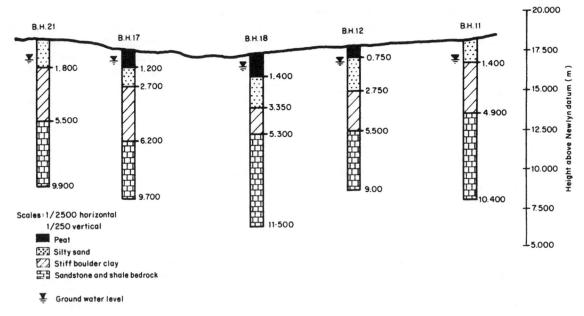

Fig. 1.14 Typical section through boreholes.

report should be broken down into a number of subheadings. First, under 'general', a description is given of the main structures and the related loadings which are to be considered, together with a general assessment of the ground conditions and the types of foundation which could be adopted, e.g. strip foundations, rafts, or piles. The remainder of the subheadings can refer either to particular structures (e.g. in the case of an electricity generating station, the boiler house, turbine house, coal or ash handling plants, switchgear, and circulating water culverts, all of which have different foundation characteristics requiring separate consideration) or they can refer to possible types of foundation design for any individual structure or structures.

The writer should come straight to the point. The reader's time should not be wasted in learning about obviously unsuitable foundation arrangements. In the case of strip foundations the required foundation depth should be stated, then the allowable bearing pressures, and then the settlements to be expected with these pressures. The advantage gained by going deeper, so enabling higher bearing pressures to be used or settlements to be reduced, should be considered.

In the case of piled foundations the writer should give the bearing stratum to which piles should be driven, the required or likely penetration of piles into this stratum, the working loads to be adopted per pile or per group of piles, and the settlements likely to occur in the individual pile or pile groups. Possible difficulties in driving or boring should be noted and any possible detrimental effects on adjacent structures should be pointed out.

In writing this section of the report care should be taken to avoid wishful thinking based on preconceived ideas on the foundation design. The problem should be studied without prejudice. For example, test results which appear to be too low should not be lightly discarded because they do not fit in with preconceived ideas on bearing pressures. The reasons why the results are low should be studied. Only if it can be established that they are due, say, to sample disturbance or are too few in number to have any practical significance to the problem as a whole can low results be neglected. Similarly, any borehole data which are unfavourable to the general ideas on foundation design must not be pushed on one side. If the results of a particular boring are unlike all the others in the vicinity, so upsetting a tidy arrangement of the foundations, the reasons for the discrepancy should be investigated. In cases of doubt a confirmatory boring or borings or check tests should be made. If it is demonstrated that the peculiar soil conditions do in fact exist the foundation design should take them into account.

The recommendations for foundation design must be based on the facts stated in the report, i.e. on the borehole records and test data. They must not be based on conjecture.

It is desirable that the engineer in charge of the site investigation and the writer of the report should work in close liaison with the designer of the project at all stages of the work. The effects of total and differential foundation settlements on the structure can be assessed and the appropriate measures recommended in the report.

Eurocode 7 requires the report to include recommendations for safety measures during construction, for monitoring behaviour during and after construction, and for post-construction maintenance.

1.13.7 Conclusion

If the preceding 'discussion' section of the report is lengthy or involved it may be convenient to summarize the main findings in itemized form. This is of help to the busy engineer who may not have time to read through pages of discussion. Alternatively, the report may commence with a brief summary of the investigation procedure and the main conclusions which have been drawn from it.

The last stages are the final typing and checking of the report, printing the drawings, and assembling and binding the whole. A neatly printed and bound report with good clear drawings free of typing and draughtsman's errors reflects the care with which the whole investigation has been done. Slipshod writing and careless typing and drawing may lead the client to think that the whole investigation has been carried out in a similar manner, but bulky report covers and over-elaborate presentation are not required.

1.14 Foundation properties of soil types

In the following pages, some notes are given on the engineering properties of various soils and rocks, with special reference to their bearing capacity and behaviour during construction and under foundation loading. Reference to BS 5930 should be made for visual and geological descriptions of these materials.

1.14.1 Non-cohesive soils

Gravels in the form of alluvial deposits are usually mixed with sands to a greater or lesser degree. Examples of the range of particle-size distribution are the beach gravels of the south coast of England which contain little or no sand. The sandy gravels which are widespread in the Thames Valley may contain 60 per cent or more of sand.

Gravels and sandy gravels in a medium-dense or denser state have a high bearing capacity and low compressibility. Compact gravelly soils give rise to difficulty in driving piles through them. If deep penetration is required into gravel strata it is usually necessary to adopt steel piles which have a higher penetrating ability than concrete or timber members.

Sandy gravels in a damp state but above the water table have some cohesion and can therefore be excavated to stand at very steep slopes provided that they are protected from erosion by flowing water. Loose gravels without sand binder are unstable in the slopes of excavations and require to be cut back to their angle of repose of about 30 to 35°.

Heavy pumping is required if deep excavations in open gravels are made below the water table, but the water table in sandy gravel can be lowered by well points or deep wells with only moderate pumping. As an alternative to providing large-capacity pumping plant the permeability of slightly sandy or clean gravels can be substantially reduced by injecting cement, clay slurries, or chemicals (see Section 11.3.7).

Erosion or solution of fine material from the interstices of gravel deposits can result in a very permeable and unstable formation. Open gravels caused by solution are sometimes found in the alluvial deposits derived from limestone formations.

Sandy soils have bearing capacity and compressibility characteristics similar to gravels, although very loosely deposited sands (e.g. dune sands) have a moderately high compressibility requiring correspondingly low bearing pressures in order to avoid excessive settlement of foundations.

Dense sands and cemented sands have a high resistance to the driving of piles and steel piles are required if deep penetrations are necessary.

Sands in their naturally deposited state above the water table are usually damp or cemented to a varying degree and thus will stand at a steep slope in excavations. However, support by timbering or sheet piling is necessary in deep and narrow excavations where a sudden collapse – caused by drying out of the sand or vibrations – might endanger workmen.

Excavation in sands below the water table will result in slumping of the sides or 'boiling' of the bottom, unless a properly designed ground-water lowering system is used. This instability, which is also known as the 'quick' or 'running sand' condition, is due to the erosive action of water flowing towards the excavation. By providing a ground-water lowering system to draw water away from the excavation towards filter wells or wellpoints a condition of high stability can be achieved (Section 11.3). In particular circumstances it may be necessary to stabilize the sands by the injection of chemicals (Section 11.3.7), or excavation under compressed air in caissons may be required.

Loosely deposited sands are sensitive to the effects of vibrations which induce a closer state of packing of the particles. Therefore, special consideration should be given to the design of machinery foundations on loose to medium-dense sands, and it is necessary to take precautions against the settlement of existing structures due to vibrations arising from such construction operations as blasting or pile driving (Section 3.3). Ground vibrations from earthquakes can cause large-scale liquefaction and subsidence of loose to medium-dense sands.

In some arid parts of the world the structure of loose deposits of sand is liable to collapse upon wetting, with consequent serious settlement of structures founded on these deposits. Wetting may be due to fracture of drains or leaking water pipes. Collapsing sands are found in some parts of South Africa, Zimbabwe, and Angola (Section 3.2). Sand deposits can be formed as a result of weathering and breakdown of calcareous formations. Examples of these are limestone sands which are found on the coasts and islands of the Mediterranean Sea; shelly sands and coral sands which are found on the coasts and islands of the Red Sea, the Arabian Gulf and the Pacific Ocean, and on the south-eastern seaboard of the USA, and the southern and western coasts of Australia. Other deposits include gypsum sands which are found in Iraq and Arabian Gulf territories. These deposits which are formed from weathering are nearly always in a loose state except at the surface where they may be weakly cemented by silt or salt spray. It should not be assumed that the relationships between SPT or CPT values and the angle of shearing resistance and deformation moduli of sands (see Sections 2.3.2 and 2.6.5) will apply to coralliferous or gypsum sands. These relationships were established for silica sands which have much higher particle strengths than the calcareous materials. Low foundation bearing pressures are required unless the loose deposits can be compacted by vibration or by the dropping weight method (Sections 11.6.4 and 11.6.5). These processes may be ineffective if the loose layers are interbedded with cemented material which cannot be broken down by vibration or impact. Piling problems can occur in very deep deposits of coralliferous shelly sands where isostatic changes in sea-level have resulted in successive layers of cemented sands interbedded with loose uncemented material. The cemented layers can cause difficulty in the penetration of piles, but after they have been crushed they provide very little resistance in skin friction or end-bearing to the loaded piles. Gerwick[1.20] quoted an extreme case of a pile driven to 60 m below the sea bed into calcareous sand where the force required to extract the pile was measured at little more than its weight.

Sands loosely deposited in the river flood plains cause problems to bridge foundations because of their susceptibility to deep erosion at times of major floods. Hinch *et al.*[1.21] described the erosion studies for the foundations of overhead power cable towers crossing the Jamuna River in Bangladesh where deposits of sands and gravels extend to depths of more than 100 m. A maximum scour depth of 60 m below bank level was predicted for the bank-full flood discharge of 70 700 m^3/s.

Distinct from these products of weathering are the cemented calcareous sands or sandstones which are formed by saline and lime-rich waters being drawn up by temperature effects and evaporation in the surface layers to form a hard crust. These soils or weak rocks are known by their geological classification of 'gypcrust' or 'salt crust' depending on the cementing medium. 'Silicretes' are formed by the evaporation of lime-rich waters in sandy sediments or sandstones, and are known by local names in various parts of the world, such as 'gatch' (Iran and Kuwait) and 'kurkar' (Israel and Jordan). The deposits occur widely in Australia where they are known as 'limestone rubble'. A feature of their formation is the irregularity in thickness and distribution of the hardened crust. It may exist in several distinct layers of varying thickness separated by loose sands or soft clay, or in irregular masses of varying degrees of cementation. Thus it is difficult to design foundations to take full advantage of the high bearing capacity of the cemented material. Disturbance of the cemented sands by excavating machinery, construction traffic, or flowing water results in rapid breakdown to a material having the texture of a sandy silt which is highly unstable when wet. Cemented sands are highly abrasive to excavation machinery.

1.14.2 Cohesive soils

The foundation characteristics of cohesive soils vary widely with their geological formation, moisture content, and mineral composition. It is impossible in this chapter to cover all the types and combinations which exist in nature, and the following notes are restricted to the characteristics of some of the well-known types.

Glacial till is generally a stiff to hard clay. Because of its heavy overconsolidation in glacial times only small consolidation settlements will occur under heavy foundation bearing pressures. Some tills are highly variable containing lenses (often water-bearing) of gravels, sands, and silts. In such conditions, foundation design must take account of the variable bearing capacity and compressibility at any particular locality. When carrying out deep excavations in variable glacial deposits, precautions should be taken against inrushes from water-bearing pockets. Excavations in stiff to hard glacial tills will stand vertically without support for long periods (Fig. 9.1), but slips may occur if steep-sided excavations are made on

sloping sites with a history of instability due to the presence of weak slicken-sided fissures or failure planes in the clay. The presence of random boulders or pockets of large gravel and cobbles can cause difficulties in driving sheet piles or bearing piles into glacial till. Another type of glacial deposit is *fluvio-glacial* or *varved clay*, which comprises layers of silty clay separated by thinner layers of sand or silt. These intervening layers are often water-bearing, which causes difficulties in 'bleeding' of sand or silt into excavations. Varved clays are usually softer in consistency and more compressible than boulder clays. Driving piles into varved clays may weaken the strength of the clay layer to that of a soft slurry. Also, where varved clays are bordering a lake or river, fluctuations in the water level may be communicated to the sand layers with a detrimental effect on their bearing capacity. For these reasons varved clays are generally held to be troublesome soils in foundation engineering.

Stiff-fissured clays such as London Clay, Barton Clay (in Hampshire), the Lias Clays of the Midlands, and the Weald and Gault Clays of south-eastern England have a relatively high bearing capacity below their softened weathered surface. Also, since they are overconsolidated clays, they have a moderate to low compressibility. They are highly plastic clays and heavy structures founded on them show slow settlement over a very long period of years. Stiff-fissured clays show marked volume changes with varying moisture content. Thus foundations need to be taken down to a depth where there will be little or no appreciable movement resulting from swelling and shrinkage of the clay in alternating wet and dry seasons (see Section 3.1.1). For the same reason it is necessary to avoid accumulation of water at the bottom of excavations in order to avoid swelling and softening of the soil. Fissuring in these soils can cause a wide variation in shear strength determined by laboratory tests on samples taken by drive tubes, due to random distribution of fissures and their partial opening during sampling. Thus it is difficult to assess the results when assessing bearing capacity (Section 4.1.2).

The fissured structure of these clays causes difficulties, mainly unpredictable, in the stability of slopes of excavations (Section 9.3), the stability of the walls of unlined holes sunk by mechanical boring methods for deep piers or piles (Section 8.14) and in the design of timbering or sheet piling to excavations (Section 9.6.2). Clays having similar characteristics to the British stiff fissured clays include the Fort Union 'shale' of Montana and the Bearpaw 'shale' of Saskatchewan. Stiff-fissured clay also occurs in northern France, Denmark, Germany, and Trinidad.

Tropical red clays are principally residual soils resulting from physical and chemical weathering of igneous rocks. They are widespread in India, Africa, South America, Hawaii, the West Indies, and Far Eastern countries. They are usually of low plasticity clays with a relatively high bearing capacity and low compressibility. However, in certain tropical conditions, leaching of the clays can occur at shallow depths, leaving a porous material with a fairly high compressibility.

In an extensive study of the formation and engineering characteristics of tropical residual soils the Engineering Group of the Geological Society of London[1.22] noted the following features of these soils which are significant to engineering problems. These are as follows:

(a) The soils may have a hard 'duricrust' capping overlying weaker material. The capping may be up to 10 m thick.

(b) Residual clay horizons may be present within saprolites.

(c) Variations occur in depth and lateral extent of weathering. Fresh rock outcrops may give way to troughs of weathering 15–20 m deep over horizontal distances of 100–200 m.

(d) The soils are often only partly saturated. The bonding between particles can be broken down by the intrusion of water or by mechanical disturbance such as by earth-moving machinery.

(e) Site investigation techniques and interpretation of soil test results can be different from those used for saturated soils in temperate climates. Soil sampling can be difficult because of the brittleness of the soils and the introduction of drilling water can weaken the soil structure.

(f) Conventional relationships between moisture content, plasticity index, and properties such as strength or stiffness or relationships with the standard penetration test do not necessarily apply to tropical residual soils.

(g) The important characteristic of the tropical clays is the yield stress beyond which their behaviour under compression loading changes from elastic to plastic. The yield stress can be determined by laboratory tests on large block samples, and in the field by large plate loading tests or pressuremeter tests. Accurate geological classification is essential for correlation with experience of engineering behaviour in similar conditions.

Laterite is a term given to a ferruginous soil of clayey texture, which has a concretionary appearance. It is essentially a product of tropical weathering and occurs widely in Central and South America, West and Central Africa, India, Malaysia, the East Indies, and Northern Australia.[1.23] Laterites are characteristically reddish-brown or yellow in colour. They exist in the form of a stiff to hard crust 6 m or more thick, overlying rather softer clayey materials followed by the parent rock. Laterites have a high bearing capacity and low compressibility. They do not present any difficult foundation engineering problems.

Tropical black clays are also developed on igneous rocks, examples being the 'black cotton soils' of the Sudan and Kenya, the 'vlei' soils of southern Zimbabwe, and the 'adobe' clays of the south-western USA.[1.24] Black clays are also found in India, Nigeria, and Australia. They are generally found in poorly drained topography. Unlike the tropical red clays, black clays are very troublesome in foundation engineering in that they show marked volume changes with changes in moisture content, and because of their poor drainage characteristics they become impassable to construction traffic in the wet season. Because these clays exist in countries where there are marked wet and dry seasons, the soil movements brought about by alternate wetting and drying are severe and extend to considerable depths. In the Sudan, seasonal swelling and shrinkage occurs to depths of 4–5 m. In many cases it has been found necessary to construct even light buildings on piled foundations to get below the zones of soil movement.

Saline calcareous clays are widely distributed in the Near and Middle East. They are found in the Mesopotamian plain of Iraq, the coastal plains of the Levant and south-west Iran, the coast of North Africa, the islands of the Mediterranean, the limestone plateaux of Jordan, and in Utah and Nevada, USA.[1.25] These soils were formed by the deposition of clay minerals in saline or lime-rich waters. The deposits are augmented by wind-blown sand and dust. The profile of calcareous silty clays is similar throughout the arid and semi-arid countries of the Near and Middle East. It comprises a surface crust about 2 m thick of hard to stiff desiccated clay overlying soft moist clay. The surface crust is not softened to any appreciable depth by the winter rains.

The stiff crust has adequate bearing capacity to support light structures, but heavy structures requiring wide foundations which transmit pressures to the underlying soft and compressible layers may suffer serious settlement unless supported by piles driven to less compressible strata. Calcareous clays show marked volume changes with varying moisture content, and where there are marked seasonal changes, as in the wet winter and dry summer of the countries bordering the Mediterranean, the soil movements extend to a depth of 5 m or more below ground level, and special precautions in foundation design are required. In regions where there are no marked differences in seasonal rainfall, as in southern Iraq, soil movement is not a serious problem. In some regions the stiff crust is a weakly cemented agglomeration of sand or gravel-size particles of clayey material, probably resulting from deposition by winds. These soils may suffer collapse on inundation combined with foundation loading.

Alluvial (including marine) clays are geologically recent materials formed by the deposition of silty and clayey material in river valleys, estuaries, and on the bed of the sea. They are 'normally consolidated', i.e. they have consolidated under their own weight and have not been subjected in their geological history to an overconsolidation load as in the case of glacial till and stiff-fissured clays. Since they are normally consolidated they show a progressive increase in shear strength with increasing depth ranging from very soft near the ground surface to firm or stiff at depth. The normally consolidated clays of the Chao Praya plain in Thailand extend to depths of 200–300 m.[1.26] Atmospheric drying and the effects of vegetation produce a stiff surface crust on alluvial clays. The thickness of this crust is generally 1–1.2 m in Great Britain, but it is likely to be much greater and liable to vary erratically in thickness in arid climates. Some regions show several layers of desiccation separated by soft, normally consolidated clayey layers. Moderately high bearing pressures, with little or no accompanying settlement, can be adopted for narrow foundations in the surface crust which do not transmit stresses to the underlying soft and highly compressible deposits. In the case of wide or deep foundations it is necessary to adopt very low bearing pressures, or to use a special type known as the buoyancy raft (Section 5.3), or to support the structure on piles driven through the soft and firm alluvial clays to a satisfactory bearing stratum.

Alluvial clays, especially marine clays, are 'sensitive' to disturbance, i.e. if they are disturbed in sampling or in construction operations they show a marked loss in shear strength. The sensitivity* can range from 2 or 3 in the case of estuarine clays of the Thames and the Firth of Forth to as much as 150 in the post-glacial clays of eastern Canada. The marine clays of Norway and Sweden are also highly sensitive due to leaching of salts from the pore water of the soil by the percolation of fresh water, leaving an open lattice structure which is readily broken down on disturbance.

Excavations below the dried-out surface crust require support by timbering or sheetpiling; open excavations require to be cut back to shallow slopes to avoid massive rotational slips. Excavations in soft clays exceeding a certain depth–width ratio are subject to failure by heaving of the bottom or appreciable inward yielding of the side supports (Sections 9.7 and 9.8).

As in the case of stiff-fissured clays, precautions must be taken against the effects on foundations of seasonal

swelling and shrinkage and the drying action of the roots of vegetation (Section 3.1.1).

Alluvial clays are frequently varved or laminated clays interbedded with layers of peat, sand, and silt as in the Fens of East Anglia, and in major river deltas.

Silts occur as glacial or alluvial deposits, or as wind-blown deposits. Examples of the latter are the 'brickearth' of south-eastern England, and 'loess' which is found in widespread tracts in the Midwest and north-western USA, China, India, Russia, and Israel. Glacial and alluvial silts are generally water bearing and soft in consistency. They are among the most troublesome soils in excavation work, since they are readily susceptible to slumping and 'boiling'. Being retentive of water they cannot readily be dewatered by conventional ground-water lowering systems. Silts are liable to frost heave.

Brickearths are generally firm to stiff and do not normally present any difficult problems in foundation work. Similarly, loess soils are slightly cemented and have a high bearing capacity. However, they are liable to collapse of their structure on wetting which may occur as a result of flooding or even broken water mains (Section 3.2). Loess soils can stand with vertical faces to a great height provided they are protected from erosion by flowing water.

Peat consists of dead and fossilized organic matter. It is found in many parts of the world. Extensive deposits occur in northern Europe, North America, and the former USSR, where it is overlain by living vegetable matter in 'muskeg' terrain. Peat is a permeable fibrous material and is highly compressible. The usual procedure is to take foundations of structures below peat to less compressible strata unless heavy settlements can be tolerated. It has the characteristic of undergoing large secondary consolidation settlements which can continue with little diminution over a very long period of years. Berry et al.[1.27] described the detailed geotechnical and structural investigations at two housing estates in north-west England where a number of blocks of two-storey dwellings had undergone large settlements due to compression and consolidation of a near-surface stratum of peat varying in thickness from 1.5 to 2.6 m. Total settlements of the buildings ranged from 60 to 258 mm over periods of about 25 years. On one of the estates the maximum settlement of 258 mm had occurred at the centre of a two-storey block and it was estimated that the secondary consolidation would continue at the rate of about 2 mm/year for the following 10 years and thereafter at only a slightly diminishing rate. The average differential settlement between the centre and corner of three blocks was 74 mm in 25 years. Another undesirable characteristic of peat is its 'wasting'. The ground surface of the peat in the Fen districts of East Anglia is slowly sinking through the years due to consolidation of the

* Sensitivity = $\dfrac{\text{Undisturbed shear strength}}{\text{Remoulded shear strength}}$

ground under its own weight (accelerated by drainage), to the fibrous peat blowing away in the wind, and to accidental or deliberate burning of the peat. Thus if foundations are taken below peat layers by means of piles or piers the surrounding ground surface sinks relative to the structure with the result that over a long period of years the foundations become exposed. Many instances of this can be seen in the Fens. Peats may contain organic acids which are aggressive to concrete (Section 13.5.4).

1.15 Foundation properties of rock types

The engineer is not usually concerned with the physical properties of the material forming the intact unweathered rock since its strength and compressibility are normally adequate for most foundation structures. However, it is essential to investigate the properties of the rock mass since these govern foundation behaviour and are relevant to problems of foundation construction. The various properties of the rock mass which should be investigated are listed below.

1.15.1 Weathering

Weathering of rocks results from the action of environmental conditions such as rainfall and temperature variation. An important influence on weathering was the deep freezing and thawing which occurred in the various glacial periods. Weathering also occurs as a result of chemical action or leaching of ground water seeping through the rocks. The effect of weathering is to degrade the physical characteristics of the intact rock, reducing it, in the ultimate state, to a mass of uncemented mineral particles or to a material of soft clay consistency. The products of weathering may be removed by solution or erosion, leaving open fissures or cavities. These may later be filled by soil washed in from the overlying superficial deposits. Weathering can occur to a highly irregular extent in a rock mass usually following joints or other structural weaknesses.

In tropical countries, weathering can extend to great depths. The *granites* of Hong Kong are weathered to depths of 60 m or more, the strong rock being degraded to a compressible porous mass of quartz particles in a clayey matrix of decomposed feldspar and biotite.[1.28]

Sandstones weather to a condition of uncemented granular particles. Meigh[1.29] reported weathering of the Bunter Sandstones in north-eastern England to a depth of 20 m below rockhead. Care is necessary in exploratory drilling in sandstones since cable percussion or small-diameter core drilling in the partly weathered rock produces loose debris which can give a false impression of the bearing characteristics of the rock. In Singapore the interbedded

sandstones, siltstones, and mudstones of the Jurong Formation have weathered to depths of more than 100 m below the surface.[1.30]

Weathering of the *mudstones* and *siltstones* of the Keuper Marls of the Midlands and north-eastern England can extend to depths of 60 m below rockhead. The weathering can degrade the rock to an uncemented silt. Leaching of calcareous materials in the marl leaves numerous cavities and open fissures which can increase the compressibility and permeability of the mass.[1.29] *Shales* or *shaly slates* are weathered by deep frost action to a clayey consistency or to a crumbly mass of crushed flat particles.

Some mudstones and shales contain pyrite. In a weathering environment removal of the pyrite causes chemical and mineralogical changes leading to degradation and weakening of the rock mass. A reduction in ground-water level or exposure in excavations can cause oxidation of the pyrite and the production of sulphuric acid. If calcite is present the reaction with sulphuric acid could cause the growth of gypsum crystals in fissures and laminations with consequent massive expansion of the rock mass. The reactions are partly chemical and partly bacterial.[1.31] They can be accelerated by warmth in the presence of moisture, e.g. beneath a heated building as described in Section 13.5.4.

Limestones (other than chalk) are generally strong and massive and do not show the effects of weathering other than those of solution which can form very large caverns or wide drift-filled fissures. Those karstic conditions occur extensively in Greece, Malaysia, and China. The features are sometimes indicated by depressions in the ground surface, but where the limestones are overlain by younger rocks or deep drift deposits it may be very difficult to explore the extent of cavities. Geophysical methods have been attempted with mixed success. On a site in Derbyshire[1.32] solution had taken place in numerous vertical joints in the limestone producing a karstic formation of pinnacles and promontories of strong limestone wholly or partially surrounded by wide sand and clay-filled joints, the infilling extending to depths of 15 m or more below the surface. These conditions are best investigated by trenching supplemented by probing with small-diameter non-coring percussion or rotary drilling.

Material infilling solution cavities in limestone formations are frequently in a loose condition. Spread or piled foundations bearing on these materials cannot be adopted because of the risks of slumping after periods of heavy rainfall. Foundations must be designed to bridge across the infilling materials.

Chalk is a form of limestone which is frequently found in a deeply weathered condition due to the effects of freezing and thawing in the glacial periods. The effect of freezing was to produce a mass of crumbly frost-shattered rock weathered to a soft or firm clayey consistency adjacent to joints and fissures. The weathering effects can extend to depths of 15 m. The heavily weathered chalks near the surface are prone to erosion and creep on hillside slopes, producing a transported very weak rock known as *solifluction chalk*, *combe deposits*, or *combe rock*.

Chalk has a porous cellular structure and mechanical disturbance such as penetration by boring tools releases water from the cells to form a soft putty-like material around the borehole. Similarly, driving piles into chalk produces a soft crushed zone around the pile. Samples of weak chalk taken by driving an open-end tube are usually crushed and softened and unsuitable for testing. The degree to which these effects occur depends on the particular geological formation (in Britain the Upper, Middle and Lower Chalk), the stratigraphic level within that formation and the weathering grade.[1.33] Hence it has proved to be impracticable to establish, for general use, simple and reliable correlations between the standard penetration test N-value and the compressibility of chalk

or the skin friction resistance of piles.[1.34] These aspects are discussed further in Sections 2.4, 2.7, and 7.12.

Chalk formations are subject to solution by water seeping through joints forming 'pipes' or cavities known variously as swallow holes, sink holes, or dene holes, similar to those described in limestone formations (see Fig. 1.15).

From the results of plate-bearing tests, pressuremeter tests or the observations of settlement of full-scale structures, deformation modulus values have been assigned to each weathering grade for various types of rock – as listed in Section 2.7. It is rarely possible to map the lateral and vertical extent of irregularly weathered rock formations, and the engineer should base allowable bearing pressures and estimate settlement on the assumptions of values for the weakest rocks at or below the design foundation level. It is usually uneconomical to attempt to use higher bearing pressures appropriate to stronger less-weathered rocks with the idea of stepping down the foundation level to get below layers or pockets of weaker material. This can lead

Fig. 1.15 Swallow holes in chalk formation filled with clay.

to deep and difficult excavation for which the contractor cannot quote firm prices since the required volume and depth of excavation will not be known until all the weak material has been removed.

1.15.2 Faulting

Faults may not be significant to foundation design since all movements along the fault plane may have ceased in geological times. However, faults are of significance in earthquake zones, and in districts subject to mining subsidence (see Section 3.5).

1.15.3 Jointing

Joints are fractures in the rock mass with little or no relative displacement of the rock on either side of the joint plane. The spacing and width of open joints is of significance to the compressibility of the rock in the stressed zone beneath foundations. Inclined joints dipping towards an excavation may be a cause of instability, particularly if the joints are filled with weak weathering products forming a plane of sliding (see Section 9.3.4). Similarly, steeply inclined clay-filled joints may result in failure of a foundation bearing on an irregular rock surface (see Section 2.3.6).

The spacing, inclination, and width of joint systems should be established by examination of rock cores, and preferably by geological mapping of the exposures in outcrops, pits, or deep shafts.

1.15.4 Strength of intact rock

The strength of intact rock can be related to the ultimate bearing capacity and compressibility of a jointed rock mass (see Section 1.12). The results of uniaxial compression and splitting tests are helpful in connection with the use of explosives or rock-boring tools for foundation excavations, and for possible uses of the rock as a constructional material.

References

1.1   Eurocode No. 7, *Geotechnics*, European Committee for Standardisation, published as a draft in 1989 by the British Standards Institution, London.

1.2   Rowe, P. W., The relevance of soil fabric to site investigation practice, *Géotechnique*, **22**, 193–300 (1972).

1.3   La Rochelle, P., Sarrailh, J., Tavenas, F., Roy, M. and Leroueil, S., Causes of sampling disturbance and design of a new sampler for sensitive soils, *Canadian Geotechnical Journal*, **18**(1), 52–66 (1981).

1.4   Chandler, R. J., Harwood, A. H. and Skinner, P. J., Sample disturbance in London Clay, *Géotechnique*, **42**(4), 577–585 (1992).

1.5   Bjerrum, L., Problems of soil mechanics and construction on soft clays and structurally unstable soils, General Report, in *Proceedings of the 8th International Conference on Soil Mechanics*, Moscow, 1973, **3**, 111–159.

1.6   Stroud, M. A., The standard penetration test in insensitive clays and soft rocks, *Proceedings of the European Symposium on Penetration Testing*, **2**, 367–375 (1975).

1.7   Cearns, P. J. and McKenzie, A., Application of dynamic cone penetrometer test in East Anglia, *Proceedings of the Symposium on Penetration Testing in the UK*, Thomas Telford, London, 1988, pp. 123–127.

1.8   Meigh, A. C., *Cone penetration testing, Methods and Interpretation*, CIRIA/Butterworths, 1987.

1.9   Burland, J. B. and Burbidge, M. C., Settlement of foundations on sand and gravel, *Proceedings of the Institution of Civil Engineers*, **78**(1), 1325–1381 (1981).

1.10   Mair, R. J. and Wood, D. M., *Pressuremeter Testing, Methods and Interpretation*, CIRIA/Butterworths, 1987.

1.11   Clarke, B. G. and Smith, A., A model specification for radial displacement measuring pressuremeters, *Ground Engineering*, Thomas Telford, London, March 1992, pp. 28–37.

1.12   Marsland, A., Large *in-situ* tests to measure the properties of stiff fissured clays, *Proceedings of the 1st Australian–New Zealand Conference on Geomechanics*, Melbourne, **1**, 180–189 (1971).

1.13   Marchetti, S., *In-situ* tests by flat dilatometer, *Journal of the Geotechnical Engineering Division*, ASCE, **106**(GT3), 299–321 (1980).

1.14   Point load test – suggested method for determining point load strength, *International Journal of Rock Mechanics and Mining Sciences*, **22**(2), 53–60 (1985).

1.15   Solomon, C. J. and Powrie, W., Assessment of health risks in the redevelopment of contaminated land, *Géotechnique*, **42**(1), 5–12 (1992).

1.16   *British Standards Institution Code of Practice for the Identification of Potentially Contaminated Land and its Development*, Draft for Development 175, BSI, London, 1988.

1.17   Olie, J. J., Van Ree, C. C. D. F. and Bremmer, C., *In-situ* measurement by chemoprobe of ground water from *in-situ* sanitation of versatic acid spill, *Géotechnique*, **42**(1), 13–21 (1992).

1.18   Bishop, A. W. and Henkel, D. J., *The Measurement of Soil Properties in the Triaxial Test*, Edward Arnold, London, 1957.

1.19   Bromhead, E. N., A simple ring shear apparatus, *Ground Engineering*, **12**(5) (1978).

1.20   Gerwick, B. C., *Construction of Offshore Structures*, John Wiley, 1986.

1.21   Hinch, L. W., McDowall, D. M. and Rowe, P. W., Jamuna River 230 kV crossing, design of foundations, *Proceedings of the Institution of Civil Engineers*, **76**(1), 927–949 (1984).

1.22   Tropical residual soils, Report of the Working Party, Engineering Group of the Geological Society of London, *Quarterly Journal of Engineering Geology*, **23**(1), 1–101 (1990).

1.23   Nixon, I. K. and Skipp, B. O., Airfield construction on overseas soils: laterite, *Proceedings of the Institution of Civil Engineers*, **8**, 253–275 (1957).

1.24   Clare, K. E., Airfield construction on overseas soils: tropical black clays, *Proceedings of the Institution of Civil Engineers*, **8**, 223–231 (1957).

1.25   Tomlinson, M. J., Airfield construction on overseas soils: saline calcareous soils, *Proceedings of the Institution of Civil Engineers*, **8**, 232–252 (1957).

1.26   Balasubramaniam, A. S., *Contributions in Geotechnical Engineering*, Asian Institute of Technology, Bangkok, 1991.

1.27   Berry, P. L., Illsley, D. and McKay, I. R., Settlement of two housing estates at St Annes due to consolidation of a near-surface peat stratum, *Proceedings of the Institution of Civil Engineers*, **77**(1), 111–136 (1985).

1.28   Lumb, P., The residual soils of Hong Kong, *Géotechnique*, **15**, 180–194 (1965).

1.29   Meigh, A. C., The Triassic rocks, with particular reference to predicted and observed performance of some major structures, *Géotechnique*, **26**, 393–451 (1976).

1.30   Poh, Kong Beng, Buttling, S. and Hwang, R., Some MRT experiences of the soils and geology of Singapore, *Proceedings of the Singapore Mass Rapid Transport Conference*, Singapore, 1987, pp. 177–191.

1.31   Hawkins, A. B. and Pinches, G. M., Cause and significance of heave at Llandough Hospital, Cardiff – a case history of ground floor heave due to gypsum growth, *Quarterly Journal of Engineering Geology*, **20**, 4–57 (1987).

1.32   Early, K. R. and Dyer, K. R., The use of a resistivity survey on a foundation site underlain by Karst dolomite, *Géotechnique*, **14**, 341–348 (1964).

1.33   Higginbottom, I. E., Engineering behaviour of chalk – a historical perspective, *Proceedings of the International Chalk Symposium*, Brighton, Thomas Telford, London 1990, pp. 5–10.

1.34   Lord, J. A., Design and construction in chalk, *Proceedings of the International Chalk Symposium*, Brighton, Thomas Telford, London 1990.

2 The General Principles of Foundation Design

2.1 Foundation types

The foundation of a structure is defined as that part of the structure in direct contact with the ground and which transmits the load of the structure to the ground.

Pad foundations are usually provided to support structural columns. They may consist of a simple circular, square, or rectangular slab of uniform thickness, or they may be stepped or haunched to distribute the load from a heavy column. Pad foundations to heavily loaded structural steel columns are sometimes provided with a steel grillage.

Various forms of pad foundations are shown in Fig. 2.1(*a*)–(*d*).

Strip foundations are normally provided for load-bearing walls (Fig. 2.2(*a*)), and for rows of columns which are spaced so closely that pad foundations would nearly touch

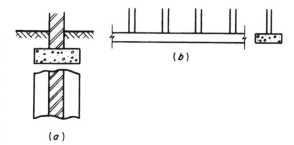

Fig. 2.2 Types of strip foundation. (*a*) Strip foundation to load-bearing wall. (*b*) Strip foundation to a row of close-spaced columns.

each other (Fig. 2.2(*b*)). In the latter case it is more economical to excavate and concrete a strip foundation than to work in a large number of individual pits. In fact, it is often thought to be more economical to provide a strip foundation whenever the distance between the adjacent square pads is less than the dimensions of the pads, and for ease in construction close-spaced pad foundations can be formed by inserting vertical joints in a continuous strip of concrete (see Section 4.2.6).

Wide strip foundations are necessary where the bearing capacity of the soils is low enough to necessitate a strip so wide that transverse bending occurs in the projecting portions of the foundation beam and reinforcement is required to prevent cracking.

Raft foundations are required on soils of low bearing capacity, or where structural columns or other loaded areas are so close in both directions that individual pad foundations would nearly touch each other. Raft foundations are useful in reducing differential settlement on variable soils or where there is a wide variation in loading between adjacent columns or other applied loads. Forms of raft foundations are shown in Fig. 2.3(*a*)–(*c*).

Bearing piles are required where the soil at normal

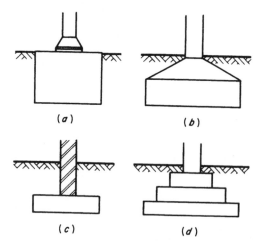

Fig. 2.1 Types of pad foundation. (*a*) Mass concrete for steel column. (*b*) Reinforced concrete with sloping upper face. (*c*) Plain reinforced concrete. (*d*) Stepped reinforced concrete.

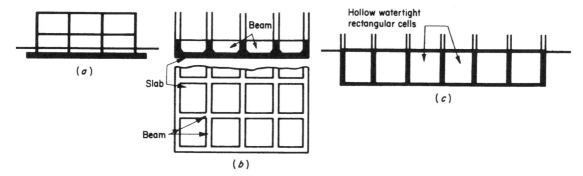

Fig. 2.3 Types of raft foundation. (*a*) Plain slab. (*b*) Slab and beam. (*c*) Cellular (or buoyancy) raft.

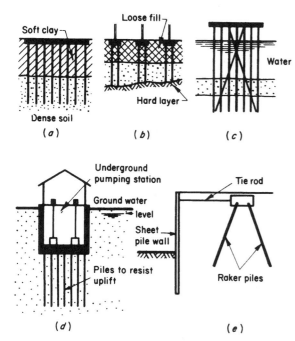

Fig. 2.4 Types of piled foundation.

foundation level cannot support ordinary pad, strip, or raft foundations (Fig. 2.4(*a*)) or where structures are sited on deep filling which is compressible and settling under its own weight (Fig 2.4(*b*)). Piled foundations are a convenient method of supporting structures built over water (Fig. 2.4(*c*)) or where uplift loads must be resisted (Fig. 2.4(*d*)). Inclined or raking piles are provided to resist lateral forces (Fig. 2.4(*e*)).

2.2 Foundation design procedures

2.2.1 General

At the time of preparing this edition the design procedure in Britain was the *permissible stress method* as followed by the British Standard Code of Practice: *Foundations* BS 8004: 1986. The same method is normal practice in USA and many other countries throughout the world. Using the permissible stress methods the actual dead loads of a structure and the most unfavourable combinations of imposed loads are assumed to be applied to the ground. The foundation design is assumed to be safe if the *permissible stress* of the soil or rock is not exceeded. The method recognizes the highly variable strength and stiffness properties of soil or rock, also the important influence of such matters as the level of the ground water and the rate of loading on strength and stiffness, and the possible uncertainty about the reliability of the analytical method used to calculate the failure conditions of the ground. To allow for these variations and uncertainties a safety factor is applied to the calculated failure stress or failure load applied to the ground to determine the allowable or design bearing pressure or loading. Because of the critical unfavourable effects of variations and uncertainties the recommended safety factors are quite high. In the case of BS 8004 a safety factor between 2 and 3 is generally adopted. It is essentially a safety factor applied to the material properties as measured in the laboratory or field, but the factor is applied globally to cover all the other uncertainties.

It is recognized when using the permissible stress method that a structure can fail due to excessive deformation caused by foundation settlement long before reaching a state of collapse due to failure in shear of the ground. In routine structures excessive total and differential settlement can often be avoided by choosing the safety factor at a sufficiently high value which has been shown by experience to avoid excessive settlement risks. In the case of heavy structures or those sensitive to differential settlement a separate calculation is made to determine the settlement. Again a realistic value of dead loading and the most favourable and unfavourable combinations of imposed loading are used in conjunction with actual (unfactored) values of the stiffness of the ground as measured by laboratory and field testing.

During the 1960s and 1970s structural codes in Europe and the UK were changing to limit state design methods. Engineering organizations in the member states of the European Community (EC) recognized the desirability of uniformity in structural and foundation codes and a drafting panel to produce a foundation code was set up under the auspices of the International Society of Soil Mechanics and Foundation Engineering. The panel produced a draft code in 1989 entitled Eurocode 7, *Geotechnics*. This code has not yet been ratified by the EC member states and it is possible that there may be some changes in the draft before an approved document is published. The comments in this and succeeding chapters will refer to the 1989 draft.

The basis of *limit state design* as applied to foundations is that a structure and its supporting foundation should not fail to satisfy the design performance requirements because of exceeding various limit states. *The ultimate limit state* is associated with safety, that is, the avoidance of the risk of collapse or any other form of failure which could endanger people. Design against the occurrence of the ultimate limit state is concerned with applying factors both to the applied loading (actions) and to the resistance of the ground (the material factors) so as to ensure that reaching the limit state is highly improbable.

Serviceability limit states are concerned with ensuring that deformations or deflections do not damage the appearance or reduce the useful life of a structure or cause damage to finishes, non-structural elements, or machinery or other installations in a structure.

Eurocode 7 also requires structures and their foundations to have sufficient *durability* to resist attack by substances in the ground or in the environment which could cause weakening and the risk of exceeding the ultimate and serviceability limit states.

In addition to a consideration of the general suitability of the ground and of the various geological and geotechnical factors set out in Sections 1.1 and 1.14 of this book, Eurocode 7 lists the following design situations which should be studied:

(a) Actions (e.g. loadings): their combinations and load cases.
(b) The nature of the environment within which the design is set including the following:
   (1) effects of scour, erosion, and excavation leading to changes in the geometry of the ground surface;
   (2) effects of chemical erosion;
   (3) effects of weathering;

(4) effects of freezing;
(5) variations in ground-water levels; including the effects of dewatering, possible flooding, failure of drainage systems, etc.;
(6) the presence of gases emerging from the ground;
(7) other effects of time and environment on the strength and other properties of materials, e.g. the effects of holes created by animal activities;
(8) earthquakes;
(9) subsidence due to mining or other causes;
(10) the tolerance of the structure to deformations;
(11) the effect of the new structure on existing structures or services.

The above items provide a useful checklist for the engineer using either permissible stress or limit state design methods.

The adoption of the limit state design procedure is advantageous in UK practice where building superstructures are designed to British Standard Codes of Practice, 8011, 'Structural concrete'; 5950, 'Structural steelwork in buildings'; and 5400, 'Bridges'. These are limit state codes and it is difficult to define the specific locations in the structure where the design procedure changes to BS 8004: *Foundations*, which is in terms of permissible stresses. In this situation the structural designer should define to the foundation engineer the loads which are to be carried by the ground. It should be clearly stated whether or not these are factored or unfactored loads, or dead loads combined with imposed loading reduced as permitted by British Standard CP 3.

2.2.2 Foundation design by calculation (Eurocode 7 procedure)

Eurocode 7 considers four methods of design. These are:

(1) By calculation
(2) By prescriptive measures
(3) From results of load tests or tests on experimental models
(4) By observation during construction

All the above are applicable both to limit state and permissible stress methods. The Eurocode methods will be described in some detail as they may be unfamiliar to many engineers. Differences with permissible stress methods will be noted where appropriate. A brief account only can be given of the Eurocode procedure. It is essential to study the draft code for a full understanding of all the relevant factors and qualifications.

Eurocode 7 refers to foundation loadings as *actions*. These can be *permanent* as in the case of weights of structures and installations, or *variable* as for imposed loading, or wind and snow loads. They can be *accidental*, e.g. vehicle impact or explosions.

Actions can vary *spatially*, e.g. self-weights are fixed (fixed actions), but imposed loads can vary in position (free actions). The *duration* of actions affects the response of the ground. It may cause strengthening such as the gain in strength of a clay by long-term loading, or weakening as in the case of excavation slopes in clay over the medium or long term. To allow for this Eurocode 7 introduces a classification related to the soil response and refers to *transient actions* (e.g. wind loads), *short-term actions* (e.g. construction loading), and *long-term actions*.

In order to allow for uncertainties in the calculation of the magnitude of actions or combinations of actions and their duration and spatial distribution, Eurocode 7 requires the design values of actions F_d to be used for the geotechnical design either to be assessed directly or to be derived from characteristic values F_k from the equation

$$F_d = F_k \times \gamma_F \qquad (2.1)$$

where γ_F is a partial safety factor. Provisional values of γ_F are shown in Table 2.1. General circulation of the draft code and experience of its use may require changes in these values. Hence they are shown in brackets to indicate that they are tentative. Eurocode 7 emphasizes groundwater level changes and drainage characteristics of the ground before and after construction as particular matters which influence the selection of design values. In the case of structures, characteristic values of actions are given in the relevant structural Eurocodes.

Favourable effects in permanent actions include such items as the buoyancy of a basement substructure. Eurocode 7 points out that the γ_F factors are lower than those used in structural design. This is because of the ability of statically indeterminate structures to redistribute the load on to the foundations. Factors as high as 1.5 are recommended for statically determinate structures, and for statically indeterminate structures but with a low degree of redistribution such as masts and towers.

Table 2.1 Partial factors on actions for conventional design situations for ultimate limit states

Action	Partial factor γ_F	
	Unfavourable effect	Favourable effect
Permanent actions including: Weight of supported structure	(1.0)	(0.95)
Soil, rock and water	(1.0)	(1.0)
Variable actions: Imposed and wind loads	(1.3)	(0)

Eurocode 7 recommends γ_F factors of unity for serviceability limit states for unfavourable variable loads, and zero for favourable variable loads. A factor of unity is recommended for accidental situations.

As noted in Section 2.2.1 the loadings (or actions) are not factored when using the *permissible stress* method. The maximum load to be carried by the ground is determined. This is the sum of the actual dead loading and the maximum credible imposed loading. Actions are resisted by the supporting ability of the ground which can be calculated from a knowledge of the geotechnical properties (referred to as material properties) to obtain design values. The partial factors are necessary to allow for uncertainties in the relation between the soil and rock properties as used in the calculation model and the properties as measured by laboratory or field tests and also uncertainties concerning the effects of brittleness, ductility, soil response to the duration of actions, and construction activities (including workmanship).

Design value of material properties, X_d, are required to be assessed directly or derived from *characteristic values* X_k from the equation

$$X_d = X_k/\gamma_m \qquad (2.2)$$

where γ_m is the partial safety factor for the material properties. Values are given in Table 2.2.

The tentative characters of the values are again denoted by brackets. Eurocode 7 provides guidance on the selection of characteristic values, defining them as those such that the probability of a more unfavourable value governing the occurrence of a limit state is not more than 5 per cent. This suggests that characteristic values should be obtained by a statistical analysis of a number of test results. However, Eurocode 7 tends to play down the statistical approach and recommends selection by noting the occurrence of particular weak layers which could influence the mechanism of failure or deformation. In contrast to structural codes, the engineer is quite unrestricted in his choice of design values and need not consider either characteristic values or the application of material factors if the design value is assessed correctly. As is the case with the permissible stress method, experienced judgement is essential to the selection process. The engineer follows the same process of considering all the available test data and the factors which govern the

Table 2.2 Partial factors on material properties for conventional design situations for ultimate limit states

Material property	Partial factor γ_m
Tan ϕ	(1.2–1.25)
c', C_u	(1.5–1.8)

selection of design parameters, no matter whether the limit state or permissible stress methods are used.

Eurocode 7 recommends a value for γ_m of unity for serviceability limit state calculations and for accidental situations. Care is needed in selecting characteristic values for the serviceability limit state where stiffness in related to strength. Later in this chapter stiffness in terms of an elastic or deformation modulus is shown to be related empirically to soil shear strength or rock unconfined compression strength by a simple multiplying factor. Hence if the soil strength is chosen conservatively, i.e. a low value, the stiffness will also be low which may give an unrealistically high estimate of foundation settlement.

Design values for *geometrical data* should be selected by reference to dimensional tolerances in the structural code or job specification. Alternatively, they can be allowed for when selecting the values of actions or material properties.

A *constraint* is defined in Eurocode 7 as an 'acceptable limiting value for a particular deformation in order to satisfy the ultimate or serviceability limit state of the structure'. Constraints are limit values of total and differential settlements, rotation, tilt, and distortion of structural members and claddings. These will be discussed in Section 2.6.2. It is evident that close liaison is essential between the designers of the superstructure and foundation in order to establish compatible constraints and to avoid the insistence by the superstructure designer of impracticable constraints such as zero settlement.

2.2.3 Foundation design by prescriptive measures

Prescriptive measures to avoid limit states are recommended in Eurocode 7 'where calculation models are not available or unnecessary'. These measures are assumed to include tables of allowable bearing pressures for various classifications of soil or rock such as those in the BS Code of Practice BS 8004: *Foundations*, and in municipal regulations in most countries of the world.

2.2.4 Design using load tests and experimental models

Loading tests, either full scale or reduced scale, on plates, piles, or on other components of a substructure are permitted by Eurocode 7 to justify a design provided that differences in the ground conditions between the test and prototype locations, and the effect of duration of loading and of scale, are taken into account. The importance of scale effects was noted in Section 1.4.5 (see Fig. 1.9). Loading tests are frequently used for the design of piles, the results being used either directly or as a check on calculated bearing capacity.

2.2.5 Design by the observational method

The observational method requires the behaviour of a structure to be monitored at intervals during construction to ensure that settlements or deflections of the substructure or superstructure comply with the constraints established before construction commences. The intervals should be sufficiently short to enable corrective measures to be taken with due allowance for time taken in recording and interpreting data from the monitoring instruments.

Although this method is frequently applied to the construction of embankments on soft ground, often with frustration and annoyance to contractors when their programme becomes uncertain or is delayed, the method has little application to structures such as bridges or buildings where design changes or delays in occupancy could have serious financial consequences.

2.2.6 The design report

Eurocode 7 requires the submission of a geotechnical report, similar in outline to that described in Section 1.13, to be part of the design process. Safety measures and requirements for maintenance and for monitoring behaviour during and after construction are items to be included in the report.

2.2.7 Stages in the design process

The stages in the process for foundation design are first to prepare a plan of the base of the structure showing the various column, wall, and distributed loadings. Dead and imposed loadings should be differentiated and any bending moments at the base of columns or walls should be noted. Second, the bearing characteristics of the ground as given by the site investigation should be studied. From this, tentative allowable bearing pressures can be allocated for the various strata below ground level. The third step is to determine the required foundation depth. This may be the minimum depth to get below surface layers of soil affected by seasonal temperature or moisture changes or by erosion. Structural requirement may involve founding at greater depths than the minimum required solely from soil considerations. For example, the building may have a basement, or there may be heating ducts or culverts below ground level. Having decided upon a minimum depth from either geotechnical or structural requirements, the dimensions of the foundations are arrived at following the procedures described in the preceding sections. This usually decides the type of foundation, i.e. pad, strip, or raft foundations. If soil of a higher bearing capacity exists at no great depth below the required minimum depth, consideration could be given to taking foundations to the deeper stratum, with consequent savings in the quantity of excavation and concrete. However, if the deeper stratum lies below the ground-water table, the cost of pumping or sheet piling might well outweigh these savings.

With the foundation depth and bearing pressures tentatively decided upon, the fourth step is to calculate or estimate the total and differential settlements of the structure, i.e. to check the serviceability limit state if the Eurocode 7 procedure is adopted. If the limits are exceeded, the bearing pressures will have to be reduced or the foundations taken to a deeper and less compressible stratum. If the bearing capacity of the shallow soil layers is inadequate, or if the estimated settlements of shallow foundations are excessive, the structure will have to be founded on piles or other special measures taken, such as providing some articulation in the structure to accommodate the estimated differential movements.

2.2.8 Definitions

It will be useful at this stage to define the various terms relating to bearing capacity and bearing pressure. These are as follows.

(a) *Total overburden pressure*, σ_{vo}, is the intensity of total pressure or total stress (σ), due to the weights of both soil and soil water, on any horizontal plane at and below foundation level before construction operations are commenced (Fig. 2.5).

(b) *Effective overburden pressure*, σ'_{vo}, is the intensity of intergranular pressure, or effective normal stress (σ_v'), on any horizontal plane at and below foundation level before construction operations are commenced. This pressure is the total overburden pressure (σ'_{vo}) less the pore-water pressure, which, in the general case, is equal to the head of water above the horizontal plane. For example, for a foundation level at a depth h below the water table,

$$\sigma'_{vo} = \sigma_{vo} - \gamma_w h. \qquad (2.3)$$

(c) *Total foundation pressure* (or gross loading intensity), q, is the intensity of total pressure on the ground beneath the foundation after the structure has been erected and fully loaded. It is inclusive of the gross load of the foundation substructure, the loading from the superstructure, and the gross loading from any

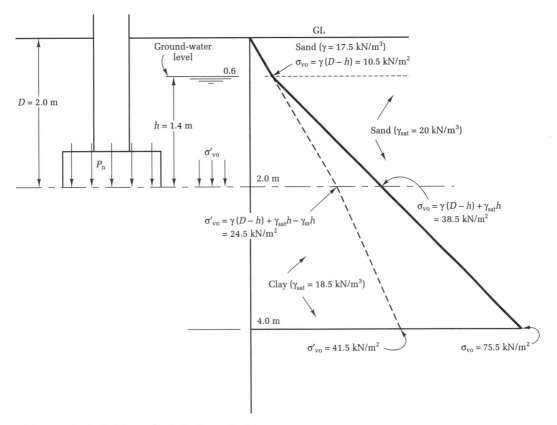

Fig. 2.5   Example of calculating total and effective overburden.

backfilled soil and soil water supported by the sub-structure.

(d) *Net foundation pressure* (or net loading intensity), p_n, is the net increase in pressure on the ground beneath the foundation due to the dead load and live load applied by the structure.

$$p_n = p - \sigma_{vo} \text{ (at foundation depth } D). \quad (2.4)$$

The term p_n is used for calculating the distribution of stress at any depth below foundation level.

(e) *Effective foundation pressure*, p'_n, is the effective increase in pressure on the ground beneath the foundation due to the dead load and live load applied by the structure.

$$p'_n = p - \sigma'_{vo} \text{ (at foundation depth } D). \quad (2.5)$$

(f) *Ultimate bearing capacity*, q, is the value of the loading intensity at which the ground fails in shear.

(g) *Net ultimate bearing capacity*, q_n, for a particular foundation is that value of the net loading intensity at which the ground fails in shear, i.e.

$$q_n = q - \sigma_v. \quad (2.6)$$

(h) *Presumed bearing value* is the net loading intensity considered appropriate to the particular type of ground for preliminary design purposes. The particular value is based either on local experience or by calculation from strength tests or field loading tests using a factor of safety against shear failure.

(i) *Allowable bearing pressure*, p_a, is the maximum allowable net loading intensity of the ground in any given case, taking into consideration the bearing capacity, the estimated amount and rate of settlement that will occur, and the ability of the structure to accommodate the settlement. It is therefore a function both of the site and of the structural conditions.

It will be seen from the above definitions that there is an important difference between the terms 'bearing capacity' and 'bearing pressure'. It is important that this distinction is clearly understood and that the terms are correctly used. The *bearing capacity* of a foundation soil is that pressure that the soil is *capable* of carrying, i.e. in the case of ultimate bearing capacity, or ultimate limit state, the pressure at which shear failure occurs, and this value, divided by a suitable safety factor, gives the safe bearing capacity. The *bearing pressure* of a foundation is the loading intensity *imposed* by the foundation on the soil.

The use of the above terms can be illustrated by the following statement: 'Calculations based on the shear strength of the soil showed that the *ultimate bearing capacity* for 1.2 m wide strip foundations at a depth of 1.5 m was 600 kN/m².' Adopting a safety factor of 3 on

this value, the *safe bearing capacity* of the soil beneath the strip foundation is 200 kN/m². However, in view of the sensitivity of the structure to the effects of small differential settlements, it was decided to limit the total and differential settlements by adopting an *allowable bearing pressure* of 150 kN/m².

In Eurocode 7 terms the statement would be expressed as follows: 'Calculations based on the shear strength of the soil gave an ultimate limit state bearing pressure of 310 kN/m² for a 1.2 m wide strip foundation at a depth of 1.5 m.' Checking the serviceability limit state for the constraints imposed for a structure sensitive to the effects of small differential settlement resulted in the reduction of the allowable bearing pressure to 150 kN/m².

2.3 Calculations for ultimate bearing capacity by analytical methods

2.3.1 General case

When a load is applied to a foundation, settlement will occur in the form shown in the load-settlement graph (Fig. 2.6). Up to a certain stage (point A), the settlement of the foundation will be comparatively small and mainly elastic (i.e. on removal of the load the foundation will rise nearly to its original level). With increased loading the settlement will increase at a disproportionate rate until ultimately (point B) the settlement will increase rapidly without further increase of loading. The ultimate bearing capacity (q_n) of the soil will have been reached, and the foundation will sink and tilt accompanied by heaving of the surrounding soil.

Sinking and tilting of the foundation will continue until the structure overturns or until a state of equilibrium is reached when the foundation reaches such a depth that the bearing capacity of the soil is sufficiently high to prevent further movement. Tilting nearly always occurs in cases of

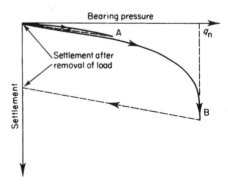

Fig. 2.6   Load–settlement relationship.

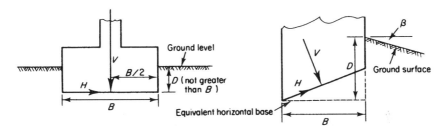

Fig. 2.7 Loading arrangements for Brinch Hansen's general equation (2.7).

foundation failure because the inevitable variation in the shear strength and compressibility of the soil from one point to another causes greater yielding on one side or another of the foundation. This throws the centre of gravity of the load towards the side where yielding has occurred, thus increasing the intensity of pressure on this side followed by further tilting.

Various analytical methods have been established for calculating the unit ultimate bearing capacity, q_n, for permissible stress calculations, or bearing pressures Q/A' for limit state calculations, of foundations. The most comprehensive ones, which take into account the shape and depth of the foundation, and the inclination of the loading, the base of the foundation and the ground surface have been derived by Brinch Hansen[2.1] and by Meyerhof.[2.2] Both use the same basic equation (2.7), but the shape, depth, inclination, and N_γ factors are computed in a different way. The Brinch Hansen factors are widely used in Western European practice while the Meyerhof method has been adopted generally in North America. The former factors give more conservative values. They have been used in this book, and they have been adopted for the purpose of calculating the design vertical bearing capacity, Q/A', in Eurocode 7. The basic equation is

$$Q/A' \text{ or } q_n = cN_c s_c d_c i_c b_c g_c + p_0 N_q s_q d_q i_q b_q g_q + \tfrac{1}{2}\gamma B N_\gamma s_\gamma d_\gamma i_\gamma b_\gamma g_\gamma, \quad (2.7)$$

where

γ = density of soil below foundation level,
B = width of foundation,
c = undrained cohesion of soil,
p_0 = effective pressure of overburden soil at foundation level,
N_γ, N_q, and N_c are bearing capacity factors,
s_r, s_q, and s_c are shape factors,
d_γ, d_q, and d_c are depth factors,
i_γ, i_q, and i_c are load inclination factors,
b_γ, b_q, and b_c are base inclination factors,
g_γ, g_q, and g_c are ground surface inclination factors.

For calculating Q/A' the value of c is the selected design

value or the characteristic value divided by the material factor γ_m. For calculating q_n for the permissible stress method the selected design value is used directly. Values of ϕ for obtaining N_q and N_γ are obtained in the same way.

The loading conditions, or spatial distribution of actions, for the Brinch Hansen equation are shown in Fig. 2.7. Values of the bearing capacity factors N_c, N_q, and N_γ and the shape factors s_c and s_γ for centrally applied vertical loading are shown in Figs 2.8 and 2.9, respectively. The shape factor s_q is calculated from the equation

$$s_q = s_c - \frac{s_c - 1}{N_q}. \quad (2.8)$$

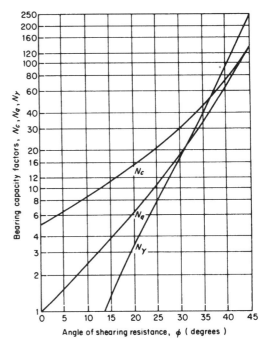

Fig. 2.8 Bearing capacity factors N_c, N_q, and N_γ (after Brinch Hansen[2.1]).

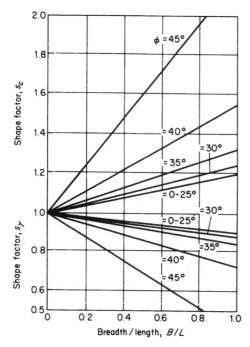

Fig. 2.9 Shape factors s_c and s_γ (after Brinch Hansen[2.1]).

Inclined loading must be considered from two directions, i.e. in the direction of the *effective* breadth B' and the *effective* length L' of the foundation. The dimensions B' and L' represent the effective foundation area for which the point of application of the load coincides with the

geometric centre of the area. A method by Meyerhof[2.2] of determining the effective foundation breadth and width is shown in Fig. 2.10.

Thus for loading in the direction of the breadth,

$$B' = B - 2e_x; \qquad (2.9)$$

and for loading in the direction of the length,

$$L' = L - 2e_\gamma. \qquad (2.9a)$$

The shape factors for inclined loading then become

$$s_{c,B} = 1 + 0.2i_{cB}B'/L', \qquad (2.10)$$

$$s_{c,L} = 1 + 0.2i_{cL}L'/B', \qquad (2.11)$$

$$s_{q,B} = 1 + \sin\phi i_{qB}B'/L', \qquad (2.12)$$

$$s_{q,L} = 1 + \sin\phi i_{qL}L'/B', \qquad (2.13)$$

$$s_{\gamma B} = 1 - 0.4_{\gamma B}B'/L', \qquad (2.14)$$

$$s_{\gamma L} = 1 - 0.4i_{\gamma L}L'/B'. \qquad (2.15)$$

The shape factors $s_{\gamma B}$ and $s_{\gamma L}$ must not be less than 0.6.

Approximate values which are sufficiently accurate for most practical purposes, for the shape factor for centrally applied vertical loading and B less than L are as follows:

Shape of base	s_c	s_q	s_γ
Continuous strip	1.0	1.0	1.0
Rectangle	$1 + 0.2B/L$	$1 + 0.2B/L$	$1 - 0.4B/L$
Square	1.3	1.2	0.8
Circle (B = diameter)	1.3	1.2	0.6

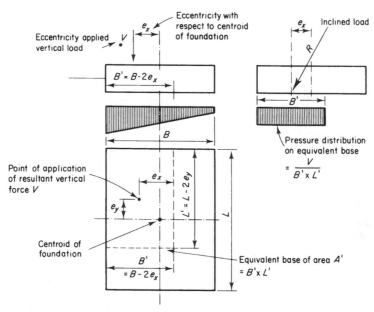

Fig. 2.10 Transformation of eccentrically loaded foundation to equivalent rectangular area carrying uniformly distributed vertical pressure.

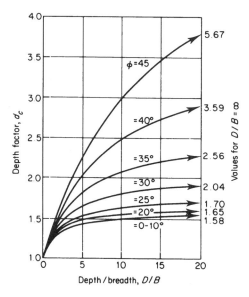

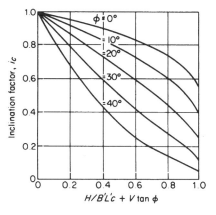

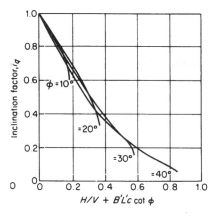

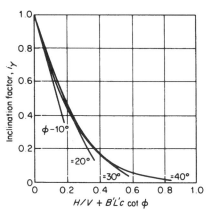

Fig. 2.11   Depth factor d_c (after Brinch Hansen[2.1]).

Values of the depth factor d_c are shown in Fig. 2.11. The final values for the base at a great depth ($D = \infty$) are shown to the right of this figure. The depth factor d_q is obtained from the equation

$$d_q = d_c - \frac{d_c - 1}{N_q}.\qquad(2.16)$$

The depth factor d_γ can be taken as unity in all cases. Also when ϕ is zero d_q is equal to unity. Where ϕ is greater than 25°, d_q can be taken as equal to d_c. A simplified value of d_c and d_q when ϕ is less than 25° is $1 + 0.35D/B$. The increase in bearing capacity provided by the use of depth factors is conditional upon the soil above foundation level being not significantly inferior in shear strength characteristics to that below this level. Where soft or loose soil exists above foundation level the depth factors should not be used.

Values for the inclination factors i_c, i_q, and i_γ are shown in Fig. 2.12 in relation to ϕ, and the effective foundation breadth and length B' and L'. Simplified values for H not greater than $V \tan \delta + cB'L'$ (where δ and c are the coefficient of friction and cohesion, respectively, between the base and the soil) are

$$i_c = 1 - \frac{H}{2cB'L'},\qquad(2.17)$$

$$i_q = 1 - \frac{1.5H}{V},\qquad(2.18)$$

$$i_\gamma = i^2{}_q.\qquad(2.19)$$

Fig. 2.12   Inclination factors i_c, i_q, and i_γ (after Brinch Hansen[2.1]).

Equation (2.19) is applicable only for $c = 0$ and $\phi = 30°$, but Brinch Hansen states that it can be used for other ϕ angles. Determination of the base inclination factors is rather complex, methods of calculating them are not given in Eurocode 7, and the reader is referred to the Brinch Hansen 1968 paper[2.1] for guidance. However, provided the load is mainly normal to the foundation surface, foundations having an inclined base can be treated as having a horizontal base at the same level as the lowest edge of the foundation and bounded by vertical planes through the other three edges (Fig. 2.7).

Where the ground surface slopes away from the foundation at an angle β to the horizontal (Fig. 2.7),

$$g_\gamma = g_q = 1 - \sin 2\beta, \qquad (2.20a)$$

$$g_c = e^{-2\beta\tan\phi}, \qquad (2.20b)$$

where e is the base 2.718 28 of the natural logarithm and β is in radians.

In a soil where the angle of shearing resistance is zero the effect of ground inclination can be allowed for by calculating the ultimate bearing capacity as for a level ground surface then reducing it by an amount equal to $2\beta c_u$ (where β is in radians). Ground inclination factors are not given in Eurocode 7.

Eurocode 7 requires the resistance to sliding to be checked where the loading is not normal to the foundation base. For safety against failure by sliding on a horizontal base Eurocode 7 requires the following inequality to be satisfied:

$$H \leqslant S + E_p \qquad (2.21)$$

where

H = horizontal component of design load including active earth force,

S = design shear resistance between the foundation and the ground,

E_p = design resistance of the ground which can be mobilized with the displacement appropriate to the limit state considered and which is available throughout the life of the structure.

Similar requirements apply to foundations on sloping bases.

For drained conditions

$$\delta = V' \tan\delta \qquad (2.22)$$

where

V' = design effective load normal to the foundation base

δ = friction angle on the foundation base.

Eurocode 7 recommends a friction angle equal to ϕ' where concrete is cast on the ground or $\frac{2}{3}\phi'$ for smooth precast concrete. Any effective cohesion c' should be neglected.

For undrained conditions

$$S \leqslant A'c_u, \qquad (2.23)$$

where A' is the effective base area.

In equation (2.21) E_p is mainly concerned with the passive resistance of the soil against the side of the base. Yielding of soil is necessary to mobilize this resistance. The reference to the life of the structure considers the possibility of trenching or reshaping the ground alongside the foundation.

Bending moments on a foundation causing a tendency to tilt introduces the possibility of ground water gaining access to a gap between the foundation base and the soil. Therefore Eurocode 7 requires the following condition to be checked:

$$S < 0.4V \qquad (2.24)$$

Checks by equations (2.23) and (2.24) can be disregarded if soil suction prevents the formation of a gap in areas where there is no positive bearing pressure.

2.3.2 Foundations on sands and gravels

Sands and gravels are highly permeable such that pore pressures induced in these soils by the applied loading are dissipated rapidly. Hence, for the calculation of bearing capacity, drained conditions can be assumed and the calculations are made in terms of *effective stresses*. The c' term in equation (2.7) is zero or very low and can be omitted except perhaps for silty sands which are sufficiently permeable to dissipate excess pore pressures during the construction period, and where the cohesion is not lost because of relief of overburden pressure by excavation or by disturbance from construction operations.

The value of the effective angle of shearing resistance, ϕ', is required to obtain the factors N_q and N_γ from Fig. 2.8. Usually the ϕ' values are obtained from standard penetration tests (SPT) or cone penetration tests (CPT) as described in Section 1.4.5. The relationship between SPT and ϕ established by Peck *et al.*[2.3] is used for ordinary foundation design (see Fig. 2.13). Values of ϕ' obtained from CP testing are more reliable because the soil does not undergo the same amount of disturbance during the test. The relationship between CPT and ϕ' established by Durgunoglu and Mitchell is shown in Fig. 2.14. Dynamic cone testing (DCPT) produces about the same degree of soil disturbance and the DCPT values can be converted to SPT values using the relationship shown in Fig. 1.5, but generally there is little published evidence of such correlations to justify using DCPT tests in preference to the SPT or CPT.

It was noted in Section 1.4.5 that ϕ' values can be obtained in the field using the self-boring pressuremeter, but the results have been shown to be higher than those given by SPTs or CPTs in the same soil deposit. However, they appear to agree with ϕ'-values obtained by triaxial testing and hence can be used with caution to obtain the

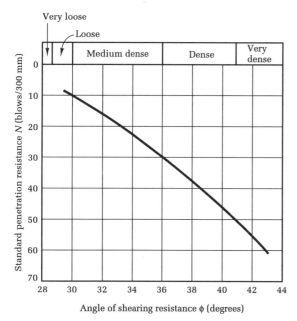

Fig. 2.13 Relationship between STP N-values and angle of shearing resistance (after Peck *et al.*[2.3]).

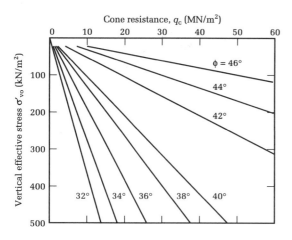

Fig. 2.14 Relationship between angle of shearing resistance and cone resistance for an uncemented, normally consolidated quartz sand (after Durgunoglu and Mitchell[2.4]).

bearing capacity factors from Fig. 2.8. The pressuremeter values must not be converted back to SPTs or CPTs for use in empirical relationships such as that used to obtain the deformation modulus of a soil (Section 2.6.5).

The pressuremeter can be used directly to calculate ultimate bearing capacity by the semi-empirical method as described in Section 2.4.1.

When calculating the bearing pressure for the ultimate limit state the values of c' obtained from triaxial testing and ϕ' from field tests should be factored (see Table 2.2) or they are used directly for the permissible stress method. In either case judgement is required to obtain design and characteristic values from the scatter of test results usually obtained from field or laboratory testing. Some illustrations of the selection procedure are given in the worked examples at the end of this chapter.

In equation (2.7) the first term is zero or very low in a sand or gravel, the second term can be high in a dense soil with a high ϕ'-value and the third term is low if B is small, as is the case for strip foundations for low-rise buildings. Hence the level of the ground-water table can be critical in determining σ'_{vo}. For example if the ground-water level were to be assumed to be at foundation level the value of the overburden pressure would be the total pressure. Then if the site were to be flooded so that the ground-water level rose to ground level the overburden pressure and consequently the ultimate bearing capacity would be approximately halved. It is for this reason that Eurocode 7 puts emphasis on the correct evaluation of ground-water levels and the need to factor the worst credible value to allow for uncertainties and accidents.

In most cases design bearing pressures for sands and gravels are governed by the serviceability limit state.

At the preliminary design stage before a site investigation report is available the ϕ-values of Terzaghi and Peck[2.5] for dry sands, sandy gravels, and silts can be used. These are as follows:

	Round grains uniform (°)	Angular grains well graded (°)	Silty sands (°)	Sandy gravels (°)	Inorganic silts (°)
Loose	27.5	33	27–33	35	27–30
Dense	34	45	30–34	50	30–35

Terzaghi and Peck state that the ϕ-values for *saturated* sands are likely to be one or two degrees below the values given in the table, except in the case of very silty sands where ϕ may be considerably lower for rapid loading conditions such as in earthquakes or machinery foundations, when ϕ must be determined from shear strength tests in the laboratory or from field loading tests.

It should be pointed out that the empirical relationship between the SPT or CPT and the angle of shearing resistance ignores the phenomenon of dilatancy in dense granular soils whereby a peak angle occurs at a relatively small shear strain followed by a lower ultimate angle at long strain. Dilatancy is of importance in the design of earth-retaining structures. It is discussed in Section 5.7.

2.3.3 Foundations on clays

Most clay soils are saturated and behave as if they are purely cohesive (angle of shearing resistance equal to zero), provided that no water is expelled from the soil as the load is applied. This is accepted as the basis for calculating ultimate bearing capacity for most normal cases of structural loading, where the load comes on to the foundations relatively quickly. Only in the case of very slow rates of loading, or with very silty soils, can the effect of decrease in water content, and hence increase in shear strength, be taken into account. Gain in shear strength with very slow rates of loading is allowed for when estimating the safe bearing pressures for high earth dams which are likely to take several years to construct. The decrease in water content of the soil is, of course, accompanied by settlement. In the case of an earth dam, such settlement would not be detrimental to the stability of the embankment. However, this procedure cannot be applied to structures which are sensitive to appreciable settlement. It can be used for flexible structures such as steel storage tanks constructed on silty clays. The tanks can be filled in small increments over long periods. The resulting settlement might dish or warp the steel plate floors without causing fracture and loss of the contents.

On the assumption that the angle of shearing resistance of the soil is equal to zero, the second and third terms in equation (2.7) are omitted. The cohesion c_u is determined from the results of 'immediate' undrained triaxial compression tests at constant volume made on soil samples from below foundation level within the zone which is stressed significantly by the foundation loading. Alternatively c can be obtained indirectly from correlations with standard penetration test N-values (Fig. 1.4), from cone penetration tests as described in Section 2.4.1, or from pressuremeter tests (Section 1.4.5). Because of the appreciable variations in strength which usually occur in natural soil deposits, the selection of a representative value of c_u for substitution in the above equation is a matter requiring some experience and judgement as discussed in Section 2.2.2. Where soils are sufficiently permeable for significant drainage to take place during the period of application of load, or where changes in the ground-water regime can cause pressures or upward flow to develop beneath the foundation, then the ultimate bearing capacity must be calculated in terms of *effective stress*. In these conditions the apparent cohesion c'_u is substituted for c in equation (2.7), and the values of N_c, N_q and N_γ are read off for the effective angle of shearing resistance ϕ'. Values of c' and ϕ' are obtained from drained triaxial compression tests (Section 1.11). It is important to note that the effective overburden pressure σ'_{v0} in equation (2.7) is equal to the total overburden pressure minus the pore-water pressure. Where changes in the pore-water pressure regime are expected during the service life of a structure it is necessary to calculate the ultimate bearing capacity both in terms of total and effective stress.

It should be noted that the bearing capacity factor N_c for undrained shear conditions ($\phi = 0$) was based on undrained shear tests made in a vertical direction on samples in the laboratory. It is sometimes the practice to obtain the undrained shear strength of soft clays by means of field vane tests (Section 1.4.5). These shear the soil in a horizontal direction and Bjerrum's correction factors related to the plasticity index of the clay (Fig. 1.3) should be applied.

The Skempton[2.6] values of N_c for vertical loading on strip or circular (and rectangular) foundations are widely used in conjunction with permissible stress methods of foundation design. However, caution is necessary when adopting them for design by the Eurocode 7 procedures. This is because the Eurocode 7 material factors in Table 2.2 have been selected by correlation of field experience with theoretical bearing capacity obtained by the Brinch Hansen equation. The Skempton factors are shown in Fig. 2.15.

Design bearing pressures for large raft foundations on uniform clays are likely to be assessed from considerations of the permissible total and differential settlements rather than calculated by dividing the calculated ultimate bearing capacity by a nominal safety factor. However, there may be a considerable variation in shear strength and compressibility of natural soil deposits beneath a large raft foundation, particularly in alluvial deposits where a stiff crust of variable thickness may be overlying soft compressible layers of varying extent both horizontally and vertically. In such conditions there is the possibility of failure of a foundation due to overstressing and squeezing out of a wedge-shaped layer of soft soil from one side of the loaded area (Section 4.1.2).

Where the shear strength and compressibility of the soil varies over the foundation area, the shear strength–depth relationship and the coefficient of volume compressibility (m_v)–depth relationship (Section 2.6.6) should be plotted at each borehole position and the information displayed on a soil profile along one or more cross-sections. Inspection of the plotted data will indicate whether or not the potential zones of overstressed soils are sufficiently large in extent to involve a risk of failure by tilting. If this is the case then the foundation should be taken below such zones to deeper and stiffer soils. If, however, the weaker soils are fairly uniformly distributed over the area, or if they are confined to a relatively small zone beneath the centre of the area, then the design bearing pressure can be calculated from considerations of settlement. A preliminary value for the design bearing pressure should be adopted

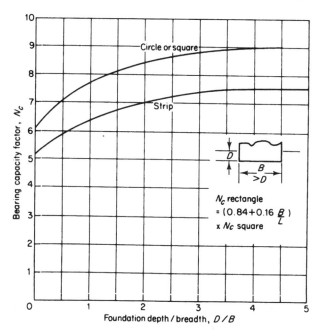

Fig. 2.15 Bearing capacity factor N_c for cohesive soil ($\phi = 0$) (after Skempton[2.6]).

based either on limit state or permissible stress methods. The immediate and consolidation settlements should be calculated corresponding to the maximum, minimum, and average compressibility of the soil layers as shown by the various plottings of m_v versus depth using the methods described in Section 2.6.6. If the differential settlements or tilting are excessive then the foundation should be taken down to deeper and less compressible soil, or if the compressible soil layers are thin, in relation to the zone of soil stressed by the raft, it may be possible to increase the plan area of the raft, so reducing the bearing pressure and hence the compression of these layers. The final stage of the settlement calculations should take into account the depth and rigidity of the raft as described later in this chapter.

The finite element method is useful for making a parametric study of the effects of variations of soil stiffness both in magnitude and spatially (see Sections 2.8.3 and 2.8.4).

2.3.4 Foundations on silts

The engineering characteristics of silts are intermediate between those of sands and clays. Calculations for ultimate bearing capacity should be made separately for undrained ($\phi = 0$) and drained (c', ϕ') conditions and judgement made on the value to be used for design. For drained conditions the value of c' obtained from drained triaxial tests should be used with caution. If it is caused by

a weak cementing medium the cement can be broken down by relief of overburden pressure or by disturbance from construction operations, when c' may fall to zero.

Soils described in borehole records as 'silts' are sometimes in the form of silty clays interbedded within layers or laminae of fine sand. In such cases the undrained shear strength of the clayey layers will govern ultimate bearing capacity for normal structural foundations.

2.3.5 Deep foundations

For economical design of deep foundations, some account must be taken of the contribution to bearing capacity provided by skin friction mobilized between the periphery of the foundation and the surrounding ground. The general equation in limit state terms is

$$Q = \frac{Q_{bk}}{\gamma_b} + \frac{Q_{sk}}{\gamma_s} \qquad (2.25a)$$

where

Q = design bearing capacity of the foundation,

Q_{bk} = characteristic value of the resistance of the soil or rock beneath the base,

Q_{sk} = characteristic value of the friction mobilized on the sides of the foundation

γ_b and γ_s are partial factors.

The axial load F applied to the foundation must be less than or equal to Q.

In permissible stress terms the equation is

$$Q_{ult} = q_b A_b + \bar{f}_s A_s \qquad (2.25b)$$

where

Q_{ult} = ultimate bearing capacity of the foundation,

q_b = unit ultimate bearing capacity of the soil or rock beneath the base,

A_b = area of base,

$\bar{f}_s$ = average unit ultimate skin friction mobilized around the foundation shaft,

A_s = area of shaft in contact with the ground.

Values of q_b can be obtained from equation (2.7) with due allowance being given to the effects of construction such as loosening of the soil by a mechanical grab or rotary auger. The average or characteristic unit skin friction depends to a great extent on the method of constructing the foundation. If concrete is dumped into an open excavation in clay the sides may swell and soften during the construction period, or if it is in the form of a pad and stem (Fig. 2.1(c)) skin friction will not be mobilized above the pad and there could be a dragdown force on the stem from the settling backfill. Low skin friction is mobilized if the deep foundation is in the form of a caisson when the surrounding soil is grossly disturbed or friction is eliminated by the use of bentonite to assist sinking.

Deep pier and pad foundations are discussed in more detail in Chapter 4, and caissons in Chapter 6. Piles are a special form of deep foundation and the calculation of bearing capacity is discussed in Chapter 7 and in Chapter 7 of Eurocode 7.

It should be noted that different partial factors are used in equation (2.25a). Partial safety factors are also applied to the base and skin friction terms in equation (2.25b) to obtain the safe bearing capacity of the foundation.

2.3.6 Shallow foundations on rock

Where rocks are completely weathered to the consistency of a sand, silt, or clay the bearing pressure for the ultimate limit state or the ultimate bearing capacity can be calculated analytically using the methods in Section

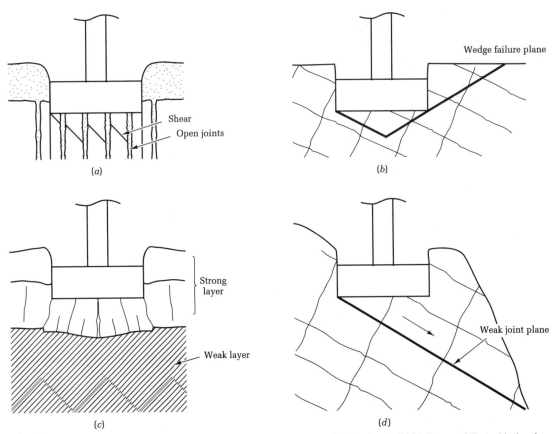

Fig. 2.16 Failure of foundations on jointed rocks. (a) Shear failure of unsupported rock columns. (b) Wedge-type failure with closed joints (horizontal or inclined). (c) Splitting and punching with strong layer over weak layer. (d) Sliding on weak inclined joint.

2.3.2–2.3.4. Failure of foundations bearing on bedded and jointed unweathered or partly weathered rocks occurs in modes different from that of soils (Fig. 2.16).

It should not be assumed that safe bearing pressures on rock are governed only from considerations of permissible settlement. It is evident from Fig. 2.16 that critical conditions causing bearing capacity failure can occur. The following methods of calculating the ultimate bearing capacity refer to foundations on level ground with horizontal or subhorizontal bedding. Stability conditions for foundations on rock slopes or on steeply bedded formations are described in detail by Wyllie[2.7].

For the case of a rock mass with wide open joints (Fig. 2.16a) the ultimate bearing capacity of the foundation is given by the unconfined compression strength of the intact rock which is measured as described in Section 1.5.4. The ultimate bearing capacity of a long strip foundation on a rock mass with closed joints is given by the equation

$$q_n = cN_c + \gamma\frac{B}{2}N_\gamma + \gamma D N_q. \tag{2.26}$$

This is basically the same as equation (2.7), but the three terms are usually arranged in a different order as shown, and the factors N_c, N_γ, and N_q are different from the Brinch Hansen factors in Fig. 2.8. The factors for rocks shown in Fig. 2.17 are appropriate to the wedge failure conditions shown in Fig. 2.16b. They are related to the friction angle, ϕ, of the jointed rock mass.

The term cN_c should be multiplied by 1.25 or 1.2 for square or circular pad foundations respectively. Also the term $\gamma(B/2)N_\gamma$ should be multiplied by 0.8 and 0.7 respec-

Table 2.3 Typical friction angles (ϕ) for clean fractures in rock (after Wyllie[2.7])

Rock type	ϕ-values (°)
Schists with high mica content Shale Marl	20–27
Sandstone Siltstone Chalk Gneiss Slate	27–34
Basalt Granite Limestone Conglomerate	34–40

tively for the two types of foundation. This term is small compared with cN_c and is often neglected.

In order to obtain c and ϕ it is necessary to make triaxial compression or shear box tests on large representative samples of the jointed rock. The samples are difficult to obtain and the tests are expensive. For ordinary foundation design it may be satisfactory to use typical values of ϕ such as those given by Wyllie[2.7] from back-analyses of failures in rock slopes. These are given in Table 2.3.

Kulhawy and Goodman[2.8, 2.8a] have shown that the c and ϕ parameters can be related to the RQD (rock quality designation) of the rock mass and they have suggested the following approximate relationships:

RQD (%)	Rock mass properties		
	q_c	c	ϕ (°)
0–70	$0.33q_{uc}$	$0.1q_{uc}$	30
70–100	0.33–$0.8q_{uc}$	$0.1q_{uc}$	30–60

where q_{uc} is the unconfined compression strength of the intact rock.

In limit state terms the characteristic values of c and ϕ are divided by the material factors γ_m to obtain the design values of q_n. For an effective foundation area A', the factored design ultimate axial load to be applied to the foundation should not exceed $q_n \times A'$.

The additional load which can be applied to a pile foundation by forming a socket in the rock will be discussed in Section 7.12.2. This method is also applicable to deep shaft foundations in rock.

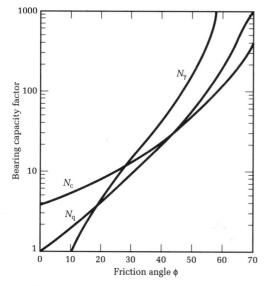

Fig. 2.17   Wedge bearing capacity factor for foundations on rock.

2.4 Calculation of ultimate bearing capacity by semi-empirical methods

2.4.1 Foundations on sands and gravels

The use of the CPT or SPT to obtain values of ϕ and hence N_q and N_γ for substituting in equation (2.7) is essentially part of the analytical method (Section 2.3.2). The direct use of the CPT values to obtain the ultimate bearing capacity of a foundation applies only to piles (Section 7.4.2). Values of SPT are not used directly either for spread or piled foundations.

The results of pressuremeter tests can be used directly. Eurocode 7 gives the equation

$$Q = A'(q + kp*_{LM}) \qquad (2.27a)$$

where

Q = ultimate limit state design load normal to the foundation base,
q = total design vertical stress at foundation level after construction due either to the embedment or a surcharge,
k = a bearing capacity factor with numerical values in the range of 0.6–1.65 depending on the type of soil, the embedment depth, and the shape of the foundation,
$p*_{LM}$ = the design net equivalent limit pressure.

Also

$$p*_{LM} = p_{LM} - \sigma_{ho} \qquad (2.27b)$$

where

p_{LM} = limit pressure measured in the field,

σ_{ho} = in-situ horizontal stress measured by the pressuremeter.

Values of k for use in conjunction with the Menard pressuremeter are given by Baguelin et al.[2.9]

2.4.2 Foundations on clays

Standard penetration tests are not used directly to calculate the ultimate bearing capacity of foundations on clays. The test results can be converted to undrained shear strengths from Fig. 1.4 and used to obtain the N_c values from Figs 2.8 or 2.15.

Cone penetration test results can be used to obtain values from the equation

$$c_u = \frac{q_c - \sigma'_{vo}}{N_k} \qquad (2.28)$$

where

$q_c - \sigma'_{vo}$ = net cone resistance,
N_k = cone factor.

Meigh[1.8] gives the following values of N_k for use with the electrical resistance cone (Fig. 1.6(c)):

Type of clay	N_k
Normally consolidated	15–19 (higher values for sensitive clays)
Overconsolidated (stiff, fissured, marine)	27–30
UK glacial	18–22

Where the mantle cone (Fig. 1.6(a)) has been used the N_k-values in normally consolidated clays are from 17.5 to 21, but the c_u-values obtained must be corrected by the Bjerrum factors for the vane test (Fig. 1.3) to give results corresponding to those obtained from plate bearing tests.

The Menard-type pressuremeter can be used directly to obtain the ultimate bearing capacity of foundations on clay from equations (2.27a) and (2.27b) using the appropriate k-values for clay given by Baguelin et al.[2.9].

2.4.3 Foundations in weak rocks

The ultimate bearing capacity of weak rocks which behave as clays can be determined from equations (2.8) or (2.15) using c_u-values obtained from SPT, CPT, or pressuremeter relationships.

Values from CPT can be used directly from equation (2.28) using the cone factor appropriate to overconsolidated clays.

Mair and Wood[1.10] state that the pressuremeter can be used only if the rock is weak enough for a significant plastic zone to develop around the cell during expansion. For a pressuremeter capable of being expanded to $20\,MN/m^2$ the maximum c_u which can be reasonably determined from the test is $7\,MN/m^2$ equivalent to an unconfined compression strength of $14\,MN/m^2$ which is in the moderately weak to moderately strong classification. However, Mair and Wood point out that many types of rock do not behave like clays when sheared by the pressuremeter, and the results cannot be used to obtain c_u-values.

2.5 Estimation of allowable bearing pressures by prescriptive methods

2.5.1 Presumed bearing values for foundations on rock

The presumed bearing values for rock stated in local or national building regulations are usually based on long experience of the behaviour of foundations on rocks in the localities covered by the regulations. They are usually conservative, but not always so. Where, for example, low bearing values are given, these may well be due to local knowledge of such factors as deep weathering or intensive fissuring of the formation. The engineer is often tempted to adopt a higher bearing value after inspecting what appears to be sound strong rock in trial pits. However, when the foundation trenches are subsequently excavated, the surface of the sound rock may be found to be highly irregular, thus necessitating deep excavation in pockets in a narrow trench to reach a suitable stratum. This delays the construction programme, and there may also be protracted delays while discussions are held to define what is the acceptable bearing stratum compatible with the adopted

bearing pressure. It is not unusual to find that the engineer who made the initial site investigation and recommendations has an opinion on an acceptable bearing stratum which is different from that of the engineer supervising the construction on site.

It is nearly always cheaper to construct wide foundations to a uniform and predetermined depth than to excavate narrow trenches or pits to a depth which cannot be forecast at the time of estimating the cost of the work. The values in Table 2.4 are based on the compressibility of rocks and soils as determined by field and laboratory tests. The adoption of these values may result in immediate and long-term settlements totalling up to 50 mm in the case of an isolated foundation. In the case of foundations on Keuper Marl or high-porosity chalk, presumed bearing

Table 2.4 Presumed bearing values for horizontal foundations under vertical static loading, settlement of individual foundation not to exceed 50 mm

(a) Strip foundations not exceeding 3 m wide, length not more than ten times width, bearing on surface of rock

Rock group	Strength grade	Discontinuity spacing (mm)	Presumed allowable bearing value (kN/m^2)
Pure limestones and dolomites, carbonate sandstones of low porosity	Strong	60 to >1000	>12500†
	Moderately strong	>600	>10000‡
		200–600	7500–10000
		60–200	3000–7500
	Moderately weak	600 to >1000	>5000†
		200–600	3000–5000
		60–200	1000–3000
	Weak	>600	>1000†
		200–600	750–1000
		60–200	250–750
	Very weak		See note‡
Igneous, oolitic, and marly limestones; well-cemented sandstones; indurated carbonate mudstones; metamorphic rocks (including slates and schists with flat cleavage/foliation)	Strong	200 to >1000	10000 to >12500†
		60–200	5000–10000
	Moderately strong	600 to >1000	8000 to >10000†
		200–600	4000–8000
		60–200	1500–4000
	Moderately weak	600 to >1000	3000 to >5000†
		200–600	1500–3000
		60–200	500–1500
	Weak	600 to >1000	750 to >1000†
		<200	See note ‡
	Very weak	All	See note ‡
Very marly limestones: poorly cemented sandstones; cemented mudstones and shales; slates and scists with steep cleavage/foliation	Strong	600 to >1000	10000 to >12 500†
		200–600	5000–10 000
		60–200	2500–5000
	Moderately strong	600 to >1000	4000 to >6000†
		200–600	2000 to >4000
		60–200	750–2000
	Moderately weak	600 to >1000	2000 to >3000†
		200–600	750–2000
		60–200	250–750
	Weak	600 to >1000	500–750
		200–600	250–500
		<200	See note ‡
	Very weak	All	See note ‡

(continued overleaf)

Table 2.4   (continued)

Rock group	Strength grade	Discontinuity spacing (mm)	Presumed allowable bearing value (kN/m²)
Uncemented mudstones and shales	Strong	200–600	250–5000
		60–200	1250–2500
	Moderately strong	200–600	1000–2000
		60–200	1300–1000
	Moderately weak	200–600	400–1000
		60–200	125–400
	Weak	200–600	150–250
		60–200	See note ‡
	Very weak	All	See note ‡

Notes: Presumed bearing values for square foundations up to 3 m wide are approximately twice the above values, or equal to the above values if settlements are to be limited to 25 mm.
† Bearing pressures must not exceed the unconfined compression strength of the rock if the joints are tight. Where the joints open the bearing pressure must not exceed half the unconfined compression strength of the rock.
‡ Bearing pressures for these weak or closely jointed rocks should be assessed after visual inspection, supplemented as necessary by field or laboratory tests to determine their strength and compressibility.

(b) Strip or pad foundations not exceeding 3 m wide, length not more than 10 times width bearing on surface of Keuper Marl or Chalk

Rock type	Weathering grade	Description	Presumed bearing value (kN/m²)
Keuper Marl	I (unweathered)	Mudstone (often fissured)	300–500
	II (lightly weathered)	Angular blocks of unweathered marl or mudstone with virtually no matrix	200–300
	III (moderately weathered)	Matrix with frequent lithorelics up to 25 mm	100–200
	IV(a) (highly weathered)	Matrix with occasional claystone pellets less than 3 mm (usually coarse sand size)	50–100
	IV(b) (completely weathered)	Matrix only (in form of silty clay)	See note ‡
Chalk	I	Blocky, moderately weak, brittle	>1000
	II	Blocky, weak, joints more than 200 mm apart and closed	500–1000
	III	Rubbly to blocky, unweathered, joints 60–200 mm apart, open to 3 mm and sometimes filled with fragments	250–500
	IV	Rubbly, partly weathered with bedding and jointing. Joints 10–60 mm apart, open to 20 mm, often filled with weak remoulded chalk and fragments	125–250
	V	Structureless, remoulded, containing lumps of intact rock	50–125
	VI	Extremely weak, structureless, containing small lumps of intact chalk	See note ‡

Notes: Presumed bearing values for square foundations up to 3 m wide are approximately twice the above values.
‡ Presumed bearing values may be assessed visually on inspection or determined by field or laboratory tests assuming that the rock has weathered to the consistency of a soil.

values have been related to the weathering grades of these rocks as shown in Table 2.4(*b*). The differential movement between adjacent foundations should be considered. If this exceeds the permissible limit for structural damage (Section 2.6.2) a detailed settlement analysis should be made following the procedure described in Section 2.6.

The values for rocks shown in Table 2.4(*a*) and (*b*) have been based on relationships between the deformation modulus and unconfined compression strength or weathering grade established by Hobbs.[2.10] All values assume

(a) That the site and adjoining sites are reasonably level;
(b) That the ground and strata are reasonably level;
(c) That there is no layer of higher compressibility below the foundation stratum;
(d) That the site is protected from deterioration.

2.5.2 Presumed bearing values for foundations on soils

Presumed bearing values for foundations on sands and gravels are shown in Table 2.5 and are related to standard penetration test N-values. They assume that the groundwater level will not rise above foundation level and that the long-term settlement should not exceed 50 mm for an isolated strip foundation. The bearing pressures shown in Table 2.5 have been calculated from the Burland and Burbidge[2.11] method described later in this chapter and

rounded up or down to take account of the average and upper bound values. An upper limiting bearing pressure of 800 kN/m² has been adopted.

Presumed bearing values for foundations on clays are shown in Table 2.6. They assume that the total of the immediate and long-term settlements should not exceed 50 mm. They have been based on published deformation modulus values of clays by routine laboratory testing on samples obtained by geotechnical category 2 methods (see Section 1.0). They are only intended as a guide and should be accepted as design values only if confirmed by experience of the settlement behaviour of foundations on similar soil types.

2.6 Settlement of foundations

2.6.1 Total and differential settlements

Settlement due to consolidation of the foundation soil is usually the most important consideration in calculating the

Table 2.5 Foundations in sands and gravels at a minimum depth of 0.75 m below ground level

Description of soil	N-value in SPT	Presumed bearing value (kN/m²) for foundation of width		
		1 m	2 m	4 m
Very dense sands and gravels	>50	800	600	500
Dense sands and gravels	30–50	500–800	400–600	300–500
Medium-dense sands and gravels	10–30	150–500	100–400	100–300
Loose sands and gravels	5–10	50–150	50–100	30–100

Note: The water table is assumed not to be above the base of foundation. Presumed bearing values for pad foundations up to 3 m wide are approximately twice the above values.

Table 2.6 Foundations on clays at a minimum depth of 1 m below ground level

Description	Undrained shear strength (kN/m²)	Presumed bearing value (kN/m²) for foundation of width		
		1 m	2 m	4 m
Hard boulder clays, hard-fissured clays (e.g. deeper London and Gault Clays)	>300	800	600	400
Very stiff boulder clay, very stiff 'blue' London Clay	150–300	400–800	300–500	150–250
Stiff-fissured clays (e.g. stiff 'blue' and brown London Clay), stiff weathered boulder clay	75–150	200–400	150–250	75–125
Firm normally consolidated clays (at depth), fluvio-glacial and lake clays, uppper weathered 'brown' London Clay	40–75	100–200	75–100	50–75
Soft normally consolidated alluvial clays (e.g. marine, river and estuarine clays)	20–40	50–100	25–50	Negligible

serviceability limit state or in assessing *allowable* bearing pressures where permissible stress methods are used. Even though sinking of foundations as a result of shear failure of the soil has been safeguarded by ultimate limit state calculations or by applying an arbitrary safety factor on the caluculated ultimate bearing capacity, it is still necessary to investigate the likelihood of settlements before the allowable bearing pressures can be fixed. In the following pages consideration will be given to the causes of settlement, the effects on the structure of total and differential movements, methods of estimating settlement, and the design of foundations to eliminate settlement or to minimize its effects.

Where the serviceability limit state is calculated by Eurocode 7 recommendations, the partial factor γ_F in Table 2.1 for actions is unity for unfavourable loads and zero for favourable loads. The material factor γ_m is unity when applied to determinations of the characteristic deformation modulus or coefficient of compressibility.

The settlement of a structural foundation consists of three parts. The *'immediate' settlement* (ρ_i) takes place during application of the loading as a result of elastic deformation of the soil without change in water content. *'Consolidation' settlement* (ρ_c) takes place as a result of volume reduction of the soil caused by extrusion of some of the pore water from the soil. *'Creep'* or *'secondary'* settlement (ρ_α) occurs over a very long period of years after completing the extrusion of excess pore water. It is caused by the viscous resistance of the soil particles to adjustment under compression. The *'final' settlement* (ρ_f) is the sum of ρ_i and ρ_α. If deep excavation is required to reach foundation level, swelling of the soil will take place as a result of removal of the pressure of the overburden. The magnitude of the swelling depends on the depth of overburden removed and the time the foundations remain unloaded. A diagram illustrating the various stages of swelling and settlement is shown in Fig. 2.18.

In the case of foundations on medium-dense to dense sands and gravels, the 'immediate' and 'consolidation' settlements are of a relatively small order. A high proportion of the total settlement is almost completed by the time the full loading comes on the foundations. Similarly, a high proportion of the settlement of foundations on loose sands takes place as the load is applied, whereas settlements on compressible clays are partly immediate and partly long-term movements. The latter is likely to account for the greater proportion of the movement and may take place over a very long period of years. Immediate settlement is defined in Eurocode 7 as 'settlement without drainage for fully saturated soil due to shear deformation at constant volume'. A reminder is given in Eurocode 7 of the need to consider causes of foundation settlement other than normal soil compression and consolidation. Some of these causes are listed in Section 2.2.1.

Settlement of foundations is not necessarily confined to very large and heavy structures. In soft and compressible silts and clays, appreciable settlement can occur under light loadings. Settlement and cracking occurred in two-storey houses founded on a soft silty clay in Scotland. The houses were of precast concrete block construction and the foundation loading was probably not more than about 3.2 kN/m run of wall. In less than 3 years from the time of completion, differential settlement and cracking of the blocks of houses were so severe that a number of the houses had to be evacuated. One block showed a relative movement of 100 mm along the wall.

The differential, or relative, settlement between one part of a structure and another is of greater significance to the stability of the superstructure than the magnitude of the total settlement. The latter is only significant in relation to neighbouring works. For example, a flood wall to a river might be constructed to a crest level at a prescribed height above maximum flood level. Excessive settlement of the wall over a long period of years might result in overtopping of the wall at flood periods.

If the whole of the foundation area of a structure settles to the same extent, there is no detrimental effect on the superstructure. If, however, there is relative movement between various parts of the foundation, stresses are set up in the structure. Serious cracking, and even collapse of the structure, may occur if the differential movements are excessive.

Differential settlement between parts of a structure may occur as a result of the following:

(a) *Variations in strata.* One part of a structure may be founded on a compressible soil and the other part on an incompressible material. Such variations are not uncommon, particularly in glacial deposits, where

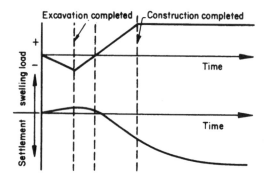

Fig. 2.18   Load-settlement time relationship for a structure.

lenses of clay may be found in predominantly sandy material or vice versa. In areas of irregular bedrock surface, parts of a structure may be founded on shallow rock and others on soil or compressible weathered rock. Wind-laid or water-laid deposits of sands and gravels can vary widely in density in a vertical and horizontal direction.

(b) *Variations in foundation loading.* For example, in a building consisting of a high central tower with low projecting wings, differential settlement between the tower and wings would be expected unless special methods of foundation design were introduced to prevent it. Similarly a factory building might have a light superstructure surrounding a very heavy item of machinery.

(c) *Large loaded areas on flexible foundations.* The settlement of large flexible raft foundations, or large loaded areas comprising the independent foundations of a number of columns, when constructed directly on a compressible soil, takes a characteristic bowl shape with the maximum settlement at the centre of the area and the minimum at the corners. The maximum differential settlement is usually about one-half of the total settlement. However, in the case of a structure consisting of a large number of close-spaced equally loaded columns, even though the maximum differential settlement between the centre and corners may be large, the relative settlement between columns may be only a fraction of the maximum. An example of the form of settlement of a large loaded area is shown in Fig. 2.19(*a*). Where the large loaded area is founded on a relatively incompressible stratum (e.g. dense gravel) overlying compressible soil (Fig. 2.19(*b*)), settlement of the structure will occur due to consolidation of the latter layer, but it will not take the form of the bowl-shaped depression. The effect of the dense layer, if thick enough, is to form a rigid raft which will largely eliminate differential settlement.

(d) *Differences in time of construction of adjacent parts of a structure.* This problem occurs when extensions

to a structure are built many years after completion of the original structure. Long-term consolidation settlements of the latter may be virtually complete, but the new structure (if of the same foundation loading as the original) will eventually settle an equal amount. Special provisions in the form of vertical joints are needed to prevent distortion and cracking between the old and new structures.

(e) *Variation in site conditions.* One part of a building area may have been occupied by a heavy structure which had been demolished; or on a sloping site it may be necessary to remove a considerable thickness of overburden to form a level site. These variations result in different stress conditions both before and after loading with consequent differential settlement or swelling.

2.6.2 The deformation of structures and their supporting foundations

In a comprehensive review of the behaviour of structures and their foundations, Burland *et al.*[2.12] stated that certain basic criteria must be satisfied when considering the limiting movements of a structure. These are:

(a) Visual appearance
(b) Serviceability or function
(c) Stability

In discussing these criteria in relation to limiting movements, it is necessary to define settlements and distortions in accordance with the accepted nomenclature shown in Fig. 2.20.

Considering first the visual appearance, a tilt or rotation greater than 1 in 250 is likely to be noticed in buildings. A local rotation of horizontal members exceeding 1 in 100 or a deflexion ratio greater than 1 in 250 is likely to be clearly visible. Cracking of load-bearing walls or of claddings to framed buildings is detrimental to their appearance. Crack widths of more than 3–5 mm at eye level are unsightly and require remedial treatment.

Cracking of structures affects their *serviceability* or

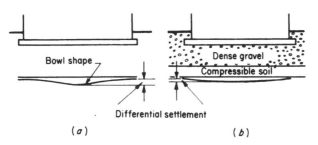

Fig. 2.19 Shape of settlement of large flexible loaded areas. (*a*) Raft on uniform compressible soil. (*b*) Raft on dense incompressible stratum overlying compressible soil.

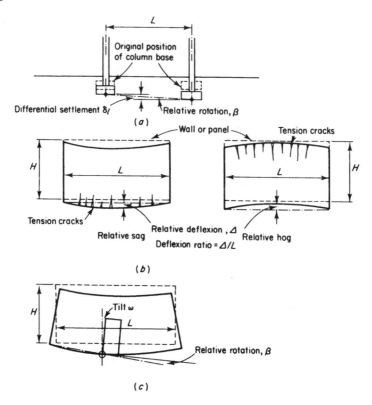

Fig. 2.20   Definitions of differential settlement and distortion for framed and load-bearing wall structures (after Burland and Wroth[2.13]).

function in that it can result in loss of weathertightness, fire-resistance, and thermal and sound insulation values. Total settlement may be critical to serviceability in relation to connections to external drains or other pipework, while deformations can affect the proper functioning of installations such as overhead cranes and precision machinery.

Relative deflexions and relative rotations may be critical to the *stability* of structures since these may cause excessive bending stresses in members. Excessive tilting may lead to the complete collapse of a structure.

The degree of damage caused by settlement is to some extent dependent on the sequence and time of construction operations. For example, in the case of a tall building founded on a deep basement in clay, the base of the excavation will initially heave to a convex shape. As the superstructure is erected, floor by floor, the foundation soil will consolidate and will eventually deform to a concave (bowl) shape, thus resulting in a complete reversal of curvature of the basement and lowermost storeys. The structural frame of multi-storey dwelling or office blocks represents the major proportion of the total dead load. Thus the greater proportion of the settlement of the

building will have taken place by the time the frame is completed. Then claddings or finishes constructed at a later stage will add to the stiffness of the structure and will suffer much less distortion than that which has already taken place in the structural frame. On the other hand the contents of a grain silo can weigh much more than their containing structure when the major proportion of the settlement does not take place until the compartments are filled for the first time.

Empirical criteria have been established for limiting the movement of structures in order to prevent or minimize cracking or other forms of structure damage. These criteria are shown in Table 2.7.

It will be noted from Table 2.7 that the critical factor for framed buildings and reinforced load-bearing walls is the relative rotation (or angular distortion), whereas the deflexion ratio is the criterion for unreinforced load-bearing walls which fail by sagging or hogging as shown in Fig. 2.20. In this respect the length to height ratio of the wall is significant, the larger the L/H ratio the higher the limiting value of Δ/L. It will also be seen that cracking by hogging occurs at deflexion ratios one-half of those which are critical for sagging movement.

Table 2.7   Limiting values of distortion and deflexion of structures

Type of structure	Type of damage	Limiting values			
		Values of relative rotation (angular distortion)			
		Skempton and MacDonald[2.14]	Meyerhof[2.15]	Polshin and Tokar[2.16]	Bjerrum[2.17]
Framed buildings and reinforced load bearing walls	Structural damage	1/150	1/250	1/200	1/150
	Cracking in walls and partitions	1/300 (but 1/500 recommended)	1/500	1/500 (0.7/1000 to 1/1000 for end bays)	1/500

		Values of deflexion ratio Δ/l		
		Meyerhof[2.15]	Polshin and Tokar[2.16]	Burland and Wroth[2.13]
Unreinforced load-bearing walls	Cracking by sagging	0.4×10^{-3}	$L/H = 3 : 0.3$ to 0.4×10^{-3}	At $L/H = 1 : 0.4 \times 10^{-3}$ At $L/H = 5 : 0.8 \times 10^{-3}$
	Cracking by hogging	–	–	At $L/H = 1 : 0.2 \times 10^{-3}$ At $L/H = 5 : 0.4 \times 10^{-3}$

Note: The limiting values for framed buildings are for structural members of average dimensions. Values may be much less for exceptionally large and stiff beams or columns for which the limiting values of angular distortion should be obtained by structural analysis.

The Building Research Establishment has classified the degree of cracking which can be regarded as acceptable or which requires various forms of remedial work as shown in Table 2.8. This table has been criticized by building owners and their technical advisers in that it underrates the seriousness of damage. The principal factor to be considered when assessing damage is the risk of future increase in crack widths. For example a 5 mm crack width can reasonably be classed as slight if there is no risk of future widening.

It is possible to examine the effects of distortion on the individual components of structures by considering the concept of a limiting tensile strain (Σ_{lim}) which can be withstood by the material of the component before cracking takes place. Burland and Wroth[2.13] defined the deflexion ratio (Δ/L) in terms of the maximum strain on a centrally loaded beam of unit thickness and their equations were later modified as follows: for maximum fibre strain

$$\frac{\Delta}{L} = \sum_{b \ max} \frac{L}{12y}\left[1 + \frac{18}{L^2} \cdot \frac{I}{H} \cdot \frac{E}{G}\right], \quad (2.29)$$

for maximum diagonal strain

$$\frac{\Delta}{L} = \sum_{d \ max} \left[1 + \frac{L^2}{18} \cdot \frac{H}{I} \cdot \frac{G}{E}\right], \quad (2.30)$$

where

y = depth of neutral axis,
I = moment of intertia,
G = shear modulus,
E = Young's modulus.

When $\Sigma_{max} = \Sigma_{lim}$, equations (2.29) and (2.30) define the limiting values of Δ/L for cracking of simple beams in bending and shear.

The relationships of Δ/L Σ_{lim} and L/H for simple rectangular beams of various E/G ratios are shown in Fig. 2.21. These curves show that even for simple beams the limiting deflexion ratios causing cracking can vary over wide limits depending on their stiffness. From a survey of case histories of cracking of buildings, Burland and Wroth[2.13] established ranges of Σ_{lim} between 0.05 and 0.1 per cent for brickwork and blockwork set in cement mortar and between 0.03 and 0.05 per cent for reinforced concrete.

2.6.3 Methods of avoiding or accommodating excessive differential settlement

Differential settlement need not be considered only in the case of structures founded on relatively incompressible bedrock. Where structures are sited on weak weathered rocks or on soils an estimate must be made of the total and differential settlements to decide whether the movements are likely to be tolerated by the design of the structure, or whether they are sufficiently large as to require special measures to avoid or accommodate them. General guidance on the approach to this study is given in a report by the Institution of Structural Engineers.[2.19]

Table 2.8 Classification of visible damage to walls with particular reference to ease of repair of plaster and brickwork or masonry (after Tomlinson *et al.*[2.18]).

Category of damage	Degree of damage	Description of typical damage† (ease of repair is underlined)	Approximate crack width‡ (mm)
		Hairline cracks of less than about 0.1 mm width are classed as negligible	≯0.1
1	Very slight	*Fine cracks which can easily be treated during normal decoration.* Perhaps isolated slight fracturing in building. Cracks rarely visible in external brickwork	≯1.0
2	Slight	*Cracks easily filled. Redecoration probably required. Recurrent cracks can be masked by suitable linings.* Cracks not necessarily visible externally; *some external repointing may be required to ensure weathertightness.* Doors and windows may stick slightly	≯5.0
3	Moderate	*The cracks require some opening up and can be patched by a mason. Repointing of external brickwork and possibly a small amount of brickwork to be replaced.* Doors and windows sticking. Service pipes may fracture. Weathertightness often impaired	5–15 (or a number of cracks ≥3.0)
4	Severe	*Extensive repair work involving breaking-out and replacing sections of walls, especially over doors and windows.* Window and door frames distorted, floor sloping noticeably.§ Walls leaning§ or bulging noticeably, some loss of bearing in beams. Service pipes disrupted	15–25 but also depends on number of cracks
5	Very severe	*This requires a major repair job involving partial or complete rebuilding.* Beams lose bearings, walls lean badly and require shoring. Windows broken with distortion. Danger of instability	Usually >25 but depends on number of cracks

† It must be emphasized that in assessing the degree of damage account must be taken of the location in the building or structure where it occurs, and also of the function of the building or structure.
‡ Crack width is one factor in assessing degree of damage and should not be used on its own as direct measure of it.
§ Local deviations of slope, from the horizontal or vertical, of more than 1/100 will normally be clearly visible. Overall deviations in excess of 1/150 are undesirable.

It is unrealistic to design foundations to prevent all cracking due to differential settlement. In most buildings with internal plaster finish cracking can be seen in walls and ceilings due to thermal and moisture movements in the structure; therefore a certain degree of cracking due to differential settlement should be accepted (see Table 2.8).

In the case of simple structures on relatively uniform compressible soils the risks of damage due to settlement can be assessed with the guidance of empirical rules based on experience. For foundations on *sands*, Terzaghi and Peck[2.5] suggested that the differential settlement is unlikely to exceed 75 per cent of the maximum movement, and since most ordinary structures can withstand 20 mm of settlement between adjacent columns a limiting maximum settlement of 25 mm was postulated. For raft foundations the limiting maximum settlement was increased to 50 mm. From a study of movements of 11 buildings, Skempton and MacDonald[2.14] concluded that for a limiting angle of distortion (β) of 1 in 500 the limiting maximum differential settlement is about 25 mm, the limiting total settlement is 40 mm for isolated foundations, and 40–65 mm for raft foundations. Studies have shown that buildings on sands rarely settle by more than 50 mm and in the majority of cases the settlement is of the order of 25 mm or less.[2.20] These rules should not be applied to sands containing silt or clay which greatly increases their compressibility.

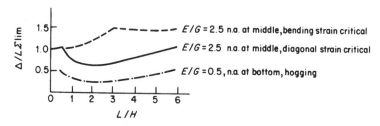

Fig. 2.21 Limit of deflection ratio for cracking in load-bearing wall or rectangular panel (after Burland *et al.*[2.12]).

For foundations on *clays*, Skempton and MacDonald[2.14] similarly prescribed a design limit for maximum *differential* settlement of 40 mm, with design limits for *total* settlement of 65 mm for isolated foundations and 65–100 mm for rafts.

If as a result of applying the above empirical rules, or of undertaking a settlement analysis of the structure based on the assumption of complete flexibility in the foundations and superstructure, it is shown that the total and differential settlements exceed the serviceability limit state, then the engineer has the choice either of avoiding settlement or of accommodating the movement by the appropriate measures in the structural design.

If the structures themselves have insufficient rigidity to prevent excessive differential movement with ordinary spread foundations, one or a combination of the following methods may be adopted in order to reduce the total and differential settlements to a tolerable figure:

(a) Provision of a rigid raft foundation either with a thick slab or with deep beams in two or three directions;
(b) Provision of deep basements to reduce the net bearing pressure on the soil;
(c) Transference of foundation loading to deeper and less compressible soil by means of basements, piers, or piles;
(d) Provision of jacking pockets, or brackets, in columns to relevel the superstructure;
(e) Provision of additional loading on lightly loaded areas in the form of kentledge or embankments.

As well as reducing maximum settlements due to relief of overburden pressure in excavating for deep basements, method (b) is useful in preventing excessive differential settlement between parts of a structure having different foundation loads. Thus the deepest basements can be provided under the heaviest parts of the structure with shallower or no basements in the areas of lighter loading.

An outstanding example of a combination of methods (b) and (c) is given by the foundations of the 40-storey Latino-Americana building in Mexico City (Fig. 2.22(a)).[2.21] A 13 m deep basement was constructed to reduce the net bearing pressure on the piled raft. The piles were driven to a depth of 34 m below street level to reach a 5 m thick stratum of sand followed by firm to stiff clays and sands. The settlement observations for this building (Fig. 2.22b)) illustrate the high compressibility of the deep volcanic clay layers in Mexico City. These deposits are settling under their own weight as shown by the surface settlement relative to the deep benchmark.

An important point to note in connection with the excavations of deep basements in clay soils is that swelling will occur to a greater or lesser degree on release of overburden pressure. This causes additional settlement as the swelled ground reconsolidates during the application of the structural load.

May[2.22] observed the heave of a basement on a site in London where the substructure did not carry any superimposed loading over the 5-year period of observation. The 11 m deep excavation was taken down through fill, sand, and sandy gravel alluvium close to the surface of the London Clay. A heave of 60 mm was observed at the centre of the basement after 5 years had elapsed from the date of completion. Movement was still continuing and it was estimated that the final heave would be 100 mm

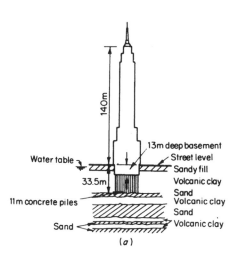

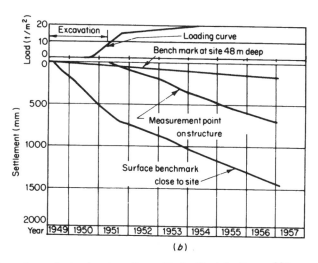

Fig. 2.22 (*a*) Foundation arrangement and (*b*) settlement observations at Latino-Americana Tower, Mexico City (after Zeevaert[2.21]).

including an upward movement of 27 mm during the period of construction of the basement. The observed heave corresponded to a coefficient of volume increase (m_v) of 0.035 m²/MN.

The effects of such swelling are eliminated by the adoption of piled basements as discussed in Section 5.5.

2.6.4 Pressure distribution beneath foundations

The first consideration in calculating the magnitude of settlements is the distribution of pressure beneath the loaded area. This depends on the rigidity of the foundation structure and the nature of the soil. The variation of *contact pressure* beneath a smooth rigid foundation on a clay (or a soil containing thick layers of soft clay) is shown in Fig. 2.23(a). A similar foundation on sand or gravel shows a very different contact pressure distribution as in Fig. 2.23(b), and the distribution for intermediate soil types takes the form shown in Fig. 2.23(c). When the bearing pressures are increased to the point of shear failure in the soil, the contact pressure is changed, tending to an increase in pressure over the centre of the loaded area in each of the above cases.

A fully flexible foundation, such as the steel plate floor of an oil storage tank, assumes the characteristic bowl shape as it deforms with the consolidation of the underlying soil. The contact pressure distribution for a fully flexible foundation on a clay soil takes the form shown in Fig. 2.23(d).

In the calculation of consolidation settlement we are concerned with the pressure distribution for a contact pressure which has a reasonable safety factor against shear failure of the soil. Also, it is impracticable to obtain complete rigidity in a normal foundation structure. Consequently the contact pressure distribution is intermediate between that of rigid and flexible foundations, and for all practicable purposes it is regarded as satisfactory to assume a *uniform pressure distribution* beneath the loaded area.

The next step is to consider the vertical stress distribution *in depth* beneath the loaded area. In the case of a concentrated load on the surface of the ground, the vertical stress (σ_z) at any point N beneath the load is given by Boussinesq's equation

$$\sigma_z = \frac{3Q}{2\pi z^2}\left[\frac{1}{(1 + (r/z)^2)^{5/2}}\right] \qquad (2.31)$$

where

Q = concentrated vertical load,
z = vertical distance between N and the underside of the foundation structure,
r = horizontal distance from N to the line of action of the load.

Boussinesq's equation is based on the assumption that the loaded material is elastic, homogeneous, and isotropic. None of this is strictly true for natural soils, but the assumptions are justifiable for practical design. Influence factors for calculating σ_z for various ratios of the diameter of a circular foundation and z have been calculated by Jurgenson[2.23] and are reproduced in Table 2.9.

In the case of *rigid foundations*, only the mean stress need be determined for settlement calculations. The coefficients for rectangular areas ranging from square to strip foundations for various depths below foundation level are shown in Fig. 2.24.

Using any of the above methods for determining stress distribution, the depth over which compression of the soil layers should be considered should be taken as the depth at which the effective vertical stress due to the foundation load decreases to 20 per cent of the effective overburden stress.

In the case of *rectangular flexible foundations* (thin plates, flexible rafts, or close-spaced pad or strip foundations), the stress is determined by Newmark's method.[2.24] The factors for determining the vertical stress at a corner of a rectangle have been tabulated by Newmark

Table 2.9 Influence factors for vertical pressure (σ_z) under centre of uniformly loaded flexible circular area of diameter D

$\dfrac{D}{z}$	Influence factor	$\dfrac{D}{z}$	Influence factor	$\dfrac{D}{z}$	Influence factor
0.00	0.0000	2.00	0.6465	4.00	0.9106
0.20	0.0148	2.20	0.6956	6.00	0.9684
0.40	0.0571	2.40	0.7376	8.00	0.9857
0.60	0.1213	2.60	0.7733	10.00	0.9925
0.80	0.1966	2.80	0.8036	12.00	0.9956
1.00	0.2845	3.00	0.8293	14.00	0.9972
1.20	0.3695	3.20	0.8511	16.00	0.9981
1.40	0.4502	3.40	0.8697	20.00	0.9990
1.60	0.5239	3.60	0.8855	40.00	0.9999
1.80	0.5893	3.80	0.8990	200.00	1.0000

Note: σ_z = influence factor × contact pressure (q).

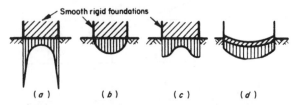

Fig. 2.23 Contact pressure distribution beneath foundations. (a) Clay. (b) Sand and gravel. (c) Intermediate soil type. (d) Fully flexible foundation on clay.

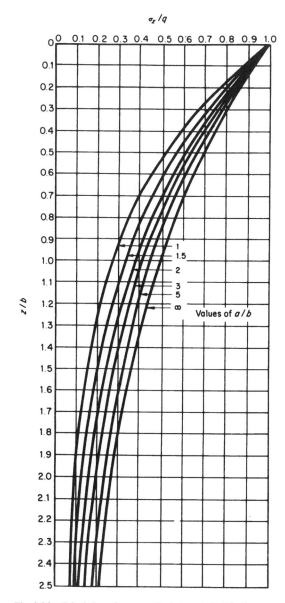

σ_z/q

Values of a/b

z/b

Fig. 2.24 Calculation of mean vertical stress (σ_z) at depth z beneath a rigid rectangular area $a \times b$ on surface, loaded at uniform pressure q.

(Table 2.10). To obtain the vertical stress at the centre of the loaded area, the area is divided into four equal rectangles. By the principle of superposition, the vertical stress at the centre of the loaded area is equal to four times the stress at any one corner. The procedure for obtaining the pressure distribution by Newmark's method beneath irregular loaded areas has been described by Terzaghi and Peck.[2.25]

In the case of *inclined loading*, the loaded area is

transformed into an equivalent rectangular area of dimensions $B' \times L'$, carrying a uniform bearing pressure, using the method shown in Fig. 2.10.

2.6.5 Estimation of settlements of foundations on sands and gravels

As already noted, settlements of foundations on sands, gravels, and granular fill materials, take place almost immediately as the foundation loading is imposed on them. Because of the difficulty of sampling these soils, there is no practicable laboratory test procedure for determining their consolidation characteristics. From a review of a number of case records of the settlement of structures founded on these soils, Sutherland[2.20] concluded that there is no reliable method for extrapolating the settlement of a standard plate to the settlement of an actual foundation at the same location. Consequently settlements of foundations on sands and gravels are estimated by semi-empirical methods based on SPT or CPT values or on the results of pressuremeter tests.

Estimation of settlements from standard penetration tests Schultze and Sharif[2.26] established an empirical relationship between the SPT N-values, foundation dimensions, and embedment depth to obtain the short-term settlement of foundations on sands. This relationship is shown as an equation with graphical values of the coefficient of settlement in Fig. 2.25. It was established from a correlation of N-values with the observed settlements of structures. It should be noted that the depth of influence over which the average N-value is taken was assumed by Schultze and Sharif to be twice the foundation width.

Burland and Burbidge[2.11] established another empirical relationship based on the standard penetration test from which the settlements of foundations on sands and gravels can be calculated from the equation

$$\text{Settlement} = \rho = f_s \cdot f_l \cdot f_t[(q'_n - \tfrac{2}{3} \cdot p'_o) \\ \times B^{0.7} \times I_c] \text{ in millimetres} \quad (2.32)$$

where

$\quad f_s$ = shape factor given by equation (2.33),
$\quad f_l$ = correction factor for the depth of the sand or gravel layer from equation (2.34),
$\quad f_t$ = time factor from equation (2.35),
$\quad q'_n$ = average net applied pressure in kN/m²,
$\quad p'_o$ = maximum previous effective overburden pressure in kN/m²,
$\quad B$ = foundation depth in metres,
$\quad I_c$ = compressibility index.

Here I_c is related to the SPT N-value as shown in Fig.

Table 2.10   Influence values (I_σ) for vertical normal stress σ_z at point N beneath corner of a uniformly loaded rectangular area

B/z	L/z													
	0.1	0.2	0.3	0.4	0.5	0.6	0.7	0.8	0.9	1.0	1.2	1.4	1.6	1.8
0.1	0.004 70	0.009 17	0.013 23	0.016 78	0.019 78	0.022 23	0.024 20	0.025 76	0.026 98	0.027 94	0.029 26	0.030 07	0.030 58	0.030 90
0.2	0.009 17	0.017 90	0.025 85	0.032 80	0.038 66	0.043 48	0.047 35	0.050 42	0.052 83	0.054 71	0.057 33	0.058 94	0.059 94	0.060 58
0.3	0.013 23	0.025 85	0.037 35	0.047 42	0.055 93	0.062 94	0.068 58	0.073 08	0.076 61	0.079 38	0.083 23	0.085 61	0.087 09	0.088 04
0.4	0.016 78	0.032 80	0.047 42	0.060 24	0.071 11	0.080 09	0.087 34	0.093 14	0.097 70	0.101 29	0.106 31	0.109 41	0.111 35	0.112 60
0.5	0.019 78	0.038 66	0.055 93	0.071 11	0.084 03	0.094 73	0.103 40	0.110 35	0.115 84	0.120 18	0.126 26	0.130 03	0.132 41	0.133 95
0.6	0.022 23	0.043 48	0.062 94	0.080 09	0.094 73	0.106 88	0.116 79	0.124 74	0.131 05	0.136 05	0.143 09	0.147 49	0.150 28	0.152 07
0.7	0.024 20	0.047 35	0.068 58	0.087 34	0.103 40	0.116 79	0.127 72	0.136 53	0.143 56	0.149 14	0.157 03	0.161 99	0.165 15	0.167 20
0.8	0.025 76	0.050 42	0.073 08	0.093 14	0.110 35	0.124 74	0.136 53	0.146 07	0.153 71	0.159 78	0.168 43	0.173 89	0.177 39	0.179 67
0.9	0.026 98	0.052 83	0.076 61	0.097 70	0.115 84	0.131 05	0.143 56	0.153 71	0.161 85	0.168 35	0.177 66	0.183 57	0.187 37	0.189 86
1.0	0.027 94	0.054 71	0.079 38	0.101 29	0.120 18	0.136 05	0.149 14	0.159 78	0.168 35	0.175 22	0.185 08	0.191 39	0.195 46	0.198 14
1.2	0.029 26	0.057 33	0.083 23	0.106 31	0.126 26	0.143 09	0.157 03	0.168 43	0.177 66	0.185 08	0.195 84	0.202 78	0.207 31	0.210 32
1.4	0.030 07	0.058 94	0.085 61	0.109 41	0.130 03	0.147 49	0.161 99	0.173 89	0.183 57	0.191 39	0.202 78	0.210 20	0.215 10	0.218 36
1.6	0.030 58	0.059 94	0.087 09	0.111 35	0.132 41	0.150 28	0.165 15	0.177 39	0.187 37	0.195 46	0.207 31	0.215 10	0.220 25	0.223 72
1.8	0.030 90	0.060 58	0.088 04	0.112 60	0.133 95	0.152 07	0.167 20	0.179 67	0.189 86	0.198 14	0.210 32	0.218 36	0.223 72	0.227 36
2.0	0.031 11	0.061 00	0.088 67	0.113 42	0.134 96	0.153 26	0.168 56	0.181 19	0.191 52	0.199 94	0.212 35	0.220 58	0.226 10	0.229 86
2.5	0.031 38	0.061 55	0.089 48	0.114 50	0.136 28	0.154 83	0.170 36	0.183 21	0.193 75	0.202 36	0.215 12	0.223 64	0.229 40	0.233 34
3.0	0.031 50	0.061 78	0.089 82	0.114 95	0.136 84	0.155 50	0.171 13	0.184 07	0.194 70	0.203 41	0.216 33	0.224 99	0.230 88	0.234 95
4.0	0.031 58	0.061 94	0.090 07	0.115 27	0.137 24	0.155 98	0.171 68	0.184 69	0.195 40	0.204 17	0.217 22	0.226 00	0.232 00	0.236 88
5.0	0.031 60	0.061 99	0.090 14	0.115 37	0.137 37	0.156 12	0.171 85	0.184 88	0.195 61	0.204 40	0.217 49	0.226 32	0.232 36	0.237 35
6.0	0.031 61	0.062 01	0.090 17	0.115 41	0.137 41	0.156 17	0.171 91	0.184 96	0.195 69	0.204 49	0.217 60	0.226 44	0.232 49	0.236 71
8.0	0.031 62	0.062 02	0.090 18	0.115 43	0.137 44	0.156 21	0.171 95	0.185 00	0.195 74	0.204 55	0.217 67	0.226 52	0.232 58	0.236 81
10.0	0.031 62	0.062 02	0.090 19	0.115 44	0.137 45	0.156 22	0.171 96	0.185 02	0.195 76	0.204 57	0.217 69	0.226 54	0.232 61	0.236 84
∞	0.031 62	0.062 02	0.090 19	0.115 44	0.137 45	0.156 23	0.171 97	0.185 02	0.195 77	0.204 58	0.217 70	0.226 56	0.232 63	0.236 86

B/z	L/z								
	2.0	2.5	3.0	4.0	5.0	6.0	8.0	10.0	∞
0.1	0.031 11	0.031 38	0.031 50	0.031 58	0.031 60	0.031 61	0.031 62	0.031 62	0.031 62
0.2	0.061 00	0.061 55	0.061 78	0.061 94	0.061 99	0.062 01	0.062 02	0.062 02	0.062 02
0.3	0.088 67	0.089 48	0.089 82	0.090 07	0.090 14	0.090 17	0.090 18	0.090 19	0.090 19
0.4	0.113 42	0.114 50	0.114 95	0.115 27	0.115 37	0.115 41	0.115 43	0.115 44	0.115 44
0.5	0.134 96	0.136 28	0.136 84	0.137 24	0.137 37	0.137 41	0.137 44	0.137 45	0.137 45
0.6	0.153 26	0.154 83	0.155 50	0.155 98	0.156 12	0.156 17	0.156 21	0.156 22	0.156 23
0.7	0.168 56	0.170 36	0.171 13	0.171 68	0.171 85	0.171 91	0.171 95	0.171 96	0.171 97
0.8	0.181 19	0.183 21	0.184 07	0.184 69	0.184 88	0.184 96	0.185 00	0.185 02	0.185 02
0.9	0.191 52	0.193 75	0.194 70	0.195 40	0.195 61	0.195 69	0.195 74	0.195 76	0.195 76
1.0	0.199 94	0.202 36	0.203 41	0.204 17	0.204 40	0.204 49	0.204 55	0.204 57	0.204 58
1.2	0.212 35	0.215 12	0.216 33	0.217 22	0.217 49	0.217 60	0.217 67	0.217 69	0.217 70
1.4	0.220 58	0.223 64	0.224 99	0.226 00	0.226 32	0.226 44	0.226 52	0.226 54	0.226 56
1.6	0.226 10	0.229 40	0.230 88	0.232 00	0.232 36	0.232 49	0.232 58	0.232 61	0.232 63
1.8	0.229 86	0.233 34	0.234 95	0.236 98	0.239 35	0.236 71	0.236 81	0.236 84	0.236 86
2.0	0.232 47	0.236 14	0.237 82	0.239 12	0.239 54	0.239 70	0.239 81	0.239 85	0.239 87
2.5	0.236 14	0.240 10	0.241 96	0.243 44	0.243 92	0.244 12	0.244 25	0.244 29	0.244 32
3.0	0.237 82	0.241 96	0.243 94	0.245 54	0.246 08	0.246 30	0.246 46	0.246 50	0.246 54
4.0	0.239 12	0.243 44	0.245 54	0.247 29	0.247 91	0.248 17	0.248 36	0.248 42	0.248 46
5.0	0.239 54	0.243 92	0.246 08	0.247 91	0.248 57	0.248 85	0.249 07	0.249 14	0.249 19
6.0	0.239 70	0.244 12	0.246 30	0.248 17	0.248 85	0.249 16	0.249 39	0.249 46	0.249 52
8.0	0.239 81	0.244 25	0.246 46	0.248 36	0.249 07	0.249 39	0.249 64	0.249 73	0.249 80
10.0	0.239 85	0.244 29	0.246 50	0.248 42	0.249 14	0.249 46	0.249 73	0.249 81	0.249 89
∞	0.239 87	0.244 32	0.246 54	0.248 46	0.249 19	0.249 52	0.249 80	0.249 89	0.250 00

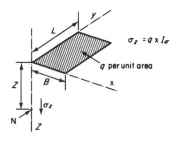

$\sigma_z = q \times I_\sigma$

q per unit area

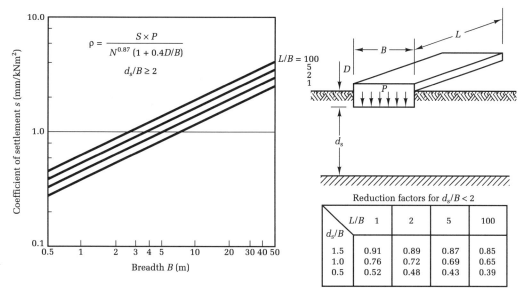

Fig. 2.25 Determining foundation settlements from results of standard penetration tests (after Schultze and Sharif[2.26]).

2.26. The average N-value is taken over the depth of influence z_I which is obtained from Fig. 2.27. The probable limits of accuracy of equation (2.32) can be assessed from the upper and lower limits of I_c in Fig. 2.31. It may be necessary to take these limits into account where total and differential settlement are a critical factor in foundation design. The term p'_o is introduced by Burland to allow for the effects of previous over-consolidation of the soil or for loading at the base of an excavation for which the maximum previous overburden pressure was p'_o. For shallow foundations on a normally consolidated sand the term within the squared brackets in equation (2.32) becomes

$$(q'_n \times B^{0.7} \times I_c)$$

The shape factor is given by the equation

$$f_s = \left(\frac{1.25 L/B}{L/B + 0.25} \right)^2. \tag{2.33}$$

Where the depth of influence of the applied pressure z_I is greater than the depth of the sand or gravel (H) the correction factor f_l is given by the equation

$$f_l = \frac{\rho_t}{\rho_i} = \frac{H}{z_I} \left(2 - \frac{H}{z_I} \right). \tag{2.34}$$

Values of z_I can be obtained from Fig. 2.27.

Burland recognizes that the settlements on sands and gravels can be time dependent. His time factor is given by the equation

$$f_t = \left(1 + R_3 + R \log\frac{t}{3} \right), \tag{2.35}$$

where

$t \geq 3$ years,

$R =$ creep ratio expressed as a proportion of the immediate settlement (ρ_i) that takes place per log cycle of time,

$R_3 =$ time-dependent settlement expressed as a proportion of the immediate settlement (ρ_i) that takes place during the first 3 years after construction.

Burland gives conservative values of R and R_3 as 0.2 and 0.3 respectively for static loading, and 0.8 and 0.7 respectively for fluctuating loads.

It is important to note that no corrections are made to the N-values to allow for the effective overburden pressure, i.e. for the level of the ground water. It is accepted that this is reflected in the N-values measured *in situ*. However, the Terzaghi and Peck correction for a fine or silty sand where N-values are greater than 15 should be applied, i.e. when the SPT value exceeds 15 it should be assumed that the density of the soil is equal to that of a sand having an N-value of $15 \pm \frac{1}{2} (N - 15)$. In gravels or sandy gravels the N-values should be increased by a factor of 1.25. Where the results of static cone tests are available they can be converted to N-values by reference to Fig. 1.7.

Where the plot of N-values increases linearly with depth, which is the case for most normally consolidated sands, the Burland and Burbidge methods tend to give higher settlements for loose soils with N-values less than 10 compared with those obtained from the Schultze and

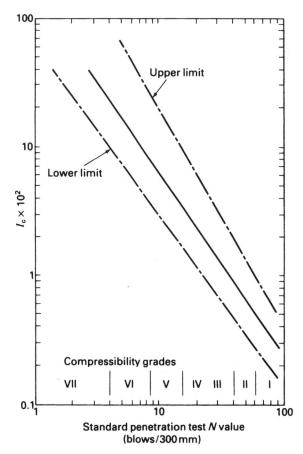

Fig. 2.26   Values of the compressibility index for sands and gravels (after Burland and Burbidge[2.11]).

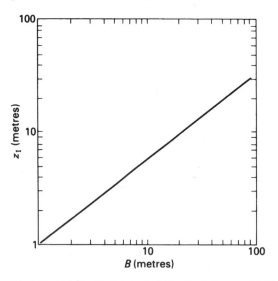

Fig. 2.27   Relationship between breadth of loaded area and depth of influence z_I (after Burland and Burbidge[2.11]).

Sharif equation (Fig. 2.25). This is because the former method is biased towards the N-value at a shallow depth below the foundation. Conversely, Burland and Burbidge give smaller settlements for N-values in excess of about 30.

The upper limit in Fig. 2.26 is likely to give over-large settlements for wide foundations on loose sands. Burland and Burbidge stated that there were very few cases recorded where buildings on sand have settled more than 75 mm.

Estimation of settlements from static cone penetration tests   The Schmertmann *et al.*[2.27] equation for calculating the settlement of foundations on cohesionless soils is

$$\rho = C_1 C_2 \Delta_\mathrm{p} \Sigma_0^{2B} \frac{I_z}{E_\mathrm{s}} \Delta_z.$$

$$(2.36)$$

where

C_1 = depth correction factor (see below),
C_2 = creep factor (see below),
Δ_p = net increase of load on soil at foundation level due to applied loading,
B = width of loaded area,
I_z = vertical strain influence factor (Fig. 2.28),
E_s = deformation modulus,
Δ_z = thickness of soil layer.

The depth correction factor is given by

$$C_1 = 1 - 0.5 \left(\frac{\sigma'_\mathrm{vo}}{\Delta_\mathrm{p}} \right),$$

$$(2.37)$$

where σ'_vo = effective overburden pressure at foundation level.

Schmertmann states that although settlements on cohesionless soils are regarded as immediate, observations frequently show long-term creep, calculated by the factor

$$C_2 = 1 + 0.2 \log_{10} \left(\frac{\mathrm{time_{years}}}{0.1} \right).$$

$$(2.38)$$

The vertical strain influence factor is obtained from one of the two curves shown in Fig. 2.28. For square foundations (axisymmetric loading) the curve for $L/B = 1$ should be used. For long strip foundations (the plane strain case), where the length is more than 10 times the breadth, use the curve for $L/B > 10$. Values for rectangular foundations for $L/B < 10$ can be obtained by interpolation.

The deformation modulus for square and long strip foundations is obtained by multiplying the static cone resistance by a factor of 2.5 and 3.5 respectively.

Where SPTs only are available the static cone resistance q_c (in kg/cm^2) can be obtained by multiplying the SPT N-values (in blows/300 mm) by an empirical factor for which Schmertmann suggests the following values:

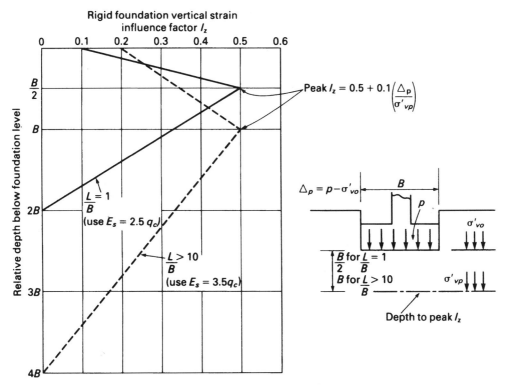

Fig. 2.28   Vertical strain influence factor diagrams (after Schmertmann *et al.*[2.27]).

Silts, sandy silts, and slightly cohesive silty sands	$N = 2$
Clean fine to medium sands, slightly silty sands	$N = 3.5$
Coarse sands and sands with a little gravel	$N = 5$
Sandy gravel and gravels	$N = 6$

The cone resistance diagram is divided into layers of approximately equal or representative values of q_c (see Worked Example 2.4) and the strain influence factor diagram is placed alongside this diagram beneath the foundation which is drawn to the same scale. The settlements in each layer resulting from the loading Δ_p are then calculated using the values of E_d and I_z appropriate to each layer. The sum of the settlements in each layer is then corrected for the depth factor and creep factor using equations (2.37) and (2.38).

It should be noted that the values for the deformation modulus E_s in Fig. 2.28 as $2.5q_c$ and $3.5q_c$ for square and strip foundations respectively assume, rather crudely, that E_s is constant irrespective of the vertical stress imposed by the foundation. This is not the case in practice. Figure 2.29 shows a typical stress–strain curve as obtained from a plate loading test. The E-value is constant only over the initial low strain portion of the curve. This value is referred to as the initial tangent modulus or Young's modulus. Hence, when attempting accuracy in settlement calculations, the stress level for the particular E-value adopted should be defined as a percentage of the stress at a strain corresponding to failure. This is illustrated in Fig. 2.30, which relates the secant modulus to the CPT cone resistance for stress levels of 25 and 50 per cent of the failure stress. It will be seen that the values of 2.5–$3.5q_c$ suggested by Schmertmann correspond to a stress level of about 25 per cent of the failure stress.

The curves in Fig. 2.30 are for uncemented normally consolidated quartz sands. Over-consolidated sands have higher modulus values for a given cone resistance as shown in Fig. 2.31. Cemented sands are likely to have a high initial modulus, but breakdown of the inter-particle bonds with increasing stress will cause drastic lowering leading to collapse settlement.

Values of the initial tangent modulus have been proposed by Lunne and Christoffersen[2.28] on the basis of a comprehensive review of field and laboratory tests as follows. For normally consolidated sands:

$$E = 4q_c (\text{MN/m}^2) \quad \text{for } q_c < 10\,\text{MN/m}^2$$
$$E = (2q_c + 20)\text{MN/m}^2 \quad \text{for } 10\,\text{MN/m}^2 < q_c$$
$$< 50\,\text{MN/m}^2$$
$$E = 120\,\text{MN/m}^2 \quad \text{for } q_c > 50\,\text{MN/m}^2.$$

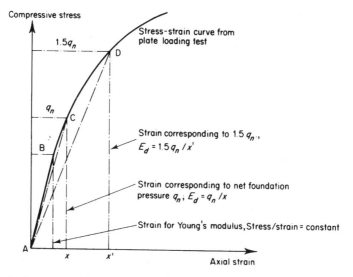

Fig. 2.29   Determination of deformation modulus from stress–strain curve obtained from plate bearing tests.

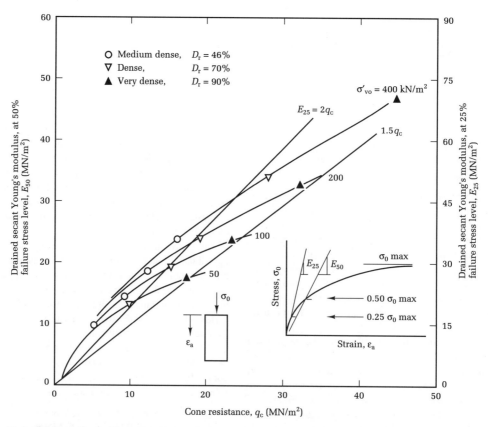

Fig. 2.30   Drained deformation modulus values for uncemented normally consolidated quartz sands in relation to cone resistance (after Robertson and Campanella[2.29] and Baldi *et al.*[2.30]).

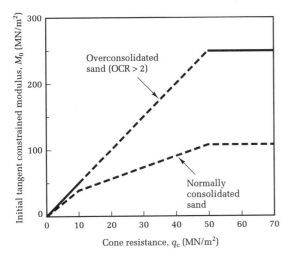

Fig. 2.31 Initial tangent constrained modulus for normally and overconsolidated sands related to cone resistance (after Lunne and Christoffersen[2.28]).

For over-consolidated sands with an over-consolidation ratio >2:

$$E = 5q_c \quad \text{for } q_c < 50 \text{ MN/m}^2$$
$$E = 250 \text{ MN/m}^2 \quad \text{for } q_c > 50 \text{ MN/m}^2$$

The deformation modulus applicable for a stress increase of Δ_p above the effective overburden pressure σ'_{vo} is given by the equation

$$E_d = E \sqrt{\frac{p_o + (\Delta_p/2)}{\sigma'_{vo}}} \qquad (2.39)$$

Estimation of settlements from pressuremeter tests
The self-boring pressuremeter, which causes the minimum disturbance, is a useful method of obtaining the shear modulus G of a sand. This is converted to the drained deformation modulus E' from the relationship

$$E' = 2G(1 + v') \qquad (2.40)$$

where v' is the Poisson's ratio for undrained conditions (typically 0.15 for coarse-grained soils and 0.25 for fine-grained soils).

The shear modulus is obtained from the load–unload loop of the pressure–volume curve at the strain level corresponding to the design stress conditions, as discussed for the derivation of E-values from the static cone resistance. The E-value is then used in conjunction with equations (2.41) or (2.42).

2.6.6 Estimation of settlements on clay soils

Very comprehensive and practical guidance on the subject of settlement of structures on clay soils is given in a publication by the Construction Industry Research and Information Association.[2.31]

The simplified procedure for estimating consolidation settlements described in the following pages is based on Skempton and Bjerrum's modification[2.32] of Terzaghi's theory of consolidation. The latter assumes that consolidation is a process of one-dimensional strain. Comparison of observed and calculated settlements indicates that the method underestimates the rate of settlement. Other theories have been published which claim to represent more accurately the consolidation process, but in most cases these are difficult to apply in practice. In any event the magnitude of settlement of engineering structures on most firm to stiff soils is relatively small. For this reason, refinement of calculations is hardly justifiable, especially as natural variations in soil compressibility can produce settlement variations of greater significance than differences in values calculated by the various methods. Only in cases of soft compressible soils carrying large heavy structures is it justifiable to adopt the more elaborate calculation methods based on three-dimensional strain. In cases where the flexural rigidity of the structure has a significant effect on the magnitude of differential settlement it can be advantageous to model the soil–structure interaction by means of a finite-element analysis as described in Section 2.8.4.

The steps in carrying out the settlement analysis are as follows:

(a) *Choice of the soil profile.* A soil profile should be drawn for the site, on which is marked the average depths of the various soil strata and the average (or characteristic) values of the undrained shear strength, the compression index (or the coefficient of volume compressibility), and coefficient of consolidation for each stratum or each division of a stratum. In the case of thick clay strata, it must not be assumed that the compressibility is constant throughout the depth of the strata. Clays usually show progressively decreasing compressibility and increase in deformation modulus with increasing depth. It frequently happens that the borehole records and soil test results show wide variations in the depths of the strata and compressibility values. In these circumstances the choice of a representative soil profile involves exercise of careful judgement. For large and important structures it is worthwhile to make settlement analyses for the highest compressibility and maximum depth of compressible strata and the lowest compressibility with the minimum depth of strata, and then to compare the two analyses to obtain an idea of the differential settlement if these two extremes of conditions occur over the area of the structure.

(b) *Assessment of the loading causing settlement.* When considering long-term consolidation settlement it is essential that the foundation loading used in the analysis should be realistic and representative of the *sustained* loading over the time period under consideration. This is a different approach to that used when calculating safe bearing pressures. In the latter case the most severe loading conditions are allowed for, with full provision for maximum imposed loading. The imposed loading used in a settlement analysis is an *average value* representing the continuous load over the time period being considered.

Wind loading is only considered in settlement analyses for high structures where it represents a considerable proportion of the total loads, and then only the wind loads representing the average of continuous wind over the full period are allowed for.

The calculation of consolidation settlement is based on the increases in effective vertical stresses induced by the loads or actions from the structure.

(c) *Calculation of pressure and stress distribution.* The distribution of *effective* vertical pressure of the overburden σ'_{vo}) and the vertical stress (σ_z) resulting from the net foundation pressure (q_n) is shown in Fig. 2.32. Values of σ_z at various depths below foundation level are obtained by the methods described in Section 2.6.4.

In the case of deep compressible soils the lowest level considered in the settlement analysis is the point where the vertical stress (σ_z) is not more than 20 per cent of σ'_{vo}.

(d) *Calculation of net immediate settlement* (ρ_i). The net immediate settlement (ρ_i), i.e. the elastic settlement beneath the corner of a flexible loaded areas, is calculated from the equation

$$\rho_i = q_n \times B \times \frac{(1 - \nu^2)}{E} \times I_p \qquad (2.41)$$

where

$\quad B$ = width of foundation,
$\quad E$ = deformation modulus,
$\quad \nu$ = Poisson's ratio,
$\quad q_n$ = *net* foundation pressure,
$\quad I_p$ = influence factor.

Eurocode 7 refers to calculations of settlement using equation (2.41) as the adjusted elasticity method. The *drained* modulus E'_d is used for calculating the total (immediate + consolidation) settlement of the foundations on gravels, sands, silts, and clays. The *undrained* modulus E_u is used to calculate the immediate settlement of foundations on clays or clayey silts. As already noted, it is rather impracticable to determine E' values of sands and gravels from laboratory tests on undisturbed samples, field tests being used instead. However, it is the general practice to obtain drained and undrained E-values of clays from laboratory tests on undisturbed samples taken from boreholes. The undrained modulus E_u of clays can be determined from relationships with the undrained shear strength, as shown in Fig. 2.33.

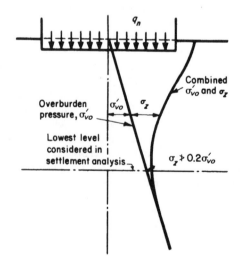

Fig. 2.32   Vertical pressure and stress distribution for deep clay layer.

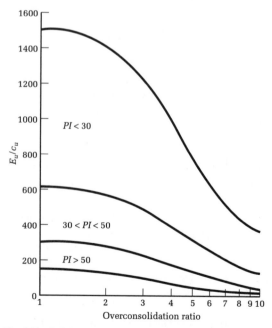

Fig. 2.33   Relationship between E_u/c_u ratio for clays with plasticity index and degree of overconsolidation (after Jamiolkowski *et al.*[2.33]).

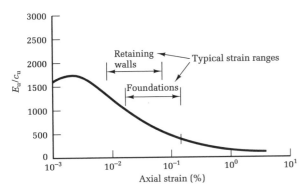

Fig. 2.34 Relationship between E_u/c_u and axial strain (after Jardine et al.[2.34]).

It should be noted that both E_u and E'_d are strain dependent. Relationships between E_u/c_u and strain for London Clay are shown in Fig. 2.34, where c_u is obtained in the laboratory from undrained triaxial tests on good-quality samples taken in thin-wall tubes. The strain applicable to normal foundations is 0.01–0.1 per cent, thus confirming the relationship $E_u/c_u = 400$ which is frequently used for intact blue London Clay.

It is important to note that swelling of the clay due to exposure in the medium or long term can reduce E_u which is the reason, when calculating foundation settlements, for adoption of E_u/c_u ratios which are lower than those obtained from triaxial or pressuremeter testing. Padfield and Sharrock[2.31] state that a ratio of 140 is often used for routine work in London Clay, but this is likely to be pessimistic by a factor of 2 or 3 for low strains. Higher ratios, as obtained by triaxial or pressuremeter testing, are used for determining ground movements around excavations, as discussed in Chapter 9. Very few E_u values have been published for other types of clay.

Marsland[2.35] obtained values of E_u/c_u from plate bearing tests of 348 for upper glacial till and 540 for a laminated glacial clay at Redcar.

If the results of drained triaxial tests are not available, the drained modulus for overconsolidated clays can be obtained approximately from the relationship $E'_d = 0.6E_u$. Alternatively, if m_v values from oedometer tests are available, E'_d is the reciprocal of m_v. This gives yet another method of obtaining E'_d from the results of SPTs using Fig. 1.4.

Values of the Poisson's ratio v in equation (2.41) are

Clays (undrained)	0.5
Clays (stiff, undrained)	0.1–0.2
Silt	0.3
Sands	0.1–0.3
Rocks	0.2

I_p is a function of the length-to-breadth ratio of the foundation, and the thickness (H) of the compressible layer. Terzaghi had given a method of calculating I_p from curves derived by Steinbrenner.

For Poisson's ratio of 0.5, $I_p = F_1$;
For Poisson's ratio of zero, $I_p = F_1 + F_2$.

Values of F_1 and F_2 for various ratios of H/B and L/B are given in Fig. 2.35. Elastic settlements should not be calculated for thickness (H) greater than $4B$.

The immediate settlement at any point N (Fig. 2.35) is given by

$$\rho_i \text{ at point N} = \frac{q_n}{E}(1 - v^2)(I_{p1}B_1 + I_{p2}B_2$$

$$+ I_{p3}B_3 + I_{p4}B_4). \qquad (2.42)$$

To obtain the settlements at the centre of the loaded area

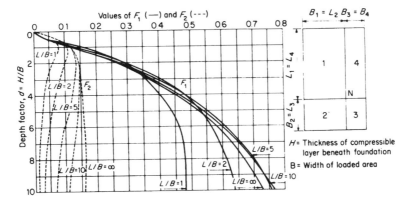

Fig. 2.35 Calculation of immediate settlements due to a flexible loaded area on the surface of an elastic layer. Note: when using this diagram to calculate ρ_i at *centre* of rectangular area, take B as half foundation width to obtain H/B and L/B.

the principle of superposition is followed in a similar way to the calculation of vertical stress (Section 2.6.4). The area is divided into four equal rectangles, and the settlement at the corner is calculated from equation (2.41). Then the settlement at the centre is equal to four times the settlement at any one corner.

The curves in Fig. 2.35 are based on the assumption that the modulus of deformation is constant with depth. However, in most natural soil and rock formations the modulus increases with depth such that calculations based on a constant modulus give exaggerated estimates of settlement. Butler[2.36] has developed a method, based on the work of Brown and Gibson[2.37], for calculating the settlement for conditions of a modulus of deformation increasing linearly with depth within a layer of finite thickness.

The value of the modulus at any depth z below foundation level is given by the equation

$$E_d = E_f(1 + kz/B), \qquad (2.43)$$

where E_f is the modulus of deformation at foundation level. The value of k is obtained by plotting measured values of E_d against depth and drawing a straight line through the plotted points to obtain values for substitution in equation (2.43). Having obtained k, the appropriate influence factor I'_p is obtained from Butler's curves shown in Fig. 2.36. These are for different ratios of L/B, and are applicable for a compressible layer of thickness not greater than nine times B. The curves are based on the assumption of a Poisson's ratio of 0.5. This ratio is

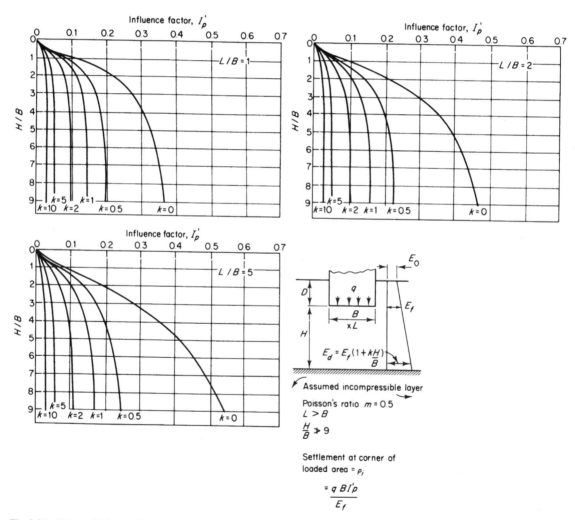

Fig. 2.36   Values of influence factor I'_p for deformation modulus increasing linearly with depth and modular ratio of 0.5 (after Butler[2.36]).

applicable to saturated clays for undrained conditions, i.e. immediate application of load.

In the case of a rigid foundation, for example a heavy beam and slab raft or a massive pier, the immediate settlement at the centre is reduced by a rigidity factor. The commonly accepted factor is 0.8, thus

$$\frac{\text{Immediate settlement of rigid foundation}}{\text{Immediate settlement at centre of flexible foundation}}$$

$$= 0.8 \text{ approx.}$$

A correction is also applied to the immediate settlements to allow for the depth of foundation by means of the 'depth factor' (see Fig. 2.39).

Generally, it will be found more convenient for the case of a constant E_d to use the method of Janbu et al.[2.38] as modified by Christian and Carrier[2.39] to obtain the aver-

age immediate settlement of a foundation where

$$\text{Average settlement} = \rho_i = \frac{\mu_1 \mu_0 q_n B}{E_d}. \qquad (2.44)$$

In the above equation Poisson's ratio is assumed to be 0.5. The factors μ_1 and μ_0 are related to the depth D of the foundation, the thickness H of the compressible layer and the length/width (L/B) ratio of the foundation. Values of these factors are shown in Fig. 2.37. Rigidity factors and depth factors are not applied to the settlements calculated from equation (2.44).

Because of the difficulty of obtaining representative values of the deformation modulus of a clay, either by correlation with the undrained shear strength or directly from field or laboratory tests, it may be preferable to determine the immediate settlement from relationships established by Burland et al.[2.12]. These are as follows.

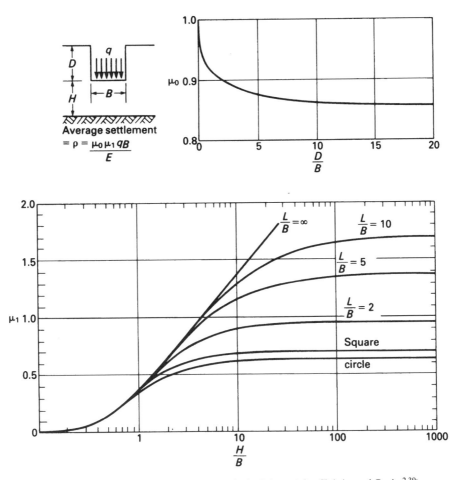

Fig. 2.37 Factors for calculating the average immediate settlement of a loaded area (after Christian and Carrier[2.39]).

For stiff over-consolidated clays:

Immediate settlement = ρ_i = 0.5 to 0.6ρ_{oed}
Consolidation settlement = ρ_c = 0.5 to 0.4ρ_{oed}
Final settlement = ρ_{oed}

For soft normally-consolidated clays:

Immediate settlement = 0.1ρ_{oed}
Consolidation settlement = ρ_{oed}
Final settlement = 1.1ρ_{oed}

(e) *Calculation of consolidation settlement* (ρ_c). If the variation in compressibility of a soil is known from the result of a number of oedometer tests, the consolidation settlement (ρ_c) is calculated preferably from the values of the coefficient of volume compressibility (m_v) as determined from the oedometer tests. Some typical values are shown in Table 2.11. Relationships between the undrained shear strength, plasticity index, N-value and modulus of volume compressibility are shown in Fig. 1.4.

Skempton and Bjerrum[2.32] have shown that the actual consolidation settlement (ρ_c) may be less than the calculated values based on oedometer tests. They give the formula

$$\rho_c = \mu_g \rho_{oed} \qquad (2.45)$$

where

μ_g = a coefficient (geological factor) which depends on the type of clay,
ρ_{oed} = settlement as calculated from oedometer tests.

Skempton and Bjerrum have related μ_g to the pore pressure coefficient of the soils as determined from undrained triaxial compression tests, and also to the dimensions of the loaded area. However, for most practical purposes it is sufficient to take the following values for μ_g:

Type of clay	μ_g
Very sensitive clays (soft alluvial, estuarine, and marine clays)	1.0–1.2
Normally consolidated clays	0.7–1.0
Overconsolidated clays (London Clay, Weald, Kimmeridge, Oxford, and Lias Clays)	0.5–0.7
Heavily overconsolidated clays (glacial till, Keuper Marl)	0.2–0.5

The μ_g value for London Clay is generally taken as 0.5.

Although the geological factor has some theoretical basis it is generally regarded as a means whereby the apparently high settlements calculated from oedometer test results can be reconciled with the much smaller settlements as measured in foundations on stiff over-consolidated clays. It is possible that oedometer tests on good samples taken by piston-driven thin-wall tubes will give lower m_v values than those obtained from conventional hammer-driven thick-wall tube samples and hence the geological factor will no longer be required. However, there is little published evidence at present to justify changing the present practice.

The oedometer settlement (ρ_{oed}) of a soil layer is calculated from the formula

$$\rho_{oed} = m_v \sigma_z H, \qquad (2.46)$$

where

m_v = average coefficient of volume compressibility obtained for the effective pressure increment in the particular layer under consideration,
σ_z = average effective vertical stress imposed on the particular layer resulting from the net foundation pressure (q_n),
H = thickness of the particular layer under consideration.

Table 2.11   Compressibility of various types of clays

Type	Qualitative description	Coefficient of volume compressibility, m_v (m²/MN)
Heavily overconsolidated boulder clays (e.g. many Scottish boulder clays) and stiff weathered rocks (e.g. weathered siltstone), hard London Clay, Gault Clay, and Oxford Clay (at depth)	Very low compressibility	Below 0.05
Boulder clays (e.g. Teesside, Cheshire) and very stiff 'blue' London Clay, Oxford Clay, Keuper Marl	Low compressibility	0.05–0.10
Upper 'blue' London Clay, weathered 'brown' London Clay, fluvio-glacial clays, Lake clays, weathered Oxford Clay, weathered Boulder Clay, weathered Keuper Marl, normally consolidated clays (at depth)	Medium compressibility	0.10–0.30
Normally consolidated alluvial clays (e.g. estuarine clays of Thames, Firth of Forth, Bristol Channel, Shatt-al-Arab, Niger Delta, Chicago Clay), Norwegian 'Quick' Clay	High compressibility	0.30–1.50
Very organic alluvial clays and peats	Very high compressibility	Above 1.50

The values of ρ_{oed} and hence ρ_c obtained for each layer are added together to obtain the total consolidation settlement beneath the loaded area.

If only one or two oedometer test results are available for a given loaded area it is more convenient to calculate ρ_{oed} directly from the *pressure–voids ratio* curves obtained in the tests. The procedure is then as follows:

For normally consolidated clays such as estuarine or marine clays, the virgin compression curve should be used for the settlement calculations. For preconsolidated or overconsolidated clays such as boulder clays and the London, Woolwich, and Reading, Gault, Weald, Kimmeridge, Oxford, and Lias Clays, the actual pressure–voids ratio curves should be used. Typical curves for normally consolidated clays and for overconsolidated clays are shown in Fig. 2,38.

The change in voids ratio due to an increase in vertical stress resulting from the foundation loading is considered at the centre of the clay layer having a thickness H. Thus for the initial (unloaded) condition, the initial voids ratio (e_1) is read off the $p–e$ curve (Fig. 2.38) corresponding to the initial overburden pressure (σ'_{vo}) at the centre of the layer. After foundation loading is applied, the initial vertical pressure (σ'_{vo}) is increased by the vertical stress (σ_z) induced at the centre of the layer by the net foundation pressure (q_n). Thus the final voids ratio (e_2) is read off the $p–e$ curve corresponding to a pressure of σ'_{vo}

$+ \sigma_z$. The decrease in thickness of the layer, i.e. the oedometer settlement ρ_{oed}, after full consolidation is then given by the equation

$$\rho_{oed} = \frac{H}{1 + e_1}(e_1 - e_2),$$

(2.47a)

or if the compression index has been used as the basis for the settlement computations, the final settlement is calculated from the equation

$$\rho_{oed} = \frac{H}{1 + e_1}C_c \log_{10}\frac{\sigma'_{vo} + \sigma_z}{\sigma'_{vo}}.$$

(2.47b)

As before,

$$\rho_c = \mu_g\rho_{oed}.$$

If the clay layer shows appreciable change in compressibility with depth, the settlement within each layer should be separately calculated. Then the total net final settlement is given by the sum of the settlements of each individual layer. For deep clay layers it is usually necessary to consider separate layers to allow for variations in compressibility with depth, and also to provide for the effect of variation in voids ratio with the rapidly decreasing vertical stress with increasing depth. The net final settlement is calculated for each layer, and the sum of the values so obtained gives the total net final settlement for the whole depth of the clay stratum affected by the foundation loading.

A correction is applied to the calculated ρ_c in the form of a 'depth factor'. This depends on the depth-to-area ratio and the length-to-breadth ratio of the foundation. Values of the depth factor,

$$\text{Depth factor} =$$

$$\frac{\text{corrected settlement for a foundation at depth } D}{\text{calculated settlement for a surface foundation}}$$

are obtained from Fox's[2.40] correction curves which are as shown in Fig. 2.39. Theoretically, Fox's curves apply only to elastic or 'immediate' settlements, but it is logical to correct consolidation settlements to allow for the depth of the foundation and Fox's curves provide a convenient method of doing this.

(f) *Estimation of final settlement* (ρ_f). The final settlement is the sum of the corrected values of the immediate settlement and the consolidation settlement, i.e.

$$\rho_f = \rho_i + \rho_c.$$

(2.48)

(g) *Estimation of the rate of consolidation settlement.* It is usually necessary to know the rate at which the foundations will settle during the long process of consolidation. This is normally calculated as the time period required for 50 per cent and 90 per cent of the final settlement.

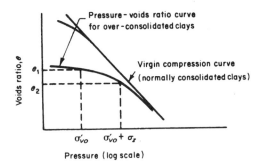

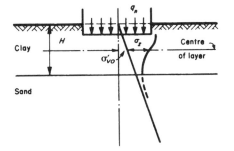

Fig. 2.38 Use of pressure–voids ratio curves in settlement analysis.

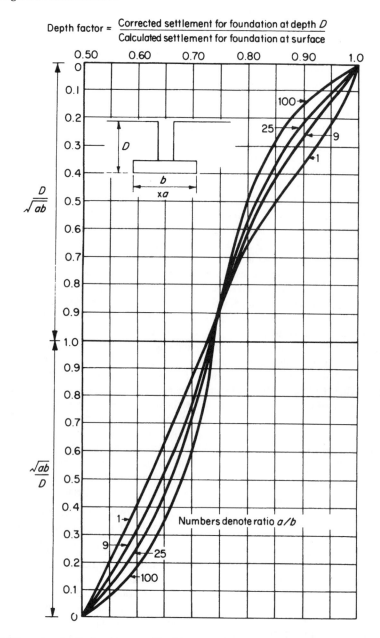

Fig. 2.39   Fox's correction curves for elastic settlement of flexible rectangular foundations at depth.

The time required is given by the equation

$$t = \frac{Td^2}{C_v} \qquad (2.49)$$

or, expressed in units of metres per year,

$$t(\text{years}) = \frac{Td^2 \times 10^{-7}\ (\text{m})}{3.154 \times C_v\ (\text{m}^2/\text{s})} \qquad (2.50)$$

where

T = time factor,

d = H (thickness of compressible stratum measured from foundation level to point where σ_z is small, say $10\text{–}20\ \text{kN/m}^2$) for drainage in one direction or $d = H/2$ for drainage at top and bottom of clay stratum,

C_v = average coefficient of consolidation over the range of pressure involved (obtained from triaxial compression or oedometer tests).

The values of the time factor (T) for various degrees of consolidation (U) are given in Fig. 2.40.

The total settlement at any time t is given by

$$\rho_t = \rho_i + U\rho_c. \qquad (2.51)$$

The type of curve to be used depends on the pressure distribution. The standard cases of pressure distribution are shown in Fig. 2.41. When there is two-way drainage,

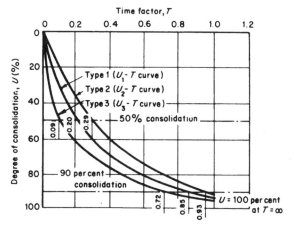

Fig. 2.40   Relation between degree of consolidation and time factor.

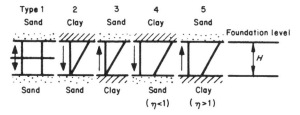

Fig. 2.41   Type of pressure distribution. Arrow indicates direction of drainage.

$$\eta = \frac{\text{Pressure at permeable surface}}{\text{Pressure at impermeable surface}}.$$

For case 4,

$$U_4 = U_1 - \left(\frac{1-\eta}{1+\eta}\right)(U_1 - U_2),$$

i.e. intermediate between U_1 and U_3. For case 5,

$$U_5 = U_1 + \left(\frac{\eta-1}{\eta+1}\right)(U_1 - U_2),$$

i.e. intermediate between U_1 and U_2.

i.e. when there is a permeable layer above and below the compressible stratum, there is only one type of curve for all types of pressure distribution.

When considering drainage of the clay layer it is always assumed that the concrete of the foundation acts as a permeable layer. In fact concrete is much more pervious than most clays. Where the foundation is constructed from impermeable material, e.g. asphalt tanking around a concrete substructure or a buried steel tank, the rate of settlement should, strictly, be based on three-dimensional theory of consolidation. However, this procedure is rarely needed in practice since asphalt tanking is nearly always laid on a mass concrete blinding layer, and similarly steel tanks are usually placed on a bed of sand or crushed stone. Therefore in nearly all cases the standard one-dimensional theory of consolidation can be followed.

Clay deposits frequently contain layers or laminations of sand or silt, and if these are continuous they form drainage layers giving a greatly increased rate of settlement compared with that predicted by oedometer tests. However, in many cases and particularly in boulder clays, the sands are in the form of isolated pockets or lenses having no drainage outlet. If there is doubt as to whether or not the sand layers are drained, the rates of settlement for the drained and undrained cases should be calculated and their relative significance considered. Fissuring in clays and the occurrence of root holes at shallow depths (Rowe[1.2]) can also accelerate the process of drainage. It is better to estimate the rate of settlement from C_v values obtained from triaxial testing,[1.18] where specimens of sufficiently large size can be obtained to incorporate these natural drainage channels. Where the results of field permeability tests are available the coefficient of consolidation can be obtained from the relationship

$$C_v = k/\gamma_w m_v. \qquad (2.52)$$

It should be noted that the imposition of foundation pressures will tend to close up natural drainage channels and hence reduce the value of the permeability coefficient k. Therefore equation (2.52) will be reliable only for low stress levels or the early stages of foundation loading. Where available, the best guide is by correlation with observed time-settlement of full-scale structures.

(h) *Estimation of settlements over the construction period.* Typical curves for the loading and settlement of a structure over the period of construction and after completion are shown in Fig. 2.42. The curve of the net settlement, assuming the final foundation loading to be instantaneously applied, is first plotted as shown by the lower curve in Fig. 2.42. The first point C on the corrected curve (allowing for progressive increase of load over the construction period) is obtained by dropping a perpendicular from a point A on the time abscissa, where OA is the time

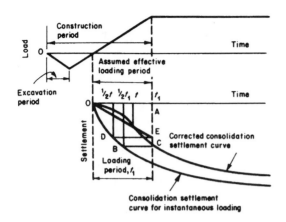

Fig. 2.42 Consolidation settlement curve during the construction period.

for completion of construction (time t_1). A perpendicular is dropped from a point equal to half t_1, to intersect the instantaneous loading curve at B. Then BC is drawn parallel to the time-scale to intersect the perpendicular from A at C. Intermediate points for any other time t are similarly obtained. A perpendicular is dropped from $\frac{1}{2}t$ to intersect the instantaneous loading curve at D. A line drawn parallel to the time-scale is extended to the perpendicular AC to intersect at the point E. Then the intersection of OE with the perpendicular from the point t gives the intermediate point on the corrected curve for the time t. Beyond C the settlement curve is assumed to be the instantaneous curve offset to the right by half the loading period, i.e. offset by the distance BC.

The corrected *total* settlement curve can be obtained by adding the immediate settlement (as calculated from equations (2.41)–(2.44)) to the corrected consolidation settlement as shown in Fig. 2.42. Assuming that the applied loading increases linearly during the construction period, the immediate settlement also increases roughly linearly.

Secondary consolidation The very long-term secondary consolidation or creep of soils was mentioned as a component of the total settlement in Section 2.6.1. It can be a significant proportion of the total in the case of soft organic alluvial clays and peats. From the results of laboratory tests and field observations Simons[2.41] derived the equation for clays:

$$\text{Coefficient of secondary consolidation} = C_\alpha$$
$$= 0.00018 \times \text{moisture content} \qquad (2.53)$$

This coefficient is used in conjunction with equation (2.47b).

Swelling of excavations Removal of the overburden when excavating to foundation level causes swelling of the soil below the excavation. After applying the loading the soil is recompressed, and after the loading has increased beyond the original overburden pressure further consolidation takes place. This swelling and recompression is usually insignificant in the case of shallow foundations and it can be neglected. It must, however, be taken into consideration for deep foundations and it is discussed further in Chapter 5.

Settlement data in an engineering report The accuracy obtainable in settlement computations does not justify quoting exact calculated figures. It is usual to give settlement to the nearest 15 mm where the settlement is 25 mm or less, or to use some such expression as 'less than 25 mm' or 'about 25 mm'. For settlements greater than 25 mm the quoted figures should be to the nearest 25 mm, stating '25–50 mm' or 'about 100 mm'. Settlements greater than 150 mm should be quoted to the nearest 50–100 mm.

2.7 Settlement of foundations on rocks

Weak to moderately weak weathered rocks and closely jointed rock formations possess a degree of compressibility such that it is necessary to make estimates of settlement wherever heavily loaded structures are founded on these formations. Meigh[1.29] described the settlement of two nuclear reactor buildings at Oldbury 'A' Power Station. The reactors and their containment structures, each weighing 680 MN, imposed net bearing pressures of 1.1 MN/m² on thinly bedded sandstones and siltstones of the Keuper Marl formation. After the end of a 2-year period when construction was nearly completed the foundations had settled 80 mm and 100 mm respectively. Thereafter slow creep settlement continued, and about 9 years after completion of loading the settlements were 100 mm and 120 mm respectively, and creep was still continuing. It was considered that a high proportion of the settlement was due to compression of a 13 m-deep leached zone of rock which contained cavities caused by the solution of gypsum nodules.

The difficulties of obtaining representative values of the deformation modulus of weak and jointed rocks have been described in Section 1.12. The most reliable method of obtaining modulus values for substitution in equation (2.41) is to make plate bearing tests at various depths within the zone stressed by the foundation loading. Alternatively, pressuremeter tests can be made using equipment appropriate to the strength of the rock (Section 1.4.5). Where the results of unconfined compression tests or point load strength tests on core samples are in sufficient numbers to be representative of the variation in strength of the rock over the depth stressed by the foundation loading, the deformation modulus of the rock mass can be obtained from the relationship

$$E_m = jM_r q_{uc} \qquad (2.54)$$

where

j = a mass factor related to the discontinuity spacing in the rock mass,

M_r = the ratio between the deformation modulus and the unconfined compression strength, q_{uc}, of the intact rock.

Values for the mass factor given by Hobbs[2.10] are shown in Table 2.12,

Table 2.12   Mass factor values

Quality classification†	RQD (%)	Fracture frequency per metre	Velocity index‡ $(V_f/V_L)^2$	Mass factor (j)
Very poor	0–25	15	0–0.2	0.2
Poor	25–50	15–8	0.2–0.4	0.2
Fair	50–75	8–5	0.4–0.6	0.2–0.8
Good	75–90	5–1	0.6–0.8	0.5–0.8
Excellent	9–100	1	0.8–1.0	0.8–1.0

† As BS 5930.
‡ V_f = wave velocity in field, V_L = wave velocity in laboratory.

Hobbs[2.10] showed that rocks of various types could be grouped together, and for practical purposes a modulus ratio could be assigned to each group as follows:

		Modulus ratio
Group 1	Pure limestones and dolomites Carbonate sandstones of low porosity	600
Group 2	Igneous Oolite and marly limestones Well-cemented sandstones Indurated carbonate mudstones Metamorphic rocks, including slates and schists (flat cleavage/foliation)	300
Group 3	Very marly limestones Poorly cemented sandstones Cemented mudstones and shales Slates and schists (steep cleavage/foliation)	150
Group 4	Uncemented mudstones and shales	75

The above values in conjunction with equation (2.54) were used in compiling the presumed bearing values for foundations on rock in Table 2.4(*a*). The high porosity chalk of south-east England and Keuper Marl are special cases. Values of the mass modulus of deformation of chalk established by Hobbs[2.10] and for Keuper Marl established by Chandler and Davis[2.42] are shown in Table 2.13. The values for chalk are considerably lower than those obtained at Mundford by Ward, Burland and Gallois[2.42a]. The latter were for relatively low stress levels where the settlement was less than about 0.01 per cent of the foundation width. The values in Table 2.13 relate to higher bearing pressures used for the foundations of structures for which a settlement of 0.5 per cent of the width is acceptable. The values of E obtained from the foregoing methods are plotted to obtain an average straight line variation of E with depth. If E is constant with depth the influence factors shown in Fig. 2.35 can be used. More likely there will be an increase in E with increasing depth, when the influence factors shown in Fig 2.36 should be used. These have been drawn from Poisson's ratio of 0.5, but most rock formations have somewhat lower ratios. Meigh[1.29] stated that the Poisson's ratio of Triassic rocks is about 0.1–0.3.

Meigh obtained curves for influence factors shown in Fig. 2.43 for various values of the constant k in equation (2.43), and for a Poisson's ratio of 0.2. He applied further corrections to the calculated settlement at the corner of the foundation (ρ_i). These corrections provide for roughness at the contact between the foundation and the rock, and the depth of embedment of the foundation (D). Thus the settlement ρ_i is given by

$$\rho_i = q(B/E_f)I_p F_B F_D, \qquad (2.55)$$

where F_B is the correction factor for roughness of base (Fig. 2.44), and F_D the correction factor for depth of embedment (Fig. 2.45). In the case of a flexible rectangular foundation, the loaded area is divided into four equal rectangles and the settlement computed at the corner of each rectangle. The settlement at the centre of the foundation is then four times the corner settlement. In the case of a relatively rigid foundation the settlement at the midpoint of a longer side is calculated. Then the average settlement of the rigid foundation is given by

$$C_{av} = \tfrac{1}{3}(C_{centre} + C_{corner} + C_{centre\ long\ edge}). \qquad (2.56)$$

Care is needed in the interpretation of the results of plate loading tests made in accordance with the procedure described in Section 1.4.5. The adoption of a modulus of elasticity (Young's modulus) corresponding to the straight line portion of the curve in Fig. 2.29 might underestimate the settlement. The usual procedure is to draw a secant AC to the stress–strain curve corresponding to a stress equal to the net foundation pressure at foundation level. More conservatively the secant can be drawn at a compressive

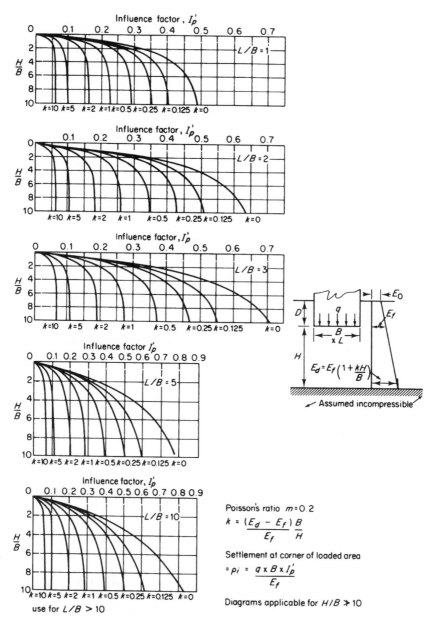

Fig. 2.43 Values of influence factor I_p for deformation modulus increasing linearly with depth and modular ratio of 0.2 (after Meigh[1.29]).

stress of 1.5 or some other suitable multiple of the foundation pressure.

It should be noted that settlements calculated on the basis of plate bearing test results do not include long-term creep settlement. At the present time this can only be estimated from experience of full-scale structures. The creep settlement observed at Oldbury[1.29] was about 20 per cent of the total. Hobbs[2.43] suggested that the long-term settlements of foundations on chalk could be between two and five times the calculated immediate settlement.

Where foundations are subjected to cyclic loading, e.g. marine structures exposed to wave impact forces, or foundations or reciprocating machinery, the plate bearing tests should also be made under conditions of cyclic

Table 2.13   Mass deformation modulus values for high-porosity chalk and Keuper Marl

Rock type	Weathering grade	Description	Deformation modulus (MN/m²)
Chalk	I	Blocky, moderately weak, brittle	>400
	II	Blocky, weak, joints more than 200 mm apart and closed	90–400
	III	Rubbly to blocky, unweathered, joints 60–200 mm apart, open to 3 mm and sometimes filled with fragments	50–90
	IV	Rubbly, partly weathered, with bedding and jointing, joints 10–60 mm apart, open to 20 mm, often filled with weak remoulded chalk and fragments	35–50
	V	Structureless, remoulded, containing lumps of intact chalk	10–35
	VI	Extremely weak, structureless, containing small lumps of intact chalk	Less than 10
Keuper Marl	I (unweathered)	Mudstone (often fissured)	>150
	II (slightly weathered)	Angular blocks of unweathered marl and mudstone with virtually no matrix	75–150
	III (moderately weathered)	Matrix with frequent lithorelicts up to 25 mm	30–75
	IV (highly to completely weathered)	Matrix with occasional claystone pellets less than 3 mm size (usually coarse sand) degrading to matrix only	Obtain from laboratory tests

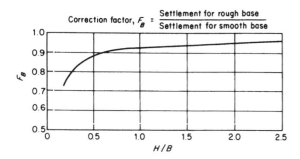

Fig. 2.44   Correction factors for roughness of base of foundation.

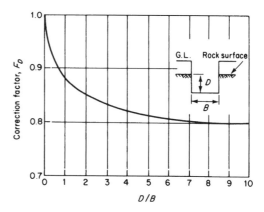

Fig. 2.45   Correction factors for depth of embedment of foundation below surface of rock.

loading at a maximum pressure corresponding to the design bearing pressure or a simple multiple, say 1.5 times this value. It may be found that the modulus for this cyclic loading is different from that of a single application of a static loading. Strong closely fissured and jointed rocks are likely to show a higher modulus value under repeated loading, i.e. the 'work-hardening' condition; whereas weak weathered rocks may show a lower modulus, or 'work-softening' condition due to breakdown of inter-particle bonds in the rock mass.

2.8 The applicability of computerized methods to foundation analysis and design

The use of computer-based methods in foundation engineering is neither as developed nor as extensive as in many other areas of civil and structural engineering. Some of the reasons for this are readily apparent; for example, most soils have stiffness properties characterized by non-linearity and inelasticity, soils are often heterogeneous and may be subject to complicated stratigraphy, and the presence of ground water and the localized pressure of that ground water have a major effect on the strength of soils. The situation is simpler in the other areas of engineering where the designer is dealing with materials that are not site dependent, are for the most part manufactured (often under factory conditions with the attendant benefits of quality control), and have critical properties that are usually far less complex than those of soils. Consequently, computer software for geotechnical engineering is very often the result of *ad hoc* development and may only be run by the program's author or that author's immediate colleagues. Frequently the software does nothing more than computerize a manual method to achieve improved speed of calculation and reliability of the arithmetic. In 1984 Smith[2.44] noted that there were no geotechnical equivalents of the many comprehensive and widely available suites of programs which existed for structural engineering analysis and design. Nor does the literature on geotechnical engineering contain details of software on a scale comparable to that for structural engineering. The reader may note that in 1974 Bowles[2.45] provided details of a number of general-purpose geotechnical programs, but beyond this there is little to be found outside specialist publications.

In later sections of this book various computer applications to foundation engineering will be noted, these are not necessarily among the latest or most sophisticated but the aim is to show applicability and accessibility where possible. However, first some background comments will be made on two of the techniques that will be mentioned most frequently: the finite difference method and the finite element method.

2.8.1 The finite difference method

This is based on the approximations to derivatives, at a specific point, of a continuous function in terms of explicit values of that function at a number of discrete points. The basic expressions can be derived by Taylor's theorem. For example, for $\phi(x)$ a function of x

$$\phi(x + h) = \phi(x) + h\frac{d\phi}{dx} + \frac{h^2}{2!}\frac{d^2\phi}{dx^2} + \frac{h^3 d^3\phi}{3!\,dx^3} + \cdots \quad (2.57)$$

where all the derivatives ($d\phi/dx$, $d^2\phi/dx^2$, etc.) are determined at point x.

Consider the sequence of five points $i - 2$ to $i + 2$ aligned in the x-direction and at equal spacing h as shown in Fig. 2.46. At point $i + 1$, where $x = x_i + h$

$$\phi_{i+1} = \phi_i + h\left(\frac{d\phi}{dx}\right)_i + \frac{h^2}{2!}\left(\frac{d^2\phi}{dx^2}\right)_i + \frac{h^3}{3!}\left(\frac{d^3\phi}{dx^3}\right)_i + \cdots \quad (2.58)$$

At point $i - 1$, where $x = x - h$

$$\phi_{i-1} = \phi_i - h\left(\frac{d\phi}{dx}\right)_i + \frac{h^2}{2!}\left(\frac{d^2\phi}{dx^2}\right)_i - \frac{h^3}{3!}\left(\frac{d^3\phi}{dx^3}\right)_i + \cdots \quad (2.59)$$

Subtracting eqn (2.59) from eqn (2.58) gives

$$\left(\frac{d\phi}{dx}\right)_i = \frac{\phi_{i+1} - \phi_{i-1}}{2h} - \frac{h^2}{6}\left(\frac{d^3\phi}{dx^3}\right)_i + \cdots \quad (2.60)$$

or, ignoring the truncation terms in h^2 and higher order

$$\left(\frac{d\phi}{dx}\right)_i \doteq \frac{\phi_{i+1} - \phi_{i-1}}{2h}. \quad (2.61)$$

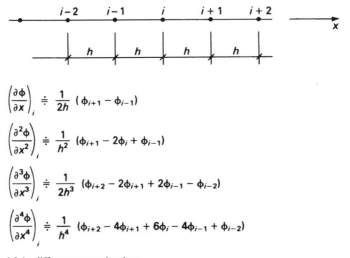

$$\left(\frac{\partial\phi}{\partial x}\right)_i \doteq \frac{1}{2h}\left(\phi_{i+1} - \phi_{i-1}\right)$$

$$\left(\frac{\partial^2\phi}{\partial x^2}\right)_i \doteq \frac{1}{h^2}\left(\phi_{i+1} - 2\phi_i + \phi_{i-1}\right)$$

$$\left(\frac{\partial^3\phi}{\partial x^3}\right)_i \doteq \frac{1}{2h^3}\left(\phi_{i+2} - 2\phi_{i+1} + 2\phi_{i-1} - \phi_{i-2}\right)$$

$$\left(\frac{\partial^4\phi}{\partial x^4}\right)_i \doteq \frac{1}{h^4}\left(\phi_{i+2} - 4\phi_{i+1} + 6\phi_i - 4\phi_{i-1} + \phi_{i-2}\right)$$

Fig. 2.46   One-dimensional finite difference approximations.

In a similar fashion the finite difference expressions for the other derivatives $(d^2\phi/dx^2)_i$, $(d^3\phi/dx^3)_i$ and $(d^4\phi/dx^4)_i$ can be established as the equations given in Fig. 2.46.

It is also possible to obtain finite difference approximations for partial differential equations in the two dimensions of x and y. These are illustrated in Fig. 2.47. It is most convenient to represent the expressions in a two-dimensional form. For instance, the form shown for $(\nabla^2\phi)_{i,j}$, that is, the approximation for $[(\partial^2\phi/\partial x^2) + (\partial^2\phi/\partial y^2)]$ at point i, j is the equivalent of

$$(\nabla^2\phi)_{i,j} \doteq \frac{1}{h^2}(\phi_{i-1,j} + \phi_{i+1,j} - 4\phi_{i,j} + \phi_{i,j-1}$$

$$+ \phi_{i,j+1}). \qquad (2.62)$$

As an example of how the technique operates, consider the case shown in Fig. 2.48 of a beam of span L, simply supported at each end and also deriving support from a Winkler spring type foundation in the span (this is a foundation where the upward reaction at a point is directly proportional to the downward displacement of the surface at that point). This could represent a beam supported by effectively rigid pile caps and additionally gaining support from the ground beneath its soffit. The governing differential equation (g.d.e.) is shown in the figure as

$$\frac{d^4y}{dx^4} = \frac{\omega}{EI} - \frac{k}{EI}y. \qquad (2.63)$$

Additionally, the boundary conditions are that both deflexion and bending moment are equal to zero at each end of the beam, i.e.

$$\left.\begin{array}{l} y = 0 \quad \text{at } x = 0 \quad \text{and} \quad x = L \\[2mm] \dfrac{d^2y}{dx^2} = 0 \quad \text{at } x = 0 \quad \text{and} \quad x = L \end{array}\right\}. \qquad (2.64)$$

Considering the simplest possible model with the beam divided into two halves gives five discrete points for the computation, one point, 1, defining deflexion at mid-span and the other four associated with the boundary conditions.

Applying the finite difference approximation to the g.d.e. gives:

$$\frac{1}{h^4}(y_{-1} - 4y_0 + 6y_1 - 4y_2 + y_3) = \frac{\omega}{EI} - \frac{k}{EI}y_1. \qquad (2.65)$$

The boundary conditions result in

$$y_0 = y_2 = 0 \qquad (2.66)$$

and

$$\left.\begin{array}{l} \dfrac{1}{h^2}(y_{-1} - 2y_0 + y_1) = \dfrac{1}{h^2}(y_1 - 2y_2 + y_3) = 0 \\[4mm] \text{or} \\[2mm] y_{-1} = y_3 = -y_1 \end{array}\right\}. \qquad (2.67)$$

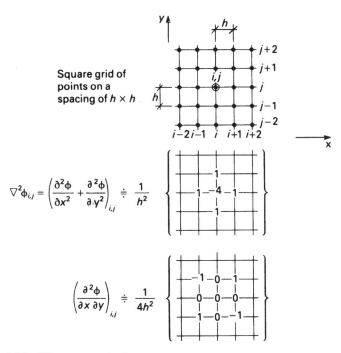

$$\nabla^2\phi_{i,j} = \left(\frac{\partial^2\phi}{\partial x^2} + \frac{\partial^2\phi}{\partial y^2}\right)_{i,j} \doteq \frac{1}{h^2}$$

$$\left(\frac{\partial^2\phi}{\partial x\,\partial y}\right)_{i,j} \doteq \frac{1}{4h^2}$$

Fig. 2.47 Two-dimensional finite difference approximations.

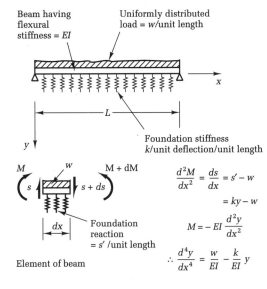

$$\frac{d^2M}{dx^2} = \frac{ds}{dx} = s' - w$$

$$= ky - w$$

$$M = -EI\frac{d^2y}{dx^2}$$

$$\therefore \frac{d^4y}{dx^4} = \frac{w}{EI} - \frac{k}{EI}y$$

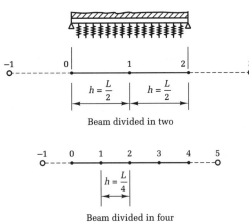

Fig. 2.48 Foundation slab carrying uniformly distributed loading.

Substituting equations (2.64) and (2.65) into equation (2.63) gives

$$y_1 = \frac{\omega L^4}{(64EI + kL^4)}. \tag{2.68}$$

If k is taken as zero then the indicated midspan deflection becomes $\omega L^4/64EI$, which is 20 per cent greater than the theoretical value of $5\omega L^4/384EI$.

To obtain improved accuracy the beam could be subdivided in four as shown in the lower part of Fig. 2.48. Application of the finite difference expressions will result in three simultaneous equations in y_2, y_3, and y_4. Solving these will result in a calculated midspan deflection of

$$y_2 = \frac{\omega L^4}{EI}\left[\frac{3584 + \dfrac{kL^4}{EI}}{\left(1536 + \dfrac{kL^4}{EI}\right)^2 - (1024 \times 2048)}\right]. \tag{2.69}$$

If k is set to zero this gives a midspan deflexion of $0.0137\omega L/EI$, which is 5 per cent greater than the exact answer. Further improvements in accuracy can be achieved by increasing the number of segments into which the beam is divided. However, this results in greater numbers of simultaneous equations and it will be appreciated that for other than comparatively simple problems, such as the example given here, use of a computer program for speed and accuracy of calculation becomes a virtual necessity. The need for computerized methods of solution becomes even more pressing in the case of two-dimensional problems.

Finite difference techniques can also be used to solve time-dependent problems such as one-dimensional consolidation (see Section 2.8.4) and heat flow (see Section 3.1), but the description of the detailed approach to these is outside the scope of this book and the reader is referred to other texts such as ref. 2.49.

2.8.2 The finite element method

In this technique the domain (one, two, or three dimensional) within which the particular problem is to be investigated is subdivided into a number of discrete areas or elements. Within each of these elements it is assumed that certain material properties (such as Young's modulus, permeability, thermal conductivity) will be constant and certain variables (such as strain, potential gradient) vary in a defined fashion; for example, depending on the type of element, strain may be defined to be constant or vary linearly or vary quadratically, etc. Adjacent elements are interconnected at nodal points. Figure 2.49 shows examples of element arrangements involving triangular, quadrilateral, and rectangular elements in representing two-dimensional problems.

In the case of an analysis to determine displacement and stresses within a domain, displacements are assumed to vary within each element according to a displacement function which satisfies the requirement for compatibility of displacements. Using this displacement function and the physical properties of the material bounded by the element it is possible to compute expressions which relate displacements and forces at the nodal points of the element. These expressions can be represented in matrix form and are then termed the element stiffness matrix. Superimposing all the individual element stiffness matrices gives the total, or global, stiffness matrix for the

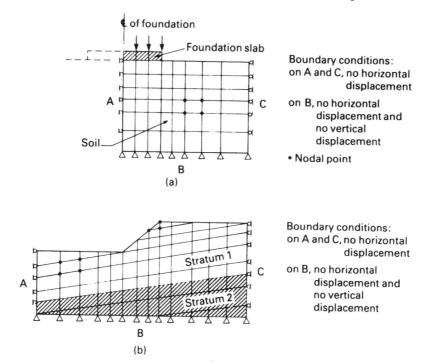

Fig. 2.49 Finite element models: (*a*) Foundation slab on soil using rectangular elements. (*b*) Slope using triangular and quadrilateral elements.

entire domain. After definition of boundary conditions and applied loads a set of simultaneous equations is obtained relating nodal point displacements, the global stiffness terms and applied loads. In matrix form the equations are represented by

$$[K]\{\delta\} = \{P\}, \qquad (2.70)$$

where

  $[K]$ = the global stiffness matrix,
  $\{\delta\}$ = the vector of nodal point displacements,
  $\{P\}$ = the vector of applied loads.

Knowing $[K]$ and $\{P\}$, solution of the equations gives $\{\delta\}$. Using the resultant nodal point displacements for an element and the element stiffness properties, the strains and stresses within the element are computed.

  Although it involves far more elaborate computations than does the finite difference method, the finite element method is consequently more versatile in that it can readily handle complex geometry and variations in material properties – both of which features present difficulties for the finite difference method. In addition, the finite element method has been extensively used to obtain solutions to problems involving non-linear behaviour of materials – a feature particularly relevant to soils – and some examples are quoted later.

  A number of good basic texts, such as that by Cheung

and Yeo,[2.46] are available for the reader who wishes to become acquainted with the details of the finite element method. A more extensive coverage and advanced treatment of the subject is provided by Zienkiewicz.[2.47] For the reader wishing not only to understand the details of the method but also to build up a library of programs capable of handling a wide range of problems in geotechnical engineering Smith[2.48] is recommended.

2.8.3 The application of finite element techniques to the delineation of over-stressed zones beneath large foundations and structures

A feature of the finite element method is that when modelling the non-linear behaviour of soils and including a yield criterion, such as von Mises or Mohr–Coulomb, it can indicate the progressive development of yielding zones and so represent the progressive onset of failure.

  Figure 2.50 illustrates this for the case of a strip footing on saturated, undrained clay. At a contact pressure equal to some 60 per cent of the ultimate value the formation of a failure zone within the clay can be seen. Subsequently, the analysis indicates a collapse at a contact pressure almost identical to the theoretical predictions of Prandtl and Hill. It should be noted that there are many situations in which strictly localized soil failure does not impair the

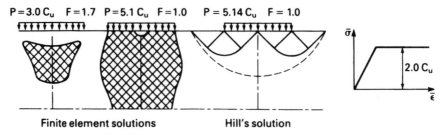

Fig. 2.50   Failure zones beneath strip foundation on saturated undrained clay (after Smith[2.50]).

efficiency of the structure supported by or retaining the soil: for example, the soil at shallow depth just in front of an anchored sheet pile retaining wall will almost always have reached a local passive failure condition, but without detriment to the wall in total.

Figure 2.51 gives an example on a larger scale and shows the development of failure as indicated in a finite element model of the soil under a large foundation subjected to both horizontal and vertical loading, in this case an offshore concrete gravity base structure.

Finite element analysis was undertaken by Potts *et al.*[2.51] in a study of the 1984 failure of the embankment of the Carsington Reservoir. The particular modelling technique adopted included representation of strain-softening behaviour of soils in the embankment cross-section. The nature of strain softening as observed in the foundation materials is illustrated in Fig. 2.52 which shows that on initial shearing of intact material the shear stress rises to a peak beyond which it progressively declines towards a

constant residual value. If there is subsequent shearing action on the previously sheared surface it is limited to the residual value following an initial modest peak. Figure 2.53 shows values of strain reached in the embankment just before failure as indicated by the finite element analysis.

2.8.4 The application of finite element techniques to the calculation of foundation settlements

The availability of computers has made possible the rapid analysis of problems concerning the settlement of foundations. General-purpose finite element programs can be used in circumstances where the soil may be considered to be reasonably represented by linear elastic behaviour. However, it must be recognized that the most successful of such applications have been in cases where the elastic properties of the soil have been derived from back analysis of field measurement for similar cases on similar strata. Many general-purpose finite element programs can also model bilinear elastic–plastic stress–strain behaviour of materials in accordance with failure criteria such as Von Mises, and this facility may be used in the total stress analysis of immediate settlement of undrained clays.

Computer-based analytical methods have been developed to cater for more generalized stress–strain behaviour and for stress-dependent strength and stiffness of soils. These particular features are recognized in the work of Duncan and Chang[2.52] in developing a non-linear stress–strain relationship which could be used readily in finite element analyses of a wide range of geotechnical problems.

Duncan and Chang extended their earlier work on the representation of the stress–strain relationship of both clay and sand by hyperbolae. The general form of curve which they proposed is shown in Fig. 2.54. The initial tangent modulus, E_i, is given by

$$E_i = Kp_a \left(\frac{\sigma_3}{p_a}\right)^n$$

(2.71)

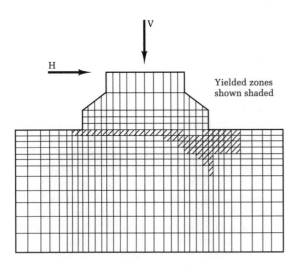

Fig. 2.51   Failure zones beneath strip foundation subjected to inclined loading (after Smith[2.50]).

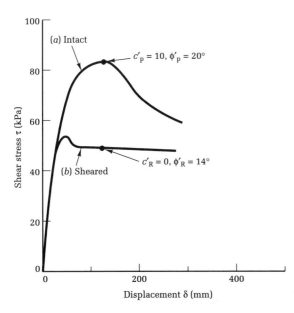

Fig. 2.52 Plotted results of shear test on foundation material at Carsington Dam (after Potts *et al.*[2.51]).

where

$$K = \text{modulus number (dimensionless)},$$
$$p_a = \text{atmospheric pressure},$$
$$\sigma_3 = \text{minor principal stress},$$
$$n = \text{exponent (dimensionless)}.$$

The Mohr–Coulomb failure criterion is used to give

$$(\sigma_1 - \sigma_3)_{ult} = \frac{2c \cos \phi + 2\sigma_3 \sin \phi}{1 - \sin \phi} \quad (2.72)$$

where c is cohesion and ϕ the angle of internal friction.

The tangent modulus, E_t, is given by

$$E_t = \left[1 - \frac{R_f(\sigma_1 - \sigma_3)}{(\sigma_1 - \sigma_3)_{ult}} \right]^2 E_i \quad (2.73)$$

where R_f is the failure ratio (typically in range 0.75–1.0) and σ_1 the major principal stress.

Finally, the stiffness for an unloading–reloading excursion from the initial loading curve is given by

$$E_{ur} = K_{ur} p_a \left(\frac{\sigma_3}{p_a} \right)^n \quad (2.74)$$

where K is the modulus number for unloading–reloading.

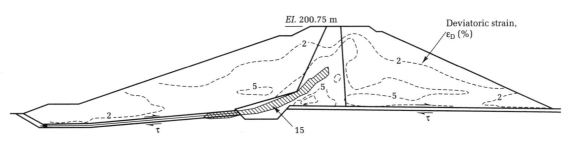

Fig. 2.53 Results of finite element analysis of Carsington Dam showing indicated shear in the embankment just before failure (after Potts *et al.*[2.51]).

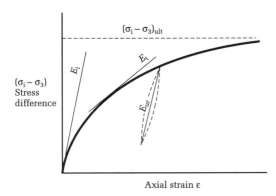

Fig. 2.54 Hyperbolic stress–strain curve for soils (after Duncan and Chang[2.52]).

Duncan and Chang[2.52] describe the use of this formulation in finite element analyses to compute the settlements for a model strip footing in sand and a circular footing on saturated, undrained clay. The analyses were conducted incrementally in the following sequence:

(a) Compute initial values of horizontal and vertical self-weight stresses σ_{x0} and σ_{y0} in the soil elements (τ_{xy0} is zero) and from these define E-values.

(b) Apply first increment of load and compute first increment deflexions and stresses. Add these to previous deflexions (nil) and previous stresses. From cumulative stress values compute σ_1 and σ_2 and hence E-value for each element.

(c) Apply second increment of load and compute second

incremental deflexions and stresses. Add these to previous cumulative deflexions and stresses, etc.

The results obtained are shown in Fig. 2.55 for the strip footing in sand and in Fig. 2.56 for the circular footing on clay. It can be seen that the values of settlements given by the finite element analyses are in good agreement with the model test results for the footing in sand and with elastic–plastic theory for the footing on clay.

In addition to the non-linear analysis to obtain values of instantaneous settlement for sands and saturated, undrained clays, computer methods have been applied to the evaluation of settlements due to consolidation. Both finite difference and finite element methods have been used for this task. CONSYST and SETTLE are two finite difference-based programs for the two-dimensional analysis of consolidation which have been available on computer bureaux in Britain since the late 1970s, CONSYST having been developed by a specialist company and SETTLE by the Highway Engineering Computer Branch

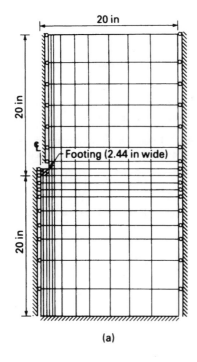

(a)

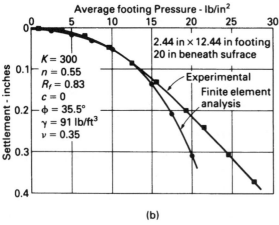

(b)

Fig. 2.55 Non-linear finite element analysis of model strip foundation in sand (after Duncan and Chang[2.52]): (a) Finite element representation; (b) Comparison of predicted and model behaviour.

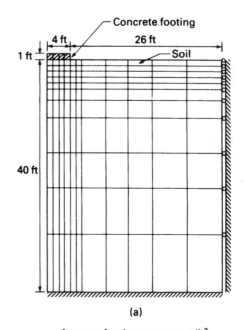

(a)

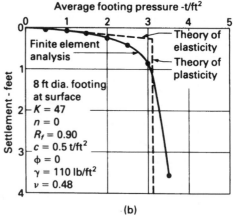

(b)

Fig. 2.56 Non-linear finite element analysis of circular foundation on clay (after Duncan and Chang[2.52]): (a) Finite element representation; (b) Comparison of predicted and field behaviour.

(HECB) of the Department of Transport. A number of finite element programs having an ability to model consolidation have also appeared. Of these, CRISTINA and subsequently CRISP were developed at Cambridge University and incorporate the Cam-Clay model and so are directed to the analysis of consolidation of soft clays. SAFE, a finite element program developed by the Ove Arup partnership,[2.53] uses the Model LC formulation and is relevant to consolidation of stiff over-consolidated clays as well as providing for calculations for other types of soil.

Although computer software for the prediction of consolidation behaviour has been in existence for several years, there has been only a limited number of published studies in which computed values of settlement, and pore water pressure, are compared with field measurements. Smith and Hobbs[2.54] reported comparisons of field measurements and computed results for consolidation beneath embankments.

The boundary element/surface element technique also has applicability to computation of settlements. This is referred to in Section 5.8.

2.9 Examples on Chapter 2

Example 2.1   A water tank 5 m wide by 20 m long is to be constructed at a depth of 0.8 m below ground level. The depth of water in the tank is 7.5 m. Borings showed a loose, becoming medium-dense, normally consolidated sand with the water table 2 m below ground level. Standard penetration tests made in the boreholes are plotted in Fig. 2.57. Check that the ultimate limit and serviceability limit states are not exceeded.

The self-weight of the structure (allowing for tolerance on dimensional inaccuracies) is 4190 kN, and the weight of the contained water is 7500 kN.

Applying the partial factors for actions from Table 2.1,

Total action normal to base slab = 4190 + 1.3 × 7500 = design F_d = 13 940 kN.

The characteristic design line from the SPT results shown in Fig. 2.57 is weighted towards the looser soils at shallow depth. Take a characteristic value of N = 6, from Fig. 2.13, ϕ = 29°.

Dividing by material factor γ_m from Table 2.2 gives

$$\text{Design } \phi' = \tan^{-1}\left(\frac{\tan 29°}{1.25}\right) = 23.9$$

Figure 2.8 gives

$$N_q = 10, \quad N_\gamma = 7.$$

Allow for a possible regrading of ground levels to give an

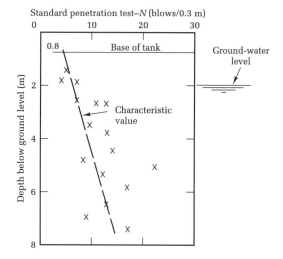

Fig. 2.57   SPT–N/depth curve.

embedment depth of 0.5 m instead of the design 0.8 m. No corrections are made for the ground-water level. Total overburden pressure

$$\sigma_{vo} = 17 \times 0.5 = 8.5 \text{ kN/m}^2.$$

For

$$B/L = 5/20 = 0.25,$$

the shape factors s_q and s_γ can be taken as unity, and for the shallow embedment the depth factors d_q and d_γ are also unity.

From equation (2.7)

$$Q/A' = 0 + 8.5 \times 10 \times 1 \times 1 + \tfrac{1}{2} \times 17 \times 5 \times 7 \times 1$$
$$= 85.0 + 297.5 = 382.5 \text{ kN/m}^2.$$

Any dimensional inaccuracies in the base slab from normal construction tolerances will have a negligible effect. Therefore take

$$A' = 20 \times 5 = 100$$
$$\text{and } Q = 382.5 \times 100 = 38\,250 \text{ kN}$$

which is greater than

$$F_d \text{ (effective)} = 13\,940 - 8.5 \times 20 \times 5 = 13\,090 \text{ kN}.$$

Hence the ultimate limit state is not exceeded. Checking by permissible stress methods

From an unfactored ϕ' = 29°, N_q = 16, N_γ = 15
From equation (2.7), = 13.6 × 16 + ½ × 17 × 5
Q/A × 15
= 217.6 + 637.5 = 855.1
At failure Q = 855.1 × 20 × 5 = 85 510 kN
Factor of safety = 85 510/11 690 = 7.3

For examining the serviceable limit state the settlement of the tank foundation can be calculated by the method of Burland and Burbidge. The depth of influence z_f from Fig. 2.27 is 3.5. The numerical average N-value (unfactored) over this depth from Fig. 2.57 is 8 and the corresponding I_c values from Fig. 2.26 are:

Minimum	4×10^{-2}
Average	9×10^{-2}
Maximum	30×10^{-2}

Average net applied pressure $= q'_u = 11\,690/100$
$$- 17 \times 0.8$$
$$= 103 \text{ kN/m}^2$$

Shape factor $= f_s = \left(\dfrac{1.25 \times 4}{4 + 0.25}\right)^2 = 1.4.$

Sand thickness factor $= f_1 = 1.$

Time factor $= f_t\left(1 + 0.3 + 0.2 \log \dfrac{30}{3}\right) = 1.5$

for 30-year period.

Average settlement at 30 years is
$1.4 \times 1 \times 1.5 \times 103 \times 5^{0.7} \times 9/100 = \underline{60\,\text{mm}}.$

From Fig. 2.26 maximum and minimum values of I_c are 30 and 4×10^{-2}.
Therefore

Maximum settlement	$= 60 \times 30/9 =$	$\underline{200\,\text{mm}}.$
Minimum settlement	$= 60 \times 4/9 =$	$\underline{25\,\text{mm}}.$

As already noted the Burland and Burbidge method tends to overestimate settlements where there are a number of low N-values at shallow depth. Checking by the Schultze and Sharif chart (Fig. 2.25), average N over depth $2B$ is 11. For $L/B = 4$, and $B = 5$, $s = 1.7$

Short-term settlement $= \dfrac{1.7 \times 103}{11^{0.87} \times (1 + 0.4 \times 0.8/5)}$

$$= 20\,\text{mm}.$$

This can be compared with the average short-term settlement by the Burland and Burbidge method of $60/1.5 = 40\,\text{mm}$. This does not exceed the limiting maximum settlement for raft foundations on sand of 40–65 mm (Section 2.6.3). Hence the serviceability limit should not be exceeded, but the tank structure should be designed to accommodate the expected differential short-term settlement. The long-term movement over a 30-year period should be capable of being accommodated by creep in the reinforced concrete structure.

Example 2.2 The retaining wall shown in Fig. 2.58

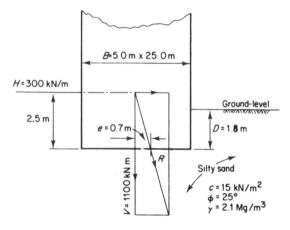

Fig. 2.58

carries a horizontal load of 300 kN/m run at a point 2.5 m above the base and a centrally applied vertical load of 1100 kN/m run. It is founded at a depth of 1.8 m in a silty sand. Using permissible stress methods, determine the factor of safety against general shear failure of the base of the wall.

For the above loading conditions we can use equation (2.4) to obtain the ultimate bearing capacity of the soil.
Resolving the vertical and horizontal forces, the resultant R cuts the base at a distance of 0.7 m from the centre of the wall. Thus, from equation (2.9),

$$B' = 5.0 - 2 \times 0.7 = 3.6\,\text{m}; \quad L' = L = 25\,\text{m}.$$

From Fig. 2.8, for $\phi = 25°$,

$$N_c = 21, \quad N_q = 11, \quad \text{and } N_\gamma = 8.$$

From Fig. 2.12, for

$H/B'L'c + V \tan \phi = 300/3.6 \times 25 \times 15 + 1100 \tan 25°$
$= 0.16$ and $H/V + B'L' \cot \phi =$
$0.08,$

$$i_c = 0.88, \quad i_q = 0.87, \quad \text{and } i_\gamma = 0.75.$$

From equation (2.10),

Shape factor $s_{c,B} = 1 + 0.2 \times 0.88 \times 3.6/25 = 1.03.$

From equation (2.12),

Shape factor $s_{q,B} = 1 + \sin 25° \times 0.87 \times 3.6/25$
$$= 1.05.$$

From equation (2.14),

Shape factor $s_{\gamma B} = 1 - 0.4 \times 0.75 \times 3.6/25 = 0.96.$

From Fig. 2.11, for $D/B' = 1.8/3.6 = 0.5.$

Depth factor, $d_c = 1.2.$

From equation (2.16),

$$d_q = 1.2 - \frac{1.2 - 1}{11} = 1.18 \quad \text{and} \quad d_\gamma = 1.$$

The base of the wall and the ground surface are level, thus the base and ground inclination factors are equal to unity. Therefore, from equation (2.7)

$$q_f = 15 \times 21 \times 1.03 \times 1.2 \times 0.88 + 1.8 \times 21 \times 11 \times$$
$$1.05 \times 1.18 \times 0.87 + \tfrac{1}{2} \times 21 \times 3.6 \times 8 \times 0.96 \times$$
$$1 \times 0.75$$
$$= 342.6 + 448.2 + 217.7 = 1008.5.$$

Bearing pressure for effective foundation width is

$$\frac{1100}{3.6} = 305.6 \text{ kN/m}^2;$$

therefore,
Factor of safety against general shear failure

$$= \frac{1008.5}{305.6} = 3.3,$$

which is satisfactory.

Example 2.3 A public building consists of a high central tower with three-storey wings projecting on each side. The central tower structure is carried by four widely spaced columns each carrying a combined dead load and representative imposed load of 2400 kN inclusive of the substructure. At the most unfavourable position the imposed load is 1200 kN on each column. Trial borings showed that below a shallow surface layer of topsoil was stiff-fissured London Clay followed by dense sand. Tests made on undisturbed samples from several boreholes gave an undrained shear strength–depth relationship shown in Fig. 2.59. Determine the required foundation depths and allowable bearing pressures for the tower and the three-storey wings.

Because of the difference in loading conditions between the tower and the wings, we shall have to investigate the possibility of differential settlement between these parts of the building.

Dealing first with the central tower, inspection of Fig. 2.59 shows that it will be desirable to take the column foundations to a depth of about 2 m to take advantage of the stiffer soil. Founding at a higher level would require excessively large foundations, thus complicating the foundations of the wings at their junction with the tower. At a depth of 2 m the fissured shear strength is about 80 kN/m².

To investigate the ultimate limit state the loads are factored from Table 2.1 to give

$$F_d = 2160 \times 1 + 1200 \times 1.3 = 3720 \text{ kN per column.}$$

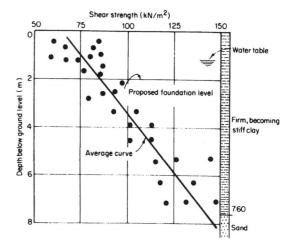

Fig. 2.59 Shear strength–depth curve.

The fissured shear strength is factored from Table 2.2 and gives $c_u = 80/1.5 = 53 \text{ kN/m}^2$.

The approximate bearing capacity factor N_c for shallow square or circular foundations is 7.5.

Ultimate bearing capacity (Fig. 2.15) is

$$\frac{Q}{A'} = 7.5 \times 53 + 19 \times 1.2 + 9 \times 0.8$$

$$= 397 + 30 = 427 \text{ kN/m}^2$$

If the ultimate limit state is not to be exceeded Q/A' must be equal to or greater than 3720 kN. Therefore

$$A' = 3720/427 = 8.7 \text{ m}^2 \quad \text{or} \quad 2.9 \times 2.9 \text{ m.}$$

To allow for dimensional inaccuracies given by machine excavation and cast in-place concrete, required size of column foundation = say 3 m square.

Calculation of net immediate settlement. To check the serviceability limit state the immediate settlement of the column foundation must be calculated for the column carrying the dead load with the imposed load in the worst position. Unfactored load = 2160 + 1200 = 3360 kN.

Net bearing pressure = 3360/9 − 30 = 343 kN/m².

The plasticity index of London Clay can be taken as 45 and the overconsolidation ratio as 3. From Fig. 2.34, $E_u/c_u = 300$. Assuming the concrete is placed fairly quickly after excavation,

$$E_u = 300 \times 85 \ (av. \ c_u) \times 10^{-3} = 25.5 \text{ kN/m}^2.$$

From equation (2.43), for $L/B = 1$, $H/B = 5.6/3 = 1.9$, and $D/B = \tfrac{2}{3} = 0.66$, $\mu_1 = 0.50$ and $\mu_0 = 0.93$ (Fig. 2.37).

Therefore

$$\rho_i = \frac{0.50 \times 0.93 \times 343 \times 3 \times 1000}{25.5 \times 1000} = 18.8\,\text{mm},$$

say <u>20 mm</u>.

Calculation of consolidation settlement ρ_c. The soil above the water table will be nearly saturated and, allowing for hydrostatic uplift in the clay below the water table, the original overburden pressure before applying foundation loading, is given by

$$\sigma_{vo} = 19 \times 1.2 + 9 \times 0.8 = 30\,\text{kN/m}^2$$

The dead load plus sustained imposed load totalling 2400 kN must be used to calculate the long-term settlements, giving a net foundation pressure of 2400/9 − 30 = 237 kN/m². We must now calculate the distribution of effective overburden pressure σ'_{vo} and the induced vertical stress σ_z throughout the full depth of the clay stratum as shown in Fig. 2.60. The vertical stress distribution is obtained from the factors shown in Fig. 2.24. The results of one typical consolidation test are shown in Fig. 2.61. Because the clay stratum is relatively thick, we must divide it into 1.4 m thick horizontal layers, and calculate separately the settlement for each layer. The calculation is shown in tabular form.

To obtain the net consolidation settlement ρ_c we must reduce the oedometer settlement ρ_{oed} by a factor which we can take as 0.5 for London Clay. Therefore,

$$\rho_c = 0.5 \times 46 = 23\,\text{mm}.$$

Here ρ_c can be reduced by a depth factor. From Fig. 2.39 the depth factor for $D/\sqrt{ab} = 0.66$ is 0.80. Therefore corrected settlement is $23 \times 0.80 = 18$ mm, say 20 mm.

Calculation of rate of consolidation settlement. From laboratory tests, coefficient of consolidation $c_v = 0.0139 \times 10^{-6}$ m²/s. The sand underlying the clay acts as a drainage layer, and also the concrete of the foundation can be taken as being more pervious than the clay; thus the drainage is in two directions. Since the sand is much more pervious than the concrete, the majority of the water expelled during the process of consolidation will travel downwards to the sand layer. However, for the purposes of calculating the rate of settlement it will be sufficiently accurate to assume equal drainage in each direction.

For two-way drainage, the time factors T_{50} and T_{90} for 50 and 90 per cent consolidation from the type 1 U_1–T curve in Fig. 2.40 are

$$T_{50} = 0.20 \qquad T_{90} = 0.85 \qquad d = \frac{H}{2} = 2.8\,\text{m}.$$

Therefore, by equation (2.50), time required for 50 per cent consolidation is

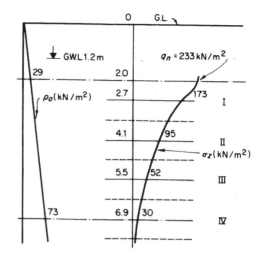

Fig. 2.60   Vertical pressure distribution.

$$t_{50} = \frac{T_{50}d^2 \times 10^{-7}}{3.154 \times c_v}$$

$$= \frac{0.20 \times 2.8^2 \times 10^{-7}}{3.154 \times 0.0139 \times 10^{-6}} = \underline{3.5\ \text{years (approx.)}}.$$

Similarly,

$$t_{90} = \frac{0.85 \times 2.8^2 \times 10^{-7}}{3.154 \times 0.0139 \times 10^{-6}} = \underline{15\ \text{years (approx.)}}.$$

Net total settlement after 4 years $= \rho_i + 50$ per cent ρ_c
$$= 10 + 0.5 \times 20$$
$$= 30\,\text{mm}$$

Net total settlement after 20 years $= 20 + \dfrac{90}{100} \times 20$

$$= 38\,\text{mm, say } 40\,\text{mm}.$$

Alternatively, from the Burland *et al.*[2.12] relationship:

Immediate settlement $= 0.5 \times \rho_{oed} = 0.5 \times 46$
$$= 23\,\text{mm}$$
Final settlement $= \rho_{oed} = 46\,\text{mm, say } 50\,\text{mm}.$

With the relatively small order of settlement there is no particular need to calculate the rate of settlement over the construction period or to draw a time–settlement curve.

If the sand stratum were not present, drainage would be one-way only (i.e. vertically upwards). We would then use the type 3 U_3–T curve in Fig. 2.40. From this curve,

$$T_{50} = 0.09 \qquad T_{90} = 0.72 \qquad d = H = 5.6\,\text{m}.$$

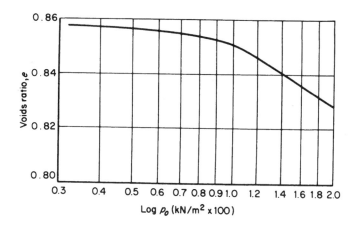

Fig. 2.61 Pressure–voids ratio curve.

Calculation of consolidation settlement for each layer

Layer no.	Depth of layer (m)	Depth to centre of layer (m)	Thickness of layer H (m)	Original effective overburden pressure, p_o at centre of layer (kN/m²)	Vertical stress σ_z due to net pressure q_n (kN/m²)	Resultant effective pressure at centre of layer, $p_o + \sigma_z$ (kN/m²)	Voids ratio at p_o, e_1	Voids ratio at $p_o + \sigma_z$ e_2	$\dfrac{e_1 - e_2}{1 + e_1}$	Settlement of layer, $\dfrac{e_1 - e_2}{1 + e_1} \times H \times 1000$ (mm)
1	2.0–3.4	2.7	1.4	36	173	209	0.858	0.827	0.0167	23
2	3.4–4.8	4.1	1.4	49	95	144	0.857	0.838	0.0102	14
3	4.8–6.2	5.5	1.4	61	52	113	0.856	0.848	0.0043	6
4	6.2–7.6	6.9	1.4	74	30	104	0.854	0.850	0.0022	3

Total oedometer settlement of clay layer, $\rho_{oed} = 46$ mm

Therefore, time required for 50 per cent consolidation

$$t_{50} = \frac{0.09 \times 5.6^2 \times 10^{-7}}{3.154 \times 0.0139 \times 10^{-6}}$$

$$= 6.5 \text{ years (approx.).}$$

Similarly,

$$t_{90} = \frac{0.72 \times 5.6^2 \times 10^{-7}}{3.154 \times 0.0139 \times 10^{-6}}$$

$$= 50 \text{ years (approx.).}$$

Strip foundation to three-storey wings. Since the structure is founded on a stiff-fissured clay it will be subject to seasonal swell and shrinkage. Since this is a public building, presumably with high standards of finish and workmanship, any cracking due to seasonal movements would be unsightly. If we take the foundations to 1.2 m depth this will get below the zone of any damaging soil movement. However, this is the water-table level and we do not want the clay in the foundation trench bottom to become puddled and soft during excavation. Therefore it is best to select a foundation level a few centimetres above the water table, i.e. at 1.15 m. From Fig. 2.59 the fissured shear strength at this level is 65 kN/m². This can be taken as the characteristic strength.

The dead loading on the wall inclusive of the strip foundation is 100 kN/m run together with an imposed loading in the most unfavourable position of 20 kN/m. Therefore to investigate the ultimate limit state

Design loading $= F_d = 100 + 1.3 \times 20 = 126$ kN/m.

For a wall thickness of 328 mm in brickwork plus 150 mm on each side to give clearance for bricklayers to work on the strip foundation, the nominal width of the strip is 0.63 m, and the effective width is 0.55 m to allow for construction inaccuracy.

$$\text{Design } c_u = 65/1.5 = 43 \text{ kN/m}^2.$$

From Fig. 2.15 for $D/B = 1.15/0.55 = 2.1$, $N_c = 7$.
Ultimate bearing capacity $= 7 \times 43 \times 0.55 = 165\,kN/m$,
which is greater than F_d. Hence the ultimate limit state is
not exceeded.

For checking the serviceability limit state, it will be
sufficiently accurate to estimate this from the immediate
settlement. Assume that the weight of soil removed in
excavation to 1.15 m roughly balances the weight of
brickwork, concrete, and clay backfill, giving a net foundation pressure for dead and unfactored imposed load of
$120/0.63 = 190\,kN/m^2$.

Take

$$E_u = 300 \times 65 \,(\text{unfactored}) \times 10^{-3} = 20\,kN/m^2.$$

From Fig. 2.23 for

$$H/B = 6.5/0.63 = 12, \quad L/B = \infty \quad \text{and} \quad D/B = 1.8.$$
$$\mu_1 = 1.4 \quad \mu_0 = 0.91$$
$$\rho_i = \frac{1.4 \times 0.91 \times 190 \times 0.63 \times 1000}{20 \times 1000} = 8\,mm,$$

$$\underline{\text{say } 10\,mm.}$$

Final settlement = oedometer settlement = 2×10
$$= \underline{20\,mm.}$$

The differential long-term settlement of $40 - 20 =$
20 mm between the tower and wings suggests the provision of vertical joints to accommodate the movement.

Example 2.4   Columns supporting a steel-framed warehouse building with sheet steel cladding carry a dead load
of 550 kN and an imposed wind load of 90 kN. The
columns are at 6 m centres. The building is located on a
river bank where medium-fine sands extend to more than
8 m below ground level. From several static cone penetration tests the loosest and densest q_c/depth diagrams are
shown in Fig 2.62(a) and (b) respectively. Design suitable
pad foundations.

A foundation depth of 1 m will be suitable to reach the
somewhat denser soil, but to remain above ground-water
level. Allowing for overbreak in excavation, net weight of
foundation for a trial size of $2.0 \times 2.0\,m$ is

$$25 \times 2.2^2 \times 1 - 18 \times 2.2^2 \times 1 = 34\,kN$$

Total dead plus imposed load is

$$(550 + 34)\,90 = 584 + 90 = 674\,kN$$

Design loading from Table 2.1 is

$$584 + 1.3 \times 90 = 701\,kN$$

The effective foundation width for calculating the ultimate
bearing capacity is 1.9 m. From Fig. 2.62(a) the characteristic q_c at 1 m bgl is 2.2 MN/m². Allowing for the ground

water to rise to ground level at the time of river flood, Fig.
2.14 gives an equivalent ϕ' value of about 34° for an
effective overburden pressure of 8 kN/m².

From Table 2.2

$$\text{Design } \phi' = \tan^{-1}\left(\frac{\tan 34°}{1.25}\right) = 28.3°$$

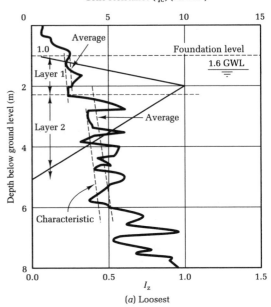

(a) Loosest

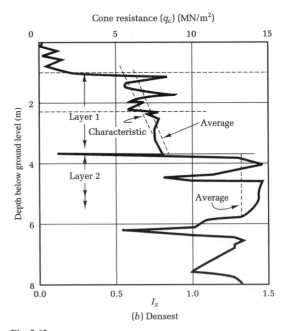

(b) Densest

Fig. 2.62

From Fig. 2.8

$$N_q = 14, N_\gamma = 13.$$

The approximate shape factors s_q and s_γ are 1.2 and 0.8 respectively, and the depth factors d_q and d_γ are unity.

From equation 2.7, ultimate bearing capacity is

$$Q/A' = 0 + 8 \times 14 \times 1.3 \times 1 + \frac{1}{2} \times 8 \times 1.9 \times 11 \times 0.8 \\ \times 1$$

$$= 225 \, kN/m^2$$

At the ultimate limit state $Q = 225 \times 1.9^2 = 812 \, kN$ which exceeds $F_d = 701 \, kN$.

For determining the serviceability limit state, the bearing pressure for the unfactored loads and the nominal foundation dimensions are $674/2^2 = 169 \, kN/m^2$.

The q_c/depth diagram is divided into two representative layers, and the drained secant modulus E' is obtained from Fig. 2.30 for the effective overburden pressure at the centre of each layer.

Layer no.	Depth (m)	Depth below FL to centre of layer (m)	σ'_{vo}	q_c (MN/m²)	E'_{50} (MN/m²)
1	1.0–2.2	0.6	29	2.3	4
2	2.2–6.0	3.1	49	4.5	9

In layer 1 from Fig. 2.37, for $H/B = 1.2/2 = 0.6$, $\mu_1 = 0.1$, for $D/B = 1.0/2 = 0.5$, $\mu_0 = 0.95$.
Short-term settlement =

$$\frac{0.1 \times 0.95 \times 169 \times 2 \times 1000}{4 \times 1000} = 8 \, mm$$

Take a 1 : 2 spread of load on to layer 2, bearing pressure at surface of layer is

$$674/3.2^2 = 66 \, kN/m^2$$

For $H/B = 3.8/3.2 = 1.2$, $\mu_1 = 0.45$, for $D/B = 2.2/3.2 = 0.7$, $\mu_0 = 0.93$.

Short-term settlement =

$$\frac{0.45 \times 0.93 \times 66 \times 3.2 \times 1000}{9 \times 1000} = 10 \, mm.$$

Total for layers 1 and 2 = 8 + 10 = 18 mm, say 20 mm.

Similarly the q_c/depth curve for the densest state is divided into two layers as shown in Figure 2.62(b) and the total short-term settlement is calculated to be 15 mm. The differential settlement between adjacent columns is therefore a maximum of 5 mm giving a relative rotation of 5 mm in 6000 mm or 1 in 1200. Table 2.7 shows that

structural damage should not occur to the steel-framed building, and brickwork cladding could be considered instead of metal sheeting.

The settlement for the loosest conditions can be checked by the Schmertmann method. The I_z/depth diagram for a peak $I_z = 0.5 + 0.1 \times 169/34 = 1.0$ (see Fig. 2.25).

From 1.0 to 2.0 m,

$$C_1 = 1 - \frac{0.5 \times 18 \times 1}{169} = 0.95 \qquad C_2 \text{ for immediate}$$

$$\text{settlement} = 1$$

$$\rho_i = \frac{0.95 \times 1 \times 169 \times 0.45 \times 1000}{2.5 \times 2.3 \times 1000} = 13 \, mm.$$

From 2.0 to 2.2

$$\rho_i = \frac{0.95 \times 1 \times 169 \times 0.97 \times 0.2 \times 1000}{2.5 \times 2.3 \times 1000}$$

$$= 5 \, mm.$$

From 2.2 to 5.0

$$\rho_i = \frac{0.95 \times 1 \times 169 \times 0.48 \times 2.8 \times 10}{2.5 \times 9 \times 1000} =$$

$$10 \, mm.$$

Total settlement = 28 mm, compared with 20 mm calculated from the charts in Fig. 2.37.

Example 2.5 Calculate the immediate and long-term settlement of a bridge pier with a base 8.50 m long by 7.50 m wide, founded at a depth of 3.0 m. The base of the pier imposes net foundation pressures of 220 kN/m² for dead loading and 360 kN/m² for combined dead and live loading. Borings showed dense sand and gravel with cobbles and boulders to a depth of 9 m below ground level, followed by very stiff over-consolidated clay to more than 25 m below ground level. Standard penetration tests gave an average N-value of 40 blows per 300 mm in the sand and gravel stratum. A number of oedometer tests were made on samples of the stiff clay. Triaxial tests on undistorted samples of the clay gave a minimum shear strength of 120 kN/m², and a modulus of deformation (E_u) of 40 MN/m².

Calculating immediate settlements in sand and gravel stratum. From Fig. 2.25, for $L/B = 8.5/7.5 = 1.1$, and $d_s/B = 6.0/7.5 = 0.8$ (reduction factor = 0.8)

Immediate settlement =

$$\frac{0.8 \times 1 \times 220}{40^{0.87}(1 + 0.4 \times 3/7.5)} = 6 \, mm.$$

Checking the settlement by the method of Burland and

Burbidge, the N-value is corrected for the gravel content to give N(corrected) $= 1.25 \times 40 = 50$, for which Fig. 2.26 gives $I_c = 0.65/100$,

$$\text{Shape factor} = f_s = \left(\frac{1.25 \times 8.5/7.5}{8.5/7.5 + 0.25}\right)^2 = 1.05.$$

For gravel thickness of 6 m, Fig. 2.27 gives $Z_I = 4.5$ m, therefore

$$f_I = 6/4.5(2 - 6/4.5) = 0.89.$$

For immediate settlement $f_t = 1$. Therefore

$$\begin{aligned}
\text{Immediate settlement} = {}&1.05 \times 0.89 \\
&\times 1[(220 - \tfrac{2}{3} \times 20 \times 3) \times \\
&7.5^{0.7} \times 0.65/100] \\
={}& 5\,\text{mm}.
\end{aligned}$$

For 360 kN/m² loading, immediate settlement is given by

$$\frac{6 \times 360}{220} = 10\,\text{mm} \quad \text{or} \quad \frac{5 \times 360}{220} = 8\,\text{mm}.$$

Bearing capacity of stiff clay. The vertical stress distribution curves (Fig. 2.63) from the curves for rigid foundations (Fig. 2.24) show that the vertical stress on the surface of the clay is 137 kN/m² for combined dead and live loading. The bearing capacity factor from Fig. 2.15 for a foundation 12 m wide at a depth of 9 m (on the surface of the clay) and $B/L = 0.80$ is 7.1. Thus the

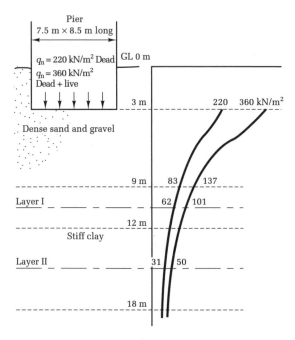

Pier
7.5 m × 8.5 m long

$q_n = 220$ kN/m² Dead   GL 0 m
$q_n = 360$ kN/m²
Dead + live

3 m   220   360 kN/m²

Dense sand and gravel

9 m   83   137
Layer I
      62   101
      12 m
Stiff clay

Layer II   31   50

18 m

Fig. 2.63 Vertical stress distribution from pier loading.

ultimate bearing capacity of the clay is $7.1 \times 120 = 852$ kN/m², which gives an ample safety factor against shear failure in the clay. However, it will be necessary to calculate the immediate and long-term consolidation settlements of the clay stratum to ensure that the total settlements of the bridge pier are not excessive.

Calculating settlements in clay stratum. From the vertical stress distribution curves (Fig. 2.63) the stresses transmitted to the surface of the clay stratum are 83 and 137 kN/m² for the 220 and 360 kN/m² loadings, respectively. These are transmitted over an area of approximately 12 m × 15 m. The vertical stress below 18 m is only 10 per cent of the applied stress. Therefore we need not consider settlements below a depth of 18 m.

The loaded area is divided into four rectangles, each 6.0 m × 7.5 m. From Fig. 2.35 for $H/B = 9/6 = 1.5$ and $L/B = 7.5/6 = 1.25$, $F_1 = I_p = 0.22$.

For 220 kN/m² loading, immediate settlement at corner of rectangle is given by equation (2.41):

$$\frac{83 \times 6 \times (1 - 0.5^2) \times 0.22 \times 1000}{40 \times 1000} = 2.0\,\text{mm}.$$

Reduce by depth factor from Fig. 2.39 of 0.7 for

$$D/\sqrt{ab} = 9/\sqrt{6 \times 7.5} = 1.3.$$

Therefore

$$\text{Reduced settlement} = 0.7 \times 1.4\,\text{mm}.$$
$$\begin{aligned}
\text{Settlement at centre of area} &= 4 \times \text{settlement at corner} \\
&= 4 \times 1.4 = 6\,\text{mm}.
\end{aligned}$$

For 360 kN/m² loading, settlement at centre of loaded area is given by

$$6 \times \frac{137}{83} = 10\,\text{mm}.$$

Calculation of net consolidation settlements. Since a number of consolidation tests were made we can use the coefficient of volume compressibility (m_v) for calculating the oedometer settlement (ρ_{oed}). The 9 m thick clay layer which is appreciably affected by the foundation pressures is divided into a 3 m thick layer from 9 to 12 m below ground level and a 6 m thick layer from 12 to 18 m below ground level. The m_v values corresponding to the respective increments of pressure in each layer are shown in the table below. Since this is an overconsolidated clay, we can take $\mu_g = 0.5$ and the depth factor for $9/\sqrt{12 \times 15} = 0.7$ is $= 0.8$. The calculations for ρ_{oed} are as in the table at the top of p. 97.

From this, total consolidation settlement for 220 kN/m² load is given by

$$0.8 \times 0.5 \times (20 + 6) = 10\,\text{mm}$$

Net foundation pressure, q_n (kN/m^2)	Depth of layer below ground level (m)	Thickness of layer (m)	m_v (m^2/kN)	Average vertical stress on layer, σ_z (kN/m^2)	Oedometer settlement ρ_{oed} (mm)
220	9–12	3	0.000 11	62	20
	12–18	6	0.000 03	31	6
360	9–12	3	0.000 20	101	61
	12–18	6	0.000 04	50	12

and total consolidation settlement for 360 kN/m^2 load by

$$0.8 \times 0.5 \times (61 + 12) = 29\,\text{mm}.$$

Summarizing total settlements. The total final settlement of the bridge pier is given by

$\rho_f =$ Immediate settlement in sand and gravel
  + Immediate settlement in clay
  + Consolidation settlement in clay.

For *dead loading* of 220 kN/m^2,

$$\rho_f = 6 + 6 + 10 = 22, \text{ say } 25\,\text{mm}.$$

For *combined dead and live loading* of 360 kN/m^2,

$$\rho_f = 10 + 10 + 29 = 49, \text{ say } 50\,\text{mm}.$$

A settlement of 15 mm will take place as the pier is constructed and this will increase to about 25 mm if and when the bridge sustains its maximum live load. The remaining settlement due to long-term consolidation of the clay under the dead load and sustained live load may take 50 years or more to attain its final value of 30 mm.

Example 2.6 A 20-storey building is founded on a stiff raft 40 m long by 20 m wide at a depth of 2 m below ground level and is sited over a weathered rock becoming

less weathered with increasing depth, until a relatively incompressible stratum is met at a depth of 45 m. Deformation modulus values obtained from plate bearing tests made in a large-diameter borehole are shown in Fig. 2.64. Estimate the settlement of the building for a net bearing pressure of 250 kN/m^2.

A straight line variation of E_d with depth is plotted in Fig. 2.64. At 30 m below ground level $E_d = 3600/\text{MN/m}^2$ and at foundation level $E_f = 200\,\text{MN/m}^2$. Therefore from the diagram in Fig. 2.43,

$$k = \frac{(3600 - 200)}{200} \times \frac{20}{28} = 12.14$$

The raft is divided into four equal rectangles each 20 m × 10 m. From Fig. 2.45 for $L/B = 20/10 = 2$, $H/B = 43/10 = 4.3$, and $k = 12.14$,

$$I_p = 0.05$$

Therefore immediate settlement at corner of rectangle is given by

$$\frac{250 \times 10 \times 0.05 \times 1000}{200 \times 1000} = 0.63\,\text{mm},$$

Settlement at centre of foundation $= 4 \times 0.63 = 2.52\,\text{mm},$

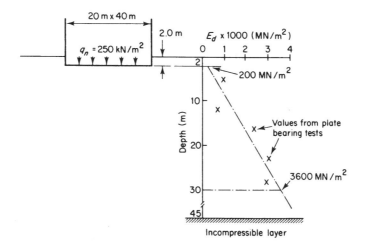

Fig. 2.64

Settlement at centre of long side = $2 \times 0.63 = 1.26$ mm,

Average settlement of 'rigid' raft = $\frac{1}{3}(2.52 + 1.26 + 0.63) = 1.47$ mm.

From Fig. 2.44, correction factor for $H/B \times 43/20 = 2.2$ is $F_B = 0.96$, and from Fig. 2.45, for $D/B = 2.0/20 = 0.1$, $F_D = 1.0$. Therefore

Corrected settlement = $1.47 \times 0.96 = 1.4$ mm,

say 2 mm.

There could, in addition, be some creep settlement in the weathered rock which might double the immediate settlement.

Example 2.7   A column carrying a dead load of 3500 kN and an imposed load of 1500 kN is to be founded at a depth of 0.6 m on a weak medium-bedded poorly cemented sandstone. Examination of rock cores showed an average joint spacing of 250 mm. Tests on the cores showed a representative uniaxial compression strength of 1.75 MN/m^2. Determine suitable dimensions for the column base.

Investigating the ultimate limit state, design load = F_d = $3.5 + 1.3 \times 1.5 = 5.45$ MN. From Section 2.3.6 for q_{uc} = 1.75 MN/m^2, $c = 0.1\ q_c = 0.175$ MN/m^2 and $\phi = 30°$.

From Fig. 2.17, $N_c = 14$, $N_\gamma = 15$ and $N_q = 9$. Take a trial width of 1.5 m.

From equation (2.26)

$$\frac{Q'}{A'} = 0.175 \times 14 + 22 \times 0.75 \times 15 \times 10^{-3}$$

$$+ 22 \times 0.6 \times 9 \times 10^{-3}$$

$$= 2.8 \text{ MN/m}^2$$

$$Q = 2.8 \times 1.5^2 = 6.3 \text{ MN}$$

which exceeds F_d = 5.45 MN. The bearing pressure for the unfactored loads is $5/1.5^2 = 2.2$ MN/m^2.

From Table 2.12 the mass factor for a joint spacing of four per metre is 0.6, and the modulus ratio for poorly cemented sandstone is 150. Therefore from equation (2.53)

Mass deformation modulus = $0.6 \times 150 \times 1.75 = 157$ MN/m^2.

Divide the base into four equal squares each of 0.75 m width.
From Fig. 2.43 for $L/B = 1$, $H/B = 10$ and $k = 0$, I'_p = 0.49.

Immediate settlement at corner of square

$$= \frac{2.2 \times 0.75 \times 0.49 \times 1000}{157} = 5.1 \text{ mm}$$

From Fig. 2.44, $F_B = 1$ and from Fig. 2.45, F_D for $D/B = 0.6/1.5 = 0.4$ is 0.92.

Corrected corner settlement = $1 \times 0.92 \times 5.1 = 4.7$ mm.

Average settlement of rigid base = $0.8 \times 4 \times 4.7 = \underline{15 \text{ mm.}}$

Creep might increase this settlement to about 25–30 mm in the long term.

References

2.1   Hansen, J. Brinch, A general formula for bearing capacity, *Danish Geotechnical Institute Bulletin No. 11* (1961); also, A revised extended formula for bearing capacity, *Danish Geotechnical Institute Bulletin No. 28* (1968); and Code of Practice for Foundation Engineering, *Danish Geotechnical Institute Bulletin No. 32* (1978).

2.2   Meyerhof, G. G., Some recent research on bearing capacity of foundations, *Canadian Geotechnical Journal*, **1**, 16–26 (1963).

2.3   Peck, R. B., Hanson, W. E., and Thornburn, T. H., *Foundation Engineering*, 2nd edn, John Wiley, New York, 1967, p. 310.

2.4   Durgunoglu, H. T. and Mitchell, J. K., Static penetration resistance of soils, *Proceedings of the Conference on In-situ Measurement of Soil Properties*, American Society of Civil Engineers, Raleigh (N. Carolina), **1**, 151–188 (1975).

2.5   Terzaghi, K. and Peck, R. B., *Soil Mechanics in Engineering Practice*, 2nd edn, John Wiley, New York, 1967.

2.6   Skempton, A. W., The bearing capacity of clays, *Building Research Congress*, Div. I, 180 (1951).

2.7   Wyllie, D. C., *Foundations in Rock*, Spon, London, 1991.

2.8   Kulhawy. F. H. and Goodman, R. E., Foundations in rock, Chapter 15 of *Ground Engineering Reference Book*, ed. F. G. Bell, Butterworths, London, 1987.

2.8a   Kulhawy, F. H. and Goodman, R. E., Design of foundations on discontinuous rock, *Structural Foundations on Rock*, Balkema, Rotterdam, 1980, pp. 209–220.

2.9   Baguelin, F., Jézéquel, J. F., and Shields, D. H., *The Pressuremeter and Foundation Engineering*, Trans. Tech. Publications, 1978, pp. 205–228.

2.10   Hobbs, N. B., General report and state-of-the-art review, in *Proceedings of a Conference on Settlement of Structures*, Pentech Press, Cambridge, 1974, pp. 579–609.

2.11   Burland, J. B. and Burbidge, M. C., Settlement of foundations on sand and gravel, *Proceedings of the Institution of Civil Engineers*, **78**(1), 1325–1381 (1985).

2.12   Burland, J. B., Broms, B. B., and De Mello, V., Behaviour of foundations and structures, in *Proceedings of the 9th International Conference on Soil Mechanics*, Tokyo, 1977, Session 2.

2.13   Burland, J. B. and Wroth, C. P., Review paper: Settlement of buildings and associated damage, in *Proceedings of the Conference on Settlement of Structures*, Pentech Press, Cambridge, 1974, pp. 611–654.

2.14   Skempton, A. W. and MacDonald, D. H. The allowable settlement of buildings, *Proceedings of the Institution of Civil Engineers, Part 3*, **5**, 727–784 (1956).

2.15   Meyerhof, G. G., The settlement analysis of building frames, *Structural Engineer*, **25**, 309 (1947).

2.16   Polshin, D. E. and Tokar, R. A., Maximum allowable non-uniform settlement of structures, in *Proceedings of the 4th International Conference on Soil Mechanics*, London, 1957, Vol 1, p. 402.

2.17   Bjerrum, L., Discussion on compressibility of soils, in *Proceedings of the European Conference on Soil Mechanics and Foundation Engineering, Wiesbaden*, 1963, Vol 2, pp. 16–17.

2.18   Tomlinson, M. J., Driscoll, R., and Burland, J. B., Foundations of low-rise buildings, *The Structural Engineer*, **56A**, 161–173 (1978).

2.19   *Structure–Soil Interaction – The Real Behaviour of Structures*, Institution of Structural Engineers, London, 1989.

2.20   Sutherland, H. B., Review paper: Granular materials, in *Proceedings of the Conference on Settlement of Structures*, Pentech Press, Cambridge, 1974, pp. 473–499.

2.21   Zeevaert, L., Foundation design and behaviour in Tower Latino Americana Building in Mexico City, *Géotechnique*, **7**, 115–133 (1962).

2.22   May, J., Heave on a deep basement in the London Clay, in *Proceedings of the Conference on Settlement of Structures*, Pentech Press, Cambridge, 1974, pp. 177–182.

2.23   Jurgenson, L., The application of theories of elasticity and plasticity to foundation problems, *Journal of the Boston Society of Civil Engineers*, **21**, 206–211 (1934).

2.24   Newmark, N. M., Simplified computations of vertical pressure in elastic foundations, *University of Illinois Engineering Experiment Station Bulletin No. 429*, (1935).

2.25   Terzaghi, K. and Peck, R. B., *Soil Mechanics in Engineering Practice*, 2nd edn, John Wiley, New York, 1967, pp. 271–276.

2.26   Schultze, E. and Sherif, G., Prediction of settlement from evaluated settlement observations for sand, *Proceedings of the 8th International Conference on Soil Mechanics*, Moscow, 1973, Vol. 1, p. 225.

2.27   Schmertmann, J. H., Hartman, J. P., and Brown, P. R., Improved strain influence factor diagrams, *Proceedings of the American Society of Civil Engineers*, 104 (GT8), 1131–1135 (1978).

2.28   Lunne, T. and Christoffersen, H. P., *Interpretation of Cone Penetrometer Data for Offshore Sands*, Norwegian Geotechnical Institute, Publication No. 156, 1–12 (1985).

2.29   Robertson, P. K., Campanella, R. G., and Wightman, A., *SPT–CPT Correlations*, University of British Columbia, Vancouver, Civ. Eng. Dept, Soil Mechanics Series, No. 62, 1982.

2.30   Baldi, G., *et al.*, Cone resistance of dry medium sand, *Proceedings of the 10th International Conference on Soil Mechanics and Foundation Engineering*, Stockholm, **2**, 427–432 (1981).

2.31   Padfield, C. J. and Sharrock, M. J., *Settlement of Structures on Clay Soils*, Construction Industry Research and Information Association, Special Publication 27 (1983).

2.32   Skempton, A. W. and Bjerrum, L., A contribution to the settlement analysis of foundations on clay, *Géotechnique*, **7**, 168–178 (1957).

2.33   Jamiolkowski, M., *et al.*, Design parameters for soft clays, *Proceedings of the 7th European Conference on Soil Mechanics and Foundation Engineering*, Brighton, **5**, 21–57 (1979).

2.34   Jardine, R., Fourie, A., Maswose, J., and Burland, J. B., Field and laboratory measurements of soil stiffness, *Proceedings of the 11th International Conference on Soil Mechanics and Foundation Engineering*, San Francisco, **2**, 511–514 (1985).

2.35   Marsland, A., *In-situ* and laboratory tests on glacial clays at Redcar, in *Proceedings of a Symposium on Behaviour of Glacial Materials*, Midlands Soil Mechanics Society, 1975, pp. 164–180.

2.36   Butler, F. G., Review paper: Heavily over-consolidated clays, in *Proceedings of the Conference on Settlement of Structures*, Pentech Press, Cambridge, 1974, pp. 531–578.

2.37   Brown, P. T. and Gibson, R. E., Rectangular loads on inhomogeneous elastic soil, *Proceedings of the American Society of Civil Engineers*, **99** (SM-10), 917–920 (1973).

2.38   Janbu, N., Bjerrum, L. and Kjaernsli, B., *Veiledning Vedløsning avfundamenteringsoppgarer*, Publication No. 16, Norwegian Geotechnical Institute (1956).

2.39   Christian, J. T. and Carrier, W. D., Janbu, Bjerrum and Kjaernsli's chart reinterpreted, *Canadian Geotechnical Journal*, **15**, 123–128 (1978), and discussion, **15**, 436–437 (1978).

2.40   Fox, E. N., The mean elastic settlement of a uniformly loaded area at a depth below the ground surface, in *Proceedings of the 2nd International Conference on Soil Mechanics, Rotterdam*, 1948, Vol. 1, pp. 129–132.

2.41   Simons, N. A., Review paper: Normally consolidated and lightly over-consolidated cohesive materials, in *Proceedings of the Conference on Settlements of Structures*, Pentech Press, Cambridge, 1974, pp. 500–530.

2.42   Chandler, R. J. and Davis, A. G., Further work on the engineering properties of Keuper Marl, *Construction Industry Research and Information Association Report No. 47* (1973).

2.42a   Ward, W. H., Burland, J. B., and Gallois, R. W. Geotechnical assessment of a site at Mundford, Norfolk, for a large proton accelerator, *Géotechnique*, **18**, 399–431 (1968).

2.43   Hobbs, N. B., General report and state-of-the-art review, in *Proceedings of the Conference on Settlement of Structures*, Pentech Press, Cambridge, 1974, pp. 579–609.

2.44   Smith, I. M., Geotechnical design and instrumentation, *Proceedings of the Conference on the Impact of Computer Technology on the Construction Industry*, Institution of Civil Engineers, London, 1984.

2.45   Bowles, J. E., *Analytical and Computer Methods in Foundation Engineering*, McGraw-Hill, Kogakusha, Tokyo, 1974.

2.46   Cheung, Y. K. and Yeo, M. F., *A practical introduction to finite element analysis*, Pitman Publishing, London, 1979.

2.47   Zienkiewicz, O. C., *The Finite Element Method*, McGraw-Hill, London, 1977.

2.48   Smith, I. M., *Programming the Finite Element Method with Application to Geomechanics*, John Wiley, Chichester, 1982.

2.49   Smith, G. D., *Numerical Solution of Partial Differential Equations*, Oxford University Press, London, 1974.

2.50   Smith, I. M. in Bell, F. G. (ed.), *Foundation Engineering in Difficult Ground*, Butterworths, London, 1978, pp. 143–160.

2.51   Potts, D. M., Dounias, G. T., and Vaughan, P. R., Finite element analysis of progressive failure of Carsington embankment, *Gétechnique*, **40**(1), 79–101 (1990).

2.52   Duncan, J. M. and Chang, C. Y., Nonlinear analysis of stress and strain in soils, *Proceedings of the American Society of Civil Engineers*, **96** (SM5), 1629–1653 (1970).

2.53   Simpson, B., O'Riordan, M. J. and Croft, D. D., A computer model for the analysis of ground movements in London Clay, *Géotechnique*, **29**(2), 149–175 (1979).

2.54   Smith, I. M. and Hobbs, R., Biot analysis of consolidation beneath embankments, *Géotechnique*, **26**(1), 149–171 (1976).

3 Foundation Design in Relation to Ground Movements

In the previous chapter we have considered foundation design in relation to bearing capacity and consolidation of the soil. However, ground movements which are independent of stresses imposed by the foundation loading can occur. Examples of these are movements due to swell and shrinkage of the soil under varying moisture and temperature conditions, frost heave, hillside creep, mining and regional subsidence, and settlements due to shock and vibration.

It is necessary to take precautions against the effects of these movements on the structure, either by deepening the foundations to place them on ground which is not susceptible to movement or, if this is not economically possible, to adopt special forms of construction which will allow appreciable movement without damaging the structure.

The various types of ground movement are described in this chapter and the foundation designs appropriate to these movements are discussed.

3.1 Soil Movements

3.1.1 Wetting and drying of clay soils due to seasonal moisture content variations

Some types of clay soil show marked swelling with increase of moisture content, followed by shrinkage after drying out. In Great Britain, the clays showing this characteristic are mainly the stiff-fissured heavy clays such as the London, Gault, Weald, Kimmeridge, Oxford and Lias Clays, and the clays of the Woolwich and Reading Beds. The leaner glacial clays and marshy clays do not show marked seasonal swell and shrinkage, except for the chalky boulder clay of East Anglia which is derived from the stiff-fissured clays mentioned above. However, these leaner clays can show substantial shrinkage if they are influenced by the roots of growing trees. As a rough guide, clays with a liquid limit of more than 50 per cent which lie above line A in the plasticity chart (Fig. 1.12) are likely to be troublesome. Local enquiry and observations of shrinkage, cracking and desiccation in trial pits are helpful. The effect of this seasonal volume change is to cause a rise and fall in the ground surface accompanied by tension cracks in the soil in drying periods and closing of the cracks in the wet season. The movements are larger in grass-covered areas than in bare ground.

Measurements of moisture content variation with depth, comparing the profiles obtained from open ground and ground close to trees, will give some indication of the depth of desiccation for these two conditions. However, because of the wide scatter in moisture content values over the range of depths and the comparatively small difference in the moisture content of desiccated and undesiccated clays a comparison of the profiles is often inconclusive, particularly on tree-covered confined sites where the profile for open ground cannot be obtained.

The pore-water suction in clays is a much more fundamental indication of the swelling and shrinkage potential. The suction is unaffected by local variations in soil properties and is much more sensitive than moisture content determination. The Building Research Establishment in UK, in association with Imperial College, London, have developed a simple test for the pore-water suction of clays[3.1]. Undisturbed samples taken from boreholes or trial pits are cut transversely into cylindrical slices and discs of filter paper are inserted between them. The assembly of discs is wrapped in clingfilm and sealed with wax. After 5–10 days of storage the water content of the filter papers is measured, from which the suction is calculated. A profile of suction versus depth is plotted showing whether it is derived from surface drying by atmospheric effects or by shallow vegetation such as grass. A deeper and irregular profile shows the effect of deep-rooted vegetation. Low or zero suction indicates the

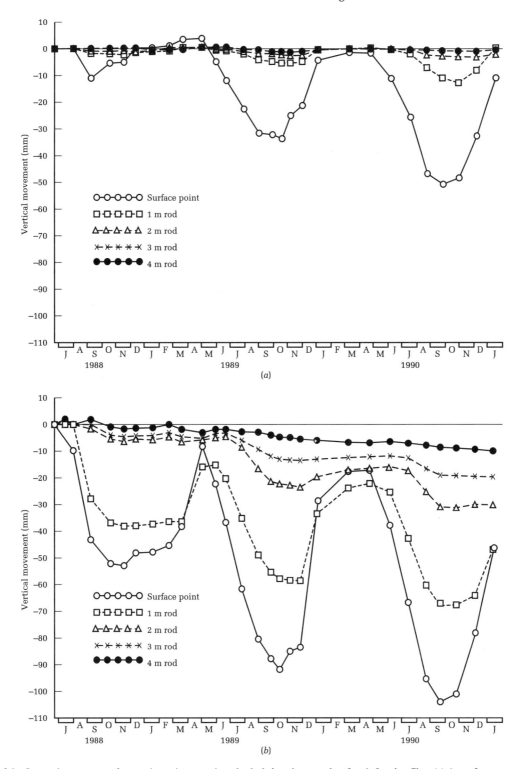

Fig. 3.1 Seasonal movements of measuring points at various depths below the ground surface in London Clay, (*a*) Away from trees. (*b*) Near trees (after Freeman *et al*[3.2]).

depth at which there is no potential for heave when the soil is wetted. The test can be made on strips cut from 38 mm tube samples, or on disturbed samples by compacting them into a mould.

Swelling and shrinkage of heavy clays can be very damaging to structures on shallow foundations. The movements are seasonal, causing subsidence of the ground surface in the summer months, particularly in grass-covered areas, and a corresponding rise during the winter. The Building Research Establishment have conducted extensive research into the effects of these movements at a London Clay site near Chattenden in Kent,[3.2] where part of the site was covered by Lombardy poplars 20–25 m high. The clay at this site has a liquid limit of 88 per cent and a plasticity index of 25, indicating a high shrinkage potential. Movements of the soil were measured in open ground and near trees at a series of measuring points between the ground surface and 4 m depth. The results for the 3 years of below-average rainfall from 1988 to 1990 are shown in Fig. 3.1. The cumulative drying of the clay over these years will be noted. The results of the research on shallow foundations showed that the long-standing recommendation for a foundation depth between 0.9 to 1.0 m below ground level for low-rise buildings on shrinkable clays was a reliable one for open-ground conditions. Many tens of thousands of houses have been constructed on sites in south-east England since that time with foundations at a depth of 0.9 m. Very few cases of damage have been reported. In most cases of reported cracking of houses on traditional strip foundations in clay, the damage can be traced to the effects of growing trees or bushes, or to building on sites where trees, hedges, or bushes had been removed.

Precautions are also necessary in the construction of ground floors on shrinkable clays. Although it is a reasonable precaution to take strip or pad foundations to a depth of 0.9–1.2 m below ground level, it will be uneconomical to excavate to this depth for the full ground-floor area of a building. The most economical procedure is to allow freedom of movement between the foundation walls and ground-floor slab, but internal non-load-bearing partitions and staircases are often supported by the ground-floor slab. A suspended ground floor with a void of at least 100 mm beneath it is the most effective way of avoiding heave damage. Alternatively the slab can be cast on to a 100 mm thick sheet of special low-density polystyrene or collapsible cardboard. Following the very dry summer of 1959, the ground floor slabs of houses in Hertfordshire were lifted some 5–20 mm above the foundation brickwork due to swelling of the clay in the autumn rains. The concrete floors were cast against the brickwork without any separating membrane, with the result that the brickwork was lifted, forming wide horizontal cracks around the foundation walls. Many other instances of such damage occurred in the dry summers of the 1970s and 1980s. Where possible, it is desirable to delay the construction of ground floors on clays which have become desiccated by exposure to summer drought conditions. In open ground away from trees the clay should have recovered to normal moisture conditions by the following spring.

Climatic factors   There are, however, two further factors which greatly increase the problem of swell and shrinkage and which may necessitate special methods of foundation design. The first factor is the effect of a wide difference in seasonal rainfall and soil temperature conditions. These conditions are met in the Sudan, the Levant coast of the Mediterranean, South Africa, south-eastern Australia, western Canada, and in the southern and southwestern parts of the USA. Measured ground movements

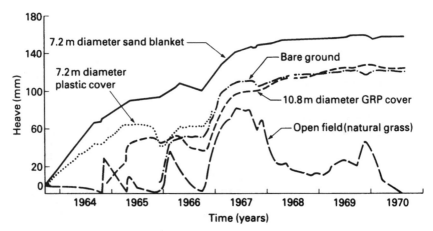

Fig. 3.2 Movements of the ground surface under various conditions of surface cover at Onderstepoort, South Africa (after Williams and Pidgeon[3.3]).

under various conditions of surface cover in a black expansive clay at the Onderstepoort test site, South Africa,[3.3] are shown in Fig. 3.2.

3.1.2 Effects of vegetation

The second factor which aggravates the swell and shrinkage problem is the effect of the roots of vegetation. The roots of trees and shrubs can extract considerable quantities of water from the soil. The root systems of isolated trees spread to a radius greater than the height of the tree, and in southern England they have caused significant drying of heavy clay soils to a depth of about 3–5 m. The

difference in the moisture content–depth relationship between open ground and tree-covered areas at the Chattenden test site is shown in Figs 3.1 and 3.3. Differential movements of 100 mm have been recorded in houses 25 m from a row of black poplars.

The problems caused by root systems are twofold. First there is the problem of heave of foundations on sites which have recently been cleared of trees and hedges, and second there is the problem of settlement in existing structures sited close to growing trees or caused by subsequent planting of trees and shrubs close to them.

Pressure exerted on the underside of foundations from heave in clay soils is much higher than that imposed by the

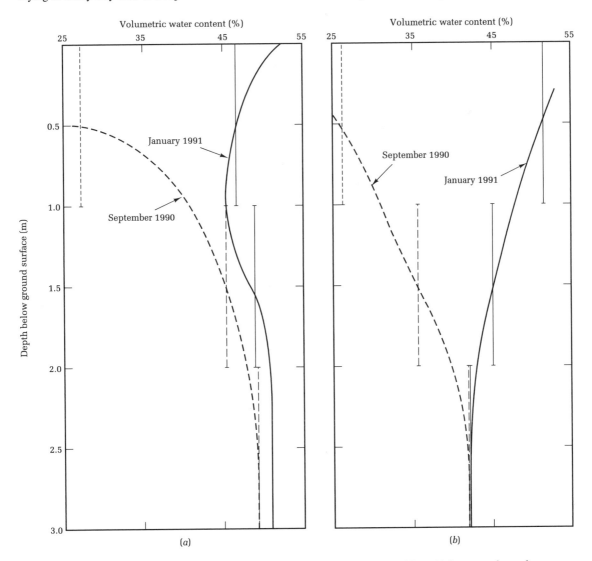

Fig. 3.3 Variation in water content versus depth in London Clay for summer and winter conditions. (a) Open ground away from trees. (b) Near trees.

superstructure of single or two-storey houses. This was demonstrated at the Chattenden site[3.2] where four 1 m square dummy foundations were set at a depth of 1.2 m and loaded to give bearing pressures of 17 kN/m² for pads 1 and 2, 34 kN/m² for pad 3, and 63 kN/m² for pad 4. The last-mentioned pressure is equivalent to that imposed by the strip foundations of a two-storey building. The pads were located in the tree-covered area. The movements shown in Fig. 3.4 are intermediate between those measured at points 1 and 2 m depth shown in Fig. 3.1(b). The cumulative downward movement due to soil shrinkage over three successive dry years will be noted.

The pronounced shrinkage which accompanies removal of water from clay soils can take place both vertically and horizontally. Thus, precautions must be taken not only against settlement but also against forces tending to tear the foundations apart. Problems with the narrow strip (trench fill) foundation in clays desiccated by vegetation are discussed in Section 4.2.1.

A severe case of cracking due to heave resulting from the removal of trees was reported by Samuels and Cheney.[3.4] Large elm trees had been cut down a few months before four cottages were built on London Clay.

Cracks developed two years after completion of building, and levelling showed that these were due to soil swelling. The movement was still not completed after 20 years from the time of cutting down the trees. The estimated heave of the soil was 150 mm. This is by no means an isolated case. The author has knowledge of several instances of heave in London Clay in the range of 100–150 mm caused by removal of mature trees, with heave movements continuing over periods of 10–20 years.

The soil beneath areas previously occupied by old buildings and paved areas must also be allowed to come to an equilibrium moisture content with the adjoining uncovered ground. General guidance on foundations affected by trees in clay soils has been given by the Building Research Establishment.[3.5]

Piled foundations Delays of 20 years or more after cutting down trees on a clay site are not usually admissible, and the only satisfactory procedure is to adopt piled foundations. A further part of the research at the London Clay site at Chattenden [3.2] included installing seven bored piles to depths between 3 and 4 m, and two 12 m deep piles. They were installed in the area of growing trees. The

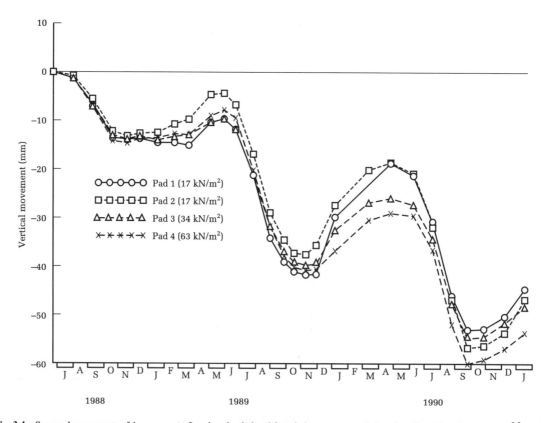

Fig. 3.4 Seasonal movements of 1 m square × 2 m deep loaded pad foundations near trees in London Clay (after Freeman et al[3.2]).

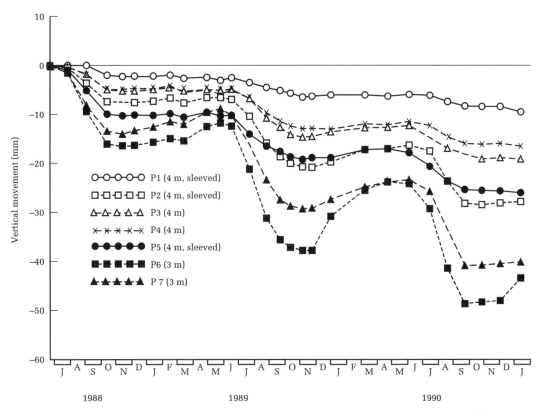

Fig. 3.5   Seasonal movements of unloaded bored and cast-in-place piles near trees in London Clay (after Freeman *et al*[3.2]).

alternate settlement and uplift movements during the summer and winter seasons are shown in Fig. 3.5. Again the cumulative downdrag which had overcome the base resistance of the piles will be noted. The two 12 m piles showed negligible movement.

The engineer should beware of advice given to provide 'short-bored' piles for house foundations on clay sites. Figure 3.5 shows that comparatively long piles are needed as a safeguard against the effects of swell or shrinkage due to trees or hedges. It must be remembered that swelling or shrinkage of the soil takes place beneath the ground floor of a building as well as beneath the foundations and to avoid damage by uplift or settlement to the interior of the building it is essential to provide a suspended floor carried by the piles.

A clear space should be left beneath the floor and beneath ground beams. In South Africa and Israel bored piles are extensively used even for light single-storey structures. The design of foundations of this type must take into consideration appreciable uplift forces which result from the adhesion of the clay to the pile shaft when the soil is swelling in the rainy season or after removal of vegetation or downdrag forces when the clay is shrinking.

Reinforcement must be provided to prevent transverse cracking and lifting of the pile shaft. Alternatively, the piles can be sleeved through the swelling zone. In this case

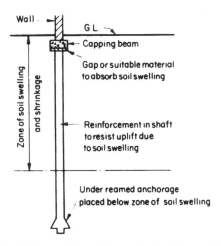

Fig. 3.6   Under-reamed bored pile used for foundations of light structures in soils subject to swelling and shrinkage to considerable depth below ground level.

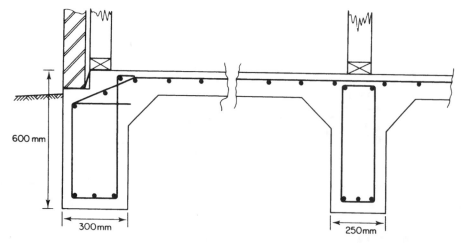

Fig. 3.7   Design of a slab and beam raft foundation as used in swelling and shrinking clays in Australia (after Walsh[3.6]).

a rigid PVC or steel sleeve surrounded by an annular void space is essential. Flexible polythene sheeting lowered down the pile borehole before placing concrete can be ineffective, either because of displacement of the sheeting in concertina form, or because of the action of the drilling auger in producing a tapering borehole. To economize in the shaft length required to resist bearing and uplift forces an under-reamed bulb end is sometimes provided on the bottom of the pile (Fig. 3.6).

In Australia[3.6] it is the usual practice to adopt stiff raft foundations as a precaution against seasonal moisture content variations. A typical design is shown in Fig. 3.7.

3.1.3 Shrinkage of clays due to high temperatures

Severe shrinkage of clay soils can be caused by the drying out of the soil beneath the foundations of boilers, kilns and furnaces. Cooling and Ward[3.7] reported that the heat from a 61 m × 30.5 m brick kiln had penetrated through 2.7 m of brick rubble filling and then for the full 19.5 m thickness of the underlying Oxford Clay. The kiln was demolished and a cold process building was erected in its place. Seven years later one corner of the new building had settled 330 mm, and elsewhere the building had risen 178 mm. In the same paper, Cooling and Ward quote the case of a battery of three Lancashire boilers 2.75 m diameter × 3 m long, where settlements up to 150 mm at the centre and 75 mm at the sides had occurred after two winter seasons of firing. The temperature and moisture conditions in the London Clay beneath the boilers are shown in Fig. 3.8.

Where furnaces, boiler houses, and the like are constructed on clay soils it is necessary to provide an

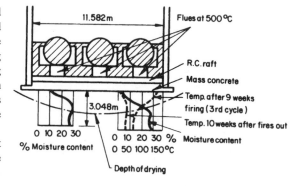

Fig. 3.8   Temperature and moisture conditions beneath Lancashire boilers founded on London Clay (after Cooling and Ward[3.7]).

insulating air-gap between the source of heat and the foundation concrete, or alternatively to provide sufficient depth of concrete or other material to ensure that the temperature at the bottom of the foundation is low enough to prevent appreciable drying of the soil.

3.1.4 Ground movements due to low temperatures

In some soils and rocks appreciable ground movement can be caused by frost. In Great Britain, chalk and chalky soils and silty soils are liable to frost heave, but the effects are not noticeable in heavy clays or sandy soils. Most of the increase in volume from frost heave is due to the formation of lenses of ice. These thaw from the surface downwards, the moisture content is high over the thawing period producing a locally weaker soil. Severe frost in the winter of 1954–55 caused a heave in chalk filling beneath the floors of partly constructed houses near London; the concrete ground floors were lifted clear of the brick footing walls by about 25 mm and frost expansion of the chalk fill caused displacement of the brickwork. In Norway and the southern regions of Canada, frost heave effects are experienced to depths of 1.2–2 m below ground level, resulting in heaving of 100–300 mm of the ground surface. Frost effects can be severe below liquefied natural gas containers and cold-storage buildings. Cooling and Ward[3.7] have described the heaving of the floor and columns of a cold-storage building over a period of 4–5 years. The temperature and ground movement are shown in Fig. 3.9.

British Standard 8103 (1986) Code of Practice for *Stability, Site Investigations, Foundations and Ground Floor Slabs for Housing*, recommends a minimum depth of 0.45 m for protection of foundations against frost heave, except for upland areas and areas known to be subject to long periods of frost where an additional depth may be desirable.

Where piled foundations are necessary to support structures sited on weak compressible soils, provision is required against damaging uplift forces on pile shafts due to expansion of the frozen soil. These are referred to in Canada as 'adfreezing' forces. Penner and Gold [3.8] measured the peak adfreezing forces on columns anchored into non-frozen soil at a site where frost penetration of 1.09 m caused a surface heave of 100 mm. The forces on the columns were

Steel: 113 kN/m^2
Concrete: 134 kN/m^2
Timber: 86 kN/m^2

Where the depth of frost is limited, the uplift forces on pile shafts, ground beams, and ground-floor slabs can be eliminated by removing the frost-susceptible soil and replacing it by clean sandy gravel or crushed and graded rock. Open gravel should not be used since silt could be washed into the interstices of the gravel during thawing periods, resulting in the formation of a silty gravel susceptible to frost expansion.

To avoid harmful effects in cold storage buildings, the floors should be constructed above ground level. If this is not practicable for structural or other reasons, a heating element may be provided below foundation level to prevent freezing of the soil.

Effects of permafrost Frost heave effects are very severe in Arctic and Antarctic regions where ground conditions known as 'permafrost' are widespread. Permafrost means permanently frozen ground, which can vary in thickness from a few metres to thousands of metres. About 50 per cent of the land mass of the former USSR is a permafrost region, generally lying north of the fiftieth parallel of latitude. Permafrost areas are also widespread in northern Canada, Alaska, and Greenland. The most difficult foundation problems occur where permafrost is overlain by soils subject to seasonal freezing and thawing. The thickness and lateral extent of this overburden do not depend solely on seasonal temperatures. Factors such as the type of soil or rock, the cover of vegetation, exposure to the sun, surface configuration, and ground-water movements, all have important effects on the zone subject to freezing and thawing. Permafrost can also occur in distinct layers, lenses, sheets, or dikes separated by thawed material. These thawed zones separating permafrost are known as 'taliks'. It is evident that severe movement will occur in the ground surface of these regions where the soils and rocks are susceptible to frost heave. Reference should be made to specialist publications for guidance on foundation design in permafrost conditions.

3.1.5 The application of finite difference or finite element methods to the calculation of temperature variations in soils

It is possible to predict both steady-state and transient (i.e. varying with time) temperature distributions in soils by application of either the finite difference method or the finite element method. Consider the one-dimensional transient heat-flow problem, the governing differential equation for which is

$$\frac{\partial T}{\partial t} = k\frac{\partial^2 T}{\partial x^2} \tag{3.1}$$

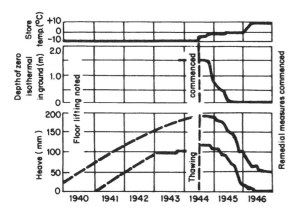

Fig. 3.9 Foundation movements and temperature conditions below a cold storage building (after Cooling and Ward[3.7]).

where

T = temperature,
t = time,
x = distance,
k = a constant ... $\lambda/\rho C$,
λ = thermal conductivity,
ρ = mass density,
C = specific heat.

Equation (3.1) can be expressed in finite difference form as

$$\frac{T_{i,j+1} - T_{i,j}}{\Delta t} = \frac{k(T_{i-1,j} - 2T_{i,j} + T_{i+1,j})}{(\Delta x)^2} \tag{3.2}$$

where

$T_{i,j}$ = temperature at point i at time step j,
Δt = time increment between time steps,
Δx = increment of distance between points.

This can be used to obtain a solution by setting all T_i to initial values and defining boundary conditions which have some time dependence. Explicit results are obtained for time step $j + 1$ as

$$T_{i,j+1} = \frac{\Delta t}{(\Delta x)^2} k(T_{i-1,j} - 2T_{i,j} + T_{i+1,j}) + T_{i,j} \tag{3.3}$$

but note that there is a stability criterion associated with the computation which is obtained by rewriting equation (3.3) as

$$T_{i,j+1} = rT_{i-1,j} + (1 - 2r)T_{i,j} + rT_{i+1,j} \tag{3.4}$$

where

$$r = \frac{\Delta t}{(\Delta x)^2} k.$$

It can be shown that $r \leqslant 0.5$ is a requirement to ensure convergence of the solution, i.e. $\Delta t \leqslant 0.5(\Delta x)^2/k$.

A solution can be obtained by an implicit method which requires solution of a set of simultaneous equations at each time step, but permits larger time steps than does the explicit method.[2.47]

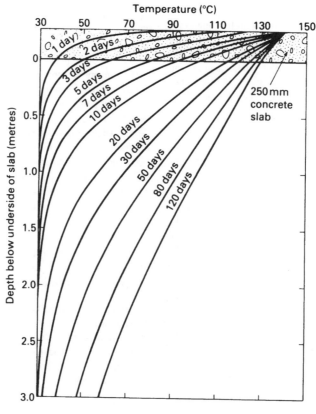

Fig. 3.10   Temperature distribution with time beneath a heated slab.

Equation (3.1) would relate, say, the rate of increase of temperature T at a level at depth x below the surface of a soil mass where a wide area of the surface of the soil is subject to some defined change in temperature with time. Such a case is illustrated in Fig. 3.10 where a 250 mm thick concrete slab rests on sand. Initially the temperature at all levels in the slab and sand was 30 °C. Due to the starting up of a smelting process the surface temperature of the concrete rose to 140 °C over a period of 7 days and thereafter remained at 140 °C. Ground water at 8 m below the slab is assumed to remain at a constant temperature of 30 °C.

A computer program was written to determine the temperature distribution at various times using equation (3.3). The results are shown in Fig. 3.10. In this instance the temperature determination was a necessary precursor to an analysis of vapour pressures in the soil beneath the slab. Diffusion of vapour through the soil and the slab was modelled by a finite difference form of equation

$$\frac{\partial g}{\partial t} = \frac{1}{\mu N} \frac{\partial P}{\partial x} \, , \qquad (3.5)$$

where

g = mass of vapour diffused per unit area,
μN = diffusion resistance,
P = vapour pressure.

The results of this second part of the analysis are shown in Fig. 3.11. The cause for concern was that vapour pressure might build up to an extent where it might lift the slab. The problem was really two-dimensional in that both heat flow and diffusion were possible in the transverse horizontal direction as well as in the vertical direction. However, the results of the simpler, and conservative, one-dimensional analysis showed that the concern was unwarranted.

Most of the widely available general-purpose finite element systems have heat flow analysis capabilities for the two-dimensional steady-state case governed by the equation

$$\frac{\partial}{\partial x}\left(k_x \frac{\partial T}{\partial x}\right) + \frac{\partial}{\partial y}\left(k_y \frac{\partial T}{\partial y}\right) = 0 \qquad (3.6)$$

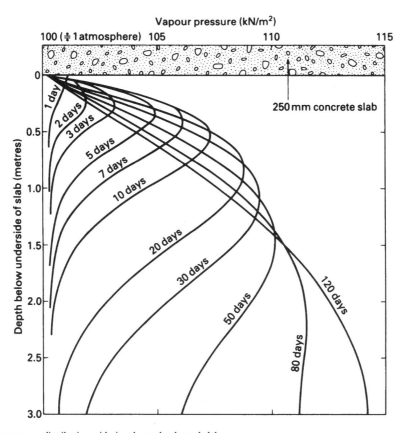

Fig. 3.11   Vapour pressure distribution with time beneath a heated slab.

and often can also solve the two-dimensional transient problem governed by

$$\frac{\partial T}{\partial t} = \frac{\partial}{\partial x}\left(k_x\frac{\partial T}{\partial x}\right) + \frac{\partial}{\partial y}\left(k_y\frac{\partial T}{\partial y}\right).\tag{3.7}$$

More complicated problems involving ground water phase changes (ice/water or water/steam) will require specialist programs.[3.9]

3.2 Ground movements due to water seepage and surface erosion

Troubles with water seepage and erosion occur mainly in sandy soils. *Internal erosion* can result from ground water seeping into fractured sewers or culverts carrying with it fine soil particles. Ground-water seepage can also cause loss or degradation of the soluble constituents of an industrial waste-fill material. The consequent loss of ground from beneath foundations may lead to collapse of structures. Trouble of this kind is liable to occur in mining subsidence areas where sewers and water mains may be broken. It can also occur as a result of careless technique in deep excavation below the water table when soil particles are carried into the excavation area by flowing water.

In South Africa, Zimbabwe, and the Luanda region of Angola dry loose sands have been known to subside as a result of seepage of water from leaking mains or drainage pipes. It is often the practice to provide special forms of foundations as a safeguard against such contingencies (Jennings and Knight[3.10]). Similar troubles have occurred in Russia, where loess soils (see Section 1.14.2) are widespread and extend to depths of hundreds of metres; a range of techniques has been described by Abelev[3.11] for treating these soils. Where the thickness of loess does not exceed 5 m, it is treated by compacting the ground surface. A heavy rammer weighing from 4–7 t is dropped from a height of 5–7 m by means of a pile-driver or crane. From 10 to 16 blows are applied to the same spot, which compacts the soil to a depth of 2–3.5 m.

Where the loess soils are more than 6 m deep they are compacted at depth by pile-driving or by blasting. Steel tubes of 280 mm outside diameter are driven by a piling hammer to depths up to 18 m. The lower end of the tube is closed by a conical shoe having a diameter 50 per cent larger than that of the tube. After full penetration of the tube it is withdrawn and the hole is filled with soil in layers, each layer being compacted by ramming.

Using the blasting method a pattern of blast holes of 75–80 mm is drilled in the area to be treated. If the loess soils are dry, water is injected to bring them to the optimum moisture content for compaction. The blasting charge consists of a string of 40–50 mm explosive cartridges suspended by cords over the full depth of the hole. The charges in each hole are fired successively at 1 min intervals. The enlarged holes resulting from the blasting are filled with soil in layers of about 1.2 m each, each layer being tamped with a rammer weighing about 1 t.

Surface erosion may take place as a result of loss of material in strong winds or erosion by flowing water. Fine sands, silts, and dry peat are liable to erosion by the wind. The possibility of undermining of foundations can readily be provided for by a minimum foundation depth of about 0.3 m, and by encouraging the growth of vegetation or by blanketing the erodible soil by gravel, crushed rock, or clay. Surface erosion by flowing water may be severe if structures are sited in the bottom of valleys, especially in regions of tropical rainstorms. Normal foundation depths (say 0.9–1.2 m) are inadequate for cases of erosion by floodwaters, but this possibility can be provided for by attention to the siting of structures, adequate drainage and paving or other forms of surface protection, of paths taken by periodical discharges of flood water. Severe erosion can take place around the foundations of bridges or other structures in waterways subjected to heavy flood discharges. The required depths of such foundations can be obtained by hydraulic calculations and local observations.

From time to time cases are reported of subsidence due to solution of minerals from the ground as a result of water seepage. Subsidence and the formation of swallow holes are quite widespread in Britain in the carboniferous limestone and chalk districts (see Section 1.15).

Where swallow holes can be located by trial pits or trenches and are found to be of limited lateral extent, they can be bridged by slabs or domes bearing on stable ground. The configuration of deep-lying swallow holes or 'karst' formations in limestone cannot be mapped reliably by exploratory drilling. It is the usual practice to provide piled foundations for heavy structures. The piles are designed to carry the load in shaft friction transferred to stable pillars of chalk or limestone below the upper weak or subsiding strata. It may be necessary to drill through several layers of rock strata to reach a formation which is judged to be stable. The piles are sleeved through the unstable upper rock layers, and the shaft friction in the underlying stable layers can be enhanced by grouting around the shaft.

3.3 Ground movements due to vibrations

The processes of using vibrators for consolidating concrete or vibratory rollers and plates for compacting sandy or gravelly soils are well known. If a poker vibrator is pushed into a mass of loosely placed concrete, the surface of the concrete will subside as its density is increased by the vibrations transmitted to it. It has been found that high frequency in vibrating plant is more effective than low frequency for consolidating concrete or soils.

The same effects of consolidation and subsidence can occur if foundations on sands or sandy gravels (or if the soils themselves) are subjected to vibrations from an external source. Thus, vibrations can be caused by out-of-balance machinery, reciprocating engines, drop hammers, pile-driving, rock blasting, or earthquakes. Damage to existing structures resulting from pile-driving vibrations is not uncommon, and it is usual to take precautions against these effects when considering schemes for piled foundations in sands adjacent to existing structures.

Experiments in the field and laboratory and records of damage have shown that the most serious settlements due to vibrations are caused by high-frequency vibrations in the range of 500–2500 impulses/min. This is also within the range of steam turbines and turbogenerators. Terzaghi and Peck[3.12] record a case of turbogenerator foundations on a fairly dense sand and gravel in Germany. The frequency of the machinery was 1500 rev/min and settlements exceeded 0.3 m within a year of putting the plant to work. Terzaghi and Peck also mention long-continued traffic vibrations and earthquakes as causes of foundation settlement. If the foundations of structures carrying vibrating machinery cannot be taken down to a stratum not sensitive to vibration (clays for example do not usually settle under vibrating loads), then special methods of mounting the machinery to damp down the vibrations must be adopted. Consolidation of sands beneath foundations by vibration processes are described in Section 11.6. Methods of designing machinery foundations to absorb or damp down vibrations are described later in this chapter.

3.4 Ground movements due to hillside creep

Certain natural hillside slopes are liable to long-term movement which usually takes the form of a mass of soil on a relatively shallow surface sliding or slipping downhill. Typical of such movements are hillsides in London Clay in inland areas with slopes of 8° or steeper, and in glacial valleys in north-east England where slopes are steeper than 15°. Their effects can be seen in parallel ridges in the ground surface; in concave scars with lobes of slip debris below them, and in leaning trees. Trial pits should be excavated on sloping sites to investigate the possible occurrence of potentially unstable slipped debris (solifluction sheet) brought down from the higher ground.

Normally, the weight of structures erected on these slopes is insignificant in relation to the mass of the slipping ground. Consequently foundation loading has little or no influence on the factor of safety against slipping. However, other construction operations may have a serious effect on the slope stability, for example regrading operations involving terracing the slopes may change the state of stress both in cut and fill areas, or the natural drainage of subsoil water may be intercepted by retaining walls. Well-grown vegetation can help to prevent shallow slides, and its clearance can often initiate renewed slipping.

Instability of slopes may occur on rocky hillsides where the strata dip with the ground surface, especially where bedding planes in shales or clayey marls are lubricated by water. Again the risk of instability is increased by regrading operations or alteration of natural drainage conditions rather than by the foundation loading.

There is little that can be done to restore the stability of hillside slopes in clays since the masses of earth involved are so large, and regrading operations on the scale required are usually impossible. The best advice is to avoid building in such areas or, if this cannot be done, to design the foundations so that the whole structure will move as one unit with provision for correcting the level as required. Suitable methods of construction are discussed in the next section of this chapter.

Local instability in rocky slopes can be corrected by grouting or by rock bolting as described in Section 9.3.4.

3.5 Ground movement due to mining subsidence

3.5.1 Forms of subsidence

The magnitude and lateral extent of subsidence due to mineral extraction depends on the method used for winning the minerals from the ground, whether by mining, pumping, or dredging. Concealed shallow caverns resulting from mining the Chalk to recover flints are quite widespread in south-east England, but the main problems in Great Britain arise from coal mine workings, and these will be discussed in some detail in the following pages. An early method of mining coal was by sinking 'bell-pits', practised in medieval times. A vertical shaft was sunk to the level of the coal seam, and mining then proceeded in all directions radially from the shaft. The bottom of the shaft was 'belled-out' as necessary to support the roof and mining continued until the roof was in imminent danger of collapse or the accumulation of ground water became too much for the primitive baling or pumping equipment. The shafts were filled with spoil from other workings or were used as rubbish tips. Most of the traces of these workings were lost over the centuries, but they still remain, notably in the Northumbrian Coalfield, as a source of trouble in foundation design. Their presence can sometimes be detected by depressions in the ground.

Similar workings were excavated in many coalfields throughout the country in the 1926 Coal Strike. These workings in the form of shafts, drifts, or deep trenches were excavated at or near the outcrops of coal seams. No records of their location were kept, but local enquiry will sometimes establish their presence. Similar workings were probably made in the 1984–85 Coal Strike.

Although geophysical methods have been used to detect concealed shafts with limited success, the presence of scrap metal, old foundations, and infilled ducts limits the usefulness of geophysical observations. The most positive method of locating the whereabouts of a suspected shaft is by trenching across the site.

3.5.2 Pillar and stall workings

As mining techniques improved, in particular with the development of steam pumping plant in the eighteenth and nineteenth centuries, workings were extended to greater distances from the shafts. Support to the roof was given by methods known variously as 'pillar and stall', 'room and pillar', or 'bord and pillar'. Galleries were driven out from the shaft with cross-galleries, leaving rectangular or diamond-shaped pillars of unworked coal to support the roof. Only 30–50 per cent of the coal was extracted in this way in the first advance of the workings from the shaft. On the return workings towards the shaft the pillars were removed either in entirety, to allow full collapse of the roof, or partially, to give continued support. Large pillars were left beneath churches and similar public buildings and sometimes beneath the colliery headworks. The patterns of pillar and stall workings adopted in various parts of Britain have been described and illustrated by Healy and Head.[3.13]

In many coalfields in Britain the presence of these old pillar and stall workings with partially worked pillars remain as a constantly recurring problem in foundation design where new structures are to be built over them. If the depth of cover of soil and rock overburden is large, the additional load of the building structure is relatively insignificant and the risk of subsidence due to the new loading is negligible (Fig. 3.12(a)). If, however, the overburden is thin, and especially if it consists of weak crumbly material, there is a risk that the additional load imposed by the new structure will cause a breakdown in an arched and partially collapsed roof, leading to local subsidence (Fig. 3.12(b)). A third possibility is the risk of collapse of the overburden due to failure of the pillars. This can be caused either by crushing of the pillar under the overburden weight, or punching of the pillar into weak material such as seatearth in the floor of the workings, or punching into weak rocks above the roof (Fig. 3.12(c)). There is an increased risk of subsidence, or renewal of past subsidence, of pillar and stall workings if underlying seams are being worked by longwall methods.

There is also a risk of subsidence if flooded workings are pumped dry, when the effective weight of the overburden will be increased as a result of removing the supporting water pressure. This will increase the load on the pillars, possibly to the point of their collapse.

3.5.3 Longwall workings

The present-day method of coalmining is by 'longwall' working whereby the coal face is continuously advanced over a long front. The roof close to the face is supported by props, and the 'goaf' or cavity left by the coal extraction is partially filled by waste material ('stowage'). As the props are removed or allowed to crush down, the roof subsides, resulting in slow settlement of the ground surface. The surface subsidence takes the form of an advancing wave moving at the same rate as the advancing coal face (Fig. 3.13). The amount of subsidence at ground level is usually less than the depth of the underground

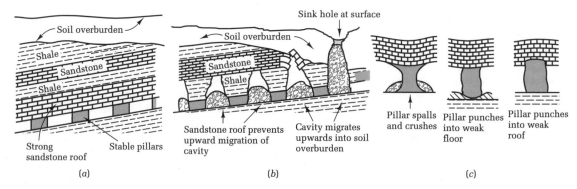

Fig. 3.12 Stable and unstable conditions in pillar and stall workings, (a) Stable. (b) Unstable due to weak shale roof. (c) Unstable due to failure of pillars.

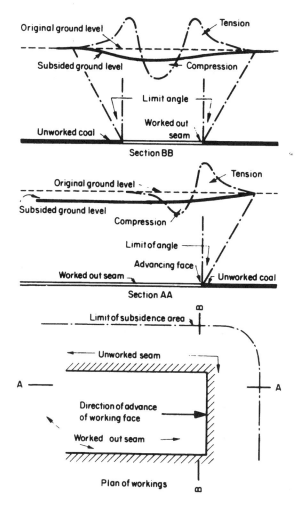

Fig. 3.14

Fig. 3.13 Form of ground subsidence given by longwall working.

gallery due to the bulking of the collapsing strata. Solid packing of the gallery with crushed mine dirt may reduce the subsidence to only one-half of that given by the unfilled gallery. The trough or basin of subsidence occurs all round the area of coal extraction, and the affected area at ground level is larger than the area of extraction. The angle between the vertical and a line drawn from the coal face to intersect the ground surface at the edge of the subsidence wave is known as the *limit angle* (Fig. 3.13); it is commonly 35°. Movement of the ground surface is not only vertical; horizontal strains are caused as the subsidence wave advances. Thus the building at A in Fig. 3.14 first experiences a tilting towards the trough accompanied by tensile strains in the ground surface which tend to pull the building apart. As the wave continues to advance, the ground becomes concave. When the concave

part of the subsidence wave reaches the building, the direction of tilting will have been reversed and the ground surface will be in a compression zone tending to crush the building. As the wave advances further the building will finally right itself and the horizontal strains will eventually die away. Vertical and horizontal movements resulting from longwall mining are severe. The amount of tilting depends on the surface slope, but it may vary from 1 in 50 or steeper over shallow workings to practically nothing over deep workings. Horizontal strains may be as much as 0.8 per cent for shallow workings, but are more commonly 0.2 per cent or less.

The protection in one form or another of structures cannot be neglected. It must be appreciated that the movements are rarely uniform. It is possible to predict with reasonable accuracy the amount of settlement and the extent of the subsidence zone if the coal is horizontal or nearly so, and if the overburden conditions are reasonably uniform. If, however, the coal seam is dipping steeply, no reasonable predictions can be made. Variations in the overburden, especially in the depth of soil cover, can cause differential vertical and horizontal movements across individual structures. Faulting can cause severe movement of the ground surface. The problem is further complicated if seams are worked at deeper levels, at different times, and in different directions of advance.

3.5.4 Other forms of subsidence

Other forms of underground mineral extraction by mining or pumping methods give rise to similar subsidence problems (see Healy and Head[3.13]). In Cheshire, brine is extracted by pumping from salt-bearing rocks. In earlier years indiscriminate pumping caused heavy long-term subsidence over a wide area due to removal of buoyant support of the overburden strata, and the formation of large cavities due to the solution of the salt in the rock. Cavities left by brine pumping or mining have coalesced due to solution effects, thus causing larger voids, and old timber-supported shafts in the salt-mining areas have collapsed. Similar subsidence is caused by the removal, by pumping, of mineral oil and natural gas.

The extent and depth of subsidence can be greatly reduced by carefully planned extraction accompanied by 'recharge' or recuperation pumping whereby water or gas is pumped in simultaneously with the extraction of the minerals. In some brine-pumping areas of the north of England the quantity pumped from a borehole is limited and the boreholes are spaced at such intervals that individual cavities are separated by pillars of salt of sufficient thickness to give support to the overburden.

3.5.5 Protection against subsidence due to longwall mining

Schemes for protection against subsidence should be drawn up in consultation with the local mine authorities. A great deal can be done to reduce the slope of the subsidence trough, and hence to reduce the tensile and compressive ground strain, by planning the extraction of the mineral in successive strips of predetermined width. The slopes can also be reduced considerably by concurrent mining of two or more seams beneath a site, advancing the working face in different directions. The opinion of a consulting mining engineer or geologist with knowledge of the area is advisable. The measures to be taken depend to a great extent on the type and function of the structure. Complete protection can be given by leaving a pillar of unworked coal beneath the structure. This involves costly payments to the mine owners for the value of the unworked coal and the subsidence effects around the fringe of the pillar are increased in severity. Therefore, protection by unworked pillars is only considered in the case of structures such as dams or historical buildings such as cathedrals where structural damage might have catastrophic effects. Measures for protection of structures and full bibliographies on the subject were given in the *Subsidence Engineers' Handbook*, published by the National Coal Board[3.14] (now British Coal).

The general principles recommended in the handbook are as follows:

(a) Structures should be completely rigid or completely flexible. Simply supported spans and flexible superstructures should be used whenever possible.
(b) The shallow raft foundation is the best method of protection against tension or compression strains in the ground surface.
(c) Large structures should be divided into independent units. The width of the gaps between the units can be calculated from a knowledge of the tensile ground strain derived from the predicted ground subsidence.
(d) Small buildings should be kept separate from one another, avoiding linkage by connecting wing walls, outbuildings, or concrete drives.

Although the orientation of a structure in relation to the direction of advance of the subsidence wave has, theoretically, an effect on the distortion of the structure (Fig.

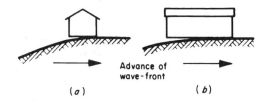

Fig. 3.15   Siting of structures in relation to advance of subsidence wave. (*a*) Building parallel to wave-front. (*b*) Building at right angles to wave-front.

3.15), the National Coal Board handbook stated that there is little point in trying to orient the structure in any particular direction relative to the mine workings unless the mining is to be undertaken in the near future to a definite plan. The handbook states that there is not enough difference between the maximum slope in a transverse profile and that in a longitudinal profile materially to affect the design of a structure. The design should allow for the maximum normal movements which are predicted for the seams to be worked. Wilson[3.15] states that with modern controlled longwall methods the amount of subsidence can be predicted to within ±10 per cent of the seam thickness.

Structures should not be sited within several metres of known geological faults, since subsidence is likely to be severe near fault planes.

Protection by raft foundations   *Raft foundations* should be as shallow as possible, preferably on the surface, so that compressive strains can take place beneath them instead of transmitting direct compressive forces to their edges, and they should be constructed on a membrane so that they will slide as ground movements occur beneath them. It is then only necessary to provide enough reinforcement in the rafts to resist tensile and compressive stresses set up by friction in the membrane. In the case of light structures such as dwelling-houses, it is not usually practicable to make the raft any smaller than the plan area of the building. However, in the case of heavy structures it is desirable to adopt the highest possible bearing pressures so that the plan dimensions of the raft are the smallest possible (Fig. 3.16). By this means the total horizontal tensile and compressive forces acting on the underside of the raft are kept to a minimum, and the lengths of raft acting as a cantilever (Fig 3.16(*a*)) at the 'hogging' stage, or as a beam (Fig. 3.16(*b*)) at the 'sagging' stage, are also a minimum. Mauntner [3.16] has analysed the conditions of support shown in Fig. 3.16 as follows:

Maximum pressure on foundation is given by

$$q_{max} = \frac{4qb}{3(b - 2l)} \quad \text{for cantilevering} \qquad (3.8)$$

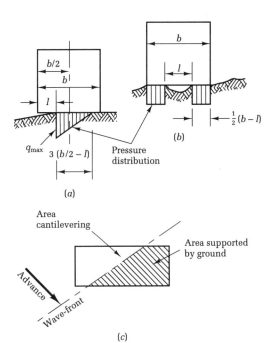

Fig. 3.16 Calculation of maximum foundation pressure (after Mauntner[3.16]). (*a*) Raft acting as cantilever. (*b*) Raft acting as beam. (*c*) Wave-front oblique to structure.

when l is greater than or equal to $b/4$; when l is less than $b/4$,

$$q_{max} = q\left(1 + \frac{3bl}{(b - l)^2}\right) \qquad (3.8a)$$

and

$$q_{max} = \frac{qb}{b - l} \quad \text{for free support,} \qquad (3.9)$$

Yielding will take place if l is greater than

$$b\left(1 - \frac{4q}{3q_f}\right) \qquad (3.9a)$$

where

b = length of the structure in the vertical plane under consideration,

q = uniformly assumed design pressure in undisturbed ground,

l = unsupported length for cantilevering or free support,

q_f = ultimate bearing capacity.

The value of q_{max} depends on the length l which in turn depends on the ratio q_{max}/q. As soon as the value of q_{max} approaches the ultimate bearing capacity of the ground, yielding of the ground will occur, causing the structure to tilt in the case of the cantilever (Fig. 3.16(*a*)) and to settle more or less uniformly in the free support case (Fig. 3.16(*b*)). In both cases the effect is to increase the area of support given to the underside of the foundation, hence reducing the length of the cantilever, or the span of the beam, and reducing the stresses in the foundation structure or superstructure. It is clear from equations (3.8) and (3.9) that the smaller the ratio of the ultimate bearing capacity to the design bearing pressure, the less will be the length of cantilever or the unsupported span length of the beam. In other words the design bearing pressure should be kept as close as possible to the ultimate bearing capacity q_{max}. The value of q can be determined from soil mechanics tests or by plate bearing tests on the ground; hence the value of l can be estimated approximately. It can be assessed only roughly because the assumed straight line pressure distribution shown in Fig. 3.16(*a*) or the uniform distribution at each end of the free support case in Fig. 3.16(*b*) is not necessarily true. It should also be noted that the alignment of the front of the subsidence wave in relation to the foundation plan is not known in advance. It is therefore necessary to analyse various positions of the subsidence wave and calculate the worst condition of support for the structure (Fig. 3.16(*c*)). Because the design bearing pressure is made close to the ultimate, the consolidation settlement may be severe if the ground is compressible (e.g. a clay or loose sand). However, the magnitude of the consolidation settlements will be small in relation to the mining subsidence movements.

A typical design of a light raft for a dwelling-house as recommended by the British Government authorities[3.17] is shown in Fig. 3.17. The design features are as follows:

(a) A 150 mm layer of compacted sand or other suitable granular material is placed on the ground surface.

(b) A layer of waterproofed paper or polythene sheeting is placed over the granular sub-base layer to act as a surface for sliding.

(c) Reinforcement is provided to resist the frictional forces acting on the underside of the slab as it slides over the sub-base.

(d) The frictional forces may be in a transverse or longitudinal direction and may be taken as the product of half the weight of the structure and the coefficient of friction between the slab and the granular material.

(e) The coefficient of friction may be taken as 2/3.

(f) The permissible tensile stress in the steel may be taken as 200 N/mm^2 and the permissible compressive stress on the concrete as 14 N/mm^2.

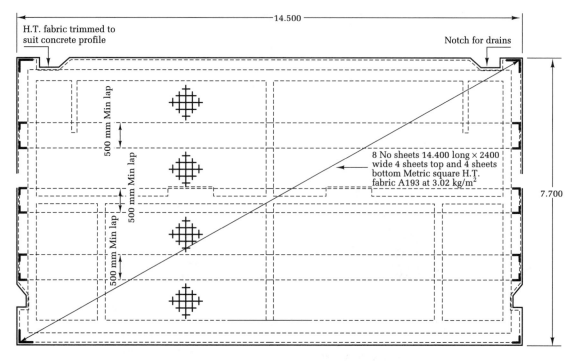

Fig. 3.17   Ministry of Works design for 150 mm slab raft foundation for light buildings subjected to mining subsidence.

(g) Snow loads and wind loads on the building may be neglected and the floor superload may be taken as 480 N/m^2.

(h) If single layer reinforcement is used it is placed in the centre of the slab to allow both for hogging and sagging of the ground surface, but the thickness of the raft and the percentage of reinforcement is such that the raft will deform under vertical movements rather than remain in one rigid plane.

(i) The design makes allowance for resistance to movement given by the superstructure, i.e. the windows and doors are arranged so as not to weaken the walls, internal load-bearing walls are tied with external walls, floors and roofs are secured to all walls, plasterboard (or fibreboard) is used for ceilings instead of plaster, and lime mortar is used for brickwork instead of cement mortar to allow movement to take place along joints instead of in wide infrequent cracks.

The National Coal Board[3.14] pointed out that the cost of providing rafts of this type in an estate, say, of hundreds of houses may not be justified since the cost of repairing the few houses damaged by mining subsidence is likely to be less than that of protecting all the houses. They recommend that where ordinary strip foundations are considered

to be satisfactory they should be laid on a bed of sand with provision at the ends of the foundation trenches for longitudinal movement.

Heavy slab and beam or cellular rafts are required for multi-storey structures or heavy plant installations. It is important that these stiffened rafts should be constructed on a membrane laid over a granular base, and that the underside of the raft should be a flat slab, i.e. the beams should be designed as upstanding beams. The stiffened rafts should be designed for the conditions of support shown in Fig. 3.16.

Because of the large movements which take place with longwall mining, appreciable deflexion of rafts cannot be avoided, and the appropriate precautions are necessary in the design of the superstructure of a suitably strengthened raft.

Protection by articulation   Articulation has been extensively used for the construction of schools in Great Britain. In the well-known 'Nottinghamshire construction', described by Lacey and Swain,[3.18] the superstructure consists of a pin-jointed steel frame designed to 'lozenge' in any direction. The cladding of hung tiles, precast concrete slabs, timber panels, or vitreous enamelled sheets is designed to move relative to the frame, as is the internal wall construction of heavy gypsum slabs.

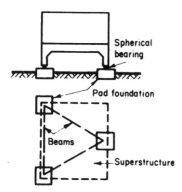

Fig. 3.18   Three-point support method.

Cracking of the lightly reinforced floor slab is expected and the floor finishes designed accordingly.

Instead of articulating the superstructure it is possible to provide articulated foundations with a rigid superstructure. This is the 'three-point support' method as used in Germany, the Netherlands, and Poland, and described by Mauntner.[3.16] The foundations consist of three piers or pads. The superstructure is constructed on columns resting on spherical bearings on the pads (Fig. 3.18). As the foundations tilt with the passing of the subsidence wave they always remain in the same plane. Therefore, although the superstructure must tilt, there is no differential movement causing racking or twisting. Another important principle of the three-point method is to use the highest possible bearing pressures in order to keep the foundation blocks as small as possible. By this means of tilting of the blocks relative to each other is minimized, and the horizontal tensile and compressive forces acting on the underside of the blocks are also kept to a minimum.

Other methods   Patented jacking systems incorporating several hydraulic jacks under walls or columns of structures have been devised. The jacks are connected to a central control system which automatically adjusts them individually as settlement takes place. In such systems it is important that they should be designed to withstand horizontal movements as well as subsidence.

Wardell[3.19] has described the use of trenches around structures to relieve horizontal compression. He states that although more cases of damage are caused by tension, the really serious cases are often irreparable damage to traditional structures caused by the compression strain, which could be twice the tensile strain.

Piled foundations should not be used under any circumstances in areas of longwall mining subsidence, since horizontal forces will either shear through the piles or else cause failure in tension of the tie beams or raft connecting the heads of the piles. Structures sited on soft ground or fill, where piling would be used in normal conditions, should be constructed on rafts designed to accommodate differential movement resulting from both consolidation of the fill and mining subsidence. The latter movements are likely to be the greater.

3.5.6 Protection against the effects of pillar and stall mining

The problems of protection against subsidence arising from the collapse of pillar and stall workings are very different from those involved in longwall mining. In the latter case the engineer is dealing with certain subsidence caused by mining in current progress or proposed for the future, and the amount and extent of the movements are to a large extent predictable.

With modern mining methods surface subsidence can also be predicted reliably for pillar and stall working. Wilson[3.15] states that these methods can be used to reduce total and differential movement, thereby giving partial protection to structures and reducing compensation paid to building owners in the event of damage. Extraction by pillar and stall working is about 30 per cent compared with 70 per cent by longwall mining. However, subsidence by the former method can be limited to $1\frac{1}{2}$ per cent of the extracted height if there is a factor of safety against crushing of the pillars of six, or 25 per cent of the height for a safety factor of unity.

The stability of the pillars is calculated from a knowledge of the ratio of the width of the pillar to its height, the average effective stress on the pillar, the unconfined compression strength of the pillar material, the weathering and jointing state of the rocks, and the percentage extraction of the seam. The influence of these factors is discussed by Cole and Statham[3.20].

It is pointed out that the ability to predict the amount of subsidence by the foregoing methods depends either on the ability to control the method of extraction to obtain pillars of the required shape and dimensions, or in the case of abandoned working it must be feasible to explore them to observe the size and condition of the pillars and to obtain cores of the rock for strength testing. This is rarely possible with former workings which may have been abandoned for 100 years or more. Mine abandonment plans, even if they are available, cannot be relied upon because they give no indication of the amount of deterioration of pillars or the roof of workings which may have taken place after cessation of mining. All subsidence may be complete or subsidence may never have taken place, but the workings are in a state of incipient instability such that the additional load of the building on the ground surface or change in underground water levels might cause the collapse of a pillar or a cave-in of an arched roof weakened by erosion or oxidation. In some cases partial subsidence may have occurred due to pushing of a pillar into the soft pavement, and the additional load of a structure may increase this form of subsidence.

The first step in considering schemes of protection is to make a detailed exploration of the workings in consultation with a geologist. All available records in the hands of the mining authorities, local museums or libraries, and the British Geological Survey, should be consulted first. Vertical boreholes using rotary core drills should be sunk to establish the depth and dip of the worked seam and the nature of the overburden. If intact coal and not cavities are found, it should not be assumed that the seam has not been worked. The depth of the seam being known, the engineer should decide whether subsidence is likely to be a serious risk. If the overburden is deep, so that the load imposed by the new structure is small in comparison, he may decide that the risk is negligible, especially if there are massive sandstone layers forming a sound roof over the cavities. If, however, the overburden is thin and consists mainly of soil or weak shales, or if there is surface evidence of past subsidence in the form of random depressions, then the engineer may decide that some form of protection is necessary. Detailed mapping of the galleries is best made by driving a heading from the outcrop if this is close at hand, or by sinking a shaft to the level of the coal seam to obtain access to the workings. However, the workings are rarely shallow enough for this purpose and there are considerable risks in entering old workings for exploratory purposes. Hence this type of mapping is only rarely undertaken. Usually, the best that can be done is to obtain a broad picture of the lateral extent and depth of the worked coal seam by means of vertical and inclined holes drilled from the ground surface. Location of individual galleries and pillars should not be attempted, and unless there is evidence to the contrary it should be assumed that the whole area beneath the structure has been worked. Jackhammer holes drilled vertically from the surface have been used to map galleries, but very close spacing is required to obtain adequate information. The method is only suitable for sites where the galleries have not collapsed and are free of heaped pit dirt. In such cases the location of the workings is revealed by the sudden drop of the drill rods. If, however, the roof of the workings has collapsed, forming numerous small cavities above the worked seam, then the vertical jackhammer holes will not give any useful information. A few rotary core-drilled holes are better than many jackhammer holes since the former give useful information on the state of collapse of the rock overburden to the coal seam, enabling the risks of future subsidence to be assessed reliably. The use of closed-circuit television cameras in small-diameter bore-

holes can often provide helpful information on the stability conditions of pillars and cavity roofs. Safety precautions against gas explosions are necessary when using television and other electrical equipment in non-water-filled voids. The advice of the mining and the Health and Safety authorities should be sought.

If the exploration shows that a large proportion of the foundation area has been mined, and the thickness or stability of the roof gives cause for concern, the engineer may decide that precautions against subsidence due to crushing of a pillar or roof collapse should be taken. Three methods have been used:

(a) The provision of a heavy raft foundation designed to bridge over a local collapsed area.
(b) Piles or piers taken down through the overburden and coal workings to the underlying 'solid' strata.
(c) Filling the workings with cement grout or other imported filling.

Raft foundations These can consist either of massive reinforced concrete slabs or stiff slab and beam cellular rafts. The latter type is suitable for the provision of jacking pockets in the upstand beams to permit the columns or walls to be relevelled if subsidence distorts the raft. A 600 mm thick slab raft was used beneath eight-storey buildings at Gateshead where workings in the 0.9–1.2 m thick seam were only 2.3 m beneath the foundations. It was decided that such rafts could bridge the cavities and any future local subsidence zones, since the 2.75–3 m wide galleries were supported by wider intact pillars and the roof to the workings consisted of massive sandstone.

Pier (deep shaft) foundations These are used for sites where the overburden is too weak to support surface foundations. Suitable types are described in Section 4.6. The piers must be taken through the overburden to a bearing stratum beneath the old workings, because founding at a higher level would involve a risk of concentrating load on potentially unstable strata above the workings. Precautions must be taken against drag-down on the piers from vertical movement of the overburden or horizontal shear forces on the piers caused by lateral ground movement. Such movements may take place at any time in the future as a result of crushing of pillars or roof collapse. A space must be provided between the piers and the overburden, and the space must be filled with a plastic material to prevent it from being filled with accumulated debris which might transmit heavy forces on to the piers.

Piled foundations If piled foundations are necessary because of weak overburden conditions, they should be installed in holes drilled through the overburden to a suitable bearing stratum beneath the workings. For reasons given in the previous paragraph, they must not be terminated above the roof of the old workings. Piles should consist of precast concrete units or steel tubes filled with concrete. Concrete cast-in-place in an unlined borehole must not be used in any circumstances, because of the possible drag-down and shear forces mentioned above. The space between the concrete units or steel tubes and the overburden should be filled with a viscous bitumen mix or bentonite. Because of the complications in providing safeguards against drag-down forces on piles, large-diameter cylinder foundations or massive piers are preferable to conventional slender piles.

Piled foundations were used for the Metropolitan-Vickers Electrical Company's factory at Sheffield. Four seams had been worked below the site. The shallowest of them outcropped beneath one end of the factory. It had been extensively worked by pillar and stall methods, and a number of shallow drifts had been made at the outcrop during the 1926 Coal Strike. Further complications were caused by the presence of deep filling over the eastern part of the site either side of the Car Brook (Fig. 3.19). A large

area of this fill was burning and it blanketed the outcrop workings. Exploration showed extensive collapsed areas near the outcrop, but as the seam dipped towards the west it was roofed by sandstone and the galleries appeared to be intact. Because of the great extent of the workings it was decided that filling would be unduly slow and expensive. The method adopted was to take the foundations of the building in the collapsed areas down to a stratum of siltstone beneath the coal seam. At and near the outcrop precast concrete piles were driven without difficulty through the overburden of glacial till and broken mudstone. Further to the west, where the roof was mainly intact, the piles were lowered down holes pre-bored through the clay, mudstone, and siltstone. Where the depth of overburden exceeded 12 m it was considered unnecessary to take any precautions because of the presence of the sandstone roof. Accordingly, the foundations to the west of the piles area consisted of piers bearing on the rock beneath the shallow clay deposits. The factory floor which carried fairly light loading was not piled, but the weak and burning fill over the east end of the factory was removed in its entirety and replaced by compacted hard filling.

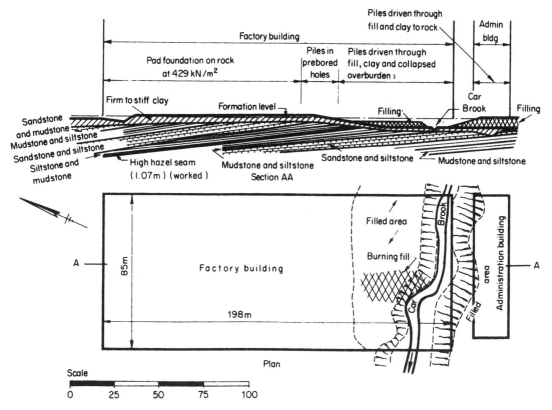

Fig. 3.19 Ground conditions and foundations of Metropolitan-Vickers Electrical Company's factory at Greenland Road, Sheffield.

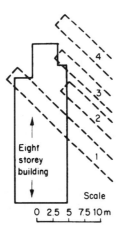

Fig. 3.20 Location of old mine-workings.

Precautions against the deeper coal seams were considered unnecessary.

Filling the workings Method (c), involving filling the workings, was used to protect an eight-storey building forming part of the College of Technology in the centre of Sheffield. The procedure described by Scott[3.21] was the basis of present-day techniques used for sites where direct

access to abandoned workings is impracticable or dangerous. Pillar and stall workings were a minimum of 10 m below ground level, and because of the width of the galleries (up to 4.5 m) and the evidence of collapse of the roof it was decided that there could be a risk of subsidence resulting in structural damage to the building. Exploration showed four wide galleries beneath the building (Fig. 3.20). Comparison with old maps of the workings indicated that there were continuous pillars of coal beneath the parallel galleries and that the workings terminated close to the building site. These conditions were favourable for filling the workings by the introduction of material from the surface. In order to prevent the filling material escaping down-dip to the north-east, dams were formed in galleries 1, 3, and 4 by dropping pea gravel down 250 mm boreholes. The gravel formed conical piles in the galleries, and the rounded material aided dispersion over the width of the cavity. Increased dispersion at the top of the pile was achieved by water and air jetting and the use of a rotating plate at the bottom of the boreholes having an action similar to a road gritter. The gravel dams were injected with cement grout using calcium chloride as an accelerator. The main filling material was an aerated sand–cement grout injected by pneumatic placer down boreholes at the up-dip ends of the galleries (Fig. 3.21). The 3.5 : 1 sand–cement mix was aerated by premixed

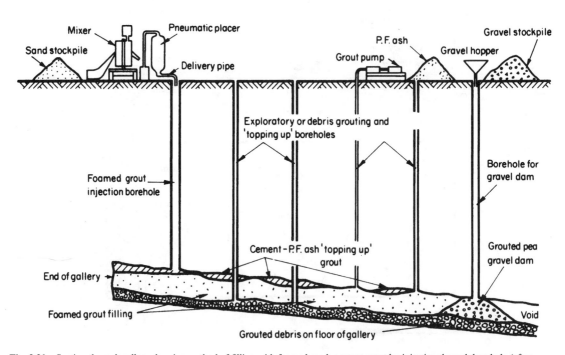

Fig. 3.21 Section through gallery showing method of filling with foamed sand–cement grout by injection through borehole (after Scott[3.21]).

foam to give $1440 \, kg/m^3$ density. This gave adequate strength with the advantage of greatly increased yield per unit weight of cement compared with non-aerated mixes. Filling of the galleries was indicated by a rise in grout level in the injection borehole. Additional jackhammer holes were drilled at intermediate points to 'top-up' the filling in roof cavities which had not been reached by the main injections. The topping-up grout was a 2 : 1 pf ash–cement mix injected by cementation pump. Exploration in gallery 2 showed extensive accumulations of debris which prevented effective formation of the pea gravel dam. Filling in this gallery was effected by washing in 35 t of sand followed by aerated sand–cement grout and finally neat cement grout.

Present-day mine-filling methods continue to follow the same techniques as described above, except that it is more usual to inject a weak pf ash–cement or pulverized rock–cement grout into the cavities. A mix of 1 : 12 cement/pf ash, or 1 : 10 : 2 cement/pf ash–sand by weight is used for the perimeter grout curtain wall, and 1 : 12 to 1 : 20 cement/pf ash for the mass of infilling grout. Healy and Head[3.13] give an outline specification for consolidating old shallow mine-workings.

It should be noted that it may not be necessary to fill completely all cavities beneath a building area. It may be sufficient to form a number of 'grout piers' over the area. The foundation would then be in the form of a stiff raft designed to span across the piers.

Where the exploratory drilling has shown partial collapse of the soil or rock overburden, attention should be given to filling cavities in these strata by injecting cement or pf ash–sand–cement mixes.

Hawkins *et al*[3.22] describes the filling of shallow workings in the 1.4 m thick Barnsley coal seam beneath the Don Valley athletics stadium in Sheffield where this form of treatment was selected as being more economical than the alternative of piled foundations. Settlement considerations were critical for the raft supporting the grandstand and for the running track which was to be constructed within a tolerance of ± 3 mm from the horizontal. In these areas the grout injections were made on a 3 m grid where the coal seam was less than 20 m from the surface. A 9 : 1 pf ash–cement mix was used with a water : solids ratio of 0.45. At the end of each injection a grouting pressure of $14 \, kN/m^2$ per metre of overburden depth was specified to be sustained for 1 min. Primary injections required between 1 and 25 t of grout per hole. Where the grout take exceeded 5 t, four secondary holes were drilled on a surrounding half-grid with tertiary injections if the secondary holes took more than 5 t. In one area, the secondary holes had a grout take between 2 and 15 t.

3.5.7 Treatment of abandoned shafts

Old mine shafts are frequently in an unstable or potentially unstable condition due to decay of the shaft lining or to collapse of the platform which was frequently built across the shaft to support fill material of limited depth. Therefore measures should be taken to stabilize the infilling or adopt other methods of strengthening the shafts where these may influence any existing structures or new structures to be built near them.

The safe distance from a shaft to a structure depends on the depth to rockhead and the character of the overburden. If the overburden soil is cohesive no precautions are necessary if the structure is beyond the limit angle or 'angle of draw' (Fig. 3.13). However, water seepage through a granular soil into the shaft can cause soil erosion and collapse of the ground surface at considerable distances from the shaft. Whether or not precautions are required from the aspect of safeguarding a new or existing structure, British Coal or other appropriate authority should be consulted whenever abandoned shafts are found on a site. Their advice on the method of plugging and sealing the shaft should be taken in all cases.

Where rockhead lies at a shallow depth and the rock or existing shaft lining is in a stable condition the treatment can be limited to providing a reinforced concrete slab over the shaft as shown in Fig. 3.22(*a*). A 'breather tube' is provided through the slab to permit the escape of gases. If the rockhead is too deep for economical excavation for a slab-type cap, and the shaft lining through the overburden is shown by inspection to be in a stable condition, the refilling to the shaft can be stabilized by the injection of cement or pf ash–cement mixes (Fig. 3.22(*b*)). If no infilling is present the shaft should be filled with stone, rubble, or sand followed by grouting the mass of fill. In the case of deep shafts it may be necessary to construct a plug or platform across the shaft to limit the amount of fill. Expert advice from a mining engineer should be sought in all such cases.

Where the overburden is deep and the existing shaft lining is in a weak condition a suitable form of treatment is to construct a ring of contiguous bored piles or a diaphragm wall (see Section 5.4.4) around the shaft, surmounted by a reinforced concrete cap (as shown in Fig. 3.22(*c*)).

Great care is necessary in conducting exploration and construction work in the vicinity of abandoned shafts. Drilling or trial pit work may weaken ground which has arched over a large cavity leading to a general collapse of the ground surface. The risks of encountering explosive or asphyxiating gases must always be considered.

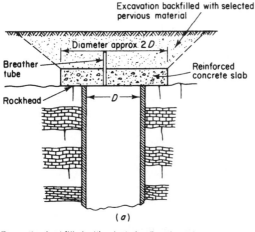

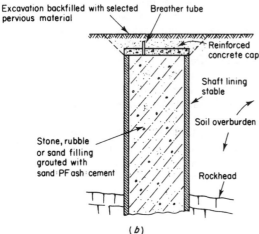

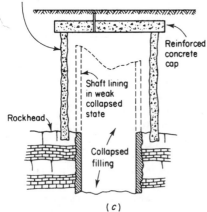

Fig. 3.22 Treatment of abandoned mine shafts. (*a*) Shallow overburden, shaft lining stable. (*b*) Deep overburden, shaft lining stable. (*c*) Deep overburden, shaft lining weak or collapsed.

3.6 Foundations on filled ground

3.6.1 The settlement of filled ground

Settlement of foundations constructed on fill material can be caused in three ways:

(a) Consolidation of compressible fill under foundation loading,

(b) Consolidation or degradation of the fill under its own weight,

(c) Consolidation of the natural ground beneath the fill under the combined weight of the fill and the structure.

These movements are illustrated in Fig. 3.23.

If the structure is light the movement due to (*a*) above will be small even in a poorly compacted fill (it is presumed that founding on very soft clayey fill would not be considered). For heavy structures the compression of fill material under foundation loading can be estimated as a result of field loading tests made on large representative areas. From the results of tests of this type and other published information the Building Research Establishment[3.23] have given values of one-dimensional compressibility expressed as a constrained modulus defined as the ratio of the increase in vertical stress to the increase in vertical strain produced by that stress increase. The values given in Table 3.1 are applicable to stress increments of about 100 kN/m^2 where the initial vertical stress was about 30 kN/m^2.

The movement due to (*b*) depends on the composition, depth and compaction of the fill layer, the conditions under which it was placed and the subsequent exposure to the environment. Chemical wastes can undergo large self-weight settlement due to volume changes resulting from continuing chemical reactions, solution due to water seepage, and the effects of mechanical disturbance on agglomerations of weakly cemented particles of materials originally dumped in slurry form. These settlements may be activated or reactivated at any time due to disturbance by bulk excavation or by the entry of surface water through

Table 3.1　Compressibility of fills

Fill type	Typical value of constrained modulus (kN/m²)
Dense well-graded sand and gravel	40000
Dense well-graded sandstone rock fill	15000
Loose well-graded sand and gravel	4000
Old urban fill	4000
Uncompacted stiff clay fill above water table	4000
Loose well-graded sandstone rockfill	2000
Poorly compacted colliery spoil	2000
Old domestic refuse	1000–2000

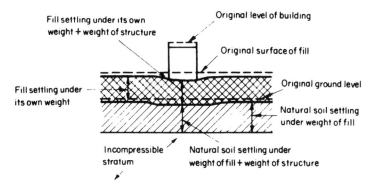

Fig. 3.23 Settlements of filled ground.

pits or trenches. Where the fill can be compacted in layers at the time of placing the settlement of granular fill materials such as gravel, sand, relatively unweathered shale and chalk, sandstone and mudstone, should not exceed 0.5 per cent of the thickness of the fill. Uncompacted fills, where the material is placed loosely by end-tipping, may show a settlement of 1–2 per cent of the thickness over a period of 10 years with continuing slow movement. Guidance on the methods and equipment for compacting fill is given in the British Standard Code of Practice, BS6031: Earthworks, and in the Department of Transport specifications for road and bridge works.

Building Research Establishment Digest 274[3.23] gives a parameter α to define the percentage vertical compression of fill that occurs during a $\log_{10}$ cycle of time, say between 1 year and 10 years after placement. Values of α in the *Digest* are:

Fill type	Creep compression rate (α) %
Well-compacted sandstone rockfill	0.2
Uncompacted opencast mining backfill	0.2–1
Domestic refuse	2–10

The large compression rate for domestic refuse will be noted. This is mainly due to degradation of organic substances with the production of methane and carbon dioxide. For this reason buildings on shallow foundations over domestic waste landfills should be avoided if alternative sites are available.

Where builders' waste and demolition debris are used for filling areas planned for future building development, care should be taken to exclude chemical wastes and organic material. Timber and other organic material can decay in anaerobic conditions and generate methane gas. Local authorities in Britain are concerned about methane gas in potentially explosive concentrations beneath buildings, and usually require tests to be made for the presence of this gas before building commences. The cost of precautionary measures such as venting beneath ground floors, sometimes with mechanical forced ventilation, can be quite high. Hence the cost of controlling incoming materials at the time of dumping will be amply repaid. Colliery wastes are liable to spontaneous combustion of carbonaceous material which can be aggravated by aeration and grinding during spreading and rolling.

Where piled foundations are required to be installed through demolition rubble the presence of masses of concrete, brickwork, and reinforcing steel can add considerably to the cost of either driven or bored piles.

Sand fills placed by pumping (hydraulic fill) show very small settlements above ground-water level due to the consolidating effect of the downward percolating water, but where the material is deposited through standing water the sand can remain in a loose compressible condition. Pumped clay in the form of lumps in a liquid slurry can settle by 10 per cent or more of the fill thickness.

Inundation of loosely placed unsaturated materials such as shale, mudstone, and dry clay lumps can cause breakdown of lumps of clay and softening at points of contact of rock particles, leading to further settlement which can be unpredictable in magnitude and time of occurrence. Charles, Naismith, and Burford[3.24] reported that local vertical settlements of more than 1 per cent of the fill thickness with surface settlements of 0.5 m, occurred when the ground-water level was allowed to rise up through a backfilled opencast mining site. There were similar observations in backfilled ironstone mine-workings at Corby[3.25] where the rate of settlement of houses built on the fill accelerated when the water level was allowed to rise. Two-storey houses built on rafts on this site showed a maximum settlement of 130 mm in 10 years with a differential movement of half this value. Cracks in the superstructure were not wider than 2 mm.

The Building Research Establishment[3.26] used the Corby site to investigate the efficacy of various types of ground treatment as a means of reducing the total and differential settlements of foundations caused by consolidation of the backfill. An area filled in 1974 was divided into four areas 50 m square. The first area was treated by surcharging with 9 m of new fill, the second by dynamic consolidation with a 15 t weight falling through 20 m, the third by flooding in trenches, and the fourth area was left untreated. The 24 m deep fill consisted mainly of loose unsaturated clay lumps deposited in the 'hill and dale' method typical of dumping by conveyor or dragline excavator. Surface settlements after 12 years of observation from 1974 to 1986 were:

Surcharge	410 mm
Dynamic consolidation	240 mm
Inundation	100 mm

Houses were built on the four areas in the 1975–76 period on conventional trench fill foundations 375 mm wide × 900 mm deep. The results of the settlement readings made in 1990 are shown in Table 3.2.

It was concluded from these experiments that surcharging was the most effective form of treatment because it gave the largest preconstruction and the lowest 15-year settlements. The dynamic consolidation and inundation methods would appear to be ineffective when compared with the settlements of the foundations on untreated ground. However, there would be a risk of quite large local settlement due to seepage of surface or shallow subsoil water into backfilled service trenches and foundation excavations. These random inundation effects could occur in areas of deep fill treated by dynamic consolidation because of the limited depth over which the treatment is effective (see Section 11.6.4.). All the foundations continued to undergo settlement after 1990 at the following rates:

Surcharge	0.66 mm/year
Dynamic consolidation	1.6 mm/year
Inundation	2 mm/year
Untreated	1 mm/year

Movement due to consolidation of the underlying natural soil again depends on the composition of the soil and the thickness of the compressible layers. Where the natural soils are highly compressible, e.g. soft clay or peat, the settlement due to consolidation of the natural ground may be high compared with that of the fill material. If the natural soil is a fairly dense sand or gravel, or a stiff to hard clay, its consolidation under the weight of filling will be of a very small order.

Buildings can be constructed on well-compacted fill with normal foundations, since the density of the fill should be in no way inferior to, or it may be better than, naturally deposited soils. Where the density of the fill is variable, it is advisable to provide some reinforcement to strip foundations to prevent the formation of stepped cracks. In areas of deep fill it may be practicable to remove the upper looser material and replace it by well-compacted quarry waste to provide a form of semi-rigid raft on which buildings can be constructed.

The presence of pyrites in ironstone shale used as filling beneath ground-floor slabs caused expansion of the shale laminae and an uplift of up to 100 mm on the slabs. Extensive damage was caused to the superstructure of 600 houses on Teesside.[3.27]

When building over poorly compacted fill, it is advisable to use raft foundations. Three-storey dwelling-houses were constructed at Bilston, on variable fill material consisting of mineral waste. The age and depth of the fill were unknown, but it was suspected to be more than 12 m deep. One block of three-storey dwellings settled 100 mm with a differential movement of 75 mm over a $1\frac{1}{4}$-year period. The ribbed raft foundation (Fig. 4.34) prevented any cracking of the superstructure. On another site in Bilston two-storey houses were constructed on ash and clinker fill which had been placed 2 years previously without any special compaction. One block of houses were founded on fill varying in thickness from 2.5 m to less than 0.25 m. There was a maximum settlement of 20 mm in a $1\frac{1}{2}$-year period. Again the raft foundation with edge beams prevented any cracking.

Preloading is a useful expedient to allow building on normal foundations in areas of deep uncompacted fills.

Table 3.2   Settlements (mm) of house foundations in areas of different ground treatment

Ground treatment	During construction				Total at end of 1990			
	Mean	Max.	Min.	Max. differential	Mean	Max.	Min.	Max. differential
Surcharge	1.4	3.0	−0.4	2.3	9.7	22.3	4.5	12.6
Dynamic consolidation	7.0	9.2	3.2	5.9	45.0	58.5	22.2	18.1
Inundation	6.1	14.3	2.8	6.8	48.0	143.0	25.2	89.5
Untreated	2.7	6.8	1.4	2.8	25.1	39.5	11.7	27.8

The efficacy of preloading of fill by surcharging was demonstrated by the Corby experiments. Another example of the method was given by the construction of a factory on a backfilled clay pit at Birtley, Co. Durham.[3.28] The 15 m deep colliery waste fill had been end-tipped through water some 35 years before, and it was still in a loose state (average N-value 10 blows per 300 mm). The overall loading on the factory floor of 86 kN/m^2 would have caused excessive settlement of foundations on untreated fill. Accordingly it was decided to adopt preloading by a mound of colliery waste some 3.5–5.5m high moved progressively across the site by scrapers and bulldozers at a rate controlled by observations on settlement plates. In areas of deep fill the surcharge mound remained in place for periods between 50 and 108 days when the average settlement of the fill surface was about 0.5 m (3.3 per cent of the fill thickness) under a surcharge of 1000 kN/m^2.

Heavy structures, including factory buildings containing plant or machinery which is sensitive to settlement, are best supported on piles taken through the fill to a relatively incompressible stratum. This procedure may not be necessary where the fill is known to be well compacted and is bearing on natural ground having a compressibility less than or not appreciably greater than that of the filling. It is common practice to construct oil storage tanks on hydraulically placed sand fill or compacted earth fill.

Where piled foundations are used in filled areas, the consolidation of the filling under its own weight or under superficial loads may cause a load to be transferred to the pile shafts due to the adhesion or skin friction of the fill material in contact with them. This phenomenon is discussed in Section 7.13.

3.7 Machinery foundations

3.7.1 General

Machinery foundations should be designed to spread the load of installed machinery on to the ground so that excessive settlement or tilting of the foundation block relative to the floor or other fixed installations will not occur; they should have sufficient rigidity to prevent fracture or excessive bending under stresses set up by heavy concentrated loads, or by unbalanced rotating or reciprocating machinery; they should absorb or damp down vibrations in order to prevent damage or nuisance to adjacent installations or property; and they should withstand chemical attack or other aggressive action resulting from manufacturing processes.

If the shear strength and the compressibility of the soil are known, the proportioning of the base of a foundation block is generally a simple matter where non-vibrating loads are carried. The allowable bearing pressures to

prevent excessive settlement can be calculated by one of the procedures described in Chapter 2. Tilting can be avoided by placing the centre of gravity of the machinery at or reasonably close to that of the foundation block.

The problems of foundation design for machinery having unbalanced moving parts are much more difficult. Unbalanced moving parts, such as large flywheels, crankshafts, or the pistons of reciprocating engines, set up vibrations in the foundation blocks which, if not absorbed by anti-vibration mountings, are transmitted to the ground. Strong ground vibrations may cause loss of bearing capacity and settlement of the soil beneath the foundations; they may also cause damage to adjacent buildings or machines.

3.7.2 Foundation vibrations

Foundation vibration problems may be considered in two categories: first, those in which the vibrations originate from external sources such as seismic activity, railway tunnels, construction activity (especially piling); and second, those where the vibrations originate from whatever the foundation is supporting; for example, machinery, or fluctuating wind or wave forces on a superstructure. The first category is very specialized and there is a considerable volume of literature on it. Within the second category the requirements of the design of offshore oil installations have stimulated a considerable amount of investigation and many sophisticated techniques, particularly those based on finite element methods, have resulted.

A frequent problem is the design of rigid foundations for machines such as compressors, fans, or turbines. The form and notation for the machine–foundation–soil system are shown in Fig. 3.24 which illustrates the case of the coupled rocking–sliding mode, that is translation in the x-direction coupled with rotation about the y-axis through the combined centre of gravity of a machine and foundation. Note that sliding and rocking will be coupled when the combined centre of gravity is not at the same elevation as the soil horizontal resistance force P_x.

For the design of a machine foundation a primary objective is to ensure that natural frequencies of the system are remote from the frequency (or possibly frequencies in the case of reciprocating machinery) at which the machine will operate. The reason for this is to avoid the greatly magnified amplitudes of movement which will occur if machine and natural frequencies coincide giving the resonance condition. However, it may not always be possible to avoid resonance. There may be constraints on the possible size of the footing, or the machine may be of the type which is required to operate over a large range of frequencies. In such circumstances it will be necessary to assess the amplitudes of movement at resonance to check that they are acceptable to the installation and that they are within acceptable limits for persons or structures in the immediate vicinity of the machine. Prediction of these amplitudes, especially of coupled modes of vibration, is complicated by the need to allow for soil damping effects. It is usual to assume that the soil can be idealized as an elastic half-space, and the method now described is based on that given by Richart et al.[3.29]

For Fig. 3.24 forces generated by the soil under dynamic displacement will be of two types: elastic forces related to the instantaneous values of translation and rotation, x and ψ, and damping forces related to the instantaneous values of translational and rotational velocities, x and ψ. Consideration of Newton's second law (force = the product of mass and acceleration) results in the following equations:

$$Q_{x(t)} - (k_x x - k_x h_0 \psi) - (c_x \dot{x} - c_x h_0 \dot{\psi}) = m\ddot{x} \quad (3.10)$$

$$T_{\psi(t)} - (k_\psi + h^2{}_0 k_x)\psi + h_0 k_x x - (c_\psi + h^2{}_0 c_x)$$
$$\dot\psi \cdot + h_0 c_x \dot{x} = I_y \ddot{\psi}, \quad (3.11)$$

where

m = mass of machine plus footing,
I_ψ = mass moment of inertia about the y-axis of machine plus footing,
k_x, k_ψ = soil spring constants for translation and rotation,
c_x, c_ψ = damping coefficients for translation and rotation,

$\ddot{x}$ = translational acceleration of machine plus foundation,
$\ddot\Psi$ = rotational acceleration of machine plus foundation,
$Q_{x(t)}, T_{\psi(t)}$ = time-dependent forces in the x-direction and about the y-axis.

For a rotary machine such as a fan or turbine the force and moment $Q_{(t)}$ and $T_{(t)}$ due to out-of-balance on the shaft will be in phase with one another and can be expressed as

$$Q_{x(t)} = Q_0 \sin \omega t, \quad T_{\psi(t)} = T_0 \sin \omega t. \quad (3.12)$$

Assuming a solution of the form

$$x = \bar{x}\sin(\omega t - \alpha), \quad \Psi = \Psi\sin(\omega t - \beta), \quad (3.13)$$

where $\bar{x}$ and Ψ are amplitudes of oscillation, α and β are phase lags, and ω is the frequency in radians/second at which the machine is operating. Substituting equations (3.12) and (3.13) into equations (3.10 and 3.11) gives

$$(-m\omega^2 + k_x)\bar{x}\sin(\omega t - \alpha) + c_x\omega\bar{x}\cos(\omega t - \alpha)$$
$$-h_0 k_x \Psi\sin(\omega t - \beta) - h_0 c_x \omega\Psi\cos(\omega t - \beta)$$
$$= Q_0\sin \omega t \quad (3.14)$$

$$(-I_y\omega^2 + k_\psi + h^2{}_0 k_x)\Psi\sin(\omega t - \beta) + (c_\psi - h^2{}_0 c_x)\omega\Psi\cos(\omega t - \beta)$$
$$-h^2{}_0 k_x \bar{x}\sin(\omega t - \alpha) - h_0 c_x \omega\bar{x}\cos(\omega t - \alpha)$$
$$= T_0\sin \omega t \quad (3.15)$$

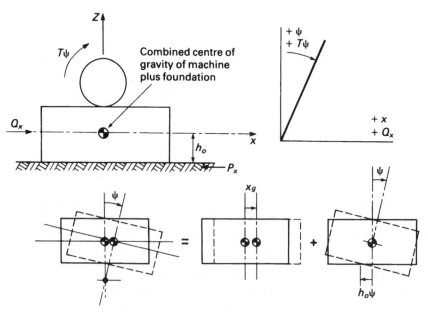

Fig. 3.24 Form and notation for the machine–foundation–soil system.

Setting ωt to zero and $\pi/2$ successively yields four simultaneous equations in

$$\bar{x}\sin(\omega t - \alpha),\ \bar{x}\cos(\omega t - \alpha),\ \Psi\sin(\omega t - \beta)\quad \text{and}$$
$$\Psi\cos(\omega t - \beta).$$

For any value of ω these equations can be solved by hand to give values of $\bar{x}$, Ψ, α and β, but this is very time-consuming when it is required to obtain answers for several values of ω in order to evaluate peak amplitudes. The computation becomes still more time-consuming when the spring stiffnesses and damping coefficients are taken to be frequency dependent,[3.30] and to achieve accuracy and reasonable speed of computation the use of a computer program is essential.

Figure 3.25 shows the results of a design analysis where a computer program was used to evaluate maximum amplitudes of movement at the bearings of a variable-speed fan operating in the range 6–16 Hz. The fan was to be set on a concrete foundation which was also to support a rigid structure around the fan. The analysis was run for a range of possible values of modulus of elasticity of the mudstone supporting the foundation.

3.7.3 Absorbing vibrations

Among the methods of absorbing vibrations, one of the simplest is to provide sufficient mass in the foundation block so that the waves are attenuated and absorbed by reflections within the block itself. A long-established rule in machinery foundations is to make the weight of the block equal to or greater than the weight of the machine. This procedure is generally satisfactory for normal machinery where there are no large out-of-balance forces. However, in the case of heavy forging hammers and presses, or large reciprocating engines, it is quite likely that the vibrations cannot be absorbed fully by the foundation block. Also, in some circumstances a large and heavy foundation block may be impracticable; for example where the machinery is carried on a suspended floor or structural framework, or where space is limited by service ducts or other foundations. In these circumstances, absorption of vibrations can be achieved by special mountings.

The aim of anti-vibration mountings for machinery or foundation blocks is to reduce the amplitude of the vibrations transmitted to the supporting foundation block or to the ground. Thus the mounting should have a much lower frequency than that of the induced vibrations of the machine. When the frequency of the mounting is very low it is said to be a 'soft mounting'. The mountings must also allow freedom for the six degrees of movement illustrated by Crockett and Hammond[3.31] in Fig. 3.26. These movements comprise translations in three dimensions and

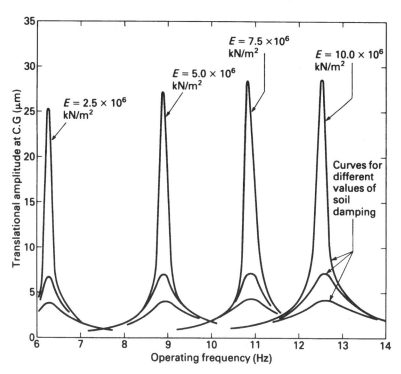

Fig. 3.25 Amplitudes of movement at the bearings of a ventilation fan at various operating speeds.

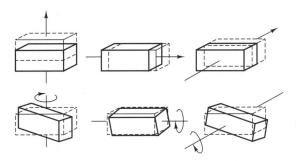

Fig. 3.26　Six degrees of freedom of movement (after Crockett and Hammond[3.31]).

rotations in planes at right angles to these dimensions. Types of anti-vibration mountings in general use include the following:

(a) *Cork slabs and rubber pads*. These are suitable for vibrating machines which do not produce severe shock or high-amplitude vibrations and where the intensity of loading on the cork slab or rubber carpet is not so high that the materials will become 'hard' under compression. Cork and rubber are reasonably durable materials and can be expected to have a life of at least 25 years.

(b) *Rubber carpet mountings*. These are designed for heavier machines such as compressors, power-hammers, presses, and generators. A type consisting of studs on either side of a rubber sheet (Fig. 3.27(*a*)) can be loaded to $36\,kN/m^2$ and a heavier type with closed-spaced ribs running at right angles on either side of a rubber sheet can be loaded to $430\,kN/m^2$ (Fig. 3.27(*b*)).

(c) *Bonded rubber mountings*. These are used for direct connection of machines to concrete or steel bases. They consist of various steel sections, e.g. plates, angles, pedestals, and plugs, which are bonded to rubber blocks. They can be used for the lightest to the heaviest machines. The sandwich type consisting of

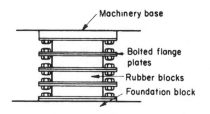

Fig. 3.28　Rubber–steel sandwich spring mounting.

layers of rubber separated by steel is illustrated in Fig. 3.28. This mounting is designed to carry $300\,kN$ at a frequency of $2.8\,Hz$. Rubber–steel sandwich mountings were used to insulate a complete five-storey building from the vibrations caused by railway trains running beneath the building. The mountings were installed beneath columns as shown in Fig. 3.29. Column loads were up to $2.7\,MN$. It was stated that the cost of the anti-vibration installation was 5 per cent of the total building cost.[3.32]

(d) *Leaf springs*. These were used in the past for forging hammers, but they have been superseded by the simpler rubber mountings.

The blow of heavy hammers is likely to cause considerable oscillation of the anvil block or base when it is supported by anti-vibration mountings. In such cases it may be necessary to provide dampers in horizontal and/or vertical positions between the anvil and base or between the base and the surrounding pit to damp down the oscillations and so bring the anvil to rest before the next blow of the hammer. Such dampers are usually a hydraulic dashpot arrangement whereby movement of the piston

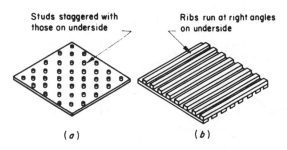

Fig. 3.27　Types of rubber carpet mounting. (*a*) Stud type. (*b*) Ribbed type.

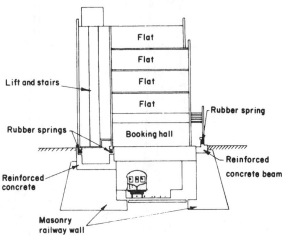

Fig. 3.29　A five-storey building mounted on rubber–steel springs.

forces oil in a cylinder through a small hole. The energy is absorbed in compression and heat in the oil. It is also the practice to provide an air-gap between the foundation block of heavy hammers and the surrounding ground by placing the block inside a lined pit. The object of this is to prevent direct transmission of shock waves to adjacent machines or enclosing buildings through the shallow soil layers. Instead the energy waves are transmitted to deeper strata and become considerably attenuated before they reach the surface; hence they are less likely to cause damage or nuisance.

It is desirable to avoid projections from machinery foundation blocks such as thin cantilever brackets in concrete or steel, since these may resonate with the machine vibrations. Similarly, any light removable steel floor-plates or decking surrounding the foundation block should be bedded on felt or rubber if there is a risk of these light units vibrating in resonance with the machinery. Problems of resonance and interference with close-spaced machinery installations can often be overcome by providing a combined foundation for a number of machines.

References

3.1 Crilly, M. S., and Chandler, R. J., A method of determining the state of desiccation in clay soils, *BRE Information Paper IP4/93*, Building Research Establishment, Watford, 1993.

3.2 Freeman, T. J., Burford, D. and Crilly, M. S., Seasonal foundation movements in London Clay, *Proceedings of the 4th International Conference on Ground Movements and Structures*, Cardiff 1991, Pentech Press, London, 1992, pp. 485–501.

3.3 Various authors, Symposium in print on the influence of vegetation on the swelling and shrinkage of clays, *Géotechnique*, **33**, 87–163 (1983).

3.4 Samuels, S. G. and Cheney, J. E., Long-term heave of a building on clay due to tree removal, in *Proceedings of the Conference on Settlement of Structures*, Pentech Press, Cambridge, 1974, pp. 212–220.

3.5 The influence of trees on house foundations in clay soils, *Building Research Establishment Digest 298* (1985).

3.6 Walsh, P. F., The design of residential slabs-on-ground, CSIRO *Division of Building Research Technical Paper No. 5*(1975).

3.7 Cooling, L. F. and Ward, W. H., Some examples of foundation movements due to causes other than structural loads, in *Proceedings of the 2nd International Conference on Soil Mechanics, Rotterdam,* 1948, Vol. 2.

3.8 Penner, E. and Gold, L. W., Transfer of heavy forces by adfreezing to columns and foundation walls in frost-susceptible soils, *Canadian Geotechnical Journal*, **8**, 514–526 (1971).

3.9 Tabb, R., Some case histories, *Finite Elements in Geotechnical Engineering*, Chapter 11, Pineridge Press, Swansea, 1981.

3.10 Jennings, J. E. and Knight, K., The additional settlement of foundations due to collapse of structure of sandy soils on wetting, in *Proceedings of the 4th International Conference on Soil Mechanics, Zurich,* 1953, Vol. 1, pp. 316–319.

3.11 Abelev, M. Y., Compacting loess soils in the USSR, in *Proceedings of a Symposium on Ground Treatment by Deep Compaction*, Institution of Civil Engineers, London, 1976, pp. 79–82.

3.12 Terzaghi, K. and Peck, R. B., *Soil Mechanics in Engineering Practice,* 2nd edn., John Wiley, New York, 1967, p. 586.

3.13 Healy, P. R. and Head, J. M., Construction Over Abandoned Mine Workings, Construction Industry Research and Information Association, Special Publication 32 (1984).

3.14 *Subsidence Engineers' Handbook*, Mining Department, National Coal Board, London, 1974.

3.15 Wilson, I., Subsidence prediction over room and pillar mining systems, *Proceedings of the 4th International Conference on Ground Movements and Structures*, Pentech Press, London, 1992, pp. 223–242.

3.16 Mauntner, K. W., Structures in areas subjected to mining subsidence, in *Proceedings of the 2nd International Conference on Soil Mechanics*, Rotterdam, 1948, Vol. 1, pp. 167–177.

3.17 *Mining Subsidence Effects on Small Houses*, National Building Studies Special Report No. 12, London, HMSO, 1951.

3.18 Lacey, W. D. and Swain, H. T., Design for mining subsidence, *Architect's Journal*, **126** (3288) (24 Oct. 1957).

3.19 Wardell, K., The protection of structures against subsidence, *Journal of the Royal Institution of Chartered Surveyors*, **90**, 573–579 (1958).

3.20 Cole, K. W. and Statham, I., General (areal) subsidence above partial extraction mine, *Ground Engineering*, Thomas Telford, London, Part I, March 1992, pp. 45–55, Part 2, April 1992, pp. 36–40.

3.21 Scott, A. C., Sheffield College of Technology: Locating and filling old mining workings, *Civil Engineering and Public Works Review* (1957).

3.22 Hawkins, A. B., Morley, S., Swainson, T. S., and Tyson, D. M. Don Valley Stadium, Sheffield, UK, in *Proceedings of the 4th International Conference on Ground Movements and Structures*, Pentech Press, London, 1992, pp. 283–301.

3.23 *Fill*, Part I: Classification and load carrying characteristics, *Building Research Establishment Digest 274* (1983).

3.24 Charles, J. A., Naismith, W. A. and Burford, D., Settlement of backfill at Horsley reported opencast coal-mining site, in *Proceedings of a Conference on Large Ground Movements and Structures*, University of Wales Institute of Science and Technology, Cardiff, 1977.

3.25 Penman, A. D. M. and Godwin, E. W., Settlement of experimental houses on land left by opencast mining at Corby, in *Proceedings of a Conference on Settlement of Structures*, Pentech Press, Cambridge, 1974, pp. 53–61.

3.26 Burford, D. and Charles, J. A., Long-term performance of houses built on opencast mining backfill at Corby, 1975–1990, in *Proceedings of the 4th International Conference on Ground Movements and Structures*, Pentech Press, London, 1992, pp. 54–67.

3.27 Chapman, D. A., Dyce, R. E. and Powell, M. J. V., NHBC structural requirements for housing, *The Structural Engineer*, **56A**, 3–10 (1978).

3.28 Tomlinson, M. J. and Wilson, D. M., Preloading of foundations by surcharge on filled ground, *Géotechnique*, **23**, 117–120 (1973).

3.29 Richart, F. E., Hall, J. R. and Woods, J. D., *Vibrations of Soils and Foundations*, Prentice-Hall, New Jersey, 1970.

3.30 Hsieh, T. K., Foundation vibrations, *Proceedings of the Institution of Civil Engineers*, **22**, 211–226 (1962).

3.31 Crockett, J. H. A. and Hammond, R. E. R., Reduction of ground vibrations into structures, *Institution of Civil Engineers Structural Paper No. 18* (1947).

3.32 Waller, R. A., A block of flats isolated from vibration, in *Proceedings of a Symposium, Society for Environmental Engineering*, London, 1966.

4 Spread and Deep Shaft Foundations

The term 'spread foundation' has been used to distinguish between foundations of the strip, pad, and raft type, and deep foundations such as basements, caissons, or piles. Terzaghi and Peck define a shallow footing as one which has a width equal to or greater than its depth. This is a reasonable definition for normal pad or strip foundations, but it is unsatisfactory for narrow or very wide foundations. In order to avoid possible misunderstandings when writing engineering reports, it is advisable to avoid using the term 'shallow foundation' unless the limiting depths are clearly defined in terms of the depth to width ratio. Definitions and general descriptions of strip, pad, and raft foundations are given in Section 2.1.

4.1 Determination of allowable bearing pressures for spread foundations

4.1.1 Foundations on sands and gravels

The allowable bearing pressures of spread foundations on sands, gravels, and other granular materials, are governed by considerations of the tolerable settlement of the structures. Thus, it is the normal practice to use semi-empirical or prescriptive methods as described in Sections 2.4 and 2.5 based on the results of *in-situ* tests or plate loading tests. The former are generally preferred since they are cheaper and rapid in execution. Only in the case of narrow foundations on waterlogged sands is it necessary to make calculations for ultimate bearing capacity by equations in Section 2.3, unless it is necessary to demonstrate compliance with limit state conditions as required by a statutory authority.

Depth of foundations If boreholes or *in-situ* test records show a marked increase in density of sands with increasing depth below ground level, it may be tempting to take the foundations deeper than normal in order to take advantage of the much higher allowable bearing pressures. This procedure is unlikely to be economical if it involves excavating below the water table. Even excavating as little as 0.5 m below the water table in a fine sand will give considerable trouble with slumping of the sides and instability of the base of the excavation. In fact uncontrolled inflow of water is likely to loosen the ground and hence decrease the allowable bearing pressures. Stability of excavations below the water table in sands can only be achieved by wellpointing or similar methods (Section 11.3.3). The cost of these measures is likely to be higher than founding on looser sands at the higher level above the water table, even though the lower bearing pressures require wider foundations.

There is not the same risk of instability when pumping from gravelly soils, but since these are usually very permeable the pumping rate will be heavy and the cost of excavation correspondingly high.

If satisfactory bearing conditions cannot be obtained above the water table, it is often more economical to pile the foundations than to excavate for spread foundations below ground water level. Alternatively, a deep compaction technique – as described in Section 11.6 – can be considered.

Settlements An example of the settlement of spread foundations on loose sand is given by three-storey flats which were built on a site at Kirkcaldy in Fifeshire. The site was near the seashore and the sand was probably a wind-blown dune deposit. In view of its looseness, as shown by the SPT which gave N-values (uncorrected for overburden pressure) varying between 1 and 8, a raft foundation was adopted. This gave an overall bearing pressure on the 61 m by 8.2 m raft of 35 kN/m². The settlements were measured on this block. They increased immediately the foundation loading increased and the average, maximum, and maximum differential values were 13, 19, and 6 mm, respectively. On an adjacent block a bearing pressure of 45 kN/m^2 was given by strip foundations 1.4 m wide in external walls and 1.8 m wide in the central longitudinal wall. The measured average, maximum, and maximum differential settlements were 25, 44 and 6 mm, respectively. Although these measurements showed that tilting had occurred, the movements were not detrimental to the structure since a rigid form of load-bearing wall construction was used. Such differential settlements would, however, have produced unsightly cracking in any form of 'hard-faced' wall panels. On the strength of these results some confidence can be placed in the use of the SPT in dry sands, and they indicate the need for caution in assessing allowable bearing pressures for structures sensitive to settlement which are founded on loose or even medium-dense sands.

Foundations on loose saturated sands and very fine or silty sands   If the sand is very loose (value of N of 5 or less) and also in a saturated state, then any form of shock loading may cause spontaneous liquefaction followed by subsidence of the foundation. A rapid change in water level, such as a sudden rise caused by severe flooding, can give the same effect. Therefore very loose sands in a saturated state should be artificially compacted by vibration (Section 11.6). If the size of the foundations and other conditions make such methods impracticable or uneconomical, then the load should be carried to underlying denser strata by means of piers or piles.

4.1.2 Foundations on clays

The ultimate bearing capacity of strip or pad foundations on clays can be calculated from the equations in Section 2.3.3. In the case of foundations of fairly light structures on firm to stiff clays it is unnecessary to make any calculations to determine consolidation settlement, especially if they are founded on boulder clays. Using permissible stress methods a factor of safety of 2.5–3 will ensure that settlements are kept within tolerable limits. It is desirable to make computations of settlements by the methods described in Section 2.6.6 in all cases of heavy structures, and in cases where there is no previous experience to guide the engineer. Calculations to determine the ultimate and service ability limit states will be necessary for Category 2 or 3 investigations (Section 1.0) to demonstrate compliance with Eurocode 7 or other statutory regulations.

Estimation of characteristic shear strengths   One of the chief difficulties in calculating the ultimate bearing capacity of spread foundations at a shallow depth is making a reasonable assessment of the characteristic shear

strength of the soil. The soil to a depth of 1.2–1.5 m in Great Britain, or deeper in tropical or subtropical conditions, is affected to a variable extent by seasonal wetting and drying and the effects of vegetation. Below this surface crust, normally consolidated clays, e.g. estuarine clays, usually show a fairly uniform variation in undrained shear strength with depth. The scatter in laboratory test values is relatively small and is mainly due to disturbance caused by sampling. There is usually a smaller scatter when shear strengths are determined by field vane tests (Section 1.4.5).

In British climatic conditions, foundations will normally be placed beneath or at the lower levels of the surface crust which is subject to seasonal wetting and drying (Section 3.1.1). Consequently, the minimum values of shear strength which are normally found immediately beneath the surface crust should be taken for the calculation of the ultimate bearing capacity. It is usually found that there is a progressive increase in shear strength with increasing depth below the surface crust, and the engineer may decide to take advantage of this by placing the foundation at a depth below that at which the minimum shear strength occurs. It is not always necessary to take the very lowest values of shear strength, since, as mentioned above, they may be random soil samples which have been affected by excessive sample disturbance.

The alternative to placing the foundations below the zone of seasonal moisture changes (and therefore on soil which has a minimum shear strength) is to place them at a fairly high level within the surface crust. This assumes that the strength of the crust is due to desiccation of the clay and that subsequent softening cannot occur, or that the ability of the crust to distribute load is not impaired by vertical fissures. The adoption of a foundation at a high level will enable relatively higher bearing pressures to be used, and because the foundations will have correspondingly smaller dimensions there will not be significant stressing of the softer soil below the surface crust.

As an example of the risks of placing foundations within the desiccated crust of a normally consolidated clay, Jarrett et al.[4.1] gave the results of settlement observations on a number of building structures at the Imperial Chemical Industries works in Grangemouth, Scotland. At this site the shear strength within the 2–2.5 m thick crust was 30–60 kN/m^2, and the minimum shear strength below the crust was about 15 MN/m^2 at a depth of 4 m. A two-storey building of brick-infilled frame construction with strip foundations loaded to 22 kN/m^2 had an average foundation settlement of about 110 mm; whereas single or one-and-a-half storey brick-infilled frame buildings of light construction and an overall loading of 12–16 kN/m^2 underwent average settlements of only 8–15 mm. These lighter buildings were bearing on continuous strip foundations at a depth of 0.65 to 1.0 m below ground level. The settlements were measured over periods of 20 years or more. The differential settlements were in the form of tilting rather than flexure and did not cause structural distress. Analysis of settlement data for these and other buildings on the site showed that there was a threshold loading intensity of 15 kN/m² below which only minor settlements occurred.

The problem of variation in shear strength is a difficult one in the case of foundations on fissured overconsolidated clays where, in addition to variations within the surface crust, a wide scatter in shear strength test results occurs throughout the full depth of the stratum owing to the effects of random fissuring on sampling and subsequent laboratory testing. It has been found that shear strengths derived from plate loading tests or tests on full-scale foundations reflect the fissured strength of the clay, i.e. failure takes place partly through the fissure planes. The scatter of triaxial compression test results on samples from boreholes and trial pits is typically shown in Fig. 4.1 for a site at Maldon in Essex. The shear strengths derived from plate loading tests in a pit on the same site are also shown. These shear strengths correspond to the lowest limits of the triaxial compression test results. Similar results have been found elsewhere in London Clay. Therefore in the case of foundations on stiff-fissured clays, characteristic values of shear strength corresponding to the lower limits should be used in conjunction with the appropriate bearing capacity factors for substitution in the equations in Section 2.3.

Glacial till also shows wide variations in shear strength due to random inclusions of sands and gravels within the test specimens. However, the *in-situ* strength as determined by full-scale foundation behaviour is usually higher than that indicated by triaxial tests in the laboratory and a characteristic strength need not necessarily be representative of lower bound values. A statistical approach could be appropriate for these soils. Nevertheless, it is advisable to make a careful study of the *in-situ* characteristics of the clayey glacial till in case low shear strengths may be the result of lenses of weak clay of appreciable extent which might cause local failure of foundations sited immediately above them.

Foundations on stiff soil overlying a soft clay stratum In the case of a foundation constructed within a stratum of stiff soil overlying a stratum of soft clay, if the foundation is close enough to the soft layer, it may break through to the latter and result in failure. This danger can arise in normally consolidated clays such as estuarine clays. Normally consolidated clays usually have a crust of stiff dry soil at the surface. Although the stiff crust may have a reasonably high shear strength enabling relatively high bearing pressures to be used, the foundation may be placed sufficiently close to the underlying soft clay for the bearing pressure transmitted through the crust to exceed the ultimate bearing capacity of the soft material.

The bearing pressure on the surface of the weaker layer can be calculated conservatively by assuming a 1 : 2 (or 30°) spread of load through the stiff layer as shown in Fig. 4.2. The bearing pressures are then given by

$$q_1 = q_n \left(\frac{B}{B + d} \right)^2 \quad \text{for square (or circular) foundations} \tag{4.1}$$

or

$$q_1 = q_n \left(\frac{B}{B + d} \right) \quad \text{for strip foundations.} \tag{4.2}$$

The above assumption of spread of load assumes a uniform bearing pressure across the width of the foundation. It tends to underestimate the vertical stress beneath the centre and overestimate it beneath the edges. The effect of the

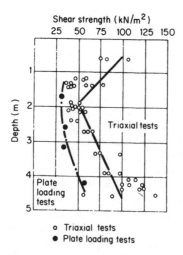

Shear strength (kN/m²)

Fig. 4.1 Variation of shear strength with depth in London Clay at Maldon.

○ Triaxial tests
● Plate loading tests

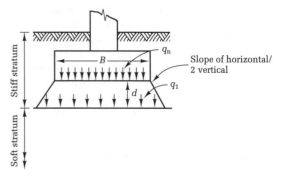

Fig. 4.2 Strip or pad foundation founded in stiff stratum overlying weak stratum.

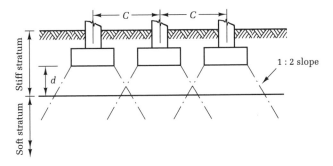

Fig. 4.3 Close-spaced foundation.

rigidity of the upper stiff stratum in reducing the stresses in the lower layer is not taken into account. These effects are discussed in more detail in Section 5.8.1.

The values of q_1, as calculated from the above equations, should not exceed the safe bearing capacity of the soft stratum. If the foundations are widely spaced (for example if the distance between them exceeds four times the width), it is possible to reduce, to some extent, the bearing pressure on the underlying stratum by increasing the size of the foundation. However, if the footings are closely spaced, the pressures transmitted to the underlying stratum will overlap, as shown in Fig. 4.3. Where they overlap, the bearing pressure on the buried stratum is increased. If the conservative 1 : 2 spread of load distribution is adopted the vertical stress at the interface with the lower layer will be twice that for an isolated foundation. The overlap stresses can be calculated more accurately by allowing for the relative rigidity of the layers.

Using Boussinesq's method, it can be shown that to ensure that the vertical stress increase due to overlapping pressure bulbs does not exceed the average value calculated from a 1 : 2 spread of load, the following minimum column spacings should be adopted for B/d less than 1.0:

B/d	0.1	0.2	0.3	0.4	0.5	0.6	0.7	0.8	0.9	1.0
C/d	0.48	0.73	0.91	1.05	1.19	1.31	1.43	1.54	1.64	1.74

when B/d is greater than 1.0, $C/d \simeq 0.8 + 0.943\,B/d$.

It may be possible to reduce the applied pressure still further by combining the footings to form a raft foundation or, in the case of a row of footings, to form a strip foundation, but if the bearing pressures are still excessive, consideration will have to be given to piling through the soft layer to a deeper bearing stratum. It should be noted that the stiff stratum in itself acts as a raft beneath the foundations. This natural raft, if thick enough, prevents the soft soil from heaving up beyond the loaded area.

Foundations constructed on a thin clay stratum

When foundations are constructed on a thin surface stratum of clay overlying a relatively rigid stratum, there may be a tendency for the thin layer to be squeezed from beneath the foundation, particularly if the soft layer is of varying thickness. Figure 4.4 shows a foundation of width B on a thin clay layer overlying a stratum of different characteristics and appreciably higher bearing capacity, for example a sand layer. The net ultimate bearing capacity of the thin clay layer is given by the formulae

$$q_{nf} = \left(\frac{B}{2d} + \pi + 1\right) c_u \quad \text{for } \frac{B}{d} \geqslant 2 \qquad (4.3)$$

for a strip foundation of width B, and

$$q_{nf} = \left(\frac{B}{3d} + \pi + 1\right) c_u \quad \text{for } \frac{B}{d} \geqslant 6 \qquad (4.4)$$

for a circular foundation of diameter B. For smaller values of B/d than those given above, q_{nf} can be calculated by the methods given for a thick clay layer below foundation level in Section 2.3.3. Alternatively, Skempton's bearing capacity factor (Fig. 2.15) can be used with the following correction factors: for strip foundations when $B/d \geqslant 2$, correction factor $= (0.1\,B/d + 0.8)$, for circular or square foundations when $B/d \geqslant 6$, correction factor $= (0.054\,B/d + 0.67)$.

It should also be noted that, with a thin clay layer, the Boussinesq theory underestimates the vertical stress near the lower boundary between the thin clay layer and the underlying stiff stratum. The underestimate is of the order of 50 per cent beneath the centre of the loaded area. This

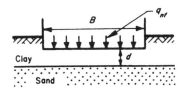

Fig. 4.4 Foundation on thin clay layer.

should be taken into account when considering factors of safety (see Section 5.8.1).

4.1.3 Spread foundations carrying eccentric loading

Examples of foundations subject to eccentric loading are column foundations to tall buildings where wind pressures cause appreciable bending moments at the base of the columns, foundations of stanchions carrying brackets supporting travelling crane girders, and the foundations of retaining walls.

The pressure distribution below eccentrically loaded foundations is assumed to be linear as shown in Fig. 4.5(a), and the maximum pressure must not exceed the maximum pressure permissible for a centrally loaded foundation. For the pad foundation shown in Fig. 4.5(a), where the resultant falls within the middle third of the base.

$$\text{Maximum pressure, } q_{max} = \frac{W}{BL} + \frac{My}{I}, \qquad (4.5)$$

which for a centrally loaded symmetrical pad becomes

$$q_{max} = \frac{W}{BL} + \frac{6M}{B^2L}. \qquad (4.5a)$$

Similarly,

$$\text{Minimum pressure, } q_{min} = \frac{W}{BL} - \frac{6M}{B^2L} \qquad (4.5b)$$

When the resultant of W and M falls outside the middle third of the base, equation (4.5b) indicates that tension will occur beneath the base. However, no tension can in fact develop and the pressure distribution is as shown in Fig. 4.5(c):

$$q_{max} = \frac{4W}{3L(B - 2e)}, \qquad (4.5c)$$

where

W = total axial load,
M = bending moment,
y = distance from centroid of pad to edge,
I = moment of inertia of plan of pad,
e = distance from centroid of pad to resultant loading.

When checking the ultimate limit state it should be noted from Table 2.1 that the partial factor for actions, including wind loading, is 1.3 but with the possibility of increasing this factor to as much as 1.5 for the case of structures such as masts and towers where redistribution of loading between different parts of the structure may not be possible.

For checking the serviceability limit state, the settlements should be calculated for uniform bearing pressure on the equivalent rectangular area (Fig. 2.10). For the case of an eccentrically loaded circular foundation the equivalent rectangular area is taken as the hatched area A in Fig. 4.6 where the dimension a' is equal to A/b. Eurocode 7 does not give any guidance on the need or otherwise to calculate settlements caused by wind loading. When using permissible stress methods the general practice is to neglect wind loading if it is less than 25 per cent of the combined dead and imposed loading. For higher percentages it is desirable to calculate immediate settlements for wind loading, but not long-term settlements assuming drained conditions. Again using permissible stress methods wind loading can be accommodated by proportioning the foundation dimensions so that the bearing pressures from the combined dead, imposed, and wind loading do not exceed the allowable bearing pressure by more than 25 per cent. Settlement is an important consideration for eccentrically loaded foundations on sands, since if it is excessive the tilting of the foundation will cause an increase in eccentricity with the higher edge pressure, followed by further yielding and possible failure. If the

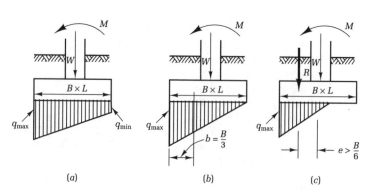

(a)   (b)   (c)

Fig. 4.5   Eccentrically loaded pad foundation.

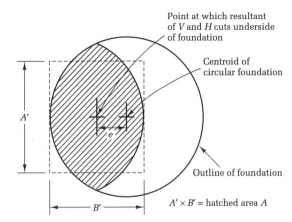

Point at which resultant of *V* and *H* cuts underside of foundation

Centroid of circular foundation

Outline of foundation

A′ × *B′* = hatched area *A*

Fig. 4.6 Equivalent uniformly loaded rectangular area for eccentrically loaded circular foundation.

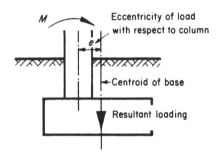

M

Eccentricity of load with respect to column

Centroid of base

Resultant loading

Fig. 4.7 Placing column eccentric to base to obtain uniform bearing pressure.

foundation is restrained against tilting it may be permissible for the maximum edge pressure to exceed that permissible for a centrally loaded foundation.

In cases where eccentric loading on a column is produced by sustained bending moments, for example dead-load bending moments, it may be advantageous to place the column off-centre of the base so that the resultant of the axial load and the bending moments passes through the centroid of the base (Fig. 4.7). Thus there is no eccentricity of loading on the foundation and the pressures are uniformly distributed.

4.2 Structural design and construction

4.2.1 Strip foundations for load-bearing walls

Unreinforced concrete strip foundations The British Standard Code of Practice, BS 8103: *Stability, Site Investigation, Foundations and Ground Floor Slabs for Housing* recommends a C20 grade mix for house foundations in unreinforced concrete. This mix has a maximum water : cement ratio of 0.8, 20 mm aggregate size, and the cement content is increased to a minimum of 220 kg/m^3 from considerations of durability in the ground. A mix as lean as this would not be used if sulphates or other aggressive substances were present in the soil in sufficient quantity to be deleterious to concrete (see Section 13.5). The loading per metre run of wall on this type of foundation is usually quite low, and in the case of strip foundations supporting brick walls, the width of the foundation is governed by the minimum width in which a bricklayer can lay the footing courses, rather than by the bearing capacity of the soil. Thus, in Fig. 4.8(*a*), a practical minimum for bricklayers to work in a shallow trench is 450 mm. The load per metre

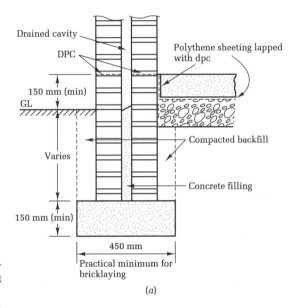

Drained cavity

DPC

Polythene sheeting lapped with dpc

150 mm (min)

GL

Varies

Compacted backfill

Concrete filling

150 mm (min)

450 mm

Practical minimum for bricklaying

(*a*)

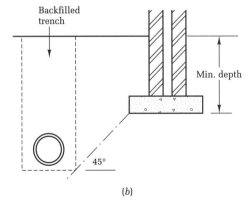

Backfilled trench

Min. depth

45°

(*b*)

Fig. 4.8 Strip foundation to 280 mm cavity brick wall. (*a*) Foundation details. (*b*) Minimum depth for foundations adjacent to service trench.

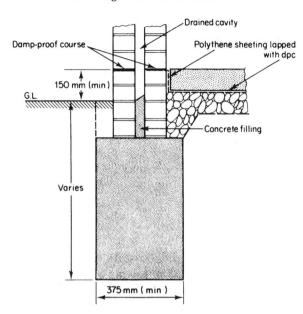

Fig. 4.9   Narrow strip foundation to 280 mm cavity brick wall.

run of the most heavily loaded wall of a two-storey dwelling house is about 50 kN, which gives a bearing pressure of little more than 110 kN/m^2 for the 450 mm wide foundation. For this value, the factor of safety on the ultimate bearing capacity of a stiff clay or a dense sand may be 6 or more. Thus the maximum safe bearing capacity of soils of good supporting values is not utilized, and in these soils it may be advantageous to adopt a narrow strip foundation as illustrated in Fig. 4.9.

BS 8103 requires a minimum depth below ground level of 1.0 m for clay sites (see Section 3.1.1). Elsewhere a minimum of 0.45 m is recommended for protection against the effects of frost heave with the possibility of an increased depth in upland areas and other areas known to be subjected to long periods of frost. A depth greater than 0.45 m may be needed if the foundation is located near a service trench (see Fig. 4.8(b)).

Narrow strip (trench fill) foundations   The essential feature of the narrow strip foundation is that the trench is too narrow to be dug by labourers working in the trench. It depends for its success on the ability of a mechanical excavator, such as a light tractor-mounted back-acter with a narrow bucket to dig the trench, which must be self-supporting until it can be backfilled with concrete. Either hand or mechanical excavation will be ineffective if the ground contains many large stones or thick roots. Also deep narrow strip foundations cannot economically be used in very soft clays or water-bearing sands which

require support by close-sheeting.

Although the narrow strip foundation is widely used in clays which are affected by swelling and shrinkage due to seasonal moisture content changes, this type is vulnerable to damage where the clay has been desiccated by vegetation to depths of more than about 1.2 m (see Section 3.1.2). Uplift of these foundations causing hogging of the superstructure can result from adhesion of the clay to the sides of the deep strip, particularly where the excavator has produced a trench wider at the top than at the bottom. Lateral movement resulting in vertical cracks in the foundation walls and lower parts of the superstructure can be caused by horizontal swelling of the mass of clay enclosed by the walls. Vertical slabs of polystyrene are sometimes used on the inside faces of trench fill foundations to relieve horizontal swelling pressures, but polystyrene, unless of a special low density, has an appreciable degree of stiffness and the pressures transmitted to the foundations can cause cracking of an unreinforced section. Volclay (bentonite-filled slabs) and Clayboard (cellular cardboard) are suitable alternatives provided that precautions are taken in the latter material against premature collapse due to wetting. The conventional strip foundation, where the soil is loosely backfilled against the deep footing walls, is less vulnerable in this type of damage.

Care is necessary to obtain accuracy in setting out and subsequent construction of narrow strip foundations and the superimposed walls. Typically the guideline for the excavator operator is a line of cement or lime dust roughly strewn on a string-line. An inexperienced operator may cut the trench 50 or 75 mm off the true centre line, so that when the brick footing courses are laid these will be eccentric to the foundation with a risk of shearing at the edge (Fig. 4.10(a)). A more serious situation, known to have occurred, is the case where the concrete placed in a trench incorrectly excavated is allowed to spill over the edges of the trench to make up low-lying ground. Then if the brickwork is also incorrectly set out in the opposite direction from the foundation the bricklayer may be unaware of the error with the result that the wall is built partly over a thin layer of concrete (Fig. 4.10(b)).

Stepped foundations   When building on sloping ground, strip foundations need not necessarily be at the same level throughout the building. It is permissible to step the foundations as shown in Fig. 4.11. Similarly, if strip foundations are taken below a surface layer of filling or weak soil on to the underlying bearing stratum, the levels of the foundation can be stepped, as required, to follow any undulations in the bearing stratum. The requirements of BS 8103 are shown in Fig. 4.12(a) and (b). The steps should not be of greater height than the

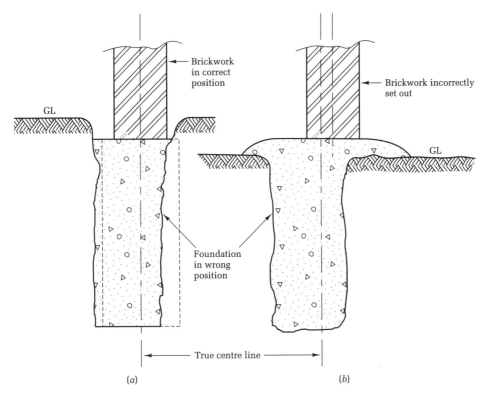

Brickwork in correct position

Brickwork incorrectly set out

GL

GL

Foundation in wrong position

True centre line

(a)   (b)

Fig. 4.10   Incorrect setting out and construction of narrow strip foundations.

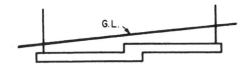

G.L.

Fig. 4.11   Foundations stepped on sloping ground.

thickness of the foundations unless special precautions are taken. Heights of steps in deep trench fill foundations (Fig. 4.12(b)), require special consideration and it might be advisable to introduce reinforcing bars to prevent cracking at the steps.

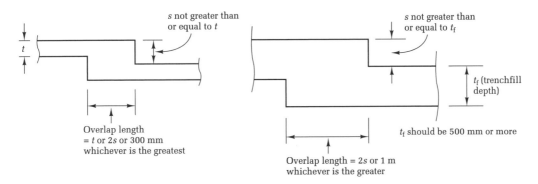

s not greater than or equal to t

t

Overlap length = t or $2s$ or 300 mm whichever is the greatest

s not greater than or equal to t_f

t_f (trenchfill depth)

t_f should be 500 mm or more

Overlap length = $2s$ or 1 m whichever is the greater

Fig. 4.12   Requirements for overlapping stepped foundations. (a) Conventional strip. (b) Narrow strip (trench fill).

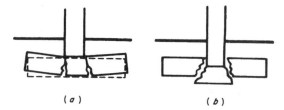

Fig. 4.13 Structural failure of unreinforced foundations. (a) Failure in bending. (b) Failure in punching shear.

Requirements for thickness Conventionally, the thickness of concrete in a lightly loaded strip foundation is equal to the projection from the face of the wall or footing with a minimum thickness of 150 mm. This minimum is necessary for rigidity, to enable the foundations to bridge over loose pockets in the soil, and to resist longitudinal forces set up by thermal expansion and contraction, and moisture movements of the footing walls. On clay soils swelling pressures can be of a fairly high order.

When strip foundations carry heavy loading, the criterion for the width of the foundations is then likely to be the allowable bearing pressure, and the thickness of the strip is governed by its strength to resist failure of the projecting portions in bending or shear. The mode of failure in bending is illustrated in Fig. 4.13(a), and in shear in Fig. 4.13(b). In an unreinforced strip foundation failure in bending will be the governing factor. This can be prevented by an adequate thickness of concrete, with or without a stepped or sloping transition from the wall to the bottom width. A conservative design rule is to proportion the thickness of the strip so that no tension is developed on the underside of the strip. This is achieved by the rule-of-thumb procedure of making the thickness equal to twice the projection. However, a distribution of loading at an angle of 45° from the base of the wall for concrete foundations, or a spread of load through brick footings of a quarter of a brick per course (51–75 mm) is generally adopted. The procedure is illustrated for plain strip, stepped, and sloping foundations in Fig. 4.14(a)–(d).

The 45° load distribution implies a small order of tension on the underside of the foundation, but its magnitude cannot be determined since it depends on the bending of the strip and the eccentricity of loading. These cannot be calculated with any pretence of accuracy; for this reason, the nominal 45° distribution is preferred to design procedures which are based on permitting a certain tensile stress to develop.

In wide and deep foundations consideration must be given to economies in concrete quantities that can be achieved by stepping or sloping the footings. It will be seen from Fig. 4.14(a) that there is a considerable area of concrete which is not contributing to the distribution of the load from the wall. The wastage is less if the concrete is stepped as shown in Fig. 4.14(b). However, the construction of such a stepped form would mean concreting in two lifts with formwork to the upper step. The cost of these operations is likely to be more than a simple addition to the thickness of the foundation. In the case of brick or stone-masoned load-bearing walls, the stepping can conveniently be carried out in the wall materials (Fig. 4.14(c)). Sloping the upper face of the projections (Fig. 4.14(d)) will give savings in concrete and will enable the concrete to be placed in one lift, but if the slope is steeper than about 1 vertical to 3 horizontal it will require formwork. Such formwork is costly to construct in a trench and requires weighting down. However, the savings given by haunching at flat slopes as in Fig. 4.14(d) are doubtful except for very long continuous foundations, and in most cases the plain strip (Fig. 4.14(a)) is the most economical.

In wide or heavily loaded strip foundations it is always worth while to keep in mind the economies that unreinforced may have over reinforced concrete foundations. This is especially the case if foundations have to be taken to a greater than nominal depth, to get below a layer of filling or weak soil for example. In such circumstances it may well be more economical to backfill the trench with a sufficient depth of lean concrete to conform to the requirement of the 45° load distribution rather than incur the expense of providing and bending reinforcement and

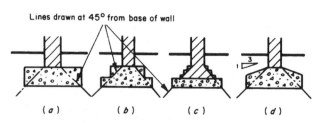

Lines drawn at 45° from base of wall

Fig. 4.14 Proportioning thickness of unreinforced strip foundations based on 45° load distribution. (a) Plain strip. (b) Stepped concrete. (c) Stepped brick footings. (d) Concrete with sloping upper face.

assembling it in a deep timbered trench, with the extra care that is necessary in placing the concrete in such conditions. Savings in cement are also given, because in non-aggressive soils 1 : 9 concrete can be used in unreinforced foundations, whereas reinforced concrete made in accordance with structural codes usually requires to be richer.

4.2.2 Reinforced concrete strip foundations

Reinforced concrete strip foundations are likely to show an advantage in cost over unreinforced concrete where weak soil and heavy wall loading require a wide strip at a comparatively shallow depth. Reinforcement in the form of *longitudinal* bars is also desirable in strip foundations on highly variable soils when the foundation is enabled to bridge over local weak or hard spots in the soil at foundation level; or when there is an abrupt change in loading.

The transverse reinforcement in a wide strip foundation is designed on the assumption that the projection behaves as a cantilever with its critical section on the face of the wall (line XX in Fig. 4.15), and with a loading on the underside of the cantilever equal to the actual net bearing pressure under the worst conditions of loading (i.e. maximum eccentricity if the loading is not wholly axial). The main reinforcement takes the form of bars at the bottom of the slab. The slab must also be designed to withstand shear and bond stresses.

In calculating bending stresses, the projecting portion b in Fig. 4.15 is taken as a cantilever, i.e. bending moment at face of wall where b is given by metres

$$M_b = \frac{q_n \times b^2}{2} \text{ kN m per metre length of wall.} \tag{4.6}$$

It is further necessary to check the shear stress at the critical section YY in Fig. 4.15 given by a vertical section at a distance of 1.5 times the effective depth of the foundation from the face of the wall:

$$\text{Shear stress} = \frac{q_n \times b'}{d} \text{ N/mm}^2, \tag{4.7}$$

where d is the effective depth at section YY and b' are in mm. If the shear stress is greater than the permissible value it should be reduced by increasing the effective depth of the strip. Shear reinforcement in the form of stirrups or inclined bars should be avoided if at all possible.

As noted for unreinforced concrete foundations, economy in concrete quantities can be made by sloping or stepping the upper face of the projections. The former shape is usually preferred, but if the foundation is short, the cost of formwork to the slope, or constructing it without formwork in a dry concrete, will exceed the cost of the extra concrete in a rectangular section. However, for long lengths of strip foundation, where travelling formwork or other methods of repetitive construction can be used, the savings in the quantity of concrete given by a sloping face may well show savings in the overall costs.

Constructional details   It is good practice to lay a blinding coat of 50–75 mm of lean-mix concrete over the bottom of the foundation trench in order to provide a clean dry surface on which to assemble the reinforcement and pack up the bars to the correct cover. The blinding concrete should be placed as quickly as possible after trimming the excavation to the specified levels: it will then serve to protect the foundation soil from deterioration by rain, sun, or frost. If the excavated material is a reasonably clean sand or sand–gravel mixture, it can be mixed with cement for the blinding layer. There is no need to use imported graded aggregates if the excavated soil will itself make a low-grade concrete.

To protect the reinforcing steel against corrosion, a

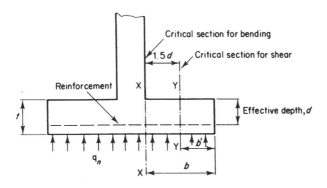

Fig. 4.15

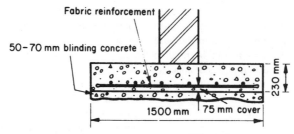

Fig. 4.16   Reinforced concrete strip foundation for 440 mm brick wall.

minimum cover of 40 mm is required by BS 8110 between the surface of the blinding concrete and the bottom layer of reinforcement. If concrete is placed against rough timber or directly against the soil the nominal cover should be increased to 75 mm.

A typical arrangement of a reinforced concrete strip foundation for a brick footing wall is shown in Fig. 4.16. The fabric reinforcement is wholly in the bottom of the slab and the 'mat' of reinforcement may be assembled on the surface, concrete spacer blocks wired on, and the whole assembly lowered on to the blinding concrete. The arrangement of reinforcement for a strip foundation to a reinforced concrete loadbearing wall is typically similar to that shown for a pad foundation in Fig. 4.19.

4.2.3   Unreinforced concrete pad foundations

The methods of design of unreinforced and reinforced concrete pad foundations are similar in principle to those described in the preceding pages for strip foundations. Thus the minimum size of the pad is given by the practical requirement of being able to excavate by hand to the required depth and level off the bottom and to lay bricks or fix steelwork for the columns. The minimum size for a 215 mm square brick pier is a base 510 mm square. The design corresponding in constructional technique to the narrow strip foundation is the circular pad in which a mechanical earth auger is used to drill a hole to the required depth which is backfilled with concrete. A brick pier may be built off the levelled surface, or a pocket may be cast into the pad to receive a steel or precast concrete column. The savings in mechanically augered excavations are more in terms of labour and time than in quantities of excavation or concrete. They are most favourable in the case of deep shaft foundations (Section 4.6).

As in the case of strip foundations, the thickness of unreinforced concrete pad foundations is given by the necessity of preventing development of tension on the underside of the base or reducing it to a small value. The former criterion is given by the rule-of-thumb proce-

dure of making the thickness twice the projection, while the latter can be determined by the normal British practice of a 45° distribution of loading (as Fig. 4.14).

The considerations for determining the choice of stepped or sloping foundations or a plain rectangular section are the same as those described for strip foundations.

4.2.4   Reinforced pad foundations

The procedure for the design of reinforced concrete pad foundations is as follows:

(a)   Determine the base area of the foundation by dividing the total load of the column and base by the allowable bearing pressure on the soil.
(b)   Determine the overall depth of the foundation required by punching shear, based on the column load only.
(c)   Select the type of foundation to be used, i.e. simple slab base, or sloping upper surface. Assume dimensions of slope.
(d)   Check dimensions by computing beam shear stress at critical sections, on the basis that diagonal shear reinforcements should not be provided.
(e)   Compute bending moments and design the reinforcement.
(f)   Check bond stresses in steel.

Calculation of required thickness of pad to resist punching shear   Punching shear occurs along a critical peripheral section at a distance of one and a half times the thickness of the pad (Fig. 4.17). Thus for the rectangular column shown the shear stress is given by

$$v = \frac{(W/bb')(bb' - A)}{(2a + 2a' + 2\pi 1.5t)d'} \qquad (4.8)$$

where A is the area within critical peripheral section, and d the effective depth of pad.

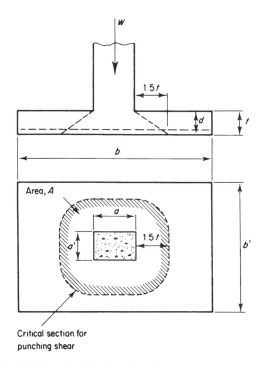

Fig. 4.17   Punching shear in a column base.

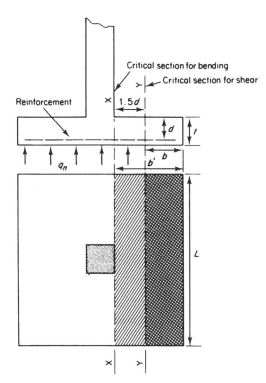

Fig. 4.18

Calculation of shear stress at critical sections   In the case of pads of uniform thickness, the critical section for shear is along a vertical section YY extending across the full width of the pad at a distance of one-and-a-half times the effective depth from the face of the column, i.e. Section XX in Fig. 4.18:

$$\text{Shear stress} = \frac{q_n \times b \times L}{L \times d}, \qquad (4.9)$$

where

q_n = net bearing pressure on area outside section under consideration,
b = width of section under consideration,
d = effective depth of section,
L = length of pad.

In the case of stepped footings, other critical sections may exist at the face of each step, and the shear stress should be checked at the reduced thickness at each step. Where pads have a sloping upper face it may be necessary to check the shear stress at two or three points. If the shear stress is greater than the permissible values given in the structural code, then it will be necessary to increase the thickness of the base. Shear reinforcement in the form of stirrups or inclined bars should be avoided if at all possible.

Calculation of bending moments in centrally loaded pad foundations   The bending moments should be taken to be the moment of the forces over the entire area on one side of the section, assuming the bearing pressure to be distributed uniformly across the section. The critical section for bending in the base should be taken at the face of the column.

Thus in Fig. 4.18 the bending moment at critical section XX is given by

$$q_n \times b' \times \frac{b'}{2} \text{ per unit width of pad.} \qquad (4.10)$$

Square or rectangular base slabs are preferred to circular slabs since the latter require reinforcing bars of varying lengths, and formwork, if required for circular steps or sloping faces, is expensive. Typical reinforcement details for a pad foundation are shown in Fig. 4.19. For square pads the reinforcement should be distributed uniformly across the section considered. Reinforcement required to resist the moment across the short span of a rectangular base should be spread over a band centred on the column and containing $2/(\beta_i + 1) = 1$ times the total area of reinforcement, and the remainder is spread over the outer parts of the section. The width of the central band is equal to the short side dimension of the base and β_i is the ratio of the longer to the shorter side.

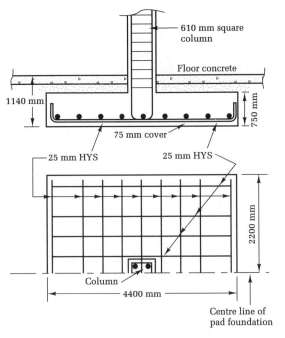

Fig. 4.19   Square pad foundation (see Example 4.4).

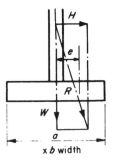

Fig. 4.20   Eccentrically loaded pad foundation.

Minimum bearing pressure at heel of slab

$$= \frac{W}{ab}\left(1 - \frac{6e}{a}\right).$$
(4.12)

When e is greater than $a/6$,

Maximum bearing pressure at toe of slab

$$= \frac{4W}{3b(a - 2e)}.$$
(4.13)

The dimensions of the base slab, a and b, are proportioned in such a way that the maximum bearing pressure at the toe does not exceed the allowable bearing pressure (see Section 4.1.3). If a column carries a permanent bending moment, for example a bracket carrying a sustained load, it may be an advantage to place the column off-centre on the pad so that the eccentricity of the resultant loading is zero, giving uniform pressure distribution on the base in a similar manner to Fig. 4.7 for the strip foundation. This eliminates the risk of tilting.

The long toe section of the slab should be designed as a cantilever about a section through the face of the column. Punching shear should be calculated on the faces of the column and the design should be checked for diagonal tension at sections outside a vertical section as shown in Fig. 4.18.

Since the bending moment at the foot of the column is likely to be large with foundations of this type, the column reinforcement should be properly tied into the base slab. Reinforcement details for an eccentrically loaded pad foundation are shown in Fig. 4.21.

In the case of large eccentricity from any direction on square or rectangular pad foundations, consideration can be given to the passive resistance of the soil in contact with the foundation block to resist overturning. For this to be taken into account it is necessary that the foundation block should be cast against the undisturbed soil, and that the soil itself should not yield appreciably under the lateral load, or that the backfilling can be rammed between the

Pad foundations for _L_-shaped columns   Columns at the corners of reinforced framed structures are often _L_-shaped for architectural reasons. The foundation slab must be made concentric with the centroid of the column. The foundation slab is designed in a similar manner to square or rectangular bases. The punching shear is small because of the large perimeter of the column in relation to the load carried.

Eccentrically loaded pad foundations   Where the lateral loads or bending moments on a column come from any direction, for example from wind loads, a square foundation slab is desirable; unless, for reasons of limitation of space, a rectangular foundation must be provided. However, where the bending moments always act in the same direction, as in columns supporting rigid framed structures, the foundation slab can be lengthened in the direction of the eccentricity. Thus in Fig. 4.20 if the resultant R of the vertical load W and the horizontal load H cuts the base at a distance e from its centroid, then when e is smaller than $a/6$,

Maximum bearing pressure at toe of slab

$$= \frac{W}{ab}\left(1 + \frac{6e}{a}\right),$$
(4.11)

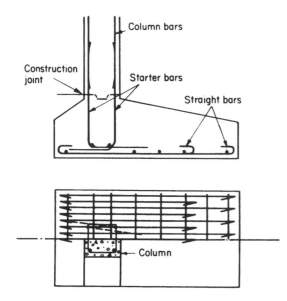

Fig. 4.21 Reinforcement details for pad foundation with eccentrically placed column.

face of the block and the excavation to a density equal to or greater than that of the surrounding soil. Thus, a dense sand or sand and gravel would have a high resistance to overturning of a foundation block. No resistance to overturning could be given by clay soil if the block were sited within the zone of seasonal moisture movements, when shrinkage might open a gap between the soil and the block.

4.2.5 Foundations close to existing structures

Where space for the base slab of foundations is restricted, for example where a strip foundation is to be built close to an existing wall (Fig. 4.22), there may not be room for a projection on both sides of the column or wall. If the projection is on one side only, eccentric loading is inevitable. This may not matter in the case of light loadings or if the foundation material is unyielding, but eccentric loading on compressible soils may lead to tilting of the foundation with consequent transmittal of dangerous horizontal forces on the walls or columns of the abutting

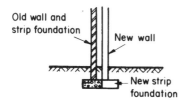

Fig. 4.22 Strip foundation close to existing structure.

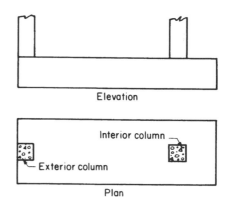

Fig. 4.23 Combined pad foundation for exterior and interior columns.

structures. Some degree of eccentricity of loading on the soil may be permitted in framed structures which are properly tied together, or in deep foundations where the tilting can be resisted by the passive resistance of the soil against the vertical surfaces of the foundation wall.

A method used to counteract tilting in column foundations of framed structures is to combine the exterior foundation with the adjoining interior foundation as shown in Fig. 4.23. Because of the eccentricity of loading on the base of such a combined foundation, there will be unequal distribution of pressure on the soil and consequently a tendency to relative settlement between the columns. This must be resisted by the structural rigidity of the base slab. Top and bottom reinforcement is required and since shear forces are likely to be large, link steel will probably be necessary.

If there are wide differences in the loading of adjacent columns in a combined foundation, the desirability of having uniform bearing pressures would theoretically

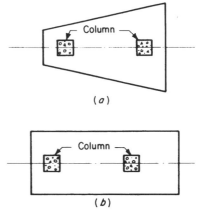

Fig. 4.24 Combined pad foundations for unequally loaded columns. (*a*) Trapezoidal base. (*b*) Rectangular base.

require a trapezoidal base slab (Fig. 4.24(*a*)). In practice the trapezoidal base slab should be avoided if possible, since it requires cutting and bending bars of differing lengths. In many cases it is possible to provide the additional bearing area for the more heavily loaded column by extending the cantilevered portion of the base slab while keeping the sides parallel (Fig. 4.24(*b*)).

4.2.6 Continuous pad and beam foundations

It may often be more economical to construct the foundations of a row of columns as a row of pad foundations with only a joint between each pad, rather than to provide individual excavations at a close spacing. Foundations of this type with individual but touching pads are more economical in reinforcing steel than continuous beam foundations, since the latter require a good deal of reinforcement to provide for the stresses due to differential settlement between adjacent columns. However, continuous beam foundations may be required to bridge over weak pockets in the soil or to prevent excessive differential settlement between adjacent columns. The advantages of the continuous pad or beam foundation are: ease of excavation by back-acter or other machines; any formwork required can be fabricated and assembled in longer lengths; there is improved continuity and ease of access for concreting the foundations. These foundations have the added advantage of providing strip foundations for panel walls of the ground floor of multi-storey framed buildings.

Structural design of continuous beam foundations
Continuous beam foundations may take the form of simple rectangular slab beams (Fig. 4.25(*a*)) or, for wider foundations with heavy loads, inverted T-beams (Fig. 4.25(*b*)). Structural design problems are complicated by factors such as varying column loads, varying live loads on columns, and variations in the compressibility of the soil. In most cases it is impossible to design the beams on a satisfactory theoretical basis. In practice, soil conditions are rarely sufficiently uniform to assume uniform settlement of the foundations, even though the column loads are equal. Inevitably there will be a tendency to greater settlement under an individual column, which will then transfer a proportion of the load to the soil beneath adjacent columns until the whole foundation eventually reaches equilibrium. The amount of load transfer and of yielding of individual parts of the foundation beam is determined by the flexural rigidity of the beam and the compressibility of the soil considered as one unit.

For reasonably uniform soil conditions and where

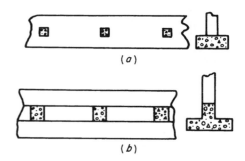

Fig. 4.25 Continuous beam foundations. (*a*) Rectangular slab. (*b*) Inverted T-beam.

maximum settlement will, in any case, be of a small order, a reasonably safe design method is to allow the maximum combined dead and live load on all columns, to assume uniform pressure distribution on the soil, and to design the foundation as an inverted beam on unyielding columns. However, if the compressibility of the soil is variable, and if the live load distribution on the columns can vary, this procedure could lead to an unsafe design. The structural engineer must then obtain from the soil mechanics engineer estimates of maximum and differential settlements for the most severe conditions of load distribution in relation to soil characteristics. The soil mechanics engineer must necessarily base his estimates on complete flexibility in the foundation, and the structural engineer then designs the foundation beam on the assumption of a beam on yielding supports. Computerized methods, as described in Section 2.8, are generally used for design purposes.

The degree of rigidity which must be given to the foundation beam is governed by the limiting differential movements which can be tolerated by the superstructure and by considerations of economies in the size and amount of reinforcement in the beams. Too great a rigidity should be avoided since it will result in high bending moments and shearing forces, and the possibility of a wide crack forming if moments and shears are underestimated (this is always a possibility since close estimate of settlements cannot be relied on from the soil mechanics engineer and it may be uneconomical to design on the worst conceivable conditions). The general aim should be a reasonable flexibility within the limits tolerated by the superstructure, and in cases of high bending moments the junctions of beams, slabs, and columns should be provided with generous splays and haunches to avoid concentrations of stress at sharp angles. When considering the effects of settlements of columns on the superstructure the structural designer should keep in mind the data on limiting distortions given in Section 2.6.2.

4.3 Foundations to structural steel columns

The traditional design of bases of structural steel columns consisted of a rectangular steel base plate, rigidly connected to the stanchion by gussets and angles, while holding-down bolts secured the plate to the concrete foundations. The size of the base plate was, more often than not, dictated by the size of the column and the space taken up by the gussets and angles, rather than by the allowable bearing pressure on the concrete beneath the base plate. However, the general adoption of welding in steel structures has led to a reduction in the size of base plates (since clearances need not be allowed for bolt holes). The use of hinged ends to columns in such structures as portal framed buildings and guyed masts permits base plates of minimum area, which are then governed in size by the allowable bearing pressure which can be imposed on the concrete. Also it is quite a usual practice to embed the feet of steel or precast concrete columns in pockets formed in the concrete foundations.

Designers often assume that the ultimate bearing capacity of the surface of the unreinforced concrete foundation is equal to its unconfined compressive cube strength. This is clearly a very conservative procedure since the confining effect of the surrounding concrete is neglected. Investigations by Meyerhof at the Building Research Station[4.2] into different sizes and thickness of blocks showed that the ultimate bearing capacity of the concrete was several times the cube strength. Indeed, for wide foundations reinforced against splitting, the ultimate bearing capacity was seven times the cube strength.

The results of these researches have not been expressed in practical rules for design. It is the usual practice to check the permissible stress on the surface of the foundation by referring to BS 5628: *Structural Use of Unreinforced Masonry*.

4.4 Grillage foundations

Where very heavy loads from structural steel columns have to be carried on a wide base, and where the overall depth of the foundation is restricted (to enable the base slab to be sited above the ground-water table, for example) a steel grillage may be required to spread the load. A typical design for a grillage base is shown in Fig. 4.26. The girders of the lowermost tier are designed to act as cantilevers carrying a distributed loading equal to the bearing pressure on the foundation slab. The upper tier distributes the column load on to the lower tier. An intermediate tier may be required. In large grillages the girders in each tier are located by tie bolts passing through holes drilled through their webs with gas barrel spacers threaded over the bolts between the webs. Attention must

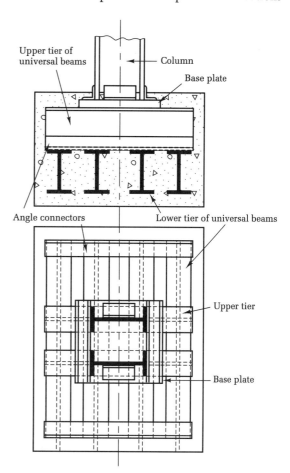

Fig. 4.26 Grillage base for heavily loaded steel column.

be given to the stability of the girder webs in bearing under concentrated loading from the column or upper tier of girders. Web stiffeners should be provided as required.

Adequate space must be provided between girder flanges to allow the concrete to flow between and underneath them. The grillage must be set very accurately in location and must be quite level before the concrete is placed, and it must not be allowed to move during concreting, since any errors due to inaccurate setting or displacement will be highly expensive to rectify once the concrete has hardened.

4.5 Raft foundations

The commonest use of raft foundations is on soils of low bearing capacity, where the foundation pressures must be spread over as wide an area as possible. They are also used for foundations on soils of varying compressibility where the partial rigidity given by stiff slab and beam construction is utilized to bridge over areas of more compressible soil, and thus differential settlement of the foundation slab is minimized. Raft foundations can be used as a matter of constructional convenience in structures supported by a grid of fairly close-spaced columns. In such cases an overall raft will avoid obstruction of the site by a number of individual excavations with their associated heaps of spoil. Some designers work on the rule that if more than 50 per cent of the area of the structure is occupied by individual pad or strip foundations it will be more economical to provide an overall raft. This is not necessarily true since the quantity of reinforcing steel required to avoid excessive deflexion and cracking of a raft carrying unequal column loads may be large. It may be more economical to excavate the site to a level formation, construct individual close-spaced pad foundations (if necessary they can touch each other), and then backfill around them.

Basements with stiff slab or slab and beam floors are forms of foundation rafts; these and the special case of buoyancy rafts are described in Chapter 5. Designs of rafts to counteract the effects of mining subsidence were described in Section 3.5.5.

4.5.1 Types of raft foundation

Plain *slab rafts* (Fig. 4.27) can be used in ground conditions where large settlements are not anticipated and hence a high degree of stiffness is not required. A layer of mesh reinforcement in the top and bottom of the slab is usually provided to resist bending moments due to hogging or sagging deflections at any point in the slab.

For architectural reasons it is usual to conceal the whole of the slab below finished ground level. Thus the top of the slab is placed at a depth of 100–150 mm to allow for minor variations in ground surface level. The slab can either be cast directly on the soil or on a thin layer of blinding concrete. The waterproof membrane is provided above the slab and is lapped with the damp-proof course. In addition a layer of polythene sheeting is provided beneath the slab as a waterproof membrane where soils aggressive to concrete are present. For a raft slab 200 mm thick suitable for a light building, the resulting foundation depth of 300–350 mm below ground level will be satisfactory for soils which are not susceptible to frost heave. In frost-susceptible soils the periphery of the slab can be constructed on a layer of lean concrete or compacted granular fill to provide an exterior foundation depth of 450 mm as a precaution against frost heave (Fig. 4.28). Somewhat greater depths are required in areas subjected to severe frost. Where a flat slab has its upper surface below ground level it is necessary to provide a separate ground-floor slab on a layer of lean concrete or compacted fill as shown in Fig. 4.27. Alternatively, a timber floor can be constructed on brick sleeper walls or timber bearers.

Where architectural considerations do not necessitate

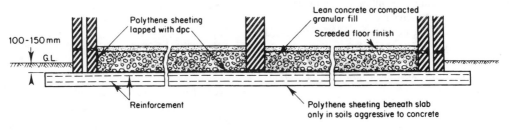

Fig. 4.27   Plain slab raft.

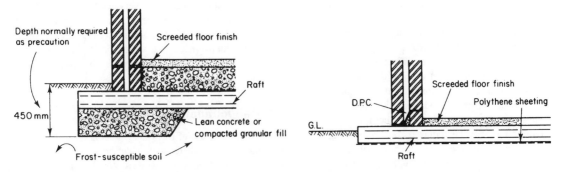

Fig. 4.28   Edge of plain slab raft in frost-susceptible soil

Fig. 4.29   Plain slab raft forming the ground floor of a building.

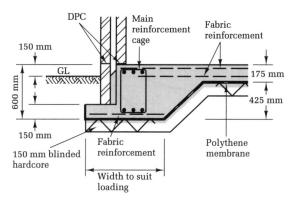

Fig. 4.30   Stiffened edge raft with integral projecting toe.

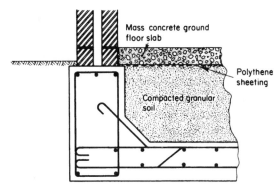

Fig. 4.32   Slab and beam raft with upstand beam.

concealing the exposed edge of the slab the raft slab can also form the ground floor (Fig. 4.29). Precautions are then necessary against rain driving beneath the walls or the external cladding of the structure. The effects of external temperature on the outer exposed edges of the raft need to be considered, particularly in cold climates. These effects may also influence the depth and amount of projection of the slab in the raft designs shown in Figs 4.27 and 4.28.

The *stiffened edge raft* (Fig. 4.30) is suitable for the foundations of houses up to four storeys on weak compressible soils or loose granular fill materials. The ground-floor slab is made an integral part of the raft, and by stepping down the peripheral part of the slab a cut-off is provided against the ingress of water to the ground floor. A stiff beam can be formed integrally with the stepped-down portion of the raft and its projecting toe.

The *slab and beam raft* is used as a foundation for heavy buildings where stiffness is the principal requirement to avoid excessive distortion of the superstructure as a result of variations in the load distribution over the raft

or in the compressibility of the supporting soil. These stiff rafts can be designed either as 'downstand-beam' (Fig. 4.31) or 'upstand-beam' types (Fig. 4.32). The downstand-beam raft has the advantage of providing a level surface slab which can form the ground floor of the structure. However, it is necessary to construct the beams in a trench which can cause difficulties in soft or loose ground when the soil requires continuous support by sheeting and strutting. Construction may also be difficult when the trenches are excavated in water-bearing soil requiring additional space in the excavation for subsoil drainage and sumps. The downstand-beam raft is suitable for firm to stiff clays which can stand for limited periods either without any lateral support or with widely spaced timbering.

The upstand-beam design ensures that the beams are constructed in clean dry conditions above the base slab. Where excavation for the base slab has to be undertaken in water-bearing soil it is easier to deal with the water in a large open excavation than in the confines of narrow trenches. However, the upstand-beam design requires the provision of an upper slab to form the ground floor of the structure. This involves the construction and removal of soffit formwork for the upper slab, or the alternative of filling the spaces between the beams with granular material to provide a surface on which the slab can be cast. Precast concrete slabs can be used for the top decking, but these require the addition of an *in-situ* concrete screed to receive the floor finish.

Basement and buoyancy rafts will be described in Chapter 5.

4.5.2 Structural design

The structural design of rafts is a problem of even greater complexity than continuous beam foundations. It is wrong in principle to assume that a raft acts as an inverted floor

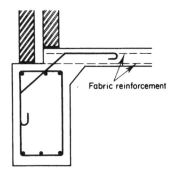

Fig. 4.31   Slab and beam raft with downstand beam.

slab on unyielding supports, and to design the slab on the assumption that its whole area is loaded to the maximum safe bearing pressure on the soil can lead to wasteful and sometimes dangerous designs. Allowance must be made for deflexions under the most unfavourable combinations of dead and imposed loading and variations in soil compressibility. Guidance is required from the soil mechanics engineer on the estimated total and differential settlements for dead and imposed loading considered separately.

The design of the raft is a compromise between the desire to keep the differential settlement of the raft, and hence the superstructure, to a minimum, and the need to avoid excessive stiffness in the raft structure. Flexibility in a raft gives minimum bending moments and hence economy in the substructure, but at the expense of relatively large differential settlement with additional costs to accommodate these movements in the superstructure, such as by joints or a flexible cladding. Stiffness in a raft minimizes differential settlement, but it redistributes load by increasing bending moments thus increasing the cost of the raft.

The stiffness of the structure relative to the raft can be expressed by the following equation:

$$K_r = \frac{4E_c(1 - v_s^2)}{3E_s(1 - v_c^2)}\left(\frac{t}{B}\right)^3 \qquad (4.14)$$

where

K_r = relative stiffness,
E_c = Young's modulus of the raft,
E_s = Young's modulus of the soil,
v_s = Poisson's ratio of the soil,
v_c = Poisson's ratio of the concrete,
t = thickness of the raft,
B = breadth of rectangular raft.

The above equation can be used to prepare design charts relating total and differential settlements and bending moments to the K_r factor. A chart for determining the settlement of a uniformly loaded square raft on a deep uniformly elastic layer is shown in Fig. 4.33. The chart gives values of the influence factor I for substituting in equation (2.41). The subscripts for I in Fig. 4.33 can be used to determine the total settlements at the points A, B, and C and differential settlements between A and B and A and C. Frazer and Wardle[4.3] have prepared additional charts giving correction factors to the calculated settlement for rectangular rafts using the ratio l/b, and for compressible layers of finite thickness from the ratio d/b.

Charts of this type are discussed in a report by the Institution of Structural Engineers.[2.19] The report points out that they are of limited practical use because they are

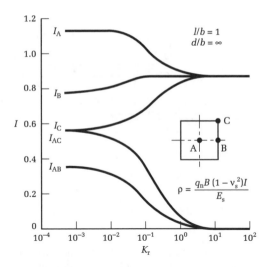

Fig. 4.33 Chart for calculating total and differential settlement of a square raft on a uniformly elastic soil of infinite depth (after Frazer and Wardle[4.3]).

generally prepared for simple circular or rectangular uniformly loaded rafts on uniform soils, whereas most practical foundation problems are concerned with irregular loadings on soils of varying compressibility. For this reason the use of finite element solutions of the types described in Section 2.8.3 are recommended.

4.5.3 Raft to withstand large settlements on made ground

A raft of a type similar to Fig. 4.34 was used to support three-storey blocks of flats of load-bearing wall construction on fill at Bilston, Staffordshire. The depth and consistency of the fill were such that piled foundations were out of the question on grounds of cost. However, since the fill varied in composition from a soft clay to a compact mass of hard slag boulders, the raft had to be designed to be stiff enough to resist a tendency towards considerable differential settlement, even though the loading at the base of the walls was only about 30–45 kN/m run. Stiffness was achieved by means of beams beneath and longitudinal and cross walls and fabric reinforcement in the top and bottom of the 254 mm thick slab. The beams were placed at the bottom of the slab to enable the latter to form the ground floor of the flats. The total dead load of the superstructure was equivalent to a uniformly distributed load over the raft of 60 kN/m², and the total design dead and live load was 64 kN/m². In one of the blocks the raft withstood a total settlement of 100 mm and a differential settlement of 65 mm without distress to the foundations or superstructure.

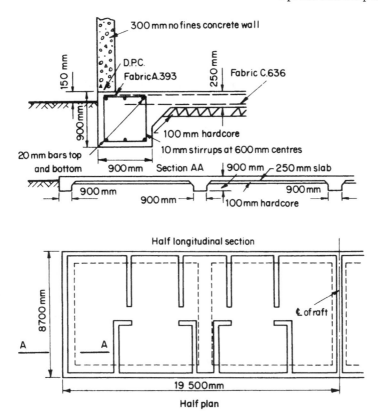

Fig. 4.34 Raft foundation for three-storey houses on filled ground.

4.6 Deep shaft foundations

4.6.1 Bearing capacity and settlements

The function of a deep shaft foundation is to enable structural loads to be taken down through deep layers of weak soil on to a stratum of stiff soil or rock which provides adequate support in end bearing, including the high edge pressures induced by lateral loads acting in combination with the bearing pressures from vertical loading. Deep shaft foundations are similar, in principle, to piled foundations, the main difference being in the method of construction.

The ultimate bearing capacity of the soil beneath the foundation base is calculated by the methods described in Sections 2.2–2.4 with particular reference to the use of partial factors in the separate evaluation of skin friction and end bearing resistance (Section 2.3.5). However, there are some situations for which the contribution of skin friction to bearing capacity should be ignored. These include:

(a) If the depth of the foundation is less than its least width.

(b) If the ground above founding level is liable to be scoured away.

(c) If backfilling between the stem of the foundation and the sides of the excavation is uncompacted and can subside causing dragdown on the stem.

(d) If loading on the ground surface or flooding can cause compression and consolidation of the upper soil layers resulting in dragdown on the surface of the foundation in contact with the soil.

(e) If the soil surrounding the sides of the foundation is liable to shrink due to drying action thus causing a gap around the foundation.

Methods of evaluating dragdown forces referred to in (d) above are discussed in Section 7.13. Where it is permissible to utilize skin friction in support of the foundation the ultimate values are calculated in the same way as for bored piles in Sections 7.6, 7.9, and 7.12. Due allowance should be made for the effects of disturbance caused by excavation, and of exposure of the soil over the excavation period which is usually much longer than with bored pile installation. The walls of a deep excavation may be exposed to water flowing down to the bottom of the shaft which can cause disintegration or complete softening of the soil or rock.

Settlements of deep shaft foundations are calculated using the methods described in Section 2.6. Where skin friction on the sides can be allowed the settlement is calculated in the same way as for piles (Section 7.11), although in most cases a high proportion of the settlement is due to compression and consolidation of the ground beneath the base. Depth correction factors are applied to settlements calculated for a foundation on the ground surface to allow for the effect of deep embedment. The depth factors of Fox (Fig. 2.39) apply to foundations where the concrete is cast against the ground, that is, when the load is applied the ground against the sides of the foundation does not deform simultaneously with deformation beneath the base. Burland[4.4] has pointed out that the Fox factors over-correct the surface settlement for the case where the foundation is at the base of an unlined shaft (or

a timbered shaft where the sides can deform). Burland's factors for various values of Poisson's ratio are shown in Fig. 4.35. The Fox correction factors for a Poisson's ratio of 0.5 are shown for comparison. The depth factors of Meigh for foundations in rock (Fig. 2.45) are roughly the same as those of Burland for a Poisson's ratio of 0.25.

Where deep shaft foundations carry combined vertical and lateral loading the magnitude of the edge pressures is calculated by the methods given in Sections 4.1.3 and 4.2.4. For the case of foundations where the concrete filling is cast against the soil up to or near the ground surface, or where the foundation consists of a lined shaft backfilled with concrete, the tendency to tilt under lateral loading is resisted by the passive resistance of the ground. This reduces the magnitude of the edge pressures. Calculations to determine the passive resistance of the soil to horizontal loading are discussed in Section 5.7.1. The degree of horizontal movement to mobilize passive resistance should be noted. This makes it impossible to eliminate tilting completely. Also the effects of exposure of the soil during excavation should be considered when selecting the parameters for soil shear strength and compressibility. The resistance to overturning of deeply embedded shafts can be determined by the Brinch Hansen method used for piled foundations (Section 7.17.1).

4.6.2 Construction by hand excavation

Hand excavation methods can be economical for constructing deep shaft foundations for building projects where relatively few heavily loaded columns are to be supported and where access conditions make it difficult or impossible to deploy equipment for drilling large-diameter bored piles. Hand excavation is also used for ground conditions where obstructions in the form of large boulders, tree trunks or man-made objects such as buried car bodies prevent the use of mechanical plant. The advantage of this system for difficult access conditions is illustrated by its use for the foundations of a new nine-storey building over Charing Cross Station in London.[4.5, 4.6] A requirement for the project was that the foundations for the new building should be completely independent of the old brick vaults supporting the rail tracks. This necessitated cutting holes through the crowns of the vaults to receive the new building columns which were in turn supported by deep shaft foundations with enlarged bases. The vaults were underpinned by raking small-diameter piles before shaft sinking commenced (Fig. 4.36).

Column loads varied between 15 and 27 MN and this was provided for by adopting a standard shaft diameter of 2.5 m and varying the base diameter from 4.5 to 6.6 m. The foundations were constructed by initially sinking the shaft as a caisson through about 10 m of fill and water-

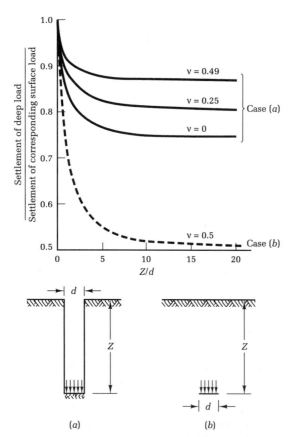

Fig. 4.35 Depth correction factors for circular foundation at base of deep unlined shaft. (*a*) Uniform circular load at base of unlined shaft. (*b*) Uniform circular load within semi-infinite solid (case treated by Fox) (after Burland[4.4]).

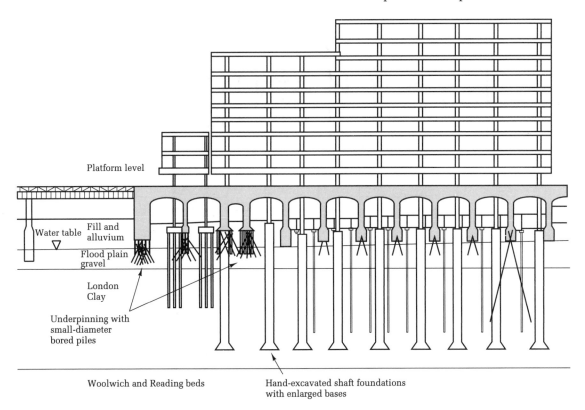

Platform level

Water table ▽

Fill and alluvium

Flood plain gravel

London Clay

Underpinning with small-diameter bored piles

Woolwich and Reading beds

Hand-excavated shaft foundations with enlarged bases

Fig. 4.36 Deep foundations and underpinning at Embankment Place, London (after Barrie and Weston[4.5]).

bearing sand and gravel. Sinking was assisted by jacking from a collar surrounding the top of the shaft with bentonite lubrication of the shaft–soil interface. The caisson was then sealed into the top of the London Clay and the remaining depth was constructed by underpinning methods using precast concrete segments as described in Section 4.6.3 (Fig. 4.37). Steel trench sheets or timber baulks were used to support the roof of the base enlargements as shown in Fig. 4.38. A single foundation required $3-3\frac{1}{2}$ weeks of double-shift working, but it should be noted that this replaced 15 or more 600 mm bored piles which were the largest which could be installed by tripod rigs working in low headroom conditions.

Hand excavation makes it possible to inspect the rock closely at foundation level, reassessments can be made of allowable bearing pressures, and where necessary plate loading tests can be made on the exposed rocks to obtain values of deformation modulus as a check on the parameters used at the initial design stage. The facility for visual examination of rock conditions at founding level was used to advantage in the lower part of the shaft foundations supporting the reactor containment vessels for the Hartlepool nuclear power station.[4.7] The vessels each

weighed 4500 MN and they were supported by 17 circular piers with a shaft diameter of 2.3 m belled out at the base to 3.9 m to limit the bearing pressure on the weak Bunter Sandstone at founding level to 2.9 MN/m². The main portions of the shaft were installed by rotary auger drilling under a bentonite slurry through 5 m of sand fill and soft alluvial clay, and 30 m of glacial till to the maximum depth into rock which could be achieved by the auger. At this stage a 6 m long lining tube of 2.7 m diameter was used to support the upper part of the shaft excavation. An inner lining tube 2.3 m in diameter was then lowered into the slurry-filled hole and driven into the rock to form a seal. This enabled the bentonite to be pumped out and the annulus between the lining tube and the excavation was filled with cement grout. The excavation was then taken out by hand and belled out at the base at a founding level 39 m below ground level. Where the flow of water into the unlined base was less than 23 litre/min the concrete was placed in the dry. Where higher inflows were measured the shaft was flooded and concrete placed by tremie pipe up to the level of the shaft liner when the remaining water was pumped out and concrete placed in dry conditions. It was necessary to drill 50 mm holes at the base of some

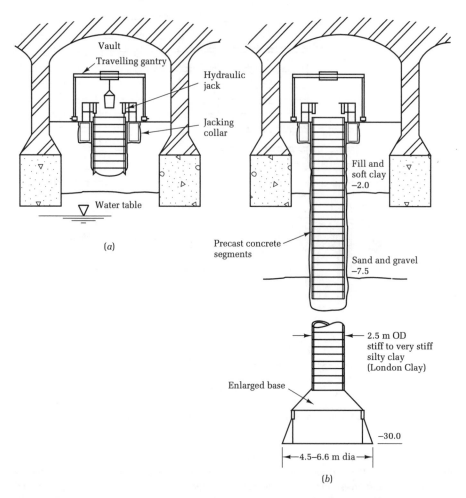

Fig. 4.37   Construction of deep shaft foundations at Embankment Place, London (after Grose[4.6]). (*a*) Upper section of lining sunk as a caisson. (*b*) Lower section below sand gravel sunk by underpinning precast concrete segments.

shafts to relieve excess water pressure in the rock, thereby preventing uplift of the base of the shaft.

Timber or steel plank sheeting is generally used for supporting shaft excavations of small to moderate depth. Runners in stages are used in loose ground, and the middle board system (Chapter 9) where the ground will stand unsupported for a metre or so of depth while the timbering is set in position and strutted. In the USA the 'Chicago' method of timbering is used for deep shafts. A circular hole is taken out for the depth to which the soil will stand unsupported, i.e. about 0.5 m for soft clays to 2 m for stiff clays. Vertical boards are set in position around the excavated face and held tightly against the soil by two or more steel rings. The shaft is then deepened for 1–2 m and another setting of boards and rings is placed. This process

is continued until founding level is reached when the base of the shaft is belled out if the soil is sufficiently stable.

Where the ground is sufficiently stable to stand unsupported to depths of 0.6–1.2 m a shaft lined with concrete cast-in-place can be an economical method of construction. The method is used quite widely in Hong Kong,[4.8] where hard granite boulders in the alluvium overlying rock or in the form of corestones in decomposed granite are common occurrences. These boulders form obstructions to piling installed by rotary mechanical augers or to diaphragm walling or barrettes installed by grabbing under a bentonite slurry. In Hong Kong shafts (referred to in that country as 'caissons') of 1–3 m diameter are excavated by hand in stages of 0.6–1.0 m. A tapered circular form is placed at the bottom of each stage of excavation and the

Fig. 4.38 Hand excavation for enlarged base to deep shaft, Embankment Place, London.

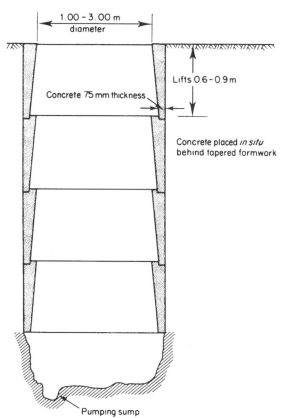

Fig. 4.39 Hand-excavated shaft foundation with cast-in-place concrete lining as used in Hong Kong.

unreinforced concrete lining is placed to give a minimum thickness of about 75 mm (Fig. 4.39). Each lift is dowelled to that above by 12 mm bars at 300 mm centres. There is a considerable risk factor (including health risks) in the method. Accidents have occurred due to objects or operatives falling down the shaft, collapses of the sides, flooding, electrocution, asphyxiation, and the constant inhaling of dust. Safety measures must be stringently observed and enforced.

The facility provided by hand-dug shafts to enable the ground at deep founding levels to be examined closely and tested is exemplified by the foundations for the 50-storey DBS building, Tower One, in Singapore. The building is located on ground which had previously been reclaimed from the waterfront in Singapore City. Preliminary boreholes showed a steeply sloping highly weathered sandstone and siltstone rock surface overlain by 5–30 m of reclamation fill and soft marine clay. The interbedded sandstones and siltstones were highly weathered and unsuitable for carrying the heavy concentrated loading from the tower superstructure which was supported by four columns each transmitting a loading of 4.5 MN to the basement at 8.5 m below street level (Fig. 4.40).

The sandstones and siltstones were underlain by partly weathered and partly unweathered closely jointed mudstone, dipping steeply across the site at angles between 45 and 50°. Examination of rock cores and compression tests on core specimens showed that the mudstone could provide a suitable founding stratum if the settlements of the heavily loaded foundations could be shown to be within acceptable limits. The consulting engineers to the Development Bank of Singapore Ltd, Steen Consultants, decided to adopt 7.3 m diameter shaft foundations taken down below the interface between the two rock formations to a level at which the mudstone could be inspected and subjected to loading tests on a 1 m diameter plate to determine the deformation modulus. The tests showed acceptable E-values of 1170 and 1460 MN/m^2. The shallowest and deepest founding levels are shown in Fig. 4.40.

The portions of the shafts within the fill and soft clay were excavated within a circular cofferdam constructed from contiguous bored piles (Fig. 4.41). Below this level

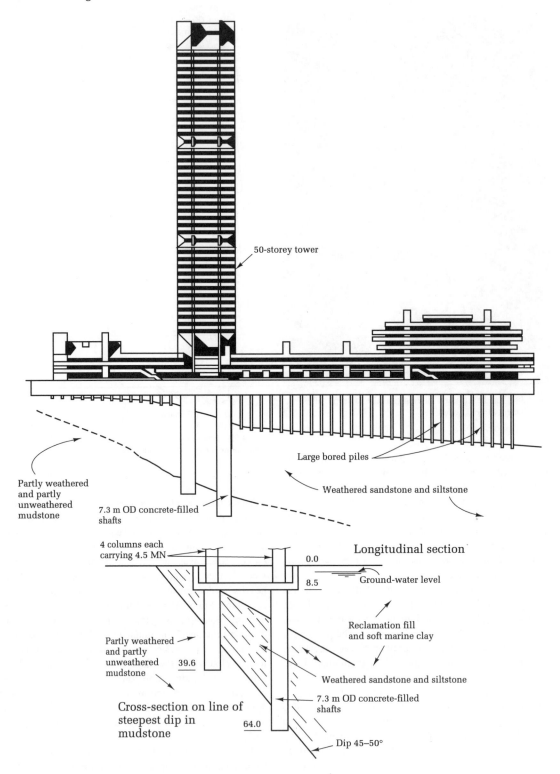

50-storey tower

Large bored piles

Weathered sandstone and siltstone

Partly weathered and partly unweathered mudstone

7.3 m OD concrete-filled shafts

4 columns each carrying 4.5 MN

0.0

Longitudinal section

8.5

Ground-water level

Reclamation fill and soft marine clay

Partly weathered and partly unweathered mudstone

39.6

Weathered sandstone and siltstone

7.3 m OD concrete-filled shafts

Cross-section on line of steepest dip in mudstone

64.0

Dip 45–50°

Fig. 4.40 Deep shaft foundations for the Development Bank of Singapore Ltd Building, Singapore.

Fig. 4.41 Upper section of the shaft foundations for the DBS building supported by contiguous bored piles.

the shaft was lined with mass concrete 450 mm thick, cast in lifts of 1.2 m using sloping formwork to produce a section similar to that shown in Fig. 4.39. The completed shafts were backfilled with mass concrete. As well as facilitating inspection and testing of the rock the shaft foundations were preferred to piling because they avoided problems of congestion when sinking a large number of piles in a confined site, the possibility of close-spaced piles moving off line when augering through variably weathered rock, and difficulties with drilling through occasional stronger bands of weathered sandstone interbedded with the weak siltstones.

A later and equally notable example of hand-excavated deep shaft foundations was their use for the support of the 47-storey headquarters building of the Hong Kong and Shanghai Banking Corporation, constructed in Hong Kong in 1981–85.[4.9] The floors of the building are suspended from two rows of four masts each consisting of four tubular steel columns. The structure was designed by

Ove Arup and Partners and their original plan of construction for the mast foundations was for each column to be supported by a single 2.5–3.5 m diameter shaft foundation sunk by hand excavation from basement level at 16–20 m below street level to rock at 25–40 m below this level. However, in order to accelerate the construction programme it was decided to sink a 10 m diameter access shaft at each mast position from a higher level within the partially excavated basement. The access shafts were sunk by a combination of hand and machine excavation to depths between 8 and 14 m. The four shafts were then commenced and taken down to the relatively fresh granite and enlarged to base diameters of 3.5–4.1 m, with rock anchors at the base to provide resistance to possible uplift conditions from wind loading on the 180 m high tower.

The ground floor and the three intermediate basement floors were supported by 58 columns bearing on 2.1 m diameter shafts excavated to rock. The management contractor, John Lok-Wimpey,[4.10] pointed out the much greater speed at which the shaft foundations were con-structed compared with what could have been achieved by machine-installed piles. It was found possible to excavate, construct linings, and backfill with concrete in 90 shafts concurrently on a congested site with a minimum of noise and with 24-hour working.

4.6.3 Segmental-lined shafts

In some of the methods of shaft sinking described above, there is a risk of settlement of the ground surface around the excavation caused by yielding or 'draw'. This is due to inward deflexion of the sheeting or of the unsupported face before the support is placed. Settlement of the ground surface may endanger adjacent structures or underground services such as water mains and sewers. The movement can be greatly reduced, if not totally eliminated, by lining the shaft with cast-iron or concrete segments. By this method only a small area of ground is exposed at any time and, because of the arching of the soil, the circular form of the segmental-lined shaft itself is resistant to inward yielding. Pressure grouting with cement behind the seg-ments is undertaken to fill in gaps and to reconsolidate any loosened ground.

In normal ground, segmental-lined shafts are sunk by the underpinning method, i.e. the first two or three rings of segments are assembled and bolted up in an open excava-tion. The excavation is backfilled and the segments anchored against sinking by grouting or by a suitable collar. Excavation then continues at the bottom of the shaft, but only sufficient ground is taken out for assembly of the complete ring before lifting and bolting to the one above. In poor ground the excavated area can be limited to the amount sufficient to place and bolt one segment to the ring above. After completion of the assembly and bolting, the dumpling at the centre is removed (Fig. 4.42), fol-lowed by pressure grouting behind the segments. Excava-tion is then commenced for the next ring and so on until

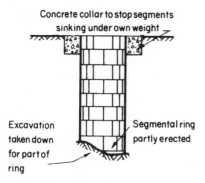

Fig. 4.42 Constructing shaft for pier with segmental lining by underpinning.

foundation level is reached. It is essential to grout behind the rings at frequent intervals during shaft sinking. If grouting is unduly delayed the dragdown forces on the back of the segments will cause excessive tension to develop around the ring leading to fracture of the bolts or circumferential cracking of the segments.

With all methods of shaft sinking by hand excavation, or where operatives are required to descend into machine-dug shafts, strict attention to safety precautions is required following the recommendations in the British Standard Code of Practice BS 5573: *Safety Precautions in the Construction of Large-diameter Boreholes for Piling or other Purposes.*

4.6.4 Barrette foundations

Deep pier foundations can be constructed using the dia-phragm walling techniques described in Section 5.4.4. The foundations are in the form of short lengths of wall which can be simple rectangular shapes or cruciform or L-shapes to suit the pattern of loading or the layout of the columns supported by the foundations. Barrettes are an alternative to the large-diameter bored and cast-in-place piles des-cribed in Chapter 8 and are useful in cases where high lateral loads or bending moments are carried by the foundations. They can be economical where the specialist equipment has to be brought to a site for the construction of basement retaining walls by diaphragm walling tech-niques. Where heavy loads are carried in end-bearing, particular care is necessary to avoid the entrapment of contaminated bentonite slurry beneath the base of the barrettes. It should also be noted that, because of the complex shapes, barrettes can take a considerably longer time to construct than simple rectangular panels. This increases the risk of collapse of the bentonite-supported soil, particularly where deep foundations are constructed in a stiff-fissured clay.

4.7 Examples on Chapter 4

Example 4.1 An isolated column carries a dead load of 400 kN and an imposed load of 200 kN. It is founded at a depth of 0.9 m in a 2 m thick stiff clay stratum having a characteristic shear strength of 130 kN/m^2 overlying a deep layer of firm clay having a characteristic shear strength of 70 kN/m^2. Determine the required size of the column foundation.

Using permissible stress methods the approximate bear-ing capacity factor N_c for the square column base is 7.5.

Therefore for a safety factor of 3, the presumed bearing value is given by

$$q_{nf} = \frac{7.5 \times 130}{3} = 325 \text{ kN/m}^2.$$

Required size of square column base $= \sqrt{600/325}$
$$= 1.36, \text{ say } \underline{1.4\,\text{m}}.$$

Checking the assumed value of N_c,

$$\frac{D}{B} = \frac{0.9}{1.4} = 0.64,$$

from Fig. 2.18. $N_c = 7.3$, which is close enough to our chosen value.

For a 1.0 m thick base, total load on soil $= 600 + 25 \times 1.4^2 \times 1 = 649\,\text{kN}$.

Net foundation pressure $= q_n = 649/1.4^2 - 19 \times 0.9 = 314\,\text{kN/m}^2$.

From equation (4.1);

$$\text{Bearing pressure on firm clay} = 314\left(\frac{1.4}{1.4 + 1.1}\right)^2$$
$$= 98\,\text{kN/m}^2.$$

For $D/B = 2/2.5 = 0.8$, Fig. 2.18 gives $N_c = 7.4$. Therefore net ultimate bearing capacity of firm clay $= 7.4 \times 70 = 518\,\text{kN/m}^2$.

Factor of safety $= 518/98 = 5.3$, which is satisfactory.

Checking by limit state methods and assuming that the nominal base dimensions of 1.4 m square might be under-dug to say 1.35 m square and over-dug by 0.05 m at the base:

Factored load from column $= 400 + 200 \times 1.3$
$$= 660\,\text{kN}$$
Load from column base $= 25 \times 1.35^2 \times 1.05 \times 1$
$$= \underline{48}\,\text{kN}$$
Total $= 660 + 48 = 708\,\text{kN}$

Weight of soil removed $= 19 \times 1.35^2 \times 1.05 \times 1$
$$= 36\,\text{kN}$$
Net design load $\quad= 708 - 36 = \underline{672}$

Design shear strength of upper stiff clay $= 130/1.8$
$$= 72\,\text{kN/m}^2$$
Design shear strength of lower firm clay $= 70/1.8$
$$= 39\,\text{kN/m}^2$$

As noted in Chapter 2 the Eurocode 7 partial factors assume that the Brinch Hansen equation is the basis for bearing capacity calculations. From Figs 2.8 and 2.11, $N_c = 5.2$, and for $D/B = 0.9/1.35$ (most unfavourable), $d_c = 1.15$, and $s_c = 1.3$. Ultimate limit state load is

$$5.2 \times 1.3 \times 1.15 \times 72 \times 1.35^2 = 1020\,\text{kN}.$$

Eurocode 7 requires the most unfavourable assumptions of geometrical data (including soil levels) where these are critical. Therefore take the shallowest level of the firm

clay to be 1.8 m below ground level, giving $d = 1.8 - 1.05 = 0.75\,\text{m}$ in Fig. 4.2, and width of equivalent foundation on firm clay $= 1.35 + 0.75 = 2.10\,\text{m}$.

For $D/B = 1.8/1.35 = 1.33$, $d_c = 1.25$ and $s_c = 1.3$.
Ultimate limit state load $= 5.2 \times 1.25 \times 1.3 \times 39$
$$\times 2.1^2$$
$$= 1453\,\text{kN}.$$

which again does not exceed the design load and a reduction of the nominal base dimensions to say 1.2 m square might be considered subject to checking the serviceability limit state.

Example 4.2   In the above example, the columns are spaced at 2.0 m centres in a single row. Determine the required size of the foundation for these conditions. Assume that the stiff and firm clay layers are part of the same formation with a plasticity index of 40.

The pressure distribution in Fig. 4.43 shows that there is overlapping of bearing pressures on the surface of the firm stratum. From equation (4.1) for a single column base, and 1 : 2 spread of load we have that the net pressure on surface of buried stratum of firm clay is given by

$$q' = 314\left(\frac{1.4}{1.4 + 1.1}\right)^2 = 98\,\text{kN/m}^2.$$

Where the pressures from adjoining columns overlap, the total pressure on the surface of the firm clay layer $= 196\,\text{kN/m}^2$. For the purpose of calculating the presumed bearing value of the firm clay, we must assume a strip footing 2.5 m wide on the surface of the clay at a depth of 2.0 m below ground level.

For $D/B = 2.0/2.5 = 0.8$, the value of N_c from Fig. 2.18 is 6.2. Therefore, for a safety factor of 3, the presumed bearing value of the firm clay stratum is

$$q_{nf} = \frac{6.2 \times 70}{3} = 145\,\text{kN/m}^2.$$

Thus where the pressure distributed by the pad foundation overlaps on the surface of the firm clay layer, the maximum safe bearing capacity is exceeded. The best expedient therefore will be to combine the column bases into a strip foundation.

Load on strip foundation $= 615/2.0 = 308\,\text{kN/m run}$.

The bearing capacity factor for the strip foundation at a depth of 0.9 m will be about 6.

Maximum net safe bearing capacity of stiff clay is given by

$$6/3 \times 130 = \underline{260\,\text{kN/m}^2}.$$

Required width of foundation $= 308/260 = \underline{1.2\,\text{m}}$

For $D/B = 0.9/1.2 = 0.75$, the value of N_c from Fig. 2.18 is 6.2. Using this higher factor, the presumed bearing value is given by

$$q_{nf} = 6.2/3 \times 130 = 269 \, \text{kN/m}^2$$

and so

Required width of foundation = $308/269$ = <u>1.15 m.</u>

From equation (4.2), the pressure on surface of firm clay stratum is

$$q_1 = \frac{308}{1.15}\left(\frac{1.15}{1.15 + 1.10}\right) = 137 \, \text{kN/m}^2.$$

From above we know that a bearing pressure of 145 kN/m² on an equivalent 2.5 m wide strip has a safety factor of 3. A bearing pressure of 137 kN/m² on a narrow strip will have a higher safety factor.

Thus a foundation width of 1.15 m should not give excessive settlement, but it will be necessary to reinforce the strip to minimize differential settlement, and if a total settlement of say 75 mm and a differential settlement of 25 mm is likely to be detrimental to the structure, it would be advisable to make a settlement analysis to obtain a closer estimate of the settlement before revising the foundation width.

Checking by limit state methods, and assuming a strip width as constructed of 1.10 m

Net design load on strip foundation = $672/2.0$
$$= 336 \, \text{kN/m run.}$$

Considering stiff clay layer:

For strip foundation $N_c = 5.2$, $s_c = 1.0$, and for D/B
$$= 0.9/1.1, d_c = 1.15.$$

Ultimate design load = $5.2 \times 1 \times 1.15 \times 72 \times 1.1 \times 1$
$$= 474 \, \text{kN/m run.}$$

Considering firm clay layer, width of equivalent strip
$$= 1.1 + 0.75 = 1.85 \, \text{m}$$

For $D/B = 1.8/1.85 = 1$, $d_c = 1.2$

Ultimate design load = $5.2 \times 1 \times 1.2 \times 39 \times 1.8 \times 1$
$$= 438 \, \text{kN/m run.}$$

Hence the design load is not exceeded either for the stiff clay or the firm clay layer, but it will be necessary to check the serviceability limit state. The factors in Fig. 2.37 will be suitable for this purpose.

From Fig. 2.33, $E_u = 450c_u$.   For $H/B = 1.1/1.1$
$$= 1 \text{ and } L/B = 10.$$

$\mu_1 = 0.35$,   for $D/B = 0.8$, $\mu_o = 0.92$.

Bearing pressure on stiff clay for unfactored load is

$$\frac{600 + 48 - 36}{2 \times 1.1} = 278 \, \text{kN/m}^2$$

Immediate settlement =

$$\frac{0.35 \times 0.92 \times 278 \times 1.1 \times 1000}{450 \times 130} = 2 \, \text{mm.}$$

Bearing pressure on firm clay = $612/2 \times 2.2$
$$= 139 \, \text{kN/m}^2$$

for H/B, say, 50 and $L/B = 10$, $\mu_1 = 1.35$, $\mu_o = 0.92$

Immediate settlement =

$$\frac{1.35 \times 0.92 \times 139 \times 2.2 \times 1000}{450 \times 70} = 12 \, \text{mm.}$$

Total immediate settlement = $2 + 12 = 14$ mm,
say 15 mm

Total long-term settlement   = 15 mm
Total immediate +
long-term settlement   = 30 mm

A maximum settlement of say 40 mm and a minimum of 20 mm should not give excessive differential settlement between the centre and ends of the strip foundation.

Example 4.3   A total of 64 columns as in Example 4.2 are spaced on a grid at 2 m centres both ways. Investigate the requirements for foundation design.

We know from Example 4.2 that pad footings cannot be used; if we use rows of strip footings running in one direction only the pressure distribution on the surface of the firm clay stratum is similar to that shown in Fig. 4.43. Thus the stiff clay stratum acts as a large raft, and we must consider the stress distribution below this raft in relation to the shear strength of the firm clay.

Approximate average loading on surface of firm clay is given by

$$\frac{600}{2^2} = 150 \, \text{kN/m}^2.$$

Considering the stiff clay stratum to act as a surface foundation, from Fig. 2.18, ultimate bearing capacity of firm clay is given by

$$= 6.1 \times 70 = 427 \, \text{kN/m}^2.$$

This gives a safety factor of 2.8 and since the firm clay layer extends to some depth, it is evident that there will be appreciable settlement. It will therefore be advisable to make a settlement analysis, and if the total and differential settlements are excessive, it will be necessary to provide piled foundations to the columns taken down to a stiffer stratum beneath the firm clay.

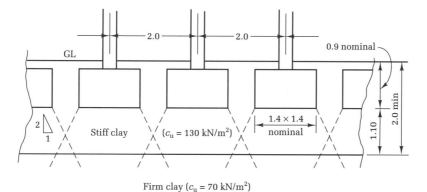

Fig. 4.43

Example 4.4   A structure is supported on widely spaced reinforced concrete columns 610 mm square, which carry design dead and imposed loads of 1100 and 700 kN, respectively, totalling 1800 kN. Borings and static cone penetration tests (Fig. 4.44(a)–(d)) show that the ground is a fairly loose becoming a medium-dense medium sand. Determine the depth and dimensions of square pad foundations and design the reinforcement.

Because of the variability in density of the sand deposits it will be necessary to design the foundations for the loosest conditions. The calculated allowable bearing pressure can then be used to compare settlements for the loosest and densest conditions, and hence to obtain the maximum differential settlements between any pair of columns. Taking the cone penetration test for the loosest conditions (Fig. 4.44), this can be simplified to give average cone resistance of 3.5, 9.0 and 15 MN/m² in three layers.

A first guide to the allowable bearing pressure is given by Table 2.5. From the relationship between q_c and the SPT N-value in Fig. 1.7, N for a medium sand can vary between about 1 and 3 times q giving N-values in the range 3.5–10. Taking an average of, say, 6, Table 2.5 gives a presumed bearing value of 45 kN/m² for a 4 m wide strip foundation or 90 kN/m² for a square pad. This requires a foundation approximately 4.40 m square for 1800 kN column load.

A foundation depth of 1.20 m below ground level will be assumed for the purpose of estimating settlements. This is deeper than is required purely from soil considerations. However, a pad of substantial thickness may be required and it is necessary to keep the pad wholly below the ground floor of the structure. Therefore, it will not alter the estimated settlements very much if the final foundation depth is made 0.5 m or so above or below the selected 1.20 m depth.

The settlements for the loosest conditions will be calculated by Schmertmann's method, for which the depth correction and creep factors by equations (2.37) and (2.38) are

$$C_1 = 1 - \frac{0.5 \times 19 \times 1.2}{95} = 0.88$$

and

$$C_2 = 1 + 0.2 \log_{10} \left(\frac{25}{0.1} \right) = 1.48 \text{ for 25 years}$$

For layer 1, from Fig. 4.44
$$I_z \text{ at centre of layer} = 0.30,$$
$$E_d = 2.5 \times 3.5 = 8.75 \text{ MN/m}^2.$$

Therefore,
$$\rho = 0.88 \times 1.48 \times 90 \times \frac{0.30}{8.75 \times 1000} \times 2.3$$
$$\times 1000 = 9.3 \text{ mm}.$$

Similarly, in layer 2,
$$I_z = 0.43,$$
$$E_d = 2.5 \times 9 = 22.5 \text{ MN/m}^2, \text{ and so } \rho = 3.4 \text{ mm}.$$

And in layer 3
$$I_z = 0.19,$$
$$E_d = 2.5 \times 15 = 37.5 \text{ MN/m}^2, \text{ and so } \rho = 3.0 \text{ mm}.$$

Total settlement for loosest layers = 9.3 + 3.4 + 3.0 = 15.7 mm.

To calculate the settlement for the densest conditions, Layer 1 is divided into two as shown.

In layer 1A
$$I_z = 0.30,$$
$$E_d = 2.5 \times 5.8 = 14.5 \text{ MN/m}^2, \text{ and so } \rho = 5.9 \text{ mm}.$$

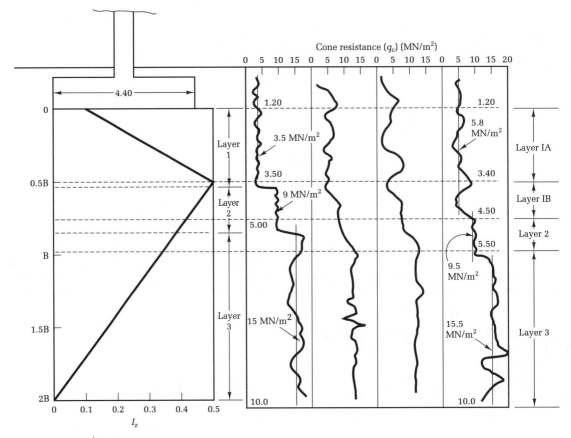

Fig. 4.44   Static cone penetration tests.

In layer 1B

$$I_z = 0.45,$$
$$E_d = 2.5 \times 5.8 = 14.5 \, \text{MN/m}^2, \text{ and so } \rho = 4.0 \, \text{mm}.$$

In layer 2

$$I_z = 0.37,$$
$$E_d = 2.5 \times 9.5 = 23.75 \, \text{MN/m}^2, \text{ and so } \rho = 1.8 \, \text{mm}.$$

In layer 3

$$I_z = 0.17,$$
$$E_d = 2.5 \times 15.5 = 38.75 \, \text{MN/m}^2, \text{ and so } \rho = 2.3 \, \text{mm}.$$

Therefore settlement for densest layers = 5.3 + 4.0 + 1.8 + 2.3 = <u>13.4 mm</u>

The differential settlement between columns founded on loosest and densest soils of 2 mm is negligible.

Structural design of pad foundations.   In view of the large dimensions of the pad it will save in quantities of excavation and concrete if they are designed in reinforced concrete.

Taking partial safety factors from BS 8110

Total ultimate load = 1.4 × 1100 + 1.6 × 700
                    = 2660 kN.

The design is set out below in the form of the output of a computer program which mimics the successive stages of a 'longhand' calculation.

(The program used is part of the SAND system developed by Fitzroy Systems of Cobham, Surrey.)

Calculations using provisions of B8110 3.11 and 3.4.4.4, fy = 460 N/mm2

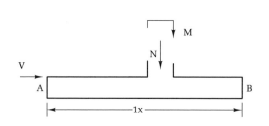

Ultimate axial force in column	N=2660 kN
Ultimate moment in column	M=0 kNm
Ultimate shear force top of base	v=0 kN
Column dimension	cx=610 mm
Column dimension	cy=610 mm
Length of base (bending dirn)	lx=4.4 m
Width of base	ly=4.4 m
Left distance to edge of column	la=1.895 m
Right distance to edge of column	lb=lx−la−cx/1000
	=4.4−1.895−610/1000
	=1.895 m
Char strength of concrete	fcu=25 M/mm2
Depth of base	h=750 mm
Cover to reinforcement	c=75 mm

Ground pressures

Taking moments about LH edge
$$Mlh= (N*(la+cx/2000)+V*h/1000+M)$$
$$= (2660*(1.895+610/2000)+0*750/1000+0)$$
$$=5852 \text{ kNm}$$

Distance of line of action from LH
edge
$$y=Mlh/N$$
$$=5852/2660$$
$$=2.2 m$$

A ——————————————— B

y = 1x/2, eccentricity   e = 0 m

p

$$p=N/(lx*LY)$$
$$=2660/(4.4*4.4)$$
$$=137.4 \text{ kN/m2}$$

Design moments and shears

Design moment (LH face)
$$Mda=p*ly*la^2/2$$
$$=137.4*4.4*1.895^2/2$$
$$=1085.5 \text{ kNm}$$

Design moment (RH face)
$$Mdb=p*ly*lb^2/2$$
$$=137.4*4.4*1.895^2.2$$
$$=1085.5 \text{ kNm}$$

For shear calc take eff depth

$$ds=0.85*h/1000$$
$$=0.85*750/1000$$
$$=0.6375\,m$$

Design shear (LH face)

$$Vda=p*ly*(la-ds)$$
$$=137.4*4.4*(1.895-0.6375)$$
$$=760.22\,kN$$

Design shear (RH face)

$$Vdb=p*ly*(lb-ds)$$
$$=137.4*4.4*(1.895-0.6375)$$
$$=760.22\,kN$$

Design for

$$Vdx=760.22\,kN$$

and

$$Mdx=1085.5\,kNm$$

Flexural reinforcement - x dirn
Diameter for x dirn steel

$$diax=25\,mm$$

Take off depth as

$$d=h-c-0.5*diax$$
$$=750-75-0.5*25$$
$$=662.5\,mm$$
$$K'=0.156$$
$$K=Mdx*10^6/(1000*ly*d^2*fcu)$$
$$=1085.5*10^6/(1000*4.4*662.5^2*25)$$
$$=0.022483 <K', OK$$

Lever arm

$$z=d*(0.5+SQR(0.25-K/0.9))$$
$$=662.5*(0.5+SQR(0.25-0.022483/0.9))$$
$$=645.51\,mm$$

but not $>0.95*d=0.95*662.5=629.38\,mm$
   i.e. take

$$z=629.38\,mm$$

Tension steel area reqd

$$Asx=Mdx*10^6/(0.87*fy*z)$$
$$=1085.5*10^6/(0.87*460*629.38)$$
$$=4309.5\,mm2$$

$750<3*d$, i.e. maximum
   clear spacing for steel
   and maximum pitch

$$maxcsp=750\,mm$$
$$pchm=750+diax\,mm$$
$$=775\,mm$$

Reqd pitch for tension steel

$$pchx=1000*ly*PI*diax^2/(4*Asx)$$
$$=1000*4.4*3.1416*25^2/(4*4309.5)$$
$$=501.18\,mm <pchm, OK$$
$$=500\,mm \text{ (rounded)}$$

Number of bars

$$Nox=INT(1000*ly/pchx)+1$$
$$=INT(1000*4.4/500)+1$$
$$=9$$

Tension steel area provided

$$Asprx=Nox*PI*diax^2/4$$
$$=9*3.1416*25^2/4$$
$$=4417.9\,mm2$$

Check percentage and spacing (see 3.12.11.2.7)
Calculate tension % provided

$$per=Asprx/(10*ly*h)$$
$$=4417.9/(10*4.4*750)$$
$$=0.13387 >.13 \ \& <4, OK$$

Percentage is <0.3, no spacing check needed.

Shear across base in x dirn
Steel % for entry to Table 3.9   pcnt=Asprx.(10*ly*d)
                                              =4417.9/(10*4.4*662.5)
                                              =0.15156
Design concrete shear stress vc from Table 3.9 for fcu=25 N/mm2
                                              =0.34093 N/mm24
Design shear stress   v=Vdx/(ly*d)
                                              =760.22/(4.4*662.5)
                                              =0.26079 N/mm2 <vc
No shear reinforcement needed across y direction faces.

Flexural reinforcement (y dirn)
Eccentricity of load within middle third, reasonable to design for uniform moment in y
   direction
Design moment   Mdy=N*(ly−cy/1000)^2/(8*ly)
                                              =2660*(4.4−610/1000)^2/(8*4.4)
                                              =1085.5 kNm
For shear calc take eff depth   ds=0.85*h/1000
                                              =0.85*750/1000
                                              =0.6375 m
Design shear   Vdy=N*(ly−cy/1000−2*ds)/(2*ly)
                                              =2660*(4.4−610/1000−2*0.6375)/(2*4.4)
                                              =760.22 kN
Diameter for y dirn steel   diay=25
                                              K'=0.156
                                               K=Mdy*10^6/(1000*lx*(d−diax)^2*fcu)
                                                =1085.5*10^6/(1000*4.4*(662.5−25)^2*25)
                                                =0.024281 <K', OK
Lever arm   z=(d−diax)*(0.5+SQR(0.25−K/0.9))
                                              =(662.5−25)*(0.5+SQR(0.25−0.024281/0.9))
                                              =619.81 mm
   but not >0.95*(d−diax)=0.95*637.5=605.63 mm
   i.e. take z=605.63 mm
Tension steel area reqd   Asy=Mdy*10^6/(0.87*fy*z)
                                              =1085.5*10^6/(0.87*460*605.63)
                                              =4478.5 mm2
750<3*(d−diax), i.e. maximum
   clear spacing for steel   maxcsp=750 mm
   and maximum pitch   pchm=750+diay mm
                                              =775 mm
Reqd pitch for tension steel   pchy=1000*lx*PI*diay^2/(4*Asy)
                                              =1000*4.4*3.1416*25^2/(4*4478.5)
                                              =482.27 mm <pchm, OK
                                              =450 mm (rounded)
Number of bars   Noy=INT(1000*lx/pchy)+1
                                              =INT(1000*4.4/450)+1
                                              =10
Tension steel area provided   Aspry=Noy*PI*diay^2/4
                                              =10*3.1416*25^2/4
                                              =4908.7 mm2

Check percentage and spacing − (see 3.12.11.2.7)
Calculate tension % provided   per=Aspry/(10*lx*h)
                                              =4098.7/(10*4.4*750)
                                              =0.14875 >.13 & <4, OK
Reinforcement percentage is <0.3, no spacing check needed.

Shear across base in y dirn
Steel % for entry, Table 3.9   pcnt=Aspry/(10*ly*(d-diax))
                                      =4908.7/(10*4.4*(662.5-25))
                                      =0.175
Design concrete shear stress vc from Table 3.9 for fcu=25 N/mm2
                                      =0.355 N/mm2
Design shear stress   v=Vdy/(lx*(d-diax))
                                      =760.22/(4.4*(662.5-25))
                                      =0.27102 N/mm2 < vc

No shear reinforcement needed across x direction faces.

Punching shear (3.7.6)
Length of side of first perimeter   x=cy+2*1.5*(d-diax/2)
                                      =610+2*1.5*(662.5-25/2)
                                      =2560 mm

Design effective shear force   Veff=N*(1+1.5*ABS(M)/(N*x/1000))
                                      =2660*(1+1.5*ABS(0)/(2660*2560/1000))
                                      =2660 kN

Average stress in punching zone   pf=pa-(pa-pb)*(la+cx/2000)/lx
                                      =137.4-(137.4-0)*(1.895+610/2000)/4.4
                                      =68.698 kN/m2

Distance out to first perimeter   lp=1.5*(d-diax/2)
                                      =1.5*(662.5-25/2)
                                      =975 mm (average of faces)

Reduction in punching load due to   Vred=pf*(cx+2*lp)*(cy+2*lp)/1E6
  direct transfer to sub-base   =68.698*(610+2*975)*(610+2*975)/1E6
                                      =450.22 kN

i.e. take Veff as   Veff=Veff-Vred
                                      =2660-450.22
                                      =2209.8 kN

fcu<39.06, max shear stress   vlim=0.8*SQR(fcu)
                                      =0.8*SQR(25)
                                      =4 N/mm2

Perimeter round loaded area   uo=2*(cx+cy)
                                      =2*(610+610)
                                      =2440 mm

Maximum design shear stress   vmax=1000*Veff/(uo*d)
                                      =1000*2209.8/(2440*662.5)
                                      =1.367 N/mm2 <4 OK

Average steel %   pcnt=(Asprx/(ly*d)+Aspry/(lx*(d
                                        -diax)))/20
                                      =(4417.9/(4.4*662.5)+4908.7/(4.4
                                        *(662.5-25)))/20
                                      =0.16328
Design concrete shear stress vc from Table 3.9 for fcu=25 N/mm2
                                      =0.34797 N/mm24
Distance out to first perimeter   lp=1.5*(d-diax/2)
                                      =1.5*(662.5-25/2)
                                      =975 mm
Distance round first perimeter   u=2*(cx+cy)+8*lp
                                      =2*(610+610)+8*975
                                      =10240 mm

Design shear stress

$$v=1000*Veff/(u*(d-diax/2))$$
$$=1000*2209.8/(10240*(662.5-25/2))$$
$$=0.332\,N/mm2 < vc, OK$$

No shear reinforcement needed for punching.

Reinforcement summary (fy=460N/mm2)
In x direction:
Area required   Asx=4309.5mm2
Area provided   Asprx=4417.9mm2
                                 (9 No 25mm dia bars @ 500mm)
Check   zoney=(3*cy/4+9*d/4)
                                       =(3*610/4+9*662.5/4)
                                       =1948.1mm, <ly/2

i.e. place 2/3 of bars in a central zone extending out 1.5*d (=993.75mm) on either side
   of the column faces. (Bars in outer zones will be lightly stressed – wider spacing
   there will be OK.)

In y direction:
Area required   Asy=4478.5mm2
Area provided   Aspry=4908.7mm2
                                 (10 No 25mm dia bars @ 450mm)
Check   zonex=(3*cx/4+9*(d-diax)/4)
                                       =(3*610/4+9*(662.5-25)/4)
                                       =1891mm, <lx/2

i.e. place 2/3 of bars in a central zone extending out 1.5*(d-diax) (=956.25mm) on
   either side of the column faces. (Bars in outer zones will be lightly stressed – wider
   spacing there will be OK.)
Shear steel;   None required

Example 4.5 A foundation wall carries a dead load of 170 kN/m run, an imposed load of 150 kN/m run, and a reversible wind load of 20 kN/m run acting at a height of 9 m above the base of the strip foundation which is 0.5 m below ground level. Borings and SPTs showed 0.2 m topsoil followed by medium-dense sand with an average SPT value of 15 blows/0.3 m. Rock was present at a depth of 3 m and ground water was at 1.1 m below ground level respectively. Design a suitable foundation.

Applying load factors from Table 2.1:

Design dead load   = 170 × 1.0 = 170 kN/m
Design imposed load = 150 × 1.3 = 195 kN/m
   Total design vertical load = 365 kN/m

Bending moment   = 20 × 9   = 180 kN/m
Design bending moment = 180 × 1.3
                                 = 234 kN/m
Eccentricity   = 234/365 = 0.64 m

Avoiding tension on the base for this condition, i.e. for e not greater than $B/6$,

$$B = 6 \times 0.64 = 3.8\,m, \underline{say, 4.0\,m}$$

Checking for conditions unfavourable to eccentric loading:

Design vertical load = 170 × 0.95 + zero imposed
                       load = 162 kN/m

Eccentricity   = 234/162 = 1.44 m

Width of equivalent strip foundation from equation (2.9) is

$$4.0 - 2 \times 1.44 = 1.1\,m$$

Corresponding uniform bearing pressure 162/1.1 = 147 kN/m².

For maximum vertical load, width of equivalent strip foundation is

$$4.0 - 2 \times 0.64 = 2.7\,m$$

Equivalent uniform bearing pressure is

$$365/2.7 = 135\,kN/m^2$$

Checking ultimate limit state (ULS) for 1.1 m width. For $\bar{N} = 15$, Fig. 2.13 gives $\phi' = 32°$. Applying material factor from Table 2.2, design $\phi' = \tan^{-1}\left(\dfrac{\tan 32°}{1.25}\right) = 26°$.

Assume ground-water level rises to base of foundation at this level, overburden pressure = 18 × 0.9 = 16 kN/m².

For $\phi' = 26°$, Fig. 2.8 gives

$$N_q = 12, N_\gamma = 9, N_c = 0.$$

For continuous strip

$$S_q = S_\gamma = 1.$$

Loading width (m)	Bearing pressure (kN/m²)	Coefficient s	D/B	ds/B	Reduction factor	Settlement (mm)
4.0	$(170 + 150)/4.0 = 80$	1.25	0.22	0.52	0.39	$\dfrac{1.25 \times 80}{150^{0.87}(1 + 4 \times 0.22)} = 5$
2.88	$(170 + 150)/2.88 = 111$	1.1	0.31	0.73	0.5	$\dfrac{1.1 \times 111}{150^{0.87}(1 + 4 \times 0.51)} = 5$

For $D/B = 0.9/1.1 = 0.8$, Fig. 2.11 gives d_c
$= d_q = 1.1$, $d_\gamma = 1$.

From Fig. 2.12 for

$$H/V = 20/162 = 0.12 \; i_q = 0.8 \quad \text{and} \quad i_\gamma = 0.7.$$

From equation (2.7)

$q_{ult} = 0 + 16 \times 12 \times 1 \times 1.1 \times 0.8 + 0.5 \times 9 \times 1.1 \times$
$9 \times 1.1 \times 1 \times 0.7$
$= 203 \, \text{kN/m}^2,$

this is not exceeded by the design pressure of $147 \, \text{kN/m}^2$, and it is evident that the ULS will not be exceeded at the design bearing pressure of $135 \, \text{kN/m}^2$ for the maximum vertical load on a 2.7 m wide equivalent foundation.

Checking the maximum edge pressure from equation (4.5c)

$$q_{max} = \frac{4 \times 162}{3 \times 1 \times (4.0 - 2 \times 1.44)} = 193 \, \text{kN/m}^2.$$

Hence the ULS bearing pressure for the 1.1 m wide strip foundation is not exceeded.

Check the serviceability limit state by using the Schultze and Sherif method (Fig. 2.25).

For the unfactored dead and imposed load plus wind loading the eccentricity is $180/320 = 0.56$ and the width of the equivalent uniformly loaded foundation is $4 - 1.12 = 2.88 \, \text{m}$.

The above tabulation gives the immediate settlement. The minimum settlement is given by the dead load without the imposed or wind loading, giving

$$\text{Minimum settlement} = 5 \times \frac{170/4}{80} = 3 \, \text{mm}$$

Hence the differential settlement is only 1 mm which could be accommodated easily by the in-plane rigidity of the wall.

Checking the dead plus imposed load from the Burland and Burbidge chart (Fig. 2.26), for $\bar{N} = 15$, average I_c
$= 3.5 \times 10^{-2}$

$$z_f = 3, f_1 = \frac{2.1}{3}\left(2 - \frac{2.1}{3}\right) = 0.9, \text{ shape factor} = 1.$$

Average immediate settlement $= 1 \times 0.9 \times 80 \times 4^{0.7}$
$\times 3.5 \times 10^{-2}$
$= \underline{7 \, \text{mm}}$

References

4.1 Jarrett, P. M., Stark, W. G. and Green, J., A settlement study within a geological investigation of the Grangemouth area, *Proceedings of the Conference on the Settlement of Structures*, Pentech Press, London, 1975, pp. 95–105.

4.2 Meyerhof, G. G., The bearing capacity of concrete and rock, *Magazine of Concrete Research* (April 1953).

4.3 Fraser, R. A. and Wardle, L. J., Numerical analysis of rectangular rafts on layered foundations, *Geotechnique*, **26**(4), 613–630 (1976).

4.4 Burland, J. B., Discussion on Session A, *Proceedings of the Conference on In-situ Investigations in Soils and Rocks*, British Geotechnical Society, London, 1969, p. 62.

4.5 Barrie, M. A. and Weston, G., Embankment Place: building over Charing Cross Station, *The Structural Engineer*, **70**(23), 405–410 (1992).

4.6 Grose, W. J., Hand-dug under-reamed piles at Embankment Place, London, in *Proceedings of the International Conference on Piling and Deep Foundations, Balkema*, **1**, 51–61 (1989).

4.7 Coates, F. N. and Taylor, R. S., Hartlepool Power Station: Major civil engineering features, *Proceedings of the Institution of Civil Engineers*, **60**(1), 95–121 (1976).

4.8 Mak, Y. W., Hand-dug caissons in Hong Kong, *The Structural Engineer*, **71**(11), 204–205 (1993).

4.9 Humpheson, C., Fitzpatrick, A. J. and Anderson, J. M. D., The basements and substructure for the new headquarters of the Hong Kong and Shanghai Banking Corporation, Hong Kong, *Proceedings of the Institution of Civil Engineers*, **80**(1), 851–883 (1986).

4.10 Archer, F. H. and Knight, D. W. M., Hong Kong and Shanghai Banking Corporation, *Proceedings of the Institution of Civil Engineers*, **84**(1), 43–65 (1988).

5 Buoyancy Rafts and Basements (Box Foundations)

5.1 General principles of design

5.1.1 Buoyant foundations

As described in the last chapter, the function of a raft foundation is to spread the load over as wide an area as possible, and to give a measure of rigidity to the substructure to enable it to bridge over local areas of weaker or more compressible soil. The degree of rigidity given to the raft also reduces differential settlement. Buoyancy rafts and basements (or box foundations) are designed on the same principles, but they have an additional and important function in that they utilize the principle of buoyancy to reduce the net load on the soil. In this way the total settlement of the foundation is reduced and it follows that the differential settlements will also be reduced. Buoyancy is achieved by providing a hollow substructure of such a depth that the weight of the soil removed in excavating for it either balances or is only a little less than the combined weight of the superstructure and substructure. In the example shown in Fig. 5.1, excavation to a depth of 4.5 m for the basement relieves the soil at foundation level of a pressure of about 80 kN/m². The substructure itself weighs about 25 kN/m²; thus a loading of 50 kN/m² can be placed on the basement before any additional loading causing settlement comes on to the soil at foundation level. A bearing pressure of 50 kN/m² is roughly equivalent to the overall loading of a four-storey block of flats or offices.† Thus, a building of this height can be supported on a basement founded in very soft and highly compressible soil, theoretically without any settlement occurring. However, in practice it is rarely possible to balance the loadings so that no additional pressure

comes on to the soil. Fluctuations in the water table affect the buoyancy of the foundation; also in most cases the intensity and distribution of live loading cannot be predicted with accuracy. Another factor causing settlement of a buoyant foundation is reconsolidation of soil which has swelled as a result of the removal of overburden pressure in excavating for the substructure. Swelling, whether by elastic or long-term movements, must be followed by reconsolidation as loading is replaced on the soil when the superstructure is built up. The order of swelling movements and measures which may be taken to reduce them are discussed later in this chapter.

For economy in the depth of foundation construction it is the usual practice to allow some net additional load to come on to the soil after the total of the dead load of the structure and its full live loading has been attained. The allowable intensity of pressure of this additional loading is determined by the maximum total and differential settlements which can be tolerated by the structure.

In limit state terms both the ultimate and serviceability limit states require consideration. Although bearing capacity failure should not occur in a properly designed buoyancy raft or basement, there is a risk of experiencing an ultimate limit state due to flotation of a completed or partly completed substructure.

The important design actions to be considered in limit state design are:

Weights of soil removed by excavation
Ground-water pressures

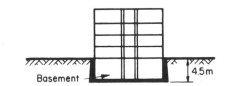

Fig. 5.1

---

† A useful approximate rule for calculating the weight of a multistorey block of flats is that a reinforced concrete framed structure with brick and concrete external walls, lightweight concrete partition walls, and plastered finishes weighs 12.5 kN/m² per storey. This figure is inclusive of 100 per cent dead load and 60 per cent maximum design live load.

Dead and imposed loads
Surcharge

An overestimate of soil density and the height of the ground-water table could lead to an underestimate of soil bearing pressure with consequent excessive settlement. Factors such as a future lowering of the ground-water table, or conversely a rise in ground-water level by flooding of the site, should be considered. The weights of constructional materials and wall thicknesses (geometrical data) can be critical in multi-cell buoyancy rafts. Surcharges such as placing fill around a semi-buoyant substructure can be critical particularly if placed on one side only causing tilting. The latter can also be caused by variations in the position of imposed loading (spatial distribution), for example by stacked containers in a warehouse.

5.1.2 Uplift on buoyant foundations

It is necessary to prevent the substructure from floating and tilting before the superstructure loads are sufficient in magnitude to prevent uplift. Floating only occurs in water-bearing ground or in a very soft silt or clay. During construction it can be prevented by keeping the water table drawn down by continuous pumping or by ballasting the substructure by flooding or other means. In some building structures it is possible to overcome the flotation problem by constructing the basement *after* the superstructure has reached a height sufficient to provide the necessary dead weight against uplift. In some underground structures there may be a net uplift because of light superstructure loading and it is necessary to provide some positive anchorage to prevent flotation. An example of this is the

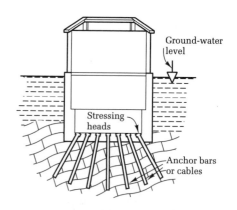

Fig. 5.3   Anchorage of buoyant structure to rock.

underground pumping station shown in Figs 5.2 and 5.3. Uplift may be resisted by providing sufficient dead weight in the structure which may result in massive and costly construction, or by means of anchor piles. If anchor piles are embedded wholly in soil, as in Fig. 5.2, they must be designed as friction piles. Where the substructure is founded on rock as in Fig. 5.3, the anchorage can take the form of steel bars or cables grouted into holes drilled into the rock and tensioned by post-stressing after completing the base slab. If rock is present at a moderate distance below the base of the substructure, a composite anchorage can be formed by driving open-ended hollow piles to rock and then grouting anchor rods into holes drilled at the bottom of the cleaned-out piles as shown in Fig. 7.30. Information in the design of piles or drilled anchorages to resist uplift is given in Section 7.16.

5.2 Drag-down effects on deep foundations

When calculating the net bearing pressure at the base of a buoyancy raft or basement the *total* foundation pressure q is reduced by the total overburden pressure σ. Any skin friction or adhesion between the walls of the substructure and the surrounding soil should not be regarded as reducing the net bearing pressure on the soil. In most cases the tendency is for the surrounding soil to produce a limited negative skin friction or drag-down effect on the substructure as a result of the construction procedure.

If a basement is constructed in a sheeted excavation, as in Fig. 5.4(*a*), the backfilling which is placed between the walls and the sheeted sides consolidates with the passage of time, thus giving a drag-down effect on the walls. In the case of a cellular buoyancy raft (Fig. 5.4(*b*)) sunk through a soft clay or silt by grabbing or hand excavation from open wells, the downward movement of the structure will cause extensive disturbance of the soil, possibly augmented by slumping of the surrounding soil consequent on

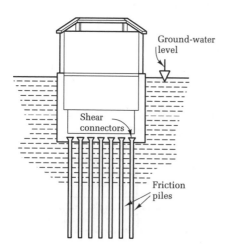

Fig. 5.2   Anchorage of buoyant structure in soil.

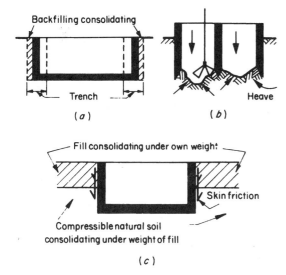

Fig. 5.4 Drag-down effects on basement structures. (*a* Basement walls constructed in sheeted trenches. (*b*) Cellular raft sunk by grabbing. (*c*) Basements in filled ground.

upward heave of the soil beneath the cells. The reconsolidation of this disturbed soil will again cause a drag-down effect on the walls. Thus in most cases the tendency will be for the soil in contact with the walls of the substructure to add to the load on the base slab rather than to relieve it of load. In practice, the normal design procedure is to ignore any support or drag-down in calculating net bearing pressures at foundation level. However, to minimize drag-down effects, the backfilling around the walls should be carefully placed and well compacted. Where an appreciable depth of filling is placed over a compressible soil (Fig. 5.4(*c*)) the drag-down effects may be marked and should be allowed for in the design.

When considering resistance against uplift, the skin friction or adhesion in the surrounding ground can be taken into account. Conservative values should be used.

5.3 Buoyancy raft foundations

5.3.1 Design

The terms 'basement' and 'buoyancy raft' have been used in the preceding pages and it is important to explain the differences between them. Although a basement is, in effect, a form of buoyancy raft, it is not necessarily designed for that purpose. The main function of a basement is to provide additional space in the building for the owner, and the fact that it reduces the net bearing pressure by the weight of the displaced soil may be quite incidental. In some cases, basements may be required for their function in reducing net bearing pressures and advantage

is taken of this to provide additional floor space in the substructure. The true buoyancy raft, however, is a foundation which is designed solely for the purpose of providing support to the structure by the buoyancy given by the displaced soil without regard to utilizing the space for any other purpose. To this end the raft is designed to be as light and rigid as possible. Lightness combined with stiffness is best achieved by cellular or 'egg-box' construction. This structural form limits the usefulness of the space within the substructure to accommodate any pipework or service ducts passing through holes in the walls of the cells. Because of the many problems inherent in the design and construction of buoyancy rafts, they have, in most cases, been supplanted by other expedients, mainly by various types of piling. Nevertheless, some account will be given in the following pages of the problems and means of overcoming them because they exemplify the true art of the foundation engineer.

Problems can arise in maintaining buoyancy in ground conditions which require the cells to be watertight. Asphalt tanking or other membrane protection (Section 5.9) is not feasible where the rafts are constructed in the form of caissons, and any water finding its way through cracks in the exterior walls or base should be removed by pumping. Openings should be provided in the interior cell walls to enable the water to drain to a sump where an automatic pump can be installed.

Where gas is used for domestic heating or industrial processes in buildings supported by buoyancy rafts, the cells should be sealed to prevent dangerous accumulations of gas within the substructure.

Buoyancy rafts can be constructed either in the form of open well caissons (Section 6.4.4) or they can be built *in situ* in an open excavation. The caisson method is suitable for soft clays where the soil within the cells can be removed by grabbing as the raft sinks down under its own weight. However, this method is unsuitable for ground conditions where heavy walls are needed to provide stiffness and weight to aid sinking through obstructions. Where the caissons are terminated within the soft clay some settlement should be expected when the soil disturbed by grabbing reconsolidates under the superstructure loading.

Construction in open excavations is suitable for site conditions where the ground-water level can be kept down by pumping without risk of 'boiling' (see Section 11.2.1), and where heave of the soil at the base of the excavation is not excessive (see Sections 2.6.3 and 9.7).

5.3.2 Buoyancy rafts constructed as caissons

The cellular structures which support the power station and other plant installations at the Grangemouth Refinery were constructed in the form of caissons.[5.1] On this site, recent alluvial silts and clays of the estuary of the River Forth extend to depths varying from 25 m to as much as 75 m, and overlie stiff glacial till. Because of the depth to the till, cellular rafts were selected as an economical alternative to long piles. Below a 1–2 m stiff clay surface crust, the shear strength of the alluvial deposits was only 10–15 kN/m^2 for a considerable depth.[5.2] Thus the ultimate bearing capacity of the soil for surface rafts was little more than 50 kN/m^2, and large loaded areas with overall bearing pressures as low as 25 kN/m^2 would have been liable to excessive settlement.

As far as possible, the cellular rafts were designed so that the weight of the displaced soil balanced the combined loading of the rafts and their superstructures. The superstructure loads were carefully positioned to ensure practically uniform bearing pressure over the foundation area. If this had not been done there would have been a risk of tilting. In practice it was found that 3 t of displaced earth gave sufficient buoyancy to support 2 t of superstructure. The loadings of the buildings and plant required excavation to a depth of 6 m, which if undertaken in open cut with sloping sides or sheetpiled supports would probably have led to difficulties with slips or base heaving. Consequently, the rafts were designed to be assembled in shallow excavations, and then sunk to the required depth by grabbing from the open cells. On reaching founding level the cells were plugged with concrete and the superstructure erected on them. By this method, the depth of open excavation was kept to a minimum and, since all the cells were not grabbed out simultaneously, there was always a surcharge of soil within the cells to prevent general heaving of the bottom of the excavation.

The largest structure on the site was the power station. This was designed to be founded on a group of four cellular rafts. Each raft was sunk independently then rigidly interconnected by *in situ* reinforced concrete panel walls. A total of 240 cells were provided in the 52 m by 52 m combined foundation. The mass of soil displaced at a founding depth of 4.72 m was 23 500 t, and the combined mass of the four cellular rafts was only 8250 t.

The standard procedure adopted for sinking the rafts was first to excavate to a depth of 2 m through the 0.6 m thick surface layer of burnt colliery shale filling to just below the stiff surface crust of the alluvial clay. The slopes remained stable at batters of about 1 to 1, and a 150 mm thick layer of shale was spread over the working area. Simultaneously with the bulk excavation, the 3 m wide by 150 mm thick precast concrete slabs for the cell walls were being made close by. A 30-cell raft required 60 slabs, and they were cast horizontally in stacks of six or seven with building paper between each two slabs.

To support the external walls, which were cast *in situ*, 1 m wide concrete strips were placed around the external walls. A 1.5 m square pad was also cast at the intersections of the cell walls as shown in Fig. 5.5(*a*). Four slabs were first set to form a cell in the centre of the raft and were strutted internally and shored externally. The subsequent slabs were placed symmetrically in turn around the first four, the object being to avoid tilting due to non-uniform loading of the soft clay. When all precast slabs had been erected, the concreting of the first 1.22 m lift of the external walls was commenced. Concreting of the succeeding 1.22 m lifts proceeded, and at the same time the intersections between the precast slabs were concreted. The concrete was placed uniformly around the structure to keep the loading as symmetrical as possible. The rate of concreting of the external walls and the cell intersections was phased so that all parts of the structure were completed at the same time.

Five days after the last lift of concrete was placed, a start was made with sinking the rafts by breaking out the concrete strips beneath the external walls. A short length of strip was left at the junction with the internal walls. The pads beneath the wall intersections were next broken out simultaneously with the remaining portions beneath the external walls. The rafts then commenced to sink through the shale layer, and movement continued until the walls had penetrated far enough to build up skin friction, thus slowing down the sinking (Fig. 5.5(*b*)). Grabbing commenced at the corners, continued with the remaining outer cells, and then proceeded towards the centre. The central row of cells was left unexcavated until the rafts had been sunk nearly to founding level. The order of grabbing from the cells is shown in Fig. 5.5(*c*). As soon as the rafts were within 1.2 m of the founding level, two of the outside cells on opposite sides were grabbed out to their full depth and the mass concrete plugs were placed. The vertically cut soil remained stable during the day or two required for the rafts to sink to their final level (Fig. 5.5(*d*)). On the day after concreting these first two plugs, two more cells were grabbed out and concreted; by this time the raft probably had another 0.6 m to sink. The corner cells were not generally selected among the first four to be plugged. It was usual to select a cell next to a corner one. For a 30-cell raft (six cells by five cells), it was necessary to grab out and concrete eight cells before the downward movement could be arrested. By the time the eighth cell had been plugged the raft was within 25–50 mm of its final level. After the raft had come to rest on the eight plugged cells, the remaining ones were grabbed out and plugged, working towards the centre row. During the plugging of the centre row the outer cells were pumped dry of any water which had seeped through the mass concrete or beneath the cutting edges. They were thoroughly cleaned out and

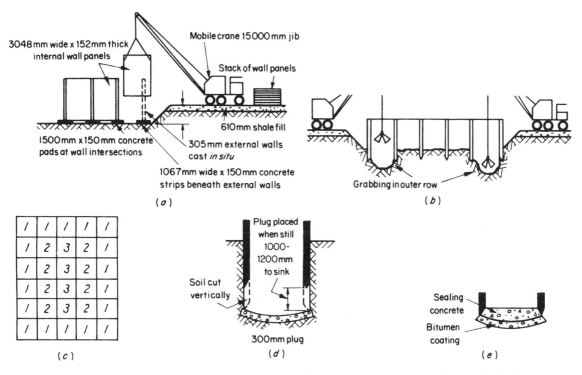

Fig. 5.5 Construction of buoyant raft foundations at Grangemouth Refinery. (*a*) Assembly of precast wall panels. (*b*) Sinking unit by grabbing. (*c*) Order of excavating cells in a 6 × 5 cell unit. (*d*) Plugging cells. (*e*) Sealing cells.

the surface of the mass concrete was given a heavy coat of bitumen. Reinforcing steel was then placed in two directions and wired to splice bars passing through holes in the internal walls of the cells. The reinforced sealing concrete, 0.6 m thick at the centre of each cell and 0.3 m thick at the walls, was then placed (Fig. 5.5(*e*)). The top surface of the concrete was given a fall towards a pumping sump at one selected cell. Any leakage through the external walls or plugs gravitated across the floor passing through holes cut through the internal walls to the sump.

Kentledge was not required to aid sinking. Control of level was achieved by grabbing from one side or one corner as required to correct any tendency to tilting. Downward movement could be slowed down or stopped altogether by clearing all men and mechanical plant or transport from the vicinity of the raft for a few hours. The 'take-up' or restoration of shear strength of the disturbed clay was sufficient to arrest the movement.

The final stage of the foundation construction was to cast the structural deck, which consisted of a 300 mm reinforced concrete slab cast *in situ* on permanent formwork provided by 76 mm thick prestressed concrete planks. The upper surface of the latter was corrugated to bond into the *in situ* slab since their combined action had

been allowed for in the design. Precautions were taken to prevent accumulations of explosive gas in the cells, resulting from leakage from the refinery plant. The structural decks were provided with airtight manhole covers set on raised plinths.

Settlements were measured on five of the structures founded on the cellular rafts. The results of these measurements were reported by Pike and Saurin[5.1] and showed settlements varying from 13 to 6 mm over a period of $1\frac{3}{4}$ years from commencement of the superstructures. Although the buoyancy raft foundations at Grangemouth have performed satisfactorily, deep piled foundations were generally preferred for most of the refinery and petrochemical installations subsequently constructed in the same area.

5.3.3 Buoyancy rafts constructed in open excavations

Buoyancy rafts were used for the foundations of six blocks of 15-storey flats in the Parkhead and Bridgeton districts of Glasgow, where the ground conditions were as shown in Fig. 5.6. Settlements of shallow raft or pad foundations would have been excessive under the gross bearing pres-

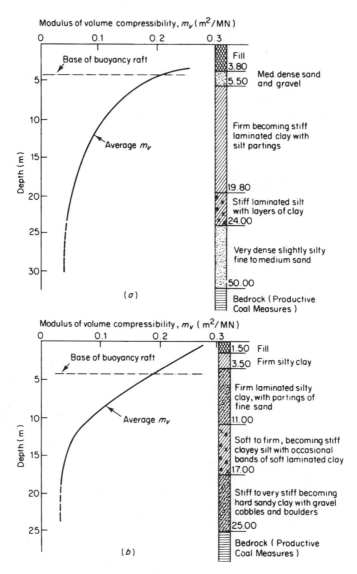

Fig. 5.6   Soil conditions beneath buoyancy raft foundations in Glasgow. (*a*) Bridgeton site. (*b*) Parkhead site.

sure of 135 kN/m². Piled foundations taken down to the glacial drift or to the bedrock of the Coal Measures were considered, but there was a risk of removal of support to the piles if collapse should occur at any time of the coal pillars left in the uncharted mine-workings. Accordingly a buoyancy raft was the only practical method of foundation design which would reduce the bearing pressures transmitted to the compressible laminated clay to a value at which the total settlement would not exceed 75 mm, stipulated by the structural designer as a maximum.

A cellular raft at a depth of 4.3 m below ground level

reduced the bearing pressure to a net value of 53.5 kN/m². Consolidation settlements were estimated to be 45 and 55 mm for the Parkhead and Bridgeton sites, respectively, and the immediate settlement about 10 mm in each case. The cellular structures were constructed in open excavations with sloped-back sides (Fig. 5.7). As a safeguard against general subsidence of the ground surface consequent on possible future collapse of the mineworkings, pockets were provided between the raft and the superstructure in which jacks could be installed to relevel the buildings.

Fig. 5.7   Construction of buoyancy raft at Bridgeton, Glasgow, in open excavation.

Settlements were reported by Somerville and Shelton[5.3] over a period of observation of 3 years from the commencement of construction. The observed movements were as follows:

	Measured total settlement at 3 years		Estimated total final settlement (mm)
	Max. (mm)	Min. (mm)	
Parkhead Block A	35	23	57
Parkhead Block B	46	23	55
Parkhead Block C	51	34	52
Bridgeton Block B	60	30	60
Bridgeton Block C	59	40	67

No cracking was observed in the superstructure of the buildings at the end of the 3-year period of observation, although the angular distortion of the raft was about 1 in 350 on one block.

5.4 Basement or box foundations

The difference between buoyancy rafts and basements is defined in Section 5.3.1. The former, which support structures by displacement of the soil without regard to the utilization of the spaces within the hollow substructure, are designed with the sole object of achieving lightness combined with rigidity. On the other hand, basements must be designed to allow the substructure to be used for various purposes such as warehouse storage or underground car parks. This requires reasonably large floor areas without close-spaced walls or columns, and the floor generally consists of a slab or slab and beams of fairly heavy construction to give the required degree of rigidity.

There are three principal methods of constructing basements. These are as follows:

(1) in excavations with sloping sides;

(2) in excavations with temporary support to the sides;

(3) in excavations supported by a permanent embedded wall constructed in advance of the main excavation.

Table 5.1   Relative construction costs of steel sheet pile and two types of concrete retaining walls (after Potts and Day[5.4])

Internal finish and other components	Diaphragm wall‡	Secant pile wall†		Larssen 6 sheet pile
		Bar reinforcement at 100 kg/m^3	Reinforcement by $914 \times 305 \times 289$ kg/m^3 universal beam	
As formed	1.00	0.81	1.24	0.77
Blockwork facing	1.05	0.85	1.29	0.82
150 mm *in-situ* concrete facing	1.08	0.88	1.32	N/A
Corrosion protection§	N/A	N/A	N/A	0.84
Fire protecant¶	N/A	N/A	N/A	0.87

† 1180 mm bored piles at 1080 mm centres.
‡ 1 m thick, 2 per cent reinforcement, 1 m × 1 m guide walls.
§ Shot-blast cleaning and coal tar epoxy pitch one-coat paint.
¶ One-hour protective coating – shot-blast cleaning, epoxy metallic zinc-rich primer and intumescent spray paint.

The essential feature of the embedded wall in method (3) is that it forms part of the permanent substructure. Embedded walls are designed to be virtually watertight, requiring only the addition of an internal facing wall which is built after completing the bulk excavation. Means of dealing with minor water seepage may be incorporated in the internal wall. Two types of embedded wall are in general use. These are (1) the reinforced concrete diaphragm wall, and (2) the contiguous or secant bored pile wall. Although steel sheet piling is widely used as a form of permanent construction for river and quay walls it is only recently that it has been considered as an economical alternative for the permanent support of basements in building works, provided that some increase in ground movements around the excavation can be permitted.

Potts and Day[5.4] used finite element analyses to assess the potential of steel sheet piling for three typical substructures in the London area. Two of these were for replacement of diaphragm walls or bored pile walls in cut-and-cover highway tunnels, and the third was for the use of sheet piling in lieu of diaphragm walling in the House of Commons underground car park (see Section 5.4.4). Cost comparisons for the three cases showed that the use of steel sheet piling could achieve savings in the range of 25 to 40 per cent of the cost of diaphragm walling for the equivalent internal finish. The cost comparison for the House of Commons car park is shown in Table 5.1.

5.4.1 Interdependence of design and construction

The choice of method depends on the depth of the basement, the ground conditions (and in particular the ground water level), and the proximity or otherwise of buildings, roads, and services which need to be safeguarded during and subsequent to construction. The

design of a deep basement requires close collaboration between the designer and contractor so that the intentions of either party are fully understood. The designer must have a knowledge of how the basement structure will be built and it is good practice to include a scheme for temporary works with the tender information drawings so that difficulties of construction are not overlooked at the design stage. The form of contract could make it optional for the designer's temporary works scheme to be adopted or not, but if the contractor should decide to adopt his own method he should be obliged to submit detailed drawings for approval after acceptance of his tender. It is essential that the responsibilities of the engineer and contractor in respect of design and construction should be clearly defined. Much useful information on the design and construction of deep basements is given in a report by the Institution of Structural Engineers.[5.5]

The design of a basement should take into account the effects of heave due to relief of overburden pressure caused by excavation (see Section 2.6.3). A classical example of the problems of heave and reconsolidation of the soil beneath basements supporting buildings of different heights and the effects of ground heave on tunnels is given by the design of the Shell Centre on the south bank of the River Thames, London.[5.6] This group of high buildings consists of two 10-storey structures separated by a railway viaduct, with a 28-storey tower block in the area upstream of the viaduct. The spaces between the buildings are occupied by underground garages in basements 9–16 m deep. The soil strata consist of made-ground followed by soft clay, sand, and gravel, a very thick stratum of London Clay, then the Woolwich and Reading beds, followed by the Thanet Sands and the Upper Chalk (Fig. 5.8). It would have been possible to found the tower block on a relatively shallow raft bearing on the sand and

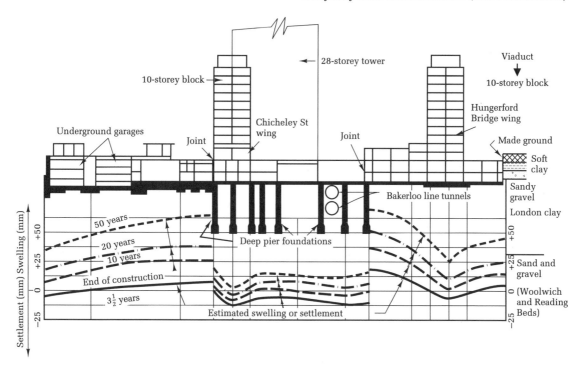

Fig. 5.8   Cross-section through Shell Centre, South Bank, London (after Measor and Williams[5.6])

gravel, but the estimated settlement of about 125 mm for a shallow raft was considered to be excessive, in view of the risk of tilting of the tall building. This could have occurred as a result of the loading on one side of the block imposed by the adjacent 10-storey wing and the relief of over-burden pressure on the other side given by the construction of the underground garage. The estimated maximum differential movement between the tower block and the underground garage was as much as 180 mm. Estimates made of swelling of the soil beneath the underground garage showed an upward movement of 75 mm in 50 years.† In view of the large differential movements and the risks of tilting it was decided to found the tower block on a deep basement. The reduction in the net bearing pressure greatly reduced the estimated settlement. The values of gross and net bearing pressures for the various basements are shown in Fig. 5.9(a) and (b), respectively.

Another complicating factor on the site was the presence of four tube railway tunnels running beneath the site

† Present-day knowledge of the effects of the soil fabric such as the fissuring of London Clay would result in estimates of the rate of settlement and heave of the structure which would be much faster than originally estimated for the Shell Centre. However, the estimate of final movements will probably be reasonably accurate.

(see Fig. 5.8). There was only 0.9 m of cover between the basement excavation and the crown of the shallowest tunnel. It was considered that the tunnels, constructed in bolted cast-iron segments, were sufficiently flexible to withstand the estimated rise in the tunnels of 40 mm due to excavation for the basements. Buildings adjacent to the tunnels were constructed on deep piers taken below tunnel level in order to prevent unequal radial pressures which might have distorted the tunnel rings. The deep piers were in the form of cylindrical shafts with belled-out bases. The bases were terminated at a sufficient height above the water-bearing Woolwich and Reading beds to avoid upheaval of the London Clay by sub-artesian water pressure in the water-bearing stratum. Vertical joints were provided between structures founded on deep piers and those on relatively shallow foundations to allow for the expected differential movement.

Extensive measurements were made by the Building Research Station[5.7] of pressures and movements in the tube railway tunnels in the course of construction of the buildings. A rise of nearly 40 mm took place without noticeable distress in the tunnel segments, and no distortion was observed as a result of excavation for the deep belled-out piers. However, driving heavy section steel sheet piling to support the trench excavation for the

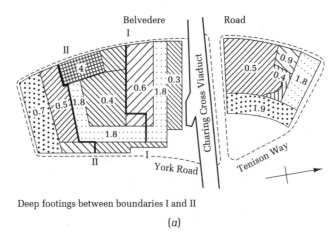

Deep footings between boundaries I and II

(a)

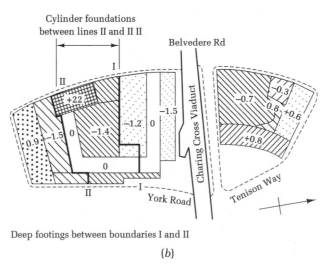

Deep footings between boundaries I and II

(b)

Fig. 5.9   Loading conditions, Shell Centre, South Bank, London. (a) Approx. average total weight per unit area (kN/m² × 100).
(b) Approx. average net weight per unit area (kN/m² × 100).

basement retaining walls caused 5 mm inward deflexion of the sides of a tunnel and a similar elongation of the vertical diameter at the soffit and invert. This movement occurred when the piles were driven parallel to and 3 m away from the tunnel lining. The movement corresponded to an increase in stress on the tunnel lining of about 110 kN/m².

If floors are very thick, as in the case of deep basements in water-bearing ground where deadweight must be provided to resist uplift, it may be necessary to excavate for a concrete floor in alternate strips in order to allow shrink-

age of the thick slabs to take place before the intervening portions are concreted. After the soil is taken out in bulk to a level at which there is sufficient overburden pressure remaining to prevent heaving or swelling, the further depth is excavated in sheet piled or timbered trenches and a section of floor slab concreted. The sheet piles are then withdrawn and the intervening excavation is taken out and the floor slab concrete poured in the space between the completed sections. This technique is used for the heavy floors of dry docks constructed in water-bearing soil or in ground liable to swelling.

5.4.2 Construction in excavations with sloping sides

This is the most economical form of construction for sites where there is sufficient space around the substructure to cut back the sides of the excavation to a stable slope, and where there are no problems of dealing with large quantities of ground water which could lead to erosion and slumping of the slopes. Guidance on stable slopes of excavations in various ground conditions is given in Chapter 9. Problems of dealing with ground water in excavations are dealt with in Chapter 11.

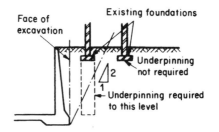

Fig. 5.10   Underpinning adjacent to basement excavations.

5.4.3 Construction in excavations with temporary support

This is a suitable method of construction for sites where insufficient space is available around the excavation to slope back the sides. If the soil conditions permit withdrawal of sheet piling for reuse elsewhere this method of ground support is very economical compared with the alternative of an embedded wall (Section 5.4.4). Support by a system of horizontal timber sheeting and vertical H-section soldiers (Section 9.5.2) is also economical for soils which will stand unsupported over a short depth for a limited period of time.

Where buildings, roads, or underground services are close to an excavation the effects on these structures of movements due to excavation and the construction of the basement need to be carefully assessed (Section 9.7). In particular the possible effects of pumping ground water from excavations in causing settlement of the ground surface around the excavation should be considered (see Section 11.4).

Excavation for deep basements may cause settlement of the surrounding ground surface. The settlement may be sufficient to cause structural damage to buildings near the excavation, and cracking of drains and other services. This settlement may be caused by the following:

(a) lateral movement of the face of the excavation due to the cumulative effects of yielding of the sheeting members, walings, and struts, or anchors which support the face;

(b) lateral movements due to elastic deflexion of the basement retaining wall after completion of backfilling;

(c) lowering of the water table surrounding the excavation due to pumping during construction;

(d) loss of ground due to slips, heave of the base, erosion, etc. as a result of ill-conceived construction methods or carelessness of execution.

Normal methods of supporting excavations by strutting or tie-back anchors, no matter how carefully done, cannot prevent inward yielding of the face of the excavation. The inward movement may be of the order of 0.3 per cent of the excavation depth in soft clays to about 0.2 per cent in dense granular soils or stiff clays accompanied by corresponding settlement of the ground surface (Section 9.8). One method of safeguarding existing structures is to underpin them, while supporting the excavations. For shallow to moderate depths of excavation (say not exceeding 5 m), it is sometimes assumed that settlement of the ground surface will not occur to any appreciable extent beyond a line drawn at a slope of 1 (horizontal) to 2 (vertical) from the base of excavation. Underpinning of structures is carried out within this line (Fig. 5.10). The possibility of significant settlements extending to a wider limit beyond the excavation is discussed in Section 12.3.2 and the use of finite element methods to predict the magnitude of movement around deep excavations is described in Section 9.9. In shallow or only moderately deep excavations it may be justifiable to accept a small amount of settlement and to repair any consequent cracking or other minor damage to existing structures. The degree of risk involved will depend on the value of the property and the effects on any activities such as manufacturing processes within the premises.

In situations where a low-rise building is constructed on a deep basement there will be a net relief of vertical pressure on the soil beneath the basement and consequent tendency to uplift of the structure. This movement may be transmitted to the ground surrounding the basement, with the result that the ground surface, which settled during the stage of excavation, will rise to its original level and then rise further to conform to the net uplift of the basement (Fig. 5.11). The convexity of the base slab will be increased where the superstructure loads are carried mainly by the basement-retaining walls. There may be net settlement of the walls and net uplift in the central areas of the base slab.

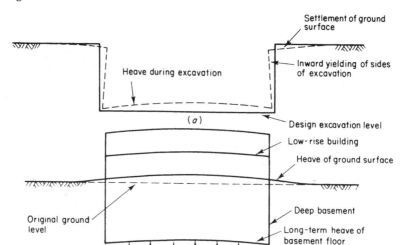

Fig. 5.11   Movements around a deep basement supporting a low-rise structure. (*a*) Movements at stage of completion of excavation. (*b*) Final movements of completed structure.

5.4.4 Construction in excavations supported by a diaphragm wall

A diaphragm wall is constructed by excavation in a trench which is temporarily supported by a bentonite slurry. On reaching founding level steel reinforcement is lowered into the trench, followed by placing concrete to displace the bentonite. This form of construction is suitable for sites where obstructions in the ground prevent sheet piles from being driven and where the occurrence of ground water is unfavourable for other methods of support (see Chapter 9). The method is also suitable for sites where considerations of noise and vibration preclude driving sheet piles and where ground heave and disturbance of the soil beneath existing foundations close to the margins of the excavation are to be avoided.

The rigidity of the diaphragm wall in combination with support by preloaded ground anchors (Section 9.5.8) followed by strutting with the floors of the permanent structure can, in comparison with support by sheet piling and timber or steel bracing, reduce the inward deflexion of the structure and hence the subsidence of the ground surrounding the excavation. However, it is not possible to eliminate completely the inward deflexion. Detailed observations of movements of the diaphragm wall supporting the 18.5 m deep substructure forming the undergound car park at the House of Commons have been described by Burland and Hancock.[5.8] Settlements around the excavations were critical at this site since Westminster Hall and the Big Ben clock tower were only 3 m and 16 m

respectively from the margins of the excavation. The observed movements are shown in Fig. 5.12. The maximum inward deflexion of the wall of 30 mm is 0.16 per cent of the depth of the excavation and the maximum subsidence of the ground surface of 20 mm is 0.11 per cent of this depth. A heave of the clay beneath the base of the excavation of 45 mm was observed. The piles supporting the seven floors of the basement heaved by 13–16 mm during the course of excavation. The inward deflexion can be compared with observed movements of other deep excavations in Fig. 9.4. The important point to note in the observations of inward yielding of the diaphragm wall is that inward movement took place below the level of the excavation at all stages. This is typical of the supports to deep excavations, and it means that inward movement cannot be prevented by strutting or anchorages installed in stages as the excavation is taken down.

The first step in diaphragm wall construction is to form a pair of reinforced concrete guide walls some 150–300 mm wide and at least 1 m deep. These walls act as guides for the excavating machinery and as a means of maintaining a head of bentonite at least 1 m above groundwater level in order to prevent collapse of the sides of the excavation. The guide walls also serve to prevent collapse of the trench sides due to the surging of the slurry caused by the excavating tools, and in soft or loose soils they must be taken deep enough or made sufficiently wide to perform this essential task. The construction of the guide walls can represent a substantial proportion of the total cost of a diaphragm wall, and in some site conditions, e.g.

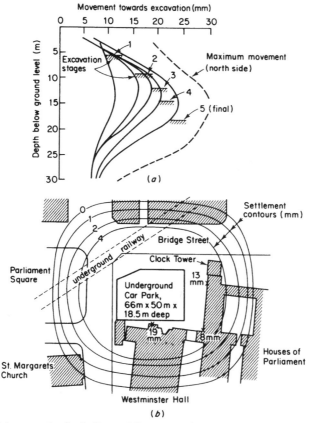

Fig. 5.12 Movements around the excavation for the House of Commons underground car park at Westminster (after Burland and Hancock[5.8]). (*a*) Observed inward yielding of diaphragm wall. (*b*) Observed settlements around excavation.

in areas of buried foundations and old cellars, they can be prohibitively costly. Problems of this kind are usually overcome by excavating generally over the site to the level of the natural undisturbed soil. The guide walls are then constructed wholly or partly within this soil. The sides of the preliminary excavation are supported temporarily (as in Fig. 5.15(*a*)) or are formed to a stable slope.

Excavation of the trench between the guide walls is undertaken in alternate panels, the length of which depends on the stability of the soil and the dimensions of the basement. A panel length of about 4–6 m is quite usual for stable ground. The excavation is performed with the support by the slurry either by grabbing or by reverse-circulation rotary drilling. Grabs operated in conjunction with a power-assisted 'kelly' are efficient in a wide range of soil types. Rotary drilling is effective in granular soils. The bentonite slurry is circulated continuously during excavation by pumping from the trench to an elevated settling tank from which it returns to the trench. The slurry can be circulated through a reconditioning plant to remove

sand and grit, or its use may be confined to a single panel and then it is pumped into a tank vehicle for disposal off-site. In either case continuous monitoring of the viscosity, density, shear strength, and pH value of the slurry is necessary to ensure that it does not become excessively diluted or contaminated by soil particles.

After completing the excavation of a panel vertical tubes are placed at each end, which form a semicircular recess at the ends of each concrete panel. The reinforcing cage with box-outs attached to it for forming junctions of floors and walls is then lowered into the slurry and suspended from the tops of the guide walls after which the concrete is placed using a tremie pipe, when the rising concrete displaces the slurry. After a period for hardening of the concrete, the intermediate panels are excavated and then concreted to form the completed wall. Increased watertightness of the semicircular joint can be achieved by welding a small-diameter pipe to the side of the stop-end tube to act as a void former. This provides a keyed vertical slot in the end of the concreted panel. The slot is grouted

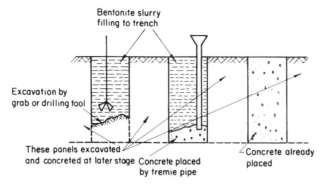

Fig. 5.13   Construction of basement retaining walls by diaphragm method.

with flexible bentonite cement after placing concrete in the abutting panel. The final stages are removal of the inner guide wall followed by bulk excavation within the enclosing wall, trimming off any protuberances caused by 'overbreak' during excavation and, finally, casting the facing wall. The model specification prepared by the UK Federation of Piling Specialists defines the face of the wall as the inside face of the inner guide wall and a tolerance of ± 15 mm in a 3 m length is allowed from this face. Also a deviation from the vertical of 1 in 80 is allowed and protuberances of more than 100 mm from the face must be cut away at the expense of the diaphragm wall contractor. According to the model specification the latter only applies in 'homogeneous clays'. In highly fissured clays, sands, gravels, and soft or loose soils the removal of the protuberances must be paid for as an extra.

The stages of excavating and concreting for a cast-in-place diaphragm wall are shown in Fig. 5.13.

Instead of placing concrete *in situ*, precast concrete panels can be lowered into the slurry-filled trench. Support to the soil behind the panels is provided by a bentonite–cement grout which also forms the seal between the grooved ends of the abutting panels. In the SIF–Bachy system the bentonite slurry is replaced by the grout immediately before placing the panels. In the Soletanche system the excavation is performed with the bentonite–cement as a supporting fluid. The advantages claimed for the precast panel systems are as follows:

(a) The interior wall surface is clean and even with no need to cut away 'overbreak' in the concrete.
(b) Better quality in the concrete and greater accuracy in positioning the reinforcement, giving savings in wall thickness compared with *in-situ* construction.
(c) The units can be set to close tolerances and openings are accurately positioned.
(d) Improved watertightness.

Both the cast-in-place or precast concrete panel systems

can achieve watertight basements. The formation of the bentonite gel in the soil behind the wall and between the panels acts as a seal against the entry of water. In rectangular or circular structures the earth pressure induces longitudinal compression in the wall which tends to close up the vertical joints. Problems with leakage can occur where there are re-entrants in the wall where the vertical joints tend to open.

Where inclined anchors are used to support a diaphragm wall the width of the wall at the toe must be sufficient to prevent failure as a strip foundation under the loading produced by the vertical component of the stress in the anchors. Also, the depth of the toe below final excavation level must be such as to provide the required passive resistance in front of the wall to the earth pressure and hydrostatic pressure behind the wall.

Consideration must be given to the capacity of diaphragm walls to carry heavy vertical loads, especially when the base of the wall is at a comparatively shallow depth and only a small proportion of the load is carried by skin friction on the rear face. As well as calculating the bearing capacity of the soil beneath the relatively slender thickness of the base, the conditions at the interface between the wall and the supporting soil or rock must be considered. It is likely that there will be a layer of

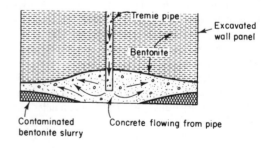

Fig. 5.14   Entrapment of contaminated slurry at base of diaphragm wall.

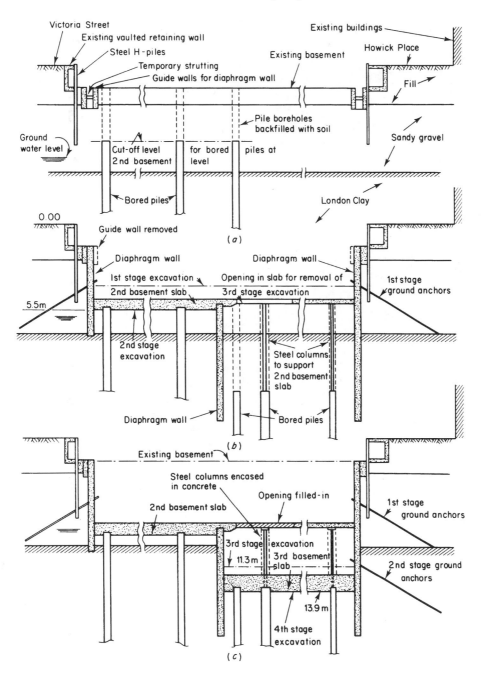

Fig. 5.15 Stages in the construction of a deep basement in Victoria Street, London. (*a*) Construction of guide walls for diaphragm walling and bored pile foundations for second basement. (*b*) Construction of second basement slab and bored pile foundations for third basement. (*c*) Construction of third basement slab.

bentonite heavily contaminated with soil particles at the base of the trench after completion of the excavation. The practice of replacing the bentonite with fresh material before placing the concrete may not remove this layer completely. The concrete in flowing laterally from the tremie pipe may then displace this layer with consequent entrapment of a thick deposit of contaminated bentonite beneath the base of the wall (Fig. 5.14). To minimize the

Fig. 5.16 Excavation completed for second basement at Victoria Street, London. (Note the diaphragm wall surrounding the basement, and the proximity of buildings in Howick Place.)

risk of entrapment of slurry the concrete should not be permitted to flow laterally along the base of the trench by a distance greater than 2.5 m from the tremie pipe.

An example of stage construction of a 14 m deep basement for an office block in Victoria Street, London, using the permanent substructure to support the diaphragm wall has been described by Hodgson.[5.8(a)]. At the first stage H-piles were driven close to the face of existing vaulted retaining walls to support these structures before trenches were excavated for the pairs of guide walls required as a preliminary to constructing the diaphragm wall. It was necessary to construct the guide walls in short lengths and to provide temporary steel struts between the walls to transfer the thrust from the existing vaulted retaining walls to the inner guide walls (Fig. 5.15(a)).

While this work was in progress the bored pile foundations beneath the second, shallower, basement were being constructed from the level of the existing basement. The concrete to the shafts of these piles was terminated at second basement level.

The diaphragm wall encircling the whole building was then constructed, followed by removal of the first stage

bulk excavation down to the level of the first stage ground anchors. After installing and stressing the anchors the second stage bulk excavation was then taken out (Fig. 5.16), followed by constructing the second basement slab which acted as a strut to the diaphragm wall at 5.5 m below ground level. The diaphragm wall enclosing three sides of the third (deep) basement was constructed from the second stage excavation level and also the bored piles which supported the second basement slab (Fig. 5.15(b)). The concrete in these piles was terminated at the level of the third basement slab and steel columns were set on this concrete to act as supports to the second basement slab. An opening was formed in this slab through which the soil excavated from the third basement was raised. The excavation was initially taken out to the level of the second stage ground anchors at 11.3 m. After installing and stressing the anchors the excavation was taken down to the final level at 13.9 m below ground level and concrete was placed in the third basement slab over the heads of the bored piles (Fig. 5.15(c)). The operations of removing the soil at this stage are shown in Fig. 5.17. Finally the steel columns were encased in concrete to form the supports for

Fig. 5.17 Excavation for the third basement at Victoria Street, London. (Note the steel columns supporting the second basement slab.)

the superstructure of the buildings, and the opening in the second basement slab was filled in.

The close proximity of roads and tall buildings surrounding the excavation can be seen in Fig. 5.16. The inward deflexions of the top of the diaphragm wall and the settlement of the ground surface were monitored throughout the construction period. Measurements at a point close to the deep basement excavation showed the following movements:

Construction stage	Horizontal deflexion of top of diaphragm wall (mm)	Settlement of ground surface at street level (mm)
1st stage excavation to 1st stage ground anchors	0	15
2nd stage excavation to 5.5 m	3.0	28
2nd basement slab constructed	0	17
Excavation to 2nd stage anchors at 11.3 m	0	17
Excavation to final level at 13.9 m	0	18
Final reading 6 months later	0	20

The ground surface settlement was considerably greater than would have been expected from the very small inward deflexion of the diaphragm wall. However, it will be noted that a large proportion of the settlement occurred at the first stage of excavation before the ground anchors were installed. Stressing of the anchors pulled back the top of the diaphragm wall and there was some recovery of the surface settlement. Thereafter the rigid strutting provided by the second basement slab prevented any further yielding of the diaphragm wall and there was little further settlement of the ground surface.

5.4.5 Construction in excavations supported by contiguous or secant bored pile wall

In the *contiguous bored pile* system of retaining wall construction, bored piles are installed in a single or double row and positioned so that they are touching or are in very close proximity to each other. Alternate piles are first drilled by power auger or grabbing rig and concreted. Then the intermediate piles are installed. The use of casing to guide the drill results in a space some 50–75 mm wide between piles. The space may be wider if the drilling deviates from the vertical (up to 1 in 80 tolerance is usual). These gaps between piles can be very troublesome in water-bearing granular soils which bleed through into the excavation. Grouting can be used to seal the gaps, but it may not be completely effective. Jet grouting as described in Section 11.3.7 provides a more positive method of sealing, but there are problems concerning the use of this method in an urban environment.

The problem of watertightness can be overcome by the *secant pile* method of interlocking bored piles. Alternate piles are first drilled and concreted at a closer spacing than the contiguous method. The intermediate piles are then installed by drilling out the soil between each pair followed by chiselling a groove down the sides of their shafts. Concrete is then placed to fill the drilled hole including the grooves, so forming a fully interlocking and watertight wall.

An alternative method of secant piling employs 'soft' primary piles using bentonite–cement or bentonite–cement–pf ash grout alternating with 'hard' concrete piles. The soft piles are readily cut into by auger drilling without the need for chiselling to form the groove for secondary piles. The primary piles have little structural strength and it is necessary to put more reinforcement with the secondary piles than is needed for the 'hard/hard' system.

Sherwood *et al*[5.9] have compared the costs of the two forms of secant piling with diaphragm walling for six different soil conditions and with secant pile diameters ranging from 650 to 1500 mm. Where the soil conditions permitted the use of low-torque CFA rigs (Section 8.14.1) for depths up to 18 m the hard/soft system was about 10–40 per cent cheaper than the hard/hard system with high torque CFA rigs (depths limited to 22 m). Using high-torque rigs with casing and in depths up to 40 m, costs were up to 50 per cent greater than the shallower hard/soft walls with the low-torque rig. The costs for diaphragm walls were shown to be roughly the same as those for the hard/hard system with casing oscillator high-torque rigs, but excluding the cost of the facing wall required for secant or contiguous pile walls (Fig. 5.18).

Other advantages and disadvantages stated by Sherwood *et al*[5.9] are:

(a) Cased secant methods provide a high degree of security when operating in granular soils adjacent to heavily loaded foundations or adjacent structures.

(b) In suitable conditions the hard/hard secant walls can approach the diaphragm wall in watertightness, but the risk of leakages due to non-intersection of piles increases with increasing depth.

(c) Only diaphragm walling can be used for depths of more than 40 m.

(d) Box-outs are not recommended for secant walls, but are readily used in diaphragm walls.

(e) More working space is required for the plant and back-up services for diaphragm walling.

(f) Secant or contiguous piling avoids the risk of loss of bentonite into abandoned sewers or cellars which can cause the collapse of the sides of a diaphragm wall trench.

Contiguous bored pile walls were used extensively in the construction of underground stations for the Hong Kong Mass Transit Railway.[5.10] For the Tsim Sha Tsui Station the piles were 450 mm in diameter drilled by continuous flight auger at 500 mm centres. Cement mortar was injected down the hollow stem of the auger while withdrawing the auger flights. Reinforcement was then lowered down the pile through the fluid mortar. Depending on the lateral loading the reinforcement consisted either of 300 × 300 mm steel H-beams or a cage of ten 25 mm steel bars. At the Choi Hung Station the piles took the form of hand-excavated shafts using the technique described in Section 4.6.2. Alternate shafts were reinforced. The intermediate ones were filled with mass concrete. In a modification of this method at Diamond Hill Station the shafts were widely spaced. The intermediate section was hand excavated in an elliptical shaft, with one side consisting of a mass concrete jack arch and the other side a temporary steel arched liner plate (Fig. 5.19).

5.5 Piled basements

Basement rafts carrying heavy buildings on weak soil are often founded on piles. The normal function of the piles is to transfer the loading to stronger and less compressible soil at greater depth, or, if economically possible, to transfer the loading to bedrock or other relatively incompressible strata. The piles also have the effect of stiffening the raft and reducing or eliminating reconsolidation of ground heave, thereby reducing differential settlements or tilting.

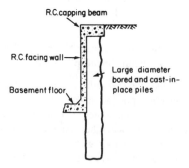

Fig. 5.18 Facing wall for contiguous or secant pile retaining wall.

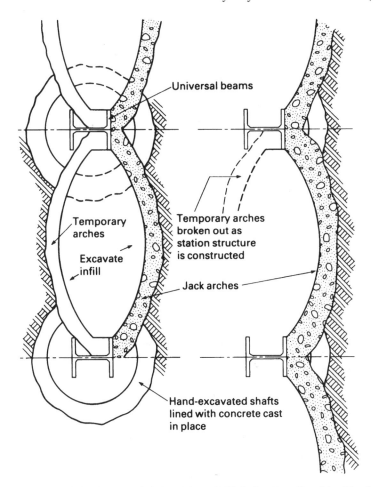

Fig. 5.19 Retaining wall constructed in hand-excavated shafts at Diamond Hill Station, Hong Kong Mass Transit Railway.

Where piles are terminated on a yielding stratum such as a stiff clay or weak weathered rock, settlement of the piles as the working load is built up will result in some of the load being carried by the underside of the slab raft or by the pile caps. The soil beneath these relatively shallow structures will then compress, causing load transfer back to the piles. The process is continuous always with some proportion of the load being carried by the piles and some by the capping structure. Therefore while the piles must be designed to carry the full superstructure loading, the slab raft which transfers the load to the piles should have sufficient strength to withstand a loading on the underside equivalent to the net load of the superstructure or to some proportionate of the net load which is assessed from a consideration of the likely yielding of the piles, the compressibility of the shallow soil layers, and the effects of basement excavation and pile installation.

Hooper[5.11] has described the measurements of the relative proportions of load carried by the underside of the basement raft and the piles supporting the 31-storey tower block of the Hyde Park Cavalry Barracks in London. The piles were installed by drilling from ground level and concreting the shaft up to the base level of the 8.8 m deep basement before commencing the bulk excavation.

The weight of the building including the imposed load but excluding wind load was 228 MN. The weight of soil removed in excavating to base of raft level was 107 MN, giving a net load to be carried by the piles of 121 MN or a net bearing pressure at raft level of 196 kN/m^2. From observations in pressure cells placed between the underside of the raft and the soil, and load cells in three of the piles, the load transfer from the superstructure to the piles was calculated from the time of commencing construction up to 3 years after completing the building. The results of these calculations are shown in Fig. 5.20, and are compared with the estimated weight of the building at various

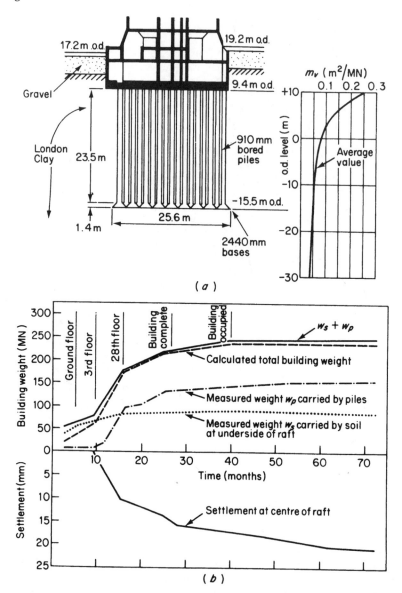

Fig. 5.20　Piled raft foundations for Hyde Park Cavalry Barracks, London. (*a*) Foundation arrangements and soil compressibility. (*b*) Transfer of load from raft to piles.

stages of construction. Hooper estimated that, at the end of construction, 60 per cent of the building load was carried by the piles and 40 per cent by the raft. In the post-construction period more load was gradually being transferred to the piles, about 6 per cent of the downward load being transferred to the piles in the 3-year period.

Some typical examples of the effects of excavation, pile installation, and pile yielding are given below.

Case A: Piles wholly in compressible clay (Fig.

5.21(*a*))　In the course of excavation for the basement, heave takes place with further upward movement caused by displacement due to pile driving, or, if bored piles are used, there may be a small reduction in the amount of heave due to inward movement of the clay around the pile boreholes. After completion of the piling, the swelled soil will be trimmed off to the specified level of the underside of the basement slab. After concreting the basement slab there will be some tendency for pressure to increase on the slab due to long-term swelling of the soil, but this will

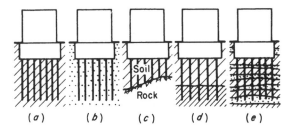

Fig. 5.21 Piled basements in various soil conditions. (*a*) Wholly in compressible clay. (*b*) Wholly in loose sand. (*c*) Bearing on rock. (*d*) Through weak clay into stiff clay. (*e*) Alternating layers of weak clay and sand.

be counteracted to some extent if driven piles are used by the soil displaced by the pile driving settling away from the slab as it reconsolidates around the piles. However, as the load on the basement increases with increasing superstructure loading, the piles themselves will settle due to consolidation of the soil in the region of the pile toes. The soil surrounding the upper part of the piles will follow the downward movement of the underlying soil and thus there will be no appreciable tendency for the full structural loading to come on to the basement slab. After completion of the building, long-term settlement due to consolidation of the soil beneath the piles will continue, but at all times the overlying soil will move downwards and will not develop appreciable pressure on the basement slab. Thus, it can be stated that the maximum load which is likely to come on the underside of the slab will be that due to soil swelling in early days after pile driving together with water pressure if the basement is below ground-water level. If, however, the working loads on the piles were to exceed their ultimate carrying capacity they would move downwards relative to the surrounding soil. The slab would then carry the full load of the building, until the consolidation of the soil throws the load back on to the piles with progressive movement continuing until equilibrium is reached.

The net downward movement resulting from the algebraic sum of heave, reconsolidation, and further consolidation will be lower for the piled basement than for an unpiled basement. This is illustrated in Fig. 5.22.

Consider first the unpiled basement. If AB represents the designed level of the underside of the basement floor slab, then as excavation proceeds the soil at level AB will rise to point 1 as the ground beneath swells due to reduction in overburden pressure. The excavation will then be trimmed back to the specified level AB (point 2) and the slab concreted. Long-term swelling will continue until point 3 is reached when the superstructure loading has increased to an extent corresponding to the original overburden pressure and reconsolidation of the swelled ground will take place. Long-term consolidation settlement will then continue until ultimate settlement is attained (point 4). The total downward movement is then Δu.

In the case of the piled basement the excavation will remain open and unconcreted for a longer period until all piles have been installed. After completion of piling (point 5) the soil is trimmed off to the specified level AB and the floor slab constructed over the piles. There will be some continuing upward movement at basement level as the soil around and beneath the piles continues to swell, but if the piles are long in relation to the width of the building such movement will be very small. When the superstructure loading reaches the original overburden pressure (point 6) reconsolidation will take place. The net downward movement (Δp) will be less, since the swelling is less and the consolidation due to net additional superstructure loading will also be less since the piles will have been terminated in soil of lower compressibility. If, however, the piles are relatively short there will be no appreciable reduction in net settlement as compared to an unpiled basement. The piles will then be wholly within the zone of swelling which may well be greater because the excavation will remain open for a longer period. To be effective in reducing net settlements, piles should be terminated below the zone of swelling.

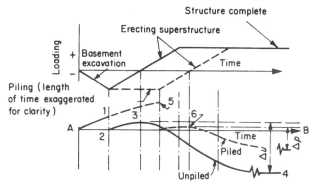

Fig. 5.22 Comparison of settlement of piled and unpiled basement walls.

It is evident from the foregoing that where piles are terminated below the zone of swelling and anchored against uplift by shaft friction or enlarged bases there will be considerable tension in the pile shafts unless measures are taken to prevent its occurrence, such as sleeving the piles within the swelling zone. Uplift on the underside of the basement slab and the consequent transfer of the uplift forces to the piles can be prevented by providing a layer of weak compressible material below the slab. However, there are drawbacks to these methods. Sleeving the pile shafts with light corrugated metal can give installation difficulties particularly if there are unstable soils within the sleeved zone as described by Findlay and Wren.[5.12] There is a risk of long-term biological decay of some types of compressible material. This can result in the accumulation of asphyxiating gases beneath the basement slabs.

A further risk with piled basements constructed by top-down methods in heaving clay is upward convexity occurring in the ground floor and upper intermediate basement slabs where these are connected to the steel columns (Fig. 5.15) at an early stage in construction. In some circumstances tension can develop at the junction between the columns and the tops of the piles, and care is necessary to ensure that the holding-down bolts to the column base plates are sufficiently long and not overstressed.

Case B: Piles driven into loose sand (Fig. 5.21(b)) In this case it is assumed that piles are required because the looseness of the sand would cause excessive settlement of an unpiled basement. It is also assumed that the density of the sand increases with depth (as is normally the case). The only swelling will be elastic movement due to removal of overburden pressure during excavation. This will take place instantaneously as the excavation is deepened, and there will be no further movement after the bottom has been reached. The total elastic movement will, in any case, be quite small and will be rapidly reversed by the consolidation of the sand by pile driving. It may even be necessary to fill up the cone-shaped depressions around the piles with sand before concreting the basement raft. As the superstructure is erected the entire loading of the superstructure and basement will be carried on the piles. This loading is thrown on to the denser sand at depth, and the settlement due to consolidation of the deeper sand strata will take place more or less instantaneously as the loading on the piles increases. The sand surrounding the piles will settle as the underlying sand compresses and thus there will be no transfer of load to the underside of the basement slab. It would, of course, be necessary to allow for water pressure.

Case C: Piles installed through compressible soil to bedrock (Fig. 5.21(c)) The soil beneath the basement excavation will swell due to removal of overburden pressure, with further swelling due to displacement by the piles if driven types are used. After completion of piling the heaved-up soil will be trimmed down and the basement floor slab concreted. There will be no further swelling because heave due to pile driving will be greater than any residual swelling movement. At some stage during erection of the superstructure, the soil heaved by driving displacement will reconsolidate and it may settle away from the underside of the floor slab. There will be no settlement of the piles other than slight elastic shortening of the pile shafts and elastic compression of the rock beneath the pile toes. Thus at no stage will any load come on to the underside of the basement slab other than that from water pressure (if any). If, however, bored piles were to be used there would be no displacement by the piles and it is quite likely that long-term swelling would cause pressure on the basement slab even with some lifting of the piles. This movement would be reversed as soon as the superstructure load exceeded the original overburden pressure.

Case D: Piles installed through soft clay into stiff clay (Fig. 5.21(d)) This case is intermediate between cases A and C. Swelling of the soil will occur because of excavation for the basement and displacement by driven pile types. After trimming off the heaved soil and concreting the floor slab there will be no further swelling. Reconsolidation of the displaced soil will proceed as the superstructure load increases. There will be a tendency for the soil to settle away from the underside of the slab since the settlement due to reconsolidation of the soft compressible soil will be greater than the consolidation of the stiff clay soil beneath the piles. Thus at no time will any soil pressure be carried by the underside of the basement floor slab, except when bored piles are used as described in case C.

Case E: Piles installed in alternating layers of soft clay and sand (Fig. 5.21(e)) This case gives heave and subsequent consolidation conditions which are intermediate between those of cases A and C.

5.6 Settlement-reducing piles in piled rafts and basements

Where piles of the same length and diameter are used to support a raft or basement slab carrying uniform loading, the slab assumes the classic dished shape when undergoing settlement. As noted above, the slab carries virtually the whole of the load at the initial stage, but as loading increases there is a progressive increase in the proportion of the load carried by the piles. In the central area of the group the piles are influenced to a greater extent by the adjacent piles than those at the periphery. Consequently the central piles are relatively less stiff and suffer a greater settlement, and load is redistributed through the slab to the stiffer outer piles which then attract a higher proportion of the load. Cooke *et al.*[5.13] measured the loads distributed over 351 piles in a rectangular group supporting a 16-storey building in London Clay. It was found that the central piles were carrying only 40 and 60 per cent of the loads carried respectively at the corners and edges of the group.

Differential settlement can be reduced and some economy in piling costs can be achieved by providing longer and hence stiffer piles in the central area and shorter more 'ductile' ones in the outer rows. The latter are allowed to yield when the resulting settlement reduces the differential settlement between the centre and the edges. The stiff slab acts as the member which redistributes the load shed from the outer piles to the stiffer central piles. If the arrangement of the superstructure loading permits, the same effect can be achieved by adopting uniform length piles but increasing the spacing of the outer rows. Alternatively, the geometry of the piles can be varied using long straight-shafted piles in the centre and short enlarged base piles around the edges, or piles with 'soft toes' can be provided by dropping a quantity of loose clay cuttings or a batch of loose sand into the pile borehole before concreting the shaft. The enlarged base pile (Section 7.9.2) undergoes a greater settlement at the same loading compared with the straight-shafted pile of equivalent bearing capacity.

Because the outer piles are permitted to yield, the overall stability of the structure must be assured. It is therefore desirable that the raft should be designed to carry the whole of the loading without structural failure or overstressing of the underlying soil. For this reason the concept of settlement-reducing piles is best applied to fairly stiff compressible soils such as London Clay. A method of designing the slab and piles to achieve uniform settlement is described by Padfield and Sharrock.[2.31]

The concept of yielding was used primarily as a means of reducing stress in a piled raft slab rather than as settlement reducers for the foundations of the Queen Elizabeth II Conference Centre in Westminster.[5.14] Five columns carrying loads between 13.2 and 25.4 MN were located along the southern edge of the basement slab and one carrying 24.1 MN on the northern edge. It was considered that these column loads when wholly supported by piles would induce excessive bending stresses in the raft with local yielding in the underlying clay. Rather than providing unyielding piles below the columns, straight-shafted piles carrying a high proportion of the load in shaft friction were used and were designed to yield under loads transmitted through the raft. In this way a higher proportion of the load would be carried by the raft,

and the edge column loads would be redistributed through the raft on to the adjacent piles. It was calculated that the particular column loads as carried by the raft were reduced by 25–46 per cent.

Monitoring of the loads carried by the piles and the stresses beneath the raft was undertaken during and after construction.[5.15] The results showed that the raft was carrying 70 per cent of the loading and the piles 30 per cent of the loading from the columns above. If unyielding piles had been used it was estimated that the raft would have carried only 20 per cent of the loads, and hence there would have been greater concentration of stress in the raft above the piles.

5.7 The structural design of basement rafts and retaining walls

5.7.1 General principles and the application of limit state design

Retaining walls around basements represent a classic example of soil–structure interaction where the degree of movement of the walls by translation or rotation under the pressure of the retained soil can modify considerably the magnitude of the pressure acting behind the walls or the resistance of the soil to movement in front of the walls. These opposing forces are known as the *active* and *passive* pressures respectively, but the author prefers the term 'passive resistance' to 'passive pressure' because it describes more accurately the condition where the soil provides *resistance* to pressure imposed on it by the structures.

The effects of movement by rotation of walls on the mobilization of active pressure and passive resistance in sand and stiff over-consolidated clay are shown in Figs 5.23(*a*) and (*b*) respectively. The values of the earth pressure coefficient K in Fig. 5.23(*a*) were obtained by measurements of a model wall rotating in sand, whereas those in Fig. 5.23(*b*) were obtained by numerical analysis and may not wholly represent actual conditions.

Two types of basement retaining walls are considered in this section, (1) free-standing walls, and (2) embedded walls, shown in Figs 5.24(*a*) and (*b*) respectively. Free-standing walls are used in shallow basements where the ground is sufficiently stable to stand without support or they are constructed within an excavation supported by timbering or sheet-piling. In a basement the free-standing condition is only temporary. At some stage the top of the wall is restrained from overturning by strutting with the permanent ground floor and the base against sliding by the abutting basement floor. Sometimes these restraints are provided temporarily by raking struts bearing on the upper part of the wall and by embedment or horizontal struts at

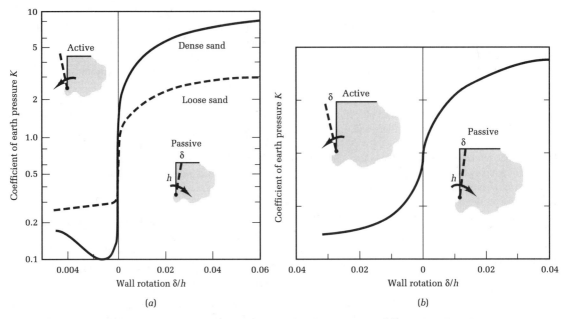

Fig. 5.23 Relationship between wall rotation and earth pressure. (*a*) In sands (after Terzaghi[5.16]). (*b*) In stiff clays (after Potts and Fourie[5.17]).

the base. Shallow walls are usually designed for the free-standing condition. Embedded walls (Fig 5.24(*b*)) are economical for deep basements and in all situations where there is insufficient space around the basement to slope back the sides of the excavation or to provide temporary support to a vertical soil face. Types of embedded walls include diaphragm, contiguous bored piles, secant piles, and sheet piling.

It is usual to assume linear distribution of active pressure and passive resistance for rigid walls. The active pressure distribution can be modified in the case of free-standing walls by the effect of compacting the backfill

causing higher pressure in the upper part of the wall. Strutting the upper part of the wall can also change the pressure distribution, but the linear distribution is not modified in normal design practice unless the struts are preloaded to push the wall back against the soil. Stressed ground anchors produce the same effect and can reverse the curvature of a flexible wall.

Three different methods are in general use to calculate the required geometry of retaining walls. Only one of these is applicable to free-standing walls. A fourth method which was introduced fairly recently is applicable only to embedded walls. All four methods are discussed in *CIRIA*

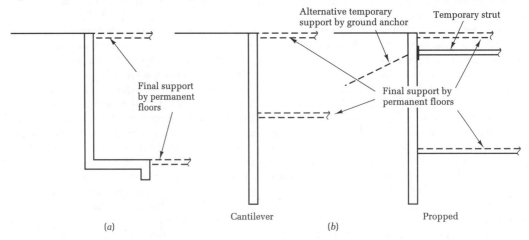

Fig. 5.24 Types of basement retaining walls. (*a*) Free-standing. (*b*) Embedded.

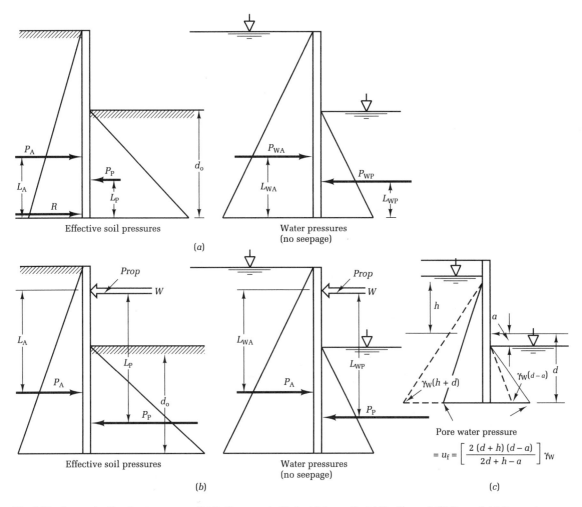

Fig. 5.25   Assumed soil and water pressure distribution on embedded retaining walls. (*a*) Cantilevered. (*b*) Propped. (*c*) Pore-water pressure distribution for seepage below toe of wall.

Report 104, 'Design of retaining walls embedded in stiff clays' (Padfield and Mair[5.18]). This report gives comprehensive and helpful advice on many aspects of the design of temporary and permanent retaining structures, and is recommended reading.

Method 1, the factor on embedment (F_d) method, is applicable only to embedded walls. Two conditions are considered. When the wall is acting as a cantilever (Fig. 5.25(*a*)) a concentrated force R is applied at the toe to represent the passive resistance mobilized at the rear due to fixity at the toe (the method is also known as the fixed earth support method as given in the British Steel handbook for the design of sheet pile walls.

No factors are applied to the limiting or fully mobilized components of active soil pressure and passive soil resistance. The factor of safety is applied empirically by factoring the depth of embedment d_o which is determined by equating the moments on each side of the wall such that the resultant is zero. For the cantilever condition (Fig. 5.25(*a*)) the depth d_o must satisfy the equation

$$P_P L_P + P_{WP} L_{WP} = P_A L_A + P_{WA} L_{WA}. \qquad (5.1)$$

Having found d_o it is increased by the factor of embedment F_d such that

Design depth $d = d_o F_d$. $\qquad (5.2)$

Then d is increased further by 20 per cent to allow for the simplifying assumption of the force R.

At the stage when the wall is propped by the ground floor slab equation (5.1) still applies, but with the different geometry of L_P, L_{WP}, L_A, and L_{WA}. The design depth d is again obtained from equation (5.2), but it is not increased by 20 per cent because free earth support conditions apply at the toe.

Method 2 is referred to as the factor of safety on shear strength (F_s) method and is applied to embedded walls. The pressure diagram is the same as Method 1 for both the cantilevered and propped stages (Figs 5.25(a) and (b)), but the active pressure and passive resistance are factored to increase the former and decrease the latter. Equations (5.1) and (5.2) apply with the design depth d being increased for the cantilevered wall, but not for the propped wall.

Method 3 is applicable to embedded and free-standing walls (Fig. 5.26). It is known as the factor of safety on moments (F_p) method, and also as the CP2 method, the latter referring to the Institution of Structural Engineers' Code of Practice: *Earth-retaining Structures*. This was published in 1951, but its revision was long delayed because of lack of agreement on the many controversial aspects of retaining wall design.*

The depth of embedment or the proportions of the wall stem and base are calculated, when taking moments about

the toe to satisfy the equation

$$F_p = \frac{\text{Restoring moments}}{\text{Overturning moments}}. \qquad (5.3)$$

Usually only net water pressures are included for overturning moments so that for embedded walls

$$F_p = \frac{P_P L_P}{P_A L_A + P_{WA} L_{WA} - P_W L_W}. \qquad (5.4)$$

The design embedment $d = d_o$ is increased by 20 per cent for cantilevered walls but not for propped walls.

For free-standing walls

$$F_p = \frac{P_P L_P + Wb}{P_A L_A + P_{WA} L_{WA} - P_{WP} L_{WP}}. \qquad (5.5)$$

Method 4 is known as the factor of safety on moments (F_r) as modified by Burland *et al*[5.19]. The method applies only to embedded walls and eliminates some of the balancing loads for the moment equilibrium equation and has the aim of using a single lumped safety factor not dependent on variations in soil strength parameters. The factor of safety is given as

$$F_r =$$

$$\frac{\text{Moment of net available passive resistance}}{\text{Moment achieved by retained material (including water) and surcharge}}. \qquad (5.6)$$

In Fig. 5.27 the net activating pressures and resistances are shown as shaded areas. For cantilever and embedded walls

$$F_r = \frac{P_{PN} L_{PN}}{P_{A1} L_{A1} + P_{A2} L_{A2} + P_{W1} L_{W1} + P_{W2} L_{W2}}. \qquad (5.7)$$

The depth of embedment d_0 which satisfies equation (5.7) is increased by 20 per cent for cantilevered walls, but not for propped walls.

Padfield and Mair[5.18] recommend safety factors for the four methods as shown in Table 5.2. Two design approaches to their use are recommended. In approach A moderately conservative values are selected for the drained soil parameters c' and ϕ', and for c_u in the total stress application. These are not average values, but would represent a cautious or 'low range' as used by engineers in bearing capacity calculations by limit equilibrium methods. The moderately conservative parameters are used in conjunction with a relatively high safety factor. 'Worst credible' values in approach B are used with correspondingly 'low safety factors. They would represent a lower bound line in a plot of c_u values against depth or the lower envelope of a plot of effective shear strength against effective stress.

A draft of the Eurocode 7, Chapter 8 ('Retaining

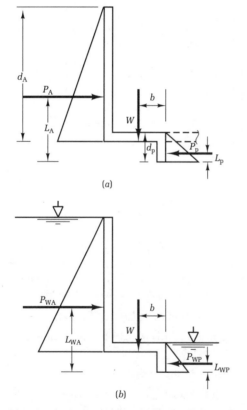

Fig. 5.26   Assumed soil and water pressure distribution on free-standing retaining walls. (*a*) Effective soil pressures. (*b*) Water pressure (no seepage).

* Published in 1994 as British Standard 8002: *Earth Retaining Structures*.

Table 5.2   Recommended factors of safety for determining a stable wall geometry in stiff clays (after Padfield and Mair[5.18])

Method		Design approach A		Design approach B		Comments
		Recommended range for moderately conservative parameters (c', ϕ', or c_u)		Recommended minimum values for worst credible parameters ($c' = 0$, ϕ')		
		Temporary works	Permanent works	Temporary works	Permanent works	
Factor on embedment F_d	Effective stress	1.1–1.2 (usually 1.2)	1.2–1.6 (usually 1.5)	Not recommended	1.2	This method is empirical. It should always be checked against one of the other methods
	Total stress†	2.0	—		—	
Strength factor method F_s	Effective stress	1.1–1.2 (usually 1.2 except for ϕ' >30° when lower value may be used)	1.2–1.5 (usually 1.5 except for ϕ' >30° when lower value may be used)	1.0	1.2	The mobilized angle of wall friction δ_m, and wall adhesion, C_{wm}, should also be reduced (see Section 7.3.2)‡
	Total stress†	1.5	—	—	—	
Factor on moments: CP2 method F_p	Effective stress	1.2–1.5	1.5–2.0	1.0	1.2–1.5	These recommended F_p values vary with ϕ' to be generally consistent with usual values of F_s and F_r
	$\phi' \geqslant 30°$	1.5	2.0	1.0	1.5	
	$\phi' = 20–30°$	1.2–1.5	1.5–2.0	1.0	1.2–1.5	
	$\phi \leqslant 20°$	1.2	1.5	1.0	1.2	
	Total stress†	2.0	—	—	—	
Factor on moments: Burland–Potts method F_r	Effective stress	1.3–1.5 (usually 1.5)	1.5–2.0 (usually 2.0)	1.0	1.5	Not yet tested for cantilevers. A relatively new method with which little design experience has been obtained
	Total stress†	2.0	—	—	—	

† Total stress factors are speculative, and they should be treated with caution. ‡ of CIRIA report 104.

Notes
(1) In any situation where significant uncertainty exists, whether design approach A or B is adopted, a sensitivity study is always recommended, so that an appreciation of the importance of various parameters can be gained.
(2) Only a few of the factors of safety recommended in Table 5.1 are based on extensive practical experience, and even this experience is recent. At present, there is no well-documented evidence of the long-term performance of walls constructed in stiff clays, particularly in relation to serviceability and movements. However, the factors recommended in the table are based on the present framework of current knowledge and good practice.
(3) Of the four factors of safety recommended, only F_p depends on the value of ϕ'.

Structures'), has not been published at the time of drafting this edition. The strength factor method represents the Eurocode 7 approach where the value of F_s corresponds to the partial factors for materials (γ_m). The characteristic soil strength could be taken as corresponding to the values in design approach A where the CIRIA F_s values are much the same as the partial factors in Eurocode 7 (see Table 2.2).

Simpson[5.20] in discussing the application of load factors or safety factors to the design of retaining structures points out the risks when walls are given only a shallow embed-ment on the passive side. The total passive resistance ($P_P = K_P\gamma'd$) in a dense granular soil is very sensitive to the embedment depth, and accidental over-excavation by only a small amount could reduce the safety factor to unity. Simpson states that Eurocode 7, Chapter 8, and the BSI committee redrafting the CP2 code are likely to suggest that the top layer of soil on the passive side should be disregarded. The layer thickness could be specified as 0.5 m or 10 per cent of the height below the lowest prop. In free-standing walls with shallow embedment the passive resistance can be ignored and stability given by the

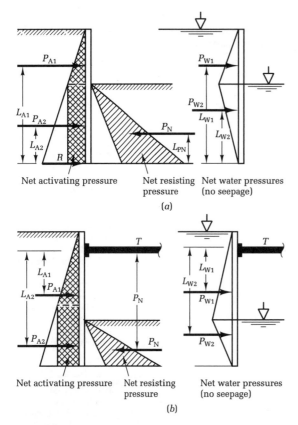

Net activating pressure   Net resisting   Net water pressures
pressure   (no seepage)

(a)

Net activating pressure   Net resisting   Net water pressures
pressure   (no seepage)

(b)

Fig. 5.27 Assumed soil and water pressures on embedded retaining walls (Burland–Potts method). (a) Cantilevered. (b) Propped.

restoring moment from gravity (*Wb* in Fig. 5.26) and from base friction until such time as the toe of the wall is propped by the basement floor slab.

CIRIA recommend that where significant uncertainty exists, whether design approach A or B is adopted, a sensitivity analysis should be made so that an appreciation of the importance of the various parameters can be gained.

The application of the four methods and the application of safety factors to the support of retaining structures in temporary works are discussed in Section 9.6.2.

5.7.2 Calculation of earth pressure on free-standing walls

Active soil pressures and passive soil resistances on permanent basement retaining walls should be calculated in terms of *effective stress*. Effective stress parameters c' and ϕ' are obtained from consolidated drained triaxial tests or consolidated undrained triaxial tests with pore-water pressure measurements (Section 1.11). It is impor-

Table 5.3 Approximate relationship between plasticity index and range of shearing resistance of clays (after Padfield and Mair[5.18])

Plasticity index	Angle of shearing resistance (ϕ)
15	30°
20	28°
25	27°
30	25°
40	22°
50	20°

tant that the stresses to which the samples are consolidated should take into account the range of stresses appropriate to the wall being designed. Where the worst credible parameters are adopted for design the cohesion intercept c' should be taken as zero.

Instead of making the special triaxial tests referred to above it may be sufficient for preliminary design purposes to use drained parameters for clays based on relationships with the plasticity index of clay as shown in Table 5.3. The relationships are empirical and are stated to be conservative. They should be used in conjunction with a c' value of zero. In clays containing layers or laminations of sand or silt it is important that the plasticity index should be representative of the clay layer.

The effective active pressure behind the retaining wall is given by the equation

$$p'_a = K_a(\gamma z - u) \tag{5.8}$$

and the effective passive resistance by the equation

$$p'_p = K_p(\gamma z - u), \tag{5.9}$$

where

K_a and K_p = the active pressure and passive resistance coefficients respectively, γ is the bulk density, and u = pore-water pressure at any depth z, below the ground or excavated surface.

Where the soil has a cohesion intercept

$$p'_a = K_a (\gamma z - u) - 2c'\sqrt{K_a} \tag{5.10}$$

or

$$p'_p = K_p (\gamma z - u) + 2c'\sqrt{K_p}. \tag{5.11}$$

The relationships between the coefficients K_a and K_p and the effective angle of shearing resistance of the soil ϕ' are shown in Tables 5.4 and 5.5.

The importance of the pore-water pressure will be noted in the above equations. The long-term values should be used in retaining wall design, and any measures to reduce it such as by drainage should be used with caution because of the possibility of drains becoming blocked. Observations of ground-water levels and seepages observed while drilling boreholes in silts and clays are likely to be

Table 5.4 Values of the coefficient K_a for active earth pressure

Angle of wall friction, δ	Effective angle of shearing resistance, φ' degrees)								
	15°	18°	21°	24°	27°	30°	33°	36°	39°
0°	0.589	0.528	0.472	0.422	0.376	0.333	0.295	0.260	0.228
10°	0.533	0.480	0.431	0.387	0.346	0.308	0.274	0.243	0.214
12°	0.526	0.474	0.426	0.382	0.342	0.305	0.271	0.240	0.212
14°	0.520	0.468	0.421	0.378	0.339	0.303	0.269	0.239	0.210
16°	0.515	0.464	0.418	0.375	0.336	0.300	0.267	0.237	0.209
18°	0.511	0.461	0.414	0.372	0.334	0.299	0.266	0.236	0.209
20°	0.508	0.458	0.412	0.370	0.332	0.297	0.265	0.235	0.208
22°	0.506	0.456	0.410	0.369	0.331	0.296	0.264	0.235	0.208
24°	0.505	0.455	0.409	0.368	0.331	0.296	0.264	0.235	0.208
26°	0.504	0.454	0.409	0.368	0.330	0.296	0.264	0.235	0.206
	Silty clays and clayey silts			Sandy silts and silts		Sands and gravels			

Note: The above values assume a horizontal plane ground surface in the retained soil.

Table 5.5 Values of the coefficient K_p for passive earth resistance

Angle of wall friction δ	Angle of shearing resistance, φ (degrees)								
	15°	18°	21°	24°	27°	30°	33°	36°	39°
0	1.70	1.89	2.11	2.37	2.66	3.00	3.40	3.86	4.40
= φ/2	1.98	2.32	2.73	3.26	3.92	4.78	5.91	7.44	9.53
	Silty clays and clayey silts			Sandy silts and silts		Sands and gravels			

Note: The above values assume a horizontal plane surface in the soil providing passive resistance.

unreliable. Standpipes or piezometers should be installed for long-term measurements, if possible over the winter months, where steady-state seepage takes place into a sheeted excavation. The pore-water pressure distribution is shown in Fig. 5.25(*d*).

Because of the movement of a retaining wall required to mobilize the full available passive resistance in clay, conservative values should be chosen for φ' for use in Table 5.5. This is particularly important in free-standing walls where the depth of embedment of the toe is likely to be small, and as noted in Section 5.7.1 much of the resistance may be lost by over-excavation, and in any case much of the resistance to horizontal sliding will be given by skin friction beneath the base slab until it is strutted by the basement floor. Padfield and Mair[5.18] state the opinion of many engineers that a minimum embedment of 2.5 m should be provided on the passive side where reliance is being placed on passive resistance for overall stability.

The angle of wall friction δ in Tables 5.4 and 5.5 should

be taken as $\frac{2}{3}$φ' in the active zone and $\frac{1}{2}$φ' in the passive zone. Mobilization of wall friction depends on relative movement between the wall and the retained soil. If the top of the wall is loaded, or if stressed ground anchors are used, the wall may move downwards relative to the soil when reduced or zero values of δ should be used.

5.7.3 Calculation of earth pressure on embedded walls

Equations (5.9)–(5.12) are used to calculate the pressure intensity on embedded walls in conjunction with the distribution of pressure shown in Figs 5.25 or 5.27. Thus at stage 1 (Fig. 5.25(a) and 5.27(a)) the wall is acting as a vertical cantilever. At stage 2, where props are used without preloading by jacking, the earth-pressure on the back of the wall will be in the active condition as in Fig. 5.25(*b*) and 5.27(*b*). In the case of an anchored wall as illustrated in Fig. 5.28, when the top level ground anchors

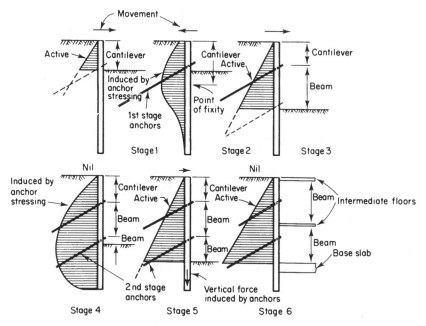

Fig. 5.28   Pressure distribution on anchored diaphragm wall at various stages of construction.

are stressed the pressure on the back of the wall will be determined by the stresses induced in the anchors and it may at some stage be intermediate between the 'at rest' and passive condition depending on the amount of reverse movement of the wall. After completion of stressing the anchors the wall acts in stage 3 as a vertical beam cantilevering above the anchors and restrained by yielding supports at anchor level and at the embedded portion below excavation level.

At stages 4 and 5 with the second stage anchors stressed, the wall acts as a vertical beam with three levels of yielding supports, and at stage 5 the construction of the basement slab and the intermediate floors provide unyielding supports and a check should be made to ensure that the wall can withstand 'at rest' earth pressure and any hydrostatic pressure on the rear face.

During construction a considerable the force will be induced in the wall by the stressing of the ground anchors, and a check should be made by the methods described in Chapter 2 to ensure that the ultimate bearing capacity of the soil beneath the toe of the wall is not exceeded.

In some situations high bending moments can be induced in the portion of the wall beneath the base slab after the latter has been completed. In fact the deeper the embedment required for the stages before constructing the base slab the higher the bending moments resulting from the unbalanced pressures, when clay on the passive side

will have been softened by the reduction in overburden pressure at the stage of bulk excavation.

For the 'at rest' earth pressure condition the coefficient K_a should be replaced by K_o. Some values of K_o are shown in Table 5.6.

Table 5.6   Values of the coefficient of earth pressure, at rest (K_o)

Loose normally consolidated sand	0.5
Medium–dense normally consolidated sand	0.45
Dense normally consolidated sand	0.35
Normally consolidated clay	0.75
Overconsolidated clay	1–2

The above values of K_o for normally consolidated sands were obtained from the relationship $K_o = (1 - \sin\phi)$. There is no simple method of obtaining K_o values for overconsolidated sands which depends on the over-consolidation ratio (OCR) of the deposit which in turn depends on its stress history. The pressuremeter (Section 1.4.5) can be used to obtain the horizontal stress in the soil and hence to obtain K_o from the relationship $K_o = \sigma'_h/\sigma'_v$, but the test may not be reliable in sands due to the sensitivity of these soils to disturbance, although reliable values can be obtained in clays.

Values of K_o for dense heavily overconsolidated sands can be as high as 2.5 which is in the region of the

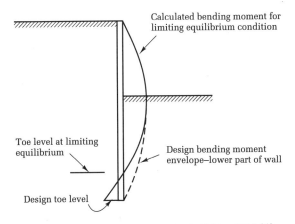

Fig. 5.29 Relationship recommended by Padfield and Mair[5.18] between calculated and design bending moment envelopes for reinforcement design of embedded retaining walls.

maximum passive state. Meigh[1.8] has given the relationship established by Schmertmann[5.21]

$$K_{o(oc)} = (OCR)^{0.42} K_{o(onc)} \qquad (5.12)$$

Padfield and Mair[5.18] recommend that the maximum bending moment in the wall stem should be calculated with unfactored soil parameters. The required design embedment depth is greater than that required to give limiting equilibrium with unfactored parameters, but this additional depth is not accounted for in the calculations. The reinforcement should not be curtailed at the point where the bending moment is zero, but it should be taken down to the bottom of the wall using a bending moment envelope shown in Fig. 5.29. Padfield and Mair[5.18] point out that the seepage path for water around the wall for the embedment depth assumed for bending moment calculations is shorter than that corresponding to working conditions, but state that the error due to neglecting the additional seepage length for calculating water pressure is small.

Calculation of bending moments using unfactored soil strengths may be adopted in Eurocode 7 where partial factors are applied to soil strengths. Simpson[5.20] points out that the bending moments calculated from unfactored soil strengths are multiplied by a safety factor of 1.4 or 1.5 to determine the required ultimate limit state of the wall section. Whereas if the factors F_p for moments or F_s for soil strengths are taken as greater than unity, the resulting bending moments can be used directly for ultimate limit state design of the wall stem. This often gives higher bending moments than the CIRIA recommended method (Padfield and Mair[5.18]). Simpson states that if the length of the wall required by the stability calculations (Section 5.7.1) is actually needed, pressure acting near its base will have large lever arms so causing bending moments. Use of

the CIRIA method will lead to walls which are not strong enough to use their length. They are either unnecessarily long or not sufficiently strong. Length in this connection refers to the sum of the retained height and embedment depth.

5.7.4 Structural design

Sometimes, for economy in the quantity of concrete, steps are provided in the retaining wall to give a progressive increase in thickness from top to bottom. However, because of the increased cost of formwork, and complications in the arrangement of reinforcing steel, stepping does not normally show economies in overall costs. In the case of backfilled retaining walls a sloping rear face may be an economical form of construction for long walls.

Free-standing retaining walls should not depend on their connection to the basement floor slab for stability against overturning or sliding. The width of the base constructed within the trench should be sufficient to keep the bearing pressure at the toe of the base slab within safe limits and to ensure that the resultant of the pressure on the back of the wall and the weight of the wall falls within the

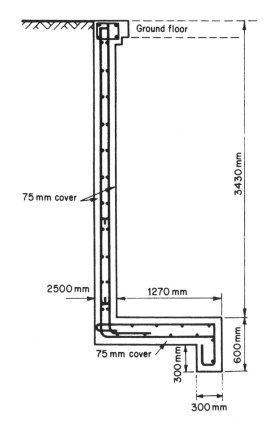

Fig. 5.30 Cantilever retaining wall for basement.

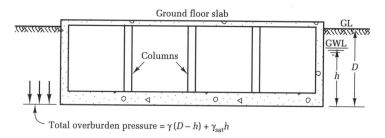

Fig. 5.31   Design of basement floor slab where basement acts as buoyancy raft.

middle third of the base. If necessary a projection should be provided beneath the wall to prevent sliding, as shown in Fig. 5.30. This illustration shows a simple design for a single-storey cantilevered basement wall. The uniformity in thickness provides economy in fabrication and erection of formwork, and there is a minimum of laps and bends in the reinforcing steel, giving ease of fixing in the confines of an excavation.

Counterforted or tee-section retaining walls are not often used for basements, since the additional excavation outside the building line to accommodate counterforts and base slabs has to be backfilled, thus increasing the cost of excavation without gaining space in the basement. Also, in built-up areas the counterforts and base slabs might encroach on adjacent property.

Basement floor slabs are designed as rafts on the lines described in Chapter 4. They must be able to withstand pressures on the underside of the slab together with stresses caused by differential settlement, non-uniform column loads, and reactions from the retaining walls. An important factor is the need to provide continuity in the base slab if the basement acts as a buoyancy raft (Fig. 5.31). If the columns are provided with independent bases with only a light slab between them, there would be likelihood of failure of the slabs from the pressure of the underlying soil.

For light or moderate bearing pressures the floor slab may consist of a slab of uniform thickness or, where columns are widely spaced, as a 'flat slab' with thickened areas beneath columns (Fig. 5.32). In order to avoid cracking of the floor slab there should not be abrupt changes from thin to thick parts of the slab and the junction of the floor slab with the column bases and wall foundations should be provided with a generous splay as

shown in Fig. 5.32. This will avoid concentrations of stress at the junction. For high bearing pressures it is usually necessary to provide a grid of heavily reinforced upstanding beams of the type shown in Fig. 4.32.

The floor slabs of piled basements should also have structural continuity where pressure develops on the underside due to soil swelling or yielding of the piles (see Section 5.5) or where water pressure is acting on the slab. Continuity is given by bonding the floor slab into the pile caps as shown in Fig. 5.33(a). Where there is no risk of any load transfer to the slab, for example where the piles are taken down to an unyielding stratum, it can be designed without structural connection to the pile caps or to the capping beam supporting the retaining wall (Fig. 5.33(b)).

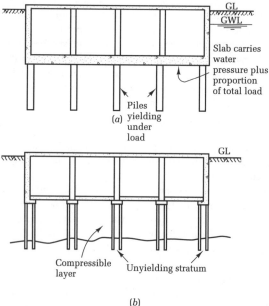

Fig. 5.33   Design of floors for piled basements. (a) Where load is partly carried by underside of floor. (b) Where no load is carried by underside of floor.

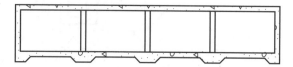

Fig. 5.32   Basement floor slab with thickened bases to columns.

5.8 The application of computer-based methods to the design of rafts and piled rafts

5.8.1 Rafts

Settlement profiles and soil contact pressures for raft foundations are dependent on the relative stiffnesses of raft and supporting soil. Use of computer-based methods is a necessity for the detailed analysis of those cases where the soil–structure interaction aspects of raft foundations are significant; that is to say, in cases where it is unrealistic to assume a raft as being either infinitely rigid or infinitely flexible. Development of a suitable representation of the soil behaviour has been the major difficulty to overcome in improving the methods for undertaking such analyses.

Winkler spring model   Initially this approach was the most widely adopted. It consists of representing the soil which supports a raft by discrete vertical springs. Usually the raft was modelled as an equivalent grillage, or grid, of rigidly connected beams (Fig. 5.34). Modelling of a slab subjected to flexure as a grillage has a long history, particularly in the analysis of bridge decks. Standard computer programs able to model grillages supported on springs have been available since the 1960s, and so it was to be expected that engineers would apply the technique to the design of rafts. Later, the use of the grillage method was largely replaced by the use of plate bending finite elements (Fig. 5.34) which provided a far more versatile means of modelling such features as irregular shapes, openings in plan, and variations in raft thickness. General-

purpose finite element programs typically include both plate bending elements and spring supports and offer the next stage of raft modelling beyond the standard grillage on springs.

However, although either grillages or plate bending elements are acceptable for representing the flexural action of a raft, the use of Winkler springs to represent the soil is questionable. The principal objection to Winkler springs is their inability to replicate the interaction, due to stress transfer within the soil mass, of settlements at separate points on the soil surface. Since the mid-1970s a number of authorities have expressed caution on the use of Winkler springs in modelling rafts. For example, Brown[5.22] remarked that 'the Winkler assumption remains what it has always been, one of mathematical convenience rather than physical reality'. A report produced by the Institution of Structural Engineers[5.23] commented that the Winkler spring model could not be recommended for the analysis of rafts, that it was a poor physical model, and could give rise to erroneous results. More recently, Hooper[5.24] stated that the method could give good answers in certain cases, but that its performance was problem-dependent. The method was not sufficiently general for practical purposes and could lead to grossly inaccurate results on the unsafe side. With regard to problem dependency, Mawditt[5.25] concluded that the Winkler spring model would give satisfactory results for structural design of a raft where, among other requirements, the loading could be approximated to a single point load. This is a severe limitation when compared with the loading configurations which need to be considered in most practical cases of raft design.

The limits to safe applicability of the Winkler spring model may well not be recognized by many design engineers, and so its use cannot be recommended as a general practice. It is much better to make use of techniques which can model soil behaviour and soil–raft interaction with more realism over a wide range of practical conditions.

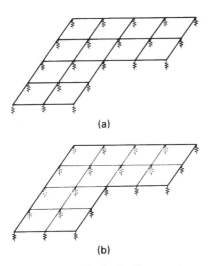

(a)

(b)

Fig. 5.34   Simple models of raft using discrete springs to represent soils. (*a*) Raft modelled as grillage of beams on Winkler springs. (*b*) Raft modelled as plate-bending finite elements on Winkler springs.

Boundary element/surface element methods   The area of raft analysis and design is one in which boundary element or quasi-boundary element methods have achieved widespread usage. The usual manner of carrying out an analysis involves use of plate bending finite elements to model the raft, and boundary elements to model the soil. It should be noted that the use of the term 'surface element' is often preferred to 'boundary element' in these applications since the methods adopted rarely involve the integral equation type of formulation which is usually associated with the boundary element method.

The first soil idealization considered is that of a linear elastic, isotropic half-space; that is, the soil is considered to have an infinite depth and to have constant linear elastic properties throughout that depth. The raft plate bending elements are defined by a two-dimensional mesh with specific node points. The surface of the soil under the raft is defined by an equivalent mesh and node points. Using the Boussinesq formulation (see Section 2.6.4), it is possible to generate values for vertical displacements at any node point on the surface of the linear elastic half-space due to the application of a vertical force at any other node point, i.e.

$$\delta_j = P_i f_{ij}. \tag{5.13}$$

where

δ = vertical displacement at point j,
P_i = vertical force at point i (equal to the product of pressure at point i and mesh area associated with node i),
f_{ij} = flexibility term dependent on elastic properties of the half space and distance between points i and j.

Considering a total of n node points the cumulative vertical displacement at j is given by

$$\delta_j = \sum_{i=1}^{n} P_i f_{ij}. \tag{5.14}$$

The relationship between displacements and forces for all nodes can be expressed in matrix form as

$$\delta = [F_s]\{P\} \tag{5.15}$$

where $[F_s]$ is the soil flexibility matrix and $\{\delta\}$ and $\{P\}$ are vectors of displacements and forces respectively. The inverse of the soil flexibility matrix is the soil stiffness matrix $[K_s]$ and

$$[K_s]\{\delta\} = \{P\}. \tag{5.16}$$

The relationship between applied loads $\{Q\}$, raft–soil contact forces $\{P\}$, and displacements $\{\delta\}$ for the plate bending elements of the raft is

$$[K_r]\{\delta\} = \{Q - P\}, \tag{5.17}$$

where $[K_r]$ is the raft stiffness matrix.

Substituting equation (5.16) into equation (5.17) gives

$$[K_r + K_s]\{\delta\} = \{Q\}. \tag{5.18}$$

Knowing $\{Q\}$ and $[K_r + K_s]$, this set of linear simultaneous equations can be solved to give nodal point values of displacements $\{\delta\}$ and hence, by back substitution, nodal point values of raft–soil contact pressure and raft bending moments.

Although the elastic half space idealization is amenable to classical elastic theory, there are few practical situations to which it approximates. The next stage in development

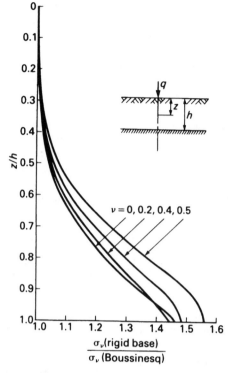

Fig. 5.35   Under-prediction by Boussinesq theory of vertical stress near lower boundary of finite layer (after Padfield and Sharrock[2.24]).

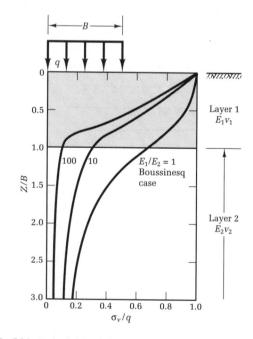

Fig. 5.36   Reduction in vertical stress as a result of stiff overlying layer (after Padfield and Sharrock[2.31]).

of boundary–surface element techniques involved direct consideration of soil layering and variations of soil stiffness with depth within layers. The basic assumption is that, within the heterogeneous soil mass beneath the raft, the stress distribution due to vertical loads applied at the surface is the same as that within a semi-infinite elastic medium, but that the vertical strain at any elevation is then governed by the defined elastic properties of the soil at that elevation.

The layered soil mass has a defined total depth at the base of which zero vertical displacement is assumed. The soil mass is further divided into a number of horizontal sublayers, and the computed vertical strain at the mid-level of each sublayer is assumed to apply throughout its thickness. Total surface displacements at any point are obtained by summation of vertical deflexions within sublayers below that point – hence flexibility matrix terms can be computed. This approach is an equivalent of the Steinbrenner method of calculating settlements (Section 2.6.6).

The assumption that a stress pattern for a semi-infinite continuum is applicable to a layered subsoil is clearly an approximation. There will be an underestimate of vertical stresses near the assumed level of zero displacement (Fig. 5.35) and, in the case of a stiff surface layer overlying softer soils, an overestimate of vertical stress in the upper levels of the softer soils (Fig. 5.36). However, the approach is an advance on the ideal semi-infinite continuum model and will give very satisfactory results over a wide range of circumstances, as has been demonstrated in a number of published case histories. The method has also been extended to include for such features as local yielding of the soil when raft contact pressures reach a limiting value and loss of contact between raft and soil in zones where an apparent vertical tension would otherwise develop at the surface of the soil. It is normal to assume a

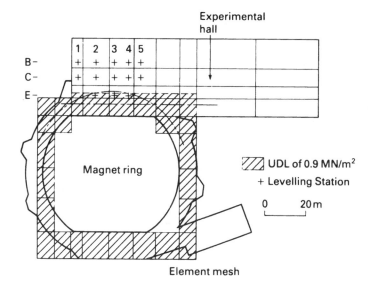

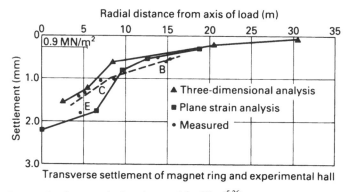

Fig. 5.37 Comparison of measured and computed raft settlements (after Wood[5.26]).

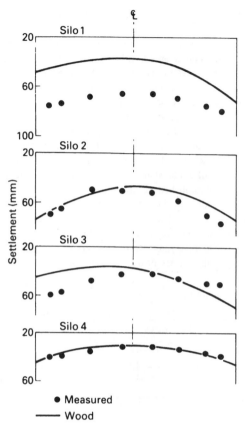

Fig. 5.38 Comparison of measured and computed settlements of silo raft foundations on Chalk (after Wood[5.28]).

smooth interface between raft and soil. This is a conservative assumption in respect of computed settlements and raft bending moments, and in any event the extent to which horizontal shear can be sustained at the raft–soil interface is difficult to quantify.

Figure 5.37 shows results obtained by Wood[5.26] when applying the method to compute settlements of an irregular plan form structure. The results obtained compare favourably with site measurements and can be compared with results of a plane strain finite element analysis[5.27]. Wood also obtained computed values for comparison with measured settlements of circular raft foundations on chalk, for a complex of four grain silos. In this instance allowance was made for local yielding of the chalk and measured and computed values are shown in Fig. 5.38. Hooper and West[5.29] stated that the simplified layered continuum method of analysis described above was highly satisfactory in most practical cases. Hooper[5.24] described eight raft analyses involving its use. He also modified the method in order to obtain results for rafts on strata whose thickness or material properties vary in the horizontal direction as well as in the vertical direction[5.30].

In the minority of cases where the simplified layered continuum method is considered to be inadequate a more rigorous method described by Hooper and West[5.29] must be adopted. This method involves obtaining explicit solutions for the layered strata by means of the integral transform method. They give various examples involving its use. Figures 5.39, 5.40 and 5.41 illustrate its application to a complicated circular plan form raft on shallow layered strata. Figure 5.41 is interesting in that it also shows results for the simplified layer method, for the isotropic half-space method, and for the Winkler spring

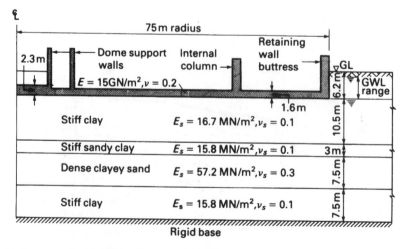

Fig. 5.39 Data used in basement raft analysis (after Hooper and West[5.29]).

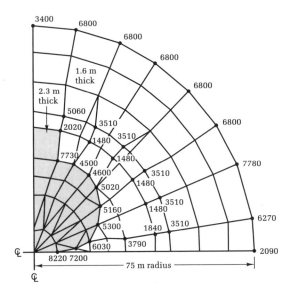

method. It can be seen that for the conditions represented in Fig. 5.41 the simplified layer and rigorous layer methods give very similar results, but that the other methods show marked differences. Also shown in Fig. 5.41 are results for an analysis ignoring the structural effect of the raft, these can be compared with traditional predictions for settlement of flexible foundations.

Finite element methods By comparison with other areas of foundation engineering, finite element methods have been applied very little to the soil model in the analysis and design of raft foundations. The principal reason for this is one of cost. For an irregular shaped raft,

Fig. 5.40 Vertical applied loads (kN) for analysis of basement quarter-raft (after Hooper and West[5.29]).

Fig. 5.41 Results using different soil models in elastic analysis of basement quarter-raft (after Hooper and West[5.29]).

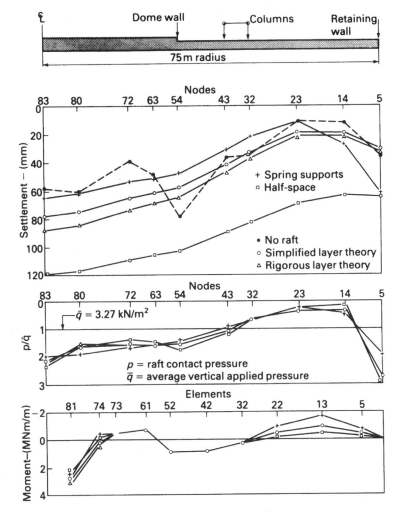

and even for a rectangular raft, a comprehensive finite element representation of the soil requires the use of three-dimensional elements which typically involve markedly more computing time than the equivalent boundary/surface element methods described above. The associated increased cost cannot be justified in a commercial environment. Two-dimensional finite element analyses are restricted either to the few cases which might be represented in plane strain (high aspect ratio rafts with loadings effectively continuous in the longitudinal direction) or to axisymmetric cases (symmetrical about a vertical axis of rotation) which are effectively limited to circular rafts subjected to axisymmetric loading, and for which a number of standard elastic solutions are already available; for example, from Poulos and Davis[5.31].

In general, it is recommended that where raft design involves computer-based methods use is made of programs based on the simplified layer method described above, including representation of soil yielding and raft–soil separation. Such programs have been generally available in Britain since the late 1970s,[5.32] and a number of published case studies exist which can be used for verification exercises.

5.8.2 Piled rafts

Following from remarks made above, it is not surprising that finite element methods have been very little used in modelling all the components of raft–pile–soil systems.

The three-dimensional modelling of such piled raft systems by finite elements would be exorbitantly expensive for use as a general design technique, as will be appreciated from the details shown in Fig. 5.42 of a finite element model of a five by three pile group – and it will be noted that, on the basis of symmetry, only one-quarter of this modest size pile group has been modelled. Two-dimensional finite element modelling was used by Hooper,[5.11] who adopted an axisymmetric idealization for the analysis of the piled raft foundation of the Hyde Park Cavalry Barracks, London. Hooper's model for this analysis is shown in Fig. 5.43.

The combined plate bending finite element and simplified layer theory boundary/surface element method described above has found reasonably wide application in the analysis of rafts on closely spaced pile groups. In such a case the pile group is modelled as an equivalent raft located at a depth at or towards pile toe level. Wood[5.34] has described this form of analysis applied to a six-storey office block founded on large-diameter bored piles in London Clay. In this instance measured values of settlement were available, and Wood concluded that the best representation of the measured values was obtained by locating the equivalent raft at a depth in the range 0.5–0.8 of the pile penetration into the London Clay. A comparison of measured values and results for an equivalent raft at a depth equal to 0.7 of the pile penetration into the London Clay is given in Fig. 5.44. Hooper[5.24] describes a similar analysis for the design of the piled raft foundation to an

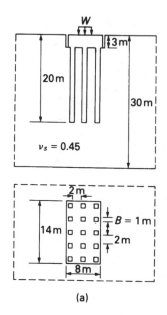

(a)

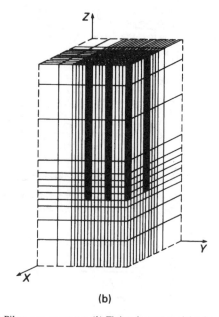

(b)

Fig. 5.42   Three-dimensional finite element model of pile group. (a) Pile group geometry. (b) Finite element model (after Ottaviani[5.33]).

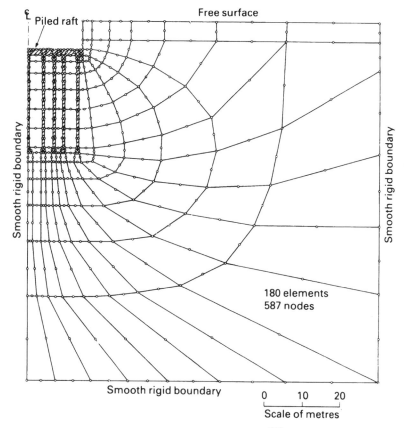

Fig. 5.43 Axisymmetric finite element model of piled raft foundation (after Hooper[5.36]).

office tower in Lagos, Nigeria: in this case the equivalent raft was modelled at pile toe level.

Hain and Lee[5.35] used a combination of plate-bending finite element model for a raft and interaction factors for pile–pile, pile–surface, surface–pile, and surface–surface effects. Approximate allowances were made for effects of soil heterogeneity, cut-off values of pile load, and localized soil yield. Hooper[5.36] commented that this particular method of analysis can be treated with a reasonable degree of confidence in cases where the piles are widely spaced and the soil stiffness is reasonably uniform from pile head level to a considerable depth below pile toe level. Figure 5.45 shows results obtained by Hain and Lee for percentage of load taken by the piles in an eight by eight group supporting a raft. In the figure K_R and K_p are relative stiffness parameters given by Hooper[5.36] in the form:

$$K_R = \frac{4E_R(1-y^2_s)}{3\pi E_s}\left(\frac{t_R}{B_R}\right)^2$$

$$K_p = \frac{E_p}{E_s}$$

where

$\quad E_R =$ Young's modulus of the raft material,
$\quad E_p =$ Young's modulus of the pile material,
$\quad E_s =$ Young's modulus of the soil mass,
$B_R, t_R =$ breadth and thickness of the raft,
$\quad v_s =$ Poisson's ratio of the soil.

An advance on the Hain and Lee approach is incorporated in a computer program PILRAFT described by Padfield and Sharrock[2.31]. Here integral transform theory is used to achieve a better representation of the effects of soil layering and variations of soil stiffness with depth. They describe the application of PILRAFT to the back analysis of an instrumental piled raft in London Clay with subsequent production of an alternative design in which pile lengths are varied across the raft with the objective of minimizing differential settlements (see Section 5.6).

Pending wider availability of, and confidence in, programs such as PILRAFT, the designer of a piled raft foundation with closely spaced piles should adopt the equivalent raft form of computer analysis described by Wood[5.32] or Hooper[5.24]. However, it must be noted that

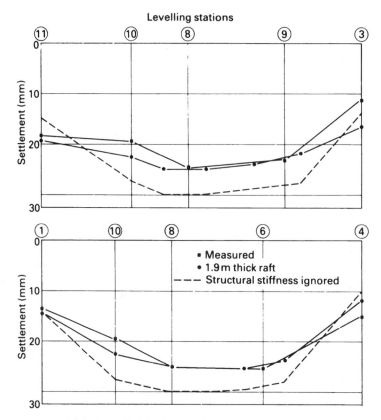

Fig. 5.44 Computed and measured settlement profiles for raft on closely spaced piles modelled as equivalent raft (after Wood[5.34]). (Transversely isotropic soil model $y/H = 0.7$ where y is depth to equivalent raft and H is the depth to pile toe level.)

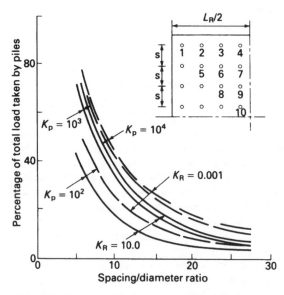

Fig. 5.45 Percentage of load taken by 8×8 pile group for a raft-pile foundation $L/d = 25 v_s = 0.5$. where L is pile penetration, d is pile diameter, and v_s is Poisson's ratio of soil (after Hain and Lee[5.35]).

the equivalent raft method is not capable of defining relative proportions of load carried by the piles and by the raft.

5.9 Waterproofing basements

The first essential in preparing a scheme for waterproofing basements is to ensure that the membrane or other water-excluding barrier is taken up to a sufficient height. Information on ground-water levels obtained from boreholes is not always a reliable guide to determining the rest level of the ground water around the walls of the completed basement. For example, the basement may be constructed on a sloping site and so form a barrier to the seepage of ground water across the site. This will result in a rise in ground-water level on the uphill side of the structure. Borings on a clay site may show only random seepages of water at depth. However, after completing the basement, water may collect in the backfilled space around the walls, especially if the backfilling has been placed in a loose state. The space may act as a sump for surface water around the walls, and this water may rise nearly to ground level.

The only satisfactory method of ensuring watertightness in an underground structure below the water table is to surround it with an impervious membrane. This process is called 'tanking' and adhesive plastic sheeting (e.g. 'Bituthene') or asphalt are the normal materials used for the membrane. An alternative to sheeting or asphalt is the use of Volclay panels. These are 1.22 m × 1.22 m by 5 mm thick fluted cardboard panels. The flutes in the panels are filled with Wyoming bentonite. Panel joints are lapped and staggered. When wetted the bentonite swells and forms a gel which provides a permanent flexible seal. Where there is a free supply of water to hydrate the Volclay panels, say by surface or underground flooding, the swelling pressure exerted by the bentonite may cause movement or structural distress to basement walls in situations where the panels are rigidly confined, for example where they are confined between an existing and a new basement wall. To deal with this possibility the suppliers of the Volclay panels recommend the use of pressure relief boards. These consist of 1.22 m × 2.44 m × 10 mm polystyrene sheets with a moulded cup surface. The sheets are placed against the existing wall and used in conjunction with Volclay Type IF (Fastseal) panels. These have one rapid degrading kraftboard face. Where they are used consideration should be given to the possibility of generation of methane gas due to biological decay of the material (see Section 5.5).

Patent additives to cement are sold for 'waterproofing' concrete. Some of these are useless, while others may be satisfactory for a time and then lose their effect. However, good-quality dense concrete may sometimes be produced for the only reason that the directions given by their manufacturers include careful proportioning of the materials and placing and compaction of the concrete, whereas when normal Portland cement is used for basement concrete such precautions are not always followed faithfully.

Some engineers believe that tanking of basements with asphalt is an unnecessary expense. It costs some 20 per cent more than an untanked basement, and if only part of this money is applied to increasing the quality of the concrete and to ensuring close supervision of placing and joint preparation then the resulting structure will be just as watertight. They claim that just as many cases of leakage occur with tanked basements as with untanked structures. This may well be true. It is certain that leaks in a tanked basement are much more difficult to seal because it is impossible to determine the point of ingress of the water, whereas in an untanked basement the point of leakages can readily be seen and sealed.

Although good-quality concrete is to all intents and purposes impervious, the construction joints are always a potential source of leakage. The joint at the base of the walls is a particular case as the reinforcement is nearly always congested at this point. Preformed water-stops at joints are often ineffective because of the difficulty in compacting the concrete around them. The best types of preformed joint are those which are placed on the outside of the structure and lie flat against the soil or formwork. Concrete is placed only against one side of the joint strip with less risk of displacing the material or forming voids in the concrete. The author would adopt tanking by plastic sheeting or asphalt to waterproof a basement in preference to an untanked structure. Imperfections in placing concrete and preparation of joints are more likely to occur with the usual standard of unskilled labour employed than imperfections in tanking laid by experienced operatives who appreciate the need for care in details of workmanship.

5.9.1 Tanking materials

Asphalt tanking can be applied either by three coats of hot mastic asphalt trowelled on to the surfaces, or by several layers of bituminous felt. The latter method is generally regarded as inferior to trowel application because of the difficulty in ensuring complete sealing at joints and laps, especially with complicated details of lapping at internal angles.

The materials for asphalt tanking are covered by BS6925, *Mastic Asphalt for Building Materials (Limestone Aggregate)*, and BS6577, *Mastic Asphalt for Building Materials (Natural Rock Asphalt Aggregate)*.

5.9.2 Application of tanking

The following points on workmanship are made in British Standard CP 102: 1973: *Protection of Buildings Against Water from the Ground*:

(1) The foundations must be kept dry, with continuous pumping, as necessary, until all asphalt is applied and protective coats have set.

(2) Horizontal asphalt should be protected by 50 mm of cement–sand mortar laid as soon as each section of the asphalt is completed, to protect the asphalt against damage by traffic, building materials, and reinforcing bars.

(3) The asphalt should be continuous and it should therefore be taken below all columns.

(4) If the asphalt is applied to a backing wall (internally applied), the space between the inside of the asphalt and the inner (structural) wall should be solidly grouted to prevent movement under external water pressure (Fig. 5.46).

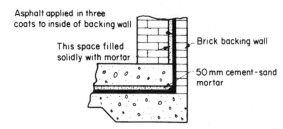

Fig. 5.46 Asphalt tanking to brick basement wall where asphalt is applied to brick backing wall (internally applied).

(5) Where the asphalt is applied to the external surface of the retaining wall (externally applied), a working space of 0.6 m should be provided, in which to execute the work, and a 100 mm thick protective wall should be built outside the asphalt to protect it from damage by sharp stones, bricks, and similar material in the backfilling.

(6) For externally applied asphalt, a horizontal set-off at least 150 mm wide should be provided to make a satisfactory double angle fillet and connection to be formed with the vertical asphalt.

(7) Internal and external angles should be suitably splayed to allow the asphalt to be carried evenly round the angle without variation in thickness.

(8) Brickwork should have all horizontal joints raked out and brushed clean to provide a good key for the asphalt; concrete should *not* have a smooth surface.

(9) Asphalt tanking should be applied in three coats to a total thickness of not less than 30 mm for horizontal work and not less than 20 mm for vertical work.

(10) Internal angles should be provided with asphalt angle fillets applied in two coats and finishing approximately 50 mm wide on the face (Fig. 5.47).

It is not always easy to decide whether the waterproof membrane should be applied to the outer face of the structural wall or to the inner face of an external protecting or backing wall. What is certain is that the membrane is useless on the inner face of the structural wall. Concrete can never be made to be completely leak-proof, and water

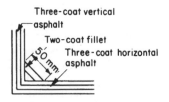

Fig. 5.47 Forming internal angle fillets.

pressure transmitted through the structural wall will readily push off the membrane, forming large water-filled blisters which eventually break, thus destroying the watertightness of the tanking.

If the tanking material is applied to the outer face of the completed structural wall, it can be easily inspected for imperfections before the protective brick wall is built. Once the wall is in position the tanking cannot be damaged unless major foundation movement causes wide cracks in the structure. This method of applying the tanking is preferred if the basement is built in an excavation with sloping sides where ample room is available for applying the membrane. However, if the method is used where the retaining wall is constructed in a trench, the excavation requires to be at least 0.6 m wider on the outer face which is wasted excavation and has to be backfilled. Therefore, for economy in trench excavation costs the alternative method of applying the membrane to the inner face of the protective wall should be adopted. When placing concrete in the structural wall great care is necessary in keeping tamping rods, shovels, or poker vibrators away from the membrane, otherwise there may be unseen damage to the material. The ground-water lowering pumps should not be shut down until the structural concrete walls have been concreted and have attained their designed strength.

The usual method of treating the waterproofing of column bases is to form pits in the tanking (Fig. 5.48). On the other hand, the membrane is usually carried over the top of piled bases (Fig. 5.49). When mastic asphalt is not fully confined to prevent extrusion, grillages to columns and pile caps should be designed to limit the pressure on the asphalt to 640 kN/m² at normal temperatures.

5.9.3 Dealing with leaking basements

Waterproofing a tanked basement which is leaking can be a lengthy, difficult, and often fruitless task. If leaks occur through imperfections in the tanking, the water creeps between the membrane and the outer face of the structural wall or floor and emerges on the inner face. The point of emergence can be a considerable distance away from the point of leakage in the tanking, making the source of leakage impossible to trace. An inner skin of waterproofing material to seal off the inflow is, in most cases, ineffective in a deep basement since the water pressure merely forces it off the wall, but such measures can be effective for low heads of water. Pressure grouting can be resorted to and is sometimes effective or partially effective if undertaken by experienced operators. The grouting should be done from the outside with the aim of forming an impervious skin of grout outside the tanking and forcing the latter into closer contact with the structural

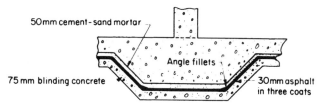

50mm cement-sand mortar

Angle fillets

75 mm blinding concrete

30mm asphalt
in three coats

Fig. 5.48   Asphalt tanking beneath column bases.

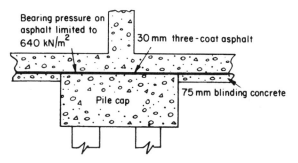

Bearing pressure on
asphalt limited to
640 kN/m^2

30 mm three-coat asphalt

75 mm blinding concrete

Pile cap

Fig. 5.49   Asphalt tanking above pile caps.

wall. If grouting is carried out on the inside of the tanking the pressure used to inject the cement may cause the membrane to burst outwards, thus worsening the situation.

In the case of thick walls and floor slabs some success may be achieved by injecting fluid chemical or resinous grouts directly into the cracks to seal them. If grouting fails the only remedy (other than reconstructing the basement) is to try and tap the source of water by chases and conduits and lead it to a pumping sump or, if surface water drains exist at a suitable level outside the basement, a drainage layer can be provided at the back of the wall connected to the piped drainage system. Even if this drainage layer cannot extend to the full depth of the wall it is often of value in reducing the head causing seepage into the basement. From this brief account of the difficulties in stopping leaks it is obvious that every care must be taken with the tanking during construction of the basement. Careful inspection of every stage of the work will be amply repaid.

5.10 Worked example of basement retaining wall

Using the four methods described in Section 5.7.1 determine the embedment depth required for the two stages of diaphragm wall construction shown in Fig. 5.50.

The required embedment depth will be governed by the forces at the propped stage. This will be calculated first and checked for the cantilever stage.

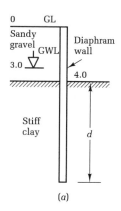

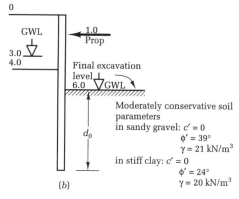

Fig. 5.50   (a) Cantilevered stage. (b) Propped stage.

Wall friction on active side = $\delta = \frac{2}{3}\phi' = 24°$ in gravel and 16° in clay.

Wall friction on passive side = $\delta = \frac{1}{2}\phi' = 12°$ in clay.

From Table 5.4, $K_a = 0.235$ in gravel and 0.375 in clay.

From Table 5.5, $K_p = 3.26$ in clay.

Pressure diagrams for Methods 1–3 in Section 5.7.1 are shown in Fig. 5.51.

Table 5.7 shows the calculation of factors of safety for embedment depths of 6.0 m and 9.0 m.

Plotting the safety factors for the two embedment

Table 5.7   Calculation of factor of safety for the two embedment depths. Take a trial depth of 6.0 m for d_0 and take moments about the prop.

	Force (kN m)		Lever arm (m)		Moment (kN m/m)
P_{A1}	$\frac{1}{2} \times 0.235 \times 21 \times 3^2 = 22.2$	L_{A1}	1.0	M_{A1}	$22.2 \times 1.0 =\ \ \ 22.2$
P_{A2}	$\frac{1}{2} \times 0.235 \times (21 \times 3 + 74) \times 1.0 = 16.1$	L_{A2}	3.67	M_{A2}	$16.1 \times 3.67 =\ \ \ 59.1$
P_{A3}	$\frac{1}{2} \times 0.375 (74 + 154) \times 8 = 342.0$	L_{A3}	$\frac{1}{2}(2 + 6) + 3 = 7.0$	M_{A3}	$342.0 \times 7.0 = 2394.0$
P_{WA}	$\frac{1}{2} \times 10 \times 9^2 = \underline{405.0}$ Total $P_A = \overline{785.3}$	L_{WA}	$\frac{2}{3}(3 + 6) + 2 = 8.0$	M_{WA}	$405.0 \times 8.0 = \underline{3240.0}$ Total $M_A = \overline{5715.3}$
P_P	$\frac{1}{2} \times 3.26 \times 10 \times 6^2 = 586.8$	L_P	$\frac{2}{3} \times 6 + 5 = 9.0$	M_P	$586.8 \times 9.0 = 5281.2$
P_{WP}	$\frac{1}{2} \times 10 \times 6^2 = \underline{180.0}$ Total $P_P = \overline{766.8}$	L_{WP}	9.0		$180.0 \times 9.0 = \underline{1620.0}$ Total $P_P = \overline{6901.2}$

$$\text{Factor of safety} = \frac{\text{Restoring moment}}{\text{Overturning moment}} = \frac{6901.2}{5915.3} = 1.2$$

Which is insufficient. Take a second trial depth of 9.0 m.

	Force (kN m)		Lever arm (m)		Moment (kN m/m)
P_{A1}	22.2	L_{A1}	1.0	M_{A1}	22.2
P_{A2}	16.1	L_{A2}	3.67	M_{A2}	59.1
P_{A3}	$\frac{1}{2} \times 0.375 (74 + 184) \times 11.0 = 532.1$	L_{A3}	$\frac{1}{2}(2 + 9) + 3 = 8.5$	M_{A3}	$532.1 \times 8.5 =\ \ \ 4522.9$
P_{WA}	$\frac{1}{2} \times 10 \times 12^2 = \underline{720.0}$ Total $P_A = \overline{1290.4}$	L_{WA}	$\frac{2}{3}(3 + 9) + 2 = 10.0$	M_{WA}	$720.0 \times 10.0 = \underline{7200.0}$ Total $M_A = \overline{11804.2}$
P_P	$\frac{1}{2} \times 3.26 \times 10 \times 9^2 = 1320.3$	L_P	$\frac{2}{3} \times 9 + 5 = 11.0$	M_P	$1320.3 \times 11.0 = 14523.3$
P_{WP}	$\frac{1}{2} \times 10 \times 9^2 = 405.0$	L_{WP}	11.0	M_{WP}	$405.0 \times 11.0 = \underline{4455.0}$ Total $M_P = \overline{18978.3}$

$$\text{Factor of safety} = \frac{18978.3}{11804.2} = 1.6.$$

This satisfies the requirement in Table 5.1 for f.o.s. between 1.5 and 2.0 where ϕ' is between 20 and 30°.

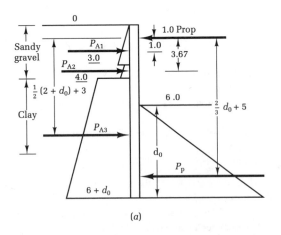

(a)

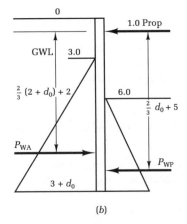

(b)

Fig. 5.51   (*a*) Soil pressures. (*b*) Water pressures (no seepage).

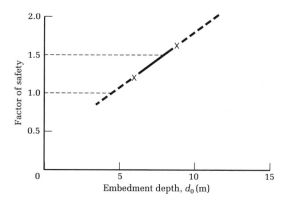

Fig. 5.52

depths in Fig. 5.52, a factor of unity is given by a depth d_o of 4.5 m.

For Method 1, the depth is increased by the factor $F_d =$ 1.2 to 1.6 for permanent works (Table 5.2). Hence embedment depth d required is

$$1.2 \times 4.5 = 5.4\,\text{m to } 1.6 \times 4.5 = 7.2\,\text{m}$$

but Method 3 governs the depth.

Checking by Method 2 (strength factor) and taking the partial factor for strength $F_s = 1.5$ giving $\phi' = \tan^{-1}$ $(39/1.5) = 28°$ in the sandy gravel and $24/1.5 = 16°$ in the stiff clay, $\delta = \frac{2}{3} \times 28° = 19°$ in gravel and $11°$ in clay. On the passive side, $\delta = \frac{1}{2} \times 16 = 8°$ in clay.

From Table 5.3, $K_a = 0.30$ in gravel and 0.51 in the clay.

From Table 5.5, $K_p = 2.1$ in clay.

For depth d_o of 9.0 m:

$$
\begin{array}{lll}
 & & \text{kN m/m} \\
M_{A1} = & 22.2 \times 0.30/0.235 \times 1.0 & = & 28.3 \\
M_{A2} = & 16.1 \times 0.30/0.235 \times 3.67 & = & 75.4 \\
M_{A3} = & 532.1 \times 0.51/0.375 \times 8.5 & = & 6151.1 \\
P_{WA} & & = & 7200.0 \\
 & \text{Total } M_A = & & 13454.8 \\
P_P = & 1320.3 \times 2.1/3.26 \times 11.0 & = & 9355.5 \\
 & & = & 4455.0 \\
P_{WP} & \text{Total } M_P = & & 13810.5
\end{array}
$$

$$\frac{\text{Restoring moment}}{\text{Overturning moment}} = \frac{13810.5}{13454.8} = \underline{\underline{1.0}}$$

which is satisfactory

Checking by Method 4, the pressure diagrams are shown in Fig. 5.53.

For simplicity assume ground-water level is at 4 m for calculating soil pressures.

$$P_{A1} = \frac{1}{2} \times 0.235 \times 21 \times 4^2 = 39.5\,\text{kN m},$$
$$P_{A2} = \frac{1}{2} \times 0.375 \times (84 + 104) \times 2 = 70.5\,\text{kN m},$$

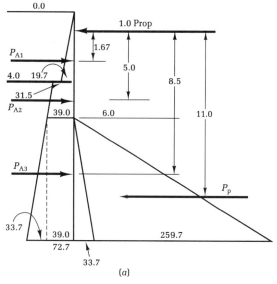

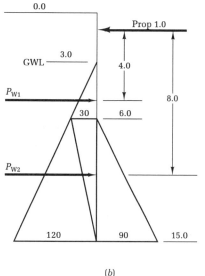

Fig. 5.53   (*a*) Soil pressures. (*b*) Net water pressures.

$$P_{A3} = 0.375\,(84 + 20) \times 9 = 351.0\,\text{kN m},$$
$$P_P = \frac{1}{2} \times 3.26 \times 10 \times 9^2 - \frac{1}{2} \times 0.375 \times 10$$
$$\qquad \times 9^2 = 1168.4$$
$$P_{W1} = \frac{1}{2} \times 10 \times 3^2 = 45\,\text{kN m},$$
$$P_{W2} = \frac{1}{2} \times 10 \times 3 \times 9 = 135\,\text{kN m}.$$

Taking moments about prop,

$$F_r =$$

$$\frac{1168.4 \times 11}{39.5 \times 1.67 + 70.5 \times 5 + 351.0 \times 8.5 + 45 \times 4.0 + 135 \times 8.0}$$

$$= \underline{\underline{2.8}}$$

From Fig. 5.51 take *a* penetration depth of 9.0 m which will give a safety factor of > 1.6 for Method 1, 1.0 for Method 2, 1.5 for Method 3 and 2.8 for Method 4.

Checking stage 1 cantilevered condition for 9 m embedment (Fig. 5.54)

$$\text{Passive resistance from soil} = P_\text{p}$$
$$= \tfrac{1}{2} \times 3.26 \times 10$$
$$\times 11^2$$
$$= 1972.3 \text{ kN m,}$$

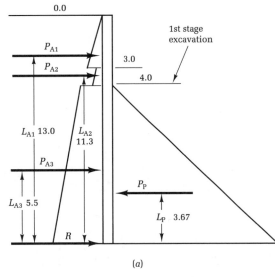

(a)

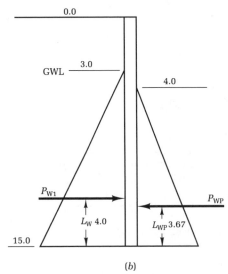

(b)

Fig. 5.54   *(a)* Soil pressures. *(b)* Water pressures.

$$\text{Passive resistance from water} = P_\text{wp}$$
$$= \tfrac{1}{2} \times 10 \times 11^2$$
$$= 605 \text{ kN m}$$

Taking moments about toe of wall:

$$\text{Overturning moment} = 22.2 \times 13.0 + 16.1 \times 11.33$$
$$+ 532.1 \times 5.5 + 720 \times 4.0$$
$$= 6277.6 \text{ kN m}$$

$$\text{Restoring moment} = 1972.3 \times 3.67 + 605 \times 3.67$$
$$= 9458.7.$$

$$\text{Factor of safety} = \frac{9458.7}{6277.6} = \underline{\underline{1.5}}$$

which is satisfactory for Method 3 – Temporary works (see Table 5.2).

References

5.1   Pike, C. W. and Saurin, B. F., Buoyant foundations in soft clay for oil refinery structures at Grangemouth, *Proceedings of the Institution of Civil Engineers*, Part 3, **1**, 301–334 (1952).

5.2   Skempton, A. W., Vane tests in the alluvial plain of the River Forth, near Grangemouth, *Géotechnique*, **1**, 111 (1948).

5.3   Somerville, S. H. and Shelton, J. C., Observed settlements of multi-storey buildings on laminated clays and silts in Glasgow, *Géotechnique*, **22**, 513–520 (1972).

5.4   Potts, D. M. and Day, R. A., Use of sheet pile retaining walls for deep excavations in stiff clay, *Proceedings of the Institution of Civil Engineers*, **88**(1), 899–927(1990), Thomas Telford, London.

5.5   *Design and Construction of Deep Basements*, Institution of Structural Engineers, London, 1975.

5.6   Measor, E. O. and Williams, G. M. J., Features in the design of the Shell Centre, London, *Proceedings of the Institution of Civil Engineers*, **21**, 475–502 (1962).

5.7   Ward, W. H., Discussion on ref. 5.6, *Proceedings of the Institution of Civil Engineers*, **24**, 411–413 (1963).

5.8   Burland, J. A. and Hancock, R. J. R., Underground car park at the House of Commons, London: geotechnical aspects, *The Structural Engineer*, **55**, 87–100 (1977).

5.8a   Hodgson, F. T., Design and construction of a diaphragm wall at Victoria Street, London, *Proceedings of the Conference on Diaphragm Walls and Anchorages*, Thomas Telford, London, pp. 51–56 (1975).

5.9   Sherwood, D. E., Haman, C. N., and Beyer, M. G., Recent developments in secant bored pile wall construction, *Proceedings of the International Conference on Piling and Deep Foundations*, Balkema, Rotterdam (1989), pp. 211–220.

5.10   McIntosh, D. F., Walker, A. J. R., Eastwood, D. J., Immamura, M. and Docherty, H., Hong Kong Mass Transit Railway, Modified Initial System, design and construction of underground stations, and cut-and-cover tunnels, *Proceedings of the Institution of Civil Engineers*, **68**(1), 599–626 (1980).

5.11   Hooper, J. A., Observations on the behaviour of a piled raft foundation on London Clay, *Proceedings of the Institution of Civil Engineers*, **55**, 855–877 (1973).

5.12   Findlay, J. D. and Wren, G. E., Review of the methods used to construct large diameter bored piles for topdown construction, *Proceedings of the International Conference on Piling and Deep Foundations*, Balkema, Rotterdam (1989), pp. 199–210.

5.13   Cooke, R. W., Bryden-Smith, D. W., Gooch, M. N. and Sillett, D. F., Some observations on the foundation loading and settlement of a multi-storey building on a piled raft foundation in London Clay, *Proceedings of the Institution of Civil Engineers*, **107**(1), 433–460 (1981).

5.14   Burland, J. B. and Kalra, V. C., Queen Elizabeth II Conference Centre: geotechnical aspects, *Proceedings of the Institution of Civil Engineers*, **80**(1), 1479–1503 (1986).

5.15   Price, G. and Wardle, I. F., Queen Elizabeth II Conference Centre: monitoring of load sharing between piles and raft, *Proceedings of the Institution of Civil Engineers*, **80**(1), 1505–1518 (1986).

5.16   Terzaghi, K., Anchored bulkheads, *Transactions of the American Society of Civil Engineers*, **119**, 1243–1281 (1954).

5.17   Potts, D. M. and Fourie, A. B., A numerical study of the effects of wall deformation on earth pressures, *International Journal for Numerical Analytical Methods* in *Geomechanics*, **10**, 383–405 (1986).

5.18   Padfield, C. J. and Mair, R. J., Design of retaining walls embedded in soft clay, *CIRIA Report 104* (1983).

5.19   Burland, J. B., Potts, D. M. and Walsh, N. M., The overall stability of free and propped embedded cantilever retaining walls, *Ground Engineering*, Thomas Telford, **14**(5), 28–38 (1981).

5.20   Simpson, B., Retaining structures: displacement and design, *Géotechnique*, **42**(4), 541–576 (1992).

5.21   Schmertmann, J. H., Measurement of *in-situ* shear strength, *Proceedings of the Conference on in-situ Measurement of Soil Properties*, American Society of Civil Engineers, North Carolina, June 1975, pp. 57–138.

5.22   Brown, P. T., Influence of soil inhomogeneity on raft behaviour, *Soils Found*, **14**(1), 61–70 (1974).

5.23   *Structure–soil Interaction, A State-of-the-Art Report*, Institution of Structural Engineers, London, 1978.

5.24   Hooper, J. A., Raft analysis and design – some practical examples, *The Structural Engineer*, **62A**(8), (1984).

5.25   Mawditt, J. M., 'The influence of discrete and continuum soil models on the structural design of raft foundations', M.Sc. dissertation, University of Surrey, UK, 1982.

5.26   Wood, L. A., The economic analysis of raft foundations, *International Journal for Numerical and Analytical Methods in Geomechanics*, **1**, 397 (1977).

5.27   Moore, J. F. A. and Jones, C. W., *In-situ* deformation in Bunter Sandstone, *Proceedings of Conference on Settlement of Structures*, Cambridge, Pentech Press, 1975, pp. 311–319.

5.28   Wood, L. A., Discussion of Majid and Rahman, Non-linear analysis of structure soil systems, *Proceedings of the Institution of Civil Engineers*, **73**(2), 527–528 (1982).

5.29   Hooper, J. A. and West, D. J., Structural analysis of a circular raft on yielding soil, *Proceedings of the Institution of Civil Engineers*, London, **75**(2), 205–242 (1983).

5.30   Hooper, J. A., Interactive analysis of foundations on horizontally variable strata, *Proceedings of the Institution of Civil Engineers*, **75**(2) 491–524 (1983).

5.31   Poulos, H. G. and Davis, E. H., *Elastic Solutions for Soil and Rock Mechanics*, Wiley, New York, 1974.

5.32   Wood L. A., RAFTS: A program for the analysis of soil – structure interaction, *Advances in Engineering Software*, **1**, 11(1978).

5.33   Ottaviani, M., Three-dimensional finite element analysis of vertically loaded pile groups, *Géotechnique*, **25**(2), 159–174 (1975).

5.34   Wood, L. A., A note on the settlement of piled structures, *Ground Engineering*, London, **11**(4), 38–42, (1978).

5.35   Hain, S. J. and Lee, I. K., The analysis of flexible raft pile systems, *Géotechnique*, **28**(1), 65–83 (1978).

5.36   Hooper, J. A., Review of the behaviour of piled raft foundations, *Construction Industry Research and Information Association*, Report No. 83, 1979.

6 Bridge Foundations

6.1 Introduction

The previous two chapters have been concerned mainly with building foundations where loadings are usually governed by architectural considerations such as the number of storeys in a building, the need or otherwise for a basement, and the spacing of columns in the superstructure. These constraints give the engineer little choice of foundation types which can be strip foundations for low-rise buildings or individual bases for framed structures. The ground conditions may require some form of piling. In contrast the bridge engineer has a much wider choice in foundation design and can often dictate the magnitude of foundation loading by choosing the span length to suit the topographical and geological conditions. A longstanding general rule is to proportion the span length so that the cost of the foundations roughly balances the cost of the superstructure of the bridge.

Usually the bridge engineer has little opportunity to choose the location of the structure to take advantage of good ground conditions. Highway or rail bridges are located to suit connections with existing roads or railways, and the sites of over-water bridges are chosen to give the shortest length of crossing or, in the case of meandering rivers, to locations where the waterway has been stabilized by natural or man-made features.

Foundation loadings for bridges are of a very different character from buildings. Imposed loads can be dominant and can be as much as half the dead load on highway bridges and two-thirds of the dead load for railway bridges. Imposed loadings from traffic are moving loads and can exert considerable longitudinal traction forces on the bridge deck. Longitudinal forces are also caused by shrinkage and temperature changes in the bridge deck, while transverse forces can be caused by wind loadings, and by current drag, wave forces, and ship collisions in the case of river or estuary crossings. Earthquake forces can be transmitted by the ground to bridge supports from any direction, and these can be critical for high-level structures or for piers in deep water where the mass of the displaced water must be added to that of the pier body. In addition to working loads from traffic, there can be rapid application of load to the foundations at the construction stage, for example when complete spans are assembled at ground level and lifted or rolled into place on to the piers.

Bridges with continuous spans can be sensitive to the effects of differential settlements between the foundations. In addition the calculated total and differential settlements must be considered in relation to the riding quality of the road surface. Critical points are the junction between the bridge and embanked approaches and joints between fixed and link spans.

6.2 Code of Practice requirements

In British practice bridge design is governed by BS 5400: *Steel, Concrete and Composite Bridges*. This standard is written in limit state forms and defines load and resistances to loads in the following terms:

$$Design\ load = Q^x = \gamma_{fL}Q_k, \qquad (6.1)$$

where Q_k *is the nominal load*

$$\gamma_{fL} = function\ (\gamma_{fi}\gamma_{fz}). \qquad (6.2)$$

The partial factor γ_{fL} *takes account of the unfavourable deviation of loads from their nominal values, and* γ_{fz} *takes account of the reduced probability that various loadings acting together will attain their nominal values simultaneously.*

Design load effects are stress resultants in a structure from its response to loading.

$$Design\ load\ effect = S^* = \gamma_{f3} \times (effects\ of\ Q^*) \quad (6.3)$$
$$= \gamma_{f3} \times (effects\ of\ \gamma_{fL}Q_k), \quad (6.4)$$

where γ_{f3} is a factor that takes into account inaccurate assessment of the effects of loading, unforeseen stress distribution in structures and variations in the dimensional inaccuracy achieved during construction.

$$\textit{The design resistance } R^* = \frac{1}{(\gamma_{m2})} \textit{ function } \frac{(f_k)}{(\gamma_{mi})}. \quad (6.5)$$

For a design to be satisfactory R^* must be equal to or greater than S^*. BS 5400 gives values of γ_{fL} and γ_{f3} for various types of loading and load combinations. The values of γ_{mi} and γ_{m2} are those given in the appropriate structure design codes.

BS 5400 dismisses foundation design in a few lines, merely stating that the foundations should be designed to BS 8004 using nominal loads as design loads (i.e. with γ_{fL} and γ_{f3} taken as unity). This is unsatisfactory. The two codes are of course incompatible, and it can often result in inappropriate and uneconomical designs for the bridge as a whole. The tendency is for the bridge engineer to design the superstructure and then to pass on the responsibility for foundation design to a geotechnical engineer who may belong to a quite separate organization and who has had no previous contact with the superstructure designer, as discussed in the following section.

6.3 Bridges on land

6.3.1 Bridge types and settlement effects

Single spans for bridges crossing minor roads or railways are supported by their abutments which are required to carry lateral pressures from the soil retained behind them in addition to the vertical and horizontal forces from the deck (Fig. 6.1(a)). Portal frames (Fig. 6.1(b)) and box structures (Fig. 6.1(c)) can also be used for these small span structures. Portal frames are not usually economical compared with simply supported beams spanning between abutments because they are sensitive to small differential settlements of the foundations which can induce bending moments at the junction between the deck and the upstand wall. Where the structure is long in a direction transverse to the span, e.g. a drainage structure beneath a dual carriageway, a box-type structure can be economical because the side walls and deck can be constructed simultaneously by travelling formwork riding on the completed base slab.

Bridges crossing dual-carriageway roads are usually designed either as a two-span structure with a central pier and conventional abutments (Fig. 6.2(a)) or with four spans supported by three piers and spill-through abutments (Fig. 6.2(b)). The spill-through abutment has the advantage of minimizing the earth pressure force to be carried by the abutment structure. The two main spans are usually designed to be continuous over the three supporting piers requiring care to avoid excessive differential settlement between them.

Arch bridges (Fig. 6.3(a)) or continuous girder bridges with inclined columns (Fig. 6.3(b)) impose an inclined thrust on the abutments when relatively unyielding soil or rock support is required. They can be economical types for deep rock cuttings, and are often preferred to slab-and-beam bridges for aesthetic reasons.

Multi-span viaducts are almost universally designed as continuous girder spans with link spans in reinforced concrete or steel and with reinforced concrete piers.

Because of the heavy moving loads on highway and rail bridges, foundation settlements are more critical to the design of the superstructure than is the case with buildings. Hambly[6.1] states that foundations for simply-supported structures are frequently designed for differential settlements of the order of 1 in 800 which corresponds to 25 mm on a 20 m span. In reasonably homogeneous soil conditions differential settlements between adjacent foundations are often assumed to be half the total settlement,

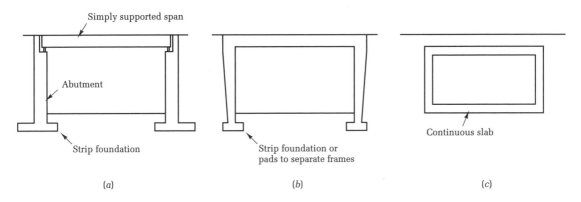

Fig. 6.1 Shallow foundations for single-span bridges. (a) With conventional abutments. (b) Portal frame. (c) Box-section.

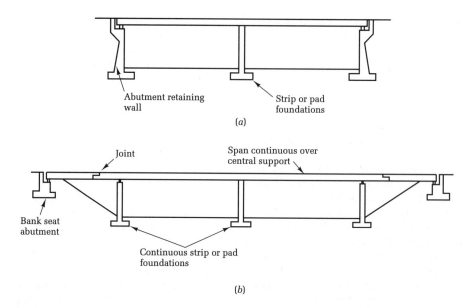

Fig. 6.2   Shallow foundation for bridges. (*a*) With conventional abutments. (*b*) With spill-through abutments.

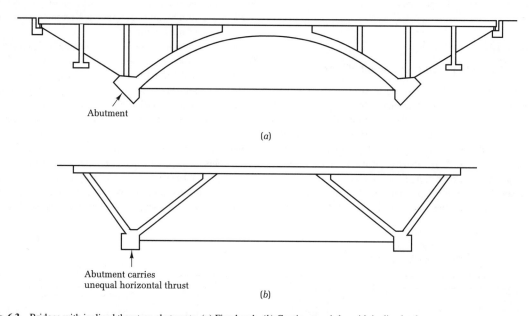

Fig. 6.3   Bridges with inclined thrust on abutments. (*a*) Fixed arch. (*b*) Continuous girder with inclined columns.

thus a total settlement of about 50 mm would be permissible. In the case of continuous deck bridges Hambly states that the 1 in 800 criterion is applied by some designers, but others expect the movement to be controlled to 1 in 4000 (5 mm on a 20 m span). Differential settlements of the order of 1 in 800 in a continuous deck are required to be treated as a load producing bending moments in the superstructure. This adds to the cost of the

structure, but it should also be noted that limitation of total settlement to 5–10 mm is difficult to achieve with spread foundations on soils of moderate to low compressibility such as medium–dense to dense sands, stiff clay, or weak rocks. Even if spread foundations are taken down through weak soils to less compressible soils at depth, the problems of heave of the underlying soil due to relief of overburden pressure and subsequent reconsolidation need to be considered. Hence the designer is often forced to adopt piled foundations to reduce settlements to tolerable limits. However, the adoption of piling should be avoided, if reasonably practicable, for the following reasons:

(a) Spread foundations are cheaper than piled foundations unless there are problems with excavations, such as inflow of ground water.
(b) The occurrence of horizontal loads on the foundations from various causes may require the use of raking piles with additional costs and construction difficulties (see Section 7.17).
(c) In main highway construction the deployment of bulky mechanical plant for installing piled foundations to bridges can cause disruption to the earth-moving programme if the piling is undertaken before the earthworks, or to the paving operation if undertaken at a later stage (see Sections 6.3.2 and 6.3.3).
(d) Railway authorities impose severe restrictions on the operation of mechanical plant close to their running lines particularly if there is overhead electrification. Piling close to the track may be restricted to periods of weekend possession.
(e) Problems of various kinds can occur in piling works causing delays and additional costs. The problems seem to be more frequent than those involving the relatively simple operations for constructing shallow spread foundations.

Before yielding to the necessity of using piles the foundation engineer should look critically at all components of the total load. It was stated in Chapter 2 that much of the settlements of foundations on sands and gravels take place immediately as the loads are applied. Hence most of the settlements due to the weight of the piers are likely to be complete before the superstructure is commenced. Where the deck and supporting girders consist of assemblies of prefabricated elements built out from the piers, much of the settlement from the superstructure dead load may have taken place before the final stressing of the main reinforcement is applied. On the other hand, where the deck is constructed *in situ* on temporary supports, nearly all the dead load is applied to the foundations as the supports are removed.

In the case of spread foundations on stiff over-consolidated clays, Chapter 2 states that the immediate settlement is about half the total, which gives scope for considering the effect of the construction programme on the amount of total and differential settlement at each stage of construction. The likely duration of the maximum traffic loading on the bridge spans should also be considered. Finally, long-term settlements due to slow consolidation of soils or creep in rocks may be capable of being accommodated by creep in the reinforced concrete or steel superstructure.

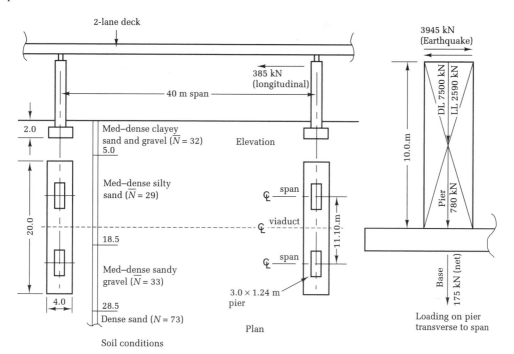

Fig. 6.4   Spread foundations for viaduct piers for dual two-lane highway.

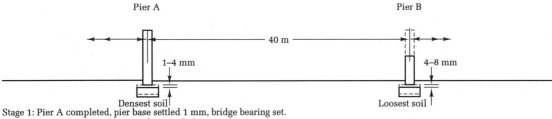

Stage 1: Pier A completed, pier base settled 1 mm, bridge bearing set.
        Pier B 50% complete, pier base settled 4 mm .
Stage 2: Balanced cantilever over pier A 50% complete, pier base settled 4 mm (total).
        Pier B completed, pier base settled 8 mm, bridge bearing set 10 mm above bearing A.

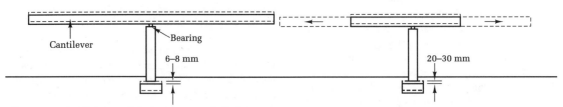

Stage 3: Balanced cantilever over pier A completed, pier base settled 6 mm (short term). Balanced cantilever over pier B 50%complete,
        pier base settled 20 mm.
Stage 4: Pier base A and bridge bearing settled 8 mm. Balanced cantilever B completed. Pier base settled 30 mm, differential settlement
        between ends of cantilevers 12 mm.

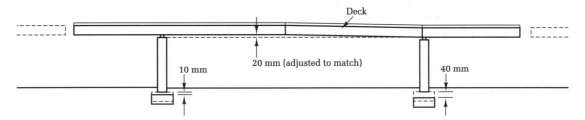

Stage 5: Deck constructed from pier A to pier B. Pier base A settled 10 mm, pier base B settled 40 mm. Differential settlement between
        bridge bearings 20 mm (1 in 2000). Differential settlement of deck between pier A and pier B is 15 mm (1 in 2700).

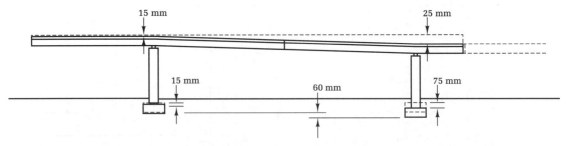

Stage 6: Long-term dead load of bridge with allowance for long-term imposed load. Deck above pier base A and pier base settlement is
        15 mm. Deck above pier base B settlement is 40 mm (total). Differential settlement of deck from A to B is 25 mm (1 in 1600).
        Final settlement of pier base B is 75 mm.

Fig. 6.5   Total and differential settlements of continuous girder bridge during construction and in long term.

6.3.2 Foundations for piers

The foundations of intermediate piers for bridges on land are required to withstand forces from the following causes:

Dead load from self-weight of pier.
Dead load from beams, deck, pavings and services.
Imposed loads from traffic in various combinations of loaded and unloaded spans.
Traction forces from traffic, longitudinal to bridge axis.
Longitudinal forces at top of pier from effects of shrinkage, temperature and creep in superstructure.
Longitudinal and transverse forces from winds and earthquakes.
Unequal horizontal earth pressure (for piers on sloping ground, or from surcharge loading by an adjacent embankment).
Impact from vehicle collision.

As an example, loadings on the piers of a dual two-lane viaduct approach to an over-water bridge are shown in Fig. 6.4. It will be noted that the unfactored imposed load is about 20 per cent of the total load on the foundations of the piers supporting the 40 m spans. A large earthquake force at the top of the pier transverse to the bridge axis necessitates a combined 20×4 m strip foundation for each pier supporting the dual carriageway structure. The 2 m deep strip is underlain by medium–dense silty clayey sands and sandy gravels, followed by dense sandy gravel (Fig. 6.4). The calculated settlements at various stages of construction and during the service life of the bridge are shown in Fig. 6.5. The feasibility of limiting the differential settlement of the bridge deck is made possible in this case by obtaining detailed information on the variations in soil density at a number of points along the viaduct location, and by precise levelling of the piers and bearings at intervals during construction. With the knowledge that pier B had settled by 8 mm while only 50 per cent complete, compared with a final settlement of 1 mm after completing pier A, the bearing on pier B is set 10 mm high after completing the pier. Then by the time that the balanced cantilevers over both piers are finished, at stage 4, the differential settlement between the bridge bearings is 20 mm compared with 30 mm between piers. The two adjacent cantilevers are stressed together at this stage. After this only the dead load of the deck in the short term and the dead and imposed load in the long term cause differential settlement of the combined deck and girders between the bearings. In the short term this differential movement is calculated to be 15 mm (stage 5) and in the long term it is 25 mm (stage 6) compared with a final differential settlement between the pier foundations of 60 mm. The final differential settlement of the deck

represents an angular distortion of 1 in 1600 which might be accommodated by creep in the superstructure of a reinforced concrete bridge.

It should be noted that the total and differential settlements were calculated by the method of Burland and Burbidge[2.11] which gives a much wider difference between the average and minimum settlements than between the average and maximum (see Section 2.6.5 and Fig. 2.26). In this example the average long-term settlement is only 30 mm. This illustrates the difficulty of making precise forecasts of settlement in variable soil conditions, but by working through the stages of construction the bridge designer can obtain a reasonably close estimate of the maximum *differential* settlement which will be experienced in the bridge spans over the life of the structure.

The procedure for minimizing differential settlement is greatly simplified if it is possible to jack the girders and adjust the bearings to level before stressing them together.

Where compressible soil conditions require the adoption of piled foundations the horizontal forces in directions parallel and transverse to the bridge axis are best resisted by a fan of raking piles (Fig. 6.6.(a)). Hambly[6.1] stated that a group of vertical piles with rakers only in the outside rows (Fig. 6.6(b)) can result in dishing at the centre of the pile cap and hence transfer of load to the outer rakers

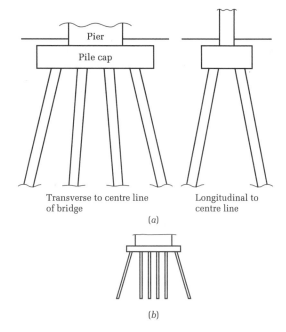

Fig. 6.6 Arrangement of raking piles for bridge pier.
(a) Preferred arrangement. (b) Unsatisfactory arrangement.

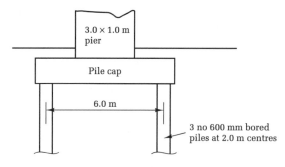

Transverse to centre line of viaduct

Fig. 6.7   Arrangement of bored piles for viaduct pier.

which could cause the latter to fail. Bored piles are very difficult, and in some cases impossible, to install on the rake. Consequently, driven displacement piles are used provided that considerations of noise levels or the effects of soil displacement do not prohibit their use.

Figure 6.7 shows an arrangement of vertical bored piles to support the pier loadings shown in Fig. 6.4 for a case where weak soil conditions exist at a shallow depth. The wide spacing of the two groups of three piles is required to resist the large overturning force.

6.3.3 Abutments

Conventional abutments of the type shown in Fig. 6.1(*a*) are designed as simple retaining walls resisting horizontal pressures from the embankment fill or natural soil on the back of the wall; the vertical load due to skin friction on the back of the abutment (or the weight of soil above the base slab); the vertical load from the bridge span; and the horizontal load from the bridge span transmitted through the bearings. The effect of friction on the back of an abutment retaining wall in reducing the active soil pressure is sometimes ignored when producing a conservative design, but the dragdown force should not be overlooked when considering vertical forces on the abutment.

Where an abutment is founded on compressible soil, settlements will occur due to the vertical forces (Fig. 6.8(*a*)). Conversely, if the bridge is in a cutting there may be uplift movement due to soil swelling (Fig. 6.8(*b*)). In both cases there is the tendency to backward rotation of the abutment.

Earth pressure is calculated in the same way as described for basement retaining walls in Section 5.7, with the important difference that there is little or no permanent restraint to horizontal movement of the top of the wall. Therefore active earth pressures apply with considerable additional pressure near the top of the wall due to heavy

compaction of the wedge of fill material between the previously placed embankment and the back of the abutment.

The amount of rotation of the top of the wall should be calculated from a knowledge of the magnitude of vertical and horizontal forces and the compressibility of the soil beneath the base of the wall. Allowance for the rotation should be made in determining the setting of the bearings.

Piled foundations may be needed for the conditions in Fig. 6.8(*a*), or to restrict the forward rotation of high retaining walls. Raking piles (Fig. 6.9) provide the most

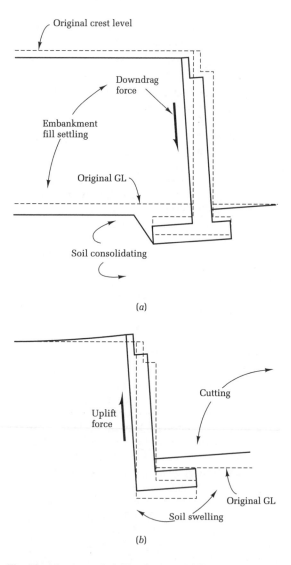

Fig. 6.8   Movement of a bridge abutment. (*a*) Due to embankment loading. (*b*) Due to soil swelling in cutting.

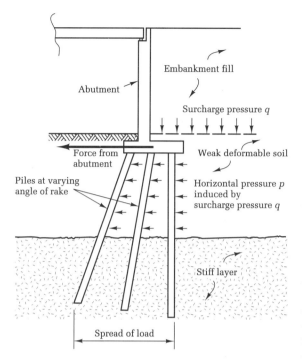

Fig. 6.9 Bridge abutment supported by raking piles.

efficient arrangement for resisting the vertical and horizontal forces. The varying angle of rake shown in Fig. 6.9 avoids the risks of excessive loading on a single outer row of rakers in combination with vertical piles directly beneath the wall. Where the rakers are bearing on rock or other unyielding material, the base of the abutment is virtually fixed against horizontal movement. Therefore a rigid type of wall should be designed for at rest (K_0) earth pressures, or higher to take account of the effects of compaction. Lateral pressures induced by the filling behind the abutment are exerted on the supporting piles where these pass through weak soils. Methods of calculating the horizontal forces on the piles are given in the following section.

Symons and Clayton[6.2] have described the results of research on the effect of compaction on earth pressures induced on backfilled retaining walls. Where walls are backfilled with free-draining granular soils the distribution of earth pressure is appreciably higher in the upper levels than that given by the normal distribution of active pressure with depth below the top of the wall. The magnitude of the additional pressure can be calculated on the assumption that the roller acts as an infinitely long line load surcharge.

Granular fill is becoming increasingly scarce in the UK

and increasing attention is being given to the use of clay backfill. If the moisture content of the clay is appreciably wet of the optimum, the horizontal earth pressure due to excess pore pressure induced by compaction will be a maximum on completion of filling, and will reduce as the pore pressures dissipate. When the clay is on the dry side swelling will take place after completion with swelling pressures approaching the passive state. Symons and Clayton report total horizontal pressures of up to 20 and 40 per cent of the undrained shear strength of medium and high plasticity clays respectively. They state that if swelling is to be avoided the moisture content of a high-plasticity clay will need to be such that the material will be barely trafficable by the compaction plant.

Wing walls to conventional abutments can cause difficult problems in foundation design because bearing pressures on shallow strip foundations vary from a maximum at the junction with the abutment to near zero at the extremity of the wing. If a joint is provided between the abutment and wing wall the differential settlement can cause inward movement at the top of the wall and consequent binding and spalling between the two components. Wing walls can be designed to cantilever from the abutments in a direction either parallel to or at right angles to the face of the abutment. The latter arrangement simplifies the foundation design, but can lead to unbalanced forces on the abutment foundations which may require temporary propping of the wing walls at the construction stage. Problems of differential settlement between abutments and wing walls can be avoided by constructing the latter from reinforced earth with suitable facing panels, or from precast concrete crib elements.

6.3.4 Spill-through abutments

It was noted in Section 6.3.1 that one of the advantages of the spill-through abutment was in the reduction in earth pressure to be carried by the structure compared with the conventional abutment retaining the full height of the soil behind it. Hambly[6.1] has described four methods of calculating the earth pressure on piers buried within the embankment (Fig. 6.10) listed below in order of magnitude of the pressure:

(1) No lateral pressure on piers buried in a stable embankment with a slope of 1 in 2 or shallower.
(2) Piers designed for active pressure immediately behind them plus an arbitrary additional allowance of up to 100 per cent (the Chettoe and Adams method[6.3]).
(3) No reduction from full active pressure over the gross width if the openings are less than twice the width of the piers, and the fill in front of the wall shall not be considered as providing resistance greater than active,

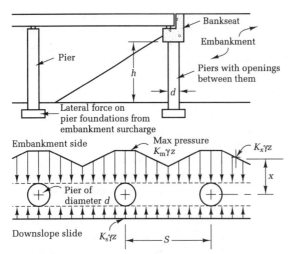

Fig. 6.10 Earth pressure distribution on piers buried in soil slope.

with reduction due to the descending slope taken into account (the Huntington method[6.4]).

(4) Full active pressure over the gross width.

The problem of assessing the magnitude of earth pressure is discussed in the Institution of Structural Engineers' report on Soil-Structure Interaction[2.19]. The report points out that if there is backward rotation of the pier due to differential settlement of the base slab then the earth pressure behind the pier could approach the passive condition. The report discussed research on the problem at Cambridge University by Ah Teck[6.5] resulting in the derivation of the earth pressure coefficient on the downslope side from the expression

$$K_s =$$

$$\frac{\cos \phi' \cos (45° + \phi'/2)}{[\tan (45° + \phi'/2) + \tan \beta] \sin (45° + 3\phi'/2)} \quad (6.6)$$

The distributions of pressure on either side of a single pier or on any row of piers are shown in Fig. 6.10. The range of values of the earth pressure coefficient K_x is given as

Maximum $K_x = K_m$, where $0 \leqslant x \leqslant d/2$,   (6.7)

$$K_x = K_m - K_s[(x/d - \tfrac{1}{2})/3.5]^{2/3},$$
where $d/2 \leqslant x \leqslant 4d$ (6.8)

Minimum $K_x = K_s$, where $x \geqslant 4d$.   (6.9)

In the above equations d is the diameter or width of the pier and x is the distance from the centre of the pier.

The horizontal stress distribution on the pile is given by

On the embankment side, $\sigma_h = K_x \gamma z$,   (6.10)

On the downslope side, $\sigma_h = K_s \gamma z$.   (6.11)

The expression for the net force W on the pier is given as

$$W = (K_m - K_s) \frac{d}{2} \gamma h^3 \left[\frac{s}{d} - 0.164 \left(\frac{s}{d} - 1 \right)^{5/3} \right], (6.12)$$

where s is the spacing between centres of the piers and h is the depth of fill at the piers. Where s/d is greater than 8 the expression becomes

$$W = 3.8 (K_m - K_s) \frac{d\gamma h^3}{2}. \quad (6.13)$$

Equations (6.7)–(6.13) were derived from centrifuge model studies where the maximum value of the coefficient K_m was taken as the at-rest value (K_0). A higher value might result from heavy compaction of embankment fill material behind the piers.

It was noted in the preceding Section 6.3.3 that lateral forces are exerted on those portions of pile shafts supporting a bridge abutment where they pass through a layer of weak deformable soil. The lateral forces are induced by the surcharge pressure of the fill material behind the abutment (Fig. 6.9). Lateral pressures are similarly exerted on the piles supporting the bank seat of a spill-through abutment (Fig. 6.11) and also on the piles at or beyond the toe of the embankment where these piles support a bridge pier. Where the embankment is placed on a soft deformable soil, such as a soft clay, some spreading within the soft layer occurs due to surcharge pressure from the fill material. If the embankment overstresses the soft layer the latter will flow and the embankment will collapse. This represents the ultimate limit state for the bridge approach. At the stage of flow of the soft layer it will pass between the piles, assuming the latter do not fall in bending

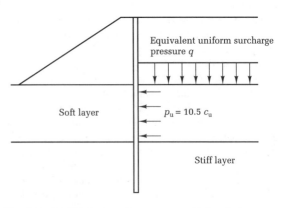

Fig. 6.11 Limiting horizontal pressure on a pile in soft clay subjected to unsymmetrical surcharge loading.

or shear. Where the surcharge pressure is low or the thickness of soft clay is small, the total force exerted on the piles will be small because the pile and soil move elastically together with little or no relative displacement between them. In the intermediate stage between the elastic and the failure condition, that is, in the elasto-plastic phase, there will be appreciable relative movement between the soft clay and the piles with pressure developed on the piles equivalent to the passive case. The heads of the piles will deflect and the movement may be sufficient to displace the bridge span bearings when the serviceability limit of the system will have been reached even though the piles have not failed in bending or shear.

Irrespective of the size of the piles or their centre-to-centre spacing the unit lateral pressure on any portion of the pile shaft will not exceed the ultimate resistance of the clay to relative movement with the pile. This has been shown by Poulos and Davis[6.6] to be given by the equation

$$\text{Limit pressure} = p_u = 10.5c_u. \qquad (6.14)$$

The above equation gives an upper bound to the lateral force on the pile, and if this pressure is assumed to act over the whole thickness of the soft clay layer (Fig. 6.11) it will give a safe but conservative value for determining the required resistance of the pile by the methods described in Sections 7.16 and 7.18. However, this approach takes no account of the effect of the sloping face of the embankment, the spacing of the piles, or the relative movement between pile and soil in reducing the overall force.

De Beer and Wallays[6.7] have established a simple empirical method of calculating the average lateral pressure on the pile due to adjacent unsymmetrical surcharge loading and for drained ($c' = 0$) conditions in the soil around the pile shaft. The surcharge is represented by a fictitious fill of height H_f with a sloping-front face as

shown for three arrangements of piles and embankment loading in Figs 6.12(a)–(c). The height of H_f is given by

$$H_f = H \times \frac{\gamma}{1.8}, \qquad (6.15)$$

where γ is the density of the fill in tonnes per cubic metre.

The fictitious fill is assumed to slope at an angle α which is drawn by one of the methods shown in Figs 6.12(a)–(c) depending on the location of the surcharge relative to the piles. The lateral pressure on the piles is then given by

$$p_z = fq, \qquad (6.16)$$

where f is a reduction factor given by

$$f = \frac{\alpha - \frac{1}{2}\phi'}{90 - \frac{1}{2}\phi'} \qquad (6.17)$$

and where q is the surcharge pressure for the height H_f and ϕ' is the effective angle of shearing resistance of the soil applying pressure to the pile.

It should be noted that when α is less than or equal to $\phi'/2$ the lateral pressure becomes negligible. De Beer and Wallays point out that the method is very approximate. It should not be used to obtain the variation in bending moments down the pile shaft, but only to obtain the maximum moment. Also the calculation method cannot be used if the safety factor for conditions of overall stability of the surcharge load is less than 1.6.

Springman and Bolton[6.8] made an extensive study of the problem as part of a research contract with the UK Department of Transport. They used finite element analyses correlated with model experiments in a centrifuge to establish the pressure distribution down the shafts of a row

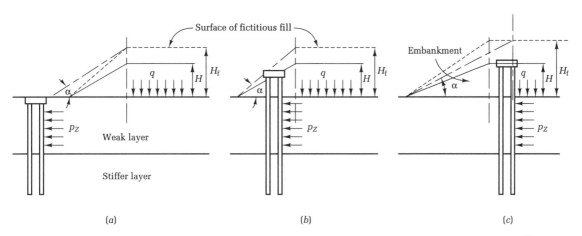

Fig. 6.12 Calculation of lateral pressure on vertical piles due to unsymmetrical surcharge loading (after De Beer and Wallays[6.7]).

of piles subjected to the one-sided surcharge pressure from an embankment where the piles extend through a weak deformable soil, such as a soft clay, into a stiffer but yielding stratum. In the case of the free-head pile, pressure at the head is low because the head is free to yield laterally as the soft clay tends to spread under the surcharge loading. Further down the pile there is less movement because of the lesser height above the point of fixity within the stiff stratum. Hence there is relative movement between the soft clay and the pile with increase in pressure on the pile surface. At deeper levels friction at the interface between the soft clay and the stiff stratum prevents significant movement of the former, and this combined with yielding of the portion of the pile embedded in the stiff clay results in the pile tending to push into the soft clay, thus reversing the direction of thrust on the pile.

The pressure distribution on a free-head pile can be idealized by the diagram in Fig. 6.13. The low pressure at the pile head is given by the equation

$$p = G_{oc}p_m/G_m, \qquad (6.18)$$

where

G_{oc} = shear modulus of the soil at the level of the pile head (Fig. 6.13(c)),

G_m = shear modulus of the soil at a depth of half the height h over which pressure is applied by the soft clay,

p_m = the mean pressure.

Over the height h the pressure distribution is parabolic with a mean pressure given by the equation below derived by Bolton et al[6.9].

$$p_m = \frac{q}{\left[\dfrac{3G_m d}{G_r h} + \dfrac{d}{s} + \dfrac{0.71G_m dh^3}{EI}\right]}, \qquad (6.19)$$

where

G_m and h are defined as above,
G_r = reduced shear modulus around the pile,
q = equivalent surcharge pressure,
d = pile diameter,
s = pile spacing centre to centre along the row parallel to the abutment.
E = Young's modulus of the pile material,
I = moment of inertia of the pile section.

Values for G_m/G_r may be taken as 1.5–2 for driven piles and 2.5–3 for bored piles respectively.

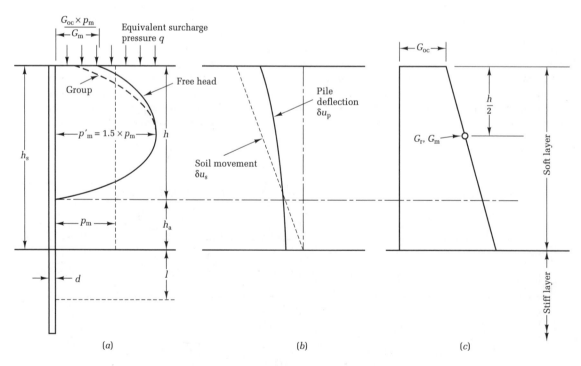

Fig. 6.13 The effect of unsymmetrical surcharge loading on a vertical pile with a free head driven through soft deformable soil into a stiff layer (after Springman and Bolton[6.8]). (a) Lateral pressure distribution. (b) Relative soil pile movement. (c) Shear modulus profile.

The maximum pressure is given by
$$p'_m = 1.5p_m. \tag{6.20}$$

Where there is a cap at the pile head which prevents relative movement with the soil there is no lateral pressure on the pile unless settlement of the soft clay layer causes it to sink below the underside of the cap.

It then remains to derive the height h_u over which there is no pressure on the lower part of the pile from the expression

$$\left[\frac{3}{2} - \frac{d}{s}\right]\frac{3\pi E_p}{128G_m} =$$

$$\left[\frac{l_e}{d} + \frac{h_u}{d}\right]^2\left[\frac{h_s}{h_u} - 1\right]\left[\frac{3h_s}{d} + \frac{4l_e}{d} - \frac{h_u}{d}\right], \tag{6.21}$$

where E_p is the equivalent Young's modulus of a solid pile, and l_e is the equivalent length of the pile which can be treated as being supported by the pile within the stiff stratum and is given by

$$l_e \simeq 0.34l_c \tag{6.22}$$

for a soil with a constant shear modulus with the stiff stratum, and

$$l_e \simeq 0.5l_c \tag{6.23}$$

where the shear module increases linearly from zero at the stiff clay surface.

The length l_c is the critical length over which the loading effects are relevant. It is dependent on the equivalent Young's modulus E_p and the shear modulus of the soil, and is derived from an expression given in Section 7.17.2.

To avoid the need for iteration in using equation (6.21) to obtain h_u, Springman and Bolton[6.8] derived charts for the non-dimensional groups s/d, E_p/G_m, h_s/d, and l_c/d as given in Fig. 6.14(a)–(c). A computer program based on the results of the research was written under the name SIMPLE which is licensed to the Cambridge University Engineering Department. The program covers surcharge loading to a single row or two rows of piles, a single 'long' flexible vertical pile, or two such piles with a rigid cap. The output includes the pile head deflections and the distribution of bending moments along the pile shaft. The SLAP program also written at Cambridge University covers the case of the single or single row of vertical free-head piles.

Springman and Bolton recommend that the embankment–pile–soil system should be designed to ensure that it lies within the pseudo-elastic zone of an interaction diagram where p_m/c_u is plotted against q/c_u (Fig. 6.15). The upper limit of the ordinate is the stage where plastic flow takes place around the piles (equation (6.14)). The limit on the abscissa at $q = (2 + \pi)c_u$ represents the

failure of the embankment on the soft layer. To avoid excessive deformation from this cause it is recommended that the embankment should have a safety factor of 1.5 against collapse. Elastic behaviour as described by equation (6.19) is represented by the limits $h/d = 4$–10. For larger values of h/d the soil tends to yield around the pile before general yield of the soil mass as the embankment loading is increased.

It is emphasized that in considering the overall stability problem the higher range of values of c_u should be used to obtain the likely maximum lateral pressure on the pile, whereas the lower range would be used to obtain the limiting height of the embankment.

Where the spill-through abutment is in the form of a bank seat at the top of a cutting (Fig. 6.2(b)) differential settlement between the abutment and the adjacent pier results from the likely greater settlement of the more heavily loaded pier foundation. The side span is jointed to accommodate the movement and rotation of the span can be minimized by delaying its construction or by adjustment in the level of the bank seat bearings.

Where the bank seat is located on the crest of an embankment, and the latter has been placed on a compressible soil, the settlement of the bank seat can be large relative to the pier. In addition there can be backward rotation of the abutment with excessive displacement of the bearings. Compaction of an embankment fill at the crest or close to the slopes is never as good as in the body of the fill due to the justifiable reluctance of plant operators to work close to the edge of the slope. Improved compaction can be given by overfilling and then cutting back the slope, but there is often insufficient working space between the abutment and the pier to perform this effectively, unless the work is done before constructing the pier. There can be continuing settlement of the bank seat due to long-term consolidation of the embankment fill, even though the embankment has been placed on relatively incompressible soil and is well compacted. Values for calculating the settlement of fills in the short and long term are given in Section 3.6.

Because of the problems of bank seat settlement and excessive rotation of the link spans or bearings it may be decided to support the ends of the side spans on buried piers with pad or combined strip foundations (Fig. 6.10), or to support the bank seats by piling (Fig. 6.16(a)). In the latter case the piles should extend through the fill as shown. If they are used to support buried piers (Fig. 6.16(b)) the loading on the piles is greatly increased because of the weight of fill carried by the pile caps. Dragdown is imposed on the pile shafts due to long-term settlement of the embankment fill and any compressible natural soil below the embankment. In this respect a driven pile is preferable to a bored pile because of the

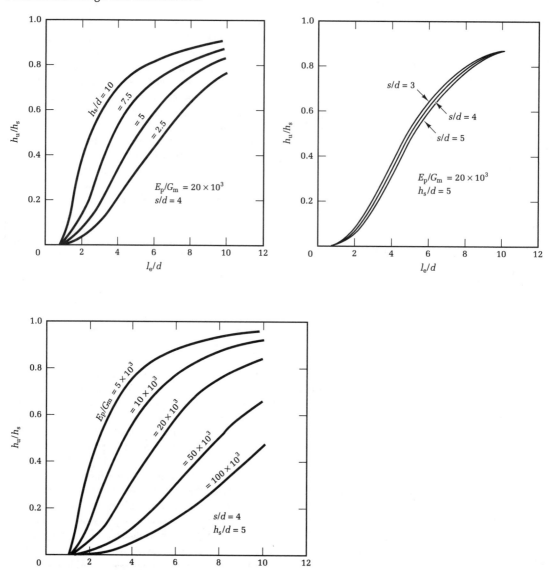

Fig. 6.14　Charts for determination of h_u/h_s (after Springman and Bolton[6.8]).

greater dragdown force imposed on enlarged and irregular shaft resulting from drilling in a soft or caving soil. The magnitude of the dragdown force is calculated by the method given in Section 7.13.

Reference was made in Section 6.3.1 to the obstruction and delays caused to the main highway construction programme by piling operations for the bridge foundations. This can to some extent be avoided by undertaking the work from the crest of the embankment or cutting. Piling for the piers can be driven by plant operating from this level (Fig. 6.17). As an alternative to conventional

bored or driven piles for supporting a spill-through abutment the bridge bearings can be set on the capping beam of a contiguous bored pile or secant pile wall (Section 5.4.5). Barrettes (Section 4.6.5) can also be used to support the spill-through abutment. They are arranged as abutting T-sections with the long leg of the T parallel to the axis of the highway and the short cross-section cast integrally forming the facing wall. The legs of the T-sections act as counterforts, thus avoiding the need for restraint to horizontal movement by ground anchors which is necessary with bored pile walls. However, abutments

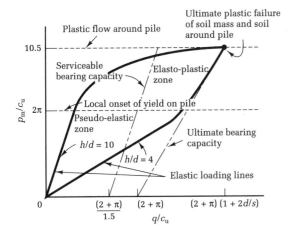

Fig. 6.15 Interaction diagram for horizontal soil pressure in a vertical pile driven through soft clay into an underlying soft stratum (after Springman and Bolton[6.8]).

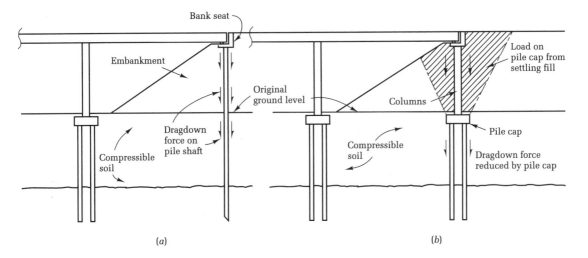

(a)   (b)

Fig. 6.16 Piling for bridges with spill-through abutments.
(a) Bank seat carried by piles driven from completed embankment.
(b) Bank seat carried by columns, pile cap at original ground level.

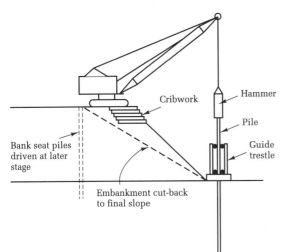

Fig. 6.17 Driving piles for bridge pier.

constructed as barrettes would need to be in sufficient numbers in a highway project to make it economical to mobilize the specialist plant and equipment required for their installation.

6.3.5 Bridge approach support piling

The problem of avoiding relative settlement between the abutment and embanked approach to a bridge is a difficult one which has not been solved satisfactorily to the present day. No matter how much compaction is given to the embankment at the time of construction long-term traf-ficking by heavy freight vehicles always produces further

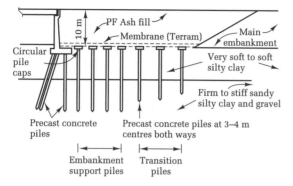

Fig. 6.18 Pile-supported approach embankment, M876 Motorway, Scotland (after Reid and Buchanan[6.10]).

settlement. Where the embankment is placed over a compressible soil as in Fig. 6.18, long-term settlement cannot be avoided. However, the relative settlement between the abutment and embankment can be minimized by giving partial support to the latter by piling.

Reid and Buchanan[6.10] used the concept of yielding piles described in Section 5.4 to provide a gradual transition from a piled bridge abutment to the full settlement of an embankment placed on soft compressible clay (Fig. 6.18). The piles close to the abutment were at close spacing and designed to carry the full load of the PF ash fill with a safety factor of 2. After the first four rows the spacing was increased to a 3–4 m grid and the piles were made progressively shorter to yield under an increasing proportion of the embankment load. Loading was transferred from the embankment to the pile caps by means of a flexible membrane made up from two layers of Terram plastics fabric reinforced with Paraweb strapping. The pile caps were 1.1–1.5 m in diameter.

6.3.6 Abutments for arch and portal frame bridges

The stability of an arch is very sensitive to movement of the abutments. If yielding of the ground causes them to spread it can cause damaging eccentricity in the line of thrust through the arch barrel or rib. Spreading can be prevented by horizontal ties between the abutments, but this is likely to be impracticable in many situations where an arch bridge is environmentally suitable. Spread foundations of the type shown in Fig. 6.3(a) can be used in a rock cutting, but groups of raking piles are usually required to restrict horizontal and vertical movements of the abutments where these are backed up soil. It was noted in Section 6.3.1 that portal frame bridges are also sensitive to

small vertical and horizontal movements at foundation level.

An example of design to prevent yielding of the foundations of a portal frame bridge is the case of the Twizel Bridge constructed in 1982–83 over the River Till in Northumberland. In order to avoid obscuring the view from the river of a closely adjacent medieval masonry arch bridge it was necessary for the 45 m span of the new bridge to have a flat and shallow profile. The vertical legs of the bridge which form the abutments were designed to have hinged bearings on strip foundations, with the base of the foundations inclined normally to the direction of the principal thrust. The direction of thrust for the most unfavourable load combination and temperature effect is shown in Fig. 6.19. These cause maximum and minimum bearing pressures at the back and the front of foundations of 540 and 220 kN/m^2 respectively. Both abutments were founded on a slightly weathered, moderately strong to strong thinly to medium bedded sandstone of the Lower Carboniferous series. Physical characteristics of the rock obtained by field and laboratory testing were as follows:

Maximum RQD	80%
Average RQD	35%
Minimum RQD (below foundation level)	25%
Unconfined compression strength (from uniaxial tests)	34 MN/m^2
Unconfined compression strength (from axial point load tests)	29 MN/m^2
Unconfined compression strength (from diametral load tests)	12 MN/m^2
Average Young's modulus of intact rock	7.6 GN/m^2
Minimum Young's modulus of intact rock	6.4 GN/m^2
Deformation modulus of rock mass (vertical)	1.3–2 MN/m^2
Deformation modulus of rock mass (horizontal)	1.3 MN/m^2

The above values of the rock mass compressibility were determined from relationships between the RQD, fracture frequency, and compression strength of the intact rock (Section 2.7). Using these values the yielding of the foundation for the average bearing pressure of 430 kN/m^2 was calculated to be 5 mm which was within the acceptable limit of 10 mm. For the maximum eccentricity of thrust the yielding was 5 mm at the back and 2 mm at the front of the strip foundation.

The calculations for yielding assumed that the rock joints were closed or filled with cementitious material. However, examination of rock outcrops and cores showed subvertical joints parallel to the abutments with many

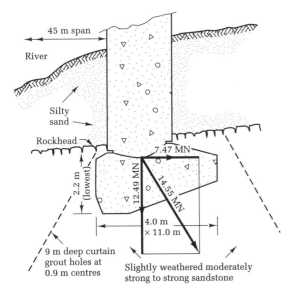

Fig. 6.19 West abutment foundation for Twizel Bridge, Northumberland (courtesy of Director of Technical Services, Northumberland County Council).

open joints, some of which having openings up to 15 mm. Air was seen to be escaping in the river when drilling exploratory boreholes on the banks. In view of this it was recommended that grout should be injected into the mass of rock influenced by the bearing pressures[6.11]. The area of each abutment was surrounded by a row of curtain grout holes 9 m deep and at 0.9 m centres. The rock mass within the curtain was then injected with cement to a depth of 9 m through primary and secondary holes on a 1.4 m grid. The average grout take was 17 kg/m^3 of treated ground.

The grout holes were drilled through the sandy overburden, and after installing the sheet pile cofferdam and excavating down to rockhead for the west abutment a fissure about 300–600 mm wide filled with loose debris was seen to lie diagonally across the area of the foundation (Fig. 6.20). The fissure was cleaned out and plugged with mass concrete to a depth of 2 m below foundation level, followed by grouting the debris below the plug. It was then 'stitched' by inclined 36 mm Dywidag bars at 400–800 mm centres arranged in a pattern to transfer thrust from the abutment to the rock behind the fissure, and to tie the rock masses on each side of the fissure together (Fig. 6.21).

Measurements of movements of the abutment were made during construction and for 5 weeks after completion of the bridge. They showed negligible settlement and an outward horizontal movement up to 2 mm on the east abutment.

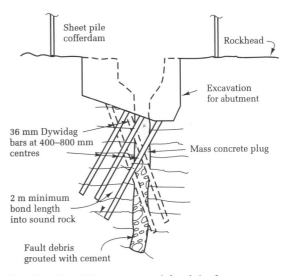

Fig. 6.20 Fissure in rock surface at foundation level of west abutment, Twizel Bridge (courtesy of Director of Technical Services, Northumberland County Council).

Fig. 6.21 Remedial treatment to rock foundation for west abutment, Twizel Bridge (courtesy of the Director of Technical Services, Northumberland County Council).

6.4 Bridges over water

6.4.1 The effect of environmental conditions on the selection of foundation types

The design of foundations for over-water bridges, jetties, and offshore marine terminals can present severe problems to the engineer. Whereas the ground conditions at the bridge location may be the dominating factor in the design of foundations for a bridge on land, in the case of over-water bridges they can be of relatively minor importance compared with the design problems posed by the environment.

The environmental conditions which govern the selection of a suitable foundation type and method of construction are given below.

Exposure conditions and water depths   Bridges in the open waters of a wide estuary or bay crossing are in a hostile environment from winds and wave action which can limit the period of operation of floating construction

plant and can cause damage to partly completed structures. This favours the use of large prefabricated elements which can be towed or transported by barge to the site of the bridge and sunk rapidly on to a prepared bed or piled platform. A box caisson is a suitable type (Fig. 6.22) where the water depth is sufficient for flotation of the unit, but weather conditions are critical at the early stages of towing the caisson to site and sinking it in position. The effects of delays due to the weather on the overall construction programme should be considered.

Open-well caissons (Fig. 6.23) are used in shallow water where the shallow draft bottom section is floated to the construction site and sunk by grabbing out the soil from the open wells as the walls are raised progressively. The construction operations for open-well sinking are more vulnerable to the weather than those for box caissons which can be sunk during a very short period of favourable weather. However, construction is feasible at exposed sites by sinking the caissons from an artificial island or by using a jack-up barge to give a stable working platform.

Cofferdams of the type described in Chapter 10 are suitable only for sheltered waters, but more robust types which can be used in moderate exposure conditions are described in Section 6.4.2.

Currents   River currents and tidal streams impose drag forces on piers or piles and create scour holes where the soil at bed level is susceptible to erosion. Scour can be

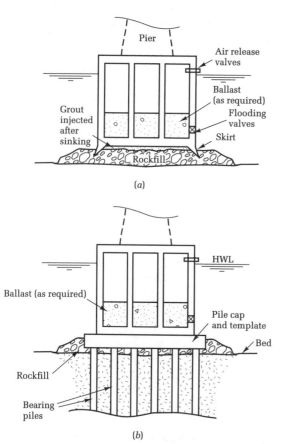

Fig. 6.22   Box caissons. (*a*) On rock blanket. (*b*) On piled raft.

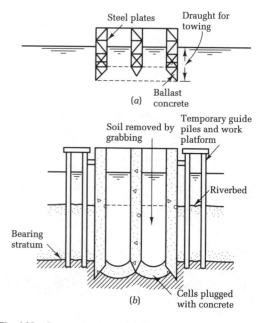

Fig. 6.23   Open-well caisson. (*a*) Shoe and lower wall section ballasted for towing to site. (*b*) Sinking and plugging completed.

critical around cofferdams where eddies are created by temporary conditions such as partly driven sheet piles. Mattressing may be required on an erodible bed to prevent erosion in the restricted flow conditions as a caisson is lowered through the last few metres of water. Current drag forces can give problems when pitching bearing piles or sheet piles. Damaging oscillations of piles can occur at certain velocities of flow before the piles are secured at the head by the pile cap or temporary girts.

Ship collisions   Precautions against the risk of collapse of bridge piers due to collision by ships can add considerably to the cost of the foundations. The risks are not confined to the designated navigation channel. Collisions are just as likely to occur from a ship straying out of control from the designated channel. In some wide estuaries the deep-water channel can swing across the river from one side to the other in a very short time. Where there is a large tidal range, as in the Severn Estuary, virtually every pier of a multi-span bridge may be at risk.

Protection can take the form of massive construction of the piers such as ballasted box or open-well caissons; a group of large-diameter piles surrounded by a ring of skirt piles to prevent ships jamming between individual units; an independent ring fender; or by surrounding the pier by an artificial island. Allowance must be made for impact at any angle to the axis of the pier. A glancing blow can induce torsional shear in the pier body.

Fender piles connected by a massive ring beam are suitable for moderately large vessels. The ring beam should be placed at a sufficient distance from the pier to allow the piles to deflect as they absorb the kinetic energy of the moving vessel and bring it to rest. The main piers of the Sungai Perak Bridge in Malaysia[6.12] are protected by a reinforced concrete ring beam fender supported by 16 1800 mm tubular steel piles. The fender beam is set about 1 m above high-water level. A horizontal load test on an individual free head pile in 22 m depth of water showed a deflection of 300 mm for a pile head load of 685 kN. Methods of calculating the ultimate resistance to hor-

izontal loads of piles embedded in soils and deflections for a given load are described in Sections 7.16 and 7.18.

Artificial islands are suitable only for shallow water because the area surrounding the pier must be sufficiently large to allow the moving vessel to ride up the slope and come to rest before the overhanging bow can strike the pier. In deep water the quantity of fill material, armouring stone for wave protection and mattressing for scour protection, become prohibitively large and the islands may obstruct the navigation channel. The artificial islands used to protect the main piers of the Penang Bridge in Malaysia[6.13] are shown in Fig. 6.24.

An upper limit to the impact force of a moving vessel on a bridge pier is the crushing strength of the vessel's hull. In this respect an angled impact from the side of a ship may be more damaging to the ship than to the pier because of the lower strength of the hull in a transverse direction.

Floating ice   The design of a bridge pier to resist impact from ice floes has some similarities with the design problem given by ship collision, but there is the additional hazard of build-up of pressure from accumulation of individual floes or from pack ice. The build-up can occur in directions transverse to the pier and vertically as pressure ridges start to form. For this reason single piers or single large piles are preferable to groups because the single member can divert the floes or concentrate the resisting force in a small area against the ice sheet causing it to rise up and fail in bending.

Gerwick[6.14] has described the characteristics of sea ice and methods used to design artificial island and platform structures in Arctic waters. Gravity base structures with a slender pier to give minimum resistance to floating ice forces and a large base to provide skin friction resistance to sliding and overturning are the favoured types. Where piling is necessary because of the ground conditions the pile group should be surrounded by a ring of close-spaced skirt piles.

In rivers where the flow is mainly parallel to the banks

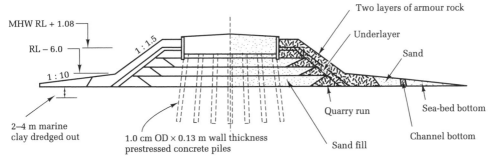

Fig. 6.24   Artificial islands protecting the piers of the Penang Island Bridge, Malaysia (after Chin Fung Kee and McCabe[6.13]).

the piers can be provided with cutwaters to break up and crush the pack ice in a manner similar to an icebreaker. Gerwick states a figure of 4–6 MN/m^2 for the maximum local pressure exerted by pack ice as measured on ice-breakers. The higher value is stated to be an extreme one.

Earthquakes   Earthquakes give severe design problems for deep-water piles because the forces exerted at high level on the bridge superstructure combine with the forces on the pier body to produce high overturning moments at base level. The mass of water displaced by the pier must be added to the mass of the pier itself. The eccentric loading on the pier base can be very large in deep water, which again favours a slender pier and a large base. A circular form is required because the earthquake forces can be aligned in any direction. Ground shaking can cause liquefaction of loose to medium-dense granular soils. The depth of liquefaction can be calculated from a knowledge of the particle-size distribution and *in-situ* density of the soil deposit[6.15, 6.16]. Piled foundations, or ground treatment to densify a loose soil deposit (Section 11.6), may be required to provide stable support to the pier.

Horizontal forces at the base of the pier may be caused by submarine flow slides initiated by liquefaction or by tsunamis. These are large-amplitude waves created by vertical shaking of the sea-floor at locations which may be hundreds of miles from the shore.

6.4.2 Pier construction in cofferdams

Construction of bridge piers within cofferdams is suitable for shallow-water sites in sheltered or moderately exposed conditions. In very shallow water or half-tide locations the piers can be constructed in simple earth bank cofferdams of the type described in Section 10.1.1.

Sheet pile cofferdams can be constructed in water depths of 15 m, but difficulties become increasingly severe at greater depths although an overall depth of 32 m from high water to base of excavation was feasible for the Thames Barrier foundations as described in Section 10.3.3. Single-skin sheet pile cofferdams are vulnerable to damage by wave action, when repeated wave impact can cause fatigue failure of welded connections. Robust cofferdams can be designed in the form of interconnected cells (Section 10.5). The ring of cells can be left in place as protection against ship collisions. The most favourable use of sheet piling is at sites where an impermeable stratum at or below excavation level provides a cut-off for inflow of ground water enabling the excavation to be pumped out and the pier foundation constructed in dry conditions. Where no cut-off is possible the excavation and concrete base construction are carried out under water.

An example of cofferdam construction in deep water and moderately exposed conditions is given by the construction of the main tower piers for the Kessock Bridge crossing the tidal waters of Moray Firth in Scotland[6.17]. Each pier leg had a separate foundation constructed in 21 m diameter cofferdams at 22 m centres. The piers were sited in 12 m depth of water with current velocities up to 3 m/s. Construction commenced by driving a group of 13 865 mm tubular steel piles to form a temporary working platform within the plan area of the cofferdam. The piles were driven from a barge and then the crane was transferred from the barge to the platform. The waling unit was prefabricated on shore. It consisted of an upper and lower ring waling formed from twin 914 mm UB sections connected by diagonal bracing (Fig. 6.25). After assembly, the waling unit was separated into four quadrants which were transported by barge, assembled on the work platform, and lowered on to the sea-bed while suspended from three guide piles threaded through collars on the inner face of the walings. The unit was finally secured by a ring of eight 750 mm tubular piles driven through collars with a grouted connection between the collars and the piles.

Larssen No. 6 sheet piles 30 m long were pitched around the waling unit and driven below the sea-bed. The excavation was taken out under water by grabbing and airlifting followed by driving 64 H-section piles (not shown in Fig. 6.25) to support the pier base, and then placing a 2.5 m thick tremie concrete slab. The water level within the cofferdam was regulated by sluices to avoid differential head during the period of concrete placing and initial curing. The pile cap and cut-water were constructed within the cofferdam which was then removed and the fender protection and scour blanket were installed.

Wastage of sheet piling due to difficulties of extraction and damage can add to the cost of pier construction in multi-span bridges and various systems for cofferdams which can be moved from pier to pier have been devised. The system used for 15 of the piers carrying the 3.3 km long bridges between the islands of Sjaelland and Falster in Denmark[6.18] is shown in Fig. 6.26. After dredging the sea-bed to formation level a blanket of crushed rock was placed as protection against scour. Then a group of 49 700 mm tubular raking piles was driven through the blanket in two concentric rings. The reinforced concrete base unit weighing 440 t was constructed on shore, brought to the pier site by barge, lowered over the projecting pile heads, and secured by three pinning piles. Concrete was then placed by tremie up to about half the height of the base unit.

The temporary cofferdam consisted of 10–11 m diameter steel plate rings, 3 m deep jointed with rubber sheeting. The assembly was lifted by floating crane and

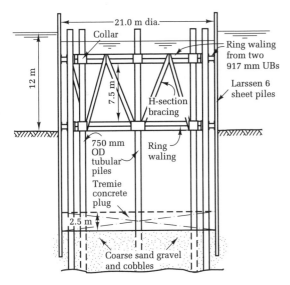

Fig. 6.25 Cofferdam for main piers of Kessock Bridge, Scotland, at stage of completing underwater excavation.

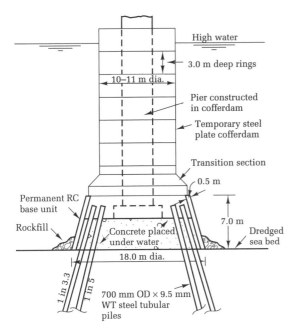

Fig. 6.26 Construction of deep-water piers for the Sjaelland–Färo–Falster Bridge, Denmark.

lowered on to the base unit to which it was locked by stressed rods. Then the cofferdam was pumped out and the pier base and stem constructed. After this the cofferdam was flooded, released from the base unit, and lifted off the pier by the crane for transport to the next pier position.

As an alternative to constructing a pile cap within a

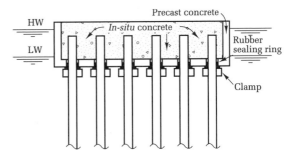

Fig. 6.27 Precast concrete shell for pile cap.

cofferdam, as in the two previous examples, the pile cap can be constructed *in situ* at or above high-water level either within formwork assembled on the pile heads or within a precast concrete shell. This is lowered by crane over the heads of the piles and rests on collars welded on to the pile shafts. Concrete bags are used to seal the annulus between the piles and the holes in the bottom of the box. After the box is pumped out, reinforcement is assembled and concrete is placed. The precast concrete shell must be designed to be light enough to be lifted by crane, and if the depth of submergence is large, as in Fig. 6.27, it must be clamped to the piles to prevent flotation until sufficient concrete is placed to overcome the buoyancy. The pile caps for the approach spans to the Penang Bridge[6.13] were constructed in this way. The caps varied in size from 6.1 to 10.0 m wide and 18.25 to 27.20 m long. The shells forming the larger caps were constructed in two sections and were handled by a 450 t floating crane. The tops of the prestressed concrete piles were cut to level by mechanical saw and metal clamps fixed to them. Rubber rings were slipped over the pile heads to seal the annulus between the holes and the piles.

6.4.3 Pier construction with box caissons

Box caissons are hollow structures with a closed bottom designed to be buoyant for towing to the bridge site and then sunk on to a prepared bed by admitting water through flooding valves. In sheltered conditions the top can be left open until sinking and ballasting is completed, or a closed top can be provided for towing in rough water. Box caissons are unsuitable for founding on weak soils, or for sites where erosion can undermine the base. They are eminently suitable for founding on compact granular soils not susceptible to erosion by scour, or on a rock surface which is dredged to remove loose material, trimmed to a level surface and covered with a blanket of crushed rock (Fig. 6.22(*a*)). Skirts are provided to allow the caisson to bed into the blanket and a cement–sand grout is injected to fill the space between the bottom of the box and the blanket.

Where the depth of mud or loose material is too deep for dredging a piled raft can be constructed to support the caisson as shown in Fig. 6.22(b).

The last few metres of lowering a large box caisson are critical. A large volume of water beneath the structure must be displaced, and if the caisson is lowered too quickly it can skid away from the intended position. In tidal conditions slack water is desirable to minimize the flow velocity in the narrowing space between the caisson bottom and the bed, thereby preventing erosion of the blanket material.

Box caissons were used for the foundations of the main tower piers of the Queen Elizabeth II Bridge crossing the Lower Thames at Dartford[6.19]. It was stipulated that the main river piers should be capable of resisting the impact of a 65 000 t ship travelling at 10 knots. This required massive construction in the form of a single box caisson 59.0 m × 28.6 m × 24.1 m deep at each pier position. The caissons shown in Fig. 6.28 were constructed in a dry dock at Rotterdam and towed across the North Sea to the bridge site. The caissons were designed by Trafalgar House Technology and sunk in place by a consortium of Cementation Construction Company and Cleveland Bridge and Engineering Company.

The size of the base was governed by the resistance to horizontal shear of the chalk on an assumed plane of weakness below the irregular interface between the 750 mm crushed rock blanket and the chalk surface. This was dredged to a depth of about 4.5 m to remove weathered chalk deemed to be too weak in shear to resist a torsional force generated by the ship striking the caisson at an angle of 57° to its axis. A design impact force of 215 MN was adopted. This was assumed to act at a height of 20 m above the base and at an eccentricity in plan of 14 m. It required an undrained shear strength of 450 kN/m² to provide the required torsional resistance to sliding. The impact force applied to the caisson was taken as the time-averaged force imparted by the crushing and buckling of the hull of the ship. Although it was accepted that the peak force could significantly exceed the average, the duration of the peak would be very small such that the mass of the caisson would not respond to it.

Considerable difficulty was experienced in achieving the design tolerance for dredging and placing the rock blanket[6.20]. It was specified that dredging should not be deeper than 0.5 m below formation level, but some areas finished 0.75 m below this level. The rock blanket consisting of 40 to 10 mm crushed stone was specified to be placed without projecting above formation level and not more than 300 mm below this level. The lower tolerance

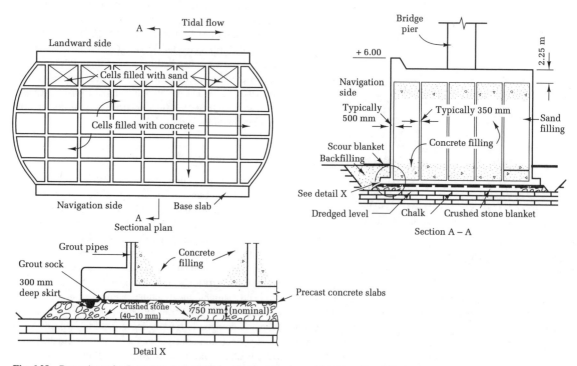

Fig. 6.28   Box caisson for Queen Elizabeth II Bridge, Dartford (courtesy of Trafalgar House Technology).

could be achieved, but some local areas were 300 mm above formation level. These areas were located and calculations were made to ensure that the structure of the caisson would be capable of withstanding the stresses which could occur when these high spots were pushed down during the stage of ballasting the caisson.

Sinking was achieved by temporarily attaching the caisson alongside a moored barge and then lowering the unit on to the rock blanket during a single tide. After ballasting grout was injected between the base of the caisson and the blanket through pipes embedded in the external and internal walls. Spread of grout beneath the base was facilitated by shallow channels formed by precast concrete slabs with geotextile 'socks' on a rectangular pattern between them. Escape of grout outside the area of the blanket was prevented by a 300 mm deep peripheral skirt. Mass concrete was used to fill the cells up to the level of the capping slab except for the row of cells alongside the shallow-water sides of the piers where impact from large ships was not possible. These cells were filled with sand.

6.4.4 Pier construction with open-well caissons

Open caissons (including monoliths) are suitable for foundations in rivers and waterways where the predominating soil consists of soft clays, silts, sands, or gravels, since these materials can be readily excavated by grabbing from the open wells, and they do not offer high resistance in skin friction to the sinking of the caissons. Open caissons are essential where the depth of sinking requires air pressures exceeding about 3.5 bar (350 kN/m²), since for physiological reasons men cannot work under compressed air at greater pressures than this. Open caissons are unsuitable for sinking through ground containing large boulders, tree trunks, and other obstructions. They can only be founded with difficulty on an irregular bedrock surface, and when sunk on to steeply sloping bedrock, they are liable to move bodily out of the vertical. Open caissons are advantageous for bridge foundations in rivers where there is a large difference in the seasonal levels such as in the major rivers in Pakistan, India, Bangladesh, and China. Caisson sinking is commenced in the low-water season and completed to the design founding level before the onset of the annual flood. The caisson can be allowed to be wholly or partly submerged by the flood water without damage during the period when no work on the bridge superstructure is possible.

On reaching founding level, open caissons are sealed by depositing a layer of concrete under water in the bottom of the wells. The wells are then pumped dry and further concrete is placed, after which the caissons can be filled with clean sand or concrete or, where their dead weight

must be kept low, by clean fresh water. Because the sealing is done under water, open caissons have the disadvantage that the soil or rock at the foundation level cannot usually be inspected before placing the sealing concrete. Only in rare cases is it possible to pump the wells dry for inspection of the bottom. Another disadvantage is that the process of grabbing under water in loose and soft materials causes surging and inflow of material beneath the cutting edge with consequent major subsidence of the ground around the caisson. Therefore open caissons are unsuitable on sites where damage might be caused by subsidence beneath adjacent structures.

Caissons should, wherever possible, be constructed as isolated units, separated by some distance from adjacent caissons or other deep structures. The effect of sinking and grabbing inevitably causes displacement of the surrounding soil, with the result that it is difficult, if not impossible, to maintain the verticality and plan location of caissons sunk close together.

Monoliths are essentially open caissons of reinforced or mass concrete construction with heavy walls. They are unsuitable, due to their weight, for sinking in deep soft deposits. Their main use is for quay walls where their massive construction and heavy weight is favourable for resisting overturning from the retained backfilling and for withstanding impact forces from berthing ships.

The principal design features of open caissons and monoliths are shown in Fig. 6.29. This illustration is of a

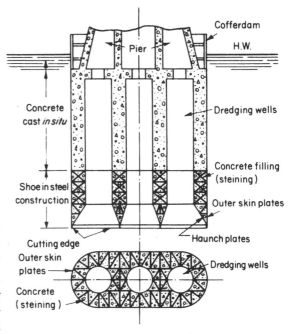

Fig. 6.29 Design features of open-well caisson.

caisson designed with a lower section to be fabricated in a dry dock or at a riverside wharf and transported as a floating unit through shallow water to the bridge site. This requires a light buoyant form of construction in steel plating. If there is sufficient depth of water for towing a deep draught unit, or if the caisson is constructed at its sinking location on staging or on a sand island, a heavier form of construction in reinforced concrete can be adopted. The *cutting edge* forms the lowermost extremity of the shoe (or curb). The latter has vertical steel *outer skin plates*, and sloping inner steel *haunch plates* (or cant plates). The skin plates are braced internally with steel trusses in vertical and horizontal planes. The trusses prevent distortion of the shoe during fabrication, towing to site, and the early stages of sinking. As soon as possible after the initial sinking the space between the skin plates is filled with concrete (*steining*). When the structure has attained sufficient rigidity by reason of the concrete filling, the skin plating can be terminated and the steining carried up in reinforced concrete placed between formwork. At or below low water level in a bridge pier the caisson proper is completed and the pier carried up in concrete masonry, or brickwork. If the water level rises above the top of the caisson at its finally sunk level, a *cofferdam* is constructed above the caisson in which to build the piers. The space within the walls may form a single *dredging well* or it may be divided by cross walls into a number of wells. Monoliths usually have only one or two small dredging wells.

In the following comments on the design of open caissons it must be realized that in most practical cases there is no ideal solution to the problem, and the final design is usually a compromise brought about by a number of conflicting requirements. For example, thick heavy walls may be desirable to provide maximum weight for sinking through stiff ground, but thick walls mean small dredging wells and the grabs cannot reach beneath the haunch plates to remove the stiff ground. Lightness of weight is desirable in the first stages of floating out the caissons, but this can only be obtained at the expense of rigidity, which is so essential at the second stage of sinking through the upper layers of soil when the caisson is semi-buoyant and may not have uniform bearing, and when stresses due to sagging of the structure consequent on differential dredging levels may be critical. Maximum height of skin plates is desirable when sinking caissons in a waterway where there is a high tidal range, but the extra height of plates may mean excessive draught for towing to site.

The *shape* of a caisson will, in most cases, be dictated by the requirements of the superstructure. The ideal shape for ease in sinking is circular in plan since this gives the minimum surface area in skin friction for a given base area. However, the structural function of the caisson is, in most cases, the deciding factor.

The *size and layout of the dredging wells* is dependent mainly on the type of soil. For sinking through dense sands or firm to stiff clays the number and thickness of cross walls, and the thickness of the outer walls, should be kept to a minimum consistent with the need for weight to aid sinking and for rigidity against distortion. Grabs can excavate close to the cutting edge in caissons with fairly thin walls. This is important in firm or stiff clays since these soils do not slump towards the centre of a dredging well; whereas in sands and soft silts, grabbing below cutting edge level will cause the ground readily to slump away from the haunch plates towards the deepest part of the excavation, especially if assisted by water jetting. However, as already noted, thin walls mean reduced sinking effort and it is inconvenient to have to take kentledge on and off the top of the steining for each lift of concrete that is placed.

Control of verticality in large caissons is facilitated by the provision of a number of wells. To give control in two directions at right angles to one another they should be disposed on both sides of the centre lines, but for a narrow caisson there may only be room for one row since sufficient width must be provided for a grab to work. Heavy monoliths sunk through soft material on to but not into a firm or hard stratum need only have small wells.

Cross walls need not extend to cutting edge level. They may be stopped at a height of, say, 8 m above the cutting edge in a deep caisson, thus reducing the end-bearing resistance of the walls and enabling the grab to be slewed beneath the cross walls to excavate over the whole base area. Such a design has the serious limitation of requiring a deep draught at the initial sinking stage. Experiences in sinking open caissons for the piers of the Lower Zambezi Bridge[6.21] showed that straight walls were preferable to circular walls when sinking through stiff clay, since with circular walls there was a tendency for the clay to arch and wedge itself around the cutting edge rather than to be forced towards the centre of the well.

If occasional obstructions to sinking are expected and are such that they cannot be broken up under water, it will be advisable to make provision for air decks in all the cells (Fig. 6.30). By such means it is possible to put individual cells under compressed air to allow obstructions to be cleared from beneath the cutting edge by men working 'in the dry'. The use of air decks or air domes also facilitates control of draught and verticality during sinking. This is the 'flotation caisson' method used by Dan Moran for the San Francisco–Oakland Bay Bridge[6.22]. Fig. 6.31 shows a section through the 40.7 m × 18.1 m caisson used for Pier 4 of the Tagus River Bridge constructed in 1960–66[6.23]. The caisson, designed by the Tudor Engineering Com-

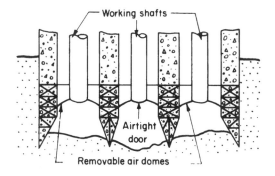

Fig. 6.30 Air domes installed in open-well caisson.

pany, was provided with a cutting edge 'tailored' to the profile of the surface of the rock at founding level. This resulted in a deep draught when in the floating position. However, the use of air domes on the dredging wells enabled the caisson to be floated with air pressure in the wells from the launching ways to the bridge site. The irregular weight distribution and submergence also gave problems in sinking but, by varying the air pressure in the different wells, it was possible to control listing and to maintain verticality in the crucial later stages of sinking.

A double skin of steel plating which is subsequently filled with concrete is used for the caisson shoe. The concrete in the lower part of the shoe should be of high

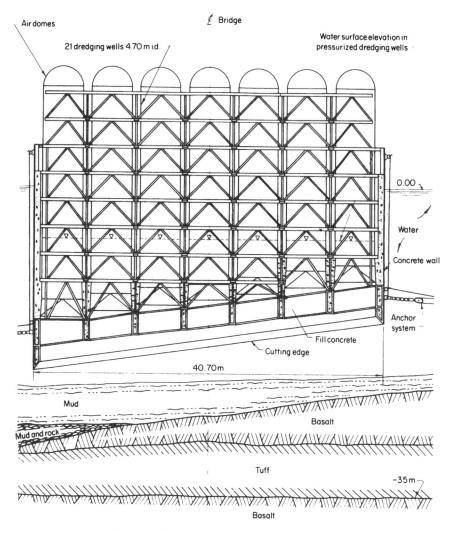

Fig. 6.31 Flotation caisson for Tagus River Bridge[6.23].

quality, since it is required to develop high early strength to resist stresses developed in the 'tender' early stages of sinking. The concrete in the upper part of the shoe and in the steining need not be of especial high quality provided that it is of massive construction. The cement content, however, should be sufficiently high to give it resistance to attack by sea or river water (see Section 13.5.3).

A design for a reinforced concrete caisson shoe conforming to the recommendations of the Indian Roads Congress (IRC)[6.24] is shown in Fig. 6.32. Indian engineers have used well-sinking methods for bridge foundations for many centuries, and it is a common method of construction at the present day. The IRC recommendations for proportioning wall thicknesses and other design details are in the form of empirical rules based on their long experience.

The design in Fig. 6.32 provides for reinforced concrete with a minimum crushing strength of 20 N/mm^2, and minimum reinforcement of 72 kg/m^3 excluding bonding rods. The cutting edge plates are required to have a weight of at least 40 kg/m. Where there are more than two compartments the cutting edges of the middle stems are kept about 300 mm above those of the outer walls to prevent rocking. Where blasting is expected to be necessary the wells are faced externally with steel plates not less than 6 mm up to half the weight of the shoe. The inner faces of the cells are protected with 10 mm plates extending to the top of the shoe continuing with 6 mm plates over a further height of 3 m. Additional hoop reinforcement in the form of 10 mm bars at 150 mm spacing is placed in the shoe and for a further height of 3 m in the walls.

Reinforced concrete was used for the shoes of two caissons forming the foundations for the river piers of the River Torridge Bridge[6.25]. The shoes were constructed in a dry dock near the site. The draught of the units was limited to 3.15 m in order to be able to float the shoes over the gate sill at a high-tide period. They were constructed to an overall height of 3.2 m, and in order to provide the required buoyancy the tops of the eight cells were covered with temporary steel plating and compressed air was introduced into the space below the plates. Further steel plates were assembled around the caisson to provide additional freeboard for towing from the dry dock to the bridge site. These plates were subsequently used as formwork for raising the walls.

The depth of the shoe (or curb) is governed by the thickness of the main walls and the angle given to the inner haunch (or cant) plates. This angle should be determined by the type of ground through which the caisson is sunk (Fig. 6.33). Generally the angle should be decided to suit the dominant factor in sinking, whether soft ground at the early stages or stiff or bouldery ground at later stages. The usual thickness of skin plating is 19 mm and bracing is provided both in horizontal and vertical planes.

The design of the cutting edge and its attachment to the shoe is an important feature. High concentrations of stress on the cutting edge are experienced when sinking through boulders, and buckling at this point would hinder sinking and might even result in the caisson splitting. Cutting edges are usually made up from 13 mm thick steel angles backed by 19 mm steel plates. The vertical plate projecting below the angle is advantageous in preventing escape of air in pneumatic caissons. Vertical stiffeners are provided on the upstanding part of the plate above the horizontal leg of the angle. Two types of cutting shoe were used on the Lower Zambezi Bridge. For founding on sand a vertical plate was used as shown in Fig. 6.34(a), but for founding

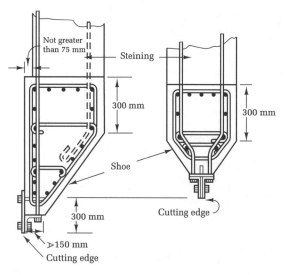

Fig. 6.32   Reinforced concrete caisson shoe (as Indian Roads Congress recommendations).

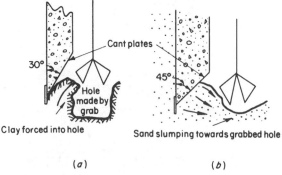

Fig. 6.33   Angles for cant plates. (a) Stiff clays. (b) Sands and weak clays.

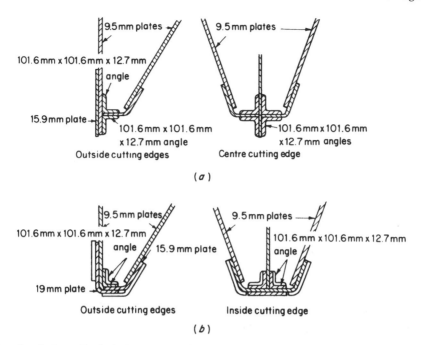

Fig. 6.34 Cutting edges for Lower Zambezi River Bridge caissons. (*a*) Cutting edges for sand. (*b*) Cutting edges for rock.

on rock the lowermost parts were stiffened plates in angles for the outer walls and a plated channel with the web horizontal for the central walls (Fig. 6.34(*b*)). It was found that the last-mentioned two designs gave difficulties in tucking poling boards behind the cutting edge while excavating beneath the cutting edge under compressed air to remove boulders. Caisson shoes are sometimes flared, either with the object of reducing skin friction or preventing adjacent caissons rubbing together and jamming when they are constructed close together. However, experience has shown that the flaring does little or nothing to reduce skin friction and there is increased difficulty in maintaining verticality while sinking.

The walls of caissons should be set back for a distance of 25–75 mm from the shoe. The thickness of the wall is dictated by the need for great rigidity to resist severe stresses which may occur during sinking, and the need to provide sufficient weight for overcoming skin friction. Lightness of the wall construction for the stages of floating-out and lowering is achieved by hollow steel plate construction, using 6–7 mm thick skin plates stiffened by vertical and horizontal trusses as provided in the caisson shoes. The plating should be arranged in strakes about 1.2–2.4 m high. These are convenient heights of lift for sinking and concreting in a 24-hour cycle.

The Indian Roads Congress recommendation for wall thickness to ensure sufficient weight to overcome skin friction is given by the empirical equation

Required thickness in meters (not less than

$$0.5 \,\text{m}) = Kd\sqrt{l} \qquad (6.24)$$

where

K = a constant (see Table 6.1),

d (in metres) = the external diameter of a circular or dumb-bell shaped well, or in the case of double D caissons the smaller dimension in plan;

l(metres) = depth of well below low-water level or ground level whichever is higher.

The IRC recommendations state that a greater thickness may be needed for sinking through boulder strata or on to rock where blasting may be needed. Lesser thickness

Table 6.1   Values of the constant K to determine wall thickness of well and caisson foundations

Type of well or caisson	Soil type	
	Predominantly sandy	Predominantly clayey
Single circular or dumb-bell shape in cement concrete	0.030	0.033
Double D caisson in cement concrete	0.039	0.043
Single circular or dumb-bell shape in brick masonry	0.047	0.052
Double D caisson in brick masonry	0.062	0.068

based on local experience may be needed when sinking through very soft clay to prevent the well from sinking uncontrollably under its own weight.

A concrete mix of 1:3:6 is recommended for mass concrete walls where not exposed to a marine or other adverse condition. Nominal steel reinforcement at 0.12 per cent of the gross cross-sectional area should be provided vertically on each face of the mass concrete walls with hoop steel at not less than 0.04 per cent of the volume per unit length of wall.

Brick masonry is frequently used in circular well foundations for bridges over minor rivers and creeks in Bangladesh where aggregates for concrete manufacture are scarce, but there is abundant labour for construction. The IRC recommend that brick masonry wells should have a diameter not greater than 6 m and a depth not greater than 20 m. Vertical bonding rods at not less than 0.1 per cent of the cross-sectional area are distributed uniformly around the circumference at the centre of the walls and encased in concrete with hoop steel similarly encased as shown in Fig. 6.35.

Vertical reinforcement in reinforced concrete or masonry caissons must extend from the shoe up through the full height of the walls to prevent the tendency for the shoe to part company from the walls and continue its downward progress while the walls are held up by skin friction at higher levels. Shear reinforcement must also be provided to withstand the racking stresses which occur during sinking.

Great care must be taken in constructing the walls and shoe of a caisson to ensure truly plane surfaces in contact with the soil. Bulging of the steel plates due to external ground or water pressure or internal pressure from air or concrete filling will greatly increase the resistance to sinking.

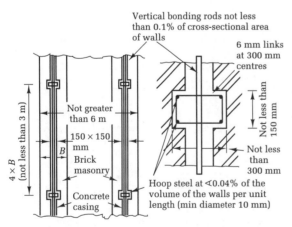

Fig. 6.35 Indian Roads Congress recommendations for reinforcement of brick masonry well foundations.

6.4.5 Pier construction with pneumatic caissons

Pneumatic caissons are used in preference to open-well caissons in situations where dredging from open wells would cause loss of ground around the caisson, resulting in settlement of adjacent structures. They are also used in sinking through variable ground or through ground containing obstructions where an open caisson would tend to tilt or refuse further sinking. Pneumatic caissons have the advantage that excavation can be carried out by hand in the 'dry' working chamber, and obstructions such as tree trunks or boulders can be broken out from beneath the cutting edge. Also the soil at foundation level can be inspected, and if necessary bearing tests made directly upon it. The foundation concrete is placed under ideal conditions in the dry, whereas with open-well caissons the final excavation and sealing concrete is almost always carried out under water.

Pneumatic caissons have the disadvantage, compared with open-well caissons, of requiring more plant and labour for their sinking, and the rate of sinking is usually slower. There is also the important limitation that men cannot work in air pressures much higher than 3.5 bar, which limits the depth of sinking to 36 m below the water table, unless some form of ground-water lowering is used externally to the caisson. If such methods are used to reduce air pressures in the working chamber they must be entirely reliable, and the dewatering wells must be placed at sufficient distance from the caisson to be unaffected by ground movement caused by caisson sinking.

The development of large-diameter cylindrical foundations installed by rotary drilling or grabbing as described in Chapter 8, and the limitation in sinking depth due to considerations of limiting air pressure, has meant that pneumatic caissons are only rarely used for foundations in the present day.

When caissons are designed to be sunk wholly under compressed air, it is usual to provide a single large *working chamber* (Fig. 6.36) instead of having a number of separate working chambers separated by cross walls. The single chamber is a convenient arrangement for minimizing resistance to sinking, since the only resistance is given by the outer walls. Control of sinking by differential excavation from a number of cells is not necessary since control of position and verticality can be readily achieved by other means, for example by the use of shores and wedges beneath the cutting edges, or by differential excavation beneath the cutting edges.

However, the location of the roof above the working chamber is a matter of some contention. If placed at a low level, as shown in Fig. 6.36, the centre of gravity of the unit is kept low, thereby assisting control of verticality. Also it is possible to place ballast weights on top of the

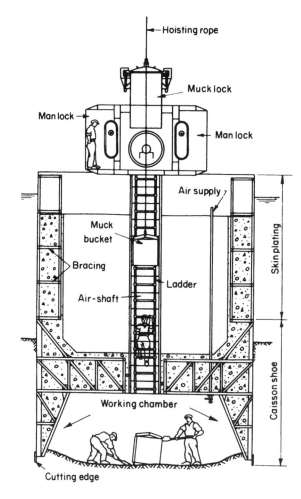

Fig. 6.36 General arrangement of pneumatic caisson.

substructure. It was necessary to underpin one of the two caissons to avoid sinking below the design depth after weak rock was encountered at the expected founding level. The paper by Pothecary and Brindle[6.25] pointed out that this possibility was one of the reasons why pneumatic caissons were adopted.

Features of pneumatic and other types of caissons have been described by Mitchell[6.26]. The working chamber is usually 2.3–2.4 m high, although where the caisson chamber is sunk to a limited penetration the height may be somewhat smaller. As already noted the roof of the working chamber must be strongly built. In large caissons, a convenient arrangement is to arch the roof of the working chamber. This is an economical arrangement since the skin plates are everywhere in tension only, instead of bending as in the case of a horizontal roof.

Access to the working chamber is through *shafts*. Since all excavated material must be lifted through the shafts or through *snorer* pipes, the shafts must have adequate capacity in size and numbers to pass the required quantity of spoil in buckets through the *air-locks* to meet the programmed rate of sinking. The air-shaft is usually oval or figure-of-eight in plan, and is divided into two compartments by a vertical ladder. One compartment is used for hoisting and lowering spoil buckets and the other is for the workmen. The shaft is built up in 1.5 or 3 m lengths to permit its heightening as the caisson sinks down. The air-locks are mounted on top of the shafts, and it is essential for the safety of the workmen to ensure that the locks are always above the highest tidal or river flood levels with sufficient safety margin to allow for unexpected rapidity in sinking of the caisson. Alternatively, the air-locks can be protected against flooding by building up the skin plating or providing a cofferdam around the top of the caisson to the required height. Details of air-shafts and air-locks are given later in this chapter.

The number of air-locks required in a caisson depends on the number of men employed in any one working chamber in the caisson. The size of the air-locks and air-shafts is governed largely by the quantity of material to be excavated, i.e. by the size of the 'muck bucket'. Reid and Sully[6.27] have given much valuable factual data on this. Their main points are as follows:

(a) For excavating in hard material, one man can be effectively employed in 3.25 m² of working area of a caisson, but in loose material such as sand or gravel one man may be allowed 6.5–7.4 m².

(b) Generally the number of men in a caisson per shaft will range from 5 to 10 except in the smallest caissons. The optimum number is 10 men per shaft.

(c) Sufficient air-locks should be provided to allow the whole shift to pass out of the caisson in reasonable

roof at the most favourable position to correct any tendency to tilt. On the other hand the massive construction of the roof required to resist internal air pressure and soil pressure when sinking through soft deposits is no longer required after the caisson has been completed. It is argued that the roof is best placed at a high level where it can be used to support the bridge pier and to strengthen the caisson against ship collisions. Against this argument it is pointed out that the caisson with a high-level roof is much more buoyant than that with a low-level roof and hence may be too light to overcome external skin friction without heavy walls, also the centre of gravity is high, and if the required founding level is deeper than expected the roof will not be in the position required to support the pier. The roof to the working chamber of the River Torridge Bridge caissons was placed at a high level on top of the walls after initial sinking in free air as an open-well time. This depends on the working pressure. For moderate pressures an air-lock should be provided for every 93–111 m² of base area. For high pressures two locks would be considered for 93–111 m² of base area because of the longer time spent in locking each shift in and out of the caisson.

(d) The size of the lock is governed by the rate of excavation and the number of men to be accommodated. Thus the main chamber (or muck-lock) has to accommodate a skip of sufficient size to pass through the excavated material at the programmed rate. For example with a base area of 93 m² per lock, a large air-lock can deal with 9.5 m³ of spoil per hour; for continuous shift working at say 2.0 bar, 188 m³ of material would be passed through in 20 h of effective working. This gives about 113 m³ of material 'in the solid' or 1.2 m of sinking per day. With a smaller air-lock under similar conditions the output would be 6.8 m³/h or 123 m³ of material in 20 h, corresponding to a sinking rate of 0.79 m per day.

The 'Gowring'-type air-lock (Fig. 6.37), originally designed by Holloway Bros., has double man-locks and a 1.12 m diameter by 1.83 m high main lock. Each man-lock can accommodate six workers, and the main lock a 1 m³ muck-skip, the same skip carrying 0.8 m³ of concrete to keep the mass within the capacity of the 2 t crane. The lock is of welded construction, and is divided into two parts for ease in transport from site to site.

In Britain the air supply to caissons and the equipment in man-locks and medical locks, and the operation of locks, is governed by the Ministry of Labour and National

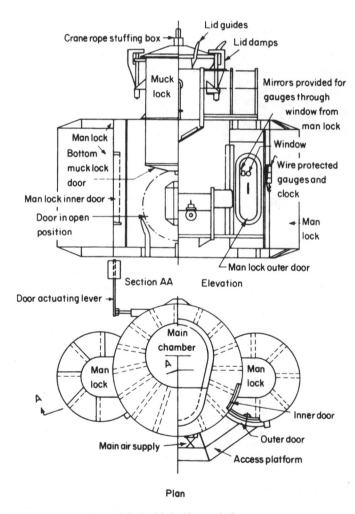

Fig. 6.37   General arrangement of 'Gowring'-type air-lock with double man-locks.

Service in the Factories Acts of 1961. The Work in Compressed Air Special Regulations, 1958, requires that the plant for the production and supply of compressed air to the working chamber shall deliver a supply sufficient to provide at the pressure in the chamber 0.30 m³ of fresh air per minute per person for the time being in the chamber. CP 2004 recommends that whenever work involving compressed air at pressures greater than 1 bar above atmospheric pressure is undertaken the Medical Research Council's Decompression Sickness Panel should be consulted for advice on decompression rates.

If the air supplied in accordance with the above rule is more than the amount lost under the cutting edge and through the air-locks, the surplus should be exhausted from the caisson through a control valve.

Compressors for air supply are usually stationary types. Ideally, they should be driven by variable-speed motors to enable the supply to be progressively increased as the caisson sinks deeper. At least 50 per cent spare compressor capacity with an alternative power supply should be provided for emergency purposes.

Improved working conditions and greater immunity to caisson sickness is given by treatment of the air supply. The air-conditioning plant should aim to remove moisture and oil, and to warm the air for cold weather working, or to cool it for working in hot climates. The need to supply cool dry air is especially important for compressed-air work in hot and humid climates. The air supply to the caissons of the Baghdad bridges described by Reid and Sully[6.27] had two stages of cooling, and a silica-gel dehumidifier which reduced the wet-bulb temperature to less than 26.6 °C at all times, even though the outside dry-bulb shade temperature was at 48.8 °C. The relative humidity in the working chamber of these caissons rarely exceeded 75 per cent. In cool climates it is advantageous to provide heating in man-locks since the cooling of the air which always takes place during decompression can cause discomfort to the occupants.

6.4.6 Caisson construction and sinking methods

(i) Construction of shoe   The normal practice in caisson construction is to build the shoe on land and slide or lower it into the water for floating out to the site, or to construct it in a dry dock which is subsequently flooded to float out the shoe. Land caissons are of course constructed directly in their final position. Caisson shoes constructed on the bank of a river or other waterway are slid down launching ways into the water, or rolled out on a horizontal track and then lowered vertically by a system of jacks and suspended links. Gently sloping banks on a waterway with a high tidal range favour construction on sloping launching ways (Fig. 6.38(a)), whereas steep banks either in tidal or non-tidal conditions usually require construction by rolling out on a horizontal track (Fig. 6.38(b)).

Care must be taken to avoid distortion of the shoe during construction. On poor ground the usual practice is

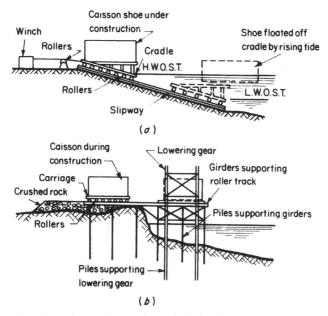

Fig. 6.38   Methods of constructing caisson shoe on shore. (a) On gently sloping river bank. (b) On steeply sloping river bank.

to lay a thick blanket of crushed stone or brick rubble over the building site and to support the launching ways on timber or steel piles.

Economy in temporary works is given by constructing caissons in their final position. This can be done for land caissons, and for river work it is sometimes possible to take advantage of low-water stages by constructing caissons on the dry river bed or on sand islands. This is only advisable when the low-water periods can be predicted reasonably accurately and there is no risk of sudden 'flash' floods.

This method is used at the present day for the foundations of bridges crossing tributary rivers and creeks in Bangladesh. A shallow excavation is made on a sandbank (or 'char'). The excavation is surrounded by a low sand and clay bund protected against erosion by driving bamboo stakes at close spacing. A ring-shaped mould is formed for the caisson shoe using clay faced with loose bricks (Fig. 6.39). Steel bar reinforcement is assembled in the mould followed by concreting the shoe up to the level of the brick masonry steining. The walls are then carried up in brickwork until sufficient weight is mobilized to

Fig. 6.39   Construction of caisson shoe, Bangladesh.

commence sinking initially by hand excavation in the well and then by grabbing using a guyed derrick pole and a hand winch (Fig. 6.40). These simple methods do not involve any heavy construction plant and there are fewer problems of logistical support over country roads and narrow bridges than would be necessary for piled foundations in sheet pile cofferdams.

(ii) Towing to sinking site   The operation of towing a caisson from the construction site to its final location must be carefully planned. Soundings must be taken along the route to ensure an adequate depth of water at the particular state of tide or river stage at which the towing is planned to take place. An essential requisite of the launching, towing, and sinking programme is a stability diagram for the caisson. This shows the draught at each stage of construction. The stability diagram used for the caissons of the Kafr-el-Zayat Bridge in Egypt[6.28] is shown in Fig. 6.41. In these diagrams the draught is plotted against the weight of the caisson for various conditions of free floating or floating with compressed air in the working chamber. The weight of each strake of skin plating and concrete in the walls to be added to give a desired draught can be read from the diagram. Also the air pressures in the working chamber required to give any desired internal water level can be read off the appropriate lines.

(iii) Bed preparation   The first operation is to take soundings over the sinking location to determine whether any dredging or filling is required to give a level bed for the caisson shoe. A study should be made of the regime of the waterway to determine whether any bed movement is caused by vagaries of current. Such movement can cause difficulty in keeping a caisson plumb when landing it on the bottom, especially at the last stages when increased velocity below the cutting edge may cause non-uniform scour. Difficulties with bed movement can be overcome by sinking flexible mattresses on the sinking site. These consist of geotextile materials or steel mesh baskets weighted with crushed stone.

(iv) Supporting structures   The various methods used to hold a caisson in position during lowering include

(a)   An enclosure formed by piling
(b)   Dolphins formed from groups of piles or circular sheet pile cells
(c)   Sinking through a sand island
(d)   Wire cables to submerged anchors

The choice of method depends on the size of caisson, the depth of water, and particularly on the stability of the bed of the waterway.

Sand islands were used for four of the caisson piers of

Fig. 6.40 Sinking brick masonry caissons, Bangladesh.

the Baton Rouge Bridge over the Mississippi River[6.29]. The fast-flowing river was known to cause deep scour, and bed protection was given to the sites of the two deepest piers in the form of 137 m × 76 m woven board mattresses. The islands were 34 and 37 m respectively in diameter, and were formed by steel plate shells filled with sand. The shells were surrounded by a double row of piles (Fig. 6.42). The sand islands narrowed the 730 m waterway to 97 m between the islands, and this caused deep scour which the mattresses did little to prevent. The scour at Pier 3 was 12 m deep, and a similar depth of scour at Pier 4 caused the whole of the sand filling in the island to disappear in 2–3 min.

The external water pressure on the shell then pushed in the 9.5 mm plating which was torn apart. The caisson, which at that time had only penetrated 4.6 m into the river bed, tilted by 2.1 m in line with the bridge and 0.6 m in the other direction, and was only plumbed with great difficulty. These experiences emphasize the hazards resulting from obstructions to flow caused by substantial temporary works in a river with an erodible bed.

Similar problems with scour around sand islands were experienced at the early stages of constructing the 1410 m span suspension bridge across the Humber Estuary com-

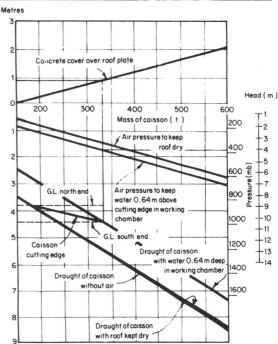

Fig. 6.41 Stability diagram for caisson no. 8 at Kafr-el-Zayat Railway Bridge (after Hyatt and Morley[6.28]).

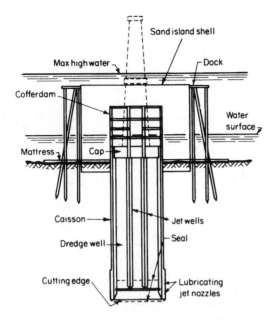

Fig. 6.42   Sand island used for caisson construction at Baton Rouge Bridge (after Blaine[6.29]).

pleted in 1983[6.30]. The main tower pier on the south side of the estuary was constructed on twin 24.4 m open well caissons sunk to a depth of 38 m below the river bed on to stiff Kimmeridge Clay. The caissons were of double wall construction divided into six circumferential compartments by radial walls (Fig. 6.43). They were sunk from two artificial sand islands enclosed by steel sheet piling (Fig. 6.44). Scour around the sheeting piling threatened to undermine the enclosures and it was necessary to dump 13 000 t of chalk and sandbags around them to check the erosion. Further problems occurred during caisson sinking when water under artesian pressure was met at 20 m below the river bed in the west side well. This caused partial collapse of the sand fill and loss of bentonite in the annulus around both caissons. Downward movement stopped and was restarted only after stacking 7000 t of

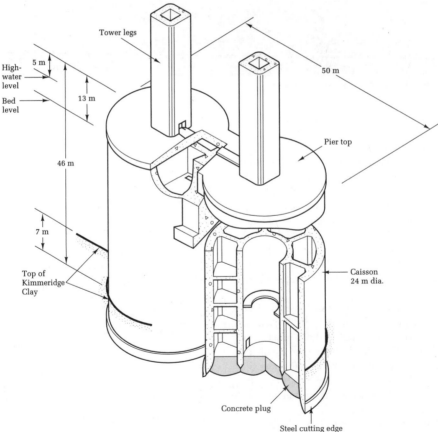

Fig. 6.43   Open-well caissons for the south tower pier, Humber Bridge.

Fig. 6.44 Artificial sand islands for constructing the caissons for the south pier, Humber Bridge (David Lee Photography).

steel ingot kentledge on top of the radial walls (Fig. 6.45) supplemented by raising the internal walls by 14 m.

The bridge was designed by Freeman Fox and Partners, and the foundations of the tower piers and the anchorages were constructed by John Howard & Co.

The minimum of temporary construction and the lowest risk of bed erosion is given by the method of securing a floating caisson to *submerged anchors*; the caisson being moored between *floating pontoons*. This method is particularly suited to a multi-span structure when the high capital cost of an elaborate pontoon-mounted sinking plant can be spread over a number of caissons; whereas if fixed stagings are provided for the piers of a multi-span structure, much time will be spent in driving and extracting piles for construction of the stagings at each pier site, with inevitable damage due to repeated reuse. Floating plant is highly mobile, and can be rapidly switched from one pier site to another to suit changing conditions of river level and accessibility at low-water stages. It is advantageous in these conditions to design the floating plant to be adaptable to working in the dry.

The arrangement of floating plant designed by Cleve-land Bridge and Engineering Ltd for the Lower Zambezi Bridge[6.21] is shown in Fig. 6.46. The two upstream moorings consisted of 10 t concrete or cast-iron sinkers, and the four crossed breast moorings were similar 5 t sinkers. The downstream moorings consisted of a single 5 t sinker. A steel staging to carry the caisson shoe was erected on the twin barges. Two 8 t steam derrick cranes were mounted on the staging, together with concreting plant, air compressors, pneumatic sinking and riveting tools, hydraulic pumps, lighting sets, and steam boilers providing steam for all purposes. Four such floating sets were provided.

Similar floating equipment was used for the foundations of two bridges over the River Tigris at Baghdad[6.27]. The concreting plant erected on one of the pontoons comprised two 0.38 m³ mixers at pontoon deck level feeding the skips running on bogie track in the bottom of the pontoon. A storage hopper was erected within the superstructure with a volume batching device. On the opposite pontoon were an electrically driven 8.5 m³/min air compressor, a standby diesel 8.5 m³/min compressor, and a portable diesel compressor for pneumatic tools. Electric power was provided

Fig. 6.45   Ballasting caisson to assist sinking, Humber Bridge (David Lee Photography).

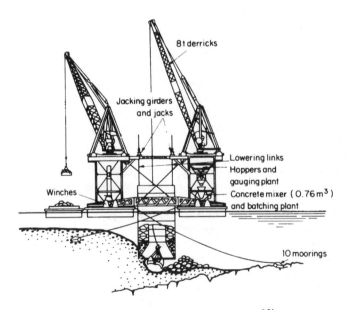

Fig. 6.46   Lowering caisson from floating plant at Lower Zambezi Bridge (after Howorth[6.21]).

by three 75 kW diesel generators, two being run at moderate load and the third a standby.

The pontoons were moored from one upstream and four breast moorings. The upstream mooring consisted of three 3 t concrete blocks (with projecting steel channels) placed in series, the 50 mm diameter mooring wire being connected to a six-part 30 t tackle leading to a 5 t hand winch on the superstructure. The breast moorings were each a 3 t concrete block connected directly to 2 t hand winches.

(v) Sinking open well caissons Four main methods are used for maintaining position and verticality of caissons during sinking. They are:

(a) Free sinking using guides between caissons and fixed stagings or floating plant;
(b) Lowering by block and tackle from piled stagings or floating plant;
(c) Lowering by suspension links and jacks from piled stagings or floating plant;
(d) Lowering without guides but controlling verticality by use of air domes.

When a caisson reaches the stage where concrete has to be added to maintain downward movement, the rate of sinking should be governed by a fixed cycle of operations. The usual procedure is to maintain a 24 h cycle comprising excavation from the wells, erecting steel plating for formwork in the walls, and concreting a 1.2–2.4 m lift of the walls. Sinking should proceed continuously, but the assembly and welding of steelwork should be scheduled for the daylight hours. The level of the concrete should be controlled by reference to stability diagrams.

The top of the skin plating should always be maintained about 1 m above water level to guard against an unexpected rise on level. However, the freeboard should not be so much that the centre of gravity is too high to give proper control of verticality.

Control of verticality can be achieved by one or a combination of the following methods:

(a) Adding concrete on one side or the other
(b) Differential dredging from beneath the cutting edge
(c) Pulling by block and tackle to anchorages
(d) Jetting under the cutting edge on the 'hanging' side
(e) Placing kentledge on one side or the other

Grabbing is the most commonly used method of excavating from the open wells although ejectors operated by compressed air or water pressure have been used. Arrangements of compressed air ejectors are shown in Fig. 6.47(a) and (b). Compressed air is injected into the riser pipe through holes in a manifold, and the aerated water, being of lighter density than the surrounding water, flows up the riser pipe to the surface. Air or water jets are

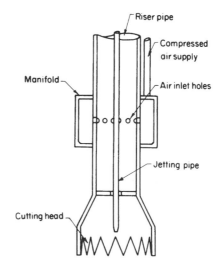

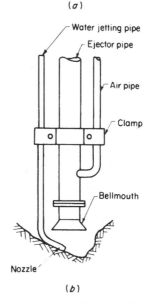

Fig. 6.47 Types of compressed-air ejector. (a) Air lift pump. (b) Air–water ejector.

used to loosen the sand at the base of the excavated area. An essential factor in the success of ejector excavation is that the material must be cohesionless, i.e. gravels, sands, or sandy silts. If ejectors are lowered into soft clays or clayey silts they merely form a vertical-sided hole and the material does not tend to slump towards the ejector unless assisted by independent water jets.

The caissons forming the foundations of the Carquinez Bridge in USA[6.31] were sunk wholly by a combination of jetting and airlifting. Each caisson was 31.2 × 16.1 m in plan divided by cross walls into three rows of six cells, the central row having 4 m square cells. A vertical 762 mm conductor tube was installed at the centre of each cell extending to the underside of timber planking which closed the bottom of the caisson just above the sloping portion of the shoe. The caisson was sunk through 27 m of water and 13.7 m of soft clay and silt to a penetration of about 9.5 m into a stratum of sand and gravel. The timber bottom was not removed until the caisson was close to founding level.

Excavation took place below the timber bottom by operating an airlift pump (of a type similar to Fig. 6.47(b)) simultaneously in four of the six cells in the central row. The airlift ejector pipe was 254 mm in diameter with air supplied through a 63.5 mm pipe at 11 m^3/min, giving an upflow velocity of 3 m/sec which was capable of lifting 60 mm gravel. The soil was loosened by a 150 mm jetting pipe operated independently of the airlift in the same 762 mm conductor tube. The four jetting pipes were supplied by two 6820 l/min pumps delivering at 34 bar pressure. They had 38 mm swivelling nozzles capable of reaching beneath the outer walls with the jet. The rate of excavation was 8 m^3/hour.

If excavation becomes difficult, a caisson can be partially pumped out. This increases its effective weight, so increasing the sinking effort. In Indian practice this is known as 'running the wells'. The procedure may be dangerous where the cutting edge has only just penetrated a clay stratum overlain by water-bearing sand (Fig. 6.48). The water in the sand may then force its way through a limited thickness of clay causing a localized 'blow' followed by tilting of the caisson.

Explosives fired in the wells can be used to cause a temporary breakdown in skin friction, but they are rarely effective in breaking down stiff material from beneath the cutting edge. Explosive charges carefully placed by divers

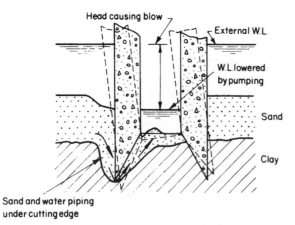

Head causing blow
External W.L.
W.L. lowered by pumping
Sand
Clay

Sand and water piping under cutting edge

Fig. 6.48

can be used to break up boulders or other obstructions to sinking.

Water jetting may not be effective in freeing hanging caissons, since 'sticky' sinking conditions usually occur in stiff clays or boulder clays which are not amenable to removal by jetting.

Open-well caissons are plugged with concrete placed by tremie pipe or by a bottom-opening skip. No attempt should be made to pump out the cells. If operating depths permit the concrete is packed under the sloping part of the shoe and bedded by divers. If the water depth is too great for diver operation the soil at the bottom of the cells is levelled by swivelling jet pipes in combination with air-lifting.

(vi) Cofferdams   In some caissons, the permanent caisson structure terminates below low-water level and the piers are constructed on the decking over the caisson. In such cases it is necessary to provide a temporary coffer-dam on top of the caisson in which the piers can be constructed in the dry. Cofferdams may be constructed in tongued and grooved timber, in timber sheeted with bituminous felt or, in cases of repetition work, by a movable steel trussed section sheeted with steel plating or interlocking sheet piling. The latter method was used for the River Tigris bridges at Baghdad[6.27] (Fig. 6.49). The lower ends of the sheet piles were made watertight by arranging them to bear against a 178 mm high raised concrete kerb cast on the top of the caisson. The kerb was faced by a steel angle bent to the shape of the cofferdam. Another angle frame was bolted to the bottom of the sheet piles to fit over the kerb angle and to complete the seal a rubber flap extended loosely below the angle frame on the cofferdam. External water pressure forced this against the kerb angle. When the cofferdam was to be removed it was flooded, so releasing the rubber flap seal. As a precaution against a sudden rise in river level a steel plate extension section was kept on hand, and if necessary it could have been bolted quickly in position on top of the sheet piling.

(vii) Sinking pneumatic caissons   Control of position and verticality of pneumatic caissons is more readily attainable than with open-well caissons. It is possible to maintain control by careful adjustments of the excavation beneath the cutting edge and, if this is insufficient, raking shores can be used in the working chamber, or the caisson can be moved bodily at early stages of sinking by placing sliding wedges or 'kickers' beneath the cutting edge.

Excavation in the working chamber is usually under-taken by men hand shovelling into crane skips, com-pressed-air tools such as clay spades or breakers being used in stiff clays or boulder clays. When excavations are in sands or gravels, hand-held water jets can be used to

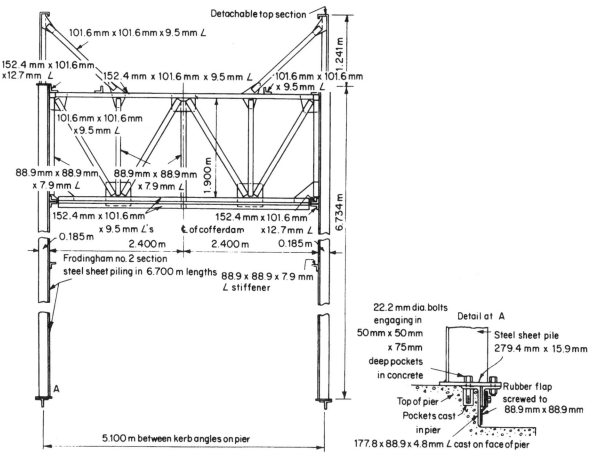

Fig. 6.49 Cofferdam erected on caissons for River Tigris bridges, Baghdad (after Reid and Sully[6.27]).

sluice the material into a sump from where it is raised to the surface by a 'snorer'. The latter is merely an open-ended pipe with its lower end dipping into water in the sump. By opening a valve on the snorer pipe, the water and soil are forced by the air pressure in the working chamber out of the caisson. The snorer also performs a useful function in clearing water from the floor of the excavation if the soil is too impermeable for the air pressure to drive the water down into it. Powerful water jets can be designed to swivel from monitors fixed to the roof of the working chamber.

The compressed-air supply must be regulated to provide adequate ventilation for the workmen. In permeable ground this is readily attained by allowing it to escape through the soil and beneath the cutting edge. However, when sinking in impermeable clays and silts, ventilation must be maintained by opening a valve to allow air to escape through the caisson roof. Careful regulation of air pressure is necessary when sinking in ground affected by changes in tidal water levels.

Smoking or naked lights should not be permitted in the working chamber because of the risk of encountering explosive gases, e.g. methane (marsh gas), during sinking. A careful watch should be maintained in neighbouring excavations. Accidents have been known to happen by compressed air passing through the beds of peat and becoming deprived of oxygen due to oxidation of the peat. The escape of this oxygen-deficient air into the confined spaces of excavations has caused asphyxiation of work-men in them.

In very permeable ground, such as open gravel or fissured rock, the escape of air may be so great as to overtax the compressor plant. The quantity escaping can be greatly reduced by pre-grouting the ground with cement, clay or chemicals as described in Section 11.3.7., or by plastering the exposed faces of the excavation with clay or by packing puddled clay beneath the cutting edge.

If a pneumatic caisson stops sinking due to build-up of skin friction, it can be induced to move by the process known as 'blowing down'. This involves reducing the air pressure to increase the effective weight of the caisson, so increasing the sinking effort. The process is ineffective if the ground is so permeable that air escapes from beneath the cutting edge at a faster rate than can be achieved by opening a valve.

Careful control should be exercised when blowing down in ground containing boulders, or when blowing down a caisson to land it on an uneven rock bed. In some circumstances, it may be necessary to excavate high spots in the rock and fill them with clay and then blow the caisson down into the clay.

If at all possible a caisson should not be blown down in soft or loose ground, as this might result in soil surging into the working chamber, so increasing the quantity to be excavated. There is also the risk of loss of ground into the working chamber causing settlement of adjacent structures. It must be remembered that pneumatic caissons are, in many instances, used as a safeguard against such settlement. Blowing down, if properly controlled, is a safe procedure in a stiff clay.

If a caisson is sinking freely without the need for blowing down, measures must be taken to arrest the sinking on reaching foundation level. This can be achieved by casting concrete blocks in pits excavated at each corner of the working chamber at such a level that the caisson comes to rest on the blocks at the desired founding level.

Concrete in pneumatic caissons is placed 'in the dry'. The procedure for small caissons as described by Wilson and Sully[6.27] is first to place a 0.6 m thick layer of fairly workable concrete over the floor of the working chamber, ramming it well under the cutting edge. Stiff concrete is then carefully packed under the haunch plates and worked back towards the air-shaft, the upper layers being well rammed against the roof of the working chamber. The space beneath the air-shaft is filled with a fairly wet concrete levelled off just below the bottom of the shaft. A 1 to 1 cement–sand grout is then placed in the shaft and allowed to fill up the lower 1.5 m section which is left in permanently. Air pressure is then raised to 0.3 or 0.6 bar above the normal working pressure to force the grout into the concrete and fill any voids underneath the roof of the chamber. The shaft is recharged with grout as necessary and it is usually found that the total amount of grout required corresponds to a volume equal to a 40 mm thick layer beneath the roof of the working chamber. After the grout, has been under air pressure for 2 days the remaining space to the top of the 1.5 m length of air-shaft is filled with concrete. About 3 or 4 days after completion of

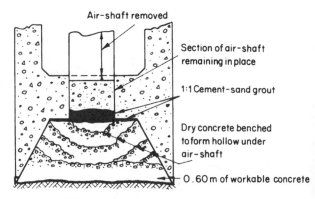

Fig. 6.50 Concreting working chamber of pneumatic caisson.

grouting, air pressure can be taken off the caisson and the remaining lengths of air-shaft and air-locks dismantled. The method of concreting is illustrated in Fig. 6.50. Concrete in large caissons is placed by pumping as described by Mitchell[6.26].

6.4.7 Skin friction in caissons

The skin friction resistance which must be overcome when sinking a caisson to the required founding level is unrelated to the *in-situ* shear strength characteristics of the soil. In the case of soft normally consolidated clays these are extensively remoulded as the caisson is sunk through them and the unit skin friction is likely to be much less than the remoulded shear strength or even the residual shear strength of the clay if the caisson is sunk rapidly. Where caissons are sunk through or into stiff clays the skin friction resistance on the external surface of the shoe is likely to be roughly equal to the residual shear strength unless undercutting below the shoe is necessary. Very little resistance will be developed above the projection at the top of the shoe.

Where sands are excavated from the cells down to or below the cutting edge they tend to flow and the consequent slumping around the caisson causes a dragdown force instead of resistance to sinking.

Present-day practice is to avoid the problems of build-up of excessive skin friction resistance by the introduction of a thixotropic clay slurry (e.g. bentonite) at the projection above the shoe. The slurry is circulated through a circumferential header pipe above the shoe with outlets spaced around the walls in case individual pipes become blocked or damaged as the caisson is sunk. Great care is necessary to avoid loss of continuity in the annulus of the bentonite from shoe level up to the ground surface, say by loss of material from a sand island such as occurred in the Humber Bridge caissons[6.30].

Table 6.2   Observed values of skin friction in caissons

Site	Type of caisson	Approx. size (m)	Soil conditions	Observed skin friction (kN/m^2)	Reference†
Lower Zambezi	Open well and pneumatic	11 × 6 × 38 m deep	Mainly sand	23	6.22
Howrah	Open well and pneumatic	55 × 25 × 32 m deep	Soft clays and sands	29	6.32
Uskmouth	Pneumatic	50 × 35 × 18.3 m deep	12.2 m soft clay 6.1 m sand	55	6.33
Grangemouth	Open well	13 × 13 × 7.3 m deep	Very soft clay	4.75	6.34
Grangemouth	Open well	19.5 × 10 × 5.5 m deep	Very soft clay	5.75–10.00	6.34
Kafr-el-Zayat	Pneumatic	15.5 × 5.5 × 25 m deep	Sand and silt	19–26	6.28
Grand Tower (Mississippi River)	Open well	19 × 8.5 × 38.1 m deep	Med. fine sand	30	6.35
Verrazana narrows	Open well	69.5 × 39 × 40 m deep	Med. dense to dense sand and fine gravel	85–95	6.36
Iskander	Open well	14.6 × 7.9 × 15.1 m deep	Sand	29	6.37
Sara	Open well	19.2 × 11.3 × 56 m deep	Silty sand	32–50	6.38
Gowtami	Open well	9 × 6 × 30.4 m deep	9.1 m sand 13.7 m stiff clay 7.6 m sand	12.6	6.39
Jamuna River	Open well (with bentonite)	12.2 m dia. × 90 m	Silty med. fine sand	260	6.40

† On 2.9 m depth of shoe at final stage of sinking.

Observed values of skin friction resistance to sinking are given in Table 6.2. It will be noted that there is no marked relationship between the depth of sinking and the unit resistance except in the case of the Jamuna River caissons where there was a linear relationship with the overburden pressure over the depth of the shoe below the bentonite annulus. The unit skin friction over this limited depth was found to be equal approximately to 0.33 times the effective overburden pressure. This relationship was expressed in a different way by Wiley[6.41] who stated:

> The skin friction on the lower section of the caisson increases directly as the depth sunk but unless the material is very unstable or practically in the liquid state the friction at any given depth on successive sections of a caisson is not as great as that exerted on the cutting edge and lower section of the caisson while at that point: or in other words the passage of the lower part of the caisson smooths, lubricates, or in some other manner tends to decrease materially the friction on the following sections. A wall thickness of 1.52 to 1.83 m gives weight enough to overcome any friction that may develop ordinarily unless the material penetrated be exceptionally difficult.

Values of the skin friction resistance obtained from sinking observations should not be used to calculate the *support* given to foundation loads after completion of a caisson. This should be calculated by the methods given for bored piling in Sections 7.7, 7.9, and 7.11. However,

there are many situations when such support cannot be allowed, for example where bentonite has been used to assist sinking, where the cutting edge has been undercut, or where the surrounding soil has slumped down and is undergoing reconsolidation and settlement.

6.4.8 Examples of the design and construction of open-well caissons

(i) Jamuna River, Bangladesh   Transmission line towers for the 250 kV east–west connection in Bangladesh are carried across the Jamuna River on piers formed from open-well caissons. Eleven single-well caissons were spaced at intervals of 1220 m across the meander belt of the river where boreholes showed alluvial silty sands and gravels extending to depths of more than 100 m. It was estimated that the river bed would scour to a depth of 38 m at a bankfull discharge of 70 700 m^3/s. To provide for this and for local scour around the caissons it was necessary to adopt a founding level of as much as 105 m below the river bed, which was the greatest founding depth anywhere in the world[6.40, 6.42].

All caissons had an outside diameter of 12.2 m with wall thicknesses increasing from 1.83 m at the top to 2.44 m above the structural steel shoe (Fig. 6.51). The soil at founding level was dense silty fine sand and gravel with an uncorrected SPT value between 50 and 125 blows/0.3 m.

The two principal problems which influenced the con-

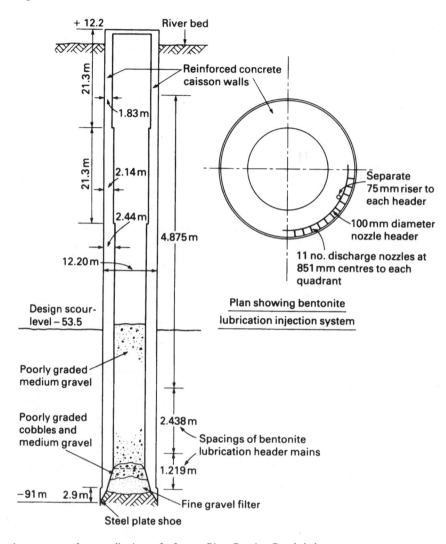

Fig. 6.51   General arrangement of open-well caissons for Jamuna River Crossing, Bangladesh.

struction method were, first, the deep scour which could occur in any annual flood season, and, second, overcoming skin friction developed in the very great depth of sinking. The deep scour (up to 10 m was recorded in the 1977–80 construction period) and the associated changes in the river channel and the formation of shallows prevented towing the caissons to the site and sinking them through water. Also it would not have been possible to recover and recirculate the bentonite slurry used for lubrication. Accordingly the six caissons located in the river channels were sunk through sand islands. These consisted of 24 m diameter sheet pile cells, using 30–37 m long piles, with sand filling. The piles were driven through a 92 m diameter scour protection blanket consisting of two layers of plastic bags filled with sand. After the sheet piles

had been driven additional sandbags and gabions filled with sandbags were stacked around the cells.

Reliance was placed on using bentonite slurry to lubricate the outer skin surface of the caissons as a means of overcoming skin friction over the great depth of sinking. The slurry was injected through 25 mm pipes spaced at 0.85 m centres around the circumference and at staggered intervals of depth varying from 1.22 to 4.87 m (Fig. 6.51). The deepest caisson had 1188 nozzles at 27 intervals of depth. A mud density of 1.2 t/m³ was used. The slurry discharging at surface level was collected in a trough, cleaned, and recirculated.

Excavation was performed by tower crane-operated grabs to a depth of about 30 m and thereafter by airlifting. The walls were concreted in 2.44 m lifts as sinking proceeded. On reaching founding level the bentonite slurry around the periphery of the caisson was replaced by a cement grout, and the interior was plugged by 30 m of medium gravel placed on a fine gravel filter.

The resistance to sinking was analysed for several of the caissons[6.40]. The unit skin friction resistance $K \tan \delta \sigma'_{vo}$ on the outer surface of the shoe was taken to be $0.33 \, \sigma'_{vo}$. The bearing resistance on the base of the shoe was calculated using the bearing capacity coefficient N_q of Caquot and Kerisel. From observations of sinking effort the angle of shearing resistance was shown to vary from 31° to 42°.

The joint consulting engineers for the project were Merz and McLellan and Rendel Palmer and Tritton. Professor P. W. Rowe advised on soil mechanics and caisson sinking. The contractors were a consortium of seven Korean companies advised by Raymond Technical Facilities.

(ii) Yokohama Bay Bridge Foundation construction in 1983 for the two main tower piers for the central 460 m span of the Yokohama Bay Bridge provided an interesting example of the use in foundation work of large-scale precasting in conjunction with heavy construction equipment. The cable-stayed spans have an overall length of 860 m and carry six lanes of traffic on each of two decks. The two main towers are each supported on a group of nine 10 m diameter open-well caissons sunk through about 50 m of silt into mudstone rock. The overall depth to founding level of the caissons was about 80–90 m below high-water level in Yokohama Bay.

The caisson groups beneath the towers are each surmounted by a $56 \times 54 \times 9.25$ m deep cap of precast concrete box construction. The construction of the first pier foundation is shown in Fig. 6.52. The work commenced by constructing the caisson cap in a dry dock. It contained three rectangular openings in each of three rows separated by U-shaped cells. At the same time as the construction work in the dry dock was proceeding, a temporary trestle supported by tubular steel piles was constructed at the pier site to act as a guide when floating in the caisson cap and as a working platform for the subsequent operations. Supports at this stage for the caisson cap consisted of a double row of tubular piles beneath the two central cells and a single row around the outside walls. These piles were driven from floating equipment and the two central rows were capped by heavy precast concrete slabs placed by floating crane at a height of about 6 m above the sea-bed.

On completion of the caisson cap in the dry dock it was towed by five tugs to the pier location, floated between the guide trestles and sunk on to the low level pile caps (Fig. 6.52(a)). The bulkheads temporarily closing the bottoms of the nine caisson openings were then removed and the 10 m diameter caissons which had been precast on shore to a height of 27 m were brought to site and pitched on to the sea-bed through the openings by a 3000 t floating crane (Fig. 6.52(b)).

The caissons were sunk to rockhead by grabbing using cranes operating from the top of the caisson cap. Further sinking to the required depth below rockhead was performed by a rotary drill mounted on a travelling carriage (Fig. 6.52(c)). An under-reaming tool enlarged the rock socket to the external diameter of the caisson. The sockets were plugged with concrete and the upper parts were capped followed by placing concrete in the box structure to form the base for the tower pier (Fig. 6.52(d)).

6.4.9 Examples of the design of pneumatic caisson foundations

(i) Pneumatic caisson for Redheugh Bridge, Newcastle upon Tyne Pneumatic caissons were used to support the two main piers of the 160 m span prestressed concrete girder bridge constructed in 1980–82 across the River Tyne. The caissons had an outside diameter of 11.0 m and they were provided to transmit the pier loads on to the mudstones, siltstones, and sandstones of the Coal Measures at a depth of 23 m below HSWT. Pneumatic caissons were selected because of the likely difficulties in sinking through boulders in the glacial deposits overlying the Coal Measures, and because of the need to inspect the rock surface at founding level and to make probe drilling to locate possible abandoned coal mine workings at depth.[6.43]

The lower part of each caisson consisting of the 4.7 m deep shoe of steel plate construction, the roof of the working chamber, and an outer cofferdam 5 m high in temporary steel tubbing plates was constructed in a dry dock 12 km downstream of the bridge. The calculated draught of 6.65 m on floating out from the dry dock gave sufficient clearance for towing at the higher tide levels to the sinking site. The required draught was achieved by air pressure in the working chamber, by the buoyancy of the section within the cofferdam above the working chamber, and by six 5 t flotation bags secured around the outside.

At the pier location the caisson was winched into position and located by three guide piles. Air was released from the working chamber to sink the unit into the alluvium below the dredged river bed. Divers using high-pressure water jets assisted by grabbing were needed to remove a large boulder from beneath the cutting edge of one caisson. To increase the sinking effort the walls were concreted in 2 m lifts using the steel tubbing plates as external formwork. These plates were later raised above the walls to maintain the top of the cofferdam above high-

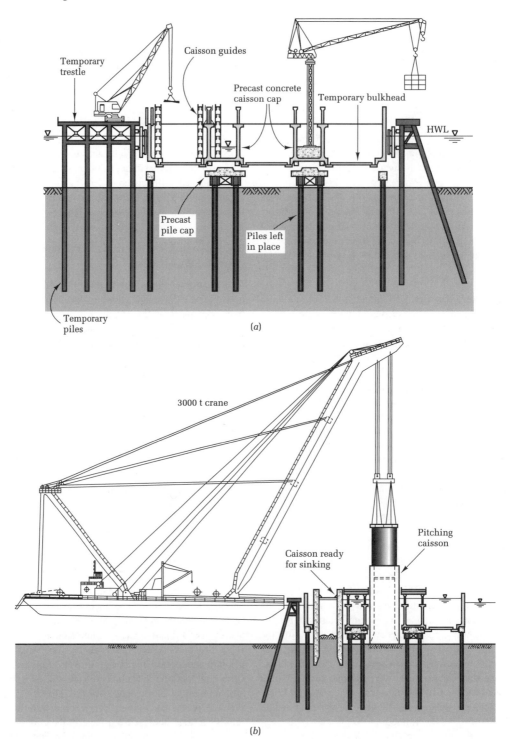

Temporary
trestle

Caisson guides

Precast concrete
caisson cap

Temporary bulkhead

HWL

Precast
pile cap

Piles left
in place

Temporary
piles

(a)

3000 t crane

Caisson ready
for sinking

Pitching
caisson

(b)

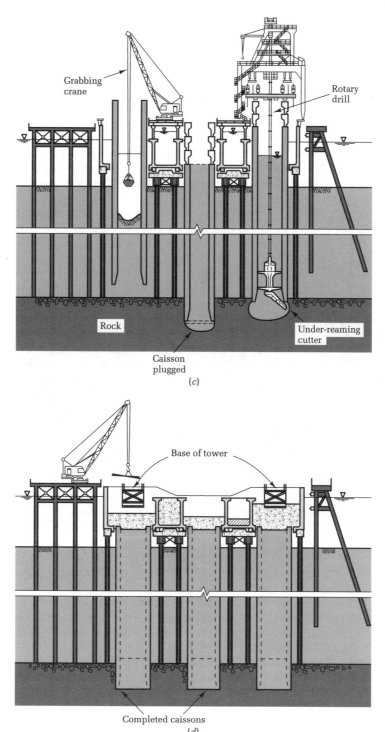

Fig. 6.52 Multi-well caisson foundations for the Yokohama Bay Bridge, Japan. (*a*) Floating in the caisson cap. (*b*) Pitching caisson. (*c*) Sinking caissons and drilling rock socket. (*d*) Constructing tower base. (Courtesy of the Metropolitan Expressway Public Authority of Yokohama).

Fig. 6.53   Sinking the pneumatic caisson for the foundations of Redheugh Bridge, Newcastle upon Tyne.

water level (Figs 6.53 and 6.54). Compressed air was supplied to the working chamber by two $28\,m^3/min$ electrically powered compressors with a third diesel-powered unit as a standby. The maximum air pressure required was 1.93 bar. The rate of sinking was controlled by varying the level of water ballast in the void above the roof of the working chamber. The outer skin was lubricated by a bentonite slurry injected above the shoe. When downward movement ceased it was restarted by the 'blowing-down' operation. Air was released rapidly from the working chamber of amounts between 0.4 and 1.0 bar to give a drop of about 0.3 m. At founding level sinking was stopped by packing concrete around the cutting edge. Probe drillings were made to a depth of 20 m followed by grouting the drill holes and the annular space around the

outer skin. The working chamber was then filled with concrete placed by pump.

The consulting engineers for the bridge project were Mott Hay and Anderson. The contractors were a consortium of Edmund Nuttall Ltd and HBM Civil Engineering Ltd.

(ii) Pneumatic caissons for the Little Belt Bridge, Denmark   Pneumatic caissons were used as combined pile caps and piers to support the towers of the 600 m main span of the suspension bridge crossing the Little Belt between Jutland and Fünen[6.44, 6.45]. The piers and the foundations of the approach spans were constructed in 1965–76 by a consortium of Christiani and Nielsen, Jespersen and Son, and Saabye and Lerche, to the designs

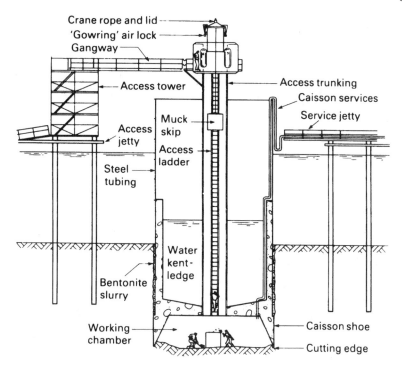

Fig. 6.54 Cross-section through pneumatic caisson for foundations of Redheugh Bridge, Newcastle upon Tyne.

of Ostenfeld and Jonson. The two main piers were located in 19–20 m of water and each of them was founded on a group of 206 vertical and raking precast concrete piles of 480 × 380 mm cross-section. The piles were driven as a first operation from staging to a depth of 31.5 m below the sea-bed. The driving was terminated when the pile heads were about 1 m above the sea-bed and a layer of gravel was placed around them.

The two caissons were constructed to a height of 10.75 m in an open excavation on the Jutland shore and were then floated to a construction berth where the walls were raised to heights between 20 and 21 m. During this time four 7 × 3.2 m bearing pads were each laid over the heads of a group of eight of the precast concrete piles in the outer two rows of vertical piles at each pier location. The gravel layer in the area of the bearing pads was grouted with cement.

The first caisson was raised from its fitting-out berth by pumping-out ballast water and it was then warped to the pier site and sunk on to concrete plinths which had been placed on the four bearing pads (Fig. 6.55). The sinking operation was performed in current velocities of 0.6–0.8 m/s. After installing the air-locks and access shafts air was pumped into the working chamber to expel the water and expose the pile heads. These were stripped down to expose the reinforcement which was bonded with a layer

of horizontal steel to make a structural connection between the pier body and the pile group. Loading tests to 1.5 times the working load of 2 MN were made on selected piles by jacking against the roof of the working chamber, after which a 0.8 m thick sealing layer of concrete was placed over all the pile heads. The concreted area was then divided into 18 compartments by steel formwork and these were concreted individually through 200 mm pipes placed at 24 positions over the area of the caisson. Concrete discharge chutes in the form of air-locks of 650 l capacity were moved from one pipe position to another as concreting progressively covered the area of the working chamber up to the roof.

The final operation was to consolidate the concrete by injecting cement grout beneath the roof of the chamber at a pressure equal to that in the chamber which was kept at about 2 bar during the work described.

6.4.10 Bridge piers supported by piling

The above example and that of the Färo Bridge (Fig. 6.26) are rather unusual methods of using piles for supporting bridge piers. The usual method is to locate the pile heads at or above high-water level and to use conventional pile caps to support the piers. Alternatively the pile caps can be constructed within a cofferdam as described for the

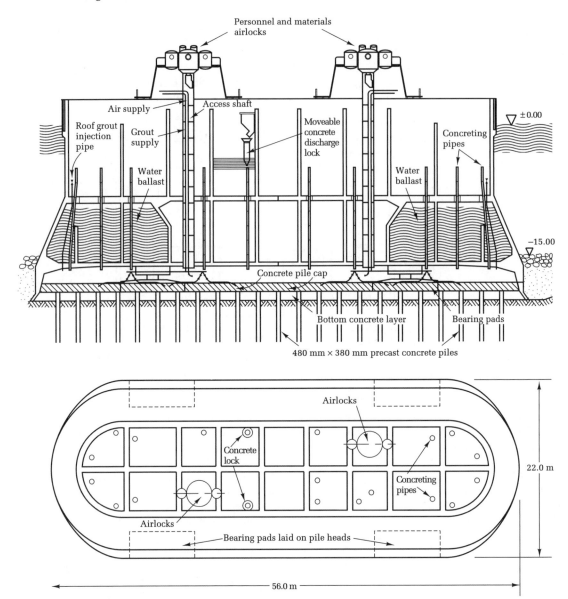

Fig. 6.55 Pneumatic caisson foundation for the Little Belt Bridge, Denmark (after Meldner[6.44]).

Kessock Bridge (Fig. 6.23). Protection against ship collision is achieved either by the combined mass and bending resistance of the pile group and cap, or by an independent fender structure[6.13].

The design of piles to resist axial and transverse loading is dealt with in the following chapter.

6.4.11 Anchorage abutments for suspension bridges

Loadings applied to the anchorages of a suspension bridge are predominantly sub-horizontal and are resisted, in the case of anchorages founded on soil, by the sliding resistance of the base and the passive resistance of the soil on the leading face. A large mass is required to provide a counterweight against overturning about a point on the bottom of the leading face. The principal features of an anchorage are:

(1)   The splay saddle where the individual strands from the suspension cables are splayed apart to the individual locking-off points.

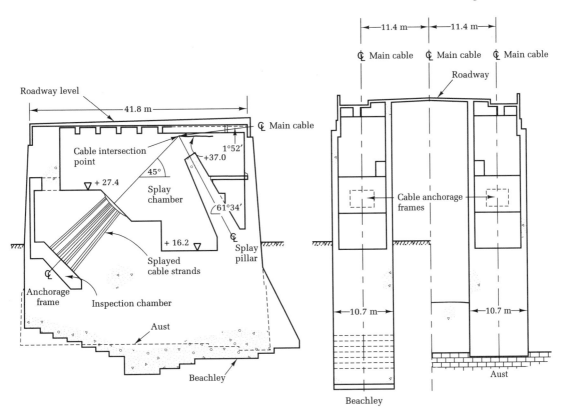

Fig. 6.56 Anchorages for the Severn Bridge (after Gowring and Hardie[6.46]).

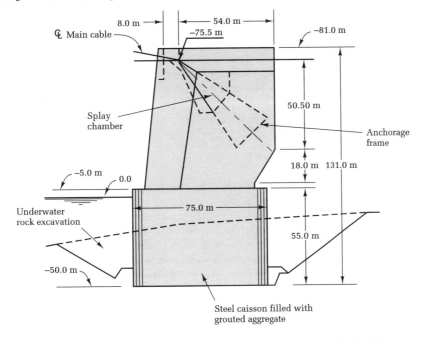

Fig. 6.57 Anchorage for South Bisan–Seto Suspension Bridge (courtesy of Honshu–Shikoku Bridge Authority).

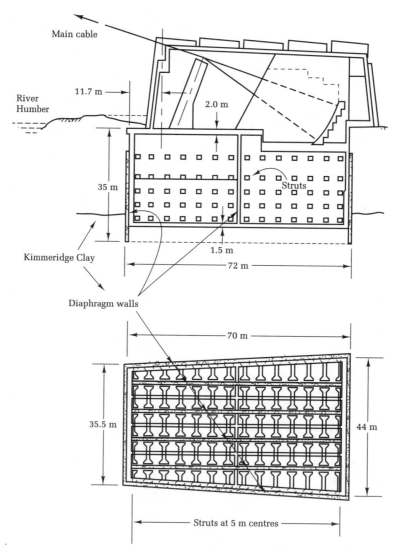

Fig. 6.58   South anchorage of the Humber Suspension Bridge.

(2) The splay pillars which are inclined columns which direct the thrust from the saddle to the base of the block.

(3) The inspection chamber where the operations of adjustment of cable tension are made.

(4) The rear gravity block which provides the resistance to overturning.

These features are shown in the arrangement of the Severn Bridge anchorages[6.46] (Fig. 6.56). A pair of cables exerted a total pull of 206 MN which was resisted by an anchorage block of about 370 MN net weight on the Aust side of the river and somewhat lighter on the Beachley side. The Aust anchorage was designed to impose a pressure of 1.3 MN/m^2 on the rock at the toe of the block and zero pressure at the heel. The gravity blocks were in two separate sections connected by a deck forming the approach roadway.

Problems of overturning may be severe where the anchorage is located in water with the splayed cables and inspection chamber set above high-water level. This was the case for the south anchorage of the 1100 m main span of the South Bisan–Seto Suspension Bridge. This bridge forms part of the Honshu–Shikoku Bridge Project and is a two-level highway–railway crossing the straits. The anchorage for a pair of cables with a combined pull of 785 MN is located in 15–20 m of water and the rock from the sea-bed downwards was removed by blasting and grab dredging with 20 m^3 buckets to a depth of −75 m to receive a 75 × 59 × 55 m steel caisson. The final trimming of the rock surface was performed by grinding with a 2.5 m diameter rotary excavator mounted on a jack-up barge. A total of 600 000 m^3 of rock was removed by these methods. After floating-in the 20 000 t caisson it was sunk on to the prepared bed and filled with coarse aggregate which was then injected by pumped-in mortar. Escape of grout from the bottom edges of the caisson was prevented by a crushable rubber cushion. The upper 76 m high section comprising the splay saddles for the cables and the inspection chamber were constructed wholly above sea-level (Fig. 6.57).

The main problem with the south anchorage of the Humber Bridge[6.47] was the extreme susceptibility to softening of the Kimmeridge Clay at the founding level of −33 m OD. This clay is of stiff to hard consistency and is fissured such that it undergoes rapid swelling and softening on relief of overburden pressure in a deep excavation. The anchorage was designed to resist a cable pull of 372 MN partly by the shear resistance of the clay at the base, partly by skin friction on the sides, and partly by passive resistance on the leading face of the substructure. In order to minimize exposure of the clay the cellular substructure was constructed by diaphragm wall methods in the form of box structure with four parallel interior walls. The exterior walls formed a trapezoidal shape in plan (Fig. 6.58). An open excavation was taken out to −4.5 m, and the 800 mm diaphragm walls were constructed from this level. The excavation within the walls was taken out in alternate 8 m wide strips with precast concrete struts installed progressively as the excavation was taken down.

The anchorage was constructed by John Howard & Co. to the designs of Freeman Fox and Partners.

References

6.1   Hambly, E. C., Bridge foundations and substructures, *Building Research Establishment Report*, HM Stationery Office, London, 1979.

6.2   Symons, I. F. and Clayton, C. R. I., Earth pressure on backfilled retaining walls, *Ground Engineering*, Thomas Telford, London, **26**(3) 26–34 (1992).

6.3   Chettoe, C. S. and Adams, H. C., *Reinforced Concrete Bridge Design*, 2nd edn, Chapman and Hall, London, 1938, p. 298.

6.4   Huntington, W. C., *Earth Pressures and Retaining Walls*, John Wiley, London, 1957.

6.5   Ah Teck, C. Y., The behaviour of spill-through abutments, M.Phil. Thesis, University of Cambridge, Dept. of Engineering, 1983.

6.6   Poulos, H. G., and Davis, E. H., *Pile Foundation Analysis and Design*, John Wiley, 1980.

6.7   De Beer, E. and Wallays, M., Forces induced in piles by unsymmetrical surcharges on the soil around the pile, *Proceedings of the 5th European Conference on Soil Mechanics and Foundation Engineering*, Madrid, **1**, 325–332 (1972).

6.8   Springman, S. M. and Bolton, M. D., The effect of surcharge loading adjacent to piling, *Contractor Report 196*, Transport and Road Research Laboratory, 1990.

6.9   Bolton, M. D., Springman, S. M., and Sun, H. W., The behaviour of bridge abutments on clay. Design and performance of earth retaining structures, *Geotech. Eng. Div. of American Society of Civil Engineers*, Specialty Conference, Cornell University, Ithaca, USA, 1990.

6.10   Reid, W. M. and Buchanan, N. W., Bridge approach support piling, *Proceedings of the Conference on Piling and Ground Treatment*, Institution of Civil Engineers, London, 1983, pp. 267–274.

6.11   Tomlinson, M. J., Unpublished report to the Northumberland County Council, 1980.

6.12   Stanley, R. P., Design and construction of the Sungai Perak Bridge, Malaysia, *Proceedings of the Institution of Civil Engineers*, **88**(1), 571–599 (1990).

6.13   Chin, Fung Kee, and McCabe, R., Penang Bridge Project: foundations and reclamation work, *Proceedings of the Institution of Civil Engineers*, **88**(1), 531–549 (1990)

6.14   Gerwick, B. C., *Construction of Offshore Structures*, John Wiley, New York, 1986.

6.15   Liquefaction of soils during earthquakes, *Report of the Committee on Earthquake Engineering*, National Research Council, National Academy Press, Washington DC, 1985.

6.16   Ishihara, K., Liquefaction and flow during earthquakes, *Géotechnique*, **43**(3), 351–415 (1993).

6.17   McGibbon, J. I. and Booth, G. W., Kessock Bridge: construction, *Proceedings of the Institution of Civil Engineers*, **76**(1), 51–66 (1984).

6.18   Levesque, M., Les fondations du Pont de Färo, en Denmark, *Travaux*, 33–36 (Nov. 1983).

6.19   Watson, R., Ship impact governs caissons at Dartford, *Ground Engineering*, 13–15 (April 1989).

6.20   Notes taken at a meeting of the London Association, Institution of Civil Engineers, on 3 May 1990.

6.21   Howorth, G. E., The construction of the Lower Zambezi Bridge, *Journal of the Institution of Civil Engineers*, **4**, 325–369 (1937).

6.22   Husband, J., The San Francisco–Oakland Bay Bridge, *The Structural Engineer*, **14**, 170–208 (1936).

6.23   Riggs, L. W., Tagus River Bridge–Tower Piers, *Civil Engineering*, 41–45 (Feb. 1965)

6.24   *Standard Specifications and Code of Practice for Roads and Bridges*, Indian Roads Congress, IRC: 78, 1983, Ministry of Shipping and Transport, New Delhi (1984).

6.25   Pothecary, C. H., and Brindle, L., Torridge Bridge, construction of substructure and deck precasting work, *Proceedings of the Institution of Civil Engineers*, **88**(1), 211–231 (1990).

6.26   Mitchell, A. J., Caisson foundations, in *Handbook of Structural Concrete*, McGraw-Hill, New York 28.1–28.43 (1983).

6.27   Reid, A. E., and Sully, F. W., The construction of the King Feisal Bridge and the King Ghazi Bridge over the River Tigris at Baghdad, *Institution of Civil Engineers Works Construction Paper No. 4* (1945–46). See also Wilson and Sully in *Institution of Civil Engineers Works Construction Paper No. 13* (1949).

6.28   Hyatt, K. E. and Morley, G. W., The construction of Kafr el-Zayat Railway Bridge, *Institution of Civil Engineers Works Construction Paper No. 19* (1952).

6.29   Blaine, E. S., Practical lessons in caisson sinking from the Baton Rouge Bridge, *Engineering News Record*, 213–215 (6 Feb. 1947).

6.30   Anon., Humber Bridge, *New Civil Engineer Supplement*, Thomas Telford, May 1991.

6.31   Anon., Hydraulic jets sink Carquinez caissons, *Engineering News Record*, **158**(16), 38–44 (1957).

6.32   Howorth, G. E. and Shirley Smith, H., The New Howrah Bridge, Calcutta: construction, *Journal of the Institution of Civil Engineers*, **28**, 211–257 (1947).

6.33   Wilson, W. S. and Sully, F. W., The construction of the caisson forming the foundations to the circulating water pumphouse for the Uskmouth Generating Station, *Proceedings of the Institution of Civil Engineers*, **1**(3), 335–356 (1952).

6.34   Murdock, L. J., Discussion on paper by Pike and Saurin (see ref. 5.1).

6.35   Anon., Piers sunk by open-well dredging for Mississippi River Bridge, Memphis, *Engineering News Record*, 449–454 (20 March 1947).

6.36   Yang, N. C., Condition of large caissons during construction, *Highway Research Record*, No. 74, 68–82 (1965).

6.37   Coales, F. G., and Clarkson, C. H., The Iskander Bridge, Perak, Federated Malay States, *Proceedings of the Institution of Civil Engineers*, **240**, 342–384 (1937).

6.38   Gales, Sir R, The Hardinge Bridge over the Lower Ganges at Sara, *Proceedings of the Institution of Civil Engineers*, **105**(1), 18–99 (1917–18).

6.39   Ramayya, T. A., Well sinking in Gowtami Bridge construction, *Journal of the Institution of Engineers (India)*, **45**, 665–686 (1965).

6.40   Chandler, J. A., Peraine, J., and Rowe, P. W., Jamuna River, 230 kV crossing, Bangladesh: Construction of foundations, *Proceedings of the Institution of Civil Engineers*, **76**(1), 965–984 (1984).

6.41   Wiley, H. L., Reply to discussion on paper 'The sinking of the piers for the Grand Trunk Pacific Bridge at Fort William, Ontario, Canada', *Transactions of the American Society of Civil Engineers*, **LXII**, 132–133 (1909).

6.42   Hinch, L. W., McDowall, D. M. and Rowe, P. W., Jamuna River 230 kV crossing, Bangladesh: Design of foundations, *Proceedings of the Institution of Civil Engineers*, **76**(1), 927–949 (1984).

6.43   Lord, J. E. D., Gill, J. M., and Murray, J., The new Redheugh Bridge, *Proceedings of the Institution of Civil Engineers*, **76**(1), 497–521 (1984).

6.44   Meldner, V., Druckluftgrundung fur die Brücke uber der Kleinen Belt, *Baumaschine und Bautecknik*, **18**(7) 289–295 (1971).

6.45   Ostenfeld, C. and Jonson, W., *Motorway Bridge across Lillebaelt*, Copenhagen, 1970.

6.46   Gowring, G. I. B. and Hardie, A., Severn Bridge, foundations and substructure, *Proceedings of the Institution of Civil Engineers*, **41**, 49–67 (1967–68).

6.47   Busbridge, J. R., Contribution to the discussion on papers 14 to 17, *Proceedings of the Conference on Diaphragm Walls and Anchorages*, Institution of Civil Engineers, London, 1975, pp. 137–138.

7 Piled Foundations 1 The Carrying Capacity of Piles and Pile Groups

7.1 Classification of piles

Piles are relatively long and slender members used to transmit foundation loads through soil strata of low bearing capacity to deeper soil or rock strata having a high bearing capacity. They are also used in normal ground conditions to resist heavy uplift forces or in poor soil conditions to resist horizontal loads. Piles are a convenient method of foundation construction for works over water, such as jetties or bridge piers. Sheet piles perform an entirely different function; they are used as supporting members to earth or water in cofferdams for foundation excavations or as retaining walls. The design and construction of foundation works in sheet piling will be described in Chapters 9 and 10.

If the bearing stratum for foundation piles is a hard and relatively impenetrable material such as rock or a very dense sand and gravel, the piles derive most of their carrying capacity from the resistance of the stratum at the toe of the piles. In these conditions they are called *end-bearing* or *point-bearing* piles (Fig. 7.1(*a*)). On the other hand, if the piles do not reach an impenetrable stratum, but are driven for some distance into a penetrable soil, their carrying capacity is derived partly from the end-bearing and partly from the skin friction between the embedded surface of the pile and the surrounding soil. Piles which obtain the greater part of their carrying capacity by skin friction or adhesion are called *friction piles* (Fig. 7.1(*b*)).

The main types of pile in general use are as follows:

(a) *Driven piles*. Preformed units, usually in timber, concrete, or steel, driven into the soil by the blows of a hammer.

(b) *Driven and cast-in-place piles*. Formed by driving a tube with a closed end into the soil, and filling the tube with concrete. The tube may or may not be withdrawn.

(c) *Jacked piles*. Steel or concrete units jacked into the soil.

(d) *Bored and cast-in-place piles*. Piles formed by boring a hole into the soil and filling it with concrete.

(e) *Composite piles*. Combinations of two or more of the preceding types, or combinations of different materials in the same type of pile.

The first three of the above types are sometimes called *displacement piles* since the soil is displaced as the pile is driven or jacked into the ground. In all forms of bored piles, and in some forms of composite piles, the soil is first removed by boring a hole into which concrete is placed or various types of precast concrete or other proprietary units are inserted. The basic difference between displacement and non-displacement piles requires a different approach to the problems of calculating carrying capacity, and the two types will therefore be discussed separately.

7.2 The behaviour of piles and pile groups under load

The load–settlement relationship for a single pile in a uniform soil when subjected to vertical loading to the point of failure is shown in Fig. 7.2(*a*). At the early stages of loading, the settlement is very small and is due almost wholly to elastic movement in the pile and the surrounding soil. When the load is removed at a point such as A in Fig. 7.2(*a*) the head of the pile will rebound almost to its original level. If strain gauges are embedded along the length of the pile shaft they will show that nearly the whole of the load is carried by skin friction on the upper part of the shaft (Fig. 7.2(*b*)). As the load is increased, the load–settlement curve steepens, and release of load from a point B will again show some elastic rebound, but the head of the pile will not return to its original level,

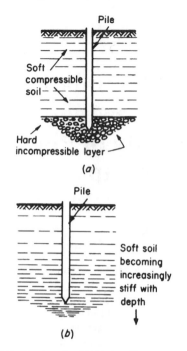

Fig. 7.1   (*a*) End-bearing pile. (*b*) Friction pile.

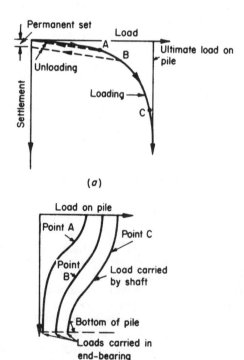

Fig. 7.2   Effects of loading a pile. (*a*) Load–settlement curve. (*b*) Strain gauge readings on pile shaft.

indicating that some 'permanent set' has taken place. The strain gauge readings will show that the shaft has taken up an increased amount of skin friction but the load carried by the shaft will not equal the total load on the pile, indicating that some proportion of the load is now being carried in the end-bearing. When the load approaches failure point C, the settlement increases rapidly with little further increase of load. The strain gauge readings show rather less load carried in skin friction than that just before failure, especially near the toe where the soil tends to flow away from the pile as failure takes place.

The relative proportions of load carried in skin friction and end-bearing depend on the shear strength and elasticity of the soil. Generally, the vertical movement of the pile which is required to mobilize full end resistance is much greater than that required to mobilize full skin friction. If the total load on the shaft and the load on the base of a pile are measured separately, the load–settlement relationships for each of these components are as shown in Fig. 7.13. It will be seen that the skin friction on the shaft increases to a peak value, then falls with increasing strain. On the other hand, the base load increases progressively until complete failure occurs.

Because of elastic movement in the pile shaft the upper part of the pile moves relatively to the soil. Thus in the case of a pile driven on to a hard and almost incompressible rock, the strain-gauge readings along the pile shaft will show some load transferred to the soil towards the top of the pile due to mobilization of skin friction as the shaft compresses elastically.

At the limit state corresponding to point C in Fig. 7.2(*a*) the ultimate bearing capacity of a pile is given by the equation

$$Q_p = Q_s + Q_b - W_p, \tag{7.1}$$

where

Q_s = ultimate shaft resistance in skin friction,
Q_b = ultimate resistance of base,
W_p = weight of pile.

Usually W_p is small in relation to Q_p and it is often neglected because it is not much greater than the weight of the displaced soil. However, it must be taken into account for marine piling where a considerable proportion of the pile length extends above the sea-bed. The draft of Eurocode 7 considers the *design bearing capacity, Q,* rather than the ultimate value. *Q* is obtained by *partial safety factors* considering shaft friction and base resistance separately from the equation:

$$Q = \frac{Q_{sk}}{\gamma_s} + \frac{Q_{bk}}{\gamma_b}, \tag{7.2}$$

where γ_s and γ_b are the partial safety factors for skin friction and base resistance respectively. These take account of the method of pile installation. In the present draft code γ_s is given as 1.3 for both driven and bored piles, but γ_b is given as 1.3, 1.6, and 1.45 for driven, bored, and continuous flight auger piles respectively. The shaft resistance is given by

$$Q_{sk} = \sum_{i=1}^{\pi} q_{sik} A_{si} \qquad (7.3)$$

and the base resistance is given by

$$Q_{bk} = q_{bk} A_b \qquad (7.4)$$

where

A_{si} = nominal surface area of the pile in soil layer i,

A_b = nominal plan area of the pile base,

q_{sik} = characteristic value of the resistance per unit of the shaft in layer i,

q_{bk} = characteristic value per unit area of base.

Eurocode 7 requires that the characteristic values q_{sk} and q_{bk} do not exceed the measured bearing capacities used to establish the correlation divided by 1.5 on average. For example if q_b is shown by correlation with loading tests to be equal to nine times the average undrained shear strength of the soil below the pile base the characteristic value should be obtained from $9 \times c_u$ (average) divided by 1.5. This additional factor is used essentially to allow for uncertainties in the calculation method or scatter in the values from which the calculation was based.

The Institution of Civil Engineers' specification for piling[7.1] defines the *allowable pile capacity* as

a load which is not less than the specified working load and which takes into account the pile's ultimate bearing capacity, the materials from which the pile is made, the required factor of safety, settlement, pile spacing, downdrag, the overall bearing capacity of the ground beneath the piles and any other relevant factors. The allowable pile capacity indicates the ability of the pile to meet the specified loading requirements.

Hence this definition is broader than the Eurocode term 'design bearing capacity' which refers only to the bearing capacity of an individual pile and takes no account of the serviceability limit state of the structure supported by the piles. The Institution of Civil Engineers' definition corresponds to that of 'allowable load' in BS 8004: *Foundations*.

When piles are arranged in close-spaced groups (Fig. 7.3) the mechanism of failure is different from that of a single pile. The piles and the soil contained within the group act together as a single unit. A slip plane occurs

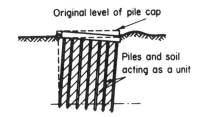

Fig. 7.3   Failure of piles by group action.

along the perimeter of the group and 'block failure' takes place when the group sinks and tilts as a unit. The failure load of a group is not necessarily that of a single pile multiplied by the number of piles in the group. In sand it may be more than this; in clays it is likely to be less. The 'efficiency' of a pile group is taken as the ratio of the average load per pile when failure of the group occurs to the load at failure of a comparable single pile.

It is evident that there must be some particular spacing at which the mode of failure changes from that of a single pile to block failure. The change is not dependent only on the spacing but also on the size and shape of the group and the length of the piles (Section 7.14).

Group effect is also important from the aspect of consolidation settlement, because in all types of soil the settlement of the pile group is greater than that of the single pile carrying the same working load as each pile in the group. The ratio of the settlement of the group to that of the single pile is proportional to the number of piles in the group, i.e. to the overall width of the group. The group action of piles in relation to carrying capacity and settlement will be discussed in greater detail later in this chapter.

7.3 Definitions of failure load on piles

In the foregoing discussion we have taken the failure load as the load causing ultimate failure of a pile, or the load at which the bearing resistance of the soil is fully mobilized. However, in the engineering sense failure may have occurred long before reaching the ultimate load since the settlement of the structure will have exceeded tolerable limits.

Terzaghi's suggestion that, for practical purposes, the ultimate load can be defined as that which causes a settlement of one-tenth of the pile diameter or width, is widely accepted by engineers. However, if this criterion is applied to piles of large diameter and a nominal safety factor of 2 is used to obtain the working load, then the settlement at the working load may be excessive.

In almost all cases where piles are acting as structural foundations, the allowable load is governed solely from considerations of tolerable settlement at the working load. An ideal method of calculating allowable loads on piles would be one which would enable the engineer to predict the load–settlement relationship up to the point of failure, for any given type and size of pile in any soil or rock conditions. In most cases a simple procedure is to calculate the ultimate bearing capacity of the isolated pile and to divide this value by a safety factor which experience has shown will limit the settlement at the working load to a value which is tolerable to the structural designer. But where settlements are critical it is necessary to evaluate separately the proportions of the applied load carried in skin friction and end bearing and then to calculate the settlement of the pile head from the interaction of the elastic compression of the pile shaft with the elasto-plastic deformation of the soil around the shaft and the compression of the soil beneath the pile base.

In all cases where piles are supported wholly by soil and are arranged in groups, the steps in calculating allowable pile loads are as follows:

(a) Determine the base level of the piles which is required to avoid excessive settlement of the pile group. This level is obtained by the methods described in Section 7.13. The practicability of attaining this level with the available methods of installing the piles must be kept in mind.
(b) Calculate the required *diameter* or width of the piles such that settlement of the individual pile at the predetermined working load will not result in excessive settlement of the *pile group*.
(c) Examine the economics of varying the numbers and diameters of the piles in the group to support the total load on the group.

The general aim should be to keep the numbers of piles in each group as small as possible, i.e. to adopt the highest possible working load on the individual pile. This will reduce the size and cost of pile caps, and will keep the settlement of the group to a minimum. However, if the safety factor on the individual pile is too low excessive settlement, leading to intolerable differential settlements between adjacent piles or pile groups, may result.

In the case of isolated piles, or piles arranged in very small groups, the diameter and length of the piles will be governed solely from consideration of the settlement of the isolated pile at the working load.

Techniques of installation have a very important effect on the carrying capacity of piles. Factors such as whether a pile is driven or cast *in situ* in a bored hole, whether it is straight-sided or tapered, whether it is of steel, concrete, or timber, all have an effect on the interrelationship between the pile and the soil. The advantage of the soil mechanics method of calculating carrying capacity is that it enables allowable loads to be assessed from considerations of the characteristics of the soil and the type of pile. These predictions do not have to wait until the actual piling or test piling work is put in hand, although confirmation of the design assumption must be made at some stage by test loading of piles. Engineers should not expect too much from formulae for calculating carrying capacity of piles and should not be disappointed when soil mechanics' methods indicate failure loads which are in error by plus or minus 50 per cent of the failure load given by test loading. It should be remembered that when a pile is subjected to test loading a full-scale foundation is being tested. Because of normal variations in ground conditions and the influence of installation techniques on ultimate resistance it is not surprising that there should be quite wide variations in failure loads on any one site. Engineers would not be surprised if such wide variations were experienced if full-scale pad or strip foundations were loaded to failure.

The alternative is to calculate allowable loads or design bearing capacities by dynamic formulae. These will give even wider variations than soil mechanics' methods and, in any case, these dynamic formulae are largely discredited by experienced foundation engineers, unless they are used in conjunction with dynamic testing and analysis using the equipment described in Section 7.20.2.

7.4 Calculating ultimate loads on isolated driven piles in cohesionless soils

When a pile is driven by hammering or jacking into a cohesionless soil it displaces the soil. A loose soil is compacted to a higher density by the pile and very little, if any, heave of the ground surface takes place. In very loose soils a depression will form in the ground surface around the pile due to the compaction of the soil by the pile driving. In dense cohesionless soils very little further compaction is possible, with the result that the pile will displace the soil, and heaving of the ground surface will result. Such displacement involves shearing of the mass of soil around the pile shaft. Resistance to this is very high in a dense cohesionless soil, so that very heavy driving is required to achieve penetration of piles in dense sands or gravels. Heavy driving may lower the shearing resistance of the soil beneath the pile toe owing to degradation of angular soil particles. Thus no advantage is gained by over-driving piles in dense cohesionless soils. In any case this is undesirable because of the possible damage to the pile itself.

The compaction of loose or medium-dense cohesionless soils by pile driving gives this type of pile a notable advantage over bored pile types. Boring operations for the latter are likely to loosen cohesionless soils and thus lower their end-bearing resistance.

The shaft frictional resistance of piles in cohesionless soils is small compared with the end resistance. This is thought to be due to the formation of a ring or 'shell' of compacted soil around the pile shaft, with an inner ring of soil particles in a relatively loose or 'live' state. The skin friction is governed by this inner shell. In the case of straight-sided piles the soil particles may stay in a loose condition on cessation of driving. In tapered piles the ring is recompacted with each blow of the pile hammer.

Because of difficulties in obtaining undisturbed samples of cohesionless soils from boreholes, it is the usual practice to calculate the ultimate bearing capacities of piles in these soils from the results of *in-situ* tests. Two methods will be described: one of these is based on the SPT and the other on the static CPT. In both methods the ultimate bearing capacity is calculated and this must be divided by a safety factor to obtain the allowable load. The safety factor is not constant. It depends on the permissible settlement at the working load, which is in turn dependent on the pile diameter and the compressibility of the soil. Experience shows that a safety factor of 2.5 will ensure that an isolated pile with a shaft diameter of not more than 600 mm driven into a cohesionless soil will not settle by more than 15 mm. Alternatively, partial safety factors can be applied in conjunction with equations (7.1)–(7.4) and the serviceability limit state checked by experience or calculation and confirmed, if necessary, by loading tests. These tests are also necessary if the 'global' safety factor of 2.5 is used unless experience provides a reliable guide to settlement behaviour.

7.4.1 Method based on standard penetration test

(i) Calculating the skin friction   The term Q_s in equation (7.1) is the sum of the components of shaft friction over the individual soil layers within the effective length of the pile shaft, i.e. the total length minus any superficial soil layers which are undergoing settlement relative to that of the pile and causing downdrag. The design ultimate unit skin friction on an individual soil layer is given by

$$q_s = K_s \sigma'_{vo} \tan \delta \qquad (7.5)$$

where

$$K_s = \text{coefficient of horizontal soil stress,}$$
$$\sigma'_{vo} = \text{average effective overburden pressure over the length of the soil layer,}$$
$$\delta = \text{angle of wall friction.}$$

The coefficient K_s is related to the coefficient of earth pressure at rest (K_0) and also to the method of installation of the piles as shown in Table 7.1. Values of K_0 for normally consolidated sands are given in Section 5.7.1. It will be seen that these decrease with increasing density of the sand. It is evident that if they were used to obtain the skin friction on a pile in a heavily over-consolidated sand, where K_0 values are likely to be greater than unity, the skin friction could be seriously underestimated. Hence an attempt should be made to obtain a reliable value for K_0 in an over-consolidated deposit for situations where the skin friction is a critical factor in design, such as in piles carrying uplift loading. It may not be critical for compression piles where the base resistance of a solid section is likely to contribute the major proportion of the bearing capacity.

Values of ϕ' for obtaining δ from Table 7.2 are obtained from SPTs using the relationship shown in Fig. 2.13, or from static CPTs using Fig. 2.14. Usually a plot of N-values against depth shows quite a wide scatter and judgement is required to obtain a characteristic value. Driving a pile tends to even out variations in soil density and the total shaft friction over the full penetration depth is, in itself, an average of all values. Hence it is usual to take an average line drawn through the plotted points to obtain ϕ' and hence the value of q_s from equation (7.5), or q_{sik} for use with equation (7.3). When applying the Eurocode 7 recommendations the tangent of the measured ϕ' is reduced by the material factor γ_m of 1.2–1.25 to obtain q_{sik}. However, when considering drivability, when the skin friction must not be underestimated, a somewhat higher mean line should be considered.

Equation (7.5) implies that in a uniform cohesionless soil the unit skin friction continues to increase linearly with increasing depth. This is not the case. Vesic[7.3] showed that at some penetration depth between 10 and 20 pile diameters, a peak value of unit skin friction is reached which is not exceeded at greater penetration depths. Thus equation (7.5) gives increasingly unsafe values as the penetration depth exceeds about 20 diameters. Research

Table 7.1 Values of the coefficient of horizontal soil stress K_s (after Kulhawy[7.2])

Installation method	K_s/K_0
Driven piles, large displacement	1–2
Driven piles, small displacement	0.75–1.75
Bored and cast-in-place piles	0.71–1
Jetted piles	0.5–0.7

Table 7.2 Values of the angle of pile to soil friction for various interface conditions (after Kulhawy[7.2])

Pile/soil interface condition	Angle of pile/soil friction(δ)
Smooth (coated) steel/sand	$0.5\phi'$ to $0.7\phi'$
Rough (corrugated) steel/sand	$0.7\phi'$ to $0.9\phi'$
Precast concrete/sand	$0.8\phi'$ to $1.0\phi'$
Cast-in-place concrete/sand	$1.0\phi'$
Timber/sand	$0.8\phi'$ to $0.9\phi'$

has not yet established whether the peak value is a constant in all conditions, or is related to factors such as soil grain size or angularity. At the present time a peak value of 110 kN/m^2 is used for straight-sided piles. However, Nordlund[7.4] has shown that a taper on the pile shaft can increase the skin friction substantially for compression loading. For example, a taper of $0.5°$ gives a 50 per cent increase in unit skin friction in a medium-dense sand, but this increase would not be permitted where uplift loads are required to be carried (see Section 7.16).

Long periods of cyclic loading on piles can degrade angular soil particles to the texture of a silt with a corresponding low ϕ'-value. Caution is required when interpreting SPT results in carbonate soils where the high N-values are due to cementing of the particles. Pile driving can break down the agglomerations to produce a soil deposit equivalent to a loose sandy silt. These and similar effects demonstrate the need to make pulling tests to check design values whenever the shaft friction is a critical component.

Calculating the base resistance The component of base resistance in equation (7.1) is obtained from the equation

$$Q_b = q_b A_b = N_q \sigma'_{vo} A_b \qquad (7.6)$$

Comparisons of observed base resistances of piles by Nordlund[7.4] and Vesic[7.5] have shown that N_q values established by Berezantsev[7.6] which take into account the depth to width ratio of the pile most nearly conform to practical criteria of pile failure. Berezantsev's values of N_q as adapted by the author are shown in Fig. 7.4. Values of ϕ' are obtained from uncorrected SPTs using Fig. 2.13 or from static CPTs using the relationships in Fig. 2.14.

The end-bearing resistance of a solid end pile driven into a cohesionless soil is usually the major component of the total resistance. Hence in order to obtain the design value of q_b from equation (7.6) or the characteristic value q_{bk} in equation (7.4) a low value of N should be selected from the scatter of results to obtain ϕ'. However, when considering pile drivability, N-values should be selected from those representative of any denser soil layers through which piles are to be driven. Eurocode 7 does not refer specifically to values of peak skin friction and base resistance, but it does state that design rules should be based on experience. When applying Eurocode 7 recommendations $\tan \delta$ should be reduced by the materials factor γ_m of 1.2–1.25 to obtain q_{sk} from equation (7.5). Although not stated in the code it will be advisable to apply the same materials factor to the characteristic ϕ'-value to obtain N_q and hence q_{bk}.

Static CPTs should not be used to obtain ϕ'-values for use with Fig. 7.4. The cone resistance should be used directly to obtain q_b as described in the following section.

It may be argued that because driving compacts the soil beneath the pile, the value of ϕ should in all cases represent the densest conditions. This is not always the case. For example, when piles are driven into loose sand the resistance is low and little compaction is given to the soil. However, when they are driven into a dense soil the resistance builds up quickly and the soil is compacted to a denser state, but because of possible weakening effects due to the crushing of particles of soil beneath the pile toe it would be unwise to assume ϕ-values higher than those represented by the *in-situ* state of the soil before pile driving. Where piles are driven into friable carbonate sands, which are found widely in Middle Eastern countries, pile driving may break down the soil particles to the texture of a silt with an angle of shearing resistance of no more than $20°$.

It will be seen from Fig. 7.4 that there is a rapid increase in N_q for high values of ϕ, giving high values of base resistance. However, research shows that at a penetration depth of 10–20 pile diameters a maximum value of base resistance is reached which is not exceeded or the rate of increase in base resistance decreases to a small value no matter how much deeper the pile is driven. Published pile test results indicate that the maximum value is 15 MN/m^2 (see Fig. 7.7). As a general rule *the allowable working load on an isolated pile driven to virtual refusal (by normal driving equipment) in a dense sand or gravel, consisting predominantly of quartz particles, is given by the allowable load on the pile considered as a structural member rather than by a consideration of failure of the supporting soil*, or, if the permissible working stress on the material of the pile is not exceeded then the pile will not fail.

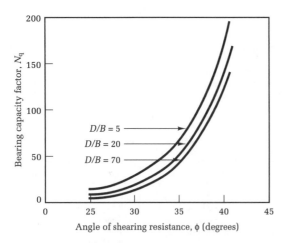

Fig. 7.4 Berezantsev's bearing capacity factor, N_q.

The reader is referred to the work of Vesic[7.3] and Kulhawy[7.2] for an account of the research on the effects of penetration depths and soil elasticity on the development of base resistance.

It is frequently necessary to drive piles to a deep penetration into cohesionless soils, for example to obtain resistance to uplift or horizontal loads. In order to avoid premature refusal of the piles to further penetration before reaching the required toe level the piles are driven with open ends. Because of radial expansion and contraction and flexure of the piles under blows from the hammer, the pile tends to slip down through the soil without the formation of a solid plug which is carried down by the pile. Measurements of the soil level inside the pile relative to the exterior show only a small drop in level due to the densifying of the interior soil. When an axial load is applied to the pile head the sand plug at the toe is pushed up, arching takes place in the soil mass and skin friction resistance corresponding to the passive coefficient of horizontal stress is developed at the interior pile–soil interface. However, it must not be assumed that the soil plug will develop a base resistance equivalent to that of a solid end pile. This is because quite appreciable movement of the pile toe is required to induce an arch to form in the soil plug and the larger the pile the greater the movement, such that pile head settlements may exceed

tolerable limits for large-diameter piles. Research by Kishida and Isemoto[7.7] has shown that the force required to push up the soil plug is related both to the pile diameter and the length to width ratio of the plug, but the work has not been taken to the stage of producing reliable design rules. Observed sand plug resistance values obtained from published and unpublished test results have been plotted by Hight[7.8] in relation to the diameter of the open-end pile as shown in Fig. 7.5. The results are very erratic with no clear trend to a reduction in resistance with increasing diameter. In most cases the piles were driven into dense or very dense soils and it was evident that failure occurred by yielding of the plug, not by failure of the soil beneath the pile toe. Also shown in Fig. 7.5 are the limiting values of the base resistance of piles in cohesionless soils set out in the American Petroleum Institute recommendations for offshore structures.[7.9] If the API values had been used for design in the cases shown the base resistance would have been grossly overestimated.

The base resistance of open-end tubular steel piles can be increased by welding a diaphragm at a predetermined height above the pile toe as described in Section 8.12.3. The base and skin resistance of tubular and H-section piles can also be increased by welding 'wings' on to the exterior of the shaft. Precast concrete piles can be provided with enlargements at the toe. However, the permissible work-

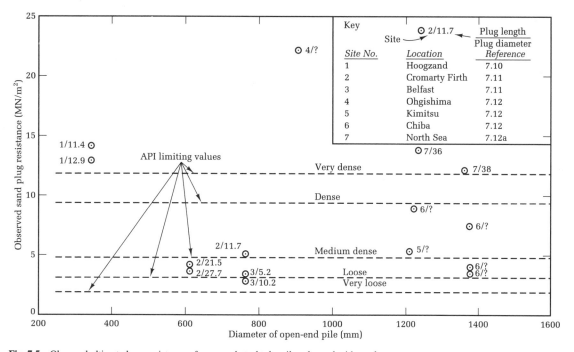

Fig. 7.5 Observed ultimate base resistance of open-end steel tube piles plugged with sand.

ing stresses on the least cross-section of pile shaft must not be exceeded and it must also be noted that driving a pile with an enlargement at the tip will reduce the skin friction on the shaft above the enlargement to that corresponding to loose soil conditions.

Equation (7.6) can be used for jacked piles in conjunction with Berezantsev's values of N_q provided it is possible to drive the piles to a penetration into the bearing stratum greater than five times the shaft diameter. For lesser penetration values of N_q as recommended for use in conjunction with Brinch Hansen's equation (Fig. 2.8) should be used. If soft clay or silt overlies the bearing stratum of sand or gravel then the D/B ratio should be calculated on the penetration into the bearing stratum only.

7.4.2 Methods based on static cone penetration test

In extensive areas of The Netherlands and Belgium, heavy structures are founded on piles driven through peats and soft clays to a suitable bearing on medium-dense to dense sands. In these countries an immense amount of experience has been gained in the interpretation of the static CPT (Section 1.4.5) in relation to piled foundations in sands.

Although most engineers in The Netherlands and others elsewhere base skin friction values on the measured local sleeve friction the author prefers to use established empirical correlations between unit shaft friction and the cone resistance q_c. This is because q_c values are more sensitive to variations in soil density than the sleeve friction and identification of the soil type from the ratio of cone resistance to sleeve friction is not always clear-cut. Empirical relationships of pile shaft friction to cone resistance are shown in Table 7.3.

A limiting value of $0.12 \, \text{MN/m}^2$ is used for the ultimate skin friction. The values in Table 7.3 are applicable to

Table 7.3 Relationships between pile shaft friction and cone resistance (after Meigh[7.13])

Pile type	Ultimate unit shaft friction
Timber	$0.012 q_c$
Precast concrete	$0.012 q_c$
Precast concrete with enlarged base†	$0.018 q_c$
Steel displacement	$0.012 q_c$
Open-end steel tube‡	$0.008 q_c$
Open-end steel tube driven into fine to medium sand	$0.0033 q_c$

† Applicable only to piles driven in dense groups otherwise use 0.003 where shaft size is less than enlarged base.
‡ Also applicable to H-section piles.

piles under static compression loading. A safety factor of 2.5 is used for q_c-values obtained from the electrical cone and 3.0 for the mechanical cone (see Section 1.4.5). These values assume that the piles are driven into a siliceous sand. These sands may undergo some abrasion of the particles under continuous cyclic compression loading. This requires the adoption of a safety factor somewhat higher than that used for static axial compression loading. The values in Table 7.3 should not be used for piles in carbonate sands.

Where the draft Eurocode 7 is used the relationships in Table 7.3 are used to obtain q_s which is divided by the factor of 1.5 to obtain q_{sik} and hence Q_{sk} from equation (7.3). The partial factor γ_s of 1.3 is applied to obtain the shaft friction component of equation (7.2). The material factor γ_m is not applied to the measured cone resistance values.

The end-bearing resistance is calculated from the equation

$$q_b = \bar{q}_c$$

where $\bar{q}_c$ is the average cone resistance within the zone influenced by stresses imposed by the toe of the pile.

A design value of $\bar{q}_c$ can be obtained by plotting on a single sheet the q_c/depth diagrams of all tests made in a given area and drawing an average curve through the readings in the vicinity of the pile toe as shown in Fig. 7.6(a). Alternatively, a statistical method can be used to obtain a mean value. The *allowable* base resistance is then obtained by dividing $\bar{q}_c A_b$ by a safety factor which depends on the scatter of results. Usually the factor is 2.5, but it is a good practice to draw a lower bound line and check that the design $\bar{q}_c$ has a small safety factor. Sharp peak depressions in the diagram should be ignored provided that they do not represent soft clay bands in a predominantly sandy deposit.

A method used in The Netherlands is to take the average cone resistance $\bar{q}_{c-1}$ over a depth of up to four diameters below the pile toe, and the average $\bar{q}_{c-2}$ eight pile diameters above the toe as described by Meigh[7.13]. The ultimate base resistance is then given by

$$q_b = \frac{\bar{q}_{c-1} + \bar{q}_{c-2}}{2} \tag{7.7}$$

The shape of the cone resistance diagram is studied before selecting the range of depth below the pile to obtain q_{c-1}. Where q_c increases continuously to a depth $4D$ below the toe the average value of q_{c-1} is obtained only over a depth of $0.7D$. If there is a sudden decrease in resistance between $0.7D$ and $4D$ the lowest value in this range should be selected for q_{c-1} as shown in Fig. 7.6(b). To obtain q_{c-2} the diagram is followed in an upward direction and an envelope is drawn only over those values

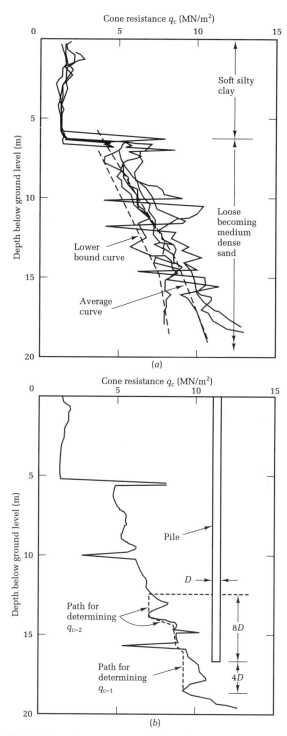

Cone resistance q_c (MN/m²)

Depth below ground level (m)

Soft silty clay

Loose becoming medium dense sand

Lower bound curve

Average curve

(a)

Cone resistance q_c (MN/m²)

Depth below ground level (m)

Pile

D

Path for determining q_{c-2}

$8D$

Path for determining q_{c-1}

$4D$

(b)

Fig. 7.6 Use of static CPTs to obtain design values of average cone resistance in cohesionless soils. (a) Determining q_c from average and lower bound q_c depth curves. (b) Method used in The Netherlands for obtaining the base resistance.

which are decreasing or remain constant at the value at the pile toe. Minor peak depressions are again ignored provided that they do not represent clay bands. Values of q_c higher than 30 MN/m² are disregarded. Safety factors generally used in The Netherlands for use with the $4D$–$8D$ method are given by te Kamp[7.14] as

Timber	1.7
Precast concrete, straight shaft	2.0
Precast concrete, enlarged shaft	2.5

An upper limit of end-bearing resistances is adopted on values obtained from either of the methods shown in Fig. 7.6. This depends on the grain size and stress history (overconsolidation ratio) of the soil as shown in Fig. 7.7. Because of the conservative assumptions applied to the measured cone resistance in both design methods, the material factor γ_m applied to the measured cone resistance values is taken as unity.

When using Eurocode 7 the value of q_c obtained from either of the methods in Fig. 7.6 should be divided by the factor of 1.5 to obtain q_{bk} and the resulting Q_{bk} should be divided by the partial factor γ_b (which is 1.3 for driven piles) to obtain the design base resistance. A further check should be made on the serviceability limit state by calculating the pile head settlement using the methods described in Section 7.11.

When considering piling into sand deposits below the base of a deep excavation or below river bed in situations where deep scour can occur, the effect of relief of overburden pressure on cone resistance must be taken into account.

In normally consolidated deposits there is a trend to a linear increase in cone resistance with increasing depth. Then if the overburden pressure at pile commencing level were to be reduced to zero by excavation or scour, it might be expected that the cone resistance would also be reduced proportionally. However, the effect of particle interlock modifies the resulting shape of the q_c/depth curve. This effect was shown by measurements of cone resistance in the very deep sandy alluvium of the Jamuna River in Bangladesh. The CPT measurements made on the river bank are compared with those made in a nearby 19 m deep scour hole in Fig. 7.8. It will be seen that the q_c-value regained the original non-scoured condition at a depth of about 20 m below scour level. Broug[7.15] has developed a method of predicting the reduction in cone resistance due to reduction in overburden pressure. The method was used to obtain skin friction and base resistance values from the results of CP testing at the site of a proposed highway and rail bridge over the Jamuna River, where scour depths up to 35 m have been predicted.

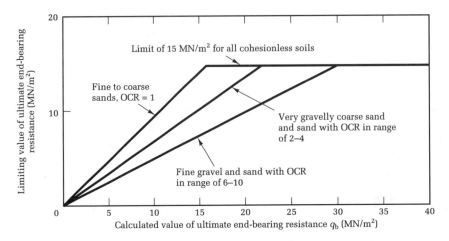

Fig. 7.7 Limiting values of pile end-bearing resistance for solid end piles in cohesionless soils (after te Kamp[7.14]).

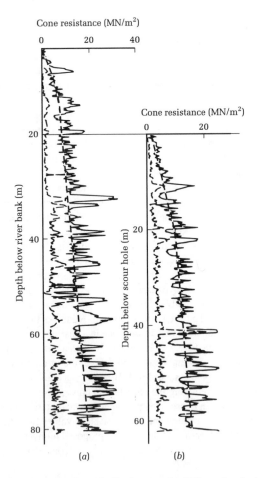

Fig. 7.8 Comparison of q_c/depth curves for CPTs on river bank and in scour hole, Jamuna River, Bangladesh (after Broug[7.15]).

7.4.3 Time effects

The carrying capacity of piles in sands and gravels does not normally show any change with time as it does in clays. However, in some circumstances, probably where piles are driven into water-bearing fine or silty sands, the time effect does seem to have some significance. Terzaghi and Peck[7.16] have noted that occasionally the carrying capacity decreases conspicuously during the first two or three days after driving. They state that it is probable, but not certain, that the high initial bearing capacity is due to a temporary state of stress that develops in the sand surrounding the point of the pile during driving. Feld[7.17] has described the case of 457 mm diameter steel pipe piles driven to depths of 15–18 m into varved silty sand overlain by about 3 m of medium sand. The piles withstood static load tests of 1200 kN and 1600 kN. However, load tests made on the same piles about a month after driving, and after further piles had been driven adjacent to them showed excessive settlement under loads of only 800 kN.

The same effect was observed when driving 2.0 m diameter steel tube piles to a penetration of about 4–5 m into dense sandy clayey silt for the foundations of the new Galata Bridge in Istanbul.[7.18] The driving resistance as expressed in blows per 250 mm penetration of the Delmag D100 hammer fell from about 700 blows to an average of 250 blows per 250 mm after waiting periods between 12 and 45 hours.

Errors made in overestimating the carrying capacity of piles in sands due to time effect can only occur if the capacity is based on calculations by dynamic formulae based on observed driving resistances. They are not likely to occur when the design has been based on static methods taking due account of the effects of pile driving on the characteristics of the soil. However, because of possible time effects, load tests or redrive tests with the dynamic analyser on piles in sands should not be made until at least 24 hours have elapsed after driving.

7.5 Calculating ultimate loads on driven and cast-in-place piles in cohesionless soils

Driven and cast-in-place piles (Section 8.13) are formed by driving a tube into the ground. On reaching founding level the tube is filled with concrete. In some types of pile the tube is withdrawn while concreting the shaft. Also it may be possible to form a bulb at the base of the pile. For pile types where a steel tube or precast concrete cylinder is left in position the shaft friction load can be calculated quite simply by the methods given in the preceding sections of this chapter. Where the drive tube has a closed end, the end resistance is calculated on the base area at the pile tip. Calculation of the base resistance is more difficult for pile types in which a bulb is formed since only the piling contractor knows the size of bulb which can be formed in any given soil conditions. Thus only the piling contractor is in a position to give reliable estimates of carrying capacity. All that the engineer can do is to obtain a preliminary idea of the range in capacity by assuming that in a dense soil the bulb will be little, if at all, larger than the shaft, and in a loose soil it may be possible to form a bulb of twice the shaft diameter. It should be kept in mind that settlement at the working load may be excessive if the pile is terminated in loose soil conditions.

In calculating the shaft friction on pile types in which the drive tube is withdrawn, it is difficult to assess whether the soil in contact with the shaft will be in a loose or dense state. This depends on the degree of compaction given to the concrete, or on the efficacy of other devices for compacting the soil while withdrawing the tube. It is also possible that the process of withdrawing the tube and compacting the concrete will enlarge the shaft, thus increasing its skin friction value. As these are essentially points of constructional technique, the engineer must be guided by the piling contractor on the proportion of the load which will be carried in skin friction. As a rough check on the contractor's estimates, the engineer can assume that for techniques in which no compaction is given to the concrete the soil will be loosened by the withdrawal of the tube. Where the concrete is compacted the soil can be assumed to be in a medium-dense state.

7.6 Calculating ultimate loads on bored and cast-in-place piles in cohesionless soils

Bored piles are installed by drilling with a mechanical auger in an unlined borehole using the equipment de-scribed in Section 8.14.1 or by cable percussion or grabbing rigs with support by temporary casing described in Section 8.14.2. Concrete is placed or injected in the drill holes as the casing, where used, is withdrawn. When using rotary mechanical augers support by bentonite slurry of the borehole below ground-water level is required, except in the case of drilling by continuous flight auger for which no temporary support is needed. In some ground conditions the casing is left in position, or precast concrete sections may be inserted in the drilled hole.

In all cases of bored piles formed in cohesionless soils by cable percussion drilling and grabbing, it must be assumed that the soil will be loosened as a result of the boring operations, even though it may initially be in a dense or medium-dense state. Equation (7.5) may be used for calculating shaft friction on the assumption that the ϕ-value will be representative of loose conditions. Similarly the ϕ-value used to obtain the bearing capacity factor N_q for calculating the base resistance from equation (7.6) must correspond to loose conditions. Where piles are installed by rotary drilling under a bentonite slurry[7.19] it may be assumed that the ϕ-value used to calculate both the skin friction and end-bearing resistance will correspond to the undisturbed soil conditions provided that allowance is made for increased working load settlements compared with driven piles, as discussed in Section 7.11.

The assumption of loose conditions for calculating skin friction and end resistance means that the ultimate carrying capacity of a bored pile in a cohesionless soil will be considerably lower than that of a pile driven into the same soil type. This is illustrated by a comparison of bored and driven piles on a factory site in the south of England. On this site 2.4–2.7 m of peat and find sand of negligible bearing value were overlying 2.7 m of dense sand and gravel (SPT N-value = 35–50 blows per 0.3 m), followed by very dense silty fine sand (N-value greater than 50 blows per 0.3 m). Bored and cast-in-place piles of 482 mm shaft diameter were installed at 4.6 and 9.1 m depth and these failed at loads of 219 and 349 kN, respectively. A pile formed from 508 mm diameter hollow precast concrete units was driven to 4.0 m below ground level, where the driving resistance was four blows of a 4 t hammer for the last 16 mm of penetration. The total settlement under a 747 kN test load was 7.5 mm and the residual settlement after removing the load was only 2.5 mm. At 747 kN the end-bearing pressure was estimated to be 3220 kN/m^2. Based on an N-value of 40 blows per 0.3 m, the base resistance was calculated by equation (7.6) to be 4830 kN/m^2. The total skin friction on the shaft of the 4.6 m bored pile and driven pile is unlikely to have been more than 50 kN in either case. Thus the loosening effect of boring operations on the base resistance of the bored pile was clearly demonstrated.

Whereas both driven and driven and cast-in-place piles can be provided with an enlarged base, forming an enlargement at the base of a bored pile with ground support by a bentonite slurry is a difficult and uncertain operation. Jet grouting as described in Section 11.6.7 can be considered as a means of providing an enlargement.

7.7 Ultimate loads on piles driven into cohesive soils

7.7.1 Skin friction on pile shaft

The carrying capacity of piles driven into cohesive materials such as silts and clays is given by the sum of the skin friction between the pile surface and the soil, and the end resistance. The skin friction is not necessarily equal to the shear strength of the soil, since driving a pile into a cohesive soil can alter the physical characteristics of the soil to a marked extent. The skin friction also depends on the material and shape of the pile.

Piles driven or jacked into a normally consolidated clay cause displacement of the soil. As a result, consolidation takes place and pore-water is squeezed out under the lateral pressures set up when the piles are forced into the ground. The effect of an increase in pore-water pressure is to decrease the effective overburden pressure on the shaft of the pile, with a corresponding decrease in skin friction. This excess pressure takes some time to dissipate. Consolidation of the soil is relatively slow and a heave of the ground surface is inevitable at the early stages after driving. As consolidation proceeds the excess pressure is dissipated into the surrounding soil or into the material of the pile and the heaved-up ground surface subsides. Dissipation of the excess pore-water pressure results in an increase in effective overburden pressure and an increase in skin friction. Figure 7.9 shows the increase in bearing capacity with time of piles driven into soft clays. It will be seen that in most cases 75 per cent of the ultimate carrying

capacity was achieved within 30 days of driving. This points to the need for delaying test loading of piles for at least 30 days in cases where a relatively high proportion of their carrying capacity is gained from skin friction in normally consolidated silts and clays.

When piles are driven into stiff over-consolidated clays such as glacial till or London Clay, little or no consolidation takes place and near the ground surface the soil cracks and is heaved up around the pile. At greater depths the radial expansion of the clay can induce negative pore pressures which provide a temporary increase in shaft friction and base resistance. It is known that a skin of clay from the softer zones near the ground surface adheres to the pile shaft following the downward movement and rebound of the pile with each blow of the hammer.[7.21] It is also known that an enlarged hole is formed in the upper part of the shaft due to lateral vibration of the pile under the hammer blows. An enlarged hole can also be caused by deviations in the cross-sectional dimensions or straightness of a pile. Ground water or exuded pore-water can collect in such an annular hole and lubricate the shaft. The possibility of 'strain softening' of the clay due to the large strain to which the soil on the contact face is subjected by the downward movement of the pile should also be considered. Tests have shown that the residual shear strength of an over-consolidated clay after high strain can be as little as 50 per cent of the peak shear strength at low strain. This is believed to be due to the effects of reorientation of the clay particles. Normally consolidated clays also show these 'strain softening' effects.

Therefore, because of the combined effects in varying degree of ground heave, the formation of an enlarged hole and strain softening, it is not surprising that the unit skin friction is often only a fraction of the undisturbed shear strength of the clay and that wide variations in the 'adhesion factor' (i.e. the ratio of the shear strength of the clay mobilized in skin friction on the pile shaft to the

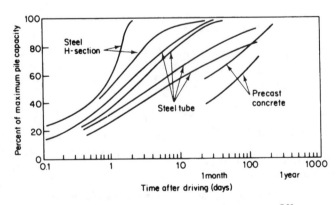

Fig. 7.9 Gain of carrying capacity with time of piles driven into soft to stiff clays (after Vesic[7.20]).

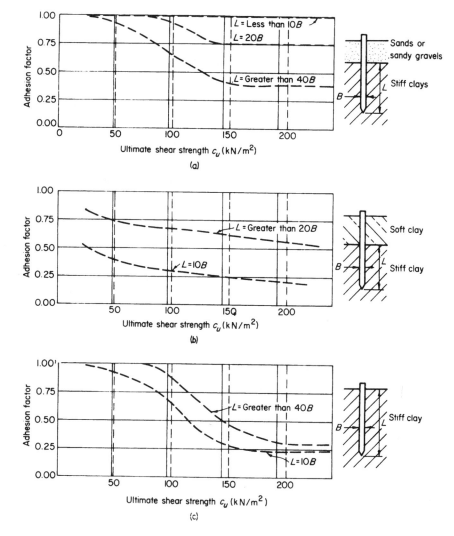

Fig. 7.10 Adhesion factors for driven piles in clay. (*a*) Case 1: piles driven through overlying sands or sandy gravels. (*b*) Case 2: piles driven through overlying weak clay. (*c*) Case 3: piles without different overlying strata.

undisturbed shear strength of the clay) can occur on any one site.

From published and unpublished records, the author has collected data which establish a relationship between the adhesion factor and the undrained shear strength of clay for three different conditions as shown in Fig. 7.10. The highest adhesion factors are obtained in Case 1 in which piles are driven through sands or sandy gravels into a clay. The gap which tends to form between the pile and clay is filled with dragged-down granular material and no skin friction is lost. The greater the penetration into the clay the less becomes the effect of the granular material with consequent reduction in adhesion factor. The reverse is true for Case 2 (soft clay over stiff clay) where the dragged-down soft clay skin has a weakening effect on skin friction. The smaller the penetration depth into the stiff clay the greater is the proportionate reduction in skin friction. For Case 3 (piles driven into a firm to stiff clay without different overlying strata) a gap forms around the upper part of the pile and no skin friction is mobilized. The smaller the penetration and the stiffer the clay the greater is the effect of the gap.

It may be asked why the adhesion factors should be based on the undrained shear strength since the clay is extensively remoulded at the time of driving the piles. The reason for adopting the undrained shear strength was that the author was seeking some basis for correlating the results of pile loading tests with some measured property of the soil. Publications giving loading test results often included some information on the undrained shear strength of the soil, and it was possible to establish a relationship, admittedly empirical, between this parameter and ultimate skin friction values obtained from the loading test data. It is claimed that effective stress methods based on the remoulded drained shear strength of the soil are more logical[7.22]. This may well be true, but published results of pile loading tests do not usually include information on drained shear strength parameters, and, more importantly, information on ground-water levels or the pore-water pressures adjacent to the pile shaft at the time of the load tests is usually lacking. Pore-water pressures are critical to the calculation of skin friction by effective stress methods and where correlations have been made between measured skin friction and drained shear strengths some very broad assumptions on the pore pressure at the time of test have had to be made.

Changes in pore-water pressure with time have an important effect on the carrying capacity of piles driven into normally consolidated clays. The increase in carrying capacity in soft to firm clays is shown in Fig. 7.9.

In the case of stiff-fissured clays there is likely to be a reduction in shear strength due to 'strain softening', but there is no clear evidence of regain in strength with time. Pile tests in London Clay made by Meyerhof and Murdock[7.23] showed that, 9 months after driving, the ultimate loads on precast concrete piles were only 80–90 per cent of the ultimate loads 1 month after driving. There was a similar decrease in the carrying capacity of piles driven into stiff-fissured clays in Denmark[7.24] after 103 days. Clayey glacial tills which have been subjected to intensive 'reworking' do not always show appreciable 'strain softening' effects. The remoulded shear strengths of clays of this type are not less than their undisturbed shear strengths, and there is little difference between peak and residual shear strengths.

The design curves in Fig. 7.10 apply essentially to piles carrying light to moderate loading driven to a relatively shallow penetration into the bearing stratum. In this condition the superficial soils overlying the bearing stratum have a significant influence on the skin friction developed below. Where heavy loads are carried, as in the case of piles in offshore petroleum production platforms, the piles may be driven very deeply into the bearing stratum. For this condition the influence of the superficial soils is negligible, but the magnitude of the overburden pressure becomes important. Considerations of settlement are also important in very long piles and these are discussed in Section 7.19.

Research, mainly in the field of pile design for offshore structures, has shown that the mobilization of skin friction is influenced principally by two factors. These are the

overconsolidation ratio of the clay and the slenderness (or aspect) ratio of the pile. The overconsolidation ratio is defined as the ratio of the maximum previous vertical effective overburden pressure σ_{vc} to the existing vertical effective overburden pressure σ_{vo}. For the purposes of pile design, Randolph and Wroth[7.25] have shown that it is convenient to represent the overconsolidation ratio by the simpler ratio of the undrained shear strength to the existing effective overburden pressure, c_u/σ'_{vo}. Randolph and Wroth showed that the c_u/σ'_{vo} ratio could be correlated with the adhesion factor α. A relationship between these two has been established by Semple and Rigden[7.26] from a review of a very large number of pile loading tests. This is shown in Fig. 7.11(a) for the case of a rigid pile, and where the skin friction is calculated from the peak value of c_u. To allow for the flexibility and slenderness ratio of the pile it is necessary to reduce the values of α_p by a length factor, F, as shown in Fig. 7.11(b). Thus total skin friction is

$$Q_s = F\alpha_p\bar{c}_u A s \qquad (7.8)$$

The slenderness ratio, L/B, influences the mobilization of skin friction in two ways. First, a slender pile can 'whip' or flutter during driving causing a gap around the pile at a shallow depth and reducing the horizontal stress

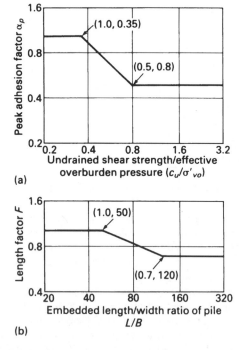

(a)

(b)

Fig. 7.11 Adhesion factors for heavily-loaded piles driven to deep penetration. (a) Peak adhesion factor versus c_u/σ'_{vo}. (b) Length factor.

at the pile–soil interface at lower levels. The second influence is the slip at the interface when the shear stress at transfer from the pile to the soil exceeds the peak value of shear strength and passes into the lower residual strength. This is illustrated by the shear stress–strain curve of the simple shear box test on a clay. The peak shear strength is reached at a relatively small strain followed by the much lower residual strength at long strain.

7.7.2 Calculating ultimate carrying capacity

The carrying capacity of piles driven into clays and clayey silts is equal to the sum of the end bearing resistance and the skin friction of that part of the shaft in contact with the soil which is giving support to the pile, i.e. not including any superficial soil layers which are undergoing consolidation and causing a dragdown force on the shaft. The end resistance is given by the equation

$$Q_b = q_b A_b = N_c \times c_{ub} \times A_b. \qquad (7.9)$$

The bearing capacity factor N_c can be taken as being equal to 9 provided that the pile is driven at least five diameters into the bearing stratum. The undrained shear strength c_{ub} at the base of the pile is taken as the *undisturbed* shear strength provided that time is given for a regain from remoulded to undisturbed shear strength conditions. In the case of stiff-fissured clays c_{ub} must be taken as representative of the *fissured* strength, i.e. the lower range of values if there is appreciable scatter in shear strength results. Reduction due to 'strain softening' should not be taken into account, since this effect is restricted to the contact face between the pile and soil. When applying the Eurocode 7 recommendations the selected c_{ub} value is divided by the materials factor γ_m of 1.5–1.8.

The skin friction on the pile shaft is given by the equation

$$Q_s = q_s A_s = \alpha \times \bar{c}_u A_s. \qquad (7.10)$$

where

$\quad q_s$ = design ultimate unit shaft friction,
$\quad \alpha$ = adhesion factor which is obtained from Fig. 7.10 or $F\alpha_p$ from Figs 7.11(a) and (b),
$\quad \bar{c}_u$ = average undrained shear strength over the depth of the pile shaft or within an individual soil layer,
$\quad A_s$ = surface area of shaft over the embedded depth within clay, giving support to the pile or within an individual soil layer.

Results of shear strength determinations from field or laboratory tests are plotted in relation to depth and an *average* line is drawn through them. When using the

results for calculations following the Eurocode recommendations, they are divided by the materials factor γ_m of 1.5–1.8.

In the case of uniform clays or clays increasing progressively in shear strength with depth, the average value of shear strength over the whole shaft length is taken for $\bar{c}_u$. Where the clay exists in layers of appreciably differing consistency, e.g. soft clay over stiff clay, the skin friction is separately calculated for each layer using the adhesion factor appropriate to the shear strength and overburden conditions. For piles in shrinkable clays (see Chapter 3) allowance must be made for the clay shrinking away from the pile shaft within the zone subjected to seasonal swelling and shrinkage. In the case of H-section piles, flexing and drag-down effects from a soft clay overburden into the stiff clay may cause a greater reduction in skin friction than would be suffered by a solid pile. Because of these uncertain effects adequate loading testing is necessary to check the validity of design calculations for H-piles. As a rough guide the skin friction is calculated conservatively on a perimeter equal to twice the flange width. For the above reasons an H-section pile is not a suitable type for use in stiff clays and tubular steel piles are preferred.

Where tubular steel piles are driven with open ends a plug of clay is carried down within the tube after the internal skin friction has built up to an amount which exceeds the base resistance. However, it must not be assumed that the bearing capacity of a clay-plugged pile will be the same as that of one with a solid end. Comparative tests on open- and closed-end piles were made by Rigden *et al.*[7.27] at a site in Yorkshire where the two types of pile with a shaft diameter of 457 mm were driven to a depth of 9 m into stiff glacial till. A clay plug formed in the open-end pile which occupied 40 per cent of the final penetration depth. The clay-plugged and solid end piles failed at 1160 kN and 1400 kN respectively. Instrumentation showed that the external skin friction on the clay-plugged pile was 20 per cent less than the other. As a result of these tests it is recommended that if measurements show that a clay plug has been carried down, the ultimate unit skin friction q_s in equation (7.10) should be reduced by a factor of 0.8 where α has been obtained from Fig. 7.10. However, where the adhesion factors of long piles are obtained from Fig. 7.11 it is likely that the reduction is already taken into account because many of the tests used to obtain these empirical factors were made on open-end piles. The unit base resistance of q_b should be multiplied by a factor of 0.5 to allow for likely yielding and compression of the plug under applied static loading. Because of the conservative assumption of shaft friction on an H-section pile by assuming that skin friction is mobilized on the outer flange surfaces only, the base resistance may be taken on the gross cross-sectional area.

The working load for all pile types is equal to the sum of the base resistance and the shaft friction divided by a suitable safety factor taking into account the range of adhesion factors. A safety factor of 2.5 is reasonable on this sum, namely

$$\text{Allowable load, } Q_a = \frac{Q_b + Q'_s}{2.5}, \qquad (7.11)$$

where Q'_s is the skin factor calculated from the adhesion factors shown in Fig. 7.10 using the *average* shear strength or calculated by equation (7.8). Also Q_a should not be more than

$$\frac{Q_b}{3} + \frac{Q''_s}{1.5}, \qquad (7.12)$$

where Q''_s is the skin friction calculated from the adhesion factors in Fig. 7.10 using the lowest range of shear strength, or from Fig. 7.11 using the average shear strength.

It is permissible to take a safety factor equal to 1.5, or even unity if conservative assumptions of shear strength have been adopted, for the skin friction because the peak value of skin friction on a pile in clay is obtained at a settlement of only 3–8 mm, whereas the base resistance requires a greater settlement for full mobilization.

A comparison can be made of allowable loads obtained from the global or partial factors in equations (7.11) and 7.12) respectively, and the design bearing capacities obtained by the Eurocode recommendations. Taking the case of a 600 mm solid end pile driven to a depth of 15 m into a firm, becoming stiff, clay with an average c_u of 75 kN/m² along the shaft and a lower bound (fissured) c_u of 100 kN/m² at the base.

From Fig. 7.11(a) for

$$c_u/\sigma_{vo} = 75/(20 \times 7.5) = 0.5, \; \alpha_p = 0.8,$$

and from Fig. 7.11(b) for $L/B = 15/0.6 = 25$, $F = 1$. Therefore from equation (7.10),

$$Q_s = 1 \times 0.8 \times 75 \times \pi \times 0.6 \times 15 = 1696 \text{ kN}.$$

From equation (7.9),

$$Q_b = 9 \times 100 \times \pi/4 \times 0.6^2 = 254 \text{ kN}$$

Hence allowable load is

$$\frac{254 + 1696}{2.5} = \underline{780 \text{ kN}} \text{ (design value)}$$

or allowable load is

$$\frac{254}{3} + \frac{1696}{1.5} = \underline{1215 \text{ kN}} \text{ (rejected)}.$$

Following the Eurocode recommendations a material factor say of 1.6 would be selected giving a factored c_u of 75/1.6 = 47 kN/m² and a factored c_{ub} of 100/1.6 = 62 kN/m². From Fig. 7.11(a) α_p for c_u/σ'_{vo} of 47/(20 × 15) is 1.0 and $F = 1$. Hence

$$q_{sik} = 1 \times 1 \times 47/1.5 = 31 \text{ kN/m}^2$$

and

$$Q_{sk} = 31 \times \pi \times 0.6 \times 15 = 876 \text{ kN}.$$

For $c_{ub} = 62$ kN/m²,

$$\begin{aligned} q_{bk} &= 9 \times 62/1.5 \\ &= 372 \text{ kN/m}^2 \end{aligned}$$

and $Q_{bk} = 372 \times \dfrac{\pi}{4} \times 0.6^2$

$$= 105 \text{ kN}.$$

Design bearing capacity from equation (7.2) is

$$\frac{876}{1.3} + \frac{105}{1.3} = \underline{755 \text{ kN}}.$$

Because the design methods are entirely empirical the comparison suggests that the Eurocode material and partial factors were selected to give results approximately equal to those obtained by permissible strength methods. Further comparisons are given in worked examples 7.1, 7.2, 7.4 and 7.8 at the end of this chapter.

7.8 Driven and cast-in-place piles in cohesive soils

For piles in which a steel tube or precast concrete shell remains in the ground the skin friction on the shaft is calculated from equation (7.10) using the adhesion factor appropriate to the undrained shear strength. For piles in which the drive tube is withdrawn, allowing the concrete to slump against the walls of the hole, the skin friction conditions are intermediate between those for a driven pile and for a bored and cast-in-place pile. Because the driving of the tube compacts the soil, and further compaction is given by tamping the concrete, there is no reason to suppose that the skin friction values will be less than those for precast concrete piles. Therefore equation (7.10) can be used to obtain a rough check on calculations submitted by the contractor for the proprietary piles concerned.

When it is known with certainty that compaction of the concrete during withdrawal of the drive tube can result in appreciable enlargement of the pile shaft in a soft or firm clay, the skin friction can be calculated on the enlarged shaft diameter.

Because of the frequently low skin friction values associated with piles driven into stiff clays, proprietary piles which incorporate an enlargement at the base formed by hammering out the concrete (see Section 8.13) have a notable advantage over straight-sided piles. In all cases the end resistance is calculated from equation (7.9) and the bearing capacity factor N_c may be taken as being equal to 9, provided that the base of the pile is driven at least five diameters into the bearing stratum.

7.9 Bored and cast-in-place piles in cohesive soils

7.9.1 Base resistance and skin friction

The base resistance and skin friction of bored piles have been extensively studied for London Clay and the adhesion and end-bearing capacity factors for this deposit have been reliably established. The researches have shown that a bearing capacity factor N_c of 9 can be used in equation (7.9) provided that c_{ub} is representative of the fissured shear strength (i.e. the lower range of values). The same value of N_c can be taken for all types of clay, provided that the base of the pile penetrates at least five diameters into the bearing stratum.

Research on bored piles in London Clay by Skempton[7.28] has shown that an adhesion factor of 0.45 may be taken on the *average* shear strength for use in equation 7.10. It is thought that the reasons for only about half the shear strength being mobilized in skin friction are due to the combined effects of swelling (and hence softening) of the clay in the walls of the borehole, the seepage of water from fissures in the clay and from the unset concrete, and 'work softening' from the boring operations. Skempton[7.28] recommended that the maximum adhesion value should be not more than 96 kN/m^2 and that in the case of short bored piles in London Clay, where the clay may be heavily fissured at a shallow depth, the adhesion factor should be taken as 0.3.

Weltman and Healy[7.29] reviewed information on the ultimate skin friction of bored piles in boulder clay and other glacial tills and produced the design curve shown in Fig. 7.12. This curve could be used for any type of clay other than London Clay, in cases where no published information or load test results are available.

A safety factor of 2.5 on the ultimate load as given by the sum of the base resistance and skin friction should ensure that the settlement at the working load will not exceed a tolerable value for piles with diameters not exceeding about 600 mm.

7.9.2 Large-diameter bored piles

A considerable volume of soil is subjected to stress beneath the base of a large-diameter pile, and elastic and consolidation settlements are thus of considerable significance in assessing the allowable load. The design methods must take into account these settlements, and it is not sufficient merely to divide the ultimate load by a nominal safety factor.

The first step is to select a base level for the piles from the aspect of overall settlement of the pile group (Section 7.14). Having thus established the required length, the engineer can carry the superstructure load on to the piles in a variety of ways. He may carry a column on a single straight-sided pile of large diameter, or he may use a shaft of smaller diameter but with an enlarged base, or he may use a group of piles of smaller dimensions with or without enlarged bases.

The governing consideration in the selection of a shaft and base diameter is, of course, the settlement at the working load, and this is determined by a rather complex interaction between the shaft and base. The load–settlement relationship of these two components for a typical pile is shown in Fig. 7.13. It will be seen from this figure that the full shaft resistance is mobilized at a settlement of only 15 mm, whereas the full base resistance, and the ultimate resistance of the entire pile, is mobilized at a settlement of 120 mm. Therefore, if the structure can

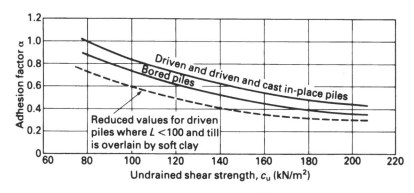

Fig. 7.12 Adhesion factors for piles in boulder clay (after Weltman and Healy[7.29]).

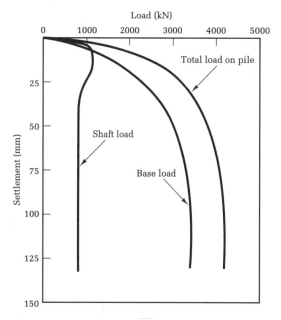

Fig. 7.13 Load–settlement relationships for large-diameter bored and cast-in-place piles.

tolerate a settlement of 15 mm at a working load of 2000 kN, the full shaft resistance, but only 57 per cent of the base resistance, will have been mobilized at the working load. The safety factor of the whole pile shown in Fig. 7.13 is 2.1. In most cases, the full shaft resistance is mobilized at a settlement less than that acceptable to the structural designer at the working load.

The separate effects of shaft and base resistance have been considered by Burland *et al.*[7.30]. They present a simple stability criterion which they say can be used to obtain the maximum safe load on any bored pile.

In general terms they state that 'in addition to an overall load factor, a factor of safety in end-bearing must be satisfied'. If an overall load factor of 2 is stipulated, together with a minimum factor of safety in end-bearing of 3, then the maximum safe load is the lesser of the two expressions,

$$\frac{Q_{ult}}{2} \text{ for the pile} \tag{7.13a}$$

and

$$Q_{ult} \text{ for the shaft } + \frac{Q_{ult}}{3} \text{ for the base.} \tag{7.13b}$$

They state that the first expression is nearly always dominant for straight-sided piles and long piles with comparatively small under-reams, whereas the second expression often controls under-reamed piles. In cases where the soil data are meagre or variable, or when loads

are indeterminate, it is thought that an overall load factor of 2.5 with a minimum factor of safety of 3.5 on the base is more suitable. The shaft and base loads are calculated using the adhesion and bearing capacity factors given in Section 7.9.1.

In the case of piles with under-reamed bases, the shaft skin friction should be ignored for a distance of two shaft diameters above the top of the under-ream. This allows for the possibility of drag-down on the pile shaft from the soil immediately above the under-ream. This drag-down might occur as a result of the normal elastic and consolidation settlement of the pile causing a small gap to open between the upper surface of the under-reamed section of the pile and the overlying soil. The latter would tend to move downward with time, resulting in drag-down on the shaft.

Because of the time required to form an under-ream by machine or hand excavation, the time spent cleaning soil debris and the delays before the under-ream is inspected and passed for concreting, it is possible that appreciable softening due to soil swelling and seepage of water from fissures will have occurred in the clay which will be in contact with the pile shaft. For this reason, the author prefers to use an adhesion factor of 0.3 for the shafts of under-reamed piles.

Using the above criterion for overall stability, the relative dimensions of the shaft and base are governed by considerations of settlement at the working load as calculated by the method described in Section 7.11.

It will be found that straight-sided piles show less elastic settlement than under-reamed piles carrying the same working load. However, the final decision as to whether or not straight-sided or under-reamed piles should be adopted and the selection of the relative shaft and base diameter of under-reamed piles is not always made from considerations of settlement alone. Economic factors often have a dominating influence. For example, considerations must be given to the relative volumes of soil and concrete involved, the practicability of forming under-reams in very silty or fissured clays, and the feasibility of forming very large under-reamed bases with the available drilling equipment. These factors will be discussed in Section 8.14.3.

The use of finite difference and finite element methods for predicting the settlement behaviour of piles under axial loadings are described in Section 7.19.1.

7.10 Calculation of the carrying capacity of piles in soils intermediate between sand and clay and layered soils

The end resistance can be calculated from Brinch Hansen's general equation (2.7) using the undisturbed values of c' and ϕ' from tests on soil samples. There are no

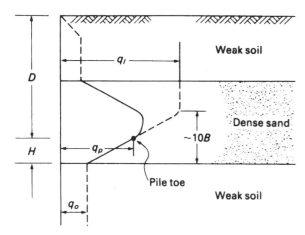

Fig. 7.14 End-bearing resistance of piles in layered soils (after Meyerhof[7.31]).

available data on the skin friction or adhesion of the pile shaft in c–ϕ soils. However, it may be taken as the sum of the adhesion and skin friction based on experience and using as a guide the values given in Fig. 7.10, the skin friction being calculated by the methods given in Sections 7.6–7.7. In the case of bored piles, due allowances should be made for likely softening or loosening effect caused by the boring operations.

Where the pile toe is terminated in a layer of stiff clay or dense sand underlain by soft clay or loose sand there is a risk of the pile punching through to the weak layer. The base resistance of the pile in the strong layer where the thickness H between the pile toe, and the top of the weak layer is less than the critical thickness of about $10B$ (Fig. 7.14) can be calculated from Meyerhof's equation.[7.31]

$$q_p = q_o + \frac{(q_1 - q_o)\,H}{10B} < q_1 \qquad (7.14)$$

where q_o and q_1 are the ultimate base resistance in the lower weak and upper strong layers respectively.

7.11 The settlement of a single pile at the working load

As already noted, a safety factor is applied to the calculated ultimate bearing capacity of a pile in order to cover natural variations in the soil properties, to provide for uncertainties in the range of applied loading and the calculation method, and to limit total and differential settlements under applied loading to amounts which can be tolerated by the superstructure. Safety factors in the range of 2–3 are normally capable of restricting settlements to acceptable amounts for piles having diameters or widths up to about 600 mm, but for larger diameters it is

often necessary to calculate the working load settlements from a knowledge of the deformation modulus of the soil or rock beneath the pile toe.

British Standard Code of Practice 8004: *Foundations*, states:

> In general an appropriate factor of safety for a single pile would be between two and three. Low values within this range may be applied where the ultimate bearing capacity has been determined by a sufficient number of loading tests, or where they may be justified by local experience; higher values should be used when there is less certainty of the value of the ultimate bearing capacity.

BS 8004 goes on to recommend the use of partial safety factors for large bored piles as described in Section 7.9.2 in this chapter.

A method of calculating pile head settlements is based on a separate evaluation of the proportion of load carried in skin friction and base resistance. The head settlement is then given by the sum of the elastic compression of the shaft and the elastic–plastic deformation of the soil or rock beneath the base using the equation

$$\rho = \frac{(W_s + 2W_b)L}{2A_sE_p} + \frac{\pi.W_b}{4A_b}\cdot\frac{B(1 - v^2)I_p}{E_b}, \qquad (7.15)$$

where

$$
\begin{aligned}
W_s \text{ and } W_b &= \text{loads on the pile shaft and base}\\
&\quad\text{respectively,}\\
L &= \text{shaft length,}\\
A_s \text{ and } A_b &= \text{cross-sectional area of the shaft and}\\
&\quad\text{base respectively,}\\
E_p &= \text{elastic modulus of the pile material,}\\
B &= \text{pile width,}\\
v &= \text{Poisson's ratio of the soil,}\\
I_p &= \text{influence factor related to the ratio}\\
&\quad\text{of } L/B,\\
E_b &= \text{deformation modulus of the soil or}\\
&\quad\text{rock beneath the pile base.}
\end{aligned}
$$

For a Poisson's ratio of 0–0.25 and $L/B > 5$, I_p is taken as 0.5 when the last term approximates to $0.5W_b/BE_b$. Values of E_b are obtained from plate loading tests at pile base level or from empirical relationships with the results of laboratory or *in-situ* soil tests given in Sections 2.6.5, 2.6.6, and 2.7. The pile shaft settlements represented by the first term in the above equation assume a uniform transfer of load down the pile shaft. However, the distribution is not uniform where a high proportion of the load is carried in skin friction on the shaft of a deeply embedded pile. Methods of simulating load transfer along the shaft of a pile are described in Section 7.19. The value of E_b for bored piles in cohesionless soils should correspond to the loose state unless the original density can be partly restored by drilling under bentonite or wholly restored by base-grouting (see p. 359).

7.12 The carrying capacity of piles founded on rock

7.12.1 Driven piles

When founding piles on rock it is usual to drive the piles to refusal in order to obtain the maximum carrying capacity from the piles. If the rock is strong at its surface, the piles will refuse further driving at a negligible penetration. In such cases the carrying capacity of the piles is governed by the strength of the pile shaft regarded as a column. When piles are driven on to strong rock through fairly stiff clays or sands, the piles can be regarded as being supported on all sides from buckling as a strut; therefore the carrying capacity is calculated from the safe load on the material of the pile at the point of minimum cross-section. In practice, it is necessary to limit the safe load on piles regarded as short columns because of the likely deviations from the vertical and the possibility of damage to the pile during driving and subsequent deterioration of the material over a long period of years. The limitations in working stresses are usually laid down in codes of practice or local by-laws. Values in general use are given in the description of various types of pile in Chapter 8.

Working loads as determined by the allowable stress on the material of the pile shaft may not be possible where piles are driven into weak rocks. It is necessary to calculate the skin friction developed over the penetration depth into the rock and the end-bearing resistance of the rock beneath the pile toe. Moderately weak to moderately strong rocks are likely to be shattered by the penetration of the pile. Hence the skin friction will be no more than that provided by a loose to medium dense gravel.

The effects of pile driving on the base resistance must also be considered. The ultimate bearing capacity of a rock mass with open or clay-filled joints may be no more than the unconfined compression strength of the intact rock. Where the joints are widely spaced (more than 600 mm) or where the joints are tightly closed and remain closed after pile driving the ultimate end-bearing resistance may be calculated from the equation

$$q_{ub} = 2N_\phi q_{uc} \qquad (7.16)$$

where $N_\phi = \tan^2(45 + \phi/2)$ and q_{uc} is the uniaxial compression strength of the rock.

For sandstone which typically has ϕ-values between 40 and 45° the end-bearing resistance is stated by Pells and Turner[7.32] to be between nine and twelve times the

Table 7.4 Friction angle values for intact rock (after Wyllie[7.33])

Classification	Type	Friction angle (degrees)
Low friction	Schists (high mica content) Shale Marl	20–27
Medium friction	Sandstone Siltstone Chalk Gneiss	27–34
High friction	Basalt Granite	34–40

uniaxial compression strength of a massive or tightly jointed rock. Wyllie[7.33] gives friction angles for intact rock as shown in Table 7.4. These should be used only as a guide because of the wide variations which can occur in the *in-situ* rock.

Where it is possible to measure the parameters c and ϕ of a jointed rock mass by laboratory tests on large-diameter core samples, Kulhawy and Goodman[7.34] state that the ultimate bearing capacity of the mass beneath a pile toe can be obtained from the equation

$$q_{ub} = cN_c + \gamma\frac{BN_\gamma}{2} + \gamma D N_q, \qquad (7.16a)$$

where

$$
\begin{aligned}
c &= \text{cohesion,}\\
B &= \text{base width,}\\
D &= \text{depth of pile base below the rock}\\
&\quad\text{surface,}\\
\gamma &= \text{effective density of rock mass,}\\
N_c, N_\gamma, \text{and } N_q &= \text{bearing capacity factors related to}\\
&\quad \phi \text{ and shown in Fig. 7.15.}
\end{aligned}
$$

The above equation represents wedge failure beneath a strip foundation. Hence cN_c should be multiplied by a factor of 1.25 for a square pile or 1.2 for a circular pile base. Also the second term in the equation should be multiplied by 0.8 or 0.7 for square or circular bases respectively. This term is small compared with cN_c and is often neglected.

It can be difficult and expensive to obtain values of c and ϕ from laboratory tests on large-diameter cores. Kulhawy and Goodman[7.34, 7.35] have shown that these

Table 7.5 Properties of a rock mass related to the unconfined compression strength (q_{uc}) and the RQD value

RQD (%)	q_c	c	ϕ
0–70	$0.33q_{uc}$	$0.1q_{uc}$	30°
70–100	0.33–$0.8q_{uc}$	$0.1q_{uc}$	30–60°

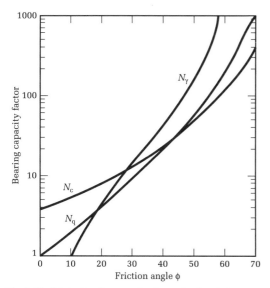

Fig. 7.15 Wedge bearing capacity factors for foundations on rock.

can be related to the RQD value (see Section 1.1) of the mass and the unconfined compression strength of the intact rock as shown in Table 7.5. It is important to note that in order to develop the maximum end-bearing resistance the settlement at the pile toe is likely to be of the order of 20 per cent of the pile diameter. Therefore an ample safety factor, at least 2.5, should be adopted to ensure that settlements at the working load are within allowable limits. Loading tests should be made whenever settlements are a critical factor in design.

Piles driven into certain types of rock such as chalk, mudstone, or shale or into weathered rocks, may penetrate to a considerable depth before a satisfactory driving resistance is obtained. Chalk is a particularly difficult material in which to estimate pile driving resistance, since the pile breaks down the cell structure of the chalk, freeing the contained water. This water forms a slurry around the pile, giving a very low frictional resistance while driving. It is probable that the slurried chalk (water and silt-size particles) settles around the pile, giving increasing frictional resistance with time after cessation of driving because piles in chalk are known to carry high working loads if a suitable period of time is allowed after driving before the pile is allowed to carry its full load.

Mudstones and shales frequently consist of bands of moderately strong partly weathered rock interbedded with weak partly to completely weathered rock of firm to stiff clay consistency. Unless rock bands can be proved to be of appreciable thickness over the piled area, it is advisable to base the end resistance of the pile on the shear strength of the clay bands.

If it is possible to obtain undisturbed samples of the cohesive completely weathered rock either by the open drive sampler or by coring, then the carrying capacity of piles can be obtained by treating them as driven into soils, using laboratory tests to determine values of the undrained or drained shear strength parameters.

Some useful information on the base resistance and skin friction values of piles driven into weak rocks can be obtained from the *Proceedings of the Symposium on Piles in Weak Rock*[7.36] and the CIRIA Report, Piling in Chalk[7.37]. Further information on the design of piles in chalk has been given by Lord.[7.38]

7.12.2 Bored piles

The action of boring a hole in rock to form a socket for a pile may weaken the bearing capacity of some types of rock. For example, the action of boring tools in chalk or mudstone causes considerable breakdown and slurrying of these rock types. Therefore, low values of skin friction should be used for bored piles in contact with the softened material. The actual value chosen will depend on the hardness and susceptibility to breakdown of these materials. In weak friable chalk or mudstone, complete softening may result and the ultimate skin friction may not be more than, say, $20\,kN/m^2$. In stronger chalk or mudstone where casing is not required to support the borehole, the walls of the hole (if drilled by percussion rather than rotary methods) will be rough. Thus, the concrete will key into the walls of the hole and the full shear strength of the rock will be developed in skin friction on the rock socket, giving values much higher than are obtainable with driven piles.

The length to diameter ratio of the rock socket is an important factor in the consideration of the bearing capacity of the pile in shaft friction and there is no point in adopting a length greater than that required to develop the full available resistance. The distribution of side wall shear stress (rock socket skin friction) for various length to shaft ratios is shown in Fig. 7.16. It will be seen that the ratio should be less than four if it is desired to mobilize base resistance in addition to socket resistance. The high interface stress over the upper part of the socket will be noted.

The ultimate skin friction value is given by the equation

$$q_s = \alpha\beta q_{uc}, \qquad (7.17)$$

where α is a reduction factor related to q_{uc} as shown in Fig. 7.17, and β a correlation factor related to the discontinuity spacing in the rock mass as shown in Fig. 7.18.

Curves relating the rock socket skin friction factor α to

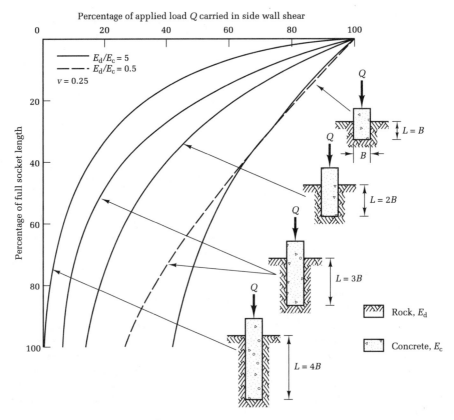

Fig. 7.16   Distribution of side wall shear stress in relation to socket length and modulus ratio (after Osterberg and Gill[7.39]).

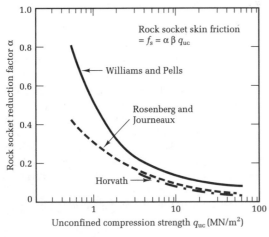

Fig. 7.17   Rock socket skin friction related to the uniaxial compression strength of intact rock.

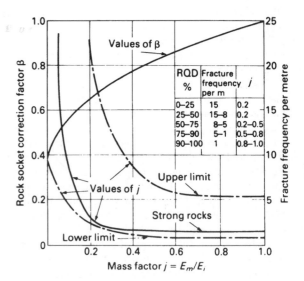

Fig. 7.18   Correction factor for rock socket skin friction allowing for discontinuities.

the uniaxial compression strength of rock established by Rosenberg and Journeaux,[7.40] Horvath[7.41] and Williams and Pells[7.42] are shown in Fig. 7.17. The Williams and

Pells curve gives values considerably higher than the other two. However, their skin friction factor should be corrected by factor β which depends on the mass factor of the rock formation as shown in Fig. 7.17. The mass factor is related to the RQD value or fracture spacing of the rock mass. Values are given in Table 2.12.

In the case of moderately weak to strong rocks where it is possible to obtain core samples for uniaxial compression or point load tests, the end-bearing resistance can be calculated by the methods given above for driven piles. However, the drilling process does not normally break up the rock to any appreciable depth below the pile toe. Hence the joint spacing can be taken as for the undisturbed rock mass. High values of end-bearing resistance should be adopted only with great caution, preferably only when it is possible to inspect the pile base to ensure that all debris and loosened rock have been removed. Again it should be noted that allowable end-bearing pressures depend on the permissible settlement at the working load. It may be necessary to make plate bearing tests (Section 1.4.5) to obtain a deformation modulus for the rock corresponding to the end-bearing pressure, or to obtain the modulus from the empirical relationship with the uniaxial compression strength (equation (2.56)). The settlement of the pile base can then be calculated from equation (7.18).

Where it is possible to drive sample tubes into a weak completely or highly weathered rock or to obtain cores from holes drilled by rotary methods, the angle of shearing resistance and cohesion of the material can be obtained by triaxial compression tests at high lateral pressure in the laboratory. These values may be substituted in equation (2.7) to determine the end resistance of the pile.

7.12.3 Working load settlements

Where piles are driven or drilled to a small penetration into a competent rock and the load is carried predominantly in end-bearing, the pile head settlements can be calculated from equation (7.15). In the case of bored piles designed to carry the applied load only in rock socket friction, the influence factors of Pells and Turner[7.43] can be used with the equation

$$\text{Settlement} = \rho = \frac{QI_p}{BE_d},\qquad(7.18)$$

where

Q = pile head load,
I_p = influence factor from Fig. 7.19,
B = diameter of socket,
E_d = deformation modulus of the rock mass surrounding the pile shaft.

Where the rock socket is recessed below the ground surface or where a layer of soil or very weak rock overlies

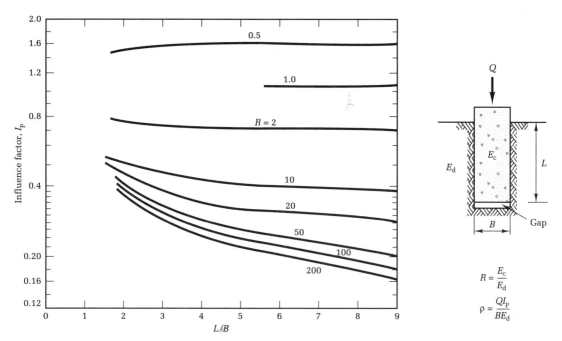

Fig. 7.19 Elastic settlement influence factors for rock socket skin friction on piles (after Pells and Turner[7.43]). (Courtesy of Research Journals, National Research Council of Canada.)

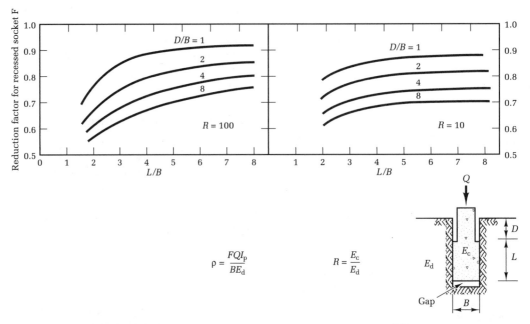

Fig. 7.20 Reduction factors for calculation of settlement of recessed sockets (after Pells and Turner[7.43]). (Courtesy of Research Journals, National Research Council of Canada.)

competent rock a reduction factor is applied to equation (7.18). Values of this factor are shown in Fig. 7.20.

Deformation modulus values for use with equations (7.15) or (7.18) are obtained from plate bearing tests at the base of trial boreholes or from the relationships between unconfined compression strength and discontinuity spacings of the rock mass described in Section 2.7. These empirical relationships are not applicable to high-porosity rocks such as detrital limestones, Chalk and Keuper Marl. Some limited information on modulus values of these rocks is given in reference 7.36.

7.13 Piles in fill–negative skin friction

If driven or bored piles are installed in compressible fill or any soil showing appreciable consolidation under its own weight, a load additional to the working load on the head of the pile is transmitted in skin friction, i.e. 'negative skin friction', to the pile surface. Additional consolidation of the fill or soil is given by superimposed loading. Negative skin friction must be allowed for when considering the safety factor on the ultimate carrying capacity of the pile.

The calculation of the total negative skin friction or drag-down force on a pile is a matter of great complexity, and the time factor is of importance. The maximum unit negative skin friction on a pile is the maximum skin friction which is mobilized on the contact face, and this

peak value can be calculated in exactly the same manner as that used for calculating the support in skin friction given to bearing piles. However, as stated in Section 7.9, the maximum skin friction will not be mobilized until the soil has moved relatively to the pile by an appreciable amount, and this may be of the order of 1 per cent of the shaft diameter. Thus if we take the simple case of a fill settling under its own weight and placed on an incompressible stratum as shown in Fig. 7.21(a), the fill immediately above the incompressible stratum will not move at all and consequently it will not cause any drag-down on the pile. In fact, the elastic compression of the pile due to the superimposed and drag-down loads may cause the pile to move downwards relatively to the fill and so cause the fill to act in *support* of the pile. The maximum settlement of the fill will be at the ground surface, but here the movement may be so large that the skin friction will have passed its peak value and the final drag-down will then be that corresponding to the residual or long strain value. Nevertheless, at some earlier stage in the settlement of the fill the movement at the ground surface will have been such as to mobilize the peak skin friction, whereas the movement at lower levels would not have been such as to mobilize the peak value. If a load is then superimposed on the pile above ground level the pile will compress elastically, the movement being a maximum at the ground surface. This will reduce the relative movement between

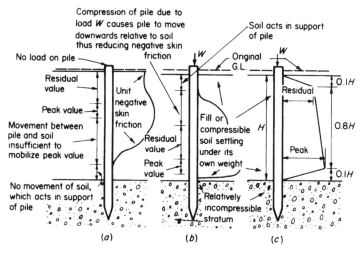

Fig. 7.21 Negative skin friction on piles driven into relatively incompressible stratum. (*a*) Early stages of settlement of compressible layer. (*b*) Late stages of settlement of compressible layer. (*c*) Average curve for design purposes.

pile and soil, which may result in mobilization of peak skin friction in support of the pile (Fig. 7.21(*b*)) or the relative movement may be so low that the peak value is not reached.

Now take the case of a pile bearing on a compressible stratum, say a stiff clay. In this case the whole pile will settle and is likely to mobilize the peak value of skin friction and a proportion of the base resistance in the bearing stratum. The pile will move downwards relatively to the fill immediately above the bearing stratum and thus the fill will act in support of the pile over this length.

Figure 7.22(*a*) and (*b*) show the distribution of drag-down forces and stresses in the pile shaft for early and late stages in consolidation of the fill.

It will be seen from the foregoing that because of the uncertainties in the magnitude of the drag-down forces mobilized in the time-dependent relative movement between pile and soil, it is impossible to obtain a close estimate of the total drag-down force. All that can be said with certainty is that peak value of skin friction will not at any time act on the whole length of shaft embedded in the fill. Figures 7.21(*c*) and 7.22(*c*) will enable the engineer to

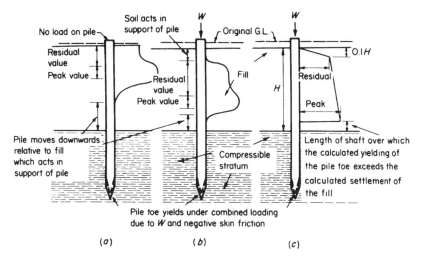

Fig. 7.22 Negative skin friction on piles driven through fills into compressible stratum. (*a*) Early stages of settlement of fill. (*b*) Late stages of settlement of fill. (*c*) Average curve for design purposes.

make an intelligent guess at a reasonable drag-down load to be allowed for design purposes.

The unit negative skin friction force at any depth can be calculated from the equation

$$f_{sneg} = \beta p_0 \qquad (7.19)$$

where p_0 is the effective overburden pressure and β is a reduction factor shown by Meyerhof[7.31] to be equal to 0.3 for piles up to 15 m long, decreasing to 0.2 and 0.1 for 40 and 60 m long piles respectively.

In cases where negative skin friction is so high that it causes difficult problems in pile design, it may be possible to eliminate the drag-down in its entirety or to control it within predetermined limits. This can be done by sleeving the pile through the compressible stratum, or by surrounding the pile by a soft asphalt membrane which has a comparatively low frictional value. However, coatings of this type can be expensive and it is usually found to be more economical to provide for the negative skin friction by increasing the penetration depth to gain additional shaft friction in support of the piles. Alternatively, the working load on the piles from the superstructure can be reduced by providing additional piles.

If the fill has been in place for a very long period of years or has otherwise been well consolidated, the negative skin friction can be ignored in estimating the total working load on the pile, but the skin friction of the fill should not be allowed to act in *support* of the pile. For cases intermediate between recently placed and old fill the safety factor, given by

$$\frac{\text{Ultimate carrying capacity}}{\text{Working load} + \text{negative skin friction}},$$

may be reduced below the value normally adopted for working loads alone.

Negative skin friction can also occur when fill is placed over peat or soft clay strata. The superimposed loading on such compressible strata causes heavy settlement of the fill with consequent drag-down on piles. The skin friction to be added to the working load includes that from the peat or soft clay as well as the fill.

In some circumstances, negative skin friction can be caused by driving piles into normally consolidated clays which causes extensive remoulding and upheaval of the ground surface but, as remarked in Section 7.7.1, the soil reconsolidates quite quickly and regains its original strength before the pile is normally required to carry its full working load. Certainly the value of driving piles wholly into soft clays without reaching any firm bearing stratum is questionable. In the case of piles driven into stiff overconsolidated clays, although remoulding takes place there is little or no loss of shear strength. However, the broken-up masses of clay reconsolidate very slowly

(possibly never within the life of the structure). Thus, the stiff clays will continue to give support to the piles throughout the long period of consolidation. This consolidation will inevitably be accompanied by some settlement.

7.14 The carrying capacity of pile groups

7.14.1 The behaviour of pile groups

As stated at the beginning of this chapter, it is important to consider the effect of driving and loading piles in groups. Only if piles are taken down to a hard incompressible stratum is the settlement of a group of piles equal to that of a single pile under the same working load as each pile in the group (the settlement of piles driven to a hard incompressible stratum is due to elastic shortening of the pile shaft and some elastic yielding of the material under the toe of the pile).

If piles are driven into a compressible bearing stratum, such as a layer of stiff clay, or if the bearing stratum is itself relatively incompressible (a dense sand, for example) but is underlain by a compressible stratum, then the carrying capacity of a group of piles may be very much less than that of the sum of the individual piles. Also, the settlement of the group of piles is likely to be many times greater than that of the individual pile under the same working load. This is illustrated by Fig. 7.23(a) and (b). In the case of the single pile (Fig. 7.23(a)) only a small zone of compressible soil around or below the pile is subjected to vertical stress; whereas with the large pile group (Fig. 7.23(b)), a considerable depth of soil around and below the group is stressed and settlement of the whole group may be large.

An example of the effects of heavy settlement due to consolidation of deep layers of soft clay beneath large pile groups is given by the case of the Campus Buildings of the Massachusetts Institute of Technology[7.44]. These buildings were constructed in 1915 and were founded on piles,

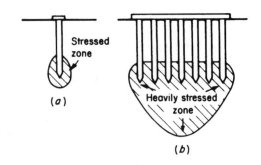

Fig. 7.23 Comparison of stressed zones beneath single pile and beneath pile group. (*a*) Single pile. (*b*) Pile group.

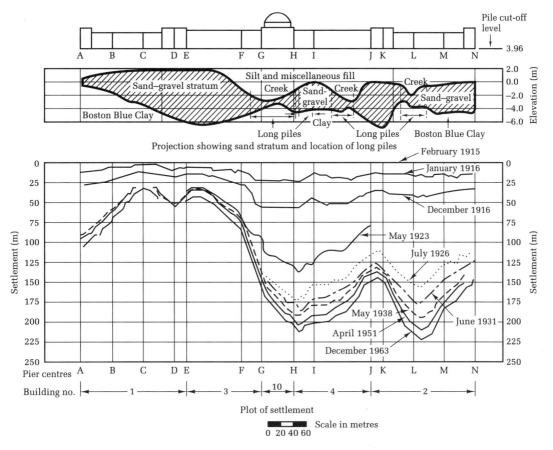

Fig. 7.24 Settlement of pile groups beneath campus buildings of the Massachusetts Institute of Technology (after Horne and Lambe[7.44]).

which in most cases were terminated in a layer of sand and gravel found beneath organic silt and peat at a depth generally of 3–6 m below ground level. The sand and gravel layer was very variable in thickness up to a maximum of about 7.6 m (Fig. 7.24). In places the layer was only about 1 m and the piles were driven through it to enter the underlying deep stratum of soft and compressible Boston Blue Clay, which was 21 m or more in thickness and was underlain by glacial till and rock. Thus the large piled areas transmitted their loads through the sand and gravel to the underlying compressible clay and this resulted in heavy settlement and extensive cracking of the buildings. The settlements varied widely over the loaded area. Between 1915 and 1943 the settlements increased progressively, the minimum being about 40 mm and the maximum 230 mm (Fig. 7.24). The variations were due mainly to the variation in thickness of the sand and gravel stratum and the lengths of the piles. The heaviest settlements occurred where the piles were driven into the Boston Blue Clay stratum.

It will be appreciated from the foregoing that the problems of the stability of pile groups are, first, the ability of the soil around as well as below the block of soil containing the pile group to support the whole load of the structure and, second, the effects of consolidation of the soil for a considerable depth below the pile group. Therefore, the manner in which the individual pile is installed, whether by driving, boring, or jacking, has little effect on these two problems. This is because the zone of soil affected by the method of installing the pile is very small compared to the very large mass of soil affected by the vertical pressures transmitted to it by piles in the group.

In order to determine the ultimate bearing capacity of a pile group and to estimate the immediate and consolidation settlements of the group, it is assumed that the piled area acts as a buried raft foundation with a degree of flexibility which depends on the rigidity of the capping system and of the superimposed structure. Transfer of load from the piles to the soil in skin friction is allowed for by assuming that the load is spread from the shafts of friction

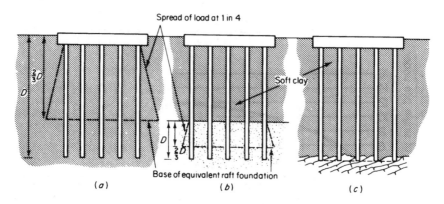

Fig. 7.25 Load transfer to soil from pile groups. (*a*) Group of piles supported mainly by skin friction. (*b*) Group of piles driven through weak clay to combined skin friction and end-bearing in stratum of dense granular soil. (*c*) Group of piles supported in end-bearing on hard incompressible stratum.

piles at an angle of 1 in 4 from the vertical as shown in Fig. 7.25(*a*)–(*c*).

The factor of safety of the equivalent raft foundation against general shear failure in the underlying soil is determined from a knowledge of the shearing resistance of the soil using the methods described in Section 2.3.

Immediate and consolidation settlements of the equivalent raft are calculated using the procedure in Section 2.6. Alternatively, elastic continuum methods can be used as described in Section 7.19.2.

7.14.2 Pile groups in cohesionless soils

The action of driving piles or pile tubes into cohesionless soils is to compact the soil around the pile to a radius of at least three times the pile diameter. Thus, when piles are driven in a group at close spacing the soil around and between them becomes highly compacted.

The spacing of piles driven in large groups in cohesionless soils requires careful consideration. If the piles are at close centres, the first few piles drive easily, but as the density of the soil increases so driving of additional piles becomes increasingly difficult. When driving piles in sands or gravels, the best procedure to avoid difficulty with 'tightening-up' of the ground is to start driving at the centre of the group and then work outwards in all directions.

The BS Code of Practice, BS 8004: *Foundations* requires a minimum spacing centre-to-centre for friction piles of not less than the perimeter of the pile or, for circular piles, three times their diameter. Piles obtaining support mainly in end-bearing may have a reduced spacing, but the distance between the surfaces of adjacent piles should not be less than their least width. Special consideration should be given to bored piles with enlarged bases where construction tolerances may cause interference between them.

For most engineering structures the load to be applied to a pile group will be calculated from considerations of settlement rather than by calculating the ultimate carrying capacity of the group and dividing it by an arbitrary safety factor of 2 or 3. Generally the settlement of the group can be calculated on the assumption that the piled area acts as a buried raft foundation (Fig. 7.25(*a*) or (*b*)).

7.14.3 Pile groups in cohesive soils

The effect of driving piles into cohesive soils (clays and silts) is very different from that in cohesionless soils. Piles driven or bored in groups into soft sensitive clays cause extensive remoulding of the soil and, in the case of driven piles, a heave of the ground surface occurs around the group. With the passage of time the soil reconsolidates and regains its original shear strength. Thus, reconsolidation causes a drag-down on the pile shaft. When piles are driven into stiff clays the same ground heave occurs, but the soil is pushed up in lumps or cracked masses and the piles may be lifted. Reconsolidation is extremely slow and the original shear strength of the whole mass of ground around and within the pile group may never be restored within the life of the structure. The drag-down effects are small and in the case of bored piles they do not occur.

When loading is applied to a group at close spacing the soil contained within the piles moves downwards with the piles and, at failure, piles and soil move together to give the typical 'block failure'. The same mechanism of failure occurs with driven or bored piles. Piles in cohesive soils generally act as friction piles, therefore the minimum centre-to-centre spacing of not less than the perimeter of the pile, as required by BS 8004, should be adopted. It may be necessary to give special consideration to the spacing of heavily loaded piles with enlarged bases. The interaction of stresses beneath the pile bases should be analysed as described in Section 7.19.2.

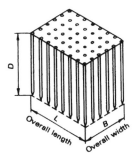

Fig. 7.26 Calculation of block failure of pile group in clay.

Block failure is unlikely to occur at a spacing of three times the perimeter, but if it is necessary to investigate this mode of failure for close-spacing the stability of a group of bored or driven piles is given by the sum of the perimeter shearing resistance and the end-bearing resistance of the block of soil contained by the piles (Fig. 7.26), as expressed in the following equation. Ultimate carrying capacity of group is

$$Q_u = 2D(B + L)\bar{c} + 1.3cN_cBL, \qquad (7.20)$$

where

D = length of piles,
B = width of group,
L = length of group,
$\bar{c}$ = average cohesion of clay around the group,
c = cohesion of clay beneath group,
N_c = bearing capacity factor (see Section 2.3).

The remoulding caused by the pile driving or boring only takes place to a relatively small distance around and beneath the pile group; thus the undisturbed shear strength beneath the group may be taken for the second term in equation (7.20). However, when considering the perimeter shearing resistance, the time effect should be taken into account allowing the fully remoulded shear strength for $\bar{c}$, if the piles are required to carry their full working load a short time after driving, or the original shear strength if full loading can be delayed for 6 months or more.

Settlements of pile groups in cohesive soils are calculated on the assumption that the group acts as a buried raft foundation (Fig. 7.25(a) or (b)) using the methods described in Section 2.6.

The effect of driving piles in groups in clays to cause ground heave has already been noted. In some circumstances this can cause lifting of piles already driven. Cases have occurred where both precast and driven and cast-in-place piles have been lifted 100 mm or more. In the case of driven and cast-in-place piles, it is not always possible to drive them down again to their original

position if they are of the type in which the tube is withdrawn. However, piles which incorporate a steel shell which remains in the ground can be redriven by reinserting the mandrel, before they are finally filled with concrete. Lifting of piles is most likely to occur when the pile shafts are wholly in firm to stiff clays. When the major portion of the shaft is in a very soft clay which is remoulded during driving, the adhesion is insufficient for the heaving ground to lift the piles. Therefore, when considering driving piles in groups in firm to stiff clays, the engineer should use precast concrete, steel, or shell piles which can be redriven if necessary, or alternatively, he should adopt bored piles, or partly bored and partly driven and cast-in-place piles. The ground displaced by the piles driven in groups can cause high lateral pressures to develop on nearby buried culverts, sewers, or tunnels, with consequent risk of serious damage.

The possible detrimental effect of driving piles wholly into soft compressible clays has been mentioned briefly in Section 7.12. It will be as well, therefore, to discuss in more detail the advantages and disadvantages of pile groups in this type of ground. It should be pointed out at this stage that *short* piles in such conditions are worse than useless. A comparison of the stress distribution in Fig. 7.27(a) and (b) between a shallow raft foundation and a short pile foundation shows that, virtually, the same volume of compressible soil is stressed in each case. In fact, the short pile group may show greater settlement than the shallow raft, due to reconsolidation of the heaved and remoulded soil. However, most normally consolidated clays show a progressive increase in shear strength and decrease in compressibility with increasing depth. Therefore, in the case of the long pile group (Fig. 7.27(c)) the stresses are transferred to deeper and less compressible

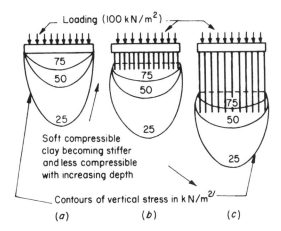

Fig. 7.27 Comparison of vertical stress distribution between (a) surface raft, (b) raft with short piles, (c) raft with long piles.

soil and the settlement of the structure is correspondingly less.

An example of the settlement of a basement foundation in conjunction with piles in stiff clays is the foundation of the Hyde Park Cavalry Barracks (see Section 5.5).

7.14.4 Ground heave and reconsolidation

In a closely-spaced pile group where soft clays or silts are overlying stiff clays, the upper part of the mass of soil enclosed by the group is heaved up during pile driving (Fig. 7.28). Therefore, when reconsolidation takes place the weight of the heaved-up soil is transferred to the piles, thus increasing the load on them. It is thus evident that the upper soft clays and silts do not contribute to the carrying capacity of the piles. The problem is to assess how much of this load should be taken as increasing the working load on the pile from the point of view of the carrying capacity of the individual pile.

The weight of the mass of soil causing drag-down should be divided up between the piles in the group and added to the working load on each pile. The total load should not exceed the ultimate carrying capacity of the individual pile as determined by calculation or loading test, but the safety factor need not necessarily be high. A safety factor less than 2 would be accepted if the settlement of the individual pile is not excessive under the total load. It should be noted that the drag-down on any individual pile in a group caused by consolidation of the soil around the pile will not be greater than can be carried in skin friction on the pile shaft, i.e. the surface area of the pile embedded in the soft clay multiplied by the adhesion corresponding to the shear strength of the soft clay (see Fig. 7.10). Also, in considering the drag-down on an individual pile it should be remembered that heave and reconsolidation sufficient to cause negative skin friction is

only likely to occur in the upper part of the pile shaft (see Fig. 7.21). The safety factor of the whole group of piles under the working load plus the weight of the consolidating mass of soft clay should also be calculated. The carrying capacity of the group is given by

$$Q_u = 2D'(B + L)\bar{c} + cN_cBL, \tag{7.21}$$

where D' is the penetration into stiff clay stratum.

Safety factor = Q_u ÷ (Total working load on pile group + Load transferred by consolidation of soft clay).

Here again the safety factor need only be low since the drag-down due to the consolidating soil will not increase the consolidation of the soil beneath the whole group because the only additional load at the base of the pile group is the working load imposed on the group from the structure plus the net weight of the piles themselves, i.e. the weight of the overburden soil on the stiff clay beneath the pile group has not been increased by the pile driving.

The procedure should therefore be as follows:

(a) For calculating the carrying capacity of individual piles in the group or the whole group, neglect any support given by the soft clay.

(b) There should be a safety factor on the expression

$$\frac{\text{Ultimate carrying capacity of individual pile}}{\text{Working load on individual pile + Load transferred to individual pile from consolidating overburden}}$$

but the safety factor can be lower than that used for

$$\frac{\text{Ultimate carrying capacity of individual pile}}{\text{Working load on an individual pile}}$$

(c) There should also be a safety factor on the expression

$$\frac{\text{Ultimate carrying capacity of pile group}}{\text{Working load on group + Load transferred to pile group from consolidating overburden}}$$

Again the safety factor can be lower than that used for

$$\frac{\text{Ultimate carrying capacity of pile group}}{\text{Working load on pile group}}.$$

7.14.5 Pile groups in filled ground

In the case of pile groups driven or bored in fill which is consolidating under its own weight, or under the weight of surface load, the weight of the whole mass of fill enclosed by the periphery of the group is transferred to the piles, as

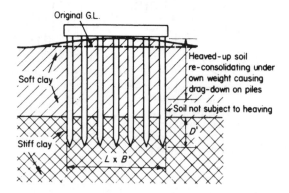

Fig. 7.28 Effects of ground heave of driven piles in clay.

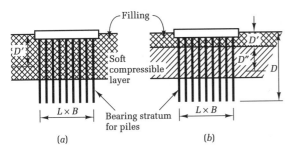

Fig. 7.29 Pile groups in filled ground.

shown in Fig. 7.29(*a*). Where the fill is underlain by a compressible stratum as shown in Fig. 7.29(*b*), the weight of that part of the soft stratum enclosed by the pile group from which the downward movement is sufficient to cause drag-down must be added to the weight of the super-imposed fill, since the latter causes the soft clay to consolidate, resulting in additional drag-down.

Thus, in Fig. 7.29(*a*), the total load on pile group at level of bearing stratum is given by

$$Q_1 = \text{Working load} + L \times B \times \gamma'D', \qquad (7.22)$$

where γ' is the density of fill and D' the depth of fill over which the movement is sufficient to cause drag-down.

In Fig. 7.29(*b*), the total load on pile group at level of bearing stratum is given by

$$\begin{aligned} Q_2 = \text{Working load} &+ L \times B \times \gamma'D' \\ &+ L \times B \times \gamma''D'', \end{aligned} \qquad (7.23a)$$

where γ'' is the density of soft stratum and D'' the depth of soft stratum over which movement is sufficient to cause drag-down. However, total load on pile group will not exceed ultimate skin friction on piles from fill and soft clay, i.e.

$$Q_2 = \not> \text{Working load} + S_1f' + S_2f'', \qquad (7.23b)$$

where

S_1 = sum of surface area of piles embedded in fill,
S_2 = sum of surface area of piles embedded in soft clay,
f' = skin friction between fill and piles,
f'' = skin friction between soft clay and piles.

The consolidation settlement of the bearing stratum due to loads imposed by the pile group should be calculated on the basis of the working load on the group plus the maximum load transferred on to the piles from the fill only. Thus, in Fig. 7.29(*b*) there is no increase in weight of the soft stratum causing additional loading on the bearing stratum. The only additional loads causing

settlement of the bearing stratum are the working loads on the group plus the weight of the fill. This assumes that the fill has been recently placed and has not had time to cause appreciable consolidation of the underlying strata.

The problem of negative skin friction from the consolidation of filling applies both to driven piles and bored piles. However, these problems do not occur if the piles are taken through fill on to an incompressible stratum such as bedrock or very compact sand and gravel. In these cases the piles cannot settle, the consolidating fill slips past the piles, and its weight is carried by the underlying incompressible stratum. However, the drag-down effect on the pile considered as a structural column should not be neglected.

7.15 The design of axially loaded piles considered as columns

Piles embedded wholly in the ground need not be considered as long columns for the purposes of structural design. Where, however, they project above the ground as in the case of jetties or piled trestles, the portion above the ground or the sea- or river-bed must be considered as a column and it is then necessary to consider their effective length and conditions of end fixity. BS 8004 makes the following recommendations:

(a) In firm ground the lower point of contraflexure can be taken as about 1 m below the ground surface.
(b) When the top stratum is soft clay or silt, this point may be taken at about half the depth of penetration into this stratum, but not necessarily more than 3 m.

Reduction factors for working stresses in piles acting as columns should be applied as recommended in the appropriate structural code of practice.

7.16 Piles resisting uplift

7.16.1 Anchoring piles in soil

In certain circumstances, piles are required to resist uplift forces, such as in foundations to structures subject to considerable overturning moment, for example tall chimneys, transmission towers, or jetty structures. Resistance to uplift is given by the friction between the pile and the surrounding soil; it may be increased in the case of bored piles by 'under-reaming' or belling-out the bottom of the piles, or by the bulb end of a driven and cast-in-place pile. A method of calculating the uplift resistance of piles with enlarged bases has been established by Meyerhof and Adams.[7.45] Various published test results have indicated that the skin frictional resistance of piles to uplift loads is appreciably lower than that mobilized in resistance to compression loading. A reduction of 50 per cent has been suggested for granular soils and the same order of reduction for long-term sustained loading for clays. The latter reduction is particularly significant for short piles due to relaxation of stresses induced in the soil during installation. Therefore the appropriate safety factor should be applied to the skin friction values calculated by the methods described earlier in this chapter, and if the uplift load on piles necessitates mobilization of skin friction approaching the ultimate values, it will be advisable to carry out uplift tests on selected full-scale piles to ensure that there is an adequate safety factor against the pile pulling out of the ground. The uplift resistance in shaft friction given by sands or gravels will be of a low order and should be neglected if piles driven to a small depth of penetration are subjected to vibration. Jetty and wharf structures are subjected to lateral forces from berthing ships and from wave action. If these forces are transferred to supporting piles the resulting lateral movement of the upper part of the embedded length of the piles may destroy most of the skin friction, for example an enlarged hole may be formed if the piles are embedded in stiff clays. This progressive deterioration in uplift resistance due to skin friction will not be reflected by a pull-out test made soon after driving; therefore, a generous safety factor on the ultimate pull-out load should be adopted for piles carrying both lateral and uplift loads.

7.16.2 Anchoring piles to rock

Piles driven to rock, although they have a high bearing resistance, may have a low uplift resistance if they are driven through soft or loose materials to a small penetration into rock. Thus, if a pile is driven only a metre or so into strong rock it will so shatter the material around the pile that uplift resistance will be negligible and due almost wholly to friction in the debris and the overburden soil. The uplift resistance may be increased by drilling a large-diameter hole deep into the rock and concreting or grouting the pile into this hole or socket or, in the case of hollow tube or box piles or bored piles, by drilling a deep hole into the rock at the bottom of the pile, followed by concreting in steel rods or cables made up from high-tensile wires lowered into the drilled hole (Fig. 7.30).

The uplift resistance of this type of anchorage is given by:

(a) The tensile strength of the anchor rods or cables;
(b) The bond resistance between the rods or cables and the concrete or grout which surrounds them;
(c) The bond resistance between the concrete or grout and the surrounding rock;
(d) The dead weight (or submerged weight if the anchorage is below water level) of a cone of rock which

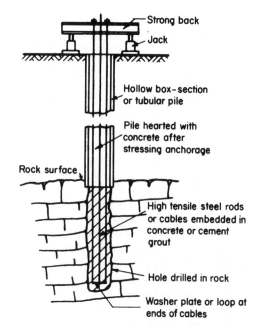

Fig. 7.30 Prestressing anchorage to tension pile.

must be lifted by the anchor if failure does not occur by (a), (b), or (c).

The uplift resistance given by (a), (b), or (c) can be determined by experiment or by calculations based on the known properties of the steel, concrete, or grout. If the surface of the anchor rod is roughened or if the rod has a hook or plate at its lower end, then anchor to grout failure should not theoretically occur. However, experiments have shown that steel to grout failure can take place under these conditions by the grout flowing plastically around the end plate or projections on the rod.

Wherever possible the allowable bond stress between steel and grout should be determined by field tests using anchors of the same diameter and the same type of end fittings as those proposed for the permanent installation. As a guide, allowable values of $0.3–0.7 \, \text{N/mm}^2$ can be used for deformed bars or cables up to 150 mm in diameter, and $0.15–0.3 \, \text{N/mm}^2$ for tube anchors up to 300 mm in diameter. Recommendations for the bond strength at the grout to anchor interface are given in BS 8081: *Code of Practice for Ground Anchors*. This Code provides a wealth of useful and practical information on the design, installation and testing of anchors in rock. The Code recommendations for *ultimate* bond strength are:

Plain bar	$\not> 1 \, \text{N/mm}^2$
Clean strand or deformed bar	$\not> 2 \, \text{N/mm}^2$
Locally noded strand	$\not> 3 \, \text{N/mm}^2$

Special consideration should be given to allowable bond stresses where high uplift loads are to be carried, for example where load is transferred from the legs of a tubular jacket structure in deep water to a large-diameter steel tube pile driven into the sea-bed. In these conditions there is an appreciable diminution of diameter in the anchor pile caused by inward radial strain under tensile load. This reduction in diameter may break the bond with the hardened grout filling the annulus between the jacket and the pile. It may be necessary to form keys on the surfaces of the pile and jacket by welding on steel strips or spiral rods. Where these are provided the steel to grout bond stress depends on the allowable stress in compression on the grout. Methods for calculating the steel to grout bond strength for the pile to jacket connections are given in the American Petroleum Institute recommendations for offshore structures[7.9] and by the UK Department of Energy[7.46].

For bars set in concrete the normal design requirements for embedment should be followed.

The allowable bond stress between the grout or concrete and the surrounding rock depends on the compressive strength of the intact rock, the size and spacing of joints and fissures, the amount of keying given to the anchor hole by the drilling bit, and the cleanness of the rock surface which can be achieved by the drilling water flush. The size of the drill-hole and of the annular space between the anchor and the wall of the hole are important. A large-diameter hole or a wide annulus can result in weakening of the grout to rock bond consequent on shrinkage of the grout, although this can be countered to some extent by using grouts which incorporate plasticizers, expanding agents, or fibrous bonding materials.

By providing a plate or special compression fitting on the bottom of the anchor, part of the grout column is put into compression. The smaller the annulus and the shorter the bonded length, the higher the compressive stress induced in the grout and consequently the higher the radial compressive stress and shearing resistance at the contact face between the grout and the rock.

The bond between grout and rock will be small if drilling softens the rock surface. This can occur with silty or clayey weathered rocks such as chalk or mudstone. Observed bond stresses at failure range from $0.15–0.3\,\text{N/mm}^2$ in weak weathered chalk, mudstone, and shale to $1.0–2.0\,\text{N/mm}^2$ in strong, relatively unweathered but jointed rocks.

With regard to (d) above, the estimation of the angle and depth of the cone for the purpose of calculating its weight is largely a matter of judgment based on experience. The shape of the cone pulled out by the anchor depends on the frequency of jointing and fissuring of the rock mass and the inclination of the bedded planes. Progressive failure of the grout to rock bond due to stretch of the anchor is prevented by sheathing the upper part of the anchor. Then if the anchor has a compression fitting at the bottom the shape of the pull-out cone is shown in Fig. 7.31(a). Where a plain bar or cable is used Wyllie[7.33] suggests that the effective length should be taken as the mid-point of the bonded length (Fig. 7.31(b)). If the rock surface has a soil overburden this is lifted with the rock cone. The failure surface is assumed to be cylindrical in

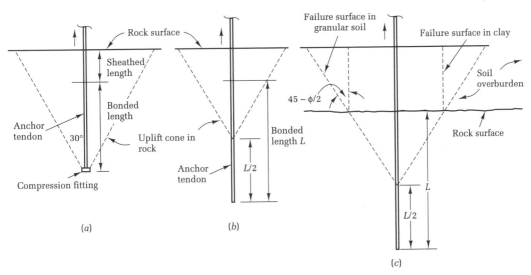

Fig. 7.31 Method of calculating uplift volume of rock cone for anchorage piles. (a) With compression fitting, wholly in rock. (b) Without compression fitting, wholly in rock. (c) Without compression fitting, in rock with soil overburden.

clay, and a truncated cone in a granular soil (Fig. 7.31(*c*)). The half-angle of 30° shown for the cone is conservative and represents a heavily jointed or shattered rock. Because shear at the interface between the cone and the surrounding rock is neglected, a safety factor of unity can be taken on the effective weight of the cone for a horizontally bedded formation or one with moderately sloping bedding planes. Where the bedding is steep the safety factor should be increased, or an attempt should be made to calculate the geometry of the failure surfaces and the frictional resistance to pull-out over these surfaces. Where uplift on anchorages is a critical factor in design, it is advisable to check the design assumptions by full-scale pull-out tests on the site. The adoption of prestressed anchorages for piles is a valuable means of checking the holding power of each individual anchorage since a prestressing load at least equal to the maximum uplift load can be applied by jacking from the top of the pile. However, jacking the anchor against the pile or against the ground resistance immediately above the pile does not test the uplift resistance of the rock cone.

A disadvantage of the rod or prestressed cable type of anchorage is the risk of corrosion particularly in marine work. If large-diameter hollow piles are used, a large hole can be drilled into the rock and a heavy steel section can be concreted into the hole. However, if the aperture in the pile is small there is likely to be insufficient clearance to enable concrete to be used to surround the anchor, which must then be grouted in. Neat cement grout of a consistency which can be pumped through a small-diameter pipe is of necessity a weak material in comparison with concrete.

If the anchorage is not prestressed, the anchor rod or cable need only extend for sufficient height up the pile to develop the required strength in bond between the inner surface of the pile and the concrete or grout. However, if

prestressing is used the cables must be carried up the pile to enable them to be stressed using the top of the pile as the reaction for the jack. It is then necessary to protect the cable with a plastic sheath or with concrete or cement grout. If space is adequate, concrete placed by tremie pipe is preferable to grouting.

If uplift piles are placed in a group, allowance should be made for the overlapping of the individual uplift cones in heavily jointed or broken rock formations, and the anchorages of the group should be deep enough to ensure that there is sufficient weight in the cone of rock encompassing the whole group (Fig. 7.32).

Uplift on piles used as foundations to light structures on swelling clay soils has been described in Section 3.1.2.

7.17 Piles subjected to horizontal or inclined loads

7.17.1 Ultimate resistance and deflexion

Foundation piles are frequently required to carry inclined loads which are the resultant of the dead load of the structure and horizontal loads from wind, water pressure, or earth pressure on the structure. Where the horizontal component of the load on the piles is small in relation to the vertical load, it can be carried safely by vertical piles. Thus, special provision is not usually made in piled foundations to buildings for the horizontal loading resulting from wind pressure.

However, in the case of piles in wharves and jetties carrying the impact forces of berthing ships, and piled foundations to bridge piers, trestles to overhead travelling cranes, tall chimneys, and retaining walls, the horizontal component is relatively large and vertical piles cannot generally be relied on to withstand the horizontal forces. Raking or batter piles have a very much higher resistance

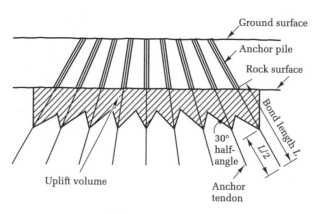

Fig. 7.32 Uplift volume of rock for a group of anchor piles.

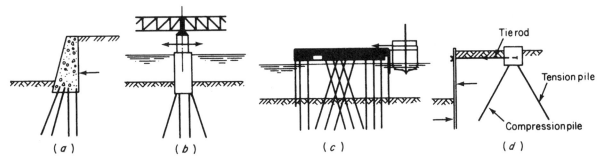

Fig. 7.33 Uses of raker piles. (*a*) Retaining wall. (*b*) Bridge pier. (*c*) Wharf. (*d*) Sheet pile retaining wall.

to horizontal loading since a large proportion of the horizontal component is carried axially by the pile. Typical arrangements of raker piles are shown in Fig. 7.33(*a*)–(*d*).

In the case of the retaining wall (Fig. 7.33(*a*)) the earth pressure always acts in the same direction and forward-raking piles only are required to carry the horizontal loading, while vertical piles are provided to support the dead weight of the wall and vertical forces transmitted to the back of the wall by the retained earth. Horizontal forces in the bridge trestle (Fig. 7.33(*b*)) are caused by braking and traction of the moving loads. These forces are reversible in direction and raking piles in two directions are required. If the bridge pier is in a deep fast-flowing river the pressure from the water may require rakers in a direction transverse to the line of the bridge. Horizontal forces in the piled wharf (Fig. 7.3(*c*)) are caused by the impact of berthing ships and by wave action and are high in relation to the dead load of the superstructure requiring a large number of raking piles. The tie rod supporting the earth forces on the sheet pile retaining wall (Fig. 7.33(*d*)) is always in tension, but a combination of forward and backward rakers is provided in the anchorage in cases where the load would be too much for a single pile. A large pile cap is provided to counteract the uplift on the backward raker.

Where the lateral loading is intermittent, as in the case of piles to wharves and jetties or in bridge trestles, the full value of skin friction on the shaft should not be allowed when calculating the resistance to axial loading along the pile. Thus, where short piles are driven into stiff clays, the deflexion and rebound under intermittent loading will result in an enlarged hole for the pile, reducing the friction to a negligible amount, and the resistance to axial loading should be calculated on end-bearing only. Where the piles are in soft clays, silts, or sands and gravels, the soil will partially close around the pile if the frequency of loading is low; thus the remoulded shear strength of clays or silts

may be used for the skin friction, and a low value of skin friction in sands and gravel should be adopted. The required value may be obtained from pulling tests when the skin friction should be calculated from the load required to maintain steady upward movement of the pile. Where short raking piles are carrying vibrating loads in soft silts and clays or cohesionless soils, skin friction will be negligible and the end-bearing resistance only should be used to carry the axial loading. Higher values of skin friction may be allowed in the case of long piles because vibrations or lower frequency movements will be, for the most part, absorbed in the upper embedded parts of the piles.

The resistance of rakers to the axial loads caused by lateral loading is often a critical factor in design since small yielding might result in serious tilting of a structure. Driving piles to batters of 1 in 3 or 1 in 4 is common, but driving to angles flatter than 1 in 2 is a difficult procedure, and is impossible for bored or driven and cast-in-place piles. It is therefore necessary to provide a larger number of rakers driven to a small batter. Wherever possible, raking piles carrying compression loads should be driven to a hard unyielding stratum.

When a horizontal load is applied to the head of a vertical pile which is free to move in any lateral direction, the load is initially carried by the soil close to the ground surface. However the soil compresses elastically and there is some transfer of load to the soil at a greater depth. When the horizontal load is increased the soil yields plastically and the load transfer extends to greater depths. A short rigid pile (length to width ratio less than 10–12) will rotate and passive resistance will develop at the toe on the same face at which the load is applied, in addition to the passive resistance of the soil near the ground surface on the opposite face (Fig. 7.34(*a*)). Failure occurs by rotation when the passive resistance at the head and toe are exceeded.

A short rigid pile restrained at the head by a pile cap or bracing fails by translation (Fig. 7.34(*b*)).

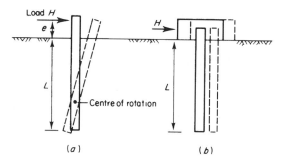

Fig. 7.34 Failure of short rigid pile under horizontal load. (*a*) Free head. (*b*) Fixed head.

The failure mechanism of a very long pile is different since the cumulative passive resistance at the lower part of the pile is very high. Hence the pile cannot rotate and failure occurs by fracture of the pile at the point of maximum bending moment (Fig. 7.35(*a*)). In the case of a long pile restrained at the head, high bending stresses occur at the point of restraint where fracture may take place (Fig. 7.35(*b*)).

The head of a laterally loaded pile may move through an appreciable distance before failure in rotation or fracture of the pile takes place, such that the movement of the structure carried by the pile may exceed tolerable limits. Therefore, in addition to ensuring that there is an adequate safety factor against failure under the lateral load, the deflexion of the pile should be calculated to ensure that it is not excessive.

The dominant factor in the interaction of the pile and the soil is the stiffness of the pile which influences the degree of deflexion and determines whether the failure mechanism is one of rotation, translation, or failure in

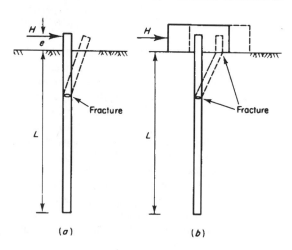

Fig. 7.35 Failure of long piles under horizontal load. (*a*) Free head. (*b*) Fixed head.

bending. The type of loading, whether sustained, alternating, or pulsating, influences the amount of yielding of the soil. External influences such as removal of soil from around the pile by scour, or shrinkage of soil away from the pile surface, affect the passive resistance of the soil against the upper part of the pile.

Brinch Hansen's method[7.47] can be used to calculate the ultimate lateral resistance of a short rigid pile. The method is a simple one applicable to cohesive or cohesionless soils either in uniform or layered state of deposition.

The resistance to rotation of the rigid pile about the point X in Fig. 7.36 is given by the sum of the moments of the soil resistance above and below this point. The passive resistance diagram is divided into a convenient number n of horizontal elements of depth L/n. The unit passive resistance of such an element at a depth z below the soil surface is then given by the equation

$$p_z = p_{oz}K_{qz} + cK_{cz}, \qquad (7.24)$$

where

p_{oz} = effective overburden pressure at depth z,
c = cohesion of soil at depth z,
K_{qz} = passive resistance coefficient for frictional component of soil at depth z,
K_{cz} = passive resistance coefficient for cohesive component of soil at depth z.

Brinch Hansen[7.47] has established values of K_q and K_c in relation to the depth (z) and the width (B) of the pile in the direction of rotation, as shown in Fig. 7.37. The total passive resistance on each horizontal element is $p_z \times (L/n) \times B$, and by taking moments about the point of application of the horizontal load:

$$\sum M = \sum_{z=0}^{z=x} p_z \frac{L}{n}(e+z)B - \sum_{z=x}^{z=L} p_z \frac{L}{n}(e+z)B. \qquad (7.25)$$

The point of rotation at depth x is chosen correctly when $\sum M = 0$; that is, when the passive resistance of the soil above the point of rotation balances that below it. Point X is thus determined by a process of trial and adjustment. If the head of the pile carries a moment M instead of a horizontal force, the moment can be replaced by a horizontal force H at a distance e above the soil surface, where M is equal to $H \times e$.

Where the pile head is fixed against rotation, the equivalent height e_1 above the soil surface of a force H acting on a pile with a free head, is given by

$$e_1 = \tfrac{1}{2}(e + z_f), \qquad (7.26)$$

where e is the height from the soil surface to the point of application of the load at the fixed head of the pile (Fig. 7.36), and z_f the depth from the soil surface to the point of virtual fixity.

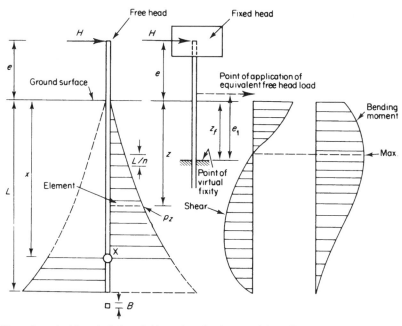

Fig. 7.36 Brinch Hansen's method for calculation of ultimate lateral resistance of short piles.

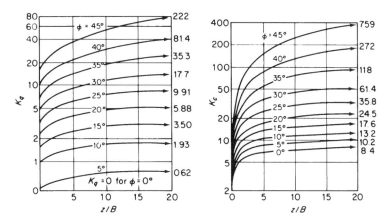

Fig. 7.37 Brinch Hansen's coefficients K_q and K_c.

The depth z_f is not known at this stage, but for trial purposes it can be taken as 1.5 m for a compact granular soil or a stiff clay (in the latter case below the zone of soil shrinkage), and 3 m for a soft clay or silt.

Having obtained the depth to the centre of rotation from equation (7.25), the ultimate lateral resistance of the pile to the horizontal force H_u can be obtained by taking moments about the point of rotation when

$$H_u (e + x) = \sum_0^x p_z \frac{L}{n} B(x - z) + \sum_x^{x+L} p_z \frac{L}{n} B(z - x).$$
(7.27)

The final steps are to construct the bending moment and shearing force diagrams (Fig. 7.36). The ultimate bending moment, which occurs at the point of zero shear, should not exceed the ultimate resistance moment M_u of the pile shaft. The appropriate load factors are applied to the design horizontal force to obtain the ultimate force H_u.

When applying the Brinch Hansen method to layered soils, assumptions must be made concerning the depth z to obtain K_q and K_c for the soft clay layer in Fig. 7.38, but z is measured from the top of the stiff clay stratum to obtain K_c for this layer.

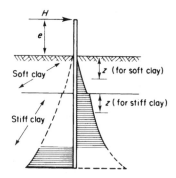

Fig. 7.38

The undrained shear strength is used for c in equation (7.24) for short-term loading conditions such as wave or ship berthing forces on a jetty. The drained effective shear strength parameters c' and ϕ' should be used for all cases of sustained loading such as earth pressure. Where the drained parameters are used a check should be made to ensure stability under undrained conditions at early stages in loading on the structure.

The deflexion of the pile head under the working load can be calculated approximately by assuming that the pile acts as a cantilever about the point of virtual fixity. Thus, from Fig. 7.36,

$$\text{Deflexion at head of free-headed pile} = \frac{H(e + z_f)^3}{3EI}, \tag{7.28}$$

$$\text{Deflexion at head of fixed headed pile} = \frac{H(e + z_f)^3}{12EI}, \tag{7.29}$$

where E is the elastic modulus of the material forming the pile shaft and I the moment of inertia of the cross-section of the pile shaft.

7.17.2 Computerized methods for analysing the behaviour of single piles under horizontal loading

There are two principal approaches to the computer-based modelling of isolated piles under lateral loading. These are (1) idealization of the soil resistance as discrete springs, and (2) consideration of the soil as an elastic continuum.

Soil spring idealization This technique involves modelling the pile as a beam supported by discrete springs to represent the soil resistance as shown in Fig. 7.39. Transfer of shear stresses through the soil is not modelled in this representation. The springs can be considered as linear or non-linear. The most usual formulation of the problem involves finite difference representations of the governing differential equation

$$EI \frac{d^4y}{dx^4} = q_{(x)}, \tag{7.30}$$

where

EI = flexural stiffness of the pile,
y = lateral displacement at x,
x = distance along pile,
$q_{(x)}$ = lateral soil force per unit length at x (note that $q_{(x)}$ will be dependent on y).

Finite difference expressions are also used to express slope, bending moment, and shear in the pile.

The linear spring model may be adopted in cases where soil strains are small and such that the soil may be assumed to be acting within the elastic range. It can be used to generate standard solutions, expressed in graphical form for example, or to solve specific problems involving conditions of complex loading and strata. Use of the technique for anything other than trivial problems involving only very small numbers of springs demands the availability of a suitable computer program.

Use of a computer program is essential for problems involving non-linear soil springs. The principles of the technique were established during the 1960s[7.48, 7.49, 7.50] and it has been used extensively since then, mainly to meet the demands of the oil industry in the design of piles to support offshore structures. Such piles are typically tubular steel and may be installed to considerable depths below the sea-bed through a wide variety of strata. Under extreme pile loading conditions due to storm waves the soils in the vicinity of the sea-bed will be subject to lateral strains beyond the elastic range, and so it is important to make use of a non-linear soil spring model as shown in Fig. 7.39, referred to as the 'p–y curve'. Considerable

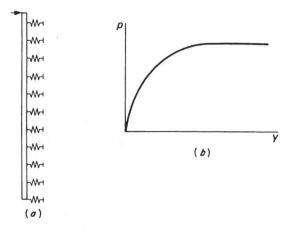

Fig. 7.39 Spring model for pile carrying lateral loading. (a) Horizontally loaded pile modelled as a vertical beam supported on horizontal springs representing soil. (b) p–y curve representing non-linear loads–deflexion response of soil spring where p is the soil resistance and y the lateral deflexion.

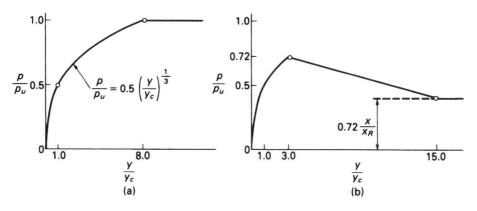

Fig. 7.40 $p–y$ curves for soft clay (after API RP2A[7.39]). (a) Short-term static loading. (b) Cyclic loading.

p = soil reaction per unit length of pile,
$p_u 1$ = ultimate soil reaction per unit length of pile,
x = depth below ground level,
x_R = transition depth (at which failure mode changes from wedge type to flow round),
y = lateral deflexion of pile,
y_c = $2.5E_{50}D$,
D = pile width,
ϵ_{50} = strain occurring at half the maximum stress on laboratory undrained compression tests.

effort has been put into the refinement of $p–y$ curve formulations on the basis of measurement of the behaviour of laterally loaded piles. As a result such formulations are widely accepted as being reliable and they are quoted in documents such as the American Petroleum Institute Code RP2A.[7.51] In the case of piling for offshore structures cyclic loading effects due to wave action are important and reference should be made to this Code for the recommended manner of constructing $p–y$ curves to be used in design for these conditions. The forms of $p–y$ curves for both short-term static loading and cyclic loading in clay are shown in Fig. 7.40.

Elastic continuum model In his development of the elastic continuum model Poulos[7.52] considered the pile as an infinitely thin strip having width and flexural stiffness equal to those of the prototype pile. Results were presented in terms of a dimensionless factor K_R, defined as

$$K_R = \frac{E_p I_p}{E_s L^4} \qquad (7.31)$$

where

$E_p I_p$ = flexural stiffness of the pile,
E_s = Young's modulus of the medium,
L = embedded length of the pile.

Poulos's method was based on the integral equation (or boundary element) method and was subsequently extended and enhanced by Poulos and by others.[7.53] An

alternative approach involving the finite element method was adopted by Randolph,[7.54] whose curves relating lateral displacement, lateral force and moment loadings are shown in Fig. 7.41. In these the following notation applies:

u = lateral displacement at ground level,
z = depth below ground level,
H = lateral load applied at ground level,
M = bending moment in pile,
r_0 = radius of pile,
E_p' = effective Young's modulus of a solid circular pile of radius r_0, i.e. $4E_p I_p / \pi r_0^4$
G_c = characteristic modulus of soil, i.e. the average value of G^* over depths l_c
$G^* = G(1 + 3v/4)$,
G = shear modulus of soil,
v = Poisson's ratio of soil,
l_c = critical length of pile

$\quad = 2r_0(E_p'/G)^{2/7}$ for homogeneous soil

$\quad = 2r_0(E_p'/m^*r_0)^{2/9}$ for soil with stiffness proportional to depth,

$m^* = m(1 + 3v/4)$,

$m = \dfrac{G}{z}$, where G varies with depth as $G = mz$,

ρ_c = homogeneity factor = $\dfrac{G^* \text{ at } l_c/4}{G^* \text{ at } l_c/2}$.

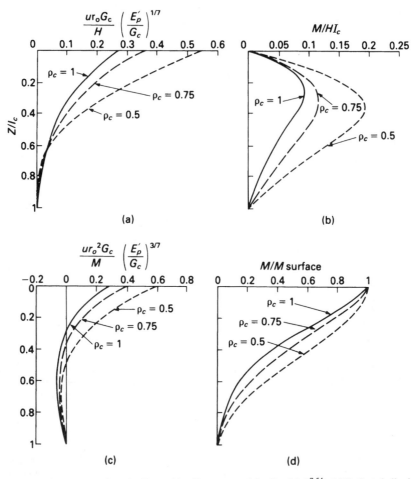

Fig. 7.41 Isolated pile subjected to lateral force loading and bending moment (after Randolph[7.54]). (*a*) Deflected pile shape for lateral force loading. (*b*) Bending moment profile for lateral force loading. (*c*) Deflected pile shape for moment loading. (*d*) Bending moment profile for moment loading.

Randolph's results are quoted for what he terms 'flexible piles' which are piles having a penetration greater than l_c. Although both Poulos and Randolph made use of computers in carrying out their analyses, their results are presented in the form of charts or generalized expressions which the engineer can apply in longhand methods of calculation for isolated vertical piles. The limitations of the elastic continuum approach must be recognized by the engineer. These are:

(a) actual soil conditions must be such as can be reasonably represented by the idealized forms of either stiffness constant with depth of stiffness proportional to depth;
(b) strain levels in the soil sufficiently low to be considered within the elastic range.

Of these, limitation (a) may not be too onerous as the behaviour of a pile under lateral loading is dominated by

soil properties within a comparatively shallow zone, say of the order of 10 pile diameters from the surface.

7.17.3 Pile groups

In the group of vertical piles shown in Fig. 7.42, the vertical component (V) of the load on any pile from the resultant load (R) on the group is given by

$$V = \frac{W}{n} + \frac{We\bar{x}}{\Sigma x^2},\qquad(7.32)$$

where

W = total vertical load on pile group,
n = number of piles in group,
e = distance between point of intersection of resultant of vertical and horizontal loading with underside of pile cap, and neutral axis of pile group,

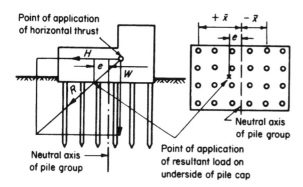

Fig. 7.42

$\bar{x}$ = distance between pile and neutral axis of
pile group; $\bar{x}$ is positive when measured in
the same direction as e and negative when
in the opposite direction.

The above equation is applicable only when the resultant
load (R) cuts the underside of the pile cap.

Computerized methods for calculating the deflexions of
pile groups subjected to horizontal loading are described
in Section 7.19.3.

7.18 The behaviour of piles under vibrating loads

Piles in clays are not liable to settlement when carrying
vibrating loads, other than that due to the normal consolidation of the soil under the working load on the piles.
However, there is a likelihood of severe settlement when
piles carrying vibrating loads are bearing on loose or
medium-dense sands. The effect of the machinery vibrations is to reduce the skin friction around the pile, thus
transferring load to the toe. The vibrating load on the toe
causes further compaction of the sand below the toe with
consequent settlement. The risks of settlement are much
greater in the case of bored and cast-in-place piles than
with driven piles, since the former method does not
compact the soil. Indeed, the effect of boring the hole for
the pile may result in loosening of the sand.

7.19 Computerized methods for predicting the load/deformation behaviour of the single pile and pile groups under axial and lateral loading

7.19.1 Axially loaded single piles

Three approaches may be considered for predicting the
settlement of the single pile under axial load. These are:

(1) Elastic methods based on Mindlin's equations for the
effects of subsurface loading in a semi-infinite elastic
medium.
(2) The t–z method.
(3) Finite element methods.

Elastic methods   These are discussed by Poulos and
Davis.[7.55] Generally the method involves dividing the pile
into a series of segments, each of which is considered to
be uniformly loaded by skin friction from the surrounding
soil. Mindlin's equations are used to express displacements in the soil mass due to loading at each of the
segments. Solutions are obtained which give compatibility
between the soil loads and displacements, and the axial
forces and displacements in the pile. Poulos and Davis
describe investigations into both the 'floating' pile and the
pile bearing on a stiffer stratum, and they also describe a
modified analysis which includes representation of pile–
soil slip occurring when shear stress at the surface of a pile
segment reaches a defined failure value.

Although the various elastic methods typically employ
computer-based techniques the results obtained are
usually presented in chart or tabular form for direct
application by the user. Subject to the applicability of
features such as pile–soil slip, elastic analyses are limited
to circumstances in which non-linear load deflexion
behaviour of the soil is not significant.

The T–z method   This is probably the most widely
adopted technique for investigation of the single axially
loaded pile in cases where non-linear soil behaviour has to
be considered and/or soil stratification is complicated. The
method involves modelling the pile as a member supported by discrete springs which represent the resistance
of the soil in skin friction and in end-bearing, see Fig.
7.43. In the usual approach the internal axial forces in the
pile are described by finite difference expressions in terms
of axial displacements at equally spaced nodes along the
pile. Likewise soil forces and/or externally applied loads
are concentrated at the nodes. Full details are given by
Meyer et al.[7.56]

An alternative approach is to use a finite element
representation of the pile; this can accommodate different
sizes of pile element and varying pile properties much
more readily than can the finite difference method. The
soil springs are non-linear representations of soil force, t,
at displacement, z, as shown in Fig. 7.43. Vijayvergiya[7.57]
proposed a form of t–z curve applicable to both side and
end springs. Vijayergiya's recommendations were based
on the results of tests on piles generally having a diameter
no greater than 0.6 m. It is considered that modifications
need to be made in defining soil springs for piles of
diameter 1.0 m and greater, otherwise the springs will be

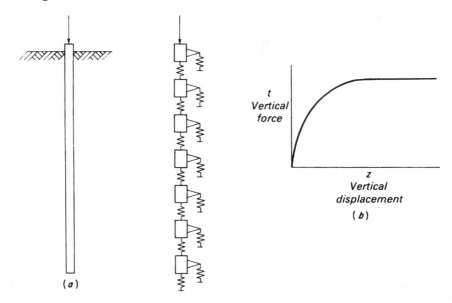

Fig. 7.43 Analysis of pile under vertical loading. (*a*) Continuous pile under vertical loading modelled as a system of rigid elements connected by springs. Soil resistance modelled as external non-linear springs. (*b*) Force-displacement curve for side-spring representing soil resistance. Curve for end-bearing spring similar (*t–z* analysis).

too stiff.[7.58] Efficient adoption of the *t–z* method for all but the most simple representations requires the use of a computer program. Typical results are shown in Fig. 7.44. The method can be extended to allow for both inelastic behaviour and strength degradation of the soil.

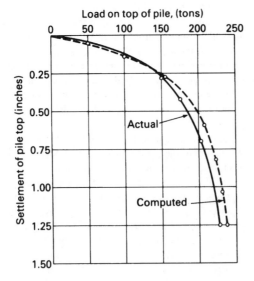

Fig. 7.44 Computed and actual load/settlement curves for 400 mm pipe pile in medium dense sand (after Vijayvergiya[7.57]).

Finite element methods   The finite element analysis of a single axially loaded pile involved idealizing not only the pile but also the soil in which it is embedded by finite elements, as shown in Fig. 7.45. The power of the method lies in its capability to model complicated conditions and to represent non-linear stress–strain behaviour of the soil over the whole zone of soil modelled. Use of computer programs is essential and the method is more suited to research or investigation of particularly complex problems than to general design.

7.19.2 Axially loaded pile groups

The elastic continuum approach previously described for the single pile under axial loading has been extended by various investigators to cater for groups of vertical piles.[7.55, 7.60] Mindlin's equations are used to obtain values for displacements experienced by one pile due to another and adjacent loaded pile. Results are typically expressed in terms of an interaction factor, α, defined

$$\alpha = \frac{\text{Additional settlement caused by adjacent pile}}{\text{Settlement of pile under its own load}},$$

where the pile and the adjacent pile both carry the same load.

Two idealized cases of pile group behaviour may be considered. First, the load on each pile is known at the outset – this applies when there is no effective structural

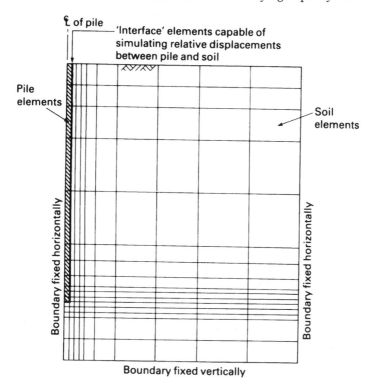

Fig. 7.45   Axisymmetric finite element model of axially loaded pile embedded in soil (after Desai and Holloway[7.59]).

connection between pile heads, and secondly a common value of settlement is imposed on all piles in the group, i.e. the case of a rigid pile cap, for which loading will vary from pile to pile (given that the group is not symmetrical with all piles equidistant from the centroid of the group).

Results are presented in chart and graphical form for a variety of conditions, e.g. piles with enlarged bases, piles of different sizes, variations in soil stiffness. Thus, although there has been extensive computer work leading to the production of the charts and tables, the user may be able to undertake design or analysis by hand.

The non-linear $t-z$ method cannot of itself be used to represent more than single pile behaviour. In circumstances where it is considered that the extent of non-linear behaviour invalidates a purely elastic solution, the most usually adopted approach is to express the displacement of any pile in a group as follows:

$$\rho_{Gi} = \rho_{Ii} + Y_L \sum_{\substack{j=1 \\ j \neq i}}^{n} P_j \, \alpha_{ij} \qquad (7.33)$$

where

    ρ_{Gi} = group displacement of pile i,
    ρ_{Ii} = isolated pile displacement of pile i (by the $t-z$ method),

    Y_L = unit elastic axial displacement of isolated pile,[7.55]
    P_j = axial load on pile j,
    α_{ij} = Poulos interaction factor for piles i and j,
    n = number of piles in group.

This approach is the direct equivalent of that proposed by Focht and Koch[7.61] for analysis of groups of laterally loaded piles and described in the subsequent section.

7.19.3 Laterally loaded pile groups

The reader is referred to Poulos and Davis[7.55] and Poulos[7.62] for descriptions of analysis of the behaviour of laterally loaded pile groups in an elastic medium. Additionally, Randolph[7.54] gives expressions for interaction factor between adjacent laterally loaded piles and compares answers obtained with Poulos's values and with the results of model tests. Where purely elastic behaviour is considered to be unrepresentative of anticipated prototype performance, a technique such as that described by Focht and Koch[7.61] is most often used. They propose that lateral displacement of a pile within a group can be expressed as

$$\rho_{Gi} = \rho_{Ii} + Y_L \sum_{\substack{j=1 \\ j \neq i}}^{n} P_j \, \alpha_{ij}, \qquad (7.34)$$

where

ρ_{Gi} = group displacement of pile i,

ρ_{Ii} = isolated pile displacement of pile i (by the $p-y$ method),

Y_L = unit elastic lateral displacement of isolated pile[7.62],

P_j = lateral load on pile j,

α_{ij} = Poulos interaction factor for piles i and j,

n = number of piles in group.

For the case of a pile group where all the pile heads are connected and undergo the same translation equation (7.34) can be rewritten as

$$\rho_{Gi} = f(P_i) + Y_L \sum_{\substack{j=1 \\ j \neq i}}^{n} P_j \, \alpha_{ij}, \qquad (7.35)$$

where $\rho_{Ii} = f(P_i)$ is the load deflexion curve for an isolated pile obtained by the $p-y$ method using a computer program of the type described by Poulos,[7.62] Randolph[7.54] or Sawko.[7.63] Hence for a prescribed displacement, n simultaneous equations result containing n unknowns (the individual pile loads). The sum of these individual loads is equal to the total group lateral load. Repetition of this

computation for a number of prescribed displacements enables a curve of total group lateral load versus lateral displacement to be drawn.

It should be noted that the value of interaction factor α_{ij} is dependent on direction of lateral loading, being greatest when the loading is along a line joining i and j and least when the loading is at right angles to this line (Fig. 7.46). The consequence of this is that group load versus group deflexion curves will differ depending on the direction of loading except for the case of a large number of piles installed at equal spacing along the perimeter of a circle. Thus far the technique has resulted in a group load versus group deflection curve and, by solution of equation (7.35), values of shear load at the heads of individual piles in the group – hence the most heavily loaded pile, or piles, in the group can be identified and the associated maximum value of shear defined for a specific total lateral load applied to the group.

To obtain the distribution of bending moment in the most heavily loaded pile the single pile computer program[7.62, 7.54, 7.63] is run a number of times, for the maximum value of shear, with the values of y in the $p-y$ curves being progressively increased through multiplication by, say, 2, 3, 4, 5, 6, etc. A curve of pile head

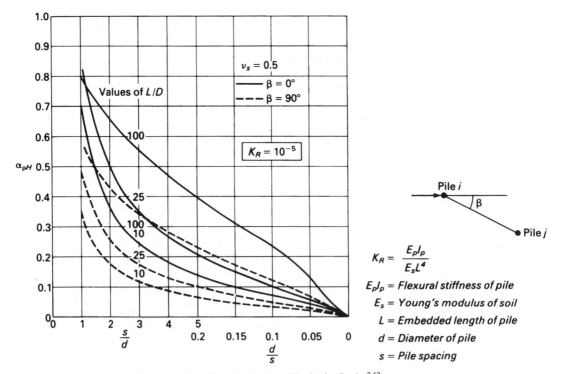

Fig. 7.46 Interaction factors for free-head piles subjected to horizontal load (after Poulos[7.62]).

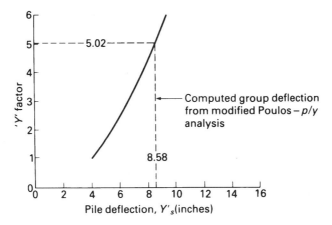

Fig. 7.47   *y*-factor assessment (after Focht and Koch[7.61]).

displacement versus the *y*-factor (also known as the *y*-modifier) can then be plotted. From this curve the value of *y*-factor to give a pile head displacement equal to the previously computed group displacement can be identified, as shown in Fig. 7.47. A further run of the program using the identified value of *y*-factor and an imposed pile head deflection equal to the group deflection will give a bending moment distribution which Focht and Koch[7.61] consider to be valid for the design of the pile (Fig. 7.48).

7.19.4 Pile groups under combined loading

A particular case in which a routine computing technique may be applied to analysis of a pile group under general

loading is shown in Fig. 7.49. This represents a pile group including both vertical and raking piles which are installed through a comparatively shallow stratum of overburden soil to develop end-bearing at rock level. The presence of the raking piles will restrict lateral pile deflections within the overburden to very small values. It is thus reasonable to analyse the group under the action of a combination of lateral, vertical, and moment forces (*H*, *V*, and *M* respectively) using a standard structural frame analysis computer program with the pile heads and toes being considered as pinned or fixed depending on the detail of pile to pile cap connection and the extent to which the piles are socketed into the rock.

The results obtained will be conservative as lateral

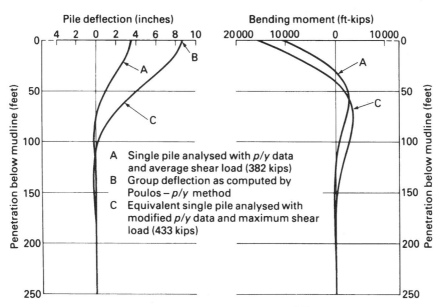

Fig. 7.48   Design bending-moment assessment for pile in group (after Focht and Koch[7.61]).

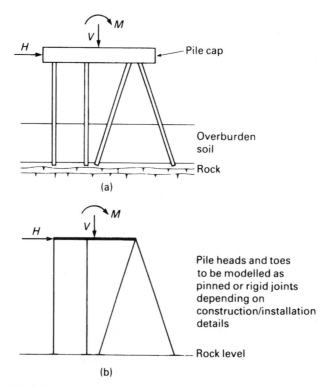

Fig. 7.49   Modelling of forces and moments on a jetty structure.

support, albeit small, which the overburden soil will provide to the piles will have been discounted. As an alternative to the use of a standard frame analysis program, Sawko[7.63] has described a program which would model the two-dimensional problem shown in Fig. 7.49 with the pile cap being considered to be fully rigid. This approach can readily be extended to the three-dimensional case.[7.64]

Most general-purpose frame analysis programs have a facility for modelling discrete linear springs attached to nodal points on the frame members. It is of course possible to introduce such springs to represent both lateral and axial resistance of the soil in an analysis of a structure such as that shown in Fig. 7.49; in such an instance these springs would be linear $p-y$ and $t-z$ curves. This option could be extended further to model cases in which pile displacements, both lateral and axial, within the soil were significant; for example, piles not bearing on an assumed rigid stratum, or vertical piles subjected to lateral loading. The limitations to such an extension are:

(a)   Soil strains should be within those which can be considered to be the elastic range;
(b)   Pile spacings should be sufficiently large to make negligible pile-to-pile interaction effects;

(c)   There may be a restriction on the number of springs that can be used in the analysis by the program.

The generalized cases of groups of piles subjected to combinations of loads in circumstances where interaction effects cannot be ignored have to be analysed by special-purpose computer programs.

A number of such programs have been developed, these include:

Program	Source	Availability
SW Pile	Midland Road Construction Unit Warwickshire County Council	Ove Arup & Partners, Warwick
Minipoint	Department of Transport Highway Engineering Computer Branch	Various bureaux
PGROUP	As above	Various bureaux
M-PILE	As above	Various bureaux
LAWPILE	Wood	London United Computing Systems
DEFPIG	Poulos	University of Sydney, Engineering Laboratory

Programs such as these can differ considerably in their governing assumptions which may lead to one or more restrictions such as:

(a) Only two-dimensional cases modelled
(b) Only elastic cases modelled
(c) Only vertical piles modelled

The reader requiring details is directed to the bureaux quoted.

7.20 Calculation of carrying capacity and pile driveability by dynamic formulae

7.20.1 Calculations of carrying capacity

All dynamic pile formulae are based on the principle that the resistance of piles to further penetration under their working load has a direct relationship to their resistance to the impact of the hammer when they are being driven. Thus, dynamic formulae take into account the weight and height of drop of the hammer, the weight of the pile, and the penetration of the pile under each blow. Refined variants of the formulae also take into account losses of energy due to elastic compression of the pile, the helmet, the packing, the ground surrounding the pile, and losses due to the inertia of the pile. The protagonists of dynamic formulae claim that they have a sound theoretical basis since they are related to Newtonian impact theory. If this were so there would not be the great multiplicity of formulae all giving different answers for the same conditions, and it would not be necessary to introduce so many empirical constants. The physical characteristics of the soil do not appear in the simple dynamic formulae. This can lead to dangerous misinterpretation of the results of dynamic formulae calculations since they represent conditions at the time of driving. They do not take into account the soil conditions which affect the long-term carrying capacity and settlement of piles, such as effects of remoulding and reconsolidation, negative skin friction, and group effects. For example, a dynamic pile formula might show a low value of resistance for a long pile driven deep into a sensitive clay. However, tests have shown that the carrying capacity in these conditions increases progressively for some months after driving. It is unusual to undertake redriving tests a month after first driving piles, and even then the results of redriving tests might not be capable of interpretation since the driving resistance again falls to a low value after only a few blows of the hammer. The time effects on piles driven into fine silty sands have been mentioned in Section 7.4.3. Again, these are not taken into account in the simple formulae, but the development of techniques based on the wave equation can simulate these effects as described below.

7.20.2 Calculations for pile drivability

Traditional dynamic formulae, such as Hiley or *Engineering News Record*, for assessing the driving resistance of piles are based on the Newtonian principles of impact between rigid bodies. However, due to the axial flexibility of a pile there will in reality be a time-lag between the occurrence of a hammer blow at the pile head and the arrival at the pile toe of the resulting compressive stress wave. To cater for this, and for various other factors difficult to represent within a manual calculation, such dynamic formulae involve a number of sweeping assumptions and empirical allowances, and the success of their application has very often depended on the subjective experience of engineers using them. The ability to analyse the behaviour of a pile during driving in a more realistic manner had to await the general availability of computers.

In 1960 Smith[7.65] proposed an idealization of the hammer–pile–soil system which would be capable of representing the passage of the stress wave down the pile (Fig. 7.50). The pile is modelled as a series of rigid masses connected by springs which can act in both compression and tension. The hammer ram and the anvil or pile cap are also modelled as rigid masses, but interfaces between ram and anvil and between anvil and top of pile are idealized by springs capable of sustaining compression but not tension. Skin friction and end-bearing actions of the soil into which the pile is being driven are represented by bilinear springs acting on the embedded rigid masses. Skin friction springs can exert forces either upward or downward while the end-bearing spring can act only in compression. Soil springs act linearly up to a limiting displacement termed the 'quake'. In addition, increased resistance of the soil due to viscous damping is represented by a series of dashpots. Smith expresses the instantaneous soil resistance force, R, acting on an adjacent rigid mass as

$$R = R_s (1 + JV), \tag{7.36}$$

where

R_s = static soil resistance,
J = a damping constant,
V = instantaneous velocity of the adjacent mass.

The analysis proceeds incrementally over a number of discrete time steps, each Δt in duration. The starting condition is defined by the velocity, V_1, of the ram, M_1, at the moment of impact with the anvil. During the first time increment it is assumed that the ram travels a distance, $V_1 \Delta t$ and compresses spring K_1 by the same amount. The resulting force in K_1 acts to decelerate the downward motion of the ram and to accelerate the anvil mass M_2 downward. From these accelerations the net velocities and

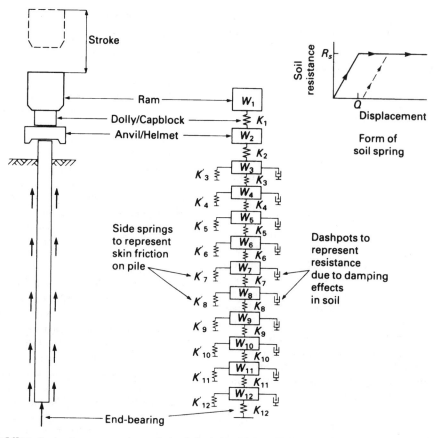

Fig. 7.50 Smith[7.65] idealization for wave equation analysis of pile during driving. (The weights W as shown above are referred to as masses M in the text.)

net displacements of M_1 and M_2 at the end of the second time increment are calculated. This process continues with velocities and displacements of all the masses being monitored at the end of each time increment. These displacements serve to define forces in the springs above and/or below individual masses. For those masses below the ground surface the displacements will define static components of force in the soil springs while the velocities are used to calculate the additional force due to the viscous damping of the soil. The net value of forces on a mass defines its acceleration for the subsequent time increment. The analysis is terminated when the toe of the pile ceases to move downward, and at this stage the predicted set is typically derived by subtracting the toe quake value from the maximum downward displacement of the toe. Permanent set will only be indicated if the maximum toe displacement exceeds the toe quake value. As a typical analysis may require consideration of the behaviour of 30–50 masses over more than 150 time increments, it will be appreciated that use of a computer is a practical necessity.

Although other forms of numerical analysis of the pile driving problem have been proposed, including the use of finite element techniques,[7.66] the approach described above remains by far the most widely used. For offshore and marine works piling it has frequently been adopted in conjunction with load testing of test piles as a means of verifying, during the course of driving, the eventual static capacity of the working piles. This technique is most successful in cases where ground conditions are such that there is no, or very little, change with time of the capacity of the pile after completion of driving, i.e. for sandy soils other than those in a very dense condition. With very dense sand there may be a reduction in capacity after completion of driving, in normally consolidated or lightly overconsolidated clays increases in resistance will occur after completion of driving (the set-up phenomenon), while for heavily overconsolidated clays there will be a tendency to reduction of resistance after driving.

As can be seen in Fig. 7.50, the basic Smith idealization is representative of a pile driven by a drop hammer or a single acting hammer. Diesel hammers have to be

Pile: 914 mm OD × 19 mm WT (Lengths > 15 metres)

Hammer: Delmag D62; Energy range 11 160 kg m – 22 320 kg m

⊙ Test pile 'A' – Predicted resistance for energy and set at completion of
   driving

⊡ Test pile 'A' – Failure load (from load test) approx. 600 tonne

▲ Test pile 'B' – Predicted resistance, based on test pile 'A' results, for energy
   and set at completion of driving (results of load test to 600
   tonne indicated resistance well above 600 tonne)

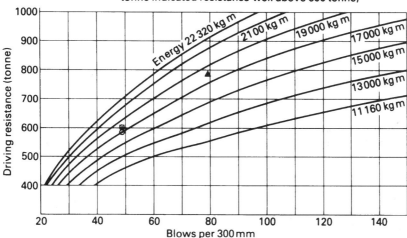

Fig. 7.51 Pile drivability analysis for diesel hammer.

considered in a rather different manner since the actual energy generated by such a hammer will vary with the resistance of the pile that is being driven. For low resistance there will be low energies per blow at a high number of blows per minute, while for high resistances the energy per blow will increase and the number of blows per minute decreases. Manufacturers can provide charts of energy versus rate of striking for diesel hammers, and when drivability predictions are being made the range of energies which a particular hammer may operate at should be considered. An example of this is shown in Fig. 7.51 in which back analyses of static load tests have been used to generate blow count versus driving resistance curves for closed-ended tubular steel piles being driven into calcareous sands and corals. In these ground conditions time-dependent effects are negligible, and so the driving resistance is considered to be the same as the static resistance. Although the original Smith idealization is incapable of modelling the interaction between driving resistance and energy generated for a diesel hammer, advances have been made. In 1976 Goble and Rausche[7.67] published details of a computer program which will model diesel hammer behaviour realistically; this program proceeds by iterations until compatibility is obtained

between the pile–soil system and the energy–blows per minute performance of the hammer.

Data required to undertake a pile drivability analysis include values for the stiffness and coefficient of restitution of the dolly–capblock and the quakes Q and damping constants J for the soil. Typical values of Q and J are given in Table 7.6. These may be used in predriving studies, but it must be remembered that they can be subject to significant variations and whenever possible they should be backfigured to fit load test results. Since the late 1970s the exact form of damping of soil for wave equation analyses has been the subject of much investigation and Heerema[7.68,7.69] and Litkouhi and Poskitt[7.70] have published findings indicating that for end-

Table 7.6 Soil property values typically used in wave equation analyses

Soil type	Damping constant		Quake side and end Q (mm)
	Skin friction J_s (s/m)	End bearing J_p (s/m)	
Clay	0.65	0.01–1.0	2.5
Sand	0.15	0.33–0.65	2.5
Silt	0.33–0.5	0.33–1.5	2.5

bearing in clays and sands and for skin friction in clays, damping is dependent on the fifth root of velocity, i.e. equation (7.36) is modified to

$$R = R_s(1 + JV^{0.2}). \qquad (7.37)$$

A method of computing driving resistances from field instrumentation readings has been described by Goble and Rausche[7.71]. Accelerometers and strain transducers are fitted near to the head of a pile and the readings from these during the course of a hammer blow are processed to give plots of force and velocity versus time as shown in Fig. 7.52. The second stage of the method involves running a wave equation analysis, but with only the pile modelled from the instrument location downward. Values of soil resistance, quake, and damping are assigned, and the recorded time-varying velocity is applied as the boundary condition at the top of the pile model. The analysis will generate a force–time plot for the instrument location and this is compared with the recorded force–time plot. Adjustments are made to the values of resistance, quake and damping, and further analyses run until agreement between the computed and recorded force–time plots is acceptable, at which stage the total soil resistance assigned in the analysis can be taken to be the resistance at time of driving. This method can be expected to give reliable assessments of static resistance for soils where time effects are negligible. It has also been claimed that the method can be used to compute the static resistance of piles in clay soils[7.72] if the analysis is made for the first few blows of a redrive after the pile has been left to set up for a time.

A frequent cause of concern in analysing the drivability of tubular piles driven with open ends is whether or not they will plug during driving. Under static loading an open-ended pile will plug if internal skin friction between the inner wall of the pile and the contained soil column exceeds the bearing resistance that the soil at the toe of the pile can exert to support the soil column. However, during driving plugging is much less likely and the soil column will either not move down at all or will move down only to a limited extent and at a slower rate than that of pile penetration. Heerema and de Jong[7.73] have proposed an extension to the basic lumped mass and spring model by including masses and springs representing the contained soil column and have used this to demonstrate the likelihood that open-ended piles will not plug or will only partially plug during driving. In assessing drivability of such piles it is advisable to consider a range of resistances at driving, catering for either full plugging or no plugging.[7.74]

The time dependency of the capacity of piles driven into clays has already been mentioned. A valuable approach to assessing the driving resistance for such soils has been given by Semple and Gemeinhardt[7.75] who relate static capacity, as defined by the American Petroleum Institute Code API RP2A (11th edition), to driving resistance at increments of depth by a factor which is dependent on the overconsolidation ratio. Figure 7.53 shows results of the technique applied to piling for a North Sea platform.

A further example is given in Fig. 7.54 which shows driving records of 762 mm OD, 19 mm wall thickness open-ended vertical tubular steel piles driven by a Delmag D46-02 diesel hammer on a marine project. Borehole logs indicated soft sea-bed sediments overlying a layer of gravel between project datum levels of −19 and −21 m approximately. Beneath the gravel stiff clay was shown. The piles were initially driven to the top of the gravel by use of a small vibrating hammer, following which

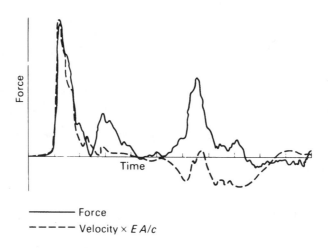

Fig. 7.52 Example of force and velocity record near to head of pile during driving (after Goble and Rausche[7.71]).

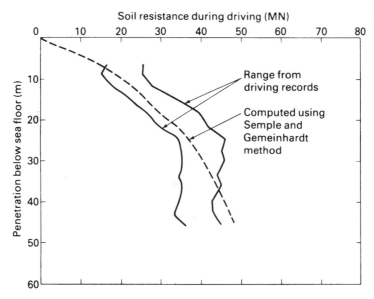

Fig. 7.53 Comparison of observed and computed driving resistance for piling to North Sea platform (after Semple and Gemeinhardt[7.75]).

installation was completed using the diesel hammer. The Semple and Gemeinhardt method[7.75] was used to obtain predictions of blow count versus depth for initial drive and redrive cases; initial drive representing behaviour during continuous driving and redrive representing behaviour immediately following an extended break in driving. The driving records show the blow count diminishing as the toe of the pile passes the −20 m level and begins to punch out of the gravel into the clay.

Figure 7.54(a) gives three records typical of piles which were installed in one continuous drive. It can be seen that there was a tendency for blow counts to be below the

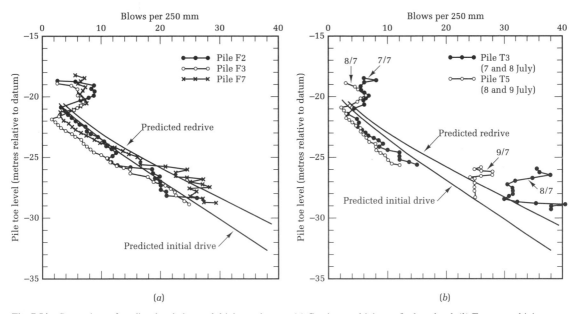

Fig. 7.54 Comparison of predicted and observed driving resistance. (a) Continuous driving to final toe level. (b) Two-stage driving to final toe level.

initial drive prediction in the upper levels of the clay down to -24 m to -26 m. However, below this penetration blow counts tended to be close to the initial drive prediction and in some cases, such as pile F7, approached or exceeded the redrive prediction. Figure 7.54(*b*) shows records for two piles which were partly driven then left until the following day. The first stage of driving, down to -25.5 m, was similar to that for the piles of Fig. 7.54(*a*); however, on start of redriving the blow counts were greater than anticipated by the predicted redrive curve – particularly so in the case of pile T3. In planning for pile driving in soils which set up, it is important to ensure that, as in this case, the hammer selected has sufficient reserve of energy to overcome the set-up resistance and start the pile penetrating once more: this is vital in the case of piles subject to tension loads which may set up before reaching a penetration adequate to ensure the necessary factor of safety against pull-out.

7.21 Procedure in correlating static methods of calculating pile resistance with driving records

The following steps should be followed at the design test piling and permanent piling stages.

(a) *Design stage.* Calculate the ultimate carrying capacity of the piles or pile group by static methods based on soil conditions as found by borings and laboratory tests, taking into account time effects, group effects, and long-term settlement.

(b) *Test piling stage.* Test piles may be driven and loaded before full-scale construction commences on a site, or the first of the permanent piles may be used as test piles and their length or working load modified in the light of the test results. The driving equipment used for the test piles should be the same as that proposed for the permanent piles. The driving resistance in blows per 0.2–0.5 m should be obtained throughout the full depth of penetration and the results should then be checked with records of adjacent boreholes. This will ensure that the required design depth of penetration into the main bearing stratum has been achieved. The stiffness or density of the bearing stratum can also be related to the driving resistance at any given depth. If possible the driving tests should be made in conjunction with the dynamic analyser incorporating accelerometers and strain transducers installed on the test piles as described in Section 7.20.2. Redrive tests should always be made at not less than 24 hours after completing the initial drive in cohesionless soils, and not less than 7 days in clays.

Loading tests are always justifiable if the number of piles to be driven is large, but the relatively high cost may not be justified for a small number of piles. If possible, test piles should be loaded to failure so that the actual safety factor of the working load may be established. The procedure for making pile driving and loading tests is described in Section 8.7. Pile loading tests are not required when piles are driven to refusal on to strong rock or other hard strata, unless the working loads are to be exceptionally heavy and approach the ultimate bearing capacity of a weak rock or unless the pile itself is suspected to be in a defective condition.

As a result of loading tests the engineer may decide to reduce or increase the depth of penetration of the pile into the bearing stratum.

(c) *Construction stage.* Unless piles are driven to refusal on a hard stratum the driving resistance for the full depth of penetration should be measured on all the permanent piles to check possible changes in level or characteristics of the bearing stratum. The measured driving resistances should be compared to those obtained for the test pile. If they are appreciably lower the piles should be driven deeper until the driving resistances reach the values obtained for the test pile.

7.22 Examples on Chapter 7

Example 7.1 An isolated 350×350 mm reinforced concrete pile in a jetty structure is required to carry a maximum compression load of 400 kN and a net uplift load of 320 kN. The soil conditions shown in Fig. 7.55 consist of a loose to medium-dense saturated fine sand

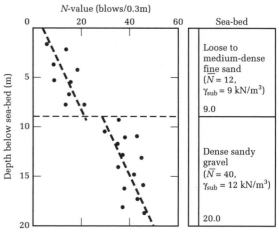

Fig. 7.55

(average $N = 12$ blows per 0.3 m) extending to a depth of 9 m below sea-bed followed by dense overconsolidated sand and gravel (average $N = 40$ blows per 0.3 m). The K_0 value of the dense stratum was established as 2.0. Determine the required depth of penetration of the pile. No erosion is expected.

From Fig. 2.13 for N in upper sand layer of 12

$$\phi' = 31° \qquad K_0 = 1 - \sin 31° = 0.5,$$

and from Table 7.1 $K_s/K_0 = 2$, giving $K_5 = 0.5 \times 2 = 1$. From Table 7.2, $\delta = 0.8 \times 31° = 24.8°$
Effective overburden pressure at 9 m = 9×9
$$= 81 \text{ kN/m}^2$$
From equation (7.5),

q_s at 9 m = $1 \times 81 \times \tan 24.8° = 37 \text{ kN/m}^2$.

Total skin friction 0–9 m is

$$\tfrac{1}{2} \times 37 \times 4 \times 0.35 \times 9 = 233 \text{ kN}.$$

Try penetration of 4 m into dense sand and gravel. For $N = 40$

$$\phi' = 38.5°, \quad K_s = 2 \times 2 = 4, \quad \delta = 0.8 \times 38.5°$$
$$= 30.8°.$$

q_s at 9 m = $4 \times 81 \times \tan 30.8° = 193 \text{ kN/m}^2$.

This exceeds the peak value of 110 kN/m^2, and total skin friction 9–13 m is

$$110 \times 4 \times 0.35 \times 4 = 616 \text{ kN}.$$

Therefore skin friction 0–12 m = $233 + 616 = \underline{849 \text{ kN}}$.

Factor of safety of uplift = $849/320 = 2.6$ which is satisfactory.
For $\phi = 38.5°$ and D/B into dense stratum = $4/0.35 = 11$, Fig. 7.4 gives $N_q = 130$. Effective overburden pressure at 13 m is

$$81 + 12 \times 4 = 129 \text{ kN/m}^2.$$

From equation (7.6),

$$q_b = 129 \times 130 = 16\,800 \text{ kN/m}^2.$$

This exceeds the peak value of 15 000 kN/m^2 (Fig. 7.7). Therefore base resistance at 13 m = $15\,000 \times 0.35^2$ = 1837
Total pile resistance is

$$849 + 1837 = \underline{2686 \text{ KN}}.$$

Factor of safety on compression load = $2686/400 = 6.7$. This is ample and shows that the penetration depth is governed by the uplift resistance.

Using the Eurocode method, the characteristic value q_{sik} for skin friction can be calculated from the average N-values in Fig. 7.55. In the upper sand layer $\tan \delta$ is

reduced by the material factor γ_m of 1.25 (Table 2.2) to give

$$q_{sik} = \frac{1 \times 81 \times \tan 0.8 \times 31°}{1.25 \times 1.5}$$
$$= 20 \text{ kN/m}^2 \text{ at 9 m}.$$

In the dense layer at 9 m

$$q_{sik} = \frac{4 \times 81 \times \tan 0.8 \times 38°}{1.25 \times 1.5} = 101 \text{ kN/m}^2.$$

In the dense layer at 13 m

$$q_{sik} = \frac{4(81 + 4 \times 12) \tan 0.8 \times 38°}{1.25 \times 1.5}$$
$$= 161 \text{ kN/m}^2.$$
Average q_{sik} 9–13 m is

$$(101 + 161)/2 = 131 \text{ kN/m}^2.$$

Therefore take the peak value of 110 kN/m^2 over this depth

$$Q_{sk} \text{ from 0–13 m}$$
$$= \tfrac{1}{2} \times 20 \times 4 \times 0.35 \times 9 + 110 \times 4 \times 0.35 \times 4$$
$$= 742 \text{ kN}.$$

Design uplift capacity is

$$Q_t = \frac{742}{1.3} = \underline{571 \text{ kN}}.$$

This exceeds the applied uplift load of 320 kN, and a smaller penetration into the dense sand could be considered subject to a check on the compression resistance. In view of the high contribution given by the base resistance to the total compression capacity it will be advisable to take a characteristic N-value as the lower bound at 13 m in Fig. 7.55. This shows $N = 34$ and Fig. 2.13 gives $\phi' = 37°$, and factored

$$\phi' = \tan^{-1}\left(\frac{\tan 37°}{1.25}\right) = 31.1°.$$

Figure 7.4 gives $N_q = 30$ for $D/B = 11$
$$q_{bk} = 30 \times (81 + 4 \times 12)/1.5 = 2580 \text{ kN/m}^2$$
and
$$Q_{bk} = 2580 \times 0.35^2 = 316.$$

Design bearing capacity in compression is

$$Q = 316/1.3 + 571 = 814 \text{ kN}$$

which exceeds the applied load of 400 kN. Hence a reduced penetration would be acceptable.

Example 7.2 A borehole and static CPT gave the results shown in Fig. 7.56. An isolated BSP cased pile is required to carry a safe working load of 500 kN in compression.

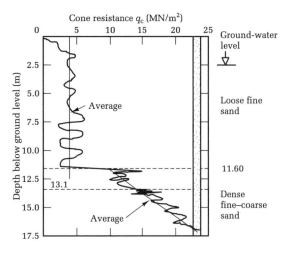

Cone resistance q_c (MN/m^2)

Fig. 7.56

Ground-water level

Loose fine sand

11.60

Dense fine–coarse sand

Determine the required diameter and penetration of the pile. We will adopt a safety factor of 2.5 to ensure that settlements are kept within tolerable limits. Therefore, required ultimate pile resistance is

$$500 \times 2.5 = 1250 \, \text{kN}.$$

A BSP cased pile of 414 mm OD is required to carry a working load of 500 kN without overstressing the concrete filling to the casing tube. The base area of this pile is 0.135 m^2 and the perimeter is 1.30 m. If as a first step the pile is regarded as wholly end-bearing, ultimate end-bearing pressure is

$$\frac{1250}{0.135} = 9300 \, \text{kN/m}^2$$

It will be seen from Fig. 7.56 that we cannot get this end resistance in the loose sand from 0 to 11.6 m below ground level. It will be necessary to drive the pile about 1.5 m into the dense sand layer to ensure that the pile has been properly seated in this layer.

From Fig. 7.56, q_{c-1} over the depth of 0.7 D below the pile toe is 13 and q_{c-2} over the height of 8 D above the toe is 7.5, giving

$$q_b = \frac{13 + 7.5}{2} = 10.25 \, \text{MN/m}^2.$$

This does not exceed the peak value (Fig. 7.7).

From Table 7.3 shaft friction from 0–9 m is

$$0.012 \times 4000 = 48 \, \text{kN/m}^2.$$

From Table 7.3 shaft friction from 11.6 to 13 m is

$$0.012 \times 11 = 132 \, \text{kN/m}^2.$$

Therefore use the upper limit of 120 kN/m^2 below 11.6 m.

Total skin friction 0–13.1 m is

$$(48 \times 11.6 + 120 \times 1.5)1.3 = 958 \, \text{kN}.$$

Total pile resistance is

$$0.135 \times 10.25 \times 10^3 + 958 = 1384 + 958$$
$$= \underline{2342 \, \text{kN}.}$$

Factor of safety on applied load = 2342/500 = 4.7. This is higher than required, but inspection of the q_c/depth curve shows that it would be undesirable to terminate the piles above 13.1 m.

Checking by the Eurocode rules:

$$q_{sik} \text{ from 0 to 11.6 m} = 48/1.5 = 32 \, \text{kN/m}^2,$$

$$q_{sik} \text{ from 11.6 to 13.1 m} = 120/1.5 = 80 \, \text{kN/m}^2.$$

Total shaft friction is

$$Q_{sk} = 1.3 \, (32 \times 11.6 + 80 \times 1.5) = 639 \, \text{kN}.$$

Base resistance is

$$Q_b = (10.25/1.5) \times 0.135 \times 10^3 = 922 \, \text{kN}$$

Design pile capacity is

$$922/1.3 + 639/1.3 = \underline{1200 \, \text{kN}.}$$

This corresponds to a factor of safety of 2.0 on the ultimate pile capacity calculated by the permissible stress method.

Example 7.3 A tubular steel pile 1.22 m diameter × 19 mm wall thickness (net steel area 0.072 m^2) carries an applied load of 7 MN. It is driven with a closed end to a depth of 4.0 m into the dense sand stratum in the soil conditions shown in Fig. 7.56. Calculate the settlement at the pile head.

Checking the factor of safety against failure in end-bearing resistance only, q_{c-1} over a depth of 0.7 × 1.22 m below the toe is 19 MN/m^2, and q_{c-2} over a height of 8 × 1.22 m above the toe is approximately 10 MN/m^2.

Ultimate unit base resistance (10 + 19)/2 = 14.5 MN/m^2. From Fig. 7.7 limiting base resistance = 14 MN/m^2. Ultimate base resistance is

$$14 \times \pi/4 \times 1.22^2 = 16.4 \, \text{MN}.$$

Factor of safety on applied load = 16.4/7 = 2.3, which is satisfactory taking into account the further contribution from shaft friction. From Example 7.2, shaft friction 0–11.6 m is

$$48 \times 11.6 \times \pi \times 1.22 \times 10^{-3} = 2.1 \, \text{MN}.$$

Shaft friction 11.6–15.6 m is

$$0.012 \times \frac{(10 + 19)}{2} \times 4 \times \pi \times 1.22 = 2.7\,\text{MN}.$$

Total shaft friction $= 2.1 + 2.7 = 4.8\,\text{MN}$. This will be fully mobilized with the expected base settlement. Hence load carried by pile base at the working load $= 7 - 4.8 = 2.2\,\text{MN}$.

Effective overburden pressure at 15.6 m is

$$19 \times 2.5 + 9 \times 9.1 + 11 \times 4 = 173\,\text{kN/m}^2.$$

From Fig. 2.30 for q_c at toe of 19 MN/m², E_{50} for soil $= 22\,\text{MN/m}^2$. From equation (7.15) pile head settlement is

$$\frac{(4.8 + 2 \times 2.2)15.6}{2 \times 0.072 \times 2.1 \times 10^5} + \frac{0.5 \times 2.2}{1.22 \times 22}$$

$$= 0.046\,\text{m}$$
$$= 46\,\text{mm}$$

say 50 mm.

Example 7.4   A jetty is sited on 4.5 m of very soft clayey silt overlying stiff to very stiff clay. Tests on undisturbed samples gave the shear strength–depth relationship shown in Fig. 7.57. For structural reasons it is desired to use 500 mm steel tube piles. Determine the required penetration to carry an allowable load of 400 kN.

Because of lateral movements in the jetty structure due to berthing forces, wind, and wave action, it will be advisable to ignore the skin friction in the very soft clayey silt. Take as a first trial a penetration length of 13 m below sea-bed level. From Fig. 7.57, the average shear strength from 5 to 13 m $= 140\,\text{kN/m}^2$. Fissured shear strength at 13 m $= 145\,\text{kN/m}^2$.

It will be seen from Fig. 7.10 that the adhesion factor

corresponding to an average shear strength of 140 kN/m², soft clay over stiff clay, and

$$\frac{L}{B} = \frac{8}{0.5} = 16 \text{ is } 0.50.$$

Therefore, from equation (7.10), total skin friction on portion of shaft in stiff clay is

$$0.50 \times 140 \times 8 \times \pi \times 0.5 = 880\,\text{kN}.$$

From equation (7.9), net end resistance in fissured stiff clay is

$$9 \times 145 \times \tfrac{1}{4}\pi \times 0.5^2 = 256\,\text{kN}.$$

Therefore, total pile resistance is

$$880 + 256 = 1136\,\text{kN}.$$

Safety factor on allowable load of 400 kN is

$$1136/400 = 2.8.$$

This is rather higher than necessary; a safety factor of 2.5 would have been adequate, but before adopting a lesser penetration we must check the safety factor on the lower range of shear strength.

From equation (7.12), and taking the average lower bound shear strength along the pile shaft as 110 kN/m², Allowable pile load

$$= \frac{256}{3} + \frac{0.52 \times 110 \times 8 \times \pi \times 0.5}{1.5}$$

$$= 564\,\text{kN}.$$

This is well in excess of the required safe working load, therefore we could consider reducing the penetration below sea-bed to about 12 m to maintain a safety factor of at least 2.5 on the combined end and skin friction resistance, subject to a loading test.

Checking by Eurocode recommendations, the c_u values are reduced by the material factor γ_m of say 1.5 (Table 2.2). On portion of shaft in stiff clay average c_u is $140/1.5 = 93\,\text{kN/m}^2$. Figure 7.10 gives an adhesion factor of 0.6, and

$$q_{sik} = \frac{0.6 \times 93}{1.5} = 37\,\text{kN/m}^2.$$

Design skin friction is

$$Q_{sk} = 37 \times 8 \times \pi \times 0.5 = 465\,\text{kN}.$$

At pile toe,

$$q_{bk} = \frac{9 \times 145}{1.5 \times 1.5} = 580\,\text{kN/m}^2, \text{ giving}$$

$$Q_{bk} = 580 \times \pi/4 \times 0.5^2 = 114\,\text{kN}.$$

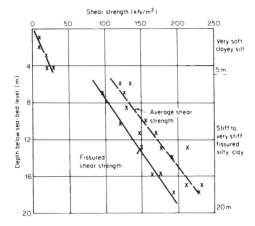

Fig. 7.57

Design pile capacity in compression is

$$114/1.3 + 465/1.3 = \underline{445 \,\text{kN}.}$$

This corresponds to a safety factor of 2.5 on the ultimate pile capacity calculated by the permissible stress method.

Example 7.5 Determine the allowable load on a 305 × 305 × 110 kg/m H-section pile driven through 4 m of loose water-bearing sand to a penetration of 2 m into weathered sandstone and terminating on strong unweathered sandstone with a joint spacing of 450 mm. Tests on cores of the unweathered sandstone showed an average uniaxial compression strength of 65 MN/m². The surface area of the pile is 2.38 m²/m and the cross-sectional area is 0.014 m².

The skin friction mobilized in the loose sand can be neglected. Driving the pile into the weathered sandstone will cause disintegration into the condition of a medium-dense sand with a ϕ' value of, say, 33°. From Section 5.7.1, K_0 is $1 - \sin 33° = 0.45$, and K_s from Table 7.1 is $1 \times 0.45 = 0.45$. At the base of the weathered sandstone, effective overburden pressure is

$$9 \times 4 + 10 \times 2 = 56 \,\text{kN/m}^2$$

q_s at 4 m = $0.45 \times 36 \times \tan(0.7 \times 33°) = 7 \,\text{kN/m}^2$

q_s at 6 m = $0.45 \times 56 \times \tan(0.7 \times 33°) = 11 \,\text{kN/m}^2$

Total Q_s in weathered sandstone is

$$\tfrac{1}{2}(7 + 11) \times 2.38 \times 2 = 43 \,\text{kN}.$$

Take the ϕ-value for the unweathered sandstone from Table 7.4 as 34° and take the cohesion as zero. From equation (7.16)

$$q_{ub} = 2 \times \tan^2(45 + 34/2) \times 65 = 460 \,\text{MN/m}^2.$$

This represents the ultimate bearing capacity of the intact rock. If pile driving were to break up the rock q_{ub} could be as low as 65 MN/m².

From Table 7.5 and an RQD between 70 and 100 per cent (i.e. joints at 450 mm spacing) $c = 0.1 \times 65 = 6.5 \,\text{MN/m}^2$. Take ϕ as before at 34°. From equation (7.16a) (ignoring the second term) and Fig. 7.15

$$q_{ub} = 6.5 \times 15 + 11 \times 2 \times 12 = 361 \,\text{MN/m}^2.$$

The above value represents closed joints in the rock mass. If the pile is not overdriven to break up the rock take q_{ub} conservatively at, say, 200 MN/m², giving a total ultimate pile resistance of

$$200 \times 10^3 \times 0.014 = 2800 \,\text{kN}$$

(the contribution from skin friction is negligible). For a safety factor of 2.5, allowable load = 2800/2.5 = 1120 kN.

The working stress on the steel is

$$1120 \times 10^3/0.014 \times 10^6 = 80 \,\text{N/mm}^2,$$

which is 0.31 × yield stress of 255 N/mm² for mild steel. It will be advisable to reduce the working stress to 0.3 × yield stress and hence the working load to

$$0.3 \times 255 \times 0.014 \times 10^3 = 1070 \,\text{kN}.$$

This complies with BS 8004.

Example 7.6 A 900 mm diameter bored and cast-in-place pile is to be installed through soft superficial soils into a weak closely jointed mudstone. Examination and testing of cores of the rock showed an RQD value of 30 per cent and an average uniaxial compression strength of 4 MN/m². Determine the required depth of the rock socket for an allowable pile load equal to the maximum allowable load on the concrete in the pile shaft.

For 25 N/mm² cube compression strength, allowable load on pile shaft is

$$0.25 \times 25 \times \pi/4 \times 0.9^2 = 4.0 \,\text{MN}.$$

For factor of safety of 2.5, required ultimate load = 10 MN.

Taking conservative value of the bearing capacity factor of 2.5 to allow for clean-out and inspection difficulties, total ultimate base resistance is

$$2.5 \times 4 \times \pi/4 \times 0.9^2 = 6.4 \,\text{MN}.$$

Therefore required ultimate skin friction = 10.0 − 6.4 = 3.6 MN. From Fig. 7.17, for q_{uc} = 4 MN/m², Williams and Pells α is 0.2. From Table 2.12 for RQD of 30 per cent, mass factor is 0.2, from Fig. 7.18, correction factor β is 0.65. Therefore ultimate rock socket skin friction is

$$0.2 \times 0.65 \times 4 = 0.52 \,\text{MN/m}^2.$$

(Checking from Horvath curve in Fig. 7.15, for q_{uc} = 4, α = 0.14.) Taking lower value, required length of rock socket is

$$3.6/0.52 \times \pi \times 0.9 = 2.4 \,\text{m}.$$

Example 7.7 Precast concrete piles of 350 mm square section driven to a penetration of 5 m below ground level are used as vertical king pile anchorages to support a retaining wall where 2 m of soft clayey silt are overlying a stiff overconsolidated clay. Ground-water level is at the base of the clayey silt. Calculate the allowable load which can be applied to the free head of the pile and the deflexion at this load.

The ratio of embedded length to width is 5/0.35 = 14. This is greater than the upper limit of 12 at which the pile no longer behaves as a short rigid unit. However, the

Brinch Hansen method is convenient to use and is applicable to layered soils.

Because of the sustained loading conditions it will be necessary to use effective shear strength parameters which can be taken as follows:

For soft clayey silt $c' = 0$, $\phi' = 20°$;

For stiff clay $c' = 10\,kN/m^2$, $\phi' = 25°$.

The embedded length of the pile is divided into five 1 m deep elements. From Fig. 7.37 the coefficients K_c and K_q and the passive resistance values per metre width of pile are tabulated below:

z (m)	For K_c	0	1	2 (soft)	0 (stiff)	1	2	3
	For K_q	0	1	2 (soft)	2 (stiff)	3	4	5
z/b	For K_c	0	2.9	5.7	0	2.9	5.7	8.6
	For K_q	0	2.9	5.7	5.7	8.6	11.4	14.3
K_c		—	—	—	0	20	25	30
$c'K_c$		0	0	0	0	200	250	300
K_q		2.5	3.5	4	7	7.5	8	9
$p_0(kN/m^2)$		0	16.7	33.3	33.3	44.1	54.9	65.7
p_0K_q		0	58.5	133.2	233.1	330.7	439.2	591.3
$c'K_c+p_0K_q$ $(kN/m^2$		0	58.5	133.2	233.1	530.7	689.2	891.3

The tabulated values of passive resistance are plotted in Fig. 7.58. Assume a centre of rotation X at 4.0 below ground level. Taking moments about the point of application of load (at ground level)

$$\Sigma M = 29.25 \times 1 \times 0.5 + 95.85 \times 1 \times 1.5 + 381.90$$
$$\times 1 \times 2.5 + 609.95 \times 1 \times 3.5 - 790.25 \times 1$$
$$\times 4.5 = -308.10\,kN\,m.$$

Taking as a second trial at centre of rotation at 4.1 m below ground level $\Sigma M = +324.8\,kN\,m$.

Therefore the actual centre of rotation can be taken as 4.05 m below ground level. Taking moments about the centre of rotation,

$$H_u \times 4.05 = 29.25 \times 1 \times 3.55 + 95.85 \times 1 \times 2.55$$
$$+ 381.9 \times 1 \times 1.55 + 609.95 \times 1 \times$$
$$0.55 + 790.25 \times 1 \times 0.45$$
$$= 1631.3\,kN$$

giving $H_u = 402.8\,kN$, which for the 350 mm pile becomes $402.8 \times 0.35 = 141.0\,kN$. This is a large figure for a 350 mm concrete pile and H_u will be governed by the ultimate moment of resistance of the pile. For one 32 mm high yield steel bar in each corner of the pile, 50 mm of cover to 8 mm link steel, concrete with a characteristic strength of 25 N/mm² and steel with a yield strength of

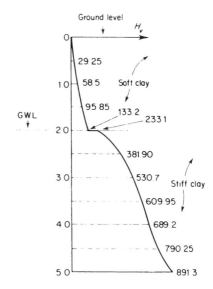

Fig. 7.58

410 N/mm², reference to Chart 22 of CP 110 (Part 2) gives an ultimate resistance moment of $125 \times 10^6\,N\,mm$.

Taking as a trial at point of zero shear at 2.8 m below ground level and resolving the horizontal forces above this point

$$402.8 - (29.25 + 95.85 + 0.8 \times 352.1) = -4\,kN.$$

This is sufficiently close to zero to give the point of zero shear and hence the point of maximum bending moment. Equating the maximum bending moment with the ultimate resistance moment of the pile,

$$H_u \times 1000 \times 2.8 \times 1000 = 125 \times 10^6,$$

therefore

$$H_u = 125/2.8 = 44.6\,kN.$$

For a safety factor of 1.6, design horizontal load at head of pile is

$$44.6/1.6 = 28\,kN.$$

The modulus of elasticity of the pile is $26 \times 10^6\,kN/m^2$, therefore from equation (7.28), deflexion of free head pile is

$$\frac{28(0 + 2.8)^3}{3 \times 26 \times 10^6 \times 0.35^4/12} = 0.006\,m = \underline{6\,mm.}$$

Example 7.8 A multi-storey building is to be constructed on a stiff to very stiff-fissured London Clay. The shear strength–depth relationship for the full depth of the clay can be taken as being the same as that shown in Fig. 7.57, assuming that the average shear strength curve for the stiff to very stiff clay extends up to ground level. Settlement calculations for the whole pile group beneath the building show that the piles must be taken to a depth of at least 15 m below ground level to keep consolidation settlements within tolerable limits. Determine the required dimensions of a large-diameter bored and cast-in-place pile to carry a column load of 2500 kN and estimate the elastic settlement of the working load.

Taking an overall load factor of 2 (equation (7.13a)), then the required ultimate pile load is 5000 kN. As a first trial take a shaft diameter of 1.0 m and a base diameter of 1.8 m. Because the piles are under-reamed the adhesion factor should not exceed 0.3. We must also deduct 3 m from the overall pile length to allow for the possible loss of adhesion over the under-reamed length and for two shaft diameters above. Ultimate shaft load is

$$0.3 \times 112 \times 12 \times \pi \times 1.0 = 1267 \, \text{kN}.$$

The fissured shear strength at 15 m depth is 162 kN/m². Ultimate base resistance is

$$9 \times 162 \times \frac{\pi}{4} \times 1.8^2 = 3710 \, \text{kN}.$$

Therefore total pile resistance is

$$1267 + 3710 = 4977 \, \text{kN}$$

From equation (7.13b), maximum safe load is

$$1267 + 3710/3 = 2500 \, \text{kN}.$$

which is satisfactory provided that settlements are not excessive. The E_u/C_u ratio for London Clay from Fig. 2.34 is about 500 for the expected order of strain. Therefore E_u at pile base is

$$500 \times 162 = 80 \times 10^3 \, \text{kN/m}^2$$

and the working load carried by the pile base is

$$2500 - 1267 = 1233 \, \text{kN}.$$

From equation (7.15) pile head settlement is

$$\frac{(1267 + 2 \times 1233)15}{2 \times 0.785 \times 20 \times 10^6} + \frac{0.5 \times 1233}{1.8 \times 80 \times 10^3}$$

$$= 0.006 \, \text{m}, \underline{\text{say between 5 and 10 mm}}.$$

Checking by the Eurocode recommendations. At 0 m

$$q_{\text{sik}} = \frac{0.3 \times 55}{1.25 \times 1.5} = 9 \, \text{kN/m}^2.$$

At 12 m

$$q_{\text{sik}} = \frac{0.3 \times 170}{1.25 \times 1.5} = 27 \, \text{kN/m}^2$$

$$Q_{\text{sk}} = \frac{1}{2}(9 + 27) \times \pi \times 1 \times 12 = 678 \, \text{kN}.$$

At 15 m

$$q_{\text{bk}} = \frac{9 \times 162}{1.25 \times 1.5} = 778 \, \text{kN}$$

$$Q_{\text{bk}} = 778 \times \pi/4 \times 1.8^2 = 1980 \, \text{kN}.$$

Design bearing capacity is

$$1980/1.6 + 678/1.3 = 1759 \, \text{kN}.$$

This is considerably less than the value of 2500 kN calculated by the permissible stress method. However, for the latter method the ultimate load of 4977 kN was divided by a safety factor of 2. If a factor of 2.5 had been used the allowable load would have been much the same as the design pile capacity given by Eurocode 7.

Example 7.9 Bored piles are required to support a wall footing carrying a uniformly distributed load of 160 kN/m run. The soil conditions below ground level consist of 3.7 m of recently placed loose ash fill overlying medium dense angular well graded sand ($N = 25$ blows per 0.3 m). The density of the ash fill is 11.0 kN/m³. No ground-water level was met in 20 m deep boreholes. Design suitable piled foundations.

It will be reasonable to place the piles in pairs with each pair at 3 m centres giving a working load on each pile of 240 kN.

The piles are too widely spaced for them to act as a group, so it is only necessary to consider the carrying capacity of a single pile. Since the ash fill is loose it is likely to consolidate under its own weight, thus causing drag-down on the pile shaft.

Since withdrawal of the temporary pile casing is likely to loosen the sand, we must take from Table 7.1, for loose conditions,

$$K_s = 0.7 \times (1 - \sin 33°) = 0.32,$$
$$\delta = 1 \times 33° = 33°.$$

Therefore, unit skin friction in sand at base of fill is given by

$$0.32 \times 11 \times 3.7 \times \tan 33° = 8 \, \text{kN/m}^2.$$

If we assume piles to be 9 m long, unit skin friction at toe of piles

$$0.32 (11 \times 3.7 + 19 \times 5.3) \times \tan 33° = 29 \, \text{kN/m}^2.$$

Assuming 400 mm diameter piles will be required, total skin friction in sand is

$$\frac{(8 + 29)}{2} \times 5.3 \times \pi \times 0.4 = 123 \, \text{kN}.$$

We must assume that the boring will loosen the sand

somewhat around the pile toe, giving ϕ say 33°, for which $N_q = 35$ (from curve for $D/B = 20$ in Fig. 7.4). Therefore, from equation (7.6), total resistance of 350 mm pile is

$$35\ (11 \times 3.7 + 19 \times 5.3) \times \pi/4 \times 0.4^2 + 123$$
$$= 622 + 123 = 745\ \text{kN}.$$

Because the fill has recently been placed it will still be settling under its own weight at the time the piles are required to carry the wall loading. Therefore negative skin friction must be added to the pile loads. Because the sand beneath the pile toe is incompressible relative to the fill (i.e. assuming that installation of the pile has not greatly loosened the soil beneath the toe) the distribution of negative skin friction on the pile shaft will be roughly the same as that shown in Fig. 7.21(c). The peak value of skin friction at the base of the fill will be roughly equal to that in the sand at this level.

Therefore if we assume from Fig. 7.21(c) that the negative skin friction acts over the upper 3.3 m of the shaft only then, at 3.3 m,

$$\text{skin friction} = (3.3/3.7) \times 8 = 7\ \text{kN/m}^2$$

and at ground level it is zero. Total negative skin friction on upper 3.3 m of pile shaft is given by

$$\frac{(0 + 7)}{2} \times \pi \times 0.4 \times 3.3 = 15\ \text{kN}.$$

Safety factor on pile is

$$\frac{745}{240 + 15} = 2.9.$$

This is above the desirable safety factor of 2.5.

Therefore, we can support the walls on pairs of 400 mm piles at 3 m centres. The piles can be spaced at three diameters centre to centre, i.e. at 1200 mm centres, requiring a pile cap 1900 mm × 700 mm × 600 mm deep.

Example 7.10 A structure weighing (inclusive of the base slab) 30 000 kN has base dimensions of 10 m square. It is to be constructed on a site where boreholes show firm intact clay (average cohesion at 1.5 m below ground level $= 40\ \text{kN/m}^2$) progressively increasing in shear strength to $240\ \text{kN/m}^2$ at a depth of 30 m. Rock is at a depth of 70 m. Design suitable foundations.

From Fig. 2.15, the ultimate bearing capacity of the firm clay for a foundation 10 m square by say 1.25 m deep is $6.2 \times 40 = 248\ \text{kN/m}^2$. This is less than the bearing pressure of $30\,000/(10 \times 10) = 300\ \text{kN/m}^2$. It will be uneconomical from the structural aspect to increase the size of the foundation, and it will be necessary to take the foundation to a depth of about 15 m before a safety factor of 3 is obtained for a foundation 10 m square. It is clearly a case for piling the foundation. Adopting a pile spacing of

1.2 m centres as a first estimate, we can obtain a pile group eight piles square, i.e. a total of 64 piles giving a working load of 469 kN per pile. It will be desirable to use driven and cast-in-place piles with an enlarged end to take advantage of their higher end resistance since there is no really good bearing stratum of rock or dense gravel within an economical piling depth.

Assume a shaft diameter of 450 mm and an enlarged end of 750 mm diameter. End resistance of piles 18 m long is

$$9 \times 160 \times \frac{\pi}{4} \times 0.75^2 = 636\ \text{kN}.$$

The pile cap will be about 1.25 m thick, giving a shaft length above the enlarged end of

$$18 - (1.25 + 0.75) = 16\ \text{m}.$$

The average shear strength along the shaft is $92\ \text{kN/m}^2$. We do not know the relationship between shear strength and skin friction for driven and cast-in-place piles, but it will certainly not be less than that for bored piles for which skin friction $= 0.45 \times 92 = 41.4\ \text{kN/m}^2$. Therefore, total skin friction on pile shaft is

$$\pi \times 0.45 \times 16 \times 41.4 = 936\ \text{kN}.$$

Total ultimate carrying capacity of a single pile is

$$636 + 936 = 1572\ \text{kN}.$$

Safety factor is

$$\frac{1572}{469} = 3.35.$$

This is somewhat higher than would be required for single piles, but we must consider the group effect. Since the piles are at a greater spacing than $2\frac{1}{4}$ diameters, block failure will not occur.

It will be advisable to check the settlement of the pile group using the charts in Fig. 2.37. From the aspect of settlement the group will behave as a virtual raft foundation at a depth of $\frac{2}{3} \times 16.75 = 11.2$ m below underside of pile cap. Allowing a spread load of 1 in 4 the virtual raft foundation is 14 m square. From Table 2.11 the m_v for the very stiff clay below pile base level will be about $0.03\ \text{m}^2/\text{MN}$.

The bearing pressure below the base of the equivalent raft is $30/14^2 = 0.15\ \text{MN/m}^2$. The drained soil modulus E_d is the reciprocal of m_v giving $E_d = 30\ \text{MN/m}^2$. From Fig. 2.37, for

$$H/B = (70 - 11)/14 = 4.2$$

and

$$D/B = 11.2/14 = 0.8,$$

$\mu_0 = 0.92$ and $\mu_1 = 0.65$, average settlement is

$$\frac{0.92 \times 0.65 \times 0.15 \times 14 \times 1000}{30}$$

$$= 42 \text{ mm, say between 30 and 50 mm.}$$

The alternative of large diameter under-reamed bored piles could be considered for this structure. Taking three rows of three piles at say 4.5 m centres both ways, the working load on each pile will be $30\,000/9 = 3333$ kN.

Take a 1.2 m diameter shaft under-reamed to, say, a 3 m diameter base at a depth of 15 m below ground level. Shear strength of clay at 15 m is 135 kN/m², end resistance of base of piles is

$$9 \times 135 \times \frac{\pi}{4} \times 3^2 = 8588 \text{ kN.}$$

In considering the shaft friction on the piles, we shall have to neglect any friction over the sloping portion, i.e. for a depth of 1.8 m at 2 : 1 slope, and for a distance up the pile shaft at least equal to twice the shaft diameter. Thus we must only consider shaft friction over a length equal to

$$15 - (1.8 + 2.4 + 1.25) = 9.55 \text{ m.}$$

Average shear strength
along shaft $= 78$ kN/m²,
Unit shaft friction on shaft $= 0.30 \times 78$
$= 23.4$ kN/m²,
Total $= 23.4 \times \pi \times 1.2 \times 9.55$
$= 842$ kN,
Total ultimate resistance of
pile $= 8589 + 842$
$= 9431$ kN,
Safety factor $= \dfrac{9431}{3333} = 2.8$,

which is satisfactory. The criterion that the working load should not be greater than the ultimate shaft load plus one-third of the ultimate base load is also satisfied.

Example 7.11 A bridge pier imposing a total load of 22 000 kN has plan dimensions at ground level of 18 × 4.5 m. It is sited on 7.5 m of recently placed loose sand filling ($N = 9$ blows per 0.3 m) overlying 4.5 m of peaty soft clay ($c = 24$ kN/m²) followed by stiff clay ($c = 60$ kN/m² at 12 m below ground level increasing to 300 kN/m² at 25 m below ground level). Ground-water level is 0.6 m below ground level.

This is obviously a case for piling, but because of the 'squeezing' ground from 7.5 to 12 m below ground level and the overlying loose water-bearing sand, this is not a clear case for an *in-situ* type pile. The clay below 12 m is not particularly stiff; therefore it will be desirable to use a pile of fairly large diameter to get the most out of the ground in end resistance and adhesion. A 508 mm OD West's Shell pile has a nominal maximum working load of 800–

1000 kN. We can space the piles at $\pi \times 0.51 = 1.6$ m centre to centre both ways, which permits a total of $13 \times 4 = 52$ piles beneath the 18 × 4.5 m base. The pile cap will be 20 × 5.8 × 1.2 m thick. It will be desirable to avoid excavating below the water table so we will allow the pile cap to project 0.6 m above ground level and we shall have to add the weight of the cap to the working load on the piles.

Calculating total structural load on piles, working load from bridge pier is

$$\frac{22\,000}{52} = 423 \text{ kN per pile.}$$

Load from pile cap is

$$\frac{20 \times 5.8 \times 1.2 \times 24}{52} = 64 \text{ kN per pile.}$$

Weight of soil displaced by pile cap is

$$\frac{20 \times 5.8 \times 0.6 \times 18}{52} = 24 \text{ kN.}$$

Net working load on top of pile shaft is

$$423 + (64 - 24) = \underline{463 \text{ kN.}}$$

Although the sand fill is recently placed it will have consolidated under its own weight, but it will be causing consolidation of the underlying soft peaty clay. Therefore there will be negative skin friction both from the sand filling and from the soft clay. Because the compression of the soft clay is large relative to the yielding of the pile toe, the negative skin friction from the sand will act over the full depth of the layer, and it will be sufficiently accurate to assume that it acts over the full depth of the soft clay layer also. We must therefore add the total weight of these two layers, which is enclosed by the perimeter of the pile group.

Gross weight of soil enclosed by pile group is given by

$$19.71 \times 5.21 \times (8 \times 6.9 + 5 \times 4.5) = 7979 \text{ kN}$$

and weight of soil displaced by piles in group is

$$52 \times \frac{\pi}{4} \times 0.508^2 \times (8 \times 6.9 + 5 \times 4.5)$$

$$= 819 \text{ kN.}$$

Hence, deducting the weight of soil displaced from the gross weight of soil, net weight causing drag-down is 7160 kN. Weight of pile shaft in fill and clay $= 55$ kN. Therefore total load on pile group is

$$52 \times (463 + 55) + 7160 = \underline{34096 \text{ kN.}}$$

From equation (7.23b), drag-down load will not exceed

$$Q_2 = 52 \times 463 + 52 \times \pi \times 0.508 \times (6.9 \times 10 + 4.5 \times 24)$$

= 38 765 kN plus weight of piles.

Total load on pile shaft at surface of stiff clay stratum is given by

$$\frac{34\,096}{52} = 157\,\text{kN}.$$

Calculate the ultimate carrying capacity of a 508 mm pile driven to 20 m below ground level, i.e. to penetration of 8 m into stiff clay:

Shear strength at 20 m below ground level $= 208\,\text{kN/m}^2$, average shear strength along pile shaft is

$$\frac{60\,+\,208}{2} = 134\,\text{kN/m}^2.$$

End resistance of pile is

$$9 \times 208 \times \frac{\pi}{4} \times 0.508^2 = 379\,\text{kN}.$$

From Fig. 7.10, shaft skin friction for c of 134 kN/m^2 and $L/B = 20$ and soft clay overburden $= 0.65 \times 134 = 87\,\text{kN/m}^2$. Total shaft skin friction is

$$\pi \times 0.508 \times 8 \times 87 = 1111\,\text{kN},$$

Ultimate carrying capacity of pile is

$$379 + 1111 = 1490\,\text{kN},$$

Safety factor on structural load is

$$\frac{1490}{463} = \underline{3.2}.$$

Safety factor on total load is

$$\frac{1490 \times 52}{34\,096} = \underline{2.3}.$$

The above safety factors should be satisfactory since the pile group is relatively narrow and the shear strength of the soil is increasing rapidly below the base of the piles; therefore the settlement of the group should be quite small.

However, we must consider the possibility of block failure of the group. The ultimate carrying capacity of a pile group considered as a block foundation is given by equation (7.20), namely

$$
\begin{aligned}
Q_{\ddot{u}} &= 2D(B + L)\bar{c} + 1.3cN_c \times B \times L \\
&= 2 \times 8(19.71 + 5.21) \times 134 + 1.3 \times 208 \times 9 \\
&\quad \times 19.71 \times 5.21 \\
&= 303\,\text{MN approx.}
\end{aligned}
$$

Safety factor against block failure $= 303/34 = 9$, which is amply safe.

Example 7.12 A piled trestle shown in Fig. 7.59 con-

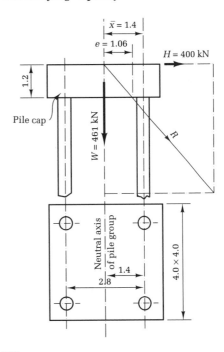

Fig. 7.59

sists of four vertical piles surmounted by a 1.200 m thick pile cap. It carries a horizontal load applied to the surface of the cap of 400 kN. Determine the loads on the piles.

For a pile cap 4.00 m square, weight of pile cap is

$$4 \times 4 \times 1.2 \times 24 = 461\,\text{kN}.$$

Resultant of vertical load of pile cap and horizontal load cuts the underside of the pile at a point 1.06 m from centre of cap. From equation (7.32), vertical component of load on piles is

$$\frac{461}{4} \pm \frac{461 \times 1.06 \times 1.4}{4 \times 1.4^2}$$

$$= 202\,\text{kN maximum and } 28\,\text{kN minimum.}$$

Any tendency to uplift will be counterbalanced by the weight of the pile and the skin friction in the ground.

Example 7.13 A steel jetty pile of hollow tubular section driven through 5.5 m of soft silt on to hard-fissured and broken rock carries an uplift load of 250 kN. Design a suitable anchorage.

We cannot obtain any uplift resistance from the soft silt; therefore we shall have to provide a positive anchorage to the rock. Since the pile is a hollow section we can drill into the rock through the bottom of the pile and provide a steel anchor rod grouted into the rock. The rod should be of deformed section and provided with an end washer.

Thus, required cross-sectional area of rod for $125 \, N/mm^2$ safe stress is

$$\frac{250 \times 1000}{125} = 2000 \, mm^2.$$

Therefore a 50 mm diameter bar will be satisfactory. (Area = $1963 \, mm^2$.)

For allowable steel-to-grout bond stress of $0.7 \, N/mm^2$, required bond length is

$$\frac{250 \times 1000}{\pi \times 50 \times 0.7 \times 1000} = 2.27 \, m.$$

The grout-to-rock bond will not require a greater bond length.

Since the rock is fissured and broken it is possible that failure would occur due to uplift of an inverted cone of rock with a half-angle of, say, 30°. Taking the submerged density of the rock to be $16 \, kN/m^3$, a 3.90 m deep cone has a weight of 265 kN. This gives a safety factor of greater than one against uplift.

Therefore the required length of rod is governed by the depth of 3.90 m required to prevent lifting the rock.

If the length of rod carried up into the pile is set in concrete having a works cube strength of $20 \, N/mm^2$, clause 303(e) of CP 114 gives

Required length

$$= \frac{Bar \ diameter \times Tensile \ stress \ in \ bar}{4 \times Permissible \ average \ bond \ stress}$$

$$= \frac{50 \times 125}{4 \times 0.80 \times 1000}$$

$$= 1.953 \, m, \text{ say } 2.000 \, m.$$

References

7.1 *Specification for Piling*, Institution of Civil Engineers, Thomas Telford, 1988.

7.2 Kulhawy, F. H., Limiting tip and side resistance, fact or fallacy, *Symposium on Analysis and Design of Pile Foundations*, American Society of Civil Engineers, San Francisco, *Proceedings*, 80–98 (1984).

7.3 Vesic, A. S., Tests on instrumented piles, Ogeechee River site, *Proceedings of the American Society of Civil Engineers*, **96** (SM-2), 561–584 (1970).

7.4 Nordlund, R. L., Bearing capacity of piles in cohesionless soils, *Proceedings of the American Society of Civil Engineers*, **89** (SM3), 1–35 (1963).

7.5 Vesic, A. S., Investigations of bearing capacity of piles in sand, *Duke University Soil Mechanics Laboratory Publication No. 3* (1964).

7.6 Berezantsev, V. G., Load-bearing capacity and deformation of piled foundations, in *Proceedings of the 5th International Conference on Soil Mechanics*, Paris, 1961, Vol. 2, pp. 11–12.

7.7 Kishida, H. and Isemoto, N., Behaviour of sand plugs in open-end steel pipe piles, *Proceedings of the 9th International Conference on Soil Mechanics*, Tokyo, 1977, Vol. 1, pp. 601–604.

7.8 Hight, D. W., Private communication.

7.9 *Recommendations for Planning, Designing and Constructing Fixed Offshore Platforms*, RP2A American Petroleum Institute, Washington, DC, 1987.

7.10 Beringen, F. L., Windle, D., and van Hooydonk, W. K., Results of loading tests on driven piles in sand, *Proceedings of the Conference on Recent Developments in the Design and Construction of Piles*, Institution of Civil Engineers, London, 1979, pp. 153–165.

7.11 Author's unpublished records and reports.

7.12 Ishihara, K., Saito, A., Shimmi, Y., Miura, Y., and Tominaga, M., Blast furnace foundations in Japan, *Proceedings of the 9th International Conference on Soil Mechanics*, Tokyo, 1977, Case History Volume, pp. 157–236.

7.12a Visser, M, van der Zwaag, G. L., and Pluimgraaf, D. J. H. M., Application of monitoring systems during pile driving in North Sea sands, *Proceedings of the Conference on the Behaviour of Offshore Structures*, Elsevier, 1985, pp. 623–631.

7.13 Meigh, A. C., *Cone Penetration Testing*, CIRIA–Butterworth, 1987.

7.14 te Kamp, W. C., Sondern end funderingen op palen in zand, *Fugro Sounding Symposium*, Utrecht, 1977.

7.15 Broug, N. W. A., The effect of vertical unloading on cone resistance q_c, a theoretical analysis and a practical confirmation, *Proceedings of the 1st International Geotechnical Seminar on Deep Foundations and Bored and Auger Piles*, Ghent, 1988, Balkema, Rotterdam, 1988, pp. 523–530.

7.16 Terzaghi, K. and Peck, R. B., *Soil Mechanics in Engineering Practice*, 2nd edn, John Wiley, New York, 1967.

7.17 Feld, J., Discussion on session 6, *Proceedings of the 4th International Conference on Soil Mechanics*, London, 1957, Vol. 3, p. 181.

7.18 Togrol, E., Aydinoglu, N., Tugcu, E. K., and Bekaroglu, O., Design and construction of large piles, *Proceedings of the 12th International Conference on Soil Mechanics*, Rio de Janeiro, 1989, Vol. 2, pp. 1067–1072.

7.19 Fleming, W. G. K. and Sliwinski, Z. J., The use and influence of bentonite in bored pile construction, *Construction Industry Research and Information Association Report No. PDP2* (1977).

7.20 Vesic, A. S., Design of pile foundations, *Transportation Research Board Publication No. 42* (1977).

7.21 Tomlinson, M. J., The adhesion of piles in stiff clay, *Construction Industry Research and Information Association Research Report No. 26* (1970).

7.22 Burland, J. B., Shaft friction of piles in clay – a simple fundamental approach, *Ground Engineering*, **6**(3), 30, 32, 37, 38, 41, 42 (1973).

7.23 Meyerhof, G. G. and Murdock, L. J., An investigation of the bearing capacity of some bored and driven piles in London Clay, *Géotechnique*, **3**, 267–282 (1953).

7.24 Ballisager, C. C., Bearing capacity of piles in Aarhus Septarian Clay, *Danish Geotechnical Institute Bulletin No. 7*, pp. 153–173 (1959).

7.25 Randolph, M. F. and Wroth, C. P., Recent developments in understanding the axial capacity of piles in clay, *Ground Engineering*, **15**(7), 17–25 (1982).

7.26 Semple, R. M. and Rigden, W. J., Shaft capacity of driven pipe piles in clay, *Proceedings of the Symposium on Analysis and Design of Pile Foundations*, American Society of Civil Engineers National Convention, San Francisco, 1984, pp. 59–79.

7.27 Rigden, W. J., Pettit, J. J., St John, H. D., and Poskitt, T. J., Developments in piling for offshore structures, *Proceedings of the 2nd International Conference on the Behaviour of Offshore Structures*, London, 1979, Vol. 2, pp. 276–296.

7.28   Skempton, A. W., Cast-in-situ bored piles in London Clay, *Géotechnique*, **9**, 153–173 (1959).

7.29   Weltman, A. J. and Healy, P. R., Piling in boulder clay and other glacial tills, *Construction Industry Research and Information Association, Report PG5* (1978).

7.30   Burland, J. B., Butler, F. G., and Dunican, P., The behaviour and design of large diameter bored piles in stiff clay, *Proceedings of the Symposium on Large Bored Piles*, Institution of Civil Engineers, Reinforced Concrete Association, London, 1966, pp. 51–71.

7.31   Meyerhof, G. G., Bearing capacity and settlement of piled foundations, *Proceedings of the American Society of Civil Engineers*, GT3, pp. 197–228, 1976.

7.32   Pells, P. J. N. and Turner, R. M., End bearing on rock with particular reference to sandstone, *Proceedings of the International Conference on Structural Foundations on Rock*, Sydney, Balkema, Rotterdam, 1980, I, pp. 181–190.

7.33   Wyllie, D. C., *Foundations on Rock*, 1st edn, Spon, 1991.

7.34   Kulhawy, F. H. and Goodman, R. E., Design of foundations on discontinuous rock, *Proceedings of the International Conference on Structural Foundations on Rock*, Sydney, 1980, Balkema, Rotterdam, 1980, Vol. 1, pp. 209–220.

7.35   Kulhawy, F. H. and Goodman, R. E., Foundations in rock, Chapter 15 of *Ground Engineering Reference Book*, ed. F. G. Bell, Butterworths, London, 1987.

7.36   Various Authors, *Proceedings of the Symposium on Piles in Weak Rock*, Institution of Civil Engineers, London, 1977.

7.37   Hobbs, N. B. and Healy, P. R., Piling in Chalk, *Construction Industry Research and Information Association, Report PG6* (1979).

7.38   Lord, J. A., Keynote address, Design and Construction in Chalk, *Proceedings of the International Chalk Symposium*, Brighton, 1989, Thomas Telford, London, 1989, pp. 301–325.

7.39   Osterberg, J. O. and Gill, S. A., Load transfer mechanisms for piers socketed in hard soils or rock, *Proceedings of the 9th Canadian Symposium on Rock Mechanics*, Montreal, pp. 235–262, 1973.

7.40   Rosenberg, P. and Journeaux, N. L., Friction and end bearing tests on bedrock for high capacity socket design, *Canadian Geotechnical Journal*, **13**, 324–333 (1976).

7.41   Horvath, R. G., Field load test data on concrete to rock bond strength, *University of Toronto, Publication No. 78-07* (1978).

7.42   Williams, A. F. and Pells, P. J. N., Side resistance rock sockets in sandstone, mudstone and shale, *Canadian Geotechnical Journal*, **18**, 502–513 (1981).

7.43   Pells, P. J. N. and Turner, R. M., Elastic solutions for design and analysis of rock socketed piles, *Canadian Geotechnical Journal*, **16**, 481–487 (1979).

7.44   Horne, H. M. and Lambe, T. W., Settlement of buildings on the MIT campus, *Proceedings of the American Society of Civil Engineers*, **90** (SM5), 181–195 (1964).

7.45   Meyerhof, G. G. and Adams, J. I., The ultimate uplift capacity of foundations, *Canadian Geotechnical Journal*, **5**, 225–244 (1968).

7.46   UK Department of Energy, *Offshore Installation: Guidance on design and construction*, HMSO, London, 1983.

7.47   Brinch Hansen, J., The ultimate resistance of rigid piles against transversal forces, *Danish Geotechnical Institute Bulletin No. 12*, pp. 5–9 (1961).

7.48   Reese, L. C. and Matlock, H., Numerical analysis of laterally loaded piles, *Proceedings of the Second Conference on Electronic Computation*, American Society of Civil Engineers, Pittsburgh, 1960.

7.49   Matlock, H. and Haliburton, T. A., BMCOL-28 – a program for numerical analysis of beam columns on non-linear supports, *Report to the Shell Group*, Austin, Texas (1964).

7.50   Haliburton, T. A., Soil–structure interaction: numerical anal-

ysis of beam-columns, *Oklahoma State University, School of Civil Engineering, Publication 14* 1971.

7.51   Recommended Practice for Planning, Designing, and Constructing Fixed Offshore Platforms, *American Petroleum Institute*, Code RP2A (15th edn), Dallas, Texas, 1984.

7.52   Poulos, H. G., Behaviour of laterally-loaded piles: single piles, *Proceedings of the American Society of Civil Engineers*, **97**(SM5), 711–731 (1971).

7.53   Elson, W. K., Design of laterally-loaded piles, *Construction Industry Research and Information Association, Report 103* (1984).

7.54   Randolph, M. F., The response of flexible piles to lateral loading, *Géotechnique*, **31**(2), 247–259 (1981).

7.55   Poulos, H. G. and Davis, E. H., *Pile Foundation Analysis and Design*, John Wiley, New York, 1980.

7.56   Meyer, P. L., Holmquist, D. V. and Matlock, H., Computer predictions for axially-loaded piles with non-linear supports, *Proceedings of the 7th Offshore Technology Conference*, Paper No. 2186, Houston, Texas, 1975.

7.57   Vijayvergiya, V. N., Load-movement characteristics of piles, *Proceedings of the Ports '77 Conference*, Long Beach, California, 1977.

7.58   Toolan, F. E. and Horsnell, M. R., Analysis of the load-deflection behaviour of offshore piles and pile groups, *Proceedings of the Conference on Numerical Methods in Offshore Piling*, Institution of Civil Engineers, London, 1979, pp. 147–155.

7.59   Desai, C. S. and Holloway, D. M., Load-deformation analysis of deep pile foundations, *Proceedings of the Conference on the Applications of the Finite Element Method in Geotechnical Engineering*, U.S. Army Engineers, Waterways Experiment Station, Vicksburg, Mississippi, 1972.

7.60   Poulos, H. G., Estimations of pile group settlements, *Ground Engineering*, **10**(2), 40–50 (1977).

7.61   Focht, J. A. and Koch, K. J., Rational analysis of the lateral performance of offshore pile groups, *Proceedings of the 5th Offshore Technology Conference*, Houston, Texas, **2**, 701–708 (1973).

7.62   Poulos, H. G., Behaviour of laterally-loaded piles: II – pile groups, *Proceedings of the American Society of Civil Engineers*, **97** (SM5), 733–751 (1971).

7.63   Sawko, F., Simplified approach to the analysis of piling systems, *The Structural Engineer*, **46**(3), 83–86 (1968).

7.64   PLATFORM – a program for the analysis of a three-dimensional group of end bearing piles supporting a rigid platform, *Wimpey Group Manual*, 1969.

7.65   Smith, E. A. L., Pile driving analysis by the wave equation, *Proceedings of the American Society of Civil Engineers*, **86** (SM4), 35–61 (1960).

7.66   Smith, I. M., *Programming the Finite Element Method with Applications to Geomechanics*, John Wiley, 1982.

7.67   Goble, G. G. and Rausche, F., Wave equation analysis of pile driving – WEAP program – *U.S. Dept. of Transportation Report, FHWA-IP-4.2* (1976).

7.68   Heerema, E. P., Relationships between wall friction, displacement velocity, and horizontal stress in clay and sand for pile driveability analysis, *Ground Engineering*, **12**(1), 55–65 (1979).

7.69   Heerema, E. P., Dynamic point resistance in sand and in clay for pile driveability analysis, *Ground Engineering*, **14**(6), 30–46 (1981).

7.70   Litkouhi, S. and Poskitt, T. J., Damping constants for pile driveability calculations, *Géotechnique*, **30**(1), 77–86 (1980).

7.71   Goble, G. G. and Rausche, R., Pile driveability predictions by CAPWAP, *Proceedings of the Conference on Numerical Methods in Offshore Piling*, Institution of Civil Engineers, London, 1979, pp. 29–36.

7.72   Mure, J. N., Kightley, M. L., Gravare, C. J. and Hermansson, I., CAPWAP – an economic and comprehensive alternative to traditional methods of load testing of piles, *Proceedings of the Conference on Piling and Ground Treatment*, Institution of Civil Engineers, London, 1983, pp. 225–232.

7.73   Heerema, E. P. and de Jong, A., An advanced wave equation computer program which simulates dynamic pile plugging through a coupled mass-spring system, *Proceedings of the Conference on Numerical Methods in Offshore Piling*, Institution of Civil Engineers, London, 1979, pp. 37–42.

7.74   Stevens, R. S., Wiltsie, E. A. and Turton, T. H., Calculating pile driveability for hard clay, very dense sand and rock, *Proceedings of the 14th Offshore Technology Conference*, Paper No. OTC 4205, Houston, Texas, 1982.

7.75   Semple, R. M. and Gemeinhardt, J. P., Stress history approach to analysis of soil resistance to pile driving, *Proceedings of the 13th Offshore Technology Conference*, Houston, Texas, 1983.

8 Piled Foundations 2 Structural Design and Construction Methods

8.1 Classification of pile types

The structural features and methods of constructing piles will be described in this chapter. The following types will be considered:

(a) *Driven piles*
   Timber (round or square sections)
   Precast concrete (solid or hollow sections)
   Prestressed concrete (solid or hollow sections)
   Steel H-section, box and tube
(b) *Driven and cast-in place piles*
   Withdrawable steel drive tube, end closed by detachable point
   Steel shells driven by withdrawable mandrel or drive tube
   Precast concrete shells driven by withdrawable mandrel
(c) *Bored piles*
   Augered
   Cable percussion drilling
   Large-diameter under-reamed
   Types incorporating precast concrete units
   Drilled-in tubes
   Minipiles
(d) *Composite piles*

Piles of the types in (a) and (b) above are referred to as *displacement piles* which are subdivided into large displacement piles for solid preformed sections and all driven and place types; and small displacement piles for hollow tubular, box or H-sections. The types listed under (c) are referred to as *replacement* piles. The types in (d) can be of any of these classifications.

The above list might at first sight present rather a bewildering choice to the engineer. However, in practice it is found that three main factors – location and type of structure, ground conditions, and durability – will narrow the choice to not more than one or two basic types. The final selection is then made from considerations of overall cost.

Dealing with the first factor, *location and type of structure*, the driven pile or the driven and cast-in-place pile in which the shell remains in position are the most favoured for works over water such as piles in wharf structures or jetties. Structures on land present a wide choice of pile type, and the bored or driven and cast-in-place types are usually the cheapest for moderate loadings and unhampered site conditions. However, the proximity of existing buildings will often necessitate the selection of a type which can be installed without ground heave or vibration, e.g. some form of bored and cast-in-place pile. Jacked piles are suitable types for underpinning existing structures. Large-diameter bored piles are normally the most economical type for very heavy structures, especially in ground which can be drilled by power augers.

The *ground conditions* influence both the choice of pile type and the technique for installing piles. For example, driven piles cannot be used economically in ground containing boulders and where ground heave would be detrimental. On the other hand, driven piles are preferred for loose water-bearing sands and gravels where compaction due to driving can develop the full potential bearing capacity for these soils. Steel H-piles, having a low ground displacement, are suitable for conditions where deep penetration is required in sands and gravels. Stiff clays favour the adoption of bored and under-reamed types. Under-reamed bases in cohesionless soils require special equipment and techniques.

Durability often affects the selection of pile type. For piles driven in marine conditions, precast concrete piles may be preferred to steel piles from the aspect of resistance to corrosion. Timber piles may be rejected for marine conditions because of the risk of attack by destructive mollusc-type borers. Where soils contain sulphates or other deleterious substances, piles incorporating high-quality precast concrete units are preferable to piles formed by placing concrete *in situ* in conditions where placing difficulties, such as the presence of ground water, may result in the concrete not being thoroughly compacted.

Having selected one or two basic types from consideration of the above factors, the final choice is made from considerations of *cost*. This does not necessarily mean the lowest quoted price per metre run of pile. The engineer must assess the overall cost of the piling work. He must take into consideration the resources of the piling contractor to achieve the speed of operations required by his construction programme; the possibility of having to take piles to a greater depth than envisaged in the piling contractor's quotation; the experience of the piling contractor in overcoming possible difficult conditions; the cost of an extensive test loading programme on sites where the engineer has insufficient experience of the behaviour of piles of the selected type in the particular ground conditions; the cost of routine check loading tests; the cost of the engineer's supervision of pile installation and test loading; and the cost of the contractor's site organization and overheads which is incurred between the time of initial site clearance and the time when he can proceed with the superstructure.

The above factors will be discussed in greater detail in the following pages in the course of describing the various types of pile. However, it will be advantageous first to describe the equipment used for pile driving, methods of driving piles including special methods for overcoming difficult conditions, observations of driving resistance, and test loading. All these are common to most forms of driven piles.

Useful general information on contract procedures and a model specification for piling work are available in an Institution of Civil Engineers' publication,[8.1] and Wynne[8.2] has reviewed the various types of proprietary and non-proprietary pile which are available in the UK.

8.2 Pile-driving equipment

8.2.1 Piling rigs

Piling rigs as used on land sites generally consist of a set of leaders mounted on a standard crane base (Fig. 8.1). The leaders consist of a stiff box or tubular member which carries and guides the hammer and the pile as it is driven into the ground. The leaders can be raked backwards and forwards by screw or hydraulic adjustment of the back stay and lower attachment to the base machine. By swivelling the base machine and adjusting the position of the leaders a group of piles can be driven without moving the machine on its tracks.

Rigs of the type shown in Fig. 8.1 can be mounted on a

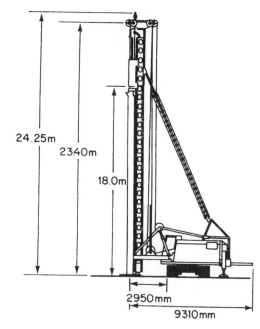

Fig. 8.1   The Ackermanns M14–5P pile frame.

pontoon for marine construction. Alternatively the leaders can be mounted in a braced framework supported by a fixed or rotating base carried by a pontoon barge, or a piled staging (Section 8.5). Extensions can be bolted to the bottom of the leaders for guiding the pile below water level.

Present-day practice is to provide high mobility, a wide range of adjustments, and a multipurpose capability of piling rigs. The Delmag MDT 0802 rig shown in Fig. 8.2 is mounted on a wheeled hydraulic excavator base machine. Hydraulic rams on the machine permit forward, backward and sideways raking of the leaders and their position relative to the machine can also be adjusted. Variation in operating height is made by telescoping the leaders, and in their closed position they can be folded back over the base machine for transport to and from the construction site. The leaders can exert pull-down and extraction forces of 40 and 60 kN respectively. A rotary drill can be clamped in the leaders or mounted at the end of the excavator bucket arm.

It is important to ensure that the leaders remain in their correct position throughout the driving of a pile. If they are allowed to move laterally or out of plumb, this may result in eccentric blows of the hammer with consequent risk of breakage and driving out of line. It is particularly necessary to check the alignment of the leaders when driving raking piles. If settlement of the equipment occurs during driving, its weight may be transmitted to a partly

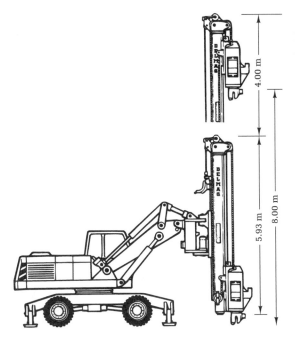

Fig. 8.2 The Delmag MDT 0802 piling rig with telescopic leader.

8.2.2 Piling winches

Winches operating with pile frames mounted on barges or fixed stagings may be powered hydraulically or by steam, diesel or petrol engines, or electric motors. Steam-powered winches are commonly used where steam is also used for the piling hammer. Diesel or petrol engines, or (less commonly) electric motors, are used in conjunction with drop-hammers or where compressed air is used to operate the hammers.

Light winches may only have a single drum, but double- and triple-drum winches which can raise hammer and pile separately are more useful where speed of handling and driving is desirable. The winches may be fitted with reversing gear so that in addition to their main purpose of lifting the hammer and pile, they can also carry out the auxiliary function of operating the raking, rotating, and travelling gear.

The reader is referred to handbooks of manufacturers of piling equipment, such as that of BSP International Foundations Ltd, for tables giving particulars of dimensions and capacities of frames, winches, and boilers for given sizes of pile and hammer.

8.2.3 Hanging leaders

Hanging leaders are designed for suspension from the jib of a crane or excavator (Fig. 8.3). A steel strut, capable of adjustment in length, forms a rigid attachment from the foot of the leaders to the bed-frame of the machine.

The winch units of the excavator or crane are used for lifting purposes with separate drums as necessary for the pile and hammer. Steam or hydraulic power or compressed air for hammer operation are supplied by a separate unit, or drop-hammers may be used with a friction winch.

Consideration should be given to the stiffness of hanging leaders, particularly when driving long raking piles. Excessive flexure can result in eccentric hammer blows, with fracture of the piles.

8.2.4 Hammer guides

In situations where it is desirable to dispense wholly with piling frames or hanging leaders, hammer guides (or rope-suspended leaders) can be attached to the piles, and the latter are guided by timber or steel frameworks. An example of a guide for a diesel hammer is shown in Fig. 8.4. These methods have the advantage of economy in plant and the piles may be driven to a very flat angle of driven pile. These conditions give rise to a serious risk of pile breakage.

Fig. 8.3 BSP crane-suspended leaders.

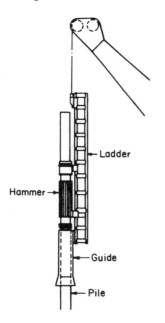

Fig. 8.4 Hammer guide and rope-suspended leaders for Delmag diesel pile hammer.

rake. However, an independent crane is necessary for pitching the pile and setting the guide and hammer. The guides for the piles have to be well constructed and rigidly secured against movement, especially for raking piles. Because of the eccentricity of thrust on the hammer guides, severe stresses and fatigue are set up in the guides and breakages of parts may be frequent. It is also necessary to check that bending stresses in the pile and guides are not excessive when long raking piles are driven without support by leaders. A heavy hammer attached to the upper end of a long pile driven at a flat angle of rake can induce high bending stresses in the pile at the point of support in the guides or 'gate'.

8.2.5 Piling hammers

The selection of a suitable hammer for a piling project depends on a number of factors including the size and weight of the pile, the driving resistance which has to be overcome to achieve the design penetration, the available space and headroom on the site, the availability of cranes, and the noise restrictions which may be in force in the locality. Some of the smaller types including drop hammers, diesel hammers, and hydraulic hammers can be adapted to operate within a sound-absorbent box which greatly reduces the noise problem.

Until recent times it was the usual practice to select a piling hammer from long experience in conjunction with the simpler types of dynamic formula for which the energy

of the blow was represented by the weight of ram, the height of drop and an efficiency factor. It is becoming more general practice to make the selection from the results of a drivability analysis using computer programs based on the Smith wave equation (see Section 7.20.2). Hammer manufacturers provide information on the energy characteristics and efficiencies of their products for use with programs for drivability, but it should be noted that the efficiency is not a constant value. In the case of diesel hammers it depends on the striking rate of the hammer, which in turn depends on the ground resistance. The operating temperature also affects the efficiency of the diesel hammer. For all types except the simple drop hammer the efficiency depends on its mechanical condition, i.e. wear and tear and maintenance. These effects can be evaluated by the output of the dynamic pile analyser as described in Section 7.20.2.

Types of piling hammer in general use are described in the following pages.

The simplest form of hammer is the *drop hammer*, which is used in conjunction with light frames and for test piling where it may be uneconomical to bring a steam boiler or compressor on to a site to drive only three or four

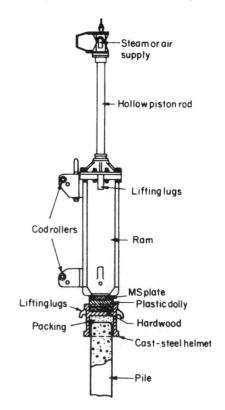

Fig. 8.5 BSP single-acting hammer driving pile with cast-steel helmet and plastic dolly.

test piles. Drop hammers are solid masses of forged steel from 1000 to 5000 kg mass fitted with a lifting eye and lugs for sliding in the leaders. They have the disadvantage that it is not easy to control the height of drop with any accuracy and there is the danger that the operator will use too great a drop when driving becomes difficult, with consequently greatly increased risk of damage to the pile.

Single-acting steam or compressed-air hammers comprise a massive weight in the form of a cylinder. Steam or compressed air admitted to the cylinder raises it up the fixed piston rod. At the top of the stroke, or at a lesser height which can be controlled by the operator, the steam is cut off and the cylinder falls freely on to the pile helmet (Fig. 8.5). The steam is admitted to the cylinder through the piston rod, the upper part of which is hollow. A valve at the top of the rod is opened by pulling on a rope attached to the lever. Releasing the rope closes the valve. Thus, the height of drop and frequency of each individual stroke is controlled by the operator. The maximum height of drop is usually about 1.37 m and the hammer can be operated at a rate of up to about 60 strokes per minute.

A useful rule to determine a suitable weight of drop-hammer or single-acting hammer is to select a hammer weighing approximately the same as the pile. This will not be possible in the case of the heavier reinforced concrete piles when the hammer is more likely to be half of the weight of the pile. However, it should not weigh less than a third of the pile. In order to avoid damage to the pile, the height of drop should be limited to 1.2 m. Concrete piles are especially liable to be shattered by a blow from too great a height. A blow delivered by a heavy hammer with a short drop is much more effective and much less damaging to the pile than a blow from a light hammer with a long drop, particularly in stiff clay soils. Special care is needed when seating a pile on to a hard bearing stratum such as rock, especially on to a steeply sloping rock surface. Single-acting hammers are made with cylinder mass varying from 2500 to 20 000 kg. The largest hammers are used for driving steel tube piles in marine structures. One of the largest is the Menck MRBS 12500 with a ram mass of 125 t.

Double-acting pile hammers are used mainly for sheet pile driving (Fig. 8.6) and are designed to impart a rapid succession of blows to the pile. The rate of driving ranges from 300 blows per minute for the light types, to 100 blows per minute for the heaviest types. The mass of the ram is generally in the range 90–2300 kg, imparting a driving energy of 165–2700 m/kg per blow. The heaviest hammer is the Vulcan 400C, with a ram mass of 18 140 kg, an energy of 15 660 m/kg, and a striking rate of 100 blows per minute.

Double-acting hammers can be driven by steam or compressed air. A piling frame is not required with this

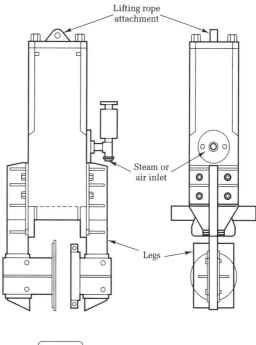

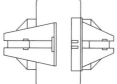

Fig. 8.6 BSP double-acting hammer.

type of hammer which can be attached to the top of the pile by leg-guides, the pile being guided by a timber framework. When used with a pile frame, back guides are bolted to the hammer to engage with the leaders, and only short leg-guides are used to prevent the hammer from moving relatively to the top of the pile. The heavy-duty hammers can drive under water to depths of 25 m.

Care is needed in their maintenance and lubrication and during driving they must be kept in alignment with the pile and prevented from bouncing. *Double-acting* hammers can be fitted with a chisel point for demolition and rock breaking or with jaws for extracting piling.

The BSP International Foundations Ltd ID17 *impulse driver* provides an air cushion between the 1130 kg ram and the driving plate with the object of achieving an appreciable noise reduction when driving steel sheet piles.

Diesel pile hammers provide an efficient means of pile driving in favourable ground conditions, but they are not effective for all types of ground. With this type of hammer (Fig. 8.7) the falling ram compresses the air in the cylinder

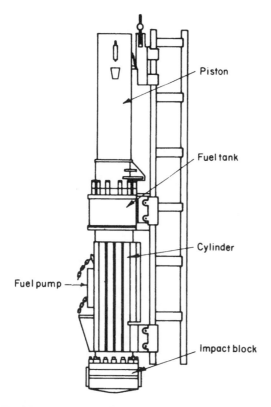

Piston

Fuel tank

Cylinder

Fuel pump

Impact block

Fig. 8.7 Delmag diesel pile hammer.

Hydraulically operated hammers produce less noise and vibration than diesel hammers and they do not emit exhaust fumes. The ram is raised by hydraulic fluid and it falls freely on to the pile. Operation of the hammer is by a solenoid valve which allows either single-stroke working by manual control or automatic striking and gives a wide range of driving energy. Hydraulic hammers can be operated in water to depths of more than 1000 m.

Types in general use for driving piles with moderate loading include the BSP HH3 to 9 series (Fig. 8.8) with ram weights in the range of 3–9 t and rated energies from 36 to 108 kN/m per blow. The IHC S.35 to S.90 series have ram weights ranging from 3.3 to 4.5 t with energies in the range of 35–80 kN/m per blow. These hammers have a power-activated downstroke. They are provided with a print-out facility giving many details of operation including the rate of striking and mean energy delivered to the pile for each 250 mm of penetration.

Hydraulic hammers are manufactured in very large sizes including the IHC S-2300 with a ram weight of 103 t and the Menck MHU 3000T with a 180 t ram.

Fig. 8.8 The BSP HH7 hydraulic hammer driving steel sheet piles.

and the impact atomizes a pool of diesel oil lying at the concave end of the cylinder. The atomized fuel ignites with the compressed air and the resulting explosion imparts an additional 'kick' to the pile, which is already moving downwards under the impact of the ram. The explosion also raises the ram ready for the next downstroke. Burnt gases are scavenged from the cylinder on the upstroke of the ram. The blow, being sustained, is more efficient than the simple impact of the conventional hammer. These hammers can be economical as they dispense with steam or compressed air plant and are entirely self-contained. Diesel hammers are ineffective in soft or yielding soils when the impact of the blow is insufficient to atomize the fuel. The largest of the various makes of diesel hammer are the BSP B45 (4500 kg), the Delmag D100–13 (10 000 kg), the Hera HD7500 (7500 kg) and the Kobe K150 (15 000 kg), the ram mass being given in each case.

These hammers may cause breakage of precast concrete piles where sharply resistant layers are encountered while driving through soft ground when the 'kick' of the explosion may result in a sharp impact on the pile and the possibility of a fracture.

8.2.6 Helmet, driving cap, dolly, and packing

A cast steel helmet is placed over the top of a concrete pile (Fig. 8.5), its purpose being to hold the resilient dolly and packing which are interposed between the hammer and the pile to prevent shattering of the latter at the head. The helmet should fit loosely around the pile, so that the latter can rotate slightly without binding on the helmet. Longer dollies or 'followers' are used when driving piles below the level of the bottom of the leaders.

The dolly is placed in a square recess in the top of the helmet. It is square at the base and rounded at the top. Elm dollies are used for easy to moderate driving, and for hard driving a hardwood, such as oak, greenheart, or pynkado, is used. Plastic dollies can withstand much heavier driving than timber dollies. They comprise a sandwich construction of a layer of resin-bonded laminated fibre between an upper metal plate and a hardwood base. In moderately hard driving conditions, plastic dollies will last for several hundred piles.

Packing is placed between the helmet and the top of the pile to cushion the blow between the two. Various types of materials are used, including hessian and paper sacking, thin timber sheets, coconut matting, sawdust in bags and wallboard.

Driving caps are used to protect the heads of steel bearing piles. They are specially shaped to receive the particular type of pile to be driven and are fitted with a recess for a hardwood or plastic dolly and with steel wedges to keep the caps tight on the pile. If the caps are allowed to work loose they will damage the pile head. A cheap form of cushioning consists of scrap wire rope in coils or in the form of short pieces laid cross-wise in two layers. Frequent renewal is necessary to maintain the required resilience.

Generally, great care is needed in the selection of materials and thicknesses for dollies and packings, since lack of resilience will lead to excessive damage to the pile head or, in severe cases, to breakage of the hammer.

8.3 Jetting piles

Water jetting may be used to aid the penetration of a pile into a sand or sandy gravel stratum. Jetting is ineffective in firm to stiff clays or any soil containing much coarse gravel, cobbles, or boulders. If the piling scheme is planned on the assumption that jetting will be used, it is preferable to cast a central jet pipe, say a 50 or 75 mm pipe terminating in a tapered nozzle into the pile. Elaborate forms of built-in jetting nozzles are sometimes used including separate nozzles leading upwards at an angle and emerging at the sides of the pile as well as the central downward hole. However, it is doubtful if these are much more effective than the single nozzle-ended vertical pipe for jetting piles up to 600 mm width. The jet pipe is led out of the side of the pile and connected by flexible armoured hose to the jetting pump. The pile should be gently 'dollied' up and down on the winch. If it is rammed hard down the nozzle is likely to be blocked by soil wedging into the tapered part. Jetting should be cut off at least 1 m above the predicted founding level and the pile driven down by hammer until the required driving resistance is achieved. Large-diameter tubular piles can be jetted down by a ring of pipes around the internal periphery with the nozzles terminating about 150 mm above the toe. Gerwick[6.14] has described the function of this arrangement as keeping the soil within the pile 'live' rather than as a means of displacing the soil to ground level.

If the jetting is only required to aid penetration of an occasional pile which 'hangs-up' in driving, a separate jet pipe is used. This is worked up and down close to the side of the pile. An angled jet may be used to ensure that the wash water flows to the pile point. In difficult conditions, two or more jet pipes may be used for a single pile. Tube or box piles driven open-ended can be jetted from within the pile, and steel H-section piles can be jetted by sinking the jet pipe down the space between the web and flanges.

An adequate quantity of water is essential for jetting. Suitable quantities of water for jetting a 250–350 mm pile are:

Fine sands	15–25 l/s
Coarse sands	25–40 l/s
Sandy gravels	45–600 l/s

A pressure of at least 5 bar or more is required. It is sometimes a difficult problem to dispose of the large quantities of water and sand flowing at ground level from around the piles, and great care is needed when jetting near existing foundations or near piles driven to depths shallower than the jetting levels. The escaping water may undermine the pile frame, causing its collapse. Jetting through sands may be impossible if the sandy strata are overlain by clays which prevent escape of the jetting water.

8.4 Pile driving by vibration

Vibratory methods of driving sheet piles or bearing piles are best suited to sandy or gravelly soils. Pile driving vibrators consist of pairs of exciters mounted on the vibrator unit, the motors of each pair rotating in opposite directions. The amplitude of vibrations is sufficient to break down the skin friction on the sides of the pile. Characteristics of some typical pile driving vibrators are as follows:

Type	Minimum power supply (kVA)	Frequency of exciter (Hz)	Mass of vibrator (kg)
Bodine	740	Up to 135	10000
BSP VF 55	179	10–120	3800
Koehring-MKT V14	104	24–31	5000
Menck 44–15/30	250	25–50	9600
Muller MS 1004	430	22	8500
Schenk DR 60	250	17–39	7200
Tomen VM2–5000	120	18–30	4887
Vibro-Mac 12	220	10–16	6100

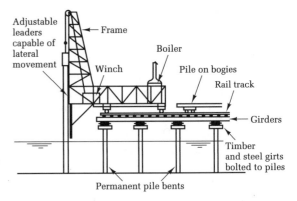

Fig. 8.9 Pile driving over water with cantilever frame.

Vibrators will drive steel piles through loose to medium-dense sands and gravels with comparative ease, but there are likely to be difficulties with dense sands, where the energy may be insufficient to displace the material to permit entry of the pile. Vibrators are little used in stiff clays, although the Bodine Resonant Driver has been used for driving tube piles in clay.

Vibrators can also be used for extracting piles and are frequently used in connection with large-diameter bored and cast-in-place piling work for sinking and extracting the pile casings.

8.5 Pile driving over water

Piles for jetty or wharf structures built over water can be driven from specially designed pile frames cantilevered out from the permanent piles already driven (Fig. 8.9), from ordinary pile frames operating from temporary piled trestles, from 'jack-up' barges or from floating plant. The first-mentioned method has the advantage of being independent of weather conditions, but progress is limited to the output of a single pile frame on narrow structures, and

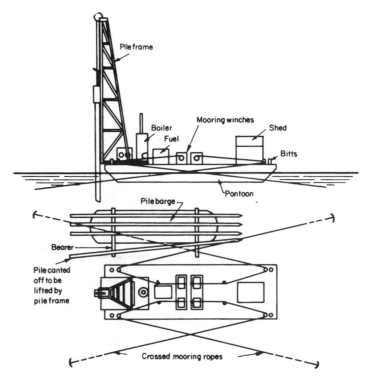

Fig. 8.10 Pile driving from pontoon.

the method cannot be used economically where bents of heavy piles are to be driven at widely spaced centres, i.e. there is a limiting distance over which a heavy frame can be designed to cantilever. Pile driving from temporary falsework trestles may be economical for wharves where piles have to be driven at fairly close spacing, a large number of them being in relatively shallow water on the inshore side of the wharf.

Floating pile-driving plant normally consists of a pile frame erected on one end of a rectangular steel pontoon (Fig. 8.10), although ordinary barges can be used for light pile frames. Pile pontoons are provided with four powered winches or two centrally placed double-drum winches for warping them into position and a very close adjustment of position can be achieved. The cells of the pontoon are ballasted, usually with water, to counteract the weight of the frame and to give the vessel the necessary stability. Piles are brought by barge alongside the pontoon and one end is lifted by the sheaves on top of the frame; the other end is canted off the deck of the barge, allowing the pile to

swing freely into a vertical position from where it is adjusted to the correct height in the leaders.

Single-acting, hydraulic, or diesel hammers are preferable to drop hammers when operating from floating plant in unsheltered waters, since the blows can be controlled with greater accuracy and there is less risk of damage to the pile.

Greatly improved operating conditions can be obtained for pile driving in unsheltered waters by mounting the equipment on a barge provided with tubular legs which can be lowered on to the sea-bed. The hull of the barge is then jacked up the legs until it is clear of the water (Fig. 8.11) to form a stable platform unaffected by wave action. It may not be feasible to use a jack-up barge in situations where the legs have to be jacked down to a deep penetration into a very soft silt or clay to obtain sufficient resistance to carry the weight of the barge and equipment. Jack-up barges specially built for marine piling are fitted with gates (pile guides) which can be adjusted in two directions using hydraulic rams and pile clamps.

Fig. 8.11 Menck piling frame mounted on jack-up barge for driving steel tube piles.

8.6 Pile driving through difficult ground

Driving piles through some types of ground, such as beds of boulders or filling containing obstructions such as large masses of old brickwork or concrete, is impossible unless special methods are used. These methods include boring a hole of sufficient diameter through the difficult ground to receive the pile. In loose ground the hole may have to be cased which sometimes leads to difficulties with extraction of the casing. Another method is to drive a heavy steel spud or joist section through the ground and to drive the pile through the loosened soil left after withdrawing the spud. This method again is liable to result in extraction difficulties and possible overstraining of the pile frame if attempts are made to extract the spud by the piling winch. In ground containing boulders, a jackhammer or rotary diamond core drills may be used to drill into boulders which are broken up by firing small explosive charges lowered down the drill-holes.

On sites where it is important to avoid noise and vibration, mechanical augers have been used to bore uncased holes to receive piles which are dropped down the hole and then driven a short distance to their required bearing level. This method would only be used where cast-*in-situ* piles are unsuitable because of ground or structural conditions.

8.7 Test piling

8.7.1 Installing test piles

Wherever possible, test piles should be of the same type and dimensions as the permanent piles which are intended to be used. This is the only way to ensure that the designed penetration will be attained and that the designer's estimate of the safe working load can be checked when the piles are subjected to test loading. Careful records should be kept of all stages of installation. The following information should be recorded:

> Contract (title and address)
> Principal dates (installation, concreting, driving, loading, etc.)
> Size, type, length (as pitched and as driven), and identification no. of pile

Age of pile (if concrete)
Reinforcement details
Weight of pile (t)
Weight of helmet, dolly, and anvil (kg)
Type of hammer
Serial no. of hammer
Mass of pile (kg)
Mass of helmet, dolly, and anvil (kg)
Type of hammer
Serial no. of hammer
Mass of ram (kg)
Condition of packing and dolly before driving
Condition of packing and dolly after driving
Condition of pile head and dolly after driving
Ground level above datum
Reduced level of pile toe
Reference to nearest borehole or other soil information
Coordinates of axis of pile after completion of driving
Observations of ground heave or subsidence around pile
Condition of pile and shoe (if extracted)
Method of load testing

The records during the actual driving should be entered on a table of the type shown in Table 8.1.

To obtain the records given below, the pile should be marked off in half metres from the toe. The driving resistance in blows per 0.2 or 0.5 m is recorded for the full depth driven, note being made of the depth to which the pile falls under its own weight or the weight of the pile and hammer. Measurements of the set in millimetres per blow need only be made when the temporary compression is being recorded and at the final stages of driving. The temporary compression is measured by clamping a piece of card or graph paper on to the pile and ruling a line across the paper on a horizontal straight edge placed just clear of the paper (Fig. 8.12(*a*)). The vertical and horizontal movements of the pile produce a pattern in the pencil stroke at each blow of the hammer as shown in Fig. 8.12(*b*), from which the temporary compression may be measured directly. Measurements of temporary compression are made at several stages after the pile has entered the stratum within which the pile may be expected to take up its bearing.

Table 8.1 Form of test pile driving record

| Time (h) | Penetration below ground level (m) | Number of blows per 200 mm | Set (mm per blow) | Actual stroke of ram (m) | Measured temporary compression† | | Remarks |
					Amount (mm)	Distance from top of pile (m)	

† Only if required in connection with dynamic formula calculations.

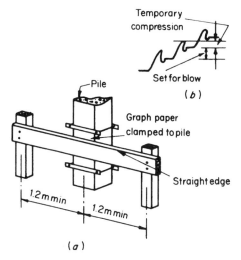

Fig. 8.12 Reading temporary compression during pile driving. (*a*) Apparatus for taking readings on pile. (*b*) Diagram of set and temporary compression.

Where piles are driven into sands, silts, or clays, the redriving characteristics should be observed. Redriving may be started after a few hours in the case of sands, after 12 hours for silts, and after 24 hours or more for clays. The redriving should continue until the resistance is similar to that previously recorded, and the final set and temporary compression should again be measured. After completion of all driving tests the data should be presented in the form of a chart (Fig. 8.13).

8.7.2 Test loading

Two types of test loading can be performed on piles. These are the constant rate of penetration (CRP) test and the maintained load (ML) test. In the latter type the loads are applied in increments.

In the *CRP test* the pile is jacked into the soil, the load being adjusted to give a constant rate of downward movement to the pile, which is maintained until failure point is reached. Failure is defined either as the load at which the pile continues to move downward without further increase in load, or the load at which the penetration reaches a value equal to one-tenth of the diameter of the pile at the base. British Standard 8004: *Foundations* states that a penetration rate of 0.75 mm/min is suitable for friction piles in clay and 1.5 mm/min for piles end-bearing on a granular soil.

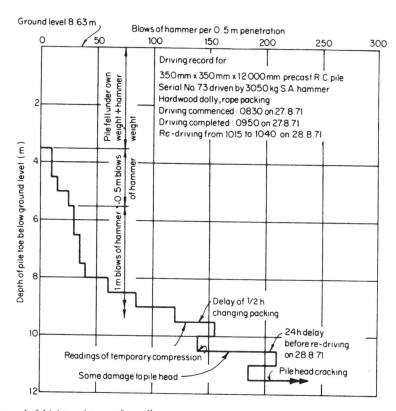

Fig. 8.13 Typical record of driving resistance of test pile.

The CRP test has the advantage that it is rapidly performed, and is thus useful for preliminary test piling when failure loads are unknown, and when the design is based on a factor of safety against ultimate failure, it is desirable to know the real safety factor. However, the method has the disadvantage that it does not give the elastic settlement under the working load (i.e. total settlement less the permanent set) which is of significance in determining whether or not there has been plastic yield of the soil at the working load. Also it has the very considerable drawback of requiring very heavy kentledge loads or high-capacity anchors when large-diameter piles are loaded to failure. For this reason the author considers that the CRP test is best suited to the research type of investigation where fundamental pile behaviour is being studied.

In the case of ground conditions where there is a reasonable amount of experience to guide the engineer on pile-carrying capacity and where pile lengths and diameters can be predicted with some degree of confidence, then a *proof load test* is all that is required. If a pile is considered in relation to its true function, i.e. to support the superstructure, all that the structural engineer needs to do is to satisfy himself that the settlement under the working load will not exceed a value which can be tolerated by the superstructure. The pile failure load has no direct bearing on the structural design. However, because of the natural variation in ground conditions, a test at the working load is insufficient. A test on one pile might just satisfy the settlement criterion, but the same load applied to an adjacent *untested* pile might cause excessive settlement. Therefore the test load must be carried to some multiple of the working load (i.e. $1\frac{1}{2}$ or 2) and a limiting settlement of all piles at the working load will not be exceeded. Proof testing is normally done by maintained load methods.

The Institution of Civil Engineers' specification for piling defines the *specified working load* (SWL) as the specified load on the head of a pile as shown on the engineer's drawings, or in the particular specification or in provided schedules. This is differentiated from the *design verification load* (DVL) which is defined as 'a load which will be substituted for the specified working load for the purpose of a test and which may be applied to an isolated or single loaded pile at the time of testing in the given conditions of the site'. The DVL can be selected as a load representative of a range of working loads when it is usually selected as the maximum of the range. It can also include an allowance for negative skin friction or variations in the pile head level where these influence the mobilization of skin friction. The ICE specification

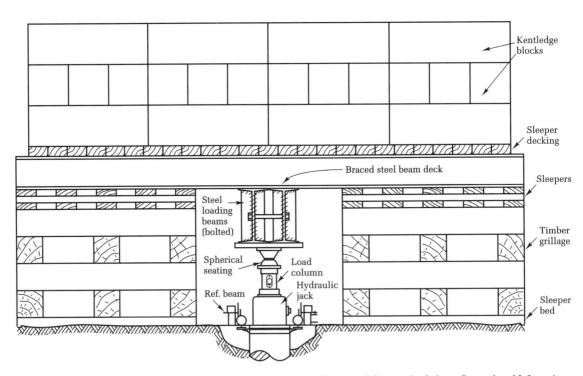

Fig. 8.14 Arrangement of loading test rig with jacking against kentledge. (Courtesy of Construction Industry Research and Information Association.)

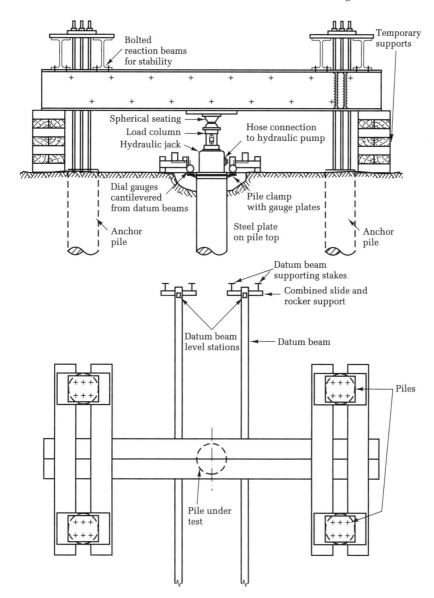

Fig. 8.15 Arrangement of loading test rig with jacking against anchor piles. (Courtesy of Construction Industry Research and Information Association.)

recommends that a proof load test should normally be the sum of DVL plus 50 per cent of the SWL applied in the sequence shown in Table 8.2.

Eurocode 7, Section 7.5.2.1, requires the load-testing procedure to be in accordance with the International Society of Soil Mechanics and Foundation Engineering Subcommittee on Field and Laboratory Testing recommended procedure, 'Axial pile loading test, suggested method', as published in the *ASTM Geotechnical Testing Journal*, June 1985, pp. 79–90.

It is usual to specify limiting total and residual (i.e. total settlement minus the elastic recovery on removal of load) settlements at the working load and at the proof load. In this way the engineer is really placing a lower limit on the early part of the load–settlement curve. The specified values are determined from considerations of the tolerable total and differential settlements of the superstructure, taking into account any group effects. Thus if piles are isolated or in isolated small groups, then the permissible settlement of the test pile will be governed solely from

Table 8.2 Loading sequence for proof load test to 100% DVL plus 50% SWL

Load	Minimum time of holding load
25% DVL	30 min
50% DVL	30 min
75% DVL	30 min
100% DVL	1 hour
75% DVL	10 min
50% DVL	10 min
25% DVL	10 min
0	1 hour
100% DVL	6 hours
100% DVL + 25% SWL	1 hour
100% DVL + 50% SWL	6 hours
100% DVL + 25% SWL	10 min
100% DVL	10 min
75% DVL	10 min
50% DVL	10 min
25% DVL	10 min
0	1 hour

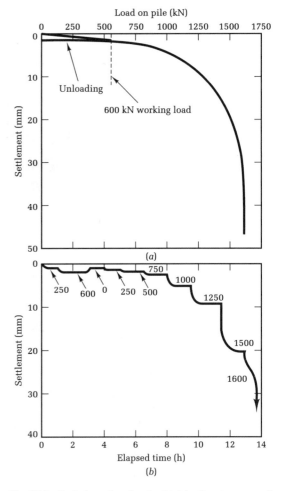

Fig. 8.16 Typical results of maintained loading test on a pile. (*a*) Load–settlement curve. (*b*) Time–settlement curve. (Loads in kN.)

considerations of settlements of the superstructure. When specifying the limiting *total* settlements, the elastic settlement should be allowed for; this will depend on the length of pile and the elastic modulus of the pile and of the soil. A proof load of one and a half times the working load is suitable for most conditions. This type of test is a useful one to perform on selected working piles as confirmation of data obtained in the preliminary test piling and as a check on the piling contractor's workmanship. It can have a salutary effect on his job supervision if he is warned at the outset that a number of the working piles may be tested at random.

In the case of piles driven into soft clays the test should not be made until at least a month has elapsed after driving to allow for the effects of increase in carrying capacity with time.

The pile should be loaded by jacking against a kentledge (Fig. 8.14) or against a beam restrained by anchor piles (Fig. 8.15) rather than by balancing the load on top of the pile. With the latter method there is a risk of a serious accident due to tilting of the platform or pile collapse. The load on the pile should be measured directly by a load cell interposed between the pile head and jack or between the jack and the platform. It is unwise to rely on the reading of the pressure gauge on the hydraulic jack, although this gauge can be used to give the increments of loading. Settlements of the pile are recorded by dial or vernier pointer gauges resting on an arm clamped to the pile head, or welded to a steel plate set on top of the pile head. The gauges are carried by beams securely supported well clear of the pile and platform supports. Alternatively, a precision level can be used to take direct readings on to a scale fixed to the head of the pile. When dial gauges are used,

levels should be taken on the supporting datum frame before and after making the loading test to ensure that no movement of the frame has taken place. The levels should be referred to a stable benchmark well clear of the pile testing area. Typical load–settlement and time–settlement curves for a maintained load test are shown in Fig. 8.16. These should be prepared on A4 size paper for ease in reference and filing.

A method of analysing the results of either CRP or ML tests to obtain an indication of the ultimate load when the tests have not been taken to failure is described by Chin[8.3]. The settlement Δ at each loading stage P is divided by the load P at that stage and plotted against Δ as shown in Fig. 8.17. For an undamaged pile a straight line plot is produced. For an end-bearing pile the plot is a single line (Fig. 8.17(*a*)). A combined friction and end-bearing pile

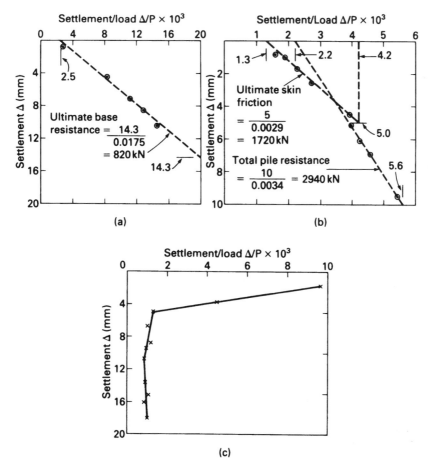

Fig. 8.17 Chin analysis for determining ultimate pile loads. (*a*) End-bearing piles; (*b*) Pile supported by skin friction and end-bearing; (*c*) Broken pile.

produces two straight lines which intersect (Fig. 8.17(*b*)). The inverse slope of the line gives the ultimate load in each case. Chin describes how a broken pile is detected by a curved plot (Fig. 8.17(*c*)).

8.7.3 Programme for test piling

On large piling projects it may be worth while to undertake a programme of CRP or ML test piling involving different types, sizes, and lengths of pile, and observing ultimate loads for the various types and for increasing depths of penetration into the bearing stratum. In piling problems involving piling through fill or other compressible strata where 'negative skin friction' may be an important factor in the design (see Section 7.13) the test piles can be driven through a casing bored through the compressible layers. This will give the carrying capacity of the pile within the bearing stratum only.

Eurocode 7, Chapter 7, for piled foundations defines the

limit load Q_L as the load causing a gross settlement of 10 per cent of the equivalent base diameter. If the load cannot be attained in a test the maximum load is to be taken as Q_L. To obtain the design bearing capacity Q_K from a loading test the value of Q_L is reduced by a factor which depends on the number of tests made. Thus:

Number of load tests	1	2	More than 2
Factor on mean Q_L	1.5	1.35	1.3
Lowest Q_L	1.5	1.25	1

Where piles are subjected to downdrag, and sleeving as suggested above is not done, Eurocode 7 recommends that the maximum load applied to a working pile test should be greater than the sum of the design external load plus twice the design downdrag force.

In the case of tension tests the same reduction factors are applied to the limit tensile load Q_{tL} to obtain the characteristic design tensile load Q_{tK}. Eurocode 7 recommends that more than one pile should be tested and in the case of large numbers of tension piles 2 per cent should be tested.

8.8 Timber piles

Timber piles are frequently used in North America, China, and Scandinavia in the form of trimmed tree trunks, driven butt uppermost. When used in foundation work their lightness gives buoyancy to the foundation. Their lightness, flexibility, and resistance to shock makes timber piles very suitable for temporary works, and in Great Britain their use is generally confined to this purpose.

If timber piles are kept permanently wet or permanently dry, i.e. driven wholly below or wholly above water level, they can have a very long life. However, they are liable to decay in the zone of a fluctuating water table. Also in the case of marine or river structures, the immersed portions of the piles are liable to severe attack by marine organisms. Timber piles are also liable to attack above the water level by fungi and ants and other wood-destroying insects. Care in selection and treatment of timber can prevent or minimize attack. Commercially available types of timber suitable for piling work are Douglas fir, parana and pitch pine, larch, and western red cedar in the softwood class, and greenheart, jarrah, opepe, teak, and European oak in the hardwood class.

Those parts of timber piles which are permanently buried in the ground and below the lowest ground water level can be left untreated. Wherever possible the concrete pile caps should be taken down so that their undersides are below ground water level, and the portion of the pile embedded in the caps should be cut off square to sound wood and liberally coated with creosote or other preservative. If the lowest ground-water table is too deep to take the pile caps down to that level the portion subjected to fluctuating water level or damp conditions can be preserved against decay to a limited extent by pressure treatment with a preservative.

When assessing the depth below ground level requiring concrete or creosote protection, due account should be taken of the possible lowering of the ground-water table by future drainage schemes.

Further information on the behaviour of timber in foundations and methods of preservation are given in Section 13.3.

British Standard 5268 recommends that timber of stress grade SS or better will be suitable for piling. Working stresses in compression should not exceed those tabulated in BS 5268 for compression parallel to the grain for the species and grade of timber being used. In calculating working stresses, the slenderness ratio and bending stresses due to transverse loading and eccentric loading

should be allowed for. Also a cautious approach should be adopted in assessing allowable loads in order to avoid unseen damage by splitting or 'brooming' of the timber under hard driving conditions. It is expected that the draft Eurocode EC5, *Design of Timber Structures*, will be published in 1994.

The *New York Foundation Code* permits a safe working stress of 5.9 N/mm^2 on the net area of cedar, Norway pine, spruce, and similar woods; and 8.3 N/mm^2 on the net area of Douglas fir, oak, southern pine, and woods of similar strength.

Where end-bearing piles are driven into dense or hard materials a shoe is required to prevent splitting or 'brooming' of the pile point. A typical shoe for timber piles is shown in Fig. 8.18. A shoe is not required where piles are driven wholly into soft ground.

It is usual in British practice to provide a steel hoop around the head of the pile to prevent damage during driving. The head of a square pile is chamfered to take the hoop as shown in Fig. 8.18. Damage at the head can be reduced by using the heaviest practicable hammer. In difficult driving conditions, jetting or pre-boring a hole for the pile should be adopted rather than risking undetected splitting or breakage of piles below ground level.

Piles are generally in 6–15 m lengths, but they may be spliced if longer lengths are required. If the splices are made before the piles are lifted and driven, it is preferable to avoid a splice near the middle of the pile since sagging and distortion are at a maximum at this point when the pile is being lifted from a horizontal position. The butting

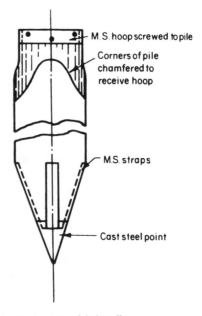

Fig. 8.18 Head and toe of timber pile.

surfaces of piles at the splices should be cut truly square to ensure even contact over the whole area of cross-section.

8.9 Precast concrete piles

8.9.1 General applications

Precast concrete piles are widely used for structures such as wharves or jetties where the pile is required to be carried above soil level in the form of a structural column. They are also used in soil conditions which may be unfavourable to cast-in-place piles, and in conditions where a high resistance to lateral forces is required; for example, in foundations to crane gantries or heavy reciprocating machinery or as anchor piles to the ties of retaining walls. Jointed types of precast pile which provide for variations in level of the bearing stratum can be an economical alternative to cast-in-place piles for ordinary foundation work.

Precast concrete piles are normally of square section for short and moderate lengths, but hexagonal, octagonal, or circular piles are usually preferred for long lengths. If soil conditions require a large cross-sectioned area, precast concrete tubes are used. They are driven hollow to save weight in handling and can be filled with concrete after driving. To avoid excessive whippiness in handling and driving, the usually accepted maximum lengths for various square-section piles are as follows:

Pile size (mm square)	Maximum length (m)
250	12
300	15
350	18
400	21
450	25

8.9.2 Design

The structural design of precast concrete piles is governed by the need to give adequate strength against the stresses caused by lifting and handling the piles and subsequently by the driving of the piles. Once the piles are driven to their final position the stresses caused by foundation loading are likely to be much lower than those induced by handling and driving.

The design of piles to resist driving stresses has been greatly influenced by the researches of Glanville et al.[8.4] at the Building Research Station. They embedded stress-recorders in piles to measure the magnitude and velocity of the stress wave, which after each blow travels from the head of the pile to the toe where it is partly reflected to return to the head. The researches showed that the stresses produced in a pile by the hammer blows far exceed those given by the driving resistance (calculated by dynamic formula) divided by the cross-sectional area of the pile.

The driving stresses were found to depend almost wholly on the fall of the hammer and the nature of the packing between the helmet and the pile. They made the following recommendations for the provision of reinforcement:

(a) The quantity of longitudinal steel should be proportional to the stresses arising in lifting and handling (the research showed that the proportion of main steel did not seem to have any effect on the resistance to driving stresses).

(b) The quantity of transverse reinforcement, where hard driving is expected, should not be less than 0.4 per cent of the gross concrete volume.

(c) The proportion of link steel in the head of the pile should be 1.0 per cent.

BS 8004: *Foundations* states:

The lateral reinforcement is of particular importance in resisting driving stresses and should be in the form of hoops, links, or spirals. For a distance of about three times the least lateral dimension of the pile from each end, the lateral reinforcement should not be less than 0.6 per cent of the gross volume except where special mechanical joints or suitable crack rings, which in themselves distribute driving stresses, are used. In the body of the pile the lateral reinforcement should be not less than 0.2 per cent, spaced at not more than half the width of the pile. Transitions in spacing of lateral reinforcement should be gradual.

The cover over all reinforcement, including binding wire, should be as specified in Table 3.4 of BS 8110 Part I: 1985.

It should be noted that the percentage of lateral reinforcement in the form of links or ties in the head and body of the pile as recommended by BS 8004 is only about half the quantity recommended by Glanville et al. However, the latter recommendations were made for hard driving conditions, whereas the figures given in BS 8004 are minimum values for easy driving conditions.

Bending moments for a variety of conditions of support in lifting and handling a pile of weight W and length L are as follows:

Condition	Maximum static bending moment
(a) Lifting by two points $\frac{1}{5} \times L$ from either end	$\pm WL/40$
(b) Lifting by two points $\frac{1}{4} \times L$ from either end	$-WL/32$
(c) Pitching by one point $\frac{3}{10} \times L$ from the head	$\pm WL/22$
(d) Pitching by one point $\frac{1}{3} \times L$ from the head	$-WL/18$
(e) Pitching by one point $\frac{1}{4} \times L$ from the head	$+WL/18$
(f) Pitching by one point $\frac{1}{5} \times L$ from the head	$+WL/14$
(g) Pitching by one end	$+WL/8$
(h) Balancing by the centre	$-WL/8$

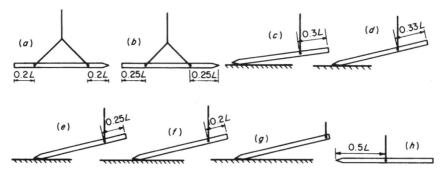

Fig. 8.19   Methods of lifting and pitching piles.

The above conditions are illustrated in Fig. 8.19(a–h).

From design charts published by the Author[8.5] maximum pile lengths for square section piles for various lifting conditions are shown in Table 8.3. The design charts also give the bending moments due to self-weight when lifting, and the ultimate resistance moments for various square and octagonal sections.

Where driving conditions are moderately hard to hard a spiral or helix has been found advantageous. The helix can be placed inside or outside the main bars. If they are placed outside the reduction in concrete cover over the helix is not detrimental since this part of the pile is normally broken down for bonding into the pile cap.

Typical details of reinforcement in precast concrete piles are shown in Fig. 8.20(a) and (b). The 787 mm octagonal pile shown in Fig. 8.20(b) was designed by

George Wimpey & Co. for the Irish Refining Company's marine terminal at Cork, Eire.

8.9.3 Pile shoes

Where piles are driven wholly in soft soils no shoe need be provided. The ends of the piles are usually cast in the shape of a blunt point as shown in Fig. 8.21(a). A sharper point (Fig. 8.21(b)) is preferred for driving into hard clays or compact sands and gravels. The metal drive shoe commonly seen on concrete piles whether driven in soft or hard conditions is based on a design used to stop timber piles from splitting or brooming, and in soft conditions no metal shoe of any kind is required. Where the piles are to be driven into soil containing large cobbles or boulders, a shoe as shown in Fig. 8.21(c) is needed to split the

Table 8.3   Maximum lengths of square section precast concrete piles for given reinforcement

Pile size (mm)	Main reinforcement (mm)	Maximum length in metres for pick up at			Transverse reinforcement	
		Head and toe	0.33 × length from head	0.2 × length from head and toe	Head and toe	Body of pile
300 × 300	4 × 20	9.0	13.5	20.5	6 mm at	6 mm at
	4 × 25	11.0	16.5	25.0	40 mm crs	130 mm crs
350 × 350	4 × 20	8.5	13.0	19.5	8 mm at	8 mm at
	4 × 25	10.5	16.0	24.0	70 mm crs	175 mm crs
	4 × 32	13.0	20.0	30.0		
400 × 400	4 × 25	10.0	15.0	22.5	8 mm at	8 mm at
	4 × 32	12.5	19.0	28.0	60 mm crs	200 mm crs
					or	
					10 mm at	
	4 × 40	15.5	23.0	34.5	100 mm crs	
450 × 450	4 × 25	9.5	14.5	22.0	8 mm at	8 mm at
	4 × 32	12.0	18.0	27.0	60 mm crs	180 mm crs
					or	or
	4 × 40	15.0	22.5	33.5	10 mm at	10 mm at
					90 mm crs	225 mm crs

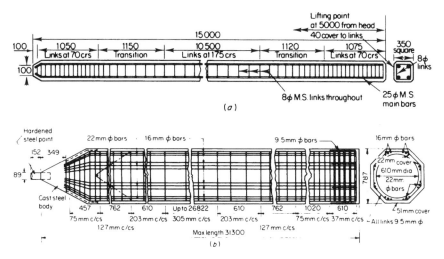

Fig. 8.20 Typical details of precast reinforced concrete piles. (*a*) RC details for 350 × 350 × 15 000 mm pile. (*b*) RC details for 787 mm hollow octagonal pile. (All distances in mm.)

boulders or to prevent breaking of the toe when the pile pushes large cobbles or boulders to one side. The area of the top of the metal shoe in contact with the concrete of the pile should be large enough to ensure that the compressive stress on the concrete is within the safe limits.

Where piles are required to penetrate rock, to obtain lateral resistance for example, a special rock point as shown in Fig. 8.21(*d*) is used. Where piles are driven on to hard rock, the 'Oslo point' (Fig. 8.21(*e*)) is recommended.

This design is particularly suited to driving on to a sloping rock surface when, under careful blows of a heavy hammer with a short drop, the sharp edge of the hollow ground point will bite into the rock, so preventing the point from slipping down the rock surface. The researches leading to the development of the Oslo point are described by Bjerrum[8.6]. This type was used for the octagonal piles as shown in Fig. 8.20(*b*). A hardened steel to BS 970:EN24 was used for the 88 mm diameter point which

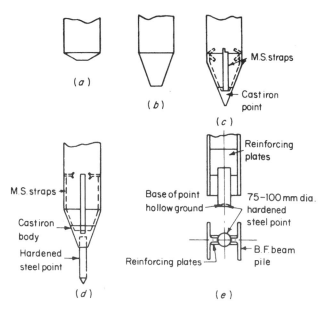

Fig. 8.21 Types of pile for various ground conditions. (*a*) Soft ground. (*b*) Stiff to hard clay, compact sands and gravels. (*c*) Ground containing cobbles or boulders. (*d*) Rock point for penetrating level bedrock surface. (*e*) Oslo point used on sloping bedrock surface.

was embedded in a chilled cast-iron shoe. The point was machined concave to 12 mm depth. The Brinell hardness of the steel was 400–600. It was found that careful flame treatment of the point was necessary after casting it into the shoe since the latter operation resulted in a loss of hardness of the steel.

The Oslo point is seated into rock with very light blows of the hammer, until it is evident that the point is wholly within rock; the hammer drop can then be increased to ensure a satisfactory penetration of the point.

Precast concrete piles cannot be expected to split large boulders when the boulders are in contact with one another or are embedded in hard or compact soil. In these cases, special precautions are taken in driving as described in Section 8.6.

8.9.4 Concrete

It is the usual practice in Britain to limit the stresses in the pile due to working loads and the handling and pitching to the maximum working stresses permitted by CP 114 or BS 8110. Considerably higher stresses may occur during driving. For this reason BS 8004 recommends for hard and very hard driving conditions for all piles and for piles in marine works concrete mix with a minimum cement content of $400\,kg/m^3$ (characteristic strength $40\,N/mm^2$ at 28 days), but for normal and easy driving conditions a cement content of $300\,kg/m^3$ (characteristic strength $25\,N/mm^2$) can be used. Where hard driving is expected it may be advantageous to adopt even richer mixes in the head and toe of the pile, say of the order of $600\,kg/m^3$ minimum cement content.

Portland cement should normally be used for piles. High alumina cement concrete is advantageous from the point of view of early release from the formwork and a reduced curing period, but there is evidence of a sub-stantial retrogression in strength of high alumina cement concrete during or subsequent to the curing period. There is no evidence of retrogression in strength of concrete made with sulphate-resisting or blast furnace slag cements after it is exposed, even to tropical sun temperatures (see Section 13.5.2).

8.9.5 Casting

Piles may be cast directly on to a concrete bed or bottom forms may be used laid on rigid timber bearers. The casting bed should be laid on sufficient depth of well-compacted hard filling to prevent settlement of the bed under the weight of the pile. This is particularly necessary when piles are cast in tiers, one on top of another. Side formwork should be provided in greater numbers than bottom forms since the side units can be struck at an earlier age. After fixing and oiling of the shutters, the pre-assembled reinforcing cage is lowered into the mould and supported by wire ties from cross-bolts as shown in Fig. 8.22. Sufficient ties should be provided to prevent sagging of the steel. If sagging occurs the cover to the bars will be reduced with consequent rusting of the steel and spalling of the concrete. Pile heads must be cast truly square to the axis.

Lifting and toggle bolt holes are formed by inserting short lengths of steel tube. Alternatively, hooks can be cast into the side of the pile. The concrete is compacted by small-diameter poker vibrators which should be inserted and withdrawn for short periods and at close-spaced intervals. The concrete between the bars and the forms should be 'sliced' or 'spaded' to avoid honeycombed patches on the face. Special care should be taken in working concrete against the upper sloping faces of hexagonal or octagonal piles. Equal care should be taken in the treatment of the exposed face by means of a wood

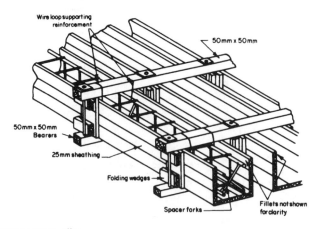

Fig. 8.22   Formwork for precast concrete piles.

float. The forms should be slightly overfilled so that concrete is pressed down by the float, so compacting it and ensuring a dense close-knit surface.

The side forms should be removed as soon as is practicable and the piles kept continuously wet for 4 days in normal (10 °C) air temperature conditions. This wet curing process should be continued until the concrete is hard enough for the piles to be lifted and transported to the stacking area. At some stage after casting, the piles should be clearly marked with a reference number, their length, and the date of casting. This will ensure that the correct piles are used for the particular location and will be a safeguard against driving too soon after casting. The allowable time between casting, lifting, and driving may be assessed by crushing tests on concrete cubes made at the same time as the piles were cast. The cubes should be cured and stored under the same conditions as the piles so that the strengths can be compared. Where piles are cast on a concrete bed, they should be slightly canted by bar and wedges before being lifted. This allows air to get beneath the pile, thus releasing suction between the pile and the bed.

Supports between piles in the stacks should be placed at the predetermined lifting-hole positions to ensure that the design bending moments are not exceeded. Figure 8.23 shows how excessive bending moments can result from carelessly placed packing. The stack should be arranged to allow air to circulate freely around the piles. The piles in the stack should be protected from over-rapid drying in hot weather by covering with tarpaulins or other forms of sheeting.

It is sometimes the practice to cast piles on to a concrete bed at a spacing, centre to centre, of twice their width. Then after striking the side forms, piles are cast in the spaces between, followed by another layer on top of the first and so on. This procedure saves space in the casting area, but has the disadvantage that the strongest piles are at the bottom of the stack, and none can be moved until the topmost pile is strong enough for handling.

Before being driven, the piles should be carefully inspected to ensure that they are free from any cracks. Fine transverse cracks can result from careless handling or lifting when they are of an immature age. Such cracks may lead to corrosion of the reinforcement, especially if the piles are in exposed situations, for example in marine structures.

Some codes of practice permit piles which are cracked during driving to remain in position. The German code (DIN 4026) does not regard cracks narrower in width than 0.15 mm to be detrimental provided that the pile is not otherwise damaged to a degree judged to be detrimental. The Swedish code (SBS S2336) also permits cracks up to 0.2 mm wide with a length not exceeding one-half of the pile circumference for transverse cracks or 100 mm for longitudinal cracks.

8.9.6 Stripping pile heads

After driving the pile to the desired level, it is usual to strip the pile head to expose the reinforcement which is then bonded into the pile cap. Tractor-mounted concrete breakers or jackhammers are the most suitable tools for this job. No method should be used which results in cracking or shattering of the concrete below the level of the underside of the pile cap.

If the pile has refused further driving at too high a level it will be necessary to cut off the surplus length. This can be done by breaking off the corners to expose the reinforcement, which is cut through by hacksaw or burning. The pile can then be broken at this point by pulling it over with ropes attached to the head. The concrete should then be broken away for sufficient depth below the cut to expose the reinforcement for bonding into the pile cap.

8.9.7 Splicing piles

If piles are required to be driven deeper than anticipated it may be necessary to extend them by splicing the reinforcement and casting on an additional length. The concrete in the head of the pile already driven should be broken away to expose the reinforcement for a distance of 40 times the diameter of the longitudinal bars. The reinforcing cage for the extension length is then set up in position and the bars welded at the joints or lapped for the full 40 diameters; after which formwork is assembled in alignment with the pile already in the ground and the extension is then concreted.

8.10 Jointed precast concrete piles

One of the principal disadvantages of the conventional precast concrete pile is that it cannot readily be extended. Thus if the bearing stratum lies locally at depths greater than indicated by borings, delays will occur while pile heads are stripped down and additional lengths cast on and allowed to mature. Furthermore, if the bearing stratum is shallower than anticipated the cut-off length of pile is wasted. This disadvantage can be overcome by using short

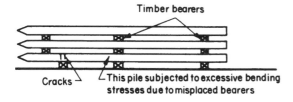

Fig. 8.23 Overstressing in pile due to careless stacking.

precast concrete units which are assembled and locked together before driving the complete pile. Additional lengths can be added as required or lengths left projecting from the ground can be removed.

The development of various proprietary types of jointed pile has done much to make the precast concrete pile competitive with driven or bored and cast-in-place types. Proprietary forms available in Britain and Europe are the Balken, Hercules, Europile, and West's (Hardrive) piles. The Balken and Hercules piles have bayonet joints while the West's pile has spigot and socket joints with metal locking pins. The pile sections are made in the factory in standard lengths of 2.5 to 12.5 m. To economize in the cost of joints, the sections are supplied to site in the longest single lengths shown to be practicable by the borehole or test pile information. Then standard short lengths are locked on to the first length driven to provide for the greater depths of driving required to accommodate variations in the level of the bearing stratum. It is claimed that the jointed piles can be driven to depths of up to 100 m.

Sizes range from 235 to 285 mm square sections with an allowable axial capacity of 700–1200 kN, and in the case of the Hercules piles, hexagonal sections 388 and 505 mm across the flats with axial load capacities ranging from 1300 to 2000 kN.

The Europile 500 is a 272 × 290 mm triangular section in unjointed lengths up to 14 m and with allowable loads up to 707 kN. However, caution is needed when using this type as a friction pile in clay. Research[8.7] has shown that the mobilized shaft friction may be less than that of a circular or hexagonal section of the same volume per unit length.

The high load capacity of the jointed pile is a factor in its competitiveness with cast-in-place types, and it is achieved by using high-grade concrete with a minimum cube crushing strength of 50 N/mm² at 28 days with high tensile steel reinforcement. However, to achieve these high working loads when driving the piles into dense granular soils or into weak jointed rocks may necessitate heavy driving. There is a tendency to concentration of driving stresses at the pile joints, with a consequent risk of unseen breakage below ground level. Occurrences of

breakage at the joints are not always apparent from an examination of driving resistance diagrams of the type shown in Fig. 8.13, which can give the appearance of a momentary sharp increase in resistance such as might be experienced when driving through a hard layer to weaker material below.

To assess the risks of pile breakage where difficult driving conditions are anticipated, it is the practice in Sweden to test drive piles having a central axial inspection hole. A proportion of the working piles are also required to have this hole. Points of breakage can be located by lowering a weighted tube down the hole. In addition excessive deviations from alignment can be located by observing whether or not the standard length tube becomes jammed in the hole as it is lowered down.

8.11 Prestressed concrete piles

Prestressed concrete piles require high-quality concrete, which in turn gives good resistance to driving stresses. As in the case of ordinary precast concrete piles, the longitudinal reinforcement is designed to resist stresses in lifting and handling, and no additional reinforcement, other than link steel, is required to resist driving stresses. Also, the pile is stressed to prevent the development of hair cracks during handling. This, together with the high-quality concrete, gives a prestressed concrete pile a good durability in corrosive soils or in marine work. The manufacturers of prestressed piles claim an advantage in that a smaller cross-section is possible than with ordinary precast piles. This is not necessarily advantageous since a large cross-section may be necessary to develop the required end-bearing and shaft friction. A typical design is shown in Fig. 8.24.

Prestressed concrete piles are usually made by the pretensioning process, i.e. the wires are placed in the moulds and stressed by jacking the ends of the wires against an abutment, after which the concrete is placed and vibrated into the moulds. After the concrete has reached the required compressive strength, the ends of the wires projecting from the concrete are cut off and the stress is transferred to the pile. The pile is then lifted and trans-

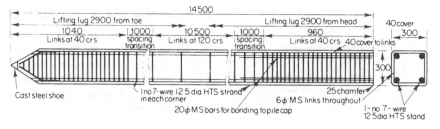

Fig. 8.24   Design for 300 × 300 × 14 500 mm prestressed concrete piles. (All distances in mm.)

Fig. 8.25   Casting yard for precast concrete piles at Drax Power Station.

ferred to the stacking area. Piles are usually made in solid square sections up to 400 mm square. Piles larger than 500 mm are more economically made with a hollow core.

The pre-tensioning method was used by Soil Mechanics Ltd for the manufacture of a large number of prestressed concrete piles for the foundations of Drax Power Station in Yorkshire. Because 18 500 piles were cast an elaborate fabrication yard was constructed on the site (Fig. 8.25). Strand reels were set up on carriers at one end of the casting beds, with winches for tensioning the strand at the other end. Each casting bed had five lines of forms constructed from concrete with heating elements in the side form upstands. The heating enabled the concrete to achieve its release strength of 27.6 N/mm^2 in 40–48 hours. An average of 300 piles per week was manufactured. The side forms had a taper of 1 in 10 to facilitate release of the piles. Whitewash was used in preference to oil as the release agent.

Another method of prestressing concrete piles is post-tensioning. This is generally used for special types such as large cylindrical sections. The precast units have holes cored longitudinally through them. After the curing and hardening period, the units are placed end to end so that the cored holes match up to give a continuous duct down the full length of the assembly. Prestressing wires or cables are threaded through the ducts and tensioned by jacking, after which the ducts are grouted under pressure, the jacks released, and the pile is then ready for lifting and stacking. By such methods, large-diameter hollow cylindrical sections can be made up in long lengths. The lengths can be readily adjusted by adding on or taking away one or more sections, although once the assemblies have been stressed and grouted-up the lengths cannot be adjusted.

Prestressed concrete piles are handled and driven in a similar manner to ordinary precast concrete piles. Since the wires are fully bonded to the concrete by the grout, there is no objection to breaking down the head for bonding into a pile cap or for splicing on an additional length in ordinary reinforced concrete. However, care is necessary in this operation because of the stress present in the wires. Hollow prestressed concrete piles are usually filled with concrete after driving, except where flexibility is required, for example to absorb the energy of berthing ships in jetties. If for structural reasons (for example if the pile is extended above ground level in the form of a column), it is necessary to provide additional reinforcement, this can be prefabricated in the form of a cage which is lowered down the interior of the hollow pile.

Prestressed concrete piles are liable to cracking and spalling if they are not handled and driven carefully.

In addition to precautions in cushioning the head and avoiding high-velocity impact, the Prestressed Concrete Institute[8.8] recommended the following precautions:

(1) Ensure that the pile driving helmet or cap fits loosely around pile top so that the pile may rotate slightly without binding within the driving head.
(2) Ensure that the pile is reasonably straight and not cambered because of uneven prestress or poor concrete placement during casting. Improper storage, handling, or hauling of piling can also cause excessive camber and even cracking. Care should be taken to support or pick up piling at the prescribed points. High flexural stresses may result during driving of a crooked pile.
(3) Ensure that the top of the pile is reasonably square or perpendicular to the longitudinal axis of the pile.
(4) Cut ends of prestressing or reinforcing steel flush with the end of the pile head to prevent their direct loading by the ram stroke.
(5) A reasonable amount of spiral reinforcing is needed at the pile head and tip.
(6) Use adequate amount of residual prestress in prestressed piles to resist reflected tensile stresses. The amounts of prestress required depend on the driving equipment, length of pile, and soil conditions as previously discussed. In general, longer piles (and batter piles) should have more prestress than shorter piles. Some experiences are as follows:

(a) short piles have been successfully driven with from 2.4 to 2.75 N/mm^2 prestress;
(b) the Joint AASHO–PCI Committee on Bridges and Structures specify a minimum of 4.8 N/mm^2 after losses;
(c) on certain occasions from 6.9 to 8.3 N/mm^2 have been used on long piles and batter piles.

(7) Chamfer top and bottom edges and corners of piles.

The Concrete Society[8.9] recommend a characteristic concrete strength of 40 N/mm^2 at 28 days. The maximum axial load capacity of the pile shaft when acting as a short column is given by the equation

$$\text{Load capacity} = \left(\frac{u_w}{3.65} - \text{prestress after losses} \right) A_c, \quad (8.1)$$

where u_w is the strength of concrete at 28 days and A_c is the cross-sectional area of pile shaft.

BS 8004 sets out the requirements for prestressing in considerable detail and this standard should be referred to for further information.

8.12 Steel piles

8.12.1 General applications

Although steel piles have their widest use in the form of sheet piling, they are used to a considerable extent as bearing piles in foundations. They have the following advantages over other types:

(a) If driven on to a hard stratum they have a high carrying capacity.
(b) They can withstand hard driving without shattering; if the pile head buckles during hard driving it can be readily cut down and reshaped for further driving.
(c) If required they can be designed to give relatively small displacement of the soil into which they are driven.
(d) They can readily be lengthened (without much delay to driving) by welding or coupling on additional lengths, thus permitting deep penetrations to be achieved without the need for a tall pile frame.
(e) They can be readily cut down if not driven to their full penetration and the cut-off portions have value as scrap.
(f) They can be roughly handled without damage.
(g) They have good resistance to lateral forces and buckling.

Steel piles are used in marine structures where their resilience and column strength are advantageous in resisting impact forces from the berthing of ships.

Corrosion need not be a drawback to the use of steel bearing piles. Methods of protection of steel piles against corrosion are described in Section 13.4.

Working stresses should be appropriate to the grade of steel (in British practice to BS 4360 grades 43A or 50B). The values adopted should allow for the higher stresses that occur while the piles are being driven to the required ultimate bearing resistance and also for the additional driving stresses at the head and possibly at the toe of the pile. BS 8004 recommends a working load stress of 30 per cent of the yield stress where the design safety factor on the driving resistance is 2, but where the piles are driven through relatively soft soils to end bearing in very dense soils or gravels or sound rock, the allowable axial stress may be increased to 50 per cent of the yield stress. A steel suitable for marine structures where piles are subjected to high impact forces from ships or waves in low temperature conditions is a high tensile alloy steel conforming to grades 55C or E of BS 4360, having minimum and average Charpy V impact values of 2.4 m/ N and 2.0 m/ N, respectively, when tested at $0\,°C$.

Where steel tube or box piles are filled with concrete (minimum cement content $300\,kg/m^3$) the load is shared between the concrete and the steel. The working stress in the concrete should not exceed the value normally used for precast concrete piles.

The New York Foundation Code permits an allowable stress of 0.35 times the minimum yield strength, the latter being not more than $248\,N/mm^2$, and places a maximum

limit on the working load to prevent excessive driving stresses.

The three main types of steel pile in general use are:

(1) H-section piles
(2) Box piles
(3) Tube piles

8.12.2 Details of H-section piles

H-section piles are usually in the form of wide-flange sections. In Great Britain they are rolled in accordance with BS 2566. Although a wide range of broad-flanged beams are rolled only the sections generally used for piling work are shown in Table 8.4.

Since they do not cause large displacement of the soil, H-section piles are useful where upheaval of the surrounding ground would damage adjoining property or where deep penetration is required through loose or medium-dense sands. In the latter case, the sand around the piles is not consolidated to any appreciable distance from them, thus facilitating driving large groups of piles without the need for hard driving and consequent vibrations which might cause settlement of adjacent property founded on the sand. Deep scour was considered likely between the piers of the Tay Road Bridge in Scotland. To achieve the required deep penetration of piles supporting the piers, an H-section was selected. Test piles of $305\,mm \times 305\,mm$ section were driven by diesel hammer to depths of up to 49 m in medium dense sands, gravels, and cobbles.

If strengthened by welding stiffening plates on to the pile toe, the H-section is a useful means of punching through thin layers of rock or boulders.

A disadvantage of the H-section pile is the tendency to bend on the weak axis during driving. Thus if piles are driven to a deep penetration, a considerable curvature may result. Measurements of curvature at Lambton, Ontario[8.10] on 310 and 352 mm H-section piles driven through 46 m of clay into shale showed up to 1.8 to 2 m deflexion from the vertical with a minimum radius of curvature of 52 m. It was evident from these measurements that the safe working stresses in the steel were exceeded even before the piles were subjected to superimposed load. The piles failed under test load. This was thought to be due to plastic deformation of the pile shaft in the region of the maximum curvature. Similar bending has been observed by Bjerrum[8.6] in H-section piles in Norway. He states that any driven H-section pile having a radius of curvature of less than 370 m after driving should be rejected. In this respect the steel tube pile filled with concrete after driving has an advantage since the concrete is not stressed until the superimposed load is applied to the pile.

The low resistance to penetration of the H-section pile in loose sandy soils may be a disadvantage in circum-

Table 8.4   Types, dimensions, and properties of steel H-section piles (as manufactured by the British Steel Corporation)

Serial size (mm)	Weight per m (kg)	Depth, D (mm)	Width, B (mm)	Thickness of web and flange, t (mm)	Area of section (cm^2)	Radius of gyration		Modulus of section	
						XX-axis (mm)	YY-axis (mm)	XX-axis (cm^3)	YY-axis (cm^3)
365 × 406	340	406.4	403.0	26.5/42.9	432.7	168	104	6027	2324
	287	393.7	399.0	22.6/36.5	366.0	165	103	5080	1940
	235	381.0	395.0	18.5/30.2	299.8	162	102	4153	1570
356 × 368	174	362	378	20.4	222.2	152	91.1	2829	976
	152	356	376	17.9	193.6	151	90.3	2464	841
	133	352	373	15.6	169.0	150	89.6	2150	727
	109	346	370	12.9	138.4	148	88.7	1762	588
305 × 305	223	338	325	30.5	285.0	136	78.0	3126	1080
	186	328	320	25.7	237.0	134	77.0	2597	881
	149	318	315	20.6	190.0	132	76.0	2075	689
	126	313	313	17.8	161.0	131	75.0	1763	577
	110	308	310	15.4	140.4	130	74.0	1532	496
	95	304	308	13.4	121.0	129	73.0	1324	424
	88	302	307	12.3	112.0	128	73.0	1220	388
	79	299	306	11.1	100.4	128	72.6	1096	346
254 × 254	85	254	260	14.3	108.1	107	62	965	323
	71	250	257	12.1	91.0	106	61	813	268
	63	247	256	10.6	79.7	105	61	711	232
203 × 203	54	204	207	11.3	68.4	85	50	489	162
	45	200	205	9.5	57.0	85	49	408	133

stances where high shaft friction and end-bearing resistance is required at an economical depth of penetration. Since little consolidation of the sand is effected around and beneath an H-section pile, the shaft friction is relatively low and the pile may not achieve satisfactory resistance until driven to a dense sand stratum or other resistant material. A tube pile, on the other hand, displaces a large volume of soil and the frictional resistance and end resistance is rapidly built up. When tube piles are driven into clays, they can be driven open-ended since a plug of clay is taken down with the pile and forms a hard compact mass. A similar plug of clay is formed within the flanges of H-section piles. However, it is necessary to check by calculation or load testing that there is adequate skin frictional resistance on the interior of the hollow pile or within the flanges of the H-section pile to yielding of the soil plug at the designed base load on the pile (see Section 7.7.2). In the case of granular soils or weak rocks shattered by pile driving, the skin friction mobilized on the interior of hollow piles or within H-section pile flanges may not be sufficient to build up any useful resistance in the form of a plug at the pile toe. Therefore the base resistance of H-section piles in granular soils or rocks should be calculated only on the net cross-sectional area of the steel. However, care is needed to avoid hard driving when terminating these pile sections on rock. There is a risk of shattering or degrading the rock to the consistency of a silt or sand with consequent loss of bearing capacity. It is a

wise precaution to make redriving tests after a waiting period of at least 24 hours to check that there has been no reduction in end-bearing resistance. Limited end-bearing resistance can be allowed for tube piles plugged with sand as discussed in Section 7.4.1. The skin frictional resistance of clays to tubular and H-section piles was discussed in Section 7.7.2. In granular soils the skin friction on hollow piles or H-section piles can be calculated on all steel surfaces provided that this can be checked by measuring the length of soil plug after driving. The interior skin friction must not exceed the plug weight where uplift loads are resisted.

Special types of steel H-section pile are manufactured which are capable of being welded or coupled together to increase their end-bearing area or resistance to lateral loads. 'Winged piles' consist of short lengths of steel H-section welded to the bottom of standard tubular and H-section piles. These 'winged piles' are specifically designed to give a high end resistance for a limited penetration into a bearing stratum of sand. However, the increase in pile resistance may not be in direct proportion to the increase in base area because of the reduction in skin friction in the portion of the shaft above the winged sections.

8.12.3 Steel tube piles

Steel tube for piles can be manufactured in seamless spirally welded, or lap-welded forms. There is nothing to choose between the latter two types of welding from the aspect of strength to resist driving stresses. Tubes are manufactured in European metric sizes from 323.9 to 2020 mm outside diameter and with wall thickness ranging from 6.3 to 12.5 m for the smallest diameters to 10–25 mm for the largest diameters. Steel tube piles can be fabricated with thicker plates than the standard sections where it is necessary to provide for high axial loads or bending moments, or to provide for wastage by corrosion. Tubes in the standard sizes are fabricated from coiled strip by spiral welding, or from plates with longitudinal and peripheral welded joints. Tubes up to 4 m diameter have been fabricated for marine structures. When driving into very stiff clays, dense granular soils, or rocks, the pile toe should be protected from buckling by a stiffening ring or proprietary cast-steel shoe. An internal ring should be used where support to the pile is provided by skin friction. In hard driving conditions the toe protection should consist of a thicker wall pile section about one to one-and-a-half pile diameters in length, butt-welded to the main pile.

Steel tube piles incorporating high tensile steel are used for marine structures where high lateral forces due to the berthing impact of ships and to wave action must be resisted. Due to their circular section these tubular piles offer less resistance to waves and currents than the rectangular H-section piles. Also the use of high tensile steel gives economy in weight of material and hence reduced shipping and handling costs. The piles can be made in three parts: the upper part and the lower part in mild steel, and the centre part in high tensile steel. This confines the more expensive high tensile steel to the highly stressed zone in the region of the sea-bed and facilitates welding of bracing steelwork to the upper section.

Tubular piles in marine structures are often driven with open ends to obtain sufficient penetration to resist lateral and uplift loading. However, the base resistance of the open-end pile can be very low in loose to medium-dense granular soils. Excessive penetration depth can be avoided by welding H- or T-sections to the outer pile surface or by providing a stiffened diaphragm plate at a calculated distance above the toe. An aperture should be provided in the diaphragm for the release of water and silt.

Tubular piles are normally installed by top driving, but in difficult ground conditions they can be installed by 'drill-and-drive' techniques in which the tube is first driven down by hammer to the maximum depth which can be achieved without the driving stresses exceeding tolerable limits. Then a grabbing rig or reverse-circulation drill (Fig. 8.26) is used to remove the plug of soil which has formed within the hollow section. Drilling is continued to a limited depth below the pile toe after which the hammer

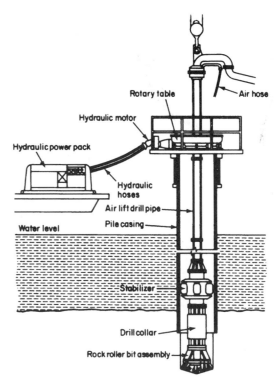

Fig. 8.26 Wirth rotary table and rotating cutter used for drilling out soil plug from within tube pile.

is replaced and the pile is driven down to its final penetration or to the next stage of 'refusal', followed by further drilling. Alternate drilling and driving can result in long delay caused by the need to dismantle and re-erect the heavy equipment. Whenever practicable the drilling should be performed as a 'once-only' operation.

Sullavan and Ehlers[8.11] recommend that the diameter of the hole drilled below pile toe level should not exceed 75 per cent of the outside diameter (and at least 150 mm less than this diameter) for clays, and should not exceed 50 per cent of the outside diameter for sands. To obtain the full design base resistance the pilot hole should be stopped 3–4.5 m above the design penetration. The maximum depth of pilot hole should not exceed 12–15 m above the design penetration for vertical piles, or 7.5 m for raking piles.

Steel tube piles are widely used in the USA, sometimes in the form of ordinary pipe-sections filled with concrete, and also in the form of specially designed fluted sections (to give rigidity for handling and driving) which are driven to the full depth by ordinary pile hammer and then filled with concrete. The Union 'monotube' pile is a tapered fluted steel section used in the USA. These piles with a tip diameter of 203 mm and butt diameters of 305, 355, 406,

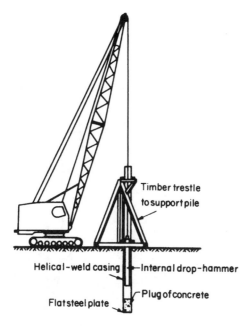

Fig. 8.27   Driving BSP cased piles with internal drop-hammer.

or 457 mm are provided in various gauges of steel depending on the driving conditions. They are driven without a mandrel. The upper part of the pile has parallel sides and the lower part is tapered.

The BSP cased pile is a spirally welded mild-steel tube driven with a closed end, either by a single- or double-acting hammer working on top of the pile or by an internal drop-hammer delivering its blow on to a plug of concrete placed in the bottom of the pile (Fig. 8.27). Care is needed to prevent splitting of the casing when driving on to the concrete plug. A dry concrete (maximum water : cement ratio 0.25) should be used for the plug which should have an initial length of about three and a half times the pile diameter. Fresh dry concrete should be added to the plug if it is more than $1\frac{1}{2}$ h old or if it has been subjected to sustained hard driving for more than $\frac{3}{4}$ h.

Standard sections are shown in Table 8.5.

Table 8.5   Dimensions of BSP concrete-filled cased piles

Internal diameter (mm)	Area of concrete (mm²)
254	50 670
305	72 960
356	99 300
406	129 700
457	164 100
508	202 700
559	245 200
610	291 800

8.12.4 Filling tube piles with concrete

If driving conditions permit, it is preferable to drive tube piles with a closed end since this avoids splitting or enlargement of the ends. Alternatively they may be driven open-ended, and if required they can then be cleaned out for their full depth and hearted with concrete. If a high end-bearing resistance is required particular care is needed to remove adhering soil coatings from the interior of the pile in the region of the base plug. Shear keys may be needed on the interior face to develop the required bond resistance between the pile and the concrete. If long piles are driven by welding on successive lengths, a convenient method is to drive the first length open-ended, then to provide the next or next-but-one length with a diaphragm so that the upper part of the pile, which may be subject to corrosion, remains empty and clean for hearting with concrete. In calculating the stresses on the cross-section resulting from the working load, the strength of the hearting material can be taken into consideration, but the cross-sectional area of steel in a hollow pile is likely to be governed by driving stresses rather than by the compressive stress set up by the working load.

8.13 Types of driven and cast-in-place pile

8.13.1 Recovered steel tube with detachable shoe types (with or without an enlarged base)

A steel tube in diameters ranging from 280 to 600 mm is driven down into the soil with its lower end closed by a precast concrete conical point or a flat steel plate. The pile tube is driven from its upper end by a drop-hammer or a diesel hammer. On reaching founding level a reinforcing cage is placed in the tube, concrete is placed in the shaft and the tube is withdrawn. Enlarged bases can be formed by using an internal drop-hammer to force out a plug of dry concrete from beneath the toe of the drive tube. A problem with driven and cast-in-place piles is ground heave caused by displacement of the soil by the drive tube. This can cause tension failure in the shafts of adjacent piles already driven, and in the worst cases to lifting of the complete piles. The enlarged base to the piles in conjunction with reinforcement in the shaft is some help in anchoring the piles against uplift and it is possible to redrive risen piles. However, the best method to overcome the problem of ground heave is to pitch the drive tube into boreholes drilled by mechanical auger in advance of the piling operations. Care is necessary when concreting the shafts of driven and cast-in-place piles. The problems are the same as those described in Section 8.14.6.

8.13.2 Concrete shells driven with mandrel (West's piles)

A proprietary type of concrete shell piles is installed by Westpile Limited. The pile consists of precast concrete shells in short lengths reinforced by polypropylene fibres which are threaded on to a straight-sided steel mandrel, the lower end of which is fitted with a precast concrete conical shoe. The shells are joined by circumferential steel bands; the inside face of these can be painted with bitumen to give a watertight joint. The mandrel and shells are driven by a drop-hammer operating inside the leaders of a pile frame. An ingeniously designed driving head allows the full weight of the hammer to strike the mandrel while a cushioned blow is transmitted to the shells. The intensity of blow delivered to the shells can be varied by adjusting the drive head to suit variations in driving resistance given by skin friction or adhesion. The sequence in installing a West's pile is illustrated in Fig. 8.28(a)–(c) as follows:

(1) The concrete shoe is set in a shallow hole and the mandrel is lowered on to the shoe (Fig. 8.28(a)).
(2) The concrete shells are threaded on to the mandrel.
(3) The pile is driven to the required level (Fig. 8.28(b)).
(4) The mandrel is withdrawn and any surplus shells removed.
(5) The interior of the pile is inspected, a steel reinforcing case set in position, and the interior of the shells is filled with concrete (Fig. 8.28(c)).

The West's pile has the advantage that the length can be readily adjusted to suit varying ground conditions by adding or taking away the shells. The longest pile driven has been 32 m. The West's system allows long piles to be driven in conditions of limited headroom since several mandrels can be coupled together to drive a long pile. The displacement of the ground given by its comparatively large diameter is advantageous in increasing the skin friction or adhesion and end-bearing resistance, except in conditions where large displacements of the ground might damage adjacent structures. Also, the transmission of the main weight of the hammer blow through the mandrel to the shoe reduces shallow ground vibrations. The shaft can be inspected to ensure that the shells are in alignment and in contact with one another before the interior is concreted.

Care must be taken in driving this type of pile through ground containing large boulders or on to steeply sloping bedrock. In such conditions the pile shaft may be deflected which causes difficulty in withdrawing the mandrel and consequent displacement of the shells. If the shells are not properly butting together the greater part of the load is transmitted to the *in-situ* concrete core which is of comparatively small diameter and, if not solidly concreted,

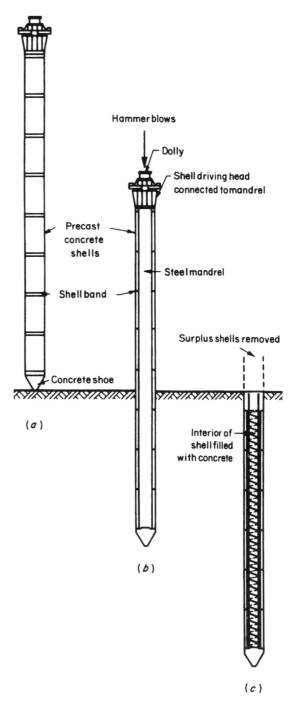

Fig. 8.28 Stages in forming a West's shell pile.

might collapse under a heavy working load. Also, if the piles are driven in groups, ground heave may cause parting of the shells unless precautions are taken such as installing the piles in a predetermined order or placing them in pre-bored holes. West's piles are installed in five sizes as follows:

External diameter (mm)	Core diameter (mm)	Shell length (mm)	Working load (kN)
280	165	914	300
380 or 405	275	915	Up to 650
445	305	915	500–800
510 or 535	380	915	750–1200
610	470	915	Up to 2000

8.13.3 Working stresses on driven and cast-in-place piles

BS 8004 limits the average compressive stress on the pile shaft at the working load to 25 per cent of the characteristic cube compression strength at 28 days. The concrete should not be leaner than 300 kg of cement per cubic metre. Where the pile has a permanent casing of adequate thickness and suitable shape the allowable compressive stress can be increased at the discretion of the engineer.

The New York Foundation Code also limits the stress on the pile shaft to 25 per cent of the cube strength, but where the metal tube or shell is thicker than 3 mm the stress on the metal is permitted to contribute to the strength of the shaft. The allowable compressive stress on the metal is limited to 0.35 times the yield strength, the latter having a maximum of 248 N/mm^2. The New York Foundation Code restricts the total load which can be carried by any single pile to not more than 1500 kN for medium-strong to strong rock, 600 kN for weak rock, and 1000 kN for hard-pan overlying rock.

8.14 Types of bored pile

8.14.1 Boring by mechanical auger

The large spiral auger or bucket auger rotary drilling machines which have been developed over the years for installing large-diameter bored piles have been brought to a stage of high efficiency and they are capable of dealing with a wide range of soil types and can drill in weak rocks. The use of a bentonite slurry in conjunction with bucket

Fig. 8.29 The Bauer BG 26 heavy-duty rotary auger. (Courtesy of Bachy Ltd.)

Fig. 8.30 The Bullivant light rotary auger.

auger drilling can eliminate some of the difficulties involved in drilling in soft silts and clays and loose granular soils without continuous support by casing tubes. Auger machines mounted on lorries or crane base machines (Fig. 8.29) when fitted with triple-telescoping 'Kelly' tubes and extension drill stems can drill to depths up to 70 m. The largest machines can drill pile shafts with diameters up to 4.57 m and with under-reaming tools can form enlarged bases with diameters up to 7.3 m. The Bullivant rotary auger shown in Fig. 8.30 can drill holes up to 400 mm in diameter to depths up to 20 m in conditions of low headroom.

For successful operation of rotary auger or bucket-type machines the soil must be reasonably free of tree roots, cobbles, and boulders, and it must be self-supporting with or without bentonite slurry. If the ground is liable to cave in, the usual practice is to place a length of casing into the augered hole by means of a crane or by the mast of the machine. The bottom of the kelly can then be locked to the top of the casing to turn and push it into the ground, so giving a seal against the entry of water. Alternatively a vibrator of the type described in Section 8.4 can be used to drive the casing to the required depth. A smaller size auger plate is then used inside the casing for the lower part of the hole. If need be a further length of casing can be used by telescoping it inside the length already set.

Boring by mechanical auger under water or a bentonite slurry can cause some loosening of the soil at pile base level. Where a high base resistance is required the soil can be compacted. In the case of the Bauer bored pile this is done by injection pressure. A flat circular steel plate is suspended from the bottom of the reinforcing cage, and a thin flexible steel sheet is attached to the underside of the plate. After concreting the pile, grout is injected around the bottom few metres of the shaft to lock the shaft to the surrounding soil. After a hardening period grout is then injected at high pressure to fill the space between the steel plate and the rubber sheet. This forces the sheet down and compresses the underlying soil. Resistance to uplift of the pile at this stage is provided by grout injections around the pile shaft. These are made in advance of the base grouting.

Other methods of base grouting after concreting the pile shaft include injecting cement grout into a gravel-filled basket with the bottom of the basket covered by a rubber sheet[8.12] and grouting through a grid of *tubes-à-manchette* (Section 11.3.7). The latter operation is undertaken at the early stage of hardening of the shaft concrete when the grout breaks through the thin concrete cover to spread across the soil at the base of the pile. The arrangement of the grid of grout pipes beneath 1200 mm bored and cast-in-place piles used to support an office building at Blackwall Yard, London.[8.13] is shown in Fig. 8.31.

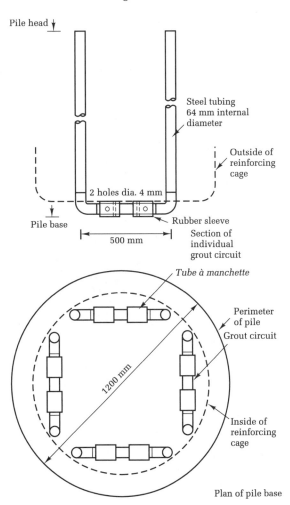

Fig. 8.31 Arrangement of grid grout pipes for base grouting of bored piles (after Yeats and O'Riordan[8.13]).

8.14.2 Boring by percussion or grab-type rigs

In ground where hand or mechanical augering is impossible, e.g. in water-bearing sands or gravels, stony or bouldery clays, or very soft clays and silts, it is the usual practice to sink the hole for small and medium diameter bored and cast-in-place piles by a conventional cable percussion boring rig (Fig. 8.32). A common type of rig for small-diameter piles is similar to that used for exploratory borings.

Specially designed boring rigs for large-diameter bored piles using grabbing methods include the Benoto machine (Fig. 8.33), the Bade machine, the Casagrande GC 1500, and the Hochstrasser–Weise. In these types the casing is given a continuous semi-rotary motion to keep it sinking as the borehole is advanced. Various types of grab are

Fig. 8.32 Installing small-diameter bored piles by cable percussion.

Fig. 8.33 Benoto grab-type rigs for large-diameter bored piles.

provided for different soil conditions. The casing may be provided with welded joints and left in position or with bolted joints and withdrawn while the shaft is concreted. The Benoto EDF 55 rig normally sinks holes up to 1 m in diameter to a depth of 30 m.

Although they can be used in all types of soil, grab types of boring rigs are best suited to difficult soils such as coarse gravel and cobbles, glacial till, marls, or thinly bedded shales and clays. The spiral auger or bucket-type rigs can usually work faster than the grab types in firm to stiff clays and uniform sands.

Grabbing rigs of the type used to construct diaphragm walls (Section 5.4.4) can also be used to construct rectangular 'piles', referred to as barrettes.

A problem in drilling for bored piles in granular soils, either by cable percussion rigs in the case of small-diameter piles, or grabbing rigs for the larger diameters, is the risk of excessive removal of soil during drilling. This is particularly liable to happen when drilling by cable percussion methods since the soil is drawn up into the 'shell' or baler by a sucking action. Water is required to induce the flow of soil into the drilling tool and if natural ground water is not present it must be poured into the pile borehole. Violent raising and lowering of the 'shell' can cause flow of soil from the surrounding ground with the risk of settlement of any adjacent structures. It is possible to minimize the risk of loss of ground by driving the casing ahead of the boring, but this can make it impossible to withdraw the casing after concreting the pile shaft since driving the casing causes the soil to tighten around it. Piling contractors like to keep the soil loose around the casing so that it follows down behind the drilling aided by occasional 'dollying'. This practice is satisfactory when piling in open ground, but can have serious consequences if buildings or services are close to the area of piling. In these circumstances the best procedure is to use rotary auger drilling (Section 8.14.1) under a bentonite slurry.

8.14.3 Under-reamed bored piles

The comments in this section are mainly concerned with the construction of enlarged bases to piles in clays where the equipment and techniques are relatively simple. Forming enlarged bases to piles in water-bearing cohesionless soils requires the use of hydraulic-powered rotary drills for operating under bentonite. The profile of the drilled holes is monitored continuously by remotely controlled sensors and ultrasonic sounding techniques are used to check the profile after completion of drilling. The equipment and techniques are described by Mori and Inamura[8.14].

The savings in cost given by under-reamed piles are mainly due to savings in material excavated from the pile borehole, and in concrete used to replace the excavated material. However, in difficult ground such as boulder clays containing lenses of silt or sand, or in any cohesion-less soil, it is very difficult to form an under-ream. Also, because of the longer time taken to form an under-ream by mechanical means or hand excavation compared with the time taken to drill the straight-sided pile, the economic advantages of the under-reamed pile are somewhat marginal even in favourable ground such as London Clay. It should be noted that London Clay, particularly in the basal beds, frequently contains layers of water-bearing fine sand and silt which can make under-reaming impracticable, and can cause the collapse of the sides of straight-shafted pile boreholes.

Excavation for the under-ream is achieved by a belling bucket rotated by the drill rods or 'kelly', Two types of belling bucket are used. The one generally favoured has arms which are hinged at the top of the bucket (Fig. 8.34) and are actuated by the drill rods or kelly. The arms are provided with cutting teeth and the excavated soil is removed by the bucket. This type cuts to a conical shape, which is an advantage in maintaining stability in fissured soils, and also the arms are forced back into the bucket when it is raised from the hole. The second type (Fig. 8.35) has arms hinged at the bottom of the bucket. This type has the advantage of being capable of cutting a larger bell than the top-hinged type and, because the cutting action is always on the base of the hole, it produces a cleaner base with less loose and softened material. However, the hemispherical upper surface of the bell is less stable than the conical surface and the bottom-hinged arms have a tendency to jam in the hole when raising the bucket.

Belling buckets normally cut to base diameters up to 3700 mm, although diameters of as much as 7300 mm are possible with special equipment. It is not usually practicable to form bells on piles having shafts of less than 762 mm diameter. Although the base of a mechanically under-reamed pile can be cleaned by specially designed mechanical tools, this is a somewhat tedious operation and it is generally preferable to clean out the base by hand,

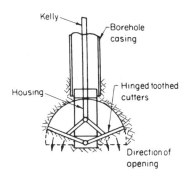

Fig. 8.35 Bottom-hinged belling bucket.

when all soil crumbs and softened material are removed.

Enlarged bases can be formed in stable and relatively dry soils or rocks by hand excavation . This requires some form of support of the roof of the bell to ensure the safety of the workmen. One method of support which has been used is in the form of a 'spider' consisting of a number of hinged steel ribs. The assembly is lowered down the borehole and the ribs are then expanded to force them into contact with the roof of the bell. If very large base areas are required, tunnels can be driven to connect the bells, and the whole base area can then be filled with concrete which is suitably reinforced to achieve the required beam action.

It is often impossible to predict from ordinary site investigation boreholes all the difficulties which may be encountered in attempting to form under-reamed bases on large-diameter piles. For this reason, it is good practice to include an item in piling contracts for drilling a trial pile borehole in advance of the main piling contract. This item can be expensive as the selected piling contractor must bring his men and equipment on to the site and take them away again, but this procedure can often save a considerable amount of money at the main piling stage since difficulties can be foreseen, any modifications to the pile design can be made and if necessary the idea of under-reaming the piles can be abandoned in favour of adopting deeper straight-sided piles. Test loading to check design assumptions can also be made at this preliminary trial stage.

8.14.4 Drilling and inspection of bored piles

It is desirable to make a close inspection of the base of all boreholes with or without bells to ensure that they are clean and free from softened material and that the walls of the shafts are in a stable condition. Large-diameter bored piles are frequently used on a 'one column–one pile' basis, and it is unusual to provide more than three piles to a column. In these conditions failure even of a single pile

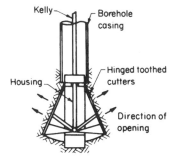

Fig. 8.34 Top-hinged belling bucket.

due to faulty construction methods would have disastrous consequences. Therefore, during the site supervision of piling contracts the engineer should treat each pile in the same way as he would an ordinary pad foundation, i.e. the piling contractor must not be permitted to place concrete until the resident engineer or clerk of works has satisfied himself that the soil is not weaker than that taken as the basis for pile design and that the hole is in a fit condition to receive concrete.

The following precautions are recommended in the construction and subsequent inspection of bored piles:

(1) The pile shaft should be supported by casing through superficial soft or loose soils to prevent the walls of the shaft from collapsing. (This casing can be extracted after concreting the shaft.)

(2) Casing should be provided to seal off water-bearing soil layers. Any soil adhering to the inside of the casing should be cleaned off before inserting the casing in the borehole, and again before placing concrete. The casing should be drilled into an impervious soil layer beneath the water-bearing layer to a sufficient depth to maintain the seal until the remainder of the borehole is completed, and until the concrete is brought up above ground-water level. Alternatively the water-bearing soil layers can be stabilized by a bentonite slurry.

(3) Soil or rock cuttings removed from the pile borehole should be compared with the descriptions stated on the site investigation borehole records.

(4) Shear-strength tests should be made when necessary on undisturbed soil samples taken from the bottom of selected piles as a check on design data.

(5) Boreholes for shallow piles, if too small to enable a man to go down and inspect the base, should be inspected by shining a light down the shaft. Any loose crumbs or lumps of soil should be cleaned out before concrete is placed.

(6) All deep pile holes should be plumbed to the bottom immediately before concreting by lowering a 'cage' to the full depth. The plumbed depth should be compared to the depth accurately measured immediately after completion of drilling. This check will ensure that no soil has collapsed into the borehole.

(7) If, for particular reasons, concrete in the bottom of large-diameter under-reamed piles is to be placed manually, any loose fallen material, or soil softened by trampling on the base, should be cleared out. If necessary, this should be done in small areas, each area being protected in turn by a layer of dry concrete. Any accumulations of water should be pumped or baled out before placing the sealing concrete or first lift of the base concrete. If the inflow

of water is such that cement would be washed from the unset concrete or instability caused to the soil beneath the base, then attempts to concrete the pile 'in the dry' should be abandoned and underwater methods should be resorted to as described in the following section.

(8) The time interval between completion of boring and placing concrete should be as short as possible, and in any case should not be longer than 6 h.

(9) If men are required to work at the base of piles, a loosely fitting safety casing should be suspended in the shaft to protect the men from material falling from the sides of the shaft. The method of suspension must be absolutely secure, as slipping might easily kill a man working in an enlarged base. The top of the casing should project above ground level as a safeguard against tools or stones being accidentally kicked down the hole.

(10) Men working down the holes should wear safety helmets and harness. The latter is required to enable them to be lifted out quickly should they be injured or overcome by gas.

(11) While men are working down the hole a 'top man' should always be in attendance to ensure that no tools or objects are lying near the hole, and that any tools required for use are properly secured before being lowered down the hole. The top of the hole should be protected by a strong well-fitting cover while it is left unattended.

(12) Safety lamps and other gas-detecting devices should be kept at hand and in proper working order.

(13) Frequent tests for gas should be made in filled ground or where pile boreholes pass through beds of peat or organic clays or when working on sites where old gas mains may be encountered.

(14) All work should conform to the requirements of BS 5573: *Safety Precautions in the Construction of Large-Diameter Boreholes for Piling and other Purposes*.

Problems associated with the installation of cast-in-place concrete piles have been reviewed by Thorburn and Thorburn[8.15].

8.14.5 Bored piles installed by continuous flight auger drilling (auger-injected piles)

Bored piles with diameters up to 1.5 m and to depths up to 35 m can be installed with continuous flight auger rigs of the type shown in Fig. 8.36. In stable ground and shallow depths the auger can be withdrawn after drilling to the required depth and the pile shaft is concreted by feeding a flexible pressure hose to the bottom of the unlined hole ever, defects with CFA piles have occurred from time to time and strict control of workmanship is required, particularly when a high proportion of the load is to be carried in end-bearing. When using conventional power augers the soil conditions at the base of the hole can be checked by examining the drill cuttings before any concrete is placed, but where the CFA rig is used the drill cuttings from the base do not reach the surface until the shaft concreting has been completed. It is possible to obtain an indication of the soil strength by measuring the torque on the drill stem over the full depth of drilling. A check can also be made on the soundness of the shaft by recording continuously the volume of concrete injected as the auger is withdrawn.

The problem of ensuring complete filling of the base of the pile with concrete was mentioned above. In this respect it is claimed that the 'Cemcore' pile installed by Cementation Piling and Foundations Ltd, which uses a side exit at the bottom of the hollow stem, is advantageous. The French 'Starsol' rig operated by Soletanche has a separate 100 mm tremie pipe within the 150 mm stem. After reaching the final depth the auger is withdrawn about 200 m and the concrete is discharged from the bottom and sides of the tremie pipe.

Massarch and Wetterling[8.16] have described a device developed in Sweden for expanding the shaft of a CFA pile. The 'expander-body' consists of a metal tube 1.25–2.5 m long. The thin metal has longitudinal concertina folds such that it can be expanded by internal pressure to enlarge the shaft of the pile after the device has been pushed down to the required depth into the fluid concrete. It was reported that the concrete in holes drilled to diameters from 87 to 160 mm was expanded to diameters between 0.4 and 0.8 m.

Derbyshire et al.[8.17] have described developments of installation techniques and instrumentation for CFA piling.

The risks of loss of ground and settlement of adjacent structures when drilling for bored piles by percussion or grabbing rigs were discussed in Section 8.14.2. These risks are reduced by drilling with CFA equipment. Thorburn et al.[8.18] stated that loss of ground is most unlikely in damp cohesive soils, but it can occur at shallow depths in loose cohesionless soils or when overbreak occurs in a sand layer causing undermining and collapse of an overlying clay layer.

Fig. 8.36 Continuous flight auger piling rig.

and withdrawing it as sand–cement mortar is pumped down. More generally, and in all cases of unstable ground, the flight auger is provided with a hollow central stem closed by a plug at the bottom. During drilling, the walls of the borehole are supported at all times by the soil rising within the flights. On reaching the required depth concrete is injected down the hollow stem which pushes out the bottom plug and the pile shaft is concreted by raising the auger with or without rotation. It is good practice to rotate the auger for a number of revolutions before raising it to ensure that the concrete has completely filled the bottom of the hole. After removing the auger the reinforcing cage is pushed down the shaft while the concrete is still fluid. Cage lengths up to 12 m are generally achievable.

The 'concrete' used in the shaft is either a fairly fluid sand–cement mortar with mix proportions of 1.5 to 1 or 2 to 1 or a concrete made with coarse aggregate not larger than 20 mm. Strengths are generally in the range of 20–30 N/mm^2 at 28 days. A plasticizing agent is used to improve the 'pumpability' and an expanding agent is added to counter shrinkage during the setting and hardening phases. The reinforcing cage for mortar-injected piles should be bound with a 6 mm spiral at 150 mm pitch. Hoops, rather than a spiral, should be used for concrete-injected piles.

The use of the continuous flight auger (CFA) rig avoids many of the problems of drilling and concreting piles experienced when using conventional power augers. How-

8.14.6 Concreting bored and driven and cast-in-place piles

In dry boreholes, or in holes where water can be removed by baling, the piles are concreted by the simple method of tipping a reasonably workable mix (not leaner than 300 kg cement per cubic metre of concrete) down the borehole from a barrow or dumper. This method is also used for the shafts of driven and cast-in-place piles where the drive tube is closed by an expendable steel plate. A hopper is provided at the mouth of the hole to receive the charge of concrete and to prevent contamination with earth and vegetable matter on the ground surface. Before any concrete is placed, the hole should be baled dry of water and loose or softened soil should be cleaned out and the bottom of the hole rammed. If the bottom of the hole is wet, a layer of dry concrete should first be placed and well rammed. Then the concrete can be placed using a readily workable mix (75–100 mm slump), which is self-compacting but does not segregate.

There are a number of practical difficulties in concreting piles in water-bearing soils, in squeezing ground, or a combination of the two. The difficulties are aggravated by the presence of the reinforcing cage. If the bottom of the casing is lifted above the concrete while the shaft is concreted and the casing is withdrawn in successive lifts, then water may surge into the hole and weaken the concrete. Also, any water which is carried up the pile shaft in the form of laitance will have a serious weakening effect on the concrete. Sand or soft clay is liable to squeeze in and cause 'waisting' or 'necking'. If, on the other hand, the concrete is kept high in the casing, then it is likely to jam inside the casing. Lifting the casing will then lift the concrete inside it and the lower part of the pile will also be lifted by the reinforcing cage. There is no simple remedy for such a mishap. The only thing to do is to remove all concrete and reinforcement, clean out the hole and start again. In these conditions concreting a pile demands the greatest of care. The concrete should be rich and easily workable. A generous space should be allowed between the vertical bars and between the hoops or spiral ties of the reinforcing cage to allow the concrete to flow easily between the bars. Wheel-type spacers should be used to maintain the clearance between the bars and the ground.

The main difficulty in concreting bored piles occurs in water-bearing ground when the hole cannot be baled or pumped dry. The entire shaft should then be concreted under water using a tremie pipe. Concrete placed through a tremie pipe should be easily workable (125–180 mm slump) and should have a minimum cement content of 400 kg/m^3. A retarder should be used if there is a risk of the concrete setting before the casing is lifted out of the hole. A finished level of concrete placed by tremie pipe should be higher than the design 'cut-off' level, to permit the removal of the thick layer of laitance which always forms on the rising surface of concrete placed by tremie methods.

The Institution of Civil Engineers specification for piling[8.1] specifies tolerances for the casting levels for three conditions of concrete placing in pile boreholes with and without the use of temporary casing. The ground surface or piling platform level is defined as the *commencing surface*. The three conditions all refer to the situation where the designated cut-off level is at a depth H below the commencing surface such that H is between 0.15 and 10 m. The latter depth is concerned with piles supporting basements where they are installed before commencing the basement excavation. The conditions are:

(a) Concrete placed in dry boreholes using temporary casing and without permanent lining. The casting tolerance in metres is specified to be $0.3 + H/12 + C/8$, where C is the length of temporary casing below the commencing surface.

(b) Concrete placed in dry boreholes within permanent tubes or permanent casings or where cut-off levels are in stable ground below the base of any casing. The casting tolerance in metres is specified to be $0.3 + H/10$.

(c) Concrete placed under water or a drilling fluid. The casting tolerance in metres is specified to be $1.0 + H/12 + C/8$.

There are qualifications to the above rules and the reader is referred to the ICE specification for details of them. It will be noted that the casing length rather than the pile diameter is a factor which influences casting tolerances. This reflects the problems of disturbing the concrete when temporary casing is extracted.

It will be noted that no mention has been made of placing concrete by means of a bottom-opening bucket. From personal experience and from several examples given in technical literature, the author is convinced that the use of such a device is unsound practice. When placing concrete in a deep borehole a crane operator does not have sufficient sensitivity in the 'feel' of the bucket to be sure that he is opening it just beneath the surface of the concrete. The bucket may lodge on the reinforcing cage at a higher level and it will then be opened to allow concrete to fall through the water; if it is lowered to too great a depth into fluid concrete there will be surging and mixing of the concrete with water as the bucket is withdrawn. The result of mishandling of a bottom-opening bucket is the formation of layers of laitance and honeycombed concrete. Nothing can be done with the defective concrete. Attempts to grout honeycombed concrete have failed because it is impossible to build up pressure in the honeycombed layer. The grout merely finds an easy way to the surface up the side of the pile.

It is frequently the practice to use bentonite mud instead of casing to support the walls of a pile borehole and then to place concrete beneath the bentonite by means of a tremie pipe. However, it is necessary to keep an adequate head of concrete in the tremie pipe and to keep it moving continuously in order to overcome the pressure of the high-density thixotropic bentonite fluid which is displaced by the outflowing concrete. Difficulties in placing concrete by tremie pipe beneath bentonite were experienced in the 18–21 m deep piled foundations of Wuya Bridge, Nigeria[8.19]. On this site a mud density of 1600 kg/m^3 was required to prevent collapsing of the soil, and the mud formed a gel due to the high ground temperatures. The concrete jammed in the tremie pipe, especially when filling was suspended to remove sections of pipe. The method finally adopted was to use a highly workable concrete with a plasticizer and retarder, and to withdraw the tremie pipe as a single unit without breaking joints.

When ground water is present in fissured rock strata, trouble may be experienced as a result of surging of this water whereby the cement is washed out of unset concrete in the pile boreholes. This surging may be due to boring operations in piles adjacent to those already concreted. If this cause is suspected the boring operations must stop until the concrete has hardened sufficiently, or the boring of all pipes in a group must be completed before concrete is allowed to be placed. Some of these difficulties can be overcome by grouting the rock through a pilot hole drilled in advance of the pile borehole.

Concreting of bored piles under water by the 'pre-packed' or 'preplaced' method is not recommended because of the likely contamination of the preplaced aggregate by silt or clay suspended in the water.

8.14.7 Integrity testing of cast-in-place piles

The soundness of the shafts of either driven or bored and cast-in-place piles can be checked by non-destructive integrity testing. Weltman[8.20] describes four methods:

(1) Acoustic
(2) Radiometric
(3) Seismic
(4) Dynamic response

In the *acoustic* and *radiometric* methods a probe is lowered down a small-diameter hole drilled or formed in the pile. The acoustic probe transmits ultrasonic pulses which pass through the pile concrete and are picked up by a receiver mounted at the bottom of the probe or in an adjacent test hole in the same pile. Radiometric probes employ gamma ray backscatter equipment. In the *seismic* method the pile head is struck by a hammer or mechanical impulse device. Reflected waves from the pile toe are picked up by an accelerometer mounted on the pile head and recorded by an oscillograph. The *dynamic response* method employs a vibrator attached to the pile head and the response is monitored by a velocity transducer.

The seismic and dynamic response methods are the ones most frequently used in the UK because these methods make it unnecessary to drill or form holes in the pile. There is a waiting period after casting of about one week before the concrete is sufficiently mature for testing. These methods are reasonably reliable for checking the overall length of a pile or for detecting gross imperfections such as local necking or large soil inclusions. They are not sufficiently reliable to give conclusive evidence on the presence of cracks across the shaft.

8.14.8 Minipiles and micropiles

Minipiles and micropiles are bored and cast-in-place piles with a diameter of less than 300 mm. They are installed by rotary or percussion drilling or a combination of the two. After completion of drilling, reinforcement is introduced followed by placing cement grout or concrete through a tremie pipe. If grouting is used the fluid grout must be consolidated by air pressure. In some applications a steel tube is used as the permanent load-bearing element with grout injected down the tube to fill the annulus. The applications of minipiles to underpinning is described in Chapter 12.

8.14.9 Working stresses for bored piles

The allowable working stresses should not exceed those given for driven and cast-in-place piles in Section 8.13.4.

8.15 Types of composite pile

Composite piles are used in ground conditions where conventional piles are unsuitable or uneconomical. They may consist of a combination of bored and driven piles, or driven piles embodying two types of material. A frequently used type of composite pile is the concrete and timber pile. This type combines the cheapness and ease of handling of the timber pile with the durability of the concrete pile. The susceptibility of the former to decay above the ground-water table has already been mentioned. Thus the timber pile is terminated below lowest ground-water level and the upper portion formed in concrete. One method of doing this is to drive a steel tube to just below water level. The tube is cleaned out, a timber pile is lowered down it and driven to the required level. The upper portion is then concreted with simultaneous withdrawal of the tube (Fig. 8.37). Alternatively, a precast concrete pile can be bolted to a timber pile using a metal sleeve connection.

Another method of composite piling is to use a hollow precast concrete driven section which remains in position after a timber pile has been driven down the interior to the required penetration. Composite steel H-section and concrete piles can be driven in a similar manner to concrete

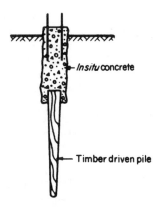

Fig. 8.37   Composite timber and concrete pile.

and timber piles. The concrete portion is used in the zone above and immediately below ground level which is most susceptible to corrosion. The design of such composite piles should provide a rigid connection between the two components.

8.16 The design of pile caps and capping beams

8.16.1 Pile caps

It is impossible to ensure that piles are driven or bored truly vertical or exactly to the prescribed rake.

The Institution of Civil Engineers' specifications for piling[8.1] permit a deviation not exceeding 75 mm in any direction from the centre of the pile as driven to the centre point shown on the setting-out drawing. This deviation is for pile heads cut off at or above ground level. An additional deviation is permitted for pile heads cut off below ground level to conform to the maximum permitted deviation of 1 in 75 for a vertical pile, and 1 in 25 for the specified rake of a raking pile for piles raking up to 1 in 6 and 1 in 15 for piles raking to more than 1 in 6. The tolerances for raking piles assume that the specified requirement to give a tolerance of 1 in 50 in setting up the piling is achieved.

Pile caps must therefore be of ample dimensions to allow them to accommodate piles which deviate from their intended position. This can be done by extending the pile cap for a distance of 100–150 mm outside the outer faces of the piles in the group. The pile cap should be deep enough to ensure full transfer of the load from the column to the cap in punching shear and from the cap to the piles.

In an isolated pile group, the pile cap must include at least three piles to ensure stability against lateral forces. A pile cap for only two piles should be connected by tie beams to adjacent caps. Typical designs of pile caps for various numbers of piles are shown in Fig. 8.38(a)–(e).

The minimum spacing of piles in a group has been referred to in Section 7.14. Using this or greater spacing

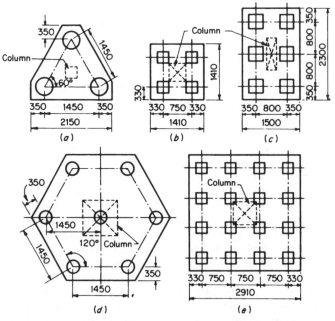

Fig. 8.38   Typical pile caps. (a) Cap for three 450 mm diameter friction piles. (b) Cap for four 350 mm square end-bearing piles. (c) Cap for six 400 mm square end-bearing piles. (d) Cap for seven 450 mm diameter friction piles. (e) Cap for sixteen 350 mm square end-bearing piles.

the piles should be arranged so that the centroid of the group coincides with the line of action of the load. This ensures that all piles carry an equal load and so avoids tilting of the group if they are bearing on compressible strata.

The heads of reinforced concrete piles should be stripped down and the projecting reinforcement bonded into the pile cap to give the required bond length. Research by the Ohio Department of Highways[8.21] showed that if the concrete cap is of adequate size and arrangement and properly reinforced for the pile reactions, there is no need to provide a bearing plate or other load transfer device at the head of a steel H-section pile. However, in the case of slender steel piles, the depth of the pile cap required for resistance to punching shear may be uneconomically large. Savings in the depth of the pile cap can be achieved by welding projections to the head of the steel pile to give increased area in bearing. Load transfer devices used for the heads of steel box piles and H-section piles are shown in Fig. 8.39(a) and (b).

Steel reinforcement should not be threaded through holes pre-drilled in the flanges or webs of steel piles. It is not always practicable to ensure that the pile heads can be driven to within 75 mm of their designed position (see tolerances, above) which means that reinforcement in pile caps or capping beams must be re-bent or moved to enable the bars to pass through the holes.

The depth of pile cap required for timber piles is very often dictated by the need to carry concrete protection to the timber below the lowest ground-water level.

For small pile caps and relatively large column bases the column load may be partly transferred directly to the piles. The part directly transferred by overlapping in a four-pile group is shown in the shaded areas in Fig. 8.40. In these conditions shear forces are negligible and only bending

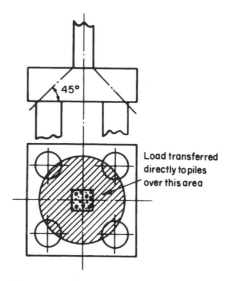

Fig. 8.40 Transfer of load through pile cap.

moments need be calculated. On the other hand, single column loads on large pile groups with widely spaced piles can cause considerable shear forces and bending moments, requiring a system of links or bent-up bars and top and bottom horizontal reinforcement in two layers. A typical arrangement of reinforcement for a small pile group is shown in Fig. 8.41. A minimum cover of 75 mm should be provided to the reinforcement in the pile cap.

Where a column is carried by a single large-diameter pile, there is no need to provide a cap. In the case of reinforced concrete columns, starter bars can be cast into the head of the pile or pockets can be left for the holding-down bolts of steel columns.

The design of reinforcement is highly indeterminate because of relative movements between piles, inequalities

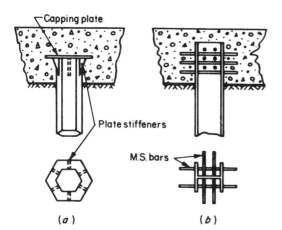

Fig. 8.39 Capping for steel piles. (a) Capping for hexagonal pile. (b) Capping for H-section pile.

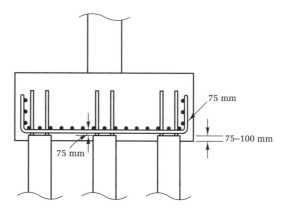

Fig. 8.41 Arrangement of reinforcement in a rectangular pile cap.

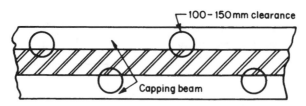

Fig. 8.42   Piles supporting load-bearing walls.

of load transfer, and the rigidity of the pile cap. A method of pile cap design by the truss analogy in which the axial thrust is taken by the concrete and the tensile force by the reinforcing steel has been described by Clarke[8.22].

8.16.2 Capping beams

Piles supporting a load-bearing wall or close-spaced columns may be tied together by a continuous capping beam. The piles can be staggered along the line of the beam to take care of small eccentricity of loading (Fig. 8.42). Where piles are used to support light structures with load-bearing walls the piles may be placed at a wide spacing beneath the centre of the wall (Fig. 8.43).

Investigations at the Building Research Station have shown that load-bearing walls themselves act as beams spanning between piles. It is therefore possible to design the capping beams for lower bending moments than those given by conventional design, which assumes that the whole of the load in a triangle of the wall above the beam is carried by the beam. The bending moment at the centre of a simply supported beam is $WL/8$, where W is the weight of the triangle of brickwork above the beam. The Building Research Station[8.23] recommends a minimum bending moment of $WL/100$, where full composite action takes place and W is the weight of the *rectangle* of brickwork above the beam plus any superimposed loads. It is stated that these 'equivalent bending moments' show considerable saving over conventional design methods, but the depth of beam to span ratio must be kept between one-fifteenth and one-twentieth. However, with a heavily loaded wall only a small degree of composite action is allowed and it may be necessary to use a beam deeper than $L/15$. With considerable composite action (bending

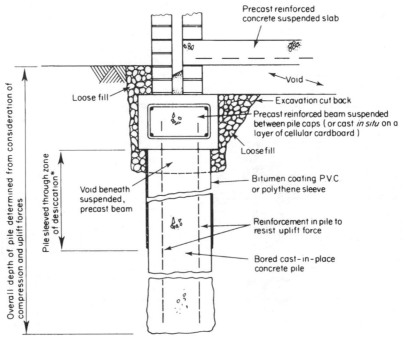

Fig. 8.43   Design for pile capping beam for house foundations where clay soil is subject to swelling or shrinkage.

moments $WL/40$ or less) the reinforcement should be calculated for a beam depth of $L/15$ if a deeper beam is required for practical reasons. A cover of 25 mm is recommended (40 mm would be preferable for good-quality work). The top steel should be placed 25 mm below the upper surface of the concrete immediately over the pile heads and should extend on each side of the pile as far as the quarter-span points. If a door is situated at the end of a span, shear reinforcement is necessary.

The design method assumes that the ground floor is carried by the soil independently of the capping beam. If a suspended floor slab is required the beam should be designed by conventional methods. It is not always desirable to design capping beams to take advantage of the composite action of wall and beam, since this could prevent openings being made for new doors or windows at some time in the future. Also, designers of such modifications or 'do-it-yourself' houseowners might not be aware of the basis of the original design for the capping beam.

Where the piles are provided as a safeguard against swelling and shrinkage of clay soils, the capping beams should be placed on 100–150 mm of specially low-density expanded polystyrene or on a collapsible void former. The purpose of this is to absorb some or all of the upthrust on the underside of the capping beam from the swelling of the soil. A suspended floor slab with a void beneath is essential for all cases where piles are required for swelling soil conditions (Fig. 8.43).

8.17 The economics of piled foundations

The engineer is frequently faced with marginal conditions when there appears to be little to choose between piling and taking conventional strip or pad foundations down to a somewhat greater depth to reach soil of satisfactory bearing capacity. For any given site there must be one particular level below which it is cheaper to pile than to adopt deep strip or pad foundations, and the engineer must make his choice on the basis of fairly detailed estimates of cost. The following notes are intended to give some information to the engineer on the various factors which influence the cost of the work.

In the first place, it is not simply a matter of adding up the cost of x cubic metres of excavation and y cubic metres of concrete for a deep pad foundation and comparing it with the cost of z metres of piling to carry the same working load. Pile foundations require a cap or a capping beam. The minimum thickness of a cap for a pair of piles is about 450 mm and caps for four piles may have to be 600–1200 mm thick, and the plan dimensions up to 2100 mm square for 550 mm diameter piles. It may also be necessary to connect pile caps together by tie beams in two or more directions. Excavation in pile caps, capping

beams, and tie beams may cost twice as much as machine excavation in column bases of fairly large dimensions. Also, piled foundations involve structural design and a higher degree of supervision over their construction. Thus the costs of excavation and concrete in pile caps and capping beams together with the costs of piling and of design and supervision weigh heavily against piled foundations. In fact, in dry soil conditions where advantage can be taken of the low cost of mechanical excavation in bulk, the total cost of excavating to a depth of as much as 4.5 m, constructing column bases in mass concrete up to ground level, and backfilling the soil around the bases, is cheaper than the cost of providing the equivalent number of 8 m long piles, together with the pile caps and capping beams, to carry the same working load. This cost comparison was made for a typical multi-storey block of dwellings. On the other hand, for small-scale work, say for individual houses where excavation is taken out by hand, conditions are suitable for the cheap, uncased, mechanically augered piles and these may be cheaper than conventional brick and concrete footings at a depth of 1.2 m, but probably not cheaper than mechanically excavated narrow strip foundations at the same depth.

When excavation must be taken down into water-bearing sands and gravels the economics may swing in favour of piled foundations, since the cost of excavation with the assistance of ground-water lowering in these conditions may be two or three times greater than excavation in dry ground.

Another justification for piled foundations on structural and economic grounds occurs in highly variable soil conditions such as random pockets or lenses of soft silt and clay in sandy glacial deposits. Ideally the foundation should be designed for the particular conditions at any given point of loading, i.e. adopting high bearing pressures with small pad foundations in areas of compact sands and low bearing pressures with large individual or combined pad foundations for the softer and more compressible soils. However, this procedure would require a very detailed soil investigation at a correspondingly high cost, or else the foundations could be designed individually as the soil conditions were recorded during the excavation work. The latter would not permit any advance design or job planning and would lead to delays due to ordering and cutting reinforcement or fabricating formwork. By adopting pile foundations for all column bases or load-bearing walls, whether in good ground or bad, the job can be designed and planned well in advance of construction, which will proceed without delay, and by using some form of *in-situ* pile the length of the pile can be varied to suit the varying soil conditions rather than by increasing or decreasing the number of piles. Settlements will be of a very small order if the piles in the bad ground conditions have an adequate safety factor; thus relative settlements between columns will be negligible. It would be difficult to achieve this with pad foundations of varying size and bearing pressure.

8.18 The choice of type of pile

Having decided that piling is necessary, the engineer must make a choice from a variety of types and sizes. It has already been noted that there is usually only one type of pile which is satisfactory for any particular site conditions. The following notes will summarize the detailed descriptions already given of the various types of pile and their particular application.

8.18.1 Driven piles

Advantages

(1) Material of pile can be inspected before it goes into the ground.
(2) Stable in 'squeezing' ground.
(3) Not damaged by ground heave when driving adjacent piles.
(4) Construction procedure unaffected by ground water.
(5) Can be readily carried above ground level, especially in marine structures.
(6) Can be driven in very long lengths.

Disadvantages

(1) May break during hard driving causing delays and replacement charges, or worse still may suffer major unseen damage in hard driving conditions.
(2) Uneconomical if amount of material in pile is governed by handling and driving stresses rather than by stresses from permanent loading.
(3) Noise and vibration during driving may cause nuisance or damage.
(4) Displacement of soil during driving piles in groups may damage adjacent structures or cause lifting by ground heave of adjacent piles.
(5) End enlargements not always advantageous.
(6) Cannot be driven in conditions of low headroom.

8.18.2 Driven and cast-in-place piles

Advantages

(1) Length can be readily adjusted to suit varying level of bearing stratum.
(2) Tube is driven with a closed end, thus excluding ground water.
(3) Possible to form an enlarged base in some types.
(4) Material in pile is not determined from handling or driving stresses.
(5) Noise and vibration can be reduced in some types.

Disadvantages

(1) 'Necking' or 'waisting' may occur in squeezing ground unless great care is taken when concreting shaft.
(2) Concrete shaft may be weakened if strong artesian water-flow pipes up outside of shaft.
(3) Concrete cannot be inspected after completion.
(4) Limitations of length of driving in most types.
(5) Displacement of ground may damage 'green' concrete of adjacent piles, or cause lifting by ground heave of adjacent piles.
(6) Noise, vibration, and ground displacement may cause a nuisance or damage adjacent structures.
(7) Cannot be used in river or marine structures without special adaptation.
(8) Cannot be driven in very large diameters.
(9) Cannot be driven in conditions of very low headroom.

8.18.3 Bored and cast-in-place piles

Advantages

(1) Length can be readily varied to suit varying ground conditions.
(2) Soil removed in boring can be inspected and if necessary sampled or *in-situ* tests made.
(3) Can be installed in very large diameters.
(4) End enlargements up to two or three diameters are possible in clays.
(5) Material of pile is not dependent on handling or driving conditions.
(6) Can be installed in very long lengths.
(7) Can be installed without appreciable noise or vibration.
(8) Can be installed in conditions of very low headroom.
(9) No risk of ground heave.

Disadvantages

(1) Susceptible to 'waisting' or 'necking' in 'squeezing' ground.
(2) Concrete is not placed under ideal conditions and cannot be subsequently inspected.
(3) Water under artesian pressure may pipe up pile shaft washing out cement.
(4) Enlarged ends cannot be formed in cohesionless materials without special techniques.
(5) Cannot be readily extended above ground level especially in river and marine structures.
(6) Boring methods may loosen sandy or gravelly soils requiring base grouting to achieve economical base resistance.
(7) Sinking piles may cause loss of ground in cohesionless soils, leading to settlement of adjacent structures.

8.18.4 Choice between types of pile in each category

Driven piles *Timber* Suitable for light loads or temporary works. Unsuitable for heavy loads. Subject to decay due to fluctuating water table. Liable to unseen splitting or brooming if driven too heavily.

Concrete Suitable for all ranges of loading. Concrete can be designed to suit corrosive soil conditions. Readily adaptable to various sizes and shapes. Disadvantages: additional reinforcement must be provided for handling and driving stresses; liable to unseen damage under heavy driving; delay between casting and driving.

Steel Suitable for all ranges of loading. Can be readily cut down or extended. Cut-off portions have scrap value and they can be used for extending other piles. Can be driven hard without damage. Can be driven in very long lengths by welding on additional lengths. Some types have small ground displacement. Structural steel bracing can be readily welded or bolted on. Resilience makes it suitable for jetty or dolphin structures. Disadvantages: subject to corrosion above the soil line in marine structures and requires elaborate paint treatment and/or cathodic protection; long and slender piles liable to go off line during driving.

Driven and cast-in-place piles Types which have a withdrawable tube are cheaper than those where a steel or concrete tube or shells are left in the ground for subsequent hearting with concrete. However, the latter types are a sounder form of construction for 'squeezing' soils or water-bearing sands or where redriving is necessary following ground heave. Generally, the recommended procedure is to approach several proprietary piling firms for quotations and let the issue be decided on the basis of costs. Reputable piling firms will not hesitate to say that their piles are unsuitable for particular ground conditions.

Bored and cast-in-place piles The cheapest forms are the simple, mechanically augered piles sunk without any casing. However, they are only suitable for reasonably firm to stiff cohesive soils. The cost increases when casing has to be installed and withdrawn and the slower conventional boring methods must be used. The use of continuous flight auger piling avoids problems with casing.

The enlarged base in favourable soils has advantages in permitting shorter piles, so bringing them within the depth range of a particular type of mechanical rig.

The use of drilling with continuous flight augers or under bentonite slurry has overcome many of the problems associated with ground water and the loosening of granular soils.

Very often headroom conditions decide the type of pile.

References

8.1 *Specification for Piling*, The Institution of Civil Engineers, Thomas Telford, London, 1988.

8.2 Wynne, C. P., *A review of bearing pile types*, Construction Industry Research and Information Association, 2nd edn *Report PG1*, 1988.

8.3 Chin, Fung Kee, Diagnosis of pile condition, *Geotechnical Engineering*, **9**, 85–104 (1978).

8.4 Glanville, W. H., Grime, G., and Davies, W. W., The behaviour of reinforced concrete piles during driving, *Journal of the Institution of Civil Engineers*, **1**, 150 (1935).

8.5 Tomlinson, M. J., *Pile Design and Construction Practice*, 4ed, Spon, London, 1977, pp. 272–275.

8.6 Bjerrum, L., Norwegian experiences with steel piles to rock, *Géotechnique*, **7**, 73–96 (1957).

8.7 Rojas, E., Static behaviour of model friction piles, *Ground Engineering*, Thomas Telford, **26**(4), 26–30 (1993).

8.8 Hirsch, T. J. Recommended practices for driving prestressed concrete piling, *Journal of the Prestressed Concrete Institute*, **2**(4), 18–27 (1966).

8.9 Foundations (piles), pretensioned, prestressed piles, *Concrete Society Data Sheet CSG1*, p. 3 (1967).

8.10 Hanna, T. H., Behaviour of long H-section piles during driving and under load, *Ontario Hydro Research Quarterly*, **18**, 17–25 (1966).

8.11 Sullavan, R. A. and Ehlers, C. J., Planning for driving offshore pipe piles, *Proceedings of the American Society of Civil Engineers*, **99** (CO-1), 57–79 (1973).

8.12 Bolognesi, A. J. L. and Moretto, O., Stage grouting preloading of large piles in sand, *Proceedings of the 8th International Conference on Sol Mechanics*, Moscow, **2**(1), 203–209 (1973).

8.13 Yeats, J. A. and O'Riordan, N. J., The design and construction of large diameter base-grouted piles in Thanet Sand, London, *Proceedings of the International Conference Piling and Deep Foundations*, London, 1, Balkema, Rotterdam, **1**, 455–461 (1989).

8.14 Mori, H. and Inamura, T., Construction of cast-in-place piles with enlarged bases, *Proceedings of the Conference on Piling and Deep Foundations*, Balkema, Rotterdam, 85–92 (1989).

8.15 Thorburn, S. and Thorburn, J. Q., *Review of Problems Associated with the Construction of Cast-in-place Concrete Piles*, Construction Industry Research and Information Association, London, 1977.

8.16 Massarch, K. R. and Wetterling, S. Improvement of auger cast pile performance by Expander Body System, *Proceedings of the Conference on Deep Foundations and Piles*, Balkema, Rotterdam, pp. 163–177 (1993)

8.17 Derbyshire, P. H., Turner, M. J., and Wain, D. E. Recent developments in continuous flight auger piling, *Proceedings of the Conference on Piling and Deep Foundations*, Balkema, Rotterdam, 33–42 (1989).

8.18 Thorburn, S., Greenwood, D. A., and Fleming, W. G. K., The response of sands to the construction of continuous flight auger piles, *Proceedings of the Conference on Deep Foundations and Piling*, Balkema, Rotterdam, 429–441 (1993).

8.19 Samuel, R. H., The construction of Wuya Bridge, Nigeria, *Proceedings of the Institution of Civil Engineers*, **33**, 353–380 (1966).

8.20 Weltman, A. J., Integrity testing of piles – a review, *Construction Industry Research and Information Association*, Report PG4 (1977).

8.21   Investigation of the strength of the connexion between a concrete cap and the embedded end of a steel H-pile, *State of Ohio Department of Highways Research Report No. 1* (1947).

8.22   Clarke, J. L., Behaviour and design of pile caps with four piles, *Cement and Concrete Association, Report, 42.489, Nov. 1973.*

8.23   Wood, R. H. and Sims, L. G., A tentative design method for the composite action of heavily loaded brick panel walls supported on reinforced concrete beams, *Building Research Establishment Current Paper 26/69* (1969).

9 Foundation Construction

9.1 Site preparation

Present-day foundation construction methods involve a high degree of mechanization. Optimum working speeds of plant are achieved only in clear working conditions giving maximum mobility for the plant and vehicles. Therefore, an efficient and well-maintained system of temporary roads should be provided on extensive sites in order to achieve and maintain a rapid tempo of construction in all weathers. Equally important is attention to site drainage to give dry working conditions and to avoid unnecessary pumping. Specialist operations such as piling and diaphragm wall construction require stable working areas for the mechanical plant.

9.1.1 Temporary roads

The form of construction of temporary roads depends, of course, on the subgrade soil conditions. On well-drained sandy or gravelly soils, no construction will be necessary other than grading to levels and rolling to give a good running surface. Sandy surfaces are liable to rutting in very dry weather or during heavy rain. If required, increased stability can be obtained by rolling in stone or burnt colliery shale or by laying some form of prefabricated metal tracking.

All forms of construction traffic can run on a clayey or silty soil when it is dry. Thus, if it is certain that all construction requiring transport over the site can be completed in the dry season, site roads are unnecessary. However, if the construction programme requires work to continue through the winter and in all periods of rainy weather, then some substantial form of temporary road construction is absolutely essential. It is also important to construct these temporary roads *before the onset of wet weather*. All too often it happens that no thought is given to the need for site roads while traffic is running unhampered over the dry sunbaked clay. Then the rains start and, within a day or two, deep wheel ruts start to form. The churned-up area gets wider and wider as drivers seek fresh routes on undisturbed ground, until eventually the whole site is a morass. Water lies in ruts, and thus aggravates the softening and slows down the drying. Drastic measures are taken when the vehicles are floundering to a standstill all over the site. Lorry loads of broken bricks and concrete are tipped on the site and bulldozed out to form a road. Half of this material is 'lost' as it sinks into the liquid mud and the remainder continues to sink down as mud is squeezed up between the interstices of the stones under the weight of vehicles. Eventually, some 1–1.5 m thickness of material has to be spread before a rough and wavy-surfaced road is obtained. This description is in no way exaggerated. It is inevitable when construction of site roads is left too late. It is also a false economy to make the site roads too narrow. On narrow roads passing vehicles are forced to run off the road, when near-side wheels become bogged and the haunches of the road are broken down.

By constructing the roads before the onset of the wet weather, advantage can be taken of the high bearing value of the dry clay to economize in thickness of road material, although some allowance must be made for softening of the subgrade due to limited percolation of water through the surfacing. The cost of providing an impervious bitumen surfacing to temporary roads cannot usually be justified, and in any case such surfacings are liable to disruption by tracked vehicles.

In very soft or peaty soils consideration should be given to spreading geotextile mattresses on to the existing cleared ground surface before laying the granular base of the site road. These plastic materials possess considerable tensile strength which allows light earth-moving vehicles to operate on them while spreading and levelling the base material.

9.1.2 Site drainage

Much can be done to improve working conditions in wet weather by attention to the surface and subsoil water drainage. This is imperative for construction in tropical or subtropical climates where heavy rainstorms can cause rapid flooding or erosion of a site with consequent damage to partly constructed works. If temporary roads cross ditches, these should be piped beneath the roads, and any ditches running towards the working area should be diverted. Any existing subsoil drains which may be exposed by excavations should be carefully inspected. If they appear to be 'live' they should be intercepted by a cross drain and led to the nearest ditch or surface-water sewer. It is highly wasteful to pump surface water from excavations which might be prevented from entering the excavation by a few well-placed cut-off ditches or stone-filled drains.

9.1.3 Site preparation in built-up areas

Site preparation in built-up areas should include tracing and clear marking of existing underground telephone and power cables, gas and water mains, and sewers. This is important since many fatal accidents have been caused by men or machines striking electric cables and gas mains, and the cost of repairing trunk telephone cables can run into many tens of thousands of pounds. The bursting of a water main can be disastrous to a partly completed excavation if flooding causes collapse of the sides. Safety precautions should be taken in consultation with the supply authorities wherever site roads or tracks are crossed by overhead cables.

If deep excavations, blasting, or pile driving are to be carried out near existing structures, careful inspections should be made jointly with the property owners to determine whether there are any signs of cracking or settlement. Any cracks should be photographed and marked with tell-tales. During the progress of the work, periodic levels should be taken on nearby buildings and measurements made of the widths of cracks at tell-tale positions as a check on possible movement. It is surprising how often property owners are quite unaware that their buildings are cracked and in all good faith make claims against the contractor if their property is shaken by blasting or pile driving.

9.2 Excavation methods

9.2.1 Bulk excavation

The choice of plant for bulk excavation is largely determined by the quantity and by the length of haul to the disposal point. If the tip area is close to the excavation, say within 100 m, then the earth can be moved by *loading shovel* or *bulldozer*.[†] Longer hauls (say 100–600 m) require crawler or rubber-tyred tractor-drawn *scrapers*. For even longer hauls of about 600–800 m, then the faster-moving rubber-tyred tractors are required for hauling the scrapers. However, the size and depth of the excavation and the soil conditions must be favourable for economical use of scrapers; they are unsuitable for deep excavations covering a small area since the ramp roads enabling the scrapers to climb in and out of the cut cannot be conveniently arranged. There must also be room for the scraper to turn in the cut. These machines are best suited to fairly large areas of shallow excavation. They can excavate all soil types except soft clays and silts. Soft or laminated rocks, such as shales or marl, can be dealt with if loosened by a ripper or rooter. Scrapers are unsuitable for haulage over public roads.

If the haul distance for the excavated material exceeds about 800–1000 m, then excavators loading into lorries or trailer wagons are required. Types of mechanical excavator available include the face shovel, backacter (backhoe), dragline, grab, and loading shovel.

Loading shovels are suitable only for shallow cuts up to a few metres deep since the method of filling the bucket is not so efficient as other types, and loading lorries requires the bucket arm to be raised thus slowing the loading cycle. Loading shovels mounted on rubber-tyred or crawler tractors are a convenient and almost universal method for shallow excavation in fairly small areas, and also in deep excavations for operating beneath the bracing frames.

The *face shovel* is the most efficient type of excavator for large quantities of bulk excavation. The height of the face is limited only by the size of the machine, but it is uneconomical for face heights lower than about 1 m. The important feature of a face shovel from the aspect of foundation excavation is that it must stand at excavation level and feed into lorries also standing at the same level. This requires a ramp road to enable the machine to dig its way into the excavation and to climb out again. For this reason face shovels are unlikely to be suitable for deep excavations in confined areas where there is no room to form the ramp road.

Backacters are rather less efficient in excavating than face shovels and their excavation depth is limited by the length of jib and bucket arm. However, they are very suitable machines for deep excavation in small areas such as for column bases or for trench excavations, and hydraulically operated machines with long arms can stand at

---

[†] The types of mechanical plant available for earthmoving are too many and diverse for detailed description and illustration in this book. The reader is referred to BS 6031: *Earthworks*, for general descriptions.

ground level and work in confined areas for basement excavation without the need for constructing ramp roads.

Draglines also have the advantage of working from ground level, and their excavation depth is limited only by the amount of wire rope that can be wound on the bucket-rope drum. Long jib machines can also dispose of the material in large piles adjacent to the cut; thus they are useful for sites where most of the excavation is to be returned around the completed foundations. They have the disadvantage that they cannot cut to a vertical face on all sides of the excavation and, because of the swinging bucket, the time of loading vehicles is slightly slower than with rigid bucket arm types.

Grabs (clam-shell buckets) suspended from mobile or derrick cranes or from excavator-type cranes are rather slow in operation due to the time required to close the grab and wind it into and out of the excavation. However, they are a suitable type of excavator for deep excavations in small areas in shafts and trenches. Powered grabs operating by rope suspension or from rigid 'kelly' masts are used for deep trench excavations in diaphragm wall construction (Section 5.4).

Belt conveyors are a useful means of transporting excavated material across ground impassable to vehicles or over water. The excavator loads into a hopper with a belt feeder beneath the hopper regulating the discharge on to the belt. The length of haul by belt conveyor is more or less unlimited and they can be set up on travelling carriages to move with the excavator.

9.2.2 Rock excavation

The use of explosives to break up the rock in advance of mechanical excavation is necessary in all but the weakest rocks such as weathered mudstones, chalk, and shales. Explosives are sometimes necessary in shales to loosen the layers, especially if, as is often the case, they are interbedded with sandstones and siltstones. However, the use of explosives involves noise and vibration and the risk of annoyance to the public and possible damage to property. Much can be done to minimize blasting vibrations by limiting the weight of charges and the use of delay detonators. Expert advice from manufacturers of explosives should be sought on these aspects of the work, and general guidance is given in British Standard Code of Practice CP 5607: *Safe Use of Explosives in the Construction Industry*.

Excessive overbreak in excavations in hard rock can be avoided by adopting 'pre-splitting' techniques in which a line of close-spaced holes are drilled along the perimeter of the excavation for its full depth. In favourable rock conditions it is unnecessary to charge these holes with

explosives since detonation of the main charges placed in the normal pattern of blast-holes within the excavated area should fracture the rock along the line of the close-spaced holes. However, to ensure effective pre-splitting, light charges should be detonated in the holes before firing the main charges in the holes within the excavated area.

If the proximity of buildings prohibits the use of conventional explosives, then there is no alternative to breaking up the rock piecemeal by hand-held pneumatic breakers or tractor mounted mechanical breakers. These types of breaker inevitably cause noise and vibration, and if these must be avoided at all costs it will be necessary to resort to special methods of loosening the rock. These include freezing water in drill-holes by liquid carbon dioxide or the use of hydraulic burster cartridges. Holes for these methods can be drilled without much noise or vibration by rotary diamond core drills or, if no noise at all can be permitted, by the thermic lance method. In the latter method drilling rates up to 9 m/h to a depth of 36 m have been observed in granite. The size of the hole is 150–250 mm diameter. However, all these special methods are necessarily costly and are only considered in special circumstances.

9.3 Stability of slopes to open excavations

9.3.1 General considerations

Three main considerations govern the determination of stable slopes for open excavations. The first of these, as would be expected, is the type of soil. The second is the length of time over which the excavation is required to remain open, and the third is the permissible degree of risk of slipping. For example, if important property is close to the top of an excavation there must be no risk of a slip, and a high safety factor must be adopted. Similarly, there must be an adequate safety factor if a slip could damage a partly completed basement retaining wall or undermine a sewer or water main. On the other hand, if slipping causes no damage to existing or new structures and if excavation plant is on hand to clear away slipped material, then it is justifiable to take some risks in the interests of economy in the total quantity of excavation, provided of course that there will be no danger to workmen. This latter factor is important in narrow excavations and will be discussed in greater detail in Section 9.4.

9.3.2 Slope stability in cohesive soils

It can be shown theoretically[9.1] that an open excavation in a normally consolidated clay soil will stand vertically without support provided that the height of the face does not exceed the critical height (H_c), where

$$H_c = \frac{4\bar{c}_u}{\gamma}, \qquad (9.1)$$

where $\bar{c}_u$ is the average undrained shear strength of clay and γ the density of clay. Values of H_c for clays of various consistencies are as follows:

	Very soft	Soft	Firm
$\bar{c}_u$ (kN/m^2)	0–17.5	17.5–35.0	35.0–70.0
Critical height (m)	4	4–8	8–16

However, the above critical heights can rarely be adopted because of changes in the stability of the ground with the passage of time as a result of changes in pore-water pressure behind the face on release of lateral pressure. This is particularly important in fissured or laminated clays, the stability of which may be provided by negative pore-water pressures which serve to keep the fissures tightly closed when the mass of clay remains undisturbed. However, on removal of lateral pressure by excavation the change from negative to positive pore pressures may take place in minutes, hours, or days, and instead of the clay being held together at the fissures it is liable to slide in a mass of small fragments falling from the face or in the form of a massive bodily slip along a well-defined fissure plane. The author conducted experiments to ascertain the 'stand-up' time of vertical-side trench excavations at Immingham where a shallow fissured desiccated crust of firm to stiff alluvial clay was overlying very soft to soft alluvial peaty clay. The following times were observed:

Depth of trench (m)	Time in minutes after commencement of excavation for	
	Partial collapse	Complete collapse
2.9	37	49
4.1	27	39
4.7	14	35
5.2	33	40
5.5	18	23
5.5	26	34

In soft or firm normally consolidated clays (which are generally unfissured), long-term drying will open up cracks in the soil. Water getting into these at times of subsequent rain is liable to force off a mass of clay from the face. Thus, in all cases the excavated face should be cut back in *normally consolidated clays*. The degree of cutting back is a matter of experienced judgement, taking into account the time over which the excavation is

required, damage to temporary or permanent work, or danger to workmen.

Where long-term stability is required in *normally consolidated soft to firm clays* the required slope for a given safety factor can be determined with fair accuracy by the normal slip circle analysis, provided that there is reasonably adequate information on the drained shear strength of the soil. The safety factor depends on the risk involved in a major slip. If this would not cause damage to property or risks to life, then a low safety factor, say 1.2–1.3, is suitable. If it is essential to avoid any risk of slipping then a safety factor of 2–3 would be required. It must be remembered that a rotational slip involves the movement of a considerable body of earth accompanied by upheaval at the toe. Therefore, in deep excavations the trouble and expense in clearing a large mass of slipped earth must be kept in mind when assessing the safety factor. The risks of slipping are greatly increased if spoil is tipped close to the top of the slope. The stability analysis should take account of this possibility and, if necessary, the spoil tip must be moved back to a safe distance beyond the top of the slope.

Stiff glacial till will stand near vertically over long periods with only minor falls from the face as a result of frost damage or erosion from sandy lenses in the clay. The main risk of instability of clayey glacial till slopes lies in the possibility of lenses or pockets of water-bearing sand and gravel in the clay and of possible massive falls from a steeply cut face where the till is intersected by fissures.

The excavation for a dry dock at South Shields, described by Stott and Ramage,[9.2] was an excellent example of how a stiff unfissured glacial till (average shear strength 120 kN/m^2) can stand at a very steep slope. The 12 m deep excavation was cut with nearly vertical sides, and the face at any point remained unsupported for periods of up to 6 months while precast concrete buttresses were propped against the face and *in-situ* concrete panel walls were cast between them (Fig. 9.1). There is no doubt that this construction method gave a very considerable saving in money over the then conventional method of constructing dock walls in a timbered trench.

Stiff-fissured clays can give difficult problems in slope stability. Because of the unpredictable effect of pore-pressure changes on removal of overburden pressure, the safe slopes over the normal construction period for foundation excavations cannot be calculated from a knowledge of shear strength of the soil. Some observations collected by the author on the stability of excavation slopes in London Clay are shown in Table 9.1. Slips in an unstable slope in this type of clay take the form of a rotational shear slide of a large mass of clay (Fig. 9.2) or minor falls caused by slipping or crumbling along a well-defined fissure plane (Fig. 9.3). If there are no risks to structures or

Fig. 9.1 Vertical face of stiff glacial till in excavations for dry dock at South Shields.

Fig. 9.2 Rotational shear slide in stiff-fissured clay (South Humberside).

Table 9.1   Stability of excavations at the construction stage in London Clay

Site	Total height (m)	Slope ver : horiz	Stability conditions	Reference
Northolt, Middx.	2.4	1 : 1	Wholly in brown London Clay; unstable	
Isleworth, Middx.	3.7–4.3	1 : $\frac{1}{2}$	Wholly in stiff blue London Clay; stable except for minor shallow slips 5 months after excavation	Tomlinson (1954)[9.3]
Muswell Hill, London	6.1	Vertical	Wholly in brown London Clay; slipped 6 weeks after excavation	Serota (1955)[9.4]
Mill Lane, London	6.1	1 : $\frac{1}{2}$	Wholly in stiff London Clay; slipping arrested by covering slopes and 1–2 m back from crest with tarpaulins	Meigh (1957)[9.5]
Bradwell-on-Sea, Essex	7.0	1 : $\frac{1}{2}$	Wholly in brown London Clay; stable 4 months after excavation	Skempton (1954)[9.6]
Paddington	7.6	Vertical	Wholly in London Clay; stable	
North London	8.5	1 : $\frac{1}{2}$	Wholly in stiff brown London Clay; stable for 6 weeks	
London Airport	7.6–9.1	1 : $1\frac{1}{2}$	3.6–4.6 m gravel overlying stiff blue London Clay; extensive slipping about 6 weeks after excavation	Tomlinson (1954)[9.3]
Euston	9.1	Vertical	Wholly in London Clay; stable	Skempton (1954)[9.6]
Bradwell-on-Sea, Essex	13.7	1 : 1 in overburden 1 : $\frac{1}{2}$ in London Clay	2.4 m fill, 2.7 m marsh clay, 7.3 m brown London Clay, 1.2 m blue London Clay, slipped 19 days after excavation	Meigh (1957)[9.5]
Bradwell-on-Sea, Essex	14.8	1 : 1 in overburden 1 : $\frac{1}{2}$ in London Clay	3.5 m fill, 2.7 m marsh clay, 7.3 m brown London Clay, 1.2 m blue London Clay; slipped 1 day after excavation	Meigh (1957)[9.5]

Fig. 9.3   Minor slipping from $1:\frac{1}{2}$ slope in stiff-fissured London Clay, Middlesex. Note smooth appearance of fissure planes.

property above the crest of the slopes they can be excavated at 1 in $\frac{1}{2}$. This batter will not by any means ensure freedom from slipping, but the mass of the slips should be small enough to avoid undue trouble in clearing them or to avoid damage to partly completed structures. If slipping is likely to undermine structures or to give risks to life, then a slope of 1 in 2 or $2\frac{1}{2}$ must be provided or else the face must be strutted.

The short-term risks of slipping of a steeply cut face can be reduced by protecting the slope with tarpaulins or polythene sheeting. This was done on a site in north London (Fig. 9.4) where tarpaulins prevented ingress of rain and water leakages from construction operations into fissures in the clay. The tarpaulins remained in place until the installation of ground anchors to restrain the sheet piles supporting the excavation had been completed allowing the clay berm to be removed.

More substantial forms of protection of steeply cut slopes in stiff-fissured clays are needed where the excavation is required to remain open for periods of weeks or months. An example of this is the 19.5 m deep excavation in completely weathered granite for Orchard Station on the Singapore Metro.[9.7] The granite had weathered to a

stiff clay consistency and a 1 to 5 vertical slope was adopted for the period of construction as shown in Fig. 9.5. The original intention was to use soil nails (unstressed ground anchors) with a length of 5 m in the upper slope and 7–10 m in the lower slope. However, movement was detected when the excavation had been taken down about 6 m. Longer soil nails were then adopted as shown in Fig. 9.5, and they were supplemented by horizontal bored drains to relieve pore pressures building up behind the slope. Shotcrete (sprayed fine concrete) was used as surface protection.

Shotcrete consists of a cement : aggregate mixture in the proportions of one part of cement to three to five parts by weight of aggregate. The maximum size of the aggregate is 5–20 mm depending on the type and size of gun being used for application. The material passing a 0.125 mm sieve should be at least 3 per cent of the total. Before application the surfaces to receive the shotcrete are cleaned with a water and air jet to remove loose material. On vertical or steeply inclined surfaces the shotcrete is then applied in successive layers of 20–60 mm. The previously sprayed layer should have commenced to bond before applying the next layer, but the time interval

Fig. 9.4 Berm slope 1 : ½ protected by tarpaulins while anchors are being installed in sheet piling driven behind berm, north London.

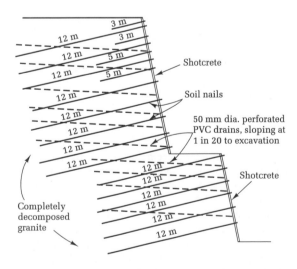

Fig. 9.5 Slope protection by soil nailing and drainage, Orchard Station, Singapore (after Hulme *et al*.[9.7]).

between each application should not be more than 24 hours. The amount of water used in the mix should be just sufficient to give good bonding properties with a minimum amount of rebound. Curing by water spray should commence as soon as drying patches are seen on the completed surface. Drainage holes are required through the shotcrete layer to prevent build-up of water pressure.

9.3.3 Slope stability in cohesionless or partly cohesive soils

Dry *sands and gravels* can stand at slopes equal to their natural angle of repose no matter what the depth. Values of the angle of repose quoted by Terzaghi and Peck are as follows:

	Angles of repose for dry sands	
	Round grains uniform	Angular grains well graded
Loose	28.5°	34°
Dense	33.0°	46°

Fig. 9.6 Stable excavated face in damp fine sand, Kent.

Damp sands and sandy gravels possess some cohesion and can stand vertically for some time. It is usual to excavate these materials to a steep batter for construction periods of up to a few weeks (Fig. 9.6). Instability of steep slopes in these soils can be caused by erosion from surface water or wind or by degradation from construction operations. Stability can be maintained by spraying the excavated face with cement mortar or fine concrete. The protective layer is reinforced with wire mesh restrained from movement by short unstressed ground anchors. *Water-bearing sands* are very troublesome in open excavations; if they are cut steeply, seepage of water from the face will result in erosion at the toe followed by collapse of the upper part of the face until a stable angle of 15–20° is eventually reached. Further information on the measures to avoid instability of slopes in water-bearing ground is given in Sections 11.2 and 11.3.

Water-bearing sandy soils are particularly troublesome when interbedded with thin layers of silt or clay. This type of ground is not uncommon in glacial deposits. The silt or silty sand is liable to bleed from the face, causing undermining and collapse of the more stable layers (Fig. 9.7). Because of the impermeable layers it may not be possible to use a ground-water lowering system; sheet piling to exclude ground water is often the only answer in this type of ground.

Dry silt will stand unsupported vertically, especially if it is slightly cemented. For example, loess soils found in the USA, the former USSR, and Middle Eastern countries are slightly cemented silts which can stand vertically to heights of more than 15 m. Indeed, excavations in such soils should preferably be cut vertically to avoid severe erosion due to water flowing over them. Uncemented dry silt is liable to flow like a liquid when subjected to vibrations and *wet silt* is one of the most troublesome materials which can be met in excavations. Seepage of water from a steeply cut face in silt causes the material to flow outwards from the toe at a very flat gradient; this is followed by slumping of the upper part and the whole face is progressively undermined until it eventually attains an overall stable angle.

9.3.4 Slope stability in rocks

It must not be taken for granted that rock excavations will stand with vertical slopes without trouble. Their stability depends on the angle of the bedding planes and the degree of shattering of an unsound rock. For example, dangerously unstable conditions can occur if bedding planes slope steeply towards an excavation (Fig. 9.8(a)), especially if there is ground water present causing lubrication along the bedding planes. Stable conditions are assured only when the bedding planes are near the horizontal or slope away from the excavation (Fig. 9.8(b)). Shattered rocks consisting of a large mass of loose fragments are liable to fall from a steeply cut face causing undermining and major collapse of overlying sound rock (Fig. 9.8(c)). Instability may occur due to toppling failures at re-entrants in steeply cut rock slopes where blasting and relief of overburden pressure cause opening of vertical fissures surrounding large blocks of rock behind the face. These conditions are shown in Fig. 9.9.

Unstable rock conditions were met in excavations for Stockholm's Underground Railway[9.8] where extensive fissures in the rock were filled with sand and clay. To prevent the rock slipping towards the excavation, prestressed anchor bolts were grouted into holes drilled deep into sound rock below excavation level (Fig. 9.10). Rock bolts can be of considerable help in preventing massive falls from a steeply cut rock face as well as in their normal application to support roofs of tunnels in rock.

Grouting of rock bolts provides a more durable anchorage with less risk of slipping than ungrouted solid rods, which rely for anchorage solely on the friction on the wedge. Long-term creep of the rock stressed by the wedge can cause relaxation of the anchorage. Retightening is

Fig. 9.7   Collapse of slopes in water-bearing fine sand interbedded with silt and clay layer, Co. Durham.

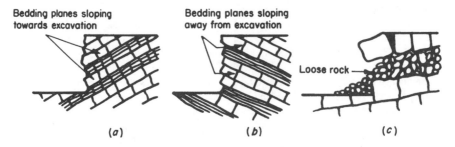

Fig. 9.8   Stability conditions in rock excavations. (*a*) Unstable conditions. (*b*) Stable conditions. (*c*) Unstable conditions.

then required to restore the original value. Washer plates, light steel sections, or steel mesh are used between the bolts to support the rock face. The type of supporting medium depends on the closeness of the joints in the rock.

Risks to operatives from rocks falling from a steeply cut face can be avoided by suspening heavy steel mesh from the top of the rock face. The mesh should be suspended well clear of the face to allow the rock to fall behind the mesh. Scrap anti-submarine netting has been used for this purpose.

Horizontally bedded mudstone or chalk can stand vertically to a considerable height, but are liable to long-term disintegration of the face by frost. The falls caused by frost action from a vertical face in chalk may be fairly extensive, but a 60° slope will remain generally stable for many years, although there will be a gradual accumulation of debris at the toe. A 45° slope in chalk will be virtually

Fig. 9.9 Re-entrant in steep rock slope showing undermining of Coal Measures sandstone by crumbling of underlying shales, South Wales. (Sandstone blocks have fallen where vertical joints emerge on face.)

trouble-free; especially if it can be covered with turf to bind the surface and prevent erosion of frost-loosened material.

Excavation in scree or boulder debris can be very difficult and adequate slopes must be given in order to avoid dangerous slides of boulders.

Further information on the stability of slopes in various types of rock formations is given in BS 6031, *Code of Practice for Earthworks*. This gives much useful information on excavation and ground support methods.

9.3.5 Computerized methods of slope stability analysis

Slope stability analysis provided one of the earliest areas of application of computer-based methods. Many programs have been developed to assist in the rapid and efficient assessment of slope stability by approaches such as that of Fellenius (Swedish circle method), and the various methods of Bishop and Janbu. Figure 9.11 shows typical plotted output from such a program, in this instance the SLOPE program developed as part of the Ove Arup & Partners OASYS system. With some programs it is also possible to analyse non-circular failure surfaces.

9.4 Trench excavation

An important factor in trench excavations for strip foundations or for retaining walls to basements is the stability against collapse of the trench sides. Such a collapse in a deep and narrow trench would be very likely to kill anyone in the trench beneath the fall. In Britain there is no definite requirement in the Factories Acts to timber trenches in all circumstances.

Regulation 8(i) of The Construction (General Provisions) Regulations, 1961, states:

> An adequate supply of timber of suitable quality or other suitable support shall where necessary be provided and used to prevent, so far as is reasonably practicable and as early as is practicable in the course

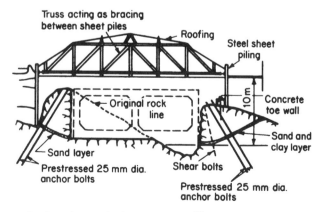

Fig. 9.10 Rock bolting in excavation for underground railway, Stockholm.[9.8]

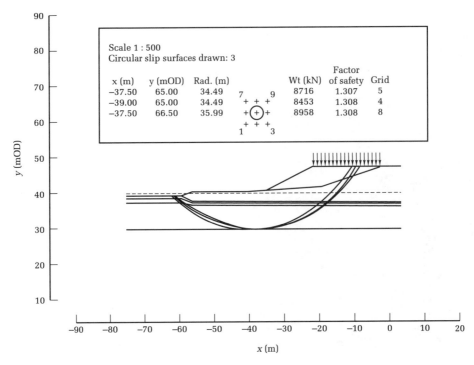

Fig. 9.11   Plotted output from computerized slope stability analysis.

of the work, danger to any person from a fall or dislodgement of earth, rock or other material forming the side or roof of or adjacent to any excavation, shaft, earthwork or tunnel:

Provided that this Regulation shall not apply

(a) to any excavation, shaft or earthwork, where, having regard to the nature and slope of the sides of the excavation, shaft or earthwork and other circumstances, no fall or dislodgement of earth or other material so as to bury or trap a person employed or so as to strike a person employed from a height of more than four feet (1.219 m) is unable to occur; or

(b) in relation to a person actually engaged in timbering or other work which is being carried out for the purpose of compliance with this Regulation, if appropriate precautions are taken to ensure his safety as far as circumstances permit.

Thus, the decision whether or not to support a trench, or the amount of support that is necessary, rests with the engineer. The usually accepted procedure is to provide some form of support, no matter what the soil conditions, whenever the trench is of such a depth that a collapse will cause death or injury to workmen. This usually requires

support to the sides of all trenches deeper than 1.2 m. Having decided to timber the trench, considerable judgement is then required on the amount of sheeting and strutting necessary. This aspect will be discussed in the next section of the chapter.

The choice between excavating a trench with supported vertical sides or with unsupported sloping sides is generally a matter of economics. However, there are circumstances where timbered trenches are unavoidable, such as in a restricted working area where space is not available for a wide trench, or in water-bearing sands which would cause slumping-in of the trench sides to a very flat angle of repose. The unsupported trench has the advantage of a clear working space unhampered by struts, but a deep trench requires the slopes to be cut well back to ensure the safety of the workmen, and the cost of this extra excavation together with the additional cost of carefully replacing and ramming the backfilled soil might well outweigh the cost of supporting a vertical-sided trench to the same depth.

It must be remembered that operatives working in excavations are at risk from causes other than the collapse of soil. Equipment or materials carelessly left on the ground close to an excavation can fall on operatives, and vehicles can run over the edge of an unfenced hole.

Therefore careful attention should be given to the provision of barriers, toe boards and walkways, with a clear space between the edge of the excavation and any stockpiled materials or equipment.

Much useful and detailed information on supporting trench excavations is given in reports by the Construction Industry Research and Information Association.[9.9, 9.10]

9.5 Support of excavations

9.5.1 General principles of design

The design of temporary supports to the sides of excavations is governed by the soil and ground-water conditions, and by the depth and width of the excavated area. In water-bearing sands and silts continuous support will have to be given to the face by means of timber *runners*, steel *trench sheets*, or *sheet piling*. With each of these methods it is necessary to drive the face support ahead of the excavation. Firm to stiff clays, compact dry or clay-bound gravels, compact or cemented sands, shales and stratified rocks can stand unsupported for varying lengths of time. Therefore, in these types of ground it is only necessary to provide support at an open spacing sufficient to withstand inward yielding of the mass of ground behind the face, and to avoid the risk of collapse of the sides due to, say, opening of fissures in a stiff clay or weakening along the bedding planes of a dipping rock stratum.

Wide excavations require heavier bracing to the frames of *walings* and *struts* than do narrow trenches or small shafts, and in the case of very wide excavations it will be necessary to support the face by *raking struts* or *shores*, or to tie it back by *ground anchors*.

Design of support schemes in advance of construction and the ordering of only sufficient material for the work will ensure safety in construction and will avoid waste and excessive cutting of the members. It must also be noted that a good design of a support scheme is one which permits easy *striking*. A design which necessitates cutting or smashing up of valuable material in heavy walings and struts in order to remove them on completion of the temporary work is thoroughly wasteful. Considerations of safety in restricted conditions also require ease of striking.

9.5.2 Support of excavations in loose sands and gravels, and soft clays and silts

In these ground conditions it is necessary to provide continuous support to the face by *close sheeting*, and it is also necessary to place the members in position as quickly as possible in order to avoid the sides slumping in. The close support can be provided either in the form of *runners*, or *sheet piles* driven ahead of the excavation or by placing horizontal *laggings* or *sheeting* behind *soldier piles* driven in advance of excavation.

Timbering with runners   The procedure in installing timber runners or steel trench sheets is shown in Fig. 9.12(a)–(d). A shallow excavation is first taken out and guide walings and struts are placed (Fig. 9.12(a)). The walings are temporarily blocked off the face by short boards (uprights) placed behind them. These enable the struts to be tightened and cleats or lip blocks are nailed or bolted to the ends of the struts. These are an important detail since they prevent the struts from falling if they work loose due to shrinkage of the timber or distortion of the walings. The struts are tightened by cutting them slightly too long and then driving one end with the other held in position, until they are at right angles to the waling. Alternatively, steel screw struts or hydraulic trench jacks can be used. Heavy struts are tightened by folding wedges at one or both ends or by jacking. Timber runners or trench sheets are then pitched behind the walings. In clays their lower ends or runners are sharpened with the bevel towards the face. This keeps them hard against the face as they are driven down. However, sharpening is not advisable in gravels because of the risk of splitting or 'brooming' as they are driven down. To facilitate pitching and driving long runners, an external guide waling is placed about 1 m above ground level. This can conveniently form part of a trestle used by workmen in driving down the runners by a heavy maul. A wedge or 'page' is placed between the runners and the waling to allow the runners to be driven down and then tightened against the face of the excavation one at a time. When the excavation reaches the level of the second bracing frame the walings are threaded through the struts of the top frame and then set in position. Longer uprights are set between the walings enabling the struts to be tightened against the face. Puncheons are placed between walings to support the top frame and lacing boards are nailed vertically between pairs of struts (Fig. 9.12(b)). The purpose of the lacing boards is to brace the two frames together, so preventing distortion resulting from uneven ground movements on either side of the excavation, and also to act as a safeguard against buckets or skips catching and lifting the struts as they are lifted out by crane. The excavation may then be taken deeper and, if necessary, a third frame is set and further puncheons and lacings are fixed between the second and third frames. The bottom frame is supported by puncheons on footblocks on the bottom of the excavation.

If the excavation has to be taken deeper than the runners, which are not usually more than 5 m long, a second setting of runners is placed within the walings of the top setting and an inner guide waling is also placed (Fig. 9.12(c)). These runners are then driven down in a similar manner to the top setting with bracing frames of walings and struts as necessary (Fig. 9.12(*d*)). There will, of course, be gaps between the runners at the ends of the struts of the top setting; these are filled with short horizontal boards or *cross poling*, tucked behind the runners as the excavation is taken down.

If still deeper excavation is required, a third setting of runners can be adopted, but this means that the excavation

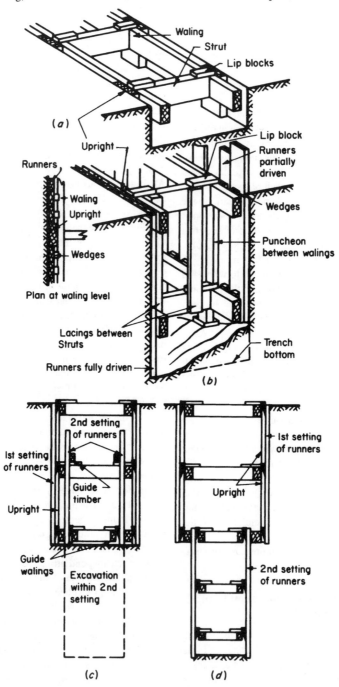

Fig. 9.12   Support of excavations in loose soil with runners. (*a*) First-stage excavation. (*b*) Second-stage excavation. (*c*) Pitching second set of runners. (*d*) Completed excavation.

at the top setting will be about 1 m wider than at the third setting with a corresponding increase in volume of excavation. Thus there is an economical limit to the number of settings that can be used in deep excavations; at some stage the use of long sheet piles or horizontal laggings will be more economical.

Support by sheet piling   Steel sheet piling driven by a piling hammer selected from the types described in Chapter 8 permits much deeper settings. In favourable ground conditions sheet piles can be driven in 15 m lengths ahead of the excavation, but in unfavourable ground such as compact gravel or soil containing cobbles or boulders it will be preferable to drive in shorter lengths and it may be necessary to excavate down and remove boulders from beneath the sheet piles as they are progressively driven deeper. Attempts to drive sheet piles too deeply in these ground conditions can lead to parting of the sheet pile clutches with consequent troublesome runs of water and sand into the excavation. The use of sheet piles also permits a much wider spacing of bracing frames. However, driving sheet piling can involve noise and vibration with annoyance to the public in populated areas and possible damage to property. The Silent pile driver operates hydraulically without noise or appreciable vibration (see Section 10.3). Also hammers can be used inside a sound-absorbent box. Further information on available sections and handling and driving sheet piling is given in the section on cofferdams in Chapter 10.

Difficulties occur from time to time where sheet piles are driven into stiff clays to some depth below excavation level, either to obtain support in passive resistance or as a cut-off against water. When the sheet piles are extracted clay wedged into the 'pans' is lifted with the piles leaving large voids below excavation level and possible disruption of permanent work. The practice of driving sheet piles to depths of more than about 0.5 m below excavation level should be avoided if at all possible in situations where the permanent work is close to the margin of the excavation.

Horizontal timbering   The other method mentioned above is to support the ground by horizontal laggings between soldier piles. This method is normally applied to deep excavations. The soldier piles are usually rolled-steel broad-flanged universal bearing pile sections driven before excavating from ground level to a level 1–2 m below the bottom of the excavation. As the excavation is taken down timber planks are inserted horizontally between the flanges of the piles (Fig. 9.13) and held against the face by wedges. Precast concrete planks are not satisfactory for the horizontal lagging since the soldier piles can rarely be installed truly vertically. Thus some

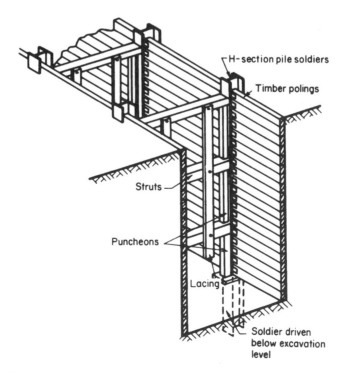

Fig. 9.13   Support by soldier piles and horizontal sheeting.

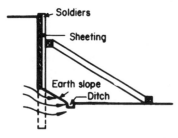

Fig. 9.14   Control of water seepage beneath timbering.

cutting of the horizontal members to make them fit between the soldiers is usually necessary. It is also necessary to expose the ground over a depth of two or three boards to provide enough space to slip one board between the flanges. In loose ground the boards can be secured to the exposed face of the inner flange of the piles by special clips. Driving the soldier piles, which are usually some 3–4 m apart, involves considerably less *duration* of vibrations than driving steel sheet piling. Where pile driving vibrations must be avoided the piles can be installed in holes drilled by a boring machine. Below excavation level the embedded lengths of the piles are surrounded by concrete or a sand–cement grout.

In water-bearing ground it is important to leave narrow gaps or louvres between the boards to allow drainage, so avoiding build-up of hydrostatic pressure behind the timbering. Where inflowing water causes erosion in the soil it can be checked by temporary earth banking and ditches to lengthen the seepage path (Fig. 9.14). The soldier pile method can also be used in conjunction with vertical runners as shown in Fig. 9.15. A waling is placed to span between the soldiers and the runners pitched behind. Further walings are set as the excavation progresses deeper. Where it is essential to avoid noise or vibration the soldiers can be set in pre-bored holes.

Horizontal timber sheeting can also be used in shallow excavations when short vertical walings are used to support the ends of the planks.

9.5.3 Support of excavations in stiff clays, compact or cohesive dry sands, or weak rock strata

These materials can usually be relied upon to stand unsupported for a varying length of time, but support is eventually needed to prevent yielding and settlement of the adjacent ground surface and to ensure the safety of workmen. Their ability to stand for a time without support greatly simplifies procedure in timbering. In trench excavations one of the simplest forms is the *open timbering* shown in Fig. 9.16.

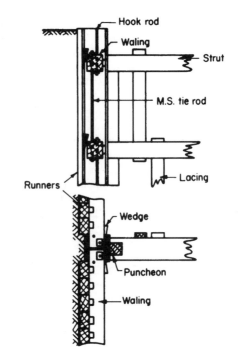

Fig. 9.15   Support by soldier piles with runners.

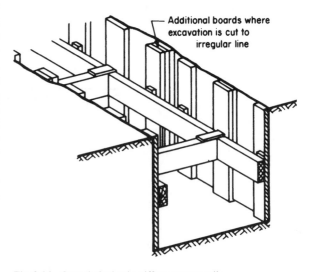

Fig. 9.16   Open timbering in stiff or compact soils.

After reaching the bottom of the excavation, 50–100 mm boards are placed against the face at each end of the walings. The walings are then placed and strutted across, further boards being placed at each intermediate strut position, and, if necessary, at intermediate positions between struts. If necessary, additional packing timbers are placed behind the boards if the face has been cut to an irregular line. If the boards are at 2 m centres the system is known as 'open timbering' and for this spacing the walings can be omitted, each pair of boards being individually strutted by 'pinchers'. Boards at 1 m centres are 'half timbering' and at 0.5 m centres 'quarter timbering', the final stage being close timbering. Open timbering can be used in stiff glacial till or sound rocky ground. Half timbering is suitable in stiff-fissured clays or compact, cohesive sandy or gravelly soils, and close timbering in dry sands and gravels or crumbly shales or in glacial till to prevent dislodgement of large stones.

9.5.4 Movable supports for trench excavations

Proprietary excavation support systems for trench excavation are widely used at the present day because of their rapidity in installation and removal, the avoidance of risks to operatives, and the lack of operatives skilled in the timbering method described in the previous sections. The principal types listed in *CIRIA Technical Note 95*[9.10] are:

Hydraulic frames (waling frames)
Shores
Boxes
Sliding panels
Shields

All the above systems can be used in reasonably stable ground which can stand unsupported for a sufficient length of time to lower the members into the trench. None of them requires the operative to work within an unsupported excavation, in fact they function more as a means of protection for the operative than as giving positive support to the ground. They are suitable for work in firm to stiff clays, damp sands and gravels, and weak weathered rocks. They can be used in water-bearing gravelly sands, sands and sandy silts if the ground water has been previously lowered by a wellpoint or bored well system (Section 11.2).

Hydraulic frame supports are shown in Fig. 9.17. The first stage consists of setting trench sheets (runners) in the excavation at a spacing which depends on the stability of the ground (Fig. 9.17(a)). The sheets are pushed into the soil below the trench bottom to a sufficient depth to make them temporarily self-supporting. The lower frame consisting of a pair of walings connected by hydraulically operated struts is then lowered into the trench with the pressure hoses attached to a pump at the ground surface (Fig. 9.17(b)). The pump is then operated to expand the struts and force the sheets against the sides of the trench. The upper frame is finally lowered into place and secured in the same way (Fig. 9.17(c)).

Shores consisting of separate pairs of steel trench sheets

forced apart by hydraulically operated struts are shown in Fig. 9.18.

Boxes (Fig. 9.19(a)) are assembled at ground level and lowered into a pre-dug trench, or lowered into a partly dug trench and then sunk as a caisson to the final level. They can be adjusted in width by selecting the appropriate strut length, and can be used in deep trenches by stacking one box above another.

Sliding panels (Fig. 9.19(b)) are assembled by pushing two pairs of slide rails (soldiers) into a partly dug trench (or from the surface in unstable ground). The sliding panels are then set between the rails and are pushed down progressively with the rails as the excavation is taken deeper. In deep trenches the panels can be stacked one above the other.

Shields (drag boxes) as shown in Fig. 9.20 consist of two metal panels permanently welded together, and towed along the trench by a crane or by the excavating machine. They are best suited as a means of protecting the operatives in stable ground when they can be set in a trench with steeply battered sides. Minor falls from the trench sides do not cause problems, but complete collapse can cause difficulty in extraction.

Systems using boxes, sliding panels or shields cannot be used economically if service pipes or cable ducts cross the line of the excavation at close intervals. Problems also occur if a heavy lifting effort is required to move them in a partly collapsed trench or if the trench has been partly backfilled and compacted in layers above the permanent work. The jerking and rocking of the units during extraction may then displace and damage the permanent work.

9.5.5 Timbering in shafts

A system of strutting different to that described above is used in timbering shafts for piers.

In the case of small shafts (Fig. 9.21(a)) no struts are needed since the ends of the walings are self-supporting where they are 'caught' by the walings or liners on the adjacent sides. The arrangement of struts for a large square shaft (Fig. 9.21(b)) is convenient, as the centre is left clear for hoisting materials by crane. In the case of rectangular shafts cross struts may be necessary (Fig. 9.21(c)). The striking piece at the end of one of the walings will be noted. This is provided because in a small excavation in swelling ground the walings can become tightly locked together. The provision of a striking piece, i.e. a piece of softwood which is readily prised out by crowbar, avoids the necessity of smashing a waling to pieces to release it.

9.5.6 Support of excavations in variable ground

It is quite the usual practice to vary the type of support for

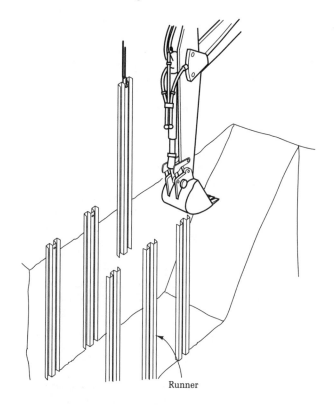

(a)

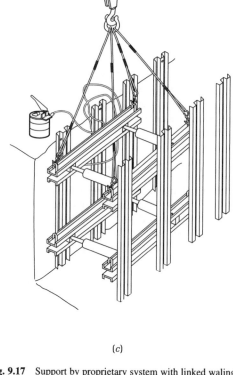

(c)

Fig. 9.17 Support by proprietary system with linked waling frames and hydraulically operated struts. (a) Runners set in place. (b) Lower waling frame lowered down. (c) Upper waling frame in place.

varying soil conditions in a given excavation. For example, Fig. 9.22 shows a method of timbering suitable for a site where loose or water-bearing sands and gravels or soft clays overlie stiff or compact material. The upper part of the excavation is supported by close-spaced timber runners or trench sheets. On reaching the stiff stratum the support can be changed to open timbering by the 'middle board' method. A 'garland drain' is placed at the base of the water-bearing stratum to intercept water seeping through the close timbering, so preventing it from accumulating at the bottom of the excavation in the stiff stratum.

The reverse method of support would be used in the case of, say, a stiff clay stratum overlying loose sand. Open timbering would be set in the stiff material; runners would then be pitched inside the walings of the upper settings and driven down with the deepening of the excavation into the loose material.

9.5.7 Support of large excavations

Very large excavations can be supported by systems of

(b)

Fig. 9.18 A movable support system for trench excavations.

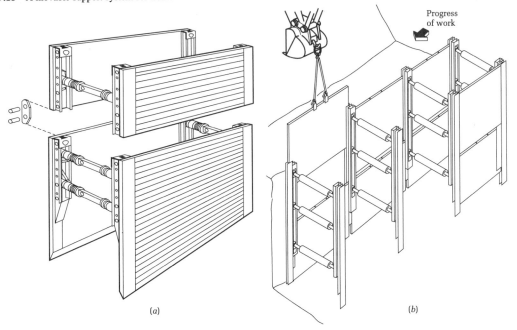

Progress
of work

(a)   (b)

Fig. 9.19 Proprietary trench support systems. (a) Box. (b) Sliding panel.

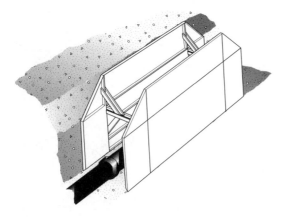

Fig. 9.20   Shield for trench support.

raking shores. In soft or cohesionless soils sheet piling can be pre-driven and the excavation is taken down, leaving a berm to support the sheet piles (Fig. 9.23(*a*)). The top waling is then placed and raking shores are fixed (Fig. 9.23(*b*)). The excavation is then taken deeper and a second waling is fixed and shored up (Fig. 9.23(*c*)). The third stage of excavation is taken out in short lengths. The third level waling is strutted horizontally and these struts are later concreted into the floor. The second waling can then be removed to allow space for constructing the bottom part of the retaining wall. The raking shores are braced together and the feet of the shores bear on king piles, on the partly completed concrete floor of the excavation, or on a timber or steel beam sill held by pins cast into a

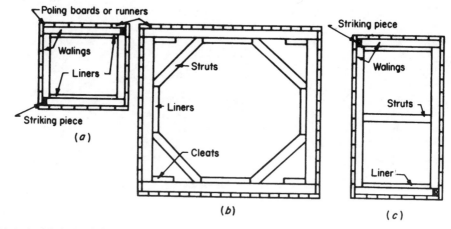

Fig. 9.21   Methods of timbering shafts. (*a*) Small shaft. (*b*) Large square shaft. (*c*) Rectangular shaft.

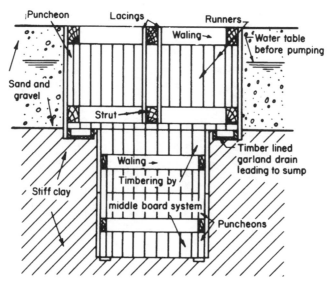

Fig. 9.22   Timbering for shaft in water-bearing sand and gravel overlying stiff clay.

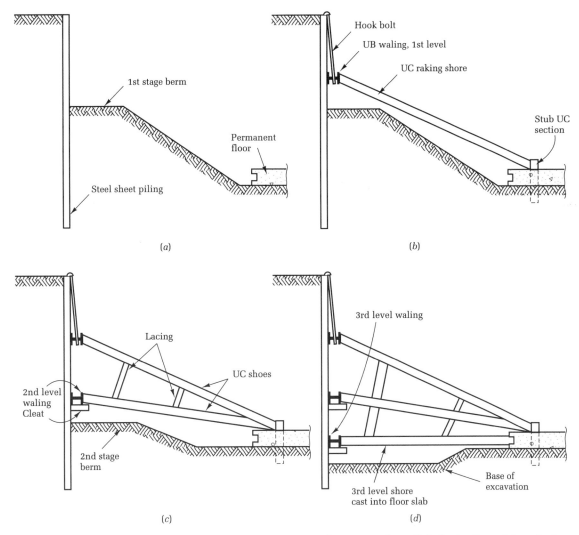

Fig. 9.23 Excavation support by raking shores. (*a*) Sheet piling cantilevers from 1st stage berm. (*b*) 1st level waling and shores fixed. (*c*) 2nd level waling and shores fixed. (*d*) 3rd level waling and shores fixed, excavation completed.

concrete floor. As an alternative to sheet piling, timber runners can be driven in one or more settings behind walings in a similar manner to that described for trench excavations. The method using soldier piles and laggings can also be used with raking shores (Fig. 9.24). In stable ground the soil between the soldier piles can be faced with *in-situ* concrete reinforced with bars or mesh or cut back to an arched profile and the exposed face covered with aprayed cement mortar or concrete. This method was used for the 24.6 m deep excavation for the four-level underground station at Raffles Place on the Singapore Metro.[9.7, 9.11] The soldier piles were set at 2 m centres in predrilled holes. The excavation was taken out in 2 m stages when the clayey completely weathered sandstone

was carved back in the form of an arc 0.45 m deep. This was sprayed with shotcrete to form a 75 mm thick layer reinforced with wire mesh. In weak collapsing soils the arch supports between the soldier piles can be formed by contiguous flight auger piles (Section 8.14.1) or jet-grouted columns (Section 11.3.7) installed in advance of the bulk excavation. The concrete in-fill methods described above are all relatively impervious and it is important to provide drainage to prevent the build-up of hydrostatic pressure behind the facing walls. Information on the use of shotcrete is given in Section 9.3.2. The layout of raking shores at the corners of excavations may become rather congested. A suitable arrangement to give improved working space is shown in Fig. 9.24.

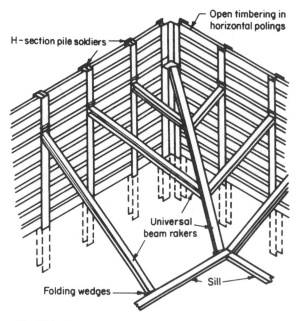

Fig. 9.24 Support by soldier piles and raking shores (view at corner of excavation).

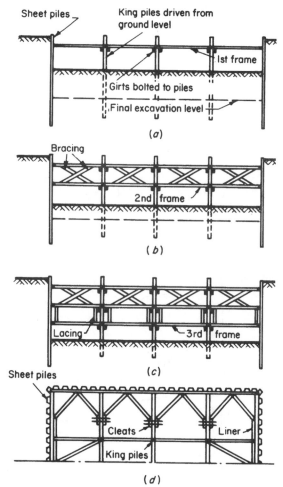

Fig. 9.25 Timbering to deep and wide excavation. (*a*) Excavate to level of second frame, fix top frame. (*b*) Excavate to level of third frame, fix second frame and bracing. (*c*) Excavate to final level, third frame fixed. (*d*) Half plan of timbering.

The methods of using raking shores, described above, have disadvantages. The shores obstruct the working area on the floor of the excavation; a flat angle is required to prevent them from sliding up on the walings or forcing up the walings; and rigid support to the lower ends is needed to avoid yielding. If the excavation is not too wide it is usually preferable to carry the struts right across. Careful design with transverse or diagonal strutting is necessary in order to prevent buckling of long struts, but at the same time the working space within the bracing frames should be as roomy as possible to avoid undue hindrance to construction operations.

A typical sequence of operations is shown in Fig. 9.25(*a*)–(*d*). The king piles, which may be universal bearing piles or column sections, are driven at the positions of joints in the struts and serve to support the bracing frames. They have a vital function in restraining the struts from buckling in a vertical direction, and they must be driven deep enough into the soil to develop satisfactory resistance to compressive or uplift forces, especially when only a single bracing frame is provided.

In wide braced excavations, careful consideration must be given to the risks of progressive collapse of the entire support system due to failure of a single member. For example, the failure of a strut dislodged by the accidental dropping of a load from a crane. Figure 9.26 shows the consequences of collapse of a 33 m × 21 m × 7.3 m deep cofferdam in which the sheet piles were supported by a

system of I-section steel walings and steel tubular struts at a single level. The struts were virtually pin-jointed at their intersections, but they were restrained from buckling over the long spans by means of steel I-section king piles driven to 1.1 m penetration into the stiff clay below the base of the excavation. This depth of embedment was sufficient to restrain the king piles against vertical movement except at one position, where an excavation was made for a sump close to one of the piles. The loss of support caused the pile to sink downwards followed by slow collapse of the entire bracing system and the sheet piles.

Fortunately there was no injury or loss of life in this case, but a more serious accident took place in 1906 in the excavation for the entrance lock to the Alexandra Dock in

Fig. 9.26   Collapse of sheet piling due to failure of a single component in the bracing system.

Newport.[9.12] The 11.3 m wide × 13.4 m deep excavation was heavily braced by frames of 300 mm × 300 mm timber walings and struts at 1.2 m vertical spacing. The struts were restrained against buckling by a row of 508 mm × 508 mm timber king piles along the centre line of the excavation. As in the previous case the king piles were designed to be restrained against vertical movement by their embedment below the base of the excavation. However, a layer of very compact gravel prevented the piles from being driven below final excavation level. At the lower stage of excavation it was necessary to excavate below the toes of the king piles which were propped by packing with timber. This prevented downward movement, but gave no restraint against upward movement. When the excavation reached a depth of 13.4 m the struts suddenly buckled upwards and the entire excavation for the lock collapsed almost instantaneously with a loss of 39 lives and many more injured.

9.5.8 Support by ground anchors

The use of tied-back anchorages instead of strutting or raking shores to support the sides of wide excavations has the considerable advantage of providing a completely clear working area. The higher cost of anchors compared with shoring is counterbalanced by the reduced cost of excavation. Ground anchors are installed by drilling holes at a downward inclination to obtain a grouted bond length beyond the zones of potential slipping or yielding of the tied-back mass of soil. The bond length depends on the friction developed between the grouted annulus around the anchor and the surrounding soil. High ultimate resistances of the order of 400–1800 kN are obtainable in cohesionless soils, the highest range being in very dense sandy gravels. Lower values are obtained in cohesive soils, depending on the diameter of the drill-hole and any local enlargements achieved by under-reaming, or cleavages formed by the action of high-pressure grout injections.

Much useful and practical information on the design and installation of ground anchors is given in BS 8081: *Code of Practice for Ground Anchors* and CIRIA report 65.[9.13] Three components of an anchor are considered in design. These are (*a*) the fixed (or bonded) anchor length, (*b*) the free (unbonded) anchor length, and (*c*) the anchor head. These components are shown in Figs 9.27 and 9.28. BS 8081 considers four types of anchor depending on the ground conditions:

(1) Type A anchors are installed in a straight shaft hole and the tendon is grouted through a tremie pipe, through a tube with a packer, or by a resin cartridge. Type A anchors are suitable for use in rocks or stiff to hard clays.

(2) Type B anchors are installed in straight shaft holes with temporary support by lining tubes. The grout is injected at low pressure through the lining tubes with a packer on the injection pipe or if secondary grouting is required through a *tube-à-manchette*. Injection pressures are less than $1000 \, \text{kN/m}^2$. Type B anchors are suitable for weak fissured rocks and fine to coarse alluvium.

(3) Type C anchors are installed in straight shaft holes supported by temporary lining tubes. Grout is injected as for type B at pressures greater than $2000 \, \text{kN/m}^2$. Type C anchors are suitable for fine-grain cohesion-less soils or stiff clays.

(4) Type D anchors are formed with enlargements on the shaft (belled anchors) as shown in Fig. 9.29(a). They are grouted through a tremie pipe and are suitable for firm to stiff clays. A single enlargement can be formed by forcing pea gravel against the walls of the borehole (Fig. 2.29(b)).

The fixed anchor length should not be less than 3 m or greater than 10 m for cohesionless soils. The area of steel in the tendon should not be more than 15 per cent of the borehole cross-sectional area for parallel multi-strand anchors, or more than 20 per cent of the area for single unit or noded multi-strand tendons.

CIRIA report 65[9.13] gives equations for determining the pull-out resistance of anchors in soils. These are as follows. For cohesionless soils:

$$T_u = Ln \tan\phi', \quad (9.2)$$

where

T_u = ultimate pull-out resistance,
L = fixed anchor length,
n = a factor,
ϕ' = effective angle of shearing resistance.

For straight anchors in clays (Fig. 9.29(a)):

$$T_u = 0.3 c_u \, \pi DL, \quad (9.3)$$

where c_u is the undrained shear strength and D the nominal borehole diameter.

For straight shaft anchors in clays with enlargement of fixed length by gravel injection (Fig. 9.28(b));

$$T_u = 0.6 c_u \, \pi DL + \frac{\pi}{4}(D^2 - d^2)c_u N_c, \quad (9.4)$$

where N_c is the bearing capacity factor.

For belled anchors in clay (Fig. 9.29(c)):

$$T_u = \pi D_u L_u c_u f_u + \frac{\pi}{4}(D_u^2 - d^2)c_u N_u + 0.3 c_u \, \pi dl, \quad (9.5)$$

where N_u is the bearing capacity factor and f_u the efficiency factor.

In equation (9.2) n is taken as 400–$600 \, \text{kN/m}$ in coarse sands and gravels and $150 \, \text{kN/m}$ in fine to medium sands. In equation (9.4) N_c is taken as 9, and in equation (9.5) N_u is taken as 0.8. The efficiency factor f_u is about 0.8.

BS 8081 recommends an ultimate shear stress between grout and rock of 10 per cent of the unconfined compression strength of the rock but not more than $4 \, \text{N/mm}^2$. BS 8081 also gives grout to tendon ultimate bond stress values of:

Not more than $1 \, \text{N/mm}^2$ for plain bars
Not more than $2 \, \text{N/mm}^2$ for clean strand or deformed bars
Not more than $3 \, \text{N/mm}^2$ for locally noded strand

Safety factors for *permanent* ground anchors are given in the code as follows:

	Minimum safety factor
Tendon	2.0
Ground to grout interface†	3.0
Grout to tendon or grout to encapsulation interface‡	3.0
Proof load factor	1.5

† This may be required to be increased to 4.0 to limit creep of the ground.
‡ A minimum of 2.0 may be used if the ultimate bond stress is obtained from full-scale field loading tests.

The above minimum safety factors relate to single isolated anchors, they do not take into consideration group effects or the pull-out resistance of a cylinder or cone of soil or rock as described in Section 7.16.

Ground anchors may be placed at a single level, or in deep excavations at several levels, installed at each successive stage of excavation. The anchors are usually inclined downwards at angles of 15–20° to the horizontal. A steep inclination induces a high vertical component of load in the anchored wall with a risk of failure at the toe, particularly if the anchored wall is of sheet piling. The 'free' or unbonded length of the anchor tendon is generally arbitrarily assumed to lie within a zone bounded by a line having an inclination of 40° to the base of the excavation (Fig. 9.27).

Earth pressure on the anchored wall is transferred to the tendons either through horizontal walings (Fig. 9.28(a)) or through bearing plates at the head of each anchor (Fig.

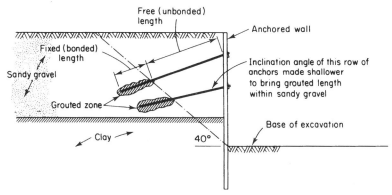

Fig. 9.27   Arrangement of ground anchors for support of excavation.

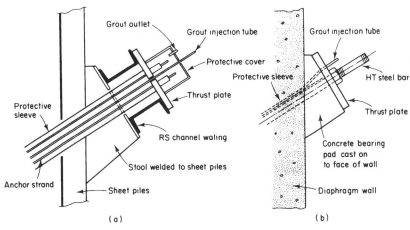

Fig. 9.28   Arrangement at head of ground anchors. (*a*) Load on cable anchors transferred to sheet piling through horizontal walings. (*b*) Load on bar anchor transferred to diaphragm wall through bearing pad.

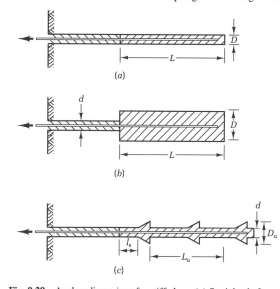

Fig. 9.29   Anchor dimensions for stiff clays. (*a*) Straight shaft. (*b*) Straight shaft enlarged by forcing out pea gravel. (*c*) Belled.

9.28(*b*)). The latter method is generally used for stiff walls capable of distributing the load over an area surrounding the anchor head. The use of walings gives greater security should one anchor fail under load.

Problems in the design of anchor support systems can occur where deep sewers or tunnels are located near the excavation, and it is necessary to vary the inclination of the anchors to avoid these obstructions. Instability can occur where ground anchors are located at re-entrant faces of excavation. Additional lengths are necessary at such locations (Fig. 9.30).

Where anchored diaphragm walls, soldier pile walls, or sheet pile walls are terminated in a rock formation above final excavation level, care is necessary to provide sufficient space between the face of the rock excavation and the face of the sheeted excavation to avoid collapse of the rock face due to the vertical compression loading induced by the anchorages (Fig. 9.31(*a*)). The inclination of joint planes in the rock can be a critical factor. Other methods of avoiding instability include installing soldier piles in

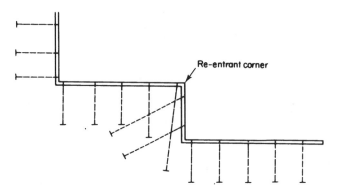

Fig. 9.30   Arrangement of anchors at re-entrant corner in excavation (plan view).

holes drilled to the base of the excavation (Fig. 9.31(*b*)) and installing an anchored reinforced concrete toe beam at the base of a sheet pile wall (Fig. 9.31(*c*)).

Installation of ground anchors may involve water flush drilling, which can cause hydrostatic pressure to develop behind the sheet piling supporting permeable soils. Also care is necessary when grouting anchors to avoid uplift of any structures sited above the anchorage zone, since the cumulative pore pressures induced in soils of low permeability can be very high. Some inward yielding of the excavation supports is likely to occur during the grouting operations, with a reversal of this movement when the anchors are stressed. Where temporary ground anchors are required for support over a long period of time some protection against corrosion should be provided. This can take the form of sheathing the unbonded length of the anchor tendon by a plastic tube or wrapping it with grease-impregnated tape, and painting the head of the tendon and bearing plate.

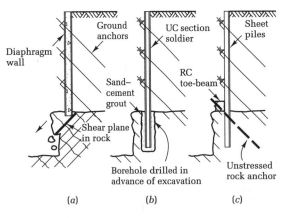

Fig. 9.31   Excavation supports close to rock face. (*a*) Failure at toe of diaphragm wall. (*b*) Soldier piles in predrilled boreholes. (*c*) Support to sheet pile wall by RC toe beam.

Problems can occur with the removal of ground anchors where these extend under adjacent property. Anchors used for excavation support at Raffles Place Station on the Singapore Mass Rapid Transport[9.11] were removed by explosive charges. Detonating cord with 25 ms delay detonators was placed in a plastic tube installed with the wire strand. The detonations broke the soil to grout bond.

The installation of a ground anchor support system should be controlled at all stages, commencing by installing trial anchors to verify the design assumptions. Preferably the trial anchors should be loaded incrementally and taken to the stage of a complete pull-out from the soil provided that the stress in the anchor is not greater than 80 per cent of the ultimate tensile strength of the steel. Where cable anchors are used the entire cable should be pulled rather than the individual strands. The load–extension curve for the trial anchors should be drawn and compared with the theoretical stress–strain curve for the unbonded length of the tendon (Fig. 9.32). (A procedure for installing and testing trial anchors is given in BS 8081.)

The working anchors should then be installed. The extension of the anchor rod or cable at the design preload and the recovery after 'locking-off' the anchors should be measured in each case, either for the complete rod or cable or for each strand in a cable. All anchors should be designed to be capable of withstanding a proof test load of 150 or 175 per cent of the working load, and the proof test load should not stress the anchor by more than 70 per cent of the ultimate tensile strength of the steel.

Random working anchors should be selected for proof tests to 150 per cent of the working load, and the load–extension graph should be compared to that of the initial trial anchors. The test loads should be maintained for at least 24 h with a check on the amount of creep during that period. Selected working anchors should also be subjected to reloading tests at various intervals of time to check whether or not there is any deterioration in the grout to soil bond resistance.

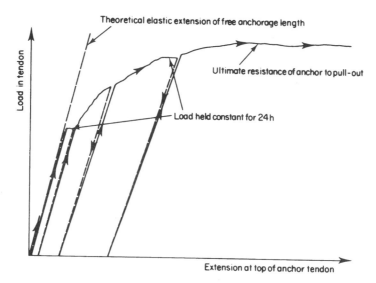

Fig. 9.32   Load-extension graph in incremental load test to failure in pull-out from soil.

It is also helpful, as a check on the assumptions made in estimating earth pressure, to monitor the development of load in selected anchors during the stages of excavation. This can be done by inserting a load cell between the anchor head and the thrust plate. Movements of the tops of the anchored wall should be recorded by reference to a datum system well clear of the excavated area.

Mitchell[9.13a] has recommended the following form of record for each working anchor:

(1) Reference number;
(2) Diameter of borehole;
(3) Direction and depth of borehole;
(4) Date of boring;
(5) Date of grouting;
(6) Date of stressing;
(7) Length and diameter of temporary lining used during boring;
(8) Consistency, colour, structure, and type of various soil strata penetrated;
(9) Details of obstructions encountered;
(10) Details of tendon installed;
(11) Length of 'free' anchorage (unbonded length);
(12) Length of 'fixed' anchorage (bonded length);
(13) Size and spacing of any enlargements in fixed anchorage length;
(14) Grout quantity injected and pressures employed;
(15) Grout test results;
(16) Tendon extension during load (including any proof test results);
(17) Maximum and minimum daily site temperatures at each day that grouting, loading, and monitoring is carried out;
(18) Measured loads.

9.6 Structural design of supports to excavations

9.6.1 Materials

The four types discussed are timber, steel, cast iron, and precast concrete.

Timber   The British Standard Code of Practice, BS 8004: *Foundations* requires timber members in cofferdams and caissons to comply with the requirements for strength classes SC3, SC4, and SC5 of BS 5268. The grade stresses should be modified in accordance with the slenderness ratio of the member.

Where the excavations have to remain open for long periods the grade stresses may have to be reduced by a factor between 1.0 and 0.7 to allow for the decay to be expected in the timber during its period of use. However, it should be noted that wetting of timber does not, by itself, cause decay and no reduction in working stresses need be made for timber subjected to wet conditions. In deep excavations where it is desired to keep the bracing frames as wide apart as possible (for ease in working), it may be advantageous to adopt a higher grade of timber for the lower levels.

Steel   Steel trench sheets can be used as runners or poling boards. Steel is also used for struts in the form of light adjustable units for trench work or for raking shores and also for walings and struts in deep excavations, either in the form of heavy universal beams or compound girders, or of lighter trussed sections built up from angles or channels. BS 8004 requires the steel to be manufactured to BS 4360 and designed in accordance with BS 449: *The*

Use of Structural Steel in Buildings. The latter code has been superseded by the limit state code BS 5950. It is possible that this new standard may be used increasingly for excavation supports after Chapter 8 (Retaining Structures) of Eurocode 7 has been published, but in its present form it is more suitable as a code for checking than for design.

Precast and cast-in-place concrete   Precast concrete is used in the form of planks or curved segments for sheeting excavations where the support has to be left permanently in place. Precast concrete can also be used for walings and struts, and on sites where sheet piles are required to be left permanently in place they are often constructed in reinforced concrete cast *in situ*. BS 8004 requires the quality of concrete to comply with BS 8110 or Clauses 201–209 of CP 114: 1949. Concrete stresses should be in accordance with BS 8110 or Clause 303 of CP 114.

Tubular steel struts were used in conjunction with reinforced concrete walings to support the 26 m wide × 13 m deep excavation for the southern approach to the River Ij Tunnel in Amsterdam constructed in 1963–64 (Fig. 9.33). In plan, the sheet piling formed a series of arches with chord lengths of 12–18 m. Each arch was supported by massive reinforced concrete arch walings at two levels. Steel H-sections were provided along the chords to resist longitudinal forces. The ends of adjacent walings formed a massive concrete thrust block fitted with bearing plates to receive the cone-shaped ends of the

1500 mm diameter tubular steel struts. The tubes were of welded plate construction without internal stiffening. They were designed to resist a thrust of 1000 kN.

An interesting feature of the design was the placing of the cone-ends eccentrically to the axis of the tube. The eccentric thrust so provided neutralized the bending stresses set up by the 5–6 t mass of the strut and ensured that the whole cross-section was more or less equally stressed in compression.

Precast bracing units consisting of centrally supported short walings were used by Edmund Nuttall Ltd in sheet piled excavations for the approaches to the Dartford–Purfleet Tunnel[9.14] (Fig. 9.34).

Cast-in-place concrete can be used instead of timber sheeting in conjunction with support by vertical steel soldier piles as described in Section 9.5.2.

The use of the permanent basement floors for strutting wide excavations was described in Section 5.4.4. A further example of this is shown in Fig. 9.35. At this site in Zürich[9.15] a 17 m deep basement excavation was surrounded by a cast-in-place concrete diaphragm wall which was supported at five levels by the permanent reinforced concrete floors. Each intermediate floor was cast on formwork propped off the ground at the four stages of excavation. Wide openings were left for removing the excavated soil and the floors were supported on the permanent steel tube piles which anchored the basement against uplift.

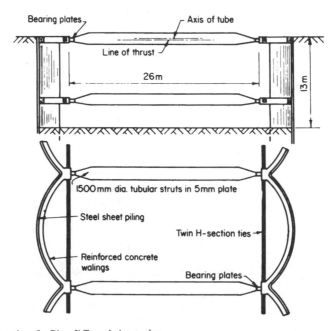

Fig. 9.33   Support of excavations for River Ij Tunnel, Amsterdam.

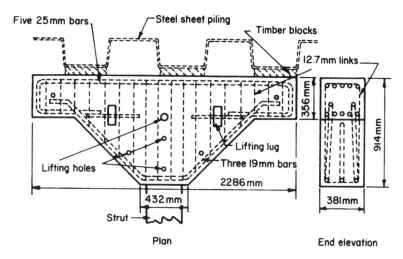

Fig. 9.34 Precast concrete waling units used for Dartford–Purfleet Tunnel.

Fig. 9.35 Permanent basement floors used to strut diaphragm wall in 17 m deep excavation at Zürich. (Note large openings.)

9.6.2 Calculation of lateral pressures on supports to excavations

Excavations up to 6 m depth It is unnecessary and unrealistic to calculate lateral pressures on supports to excavations shallower than 6 m unless hydrostatic pressure has to be taken into account. Lateral pressures at shallow depths can be highly variable in any given soil type. For example, a clay will shrink away from behind the timbering in dry weather, sometimes forming a wide gap down which crumbs of dry clay may fall. The onset of wet weather causes the clay to swell and, if the gap has become filled with clay debris, the swelling forces on the timbering may be high enough to cause crushing or buckling of the struts. In stiff or compact soil higher stresses can be caused in struts and walings by hard driving of folding wedges at the ends of struts than from earth pressure. Again, if dry timber is used to support an excavation in water-bearing ground, the seepage through the runners on to the walings and struts, and the effects of rain falling on to the struts, can cause heavy stresses in the timbers due to swelling of the dry timber. Expansion of soil at times of hard frost can also cause heavy loads on timbering.

Therefore, the usual method of designing timbering by rule of thumb for shallow excavations is fully justified. The sizes of timbers given by such methods have been based on centuries of experience and take into account the desirability of reusing the timber as many times as possible, the requirement of withstanding stresses due to swelling of dry timber, and the necessity or otherwise of driving wedges tightly at the ends of the struts (or of jacking the ends of the struts) to prevent yielding of the sides of the excavation. They bear little or no relationship to the stresses arising from earth pressure. The reader may have noticed that the sizes of timbers used in the shallower excavations are more or less the same on any job no matter what the type of soil or even the depth of excavation.

If steel struts and walings are used, the stresses in them may be more representative of earth pressure than with timber members. However, the usual procedure is to use the struts to apply pressure *to the ground* by wedging or jacking rather than to allow earth pressure to come on to the supports. The jacking and wedging forces depend on the extent to which it is desired to prestress the ground to minimize yielding. An example of the large forces required to prestress the ground to prevent settlement is given by the 10–12 m deep excavation for London's Hyde Park Underpass,[9.16] which was taken out within 1 m of the foundations of St George's Hospital. As a precaution against yielding and settlement, the 9 m long steel walings were set at three levels, each waling having three struts (twin 356 mm × 406 mm I-beams). Total loads jacked into each waling frame were 837, 3198, and 4065 kN respectively, for the three levels, i.e. a total jacking load of 8100 kN for about 100 m² of excavated face. The measured settlement of the hospital wall was less than 3 mm.

A more general practice is to preload the struts to some value less than the predicted earth pressure, bearing in mind that some permitted amount of yielding will reduce the active earth pressure on the support system. The 24.6 m deep excavation for Raffles Place Station on the Singapore MRT[9.11] had six levels of bracing frames. The struts for the upper three levels were preloaded to 30 per cent of the design load, and those for the lower three levels to 50 per cent of this load.

Sizes of timber struts for excavations to a depth of about 6 m are only nominal and calculated on 'rule of thumb' procedures. One such rule which is applicable to square struts for excavations from 1.2 to 3 m wide is

Strut thickness (mm) = 80 × Trench width (m).   (9.6)

It is convenient practice, though not always employed, to use struts of the same depth as the walings and often of the same scantling in order to avoid too many sizes of timber on the same job. Thus, 225 mm × 75 mm struts are used with 225 mm × 75 mm walings, 300 mm × 150 mm struts with 300 mm × 150 mm walings, and so on. The spacings of struts are governed by the lengths of the walings. Thus, 225 mm × 75 mm walings are usually supplied in 3600–4800 mm lengths. Three struts is a convenient number for a pair of 3600 mm walings, giving a spacing of 3400 mm for the three struts, with the pair at the abutting ends of walings at about 200 mm centres. Heavier timbers, for example 300 mm × 150 mm, 225 mm × 225 mm, and 300 mm × 300 mm, can be obtained up to 6.3 m long, or

even longer by special order. However, it is often convenient to work with much shorter lengths because of the difficulty in threading long walings for lower settings through the previously set frames. Also, short walings are preferred for convenience in striking timbering before backfilling deep excavations, when the lower frames have to be struck first and threaded back through the upper frames with the added obstruction of the completed foundation structure within the excavation.

Excavations deeper than 6 m   Economies in material can be achieved by calculating earth pressures in excavations deeper than 6 m. It is not necessary to do this in all cases. Long experience in known ground conditions is often a better guide than calculations based on soil mechanics analysis. But where deep excavations have to be carried out in soil conditions of which the engineer has no previous experience to guide him, or where a different system of ground support is employed to that usually adopted for the locality, then it will be advisable to calculate earth pressures based on information provided by borings and laboratory tests on soil samples. Earth pressure calculations also make possible an economical spacing of walings and struts and permit the development of the full flexural strength of the various components including the sheeting members.

The method of earth pressure calculation in general use for excavation supports has been developed by Terzaghi based on observations of actual loads in struts in full-scale excavations in sand in Berlin (Spilker, 1937)[9.17] and in soft clay in Chicago (Peck, 1943).[9.18] Terzaghi and Peck's empirical design procedure is described in *Soil Mechanics in Engineering Practice*.[9.19] Figure 9.36 shows their rules for calculating the pressure distribution to obtain the maximum strut loads in sands and clays.

It is important to recognize the difference between the pressure distribution on a retaining wall and on a strutted excavation. A retaining wall acts as a structural unit and fails as a unit, whereas a strutted excavation has some flexibility, and local concentrations of earth pressure can cause high loads on individual bracing members. If one strut fails it will immediately throw increased loads on to the adjoining members, thus initiating a general collapse of the system. Hence the trapezoidal distribution of Terzaghi and Peck was intended to be an *envelope* covering the maximum strut loads likely to occur at any level rather than representing the average strut loads.

Studies of earth pressure in soft clay at Shellhaven by Skempton and Ward[9.20] illustrated the effects on the strut loads of pile driving into a soft clay. The pile driving lowered the shear strength of the sensitive clay resulting in a 60 per cent increase in the strut loads in the lowest frame. The effects of frost in Norwegian stiff clay on strut

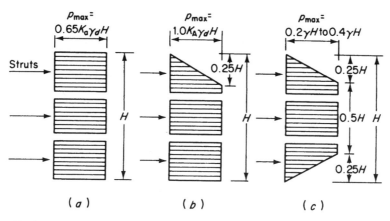

Fig. 9.36 Terzaghi and Peck's rules for pressure distribution on strutted excavations. (*a*) Sands. (*b*) Soft to firm clays. (*c*) Stiff-fissured clays.

loads were measured by Di Biagio and Bjerrum.[9.21] They caused a fivefold increase in strut loading in a 4 m deep excavation accompanied by buckling of walings and struts, necessitating the installation of additional timber bracing frames. the subsequent thaw was accompanied by a large reduction in strut loads. The two upper frames showed a marked reduction in loading over the winter period, probably due to expansion of the ground surface adjacent to the trench.

Calculation of earth pressure on strutted excavations in cohesionless soils   The empirical envelope of pressure distribution in sands is shown in Fig. 9.36(*a*). The maximum intensity of pressure is given by

$$p_{max} = 0.65K_a\gamma_a H, \qquad (9.7)$$

where

$\quad K_a$ = coefficient of active earth pressure
$\qquad = P_a/\gamma_d(H^2/2)$,
$\quad P_a$ = total pressure calculated as for a retaining wall with zero wall friction,
$\quad \gamma_d$ = bulk density of the soil,
$\quad H$ = depth of excavation.

The value of K_a varies with the angle of shearing resistance of the soil, and values are given in Table 5.4. Typical values for the bulk density of various materials can be found in Appendix A.

It should be emphasized that the empirical rules for the calculation of earth pressures are based on *dry* or *drained* soils. Hydrostatic pressure cannot be redistributed and it must be allowed for in full with the normal triangular distribution associated with hydrostatic pressure. Methods of calculation in such cases are described in Section 10.2.

Calculation of earth pressure in strutted excavations in soft to firm cohesive soils   The trapezoidal envelope of earth pressure in strutted excavations in cohesive soils is shown in Fig. 9.36(*b*). The maximum intensity of pressure is given by the formula

$$p_{max} = 1.0K_A\gamma H, \qquad (9.8)$$

where

$\quad K_A = 1 - m4\bar{c}/\gamma H$,
$\quad \gamma$ = density of the soil,
$\quad H$ = depth of cut,
$\quad \bar{c}$ = average cohesion of clay,
$\quad m$ = a coefficient.

This pressure distribution was derived empirically by Peck for soft to medium clays, and Terzaghi and Peck state that in adopting values of cohesion of the clay from field and laboratory test data the *lowest average* shear strength from any one of the borings should be taken. In the case of soft sensitive clays due allowance should be made for loss in shear strength due to remoulding or disturbance of such soils. The coefficient m can be taken as unity for most soft to firm clays, but observations in soft clays at Oslo indicated a value of 0.4 where $\gamma H/c$ was greater than 4.

Calculation of earth pressure in strutted excavations in stiff-fissured clay   The pressure distribution of Terzaghi and Peck is shown in Fig. 9.36(*c*). The maximum pressure is given by

$$p_{max} = 0.2\gamma H \text{ to } 0.4\gamma H. \qquad (9.9)$$

The lowest range is applicable to excavations which remain open only for a very short period of time.

Some engineers believe that the Terzaghi and Peck design rules overestimate earth pressures on excavation supports, but it is often forgotten that the diagrams in Fig. 9.36 represent an envelope of the maximum strut loads. Flaate and Peck[9.22] showed that the great majority of the measurements of the value of K_A in equation (9.8) were within ±30 per cent of the average calculated from this equation for excavations in soft to firm clays in Japan, UK, USA (Chicago) and Norway (Oslo). Strut load measurements made in six deep excavations for the Singapore MRT[9.23, 9.7] showed that a value of unity for m in equation (9.8) gave reasonable results for excavation supports in soft peaty clays and soft to firm marine clays. For excavations up to 30 m deep in hard clays and boulder clays the pressure envelope corresponded to a maximum of $0.25H$ in equation (9.9). In sandstone and siltstone rocks completely weathered to a hard clay consistency strut load measurements in a 16 m deep excavation gave an envelope corresponding to $0.5H$ in this equation.

A recent trend is the 'observational' method to the design and installation of excavation support. Measurement of strut loads are continuously monitored from the time of their installation and adjustments are made to the spacing of the struts depending on whether the strut loads are lower or higher than the values calculated for the initial design. This procedure is similar to the use of the observational method for the construction of embankments on soft clays where the rate of placing the fill is determined from observations of pore pressure development within the soft clay and the settlement of the base of the fill.

The method is best applied to deep excavations where the strut spacing at the lower levels can be adjusted from observations of loading at higher levels. It is also suited to cut-and-cover work over considerable lengths of excavation, for example in highway and railway tunnels. It was used for this purpose in the construction of the 1.8 km long Limehouse Link Road[9.24] where part of the excavated length was supported by steel sheet piling with tubular struts at two levels in the 18 m deep cut. The savings were not so much in the saving of materials for strutting as in the reduced costs of constructing the permanent work without the hindrance of working between close-spaced struts.

Instrumentation for monitoring strut loads when using the observational method include data loggers on the load-cells producing readings at hourly intervals. It is important that these should be reviewed by a responsible person so that timely action can be taken to install additional struts from a reserve held on site if loads become excessive, say from temperature effects on long steel members.

It should be pointed out that the application of the observational method to the design of excavation supports is not new. It has been practised for over 100 years by experienced timbermen who will strike timber struts or mine roof supports by a sledge-hammer, and will know from the sound of the blow whether or not the member is overstressed.

Water pressure on sheet piled excavations in clay Because borings show no water present in a clay soil it must not be assumed that hydrostatic pressure cannot build up outside a sheet piled excavation in clay. For example, if the sheet piles are driven completely through the clay stratum into a water-bearing sand layer, the water may creep up from this layer into the slight gap which is always likely to be present between the sheet piles and the clay in the region of the interlocks. The water pressure in this gap may then force back the clay sufficiently to cause water pressure over a wide area of the sheet piling at the sub-artesian pressure within the sand layer (Fig. 9.37(a)). Another possible source of water developing pressure is seepage down the back of the piling from ponded surface

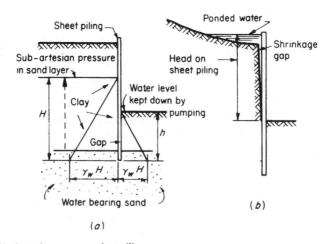

Fig. 9.37 Development of hydrostatic pressure on sheet piling.

water (Fig. 9.37(*b*)). A route for such seepage is the enlarged hole formed by whip in long sheet piles during driving or shrinkage of the clay away from the sheet piles in dry weather. This entry of water can be prevented by keeping the clay well rammed against the back of the piling. Some stiff-fissured clays contain water under pressure in random fissures or pockets. If such fissures are intersected by sheet piles they form a route for water to find its way to the back of the sheet piles. This possibility of the development of hydrostatic pressure in clays does not arise in excavations supported by *in-situ* concrete with drainage holes or by timber face boards where the gaps between the boards will allow the water to leak through, thus preventing build-up of pressure.

As a final word on methods of supporting excavations by sheet piling or timbering, the author stresses the need for accuracy in setting the timbers. Runners should be set truly vertical or at the specified rake, walings should be set horizontally using a spirit level and struts should be truly horizontal and at right angles to the walings, and also vertically over one another. This need for accuracy is not merely a whim of the old school of timbermen, it plays a vital part in the early detection of dangerous ground movements. Thus if walings are bulging or struts bowing at a particular place, the movement can be readily seen if the struts and walings have been placed in truly vertical planes in the first instance. Any tendencies to bodily ground movements resulting in displacement of one side of the excavation relative to the other will also be noticed if the bracing frames show signs of racking.

9.7 Overall stability of strutted excavations

In all types of ground except massive rock some inward yielding of the sides of a strutted or anchored excavation will take place no matter how carefully the support system is installed. The yielding is accompanied by settlement of the ground surface near the excavation. The magnitude of the yielding and its accompanying settlement depends on the type of ground and the care with which the ground is supported. In soft silts and clays there is the additional risk of upward heaving of the bottom of the excavation accompanied by major settlement of the ground surface, as shown in Fig. 9.38.

The mechanics of bottom heave have been studied by Bjerrum and Eide[9.25] who derived the following formula for the critical depth of an excavation:

$$\text{Critical depth, } D_c = N_c s/\gamma, \qquad (9.10)$$

$$\text{Factor of safety against bottom heave, } F = \frac{N_c \times s}{\gamma D + P}, \qquad (9.11)$$

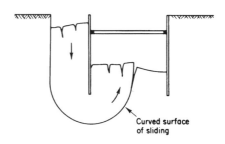

Fig. 9.38 Base heave in strutted excavation in soft clay.

where

N_c = coefficient depending on the dimensions of the excavation,

s = undrained shear strength of soil in a zone immediately around the bottom of the excavation,

γ = density of soil,

D = depth of excavation,

P = surface surcharge.

Values of N_c, as given by Bjerrum and Eide, are shown in Fig. 9.39.

Where vane tests are used to obtain the undrained shear strength of a soft or firm normally consolidated clay the vane shear strengths should be corrected by the factors shown in Fig. 1.3.

Four cases of inward yielding and bottom heave of sheet piled excavations in a soft sensitive clay at Tilbury, Essex, have been analysed by Ward.[9.26] In these cases the tops of the sheet piles came in, or if a top frame was fixed in time to stop this movement, the bottom heaved and the top frame was displaced. A typical case of a 4.6 m deep excavation is shown in Fig. 9.40. The piling on one side moved outwards and away from the top frame of timbering and rose *above* its original level. A gap was formed between the clay and the back of the sheet piling (which remained straight) and the bottom heaved. Ward

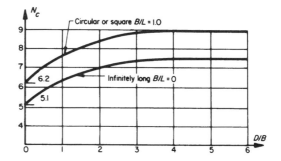

Fig. 9.39 Critical depth of excavations in clay (after Bjerrum and Eide[9.25]).

$$N_c \text{ rectangular} = (0.84 + 0.16B/L)N_c \text{ square}$$

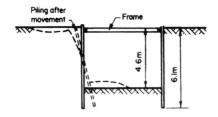

Fig. 9.40 Movement of sheet piled excavations at Tilbury Power Station (after Ward[9.26]).

concluded from this investigation that in a deep deposit of normally consolidated clay the uppermost frame of struts should be placed across a cofferdam before the depth of excavation reaches a value given by $D_c = 4s/\gamma$.

Ward also made the important point that inward yielding and bottom heave appeared to be quite unrelated to the length of the sheet piling and to the materials into which the sheet piles were driven. In one of his cases where heave and yielding occurred at an excavation depth of only 1.8–2.1 m the sheet piles were over 21 m long and were driven into sandy gravel or chalk. The type of inward yielding described by Ward is a form of shear failure of the soil and should not be confused with the smaller scale yielding resulting from soil swelling. Bjerrum and Eide state that

> experience cannot yet indicate exactly how much the danger of a base failure is reduced if sheet piles are driven below the bottom of an excavation. The stabilizing effect of sheet piling seems, however, to be small for the cases where the shear strength of the clay does not increase with depth.

Base failure of cofferdams or deep excavations due to 'piping' or 'boiling' of ground water and water under artesian pressure beneath excavations is described in Section 10.2.5.

9.8 Inward yielding and settlement of the ground surrounding excavations

The problems of inward yielding of the supported sides of strutted or anchored excavations and the accompanying settlement of the ground surface have been mentioned in the preceding pages of this chapter and also in Chapter 5. This subject was reviewed comprehensively by Peck[9.27] and by Clough and Davidson,[9.28] who stated that the amount of yielding for any given depth of excavation is a function of the characteristics of the supported soil and not of the stiffness of the supports. Steel structural members, even of heavy section, are not stiff enough to reduce yielding by any significant amount. Yielding also takes place with cast-in-place concrete diaphragm walls which

can be of the same order as that experienced with sheet pile walls.

The author has reviewed 34 case histories of yielding of the sides of excavations supported by different methods and grouped into three categories of ground conditions. The results of observations at the individual locations are shown in Table B.1. The amount of inward movement expressed as a percentage of the excavation depth is shown in Fig. 9.41. Excluding one high value in each soil category the average and range of movements are:

Soil type	Fig no.	Maximum yield/ excavation depth (%)	
		Average	Range
Soft to firm NC Clays	9.41(a)	0.30	0.08–0.58
Stiff to hard OC Clays	9.41(b)	0.16	0.06–0.30
Sands and gravels	9.41(c)	0.19	0.04–0.46

The plotted points show little difference in the amount of yielding between anchored and strutted supports and little difference between diaphragm walls and sheet piling. This reinforces the view stated above that the amount of movement is a characteristic of the soil type not the stiffness of the support system. The value of making a finite element analysis to determine the amount of yielding of excavation supports in all routine cases of basement construction is questionable. In most cases the overall excavation depth for a two-storey basement does not exceed about 10 m. Thus for the two categories of stiff to hard clays and sands and gravels the range of maximum inward yielding is 5–30 mm and 5–45 mm respectively.

An important consideration is the amount of settlement of the ground surface around the excavation, particularly if buildings or services are located close to the work. Recorded cases of surface settlement are fewer than those of horizontal movement. Figures for 13 case histories are shown in Table B.2. The numbers in each soil category are too few to draw definite conclusions on relationship with excavation depth. They can be summarized as follows:

Soil type	Surface settlement/ excavation depth (%)	
	Average	Range
Soft to firm NC Clays	0.8	0.2–1.7
Stiff to hard OC Clays	0.3	0.1–0.6
Sands and gravels	0.1	0.1–0.2

The wide range of values for soft to firm clays reflects the time-dependent character of surface settlements which are likely to be governed more by the degree of consolida-

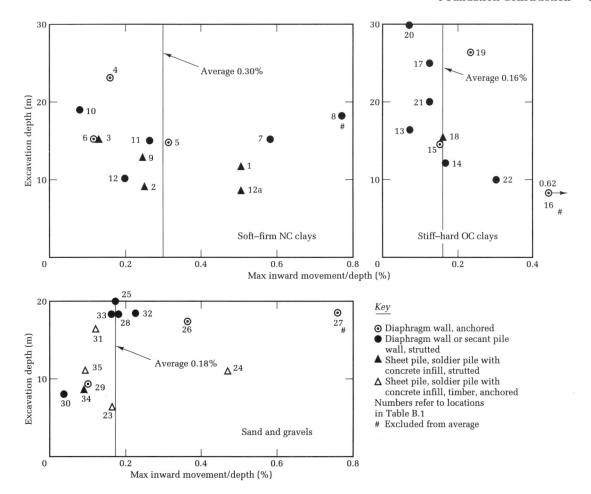

Fig. 9.41 Observed maximum inward movement of excavation supports expressed as a percentage of the excavation depth.

tion of the compressed soils caused by reduction in the pore-water pressure of the soil mass during the period of pumping from within the excavation than by the amount of horizontal movement (see Section 11.4.1). These effects are not so important in the case of stiff clays and sandy deposits which have considerably lower compressibility, and the amount of surface settlement can be related to the degree of horizontal movement. In the case histories shown in Tables B.1 and B.2 the maximum surface settlement was about 0.5–1.0 times the maximum inward yielding.

In all three soil categories the distance from the face of the excavation to the point of maximum settlement was 0.2–0.7 times the excavation depth. Other factors which influence the amount of surface settlement include surcharge loading around the excavation, and the use of ground anchors. Vertical compression induced in an

anchored diaphragm wall can increase the amount of settlement close to the wall, but grouting pressures can cause ground heave within the anchorage zone which may reverse the settlement caused by yielding of the supports.

Where sheet pile or diaphragm walls are supported by a berm of soil followed by installing raking shores at one or more levels (Fig. 9.23) the inward yielding and settlement of the surrounding ground surface can be of a larger order than that observed for strutted or anchored walls. Some observed movements are summarized in Table 9.2. Finite element analyses by Clough and Danby[9.29] showed that the degree of settlement of the ground surface was related to the undrained shear strength of the clay. They plotted the maximum ground settlement as a percentage of the excavation depth against a stability number given by

$$N = \gamma H / c_{ub}, \qquad (9.12)$$

Table 9.2   Observations of maximum inward deflexion of sheet pile walls supported by berms in clay

Location	Berm slope (horiz.:vert.)	Excavation depth (m)	Maximum inward deflexion (mm)	Deflexion Depth × 100	Soil type	Ref.
Chicago	2 : 1	7	64	0.91	2–3 m fill, then soft clay	
Chicago	2 : 1	10	100	1.00	2–3 m fill, then soft clay	
St Louis	2.2 : 1	9	45	0.50	3 m fill, then firm clay	
San Francisco	3 : 1	14	200	1.43	4 m fill, then soft clay	9.28
San Francisco	1.5 : 1	8	330	4.12	5 m fill, then soft clay	
San Francisco	1.5 : 1	11	100	0.91	5 m fill, then soft clay	
San Francisco	4 : 1	14	25	0.18	5 m fill, 7 m soft clay, then dense sand	
North London	$1 : \frac{1}{2}$	11.5	25	0.22	Stiff clay	—
London, Moorgate	1 : 1	18	57	0.32	Stiff clay	9.30

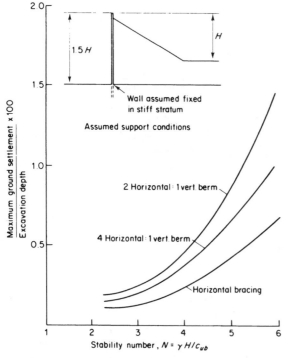

Fig. 9.42   Settlement of ground surface adjacent to excavations supported by berms (after Clough and Danby[9.29]).

where

γ = unit weight of the soil,

H = excavation depth,

c_{ub} = undrained shear strength of the soil at the base of the excavation.

It will be seen from Fig. 9.42 that flattening the berm had little effect in reducing the percentage settlement.

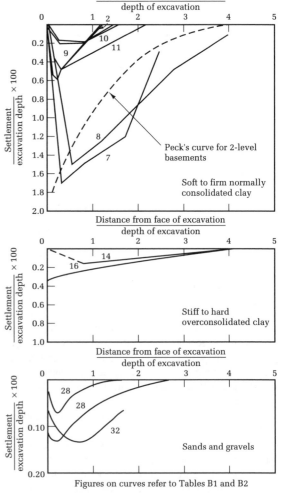

Figures on curves refer to Tables B1 and B2

Fig. 9.43   Profile of ground surface settlement in relation to distance from excavated face.

Observations of settlement agreed closely with the Clough and Danby curves.

Observed settlements of the ground surface adjacent to excavations are plotted in relation to the distance from the face of the excavation in Fig. 9.43. It will be seen that the settlements decrease to zero or to a negligible amount at distances between one and a half and four times the excavated depth. As already noted the maximum settlement occurs at distances between 0.2 and 0.7 of the depth. Also plotted in Fig. 9.43 is Peck's curve obtained from his settlement observations adjacent to strutted sheet pile supports in Chicago.[9.18] The settlement curves 7 and 8 in Fig. 9.43 fall below Peck's curve. These were influenced by long periods of pumping from the excavations. The small settlements close to the excavated face will be noted. These are due to the restraint provided by skin friction on the back of the retaining structure and also to the effect of uplift relief of overburden pressure at the base of the excavation causing uplift at that level.

9.9 The use of finite element techniques for predicting deformations around deep excavations

Prediction of ground movements likely to occur as a consequence of deep excavation can be a matter of crucial importance, particularly in cases where existing structures, which may be potentially sensitive to such movements, are near by. The finite element method provides one of the very few means of predicting with any reasonable degree of accuracy the behaviour of deep excavations both in open cut and in cases where support is provided by diaphragm walls, sheet piling, etc.

Two ways by which excavation can be represented in a two-dimensional finite element analysis are described in Section 3.5 of ref. 11.3. The more readily grasped of these is illustrated in Fig. 9.44. For excavation to the surface ABCD the sequence is as follows:

(i)   Evaluate the stresses σ_x, σ_y, τ_{xy} which exist within elements, a, at the boundary ABC. Convert these stresses to equivalent nodal point loads Px, Py.

(ii) Reduce stiffness of elements, a, to a negligible value. This is to ensure that these elements will then offer no restraint to any movements of the remaining elements.

(iii) Apply the nodal point loads Px, Py in the opposite sense and analyse the system to obtain the displacement and stress increments in the elements below ABCD.

(iv) Add the incremental displacements and stresses to those previously existing in the elements below ABCD to obtain the net values after excavation.

Supposing that it is required to model excavation to the deeper profile of EFC. If the soil is to be considered as behaving in a linear elastic manner, and if no changes in boundary conditions occur during the course of the excavation, then the analysis can be completed in one stage

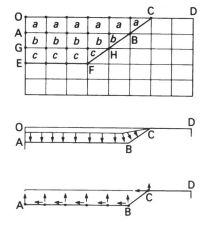

Required to represent excavation from level surface O C down to surface A B C

Evaluate stresses at excavation boundary at lower edges of elements a

(i)   Convert these stresses to equivalent nodal point forces

(ii) Reduce stiffness of elements a to negligible value

(iii) Apply nodal point forces in opposite sense and analyse system to give displacement and stress changes in elements below A B C D.

Fig. 9.44   Simulation of excavation in finite element analysis.

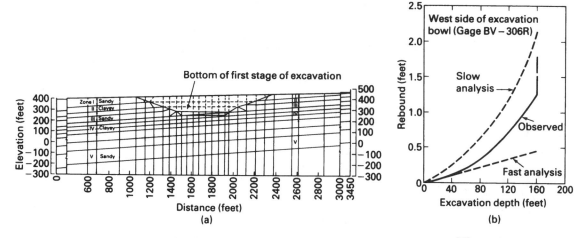

Fig. 9.45 Finite element analysis of deep open-cut excavation for pumping station (after Chang and Duncan[9.33]). (*a*) Finite element mesh. (*b*) Variation of computed and observed rounds with excavation depth.

substituting boundary EFC and all the elements above it for ABC and elements, *a*, in steps (i)–(iv) above. However, if non-linear stress–strain behaviour of the soil is to be modelled then the analysis involves the following sequence of increments:

(1) Perform the sequence (i) to (iv) above for elements, *a*, and surface ABC.
(2) From the results of (1) define new values of stiffness for elements below ABCD.
(3) Using these new values of stiffness repeat (i) to (iv) for elements, *b*, and surface GHC.
(4) From the results of (3) define new values of stiffness in the elements below GHCD.
(5) Using these new values of stiffness repeat (i) to (iv) for elements, *c*, and surface EFC.

The analysis must also be incremental if the soil is modelled as linear elastic, but boundary conditions change during the course of the excavation; for example, if temporary props are added or removed as various excavation levels are reached.

Figure 9.45(*a*) shows the finite element mesh used by Chang and Duncan[9.33] in analyses of a 49 m deep open cut excavation for a pumping station near Bakersfield, California. Non-linear stress–strain behaviour of the soils was allowed for, and so the excavation was modelled in increments. Figure 9.45(*b*) shows values of ground heave, or rebound, measured at one of a number of gauges set in the subsoil. Results of the finite element analyses are also plotted: the terms 'fast analysis' and 'slow analysis' relate respectively to undrained and fully drained soil properties.

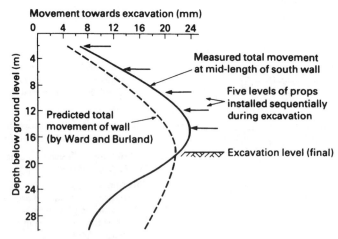

Fig. 9.46 Predicted and measured movements of diaphragm wall at House of Commons underground car park (after Burland and Hancock[9.32]).

A second example of modelling excavation by the finite element method is shown in Fig. 9.46. Here measured and predicted movements of diaphragm walls surrounding excavation for the underground car park at the House of Commons, London, are reported by Burland and Hancock.[9.32] Although the soil was considered to be linear elastic, the analysis was non-linear in the sense that the boundary conditions were changed by the sequential installation of five levels of props during the excavation: hence the analysis had to be incremental. While the analysis predicted the magnitude of maximum wall displacement quite well it indicated that the elevation at which this displacement occurred would be significantly lower than was subsequently recorded.

The House of Commons underground car park results were also used by Simpson et al.[9.34] as a test case in proving their computer-based model (Model LC) for the stress-strain behaviour of London Clay. Model LC is a non-linear elastic–plastic formulation which defines the behaviour of London Clay in terms of three ranges of strain. These are (a) very small strains, less than 0.02 per cent, the 'elastic' range associated with high stiffness values for the soil (some 10 times greater than in the 'intermediate' range); (b) the 'intermediate' range approximating to the linear range of strains normally measured in laboratory tests; and (c) the 'plastic' range associated with large strains. Results from the Model LC based analysis are shown in Fig. 9.47 for comparison with the original predictions by Ward and Burland as well as with the field measurements. It can be seen that results of the Model LC based analysis are a good fit to the measured values not only in terms of maximum displacement but also on the overall form of the wall displacement curve. It should also be noted that Simpson et al. used an axisymmetric idealization (rotational symmetry about a vertical axis), considering that this would be more appropriate for the proportions of the excavation than the plane strain idealization used by Ward and Burland.

Further finite element analysis of the House of Commons underground car park was undertaken by Potts and Day[9.35] who, in a study of retaining walls for deep excavations in stiff clay, used sheet pile sections instead of the concrete diaphragm wall provided. The principal objective of the study was to investigate the effects of variations in wall stiffness. The clay soils were assumed to remain undrained throughout all analysed stages of construction. The final stage of analysis for each of the three structures was to determine the effect on the completed structure of excess pore pressure dissipation back to the assumed initial pore pressure distribution. Figure 9.48 shows values of wall displacement and bending moment at completion of construction and long term for four different sheet pile sections, having comparative stiffnesses varying over a range of 1 to 183.

References

9.1   Terzaghi, K. and Peck, R. B., *Soil Mechanics in Engineering Practice*, 2nd edn, John Wiley, New York, 1967, pp. 236–237.
9.2   Stott, P. F. and Ramage, L. M., The design and construction of a dry dock at South Shields for Messrs Brigham and Cowan Ltd, *Proceedings of the Institution of Civil Engineers*, **8**, 161–188 (1957).
9.3   Tomlinson, M. J., Discussion, in *European Conference on the Stability of Earth Slopes, Royal Swedish Geotechnical Institute*, Stockholm, 1954, Vol. 3, pp. 119–120.

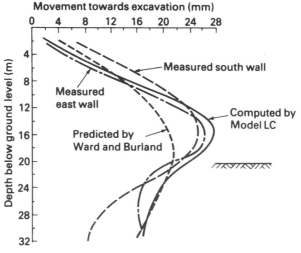

Fig. 9.47 Comparison of predicted and measured movements of diaphragm wall at House of Commons underground car park. Computer values using model LC (after Simpson *et al.*[9.34]).

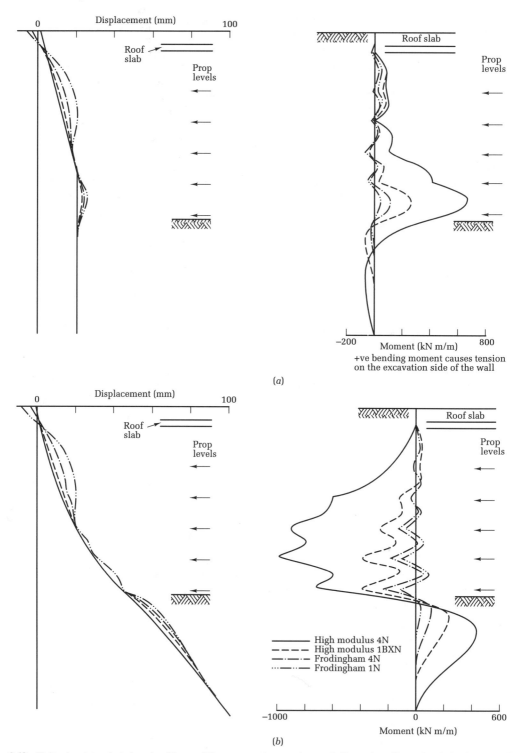

Fig. 9.48 Finite element analysis based on House of Commons underground car park illustrating effects of variation in wall stiffness on wall displacement and bending moment (after Potts and Day[9.35]). (*a*) Displacements and bending at end of construction. (*b*) Displacement and bending in long-term.

9.4   Serota, S., Discussion, in *European Conference on the Stability of Earth Slopes, Royal Swedish Geotechnical Institute*, Stockholm, 1954, Vol. 3, pp. 121–122.

9.5   Meigh, A. C., Discussion on Section 9, in *Proceedings of the 4th International Conference on Soil Mechanics*, London, 1957, Vol. 3, pp. 255–256.

9.6   Skempton, A. W., Discussion, in *European Conference on the Stability of Earth Slopes, Royal Swedish Geotechnical Institute*, Stockholm, 1954, Vol. 3, pp. 113–119 and 120–121.

9.7   Hulme, T. W., Potter, L. A. C., and Shirlaw J. N., Singapore Mass Rapid Transport System: Construction, *Proceedings of the Institution of Civil Engineers*, **86**(1), 709–770 (1989).

9.8   Subway for Stockholm built in unusual cofferdams, *Engineering News Record*, **159**(12), 52–57 (1954).

9.9   Irvine, D. J. and Smith, R. J. H., Trenching practice, *Construction Industry Research and Information Association, Report 97* (1983).

9.10   Mackay, E. B., Proprietary trench support systems, *Construction Industry Research and Information Association, Technical Note 95*, 3rd edn, 1986.

9.11   Benjamin, B. A., Whiting, P. and Kou Aik Kwang, Construction of Raffles Place Station, *Proceedings of the Singapore Mass Rapid Transport Conference*, Institution of Engineers, Singapore, 1987, pp. 111–119.

9.12   Squire, W. W., Report to the Secretary of State for the Home Department on the disaster at Alexandra Dock, Newport, which occurred on 2 July 1906, Cmnd 4921 (1909).

9.13   Hanna, T. H., *Design and Construction of Ground Anchors*. Construction Industry Research and Information Association, Report 65, 2ed, 1980.

9.13a   Mitchell, J. M., Some experiences with ground anchors in London, *Proceedings of the Conference on Diaphragm Walls and Anchorages*, Institution of Civil Engineers, London, 1975, pp. 129–133.

9.14   Kell, J., The Dartford Tunnel, *Proceedings of the Institution of Civil Engineers*, **24**, 359–372 (1963).

9.15   Huder, J., Deep braced excavation with high ground water level, in *Proceedings of the 7th International Conference on Soil Mechanics*, Mexico, 1969, Vol. 2, pp. 443–448.

9.16   Granter, E., Park Lane improvement scheme: design and construction, *Proceedings of the Institution of Civil Engineers*, **29**, 293–318 (1964).

9.17   Spilker, A., Mitteilung über die Messung der Kräfte in einer Baugrubenaussteifung, *Bautechnik*, **15**, 16 (1937).

9.18   Peck, R. B., Earth pressure measurement in open cuts, Chicago subway, *Transactions of the American Society of Civil Engineers*, **108**, 1008–1036 (1943).

9.19   Terzaghi, K. and Peck, R. B., *Soil Mechanics in Engineering Practice*, 2nd edn, John Wiley, New York, 1967, pp. 394–413.

9.20   Skempton, A. W. and Ward, W. H., Investigations concerning a deep cofferdam in the Thames Estuary clay at Shellhaven, *Géotechnique*, **3**, 119–139 (1952).

9.21   Di Biagio, E. and Bjerrum, L., Earth pressure measurements in a trench excavated in stiff marine clay, in *Proceedings of the 4th International Conference on Soil Mechanics*, London, 1957, Vol. 2, pp. 196–202.

9.22   Flaate, K. and Peck, R. B., Braced cuts in sand and clay, *Norwegian Geotechnical Institute, Publication no. 96*, 7–29 (1973).

9.23   Hwang, R., Quah Hong Pin, and Buttling, S., Measurements of strut forces in braced excavations; *Proceedings of the Singapore Mass Rapid Transport Conference*, Institution of Engineers, Singapore, 1987, pp. 239–244.

9.24   Fowler, D., Joint strides, *The New Civil Engineer Limehouse Link Supplement*, Thomas Telford, London, 1993, pp. 8–16.

9.25   Bjerrum, L. and Eide, O., Stability of strutted excavations in clay, *Géotechnique*, **6**, 32–47 (1956).

9.26   Ward, W. H., Experiences with some sheet pile cofferdams at Tilbury, *Géotechnique*, **5**, 327 (1955).

9.27   Peck, R. B., Deep excavations and tunnelling in soft ground, in *Proceedings of the 7th International Conference on Soil Mechanics*, Mexico, 1969, State-of-the-art, pp. 225–290.

9.28   Clough, G. W. and Davidson, R. R., Effects of construction on geotechnical performance, in *Proceedings of the 9th International Conference on Soil Mechanics*, Tokyo, 1977, Specialty session no. 3.

9.29   Clough, G. W. and Danby, G. M., Stabilizing berm design for temporary walls in clay (see Clough and Davidson[9.28]).

9.30   Sills, G. C., Burland, J. B., and Czechowski, M. K., Behaviour of an anchored diaphragm wall in clay, in *Proceedings of the 9th International Conference on Soil Mechanics*, Tokyo, 1977, Vol. 2, pp. 147–154.

9.31   Cole, K. W. and Burland, J. B., Observations of retaining wall movements associated with a large excavation, in *Proceedings of the 5th European Conference on Soil Mechanics*, 1972, Vol. I, pp. 445–453.

9.32   Burland, J. B. and Hancock, R. J. R., Underground car park at the House of Commons, London, *The Structural Engineer*, **55**, 87–100 (1977).

9.33   Chang, C. Y. and Duncan, J. M., Analysis of soil movement around a deep excavation, *Proceedings of the American Society of Civil Engineers*, **96** (SM5), 1655 (1970).

9.34   Simpson, B., O'Riordan, N. J. and Croft, D. D., A computer-model for the analysis of ground movements in London Clay, *Géotechnique*, **29**(2), 149–175 (1979).

9.35   Potts, D. M. and Day, R. A., Use of sheet pile retaining walls for deep excavations in stiff clay, *Proceedings of the Institution of Civil Engineers*, **88**(1), 899–927 (1990).

10 Cofferdams

10.1 Cofferdam types

A cofferdam is essentially a temporary structure designed to support the ground and to exclude water from an excavation – either ground water or water lying above ground level. A sheet piled excavation in dry ground is not a cofferdam. It should also be noted that a cofferdam does not necessarily exclude all water and it is usually uneconomical to design cofferdams to do so.

Although steel sheet piling is widely used for cofferdams because of its watertightness and structural strength, it is by no means the only material, and a wide variety of types is available. Some of these are:

(1) Earth embankments
(2) Rockfill embankments
(3) Sandbag embankments
(4) Single-wall timber sheet piling
(5) Double-wall timber sheet piling
(6) Flexible sheeting on timber or steel framing
(7) Rock or earth-filled timber cribs
(8) Single-wall steel sheet piling
(9) Double-wall steel sheet piling
(10) Cellular steel sheet piling
(11) Bored cast-in-place piling
(12) Precast concrete blockwork
(13) Precast concrete frame units
(14) Structural steel cylinders and shells (movable cofferdams)

The choice of type depends on the site conditions, for example depth of water, depth and size of excavation, soil types, velocity of flow in waterway, tide levels, and the risk of damage by floating debris or ice. The choice depends also on the availability and ease of transport to the site of heavy construction plant and such materials as timber and sheet piling. Thus, earthfill cofferdams are suitable for low heads of water and, if given surface protection, they can be used for half-tide work, or on sites

exposed to flowing water. Single-wall sheet pile cofferdams are suitable for restricted site areas where cross bracing or ground anchors can be used. Double-wall cofferdams or cellular sheet piling are used for wide excavations where self-supporting dams are required. Rock or earth-filled timber cribs would be suitable for a remote site in undeveloped territory where heavy timber in log form is available, and the cost of importing and transporting steel sheet piling and the necessary plant to handle and drive it might be prohibitively high. The foundation soil conditions are an important factor in the choice of cofferdam types. Thus, a heavy earth-filled crib or cellular cofferdam could not be carried by deep deposits of soft clay, and single-wall timber or steel sheet piling would be required in these conditions. Sheet piling is unsuitable for ground containing numerous boulders.

The design of cofferdams which obstruct, wholly or partly, the flow of rivers must take into account hydraulic problems such as bed erosion and overtopping. These problems can often be solved with the aid of scale models employing an erodible bed to enable the scour tendencies to be studied. The book on cofferdams by White and Prentis[10.1] gives an excellent account of the theoretical and practical considerations involved in the design and construction of large river cofferdams. The authors have a wide range of experience of this type of work and their account of practical difficulties and cofferdam failures is especially interesting and informative.

Generally, the choice of cofferdam type and the detailed designs should be based on a thorough knowledge of the soil conditions by an adequate number of borings and on a careful study of available materials and statistics of the flows and hydrographs of the waterway.

10.1.1 Earthfill cofferdams

Cofferdams formed by an embankment or dike of earth are

suitable for sluggish rivers, lakes, or other sheltered waters not subject to high-velocity flow or wave action. Fast-flowing water causes severe erosion of earth embankments and the cost of protection by stone or other types of blanketing may not be justifiable if other forms of construction are available. Earthfill dams are usually restricted to low or moderate heads of water. The type of soil to be used depends on availability and the site conditions. For example, clay fill is unsuitable if the bank has to be formed by dumping material under water, when the clay would soften and stable slopes could not be achieved. However, clay fill is very satisfactory if the bank can be constructed in the dry, say, at a low-water season of river flow. If the clay is excavated in a fairly dry condition and is spread and rolled in thin layers, an embankment of considerable stability and watertightness can be obtained. Sand is generally the best material for earthfill cofferdams constructed 'in the wet'. Embankments of sand or gravelly sand fill can be formed by dumping from barges, by pumping from a suction dredger, or by end tipping from the shore. Consideration can be given to composite construction such as a clay core with sand fill on both sides.

The possibility of erosion due to seepage from the inside slopes is liable to occur in sand fill embankments. The flow lines of seepage through permeable material are illustrated in Fig. 10.1(a). They emerge on the inside face in a concentrated pattern and it is necessary to provide a drainage ditch at this point to collect the seepage and convey it to a pumping sump. Unstable conditions caused by 'boils' or erosion of the sand due to excessive velocity in the outflowing water will occur if the hydraulic gradient, i.e. the head of water divided by the length of the seepage path, exceeds unity. The hydraulic gradient can be determined by drawing a flow net (Fig. 10.1(a)). If there is

no safety factor against erosion, the seepage path must be lengthened by flattening the inside slope or by blanketing the outside face and bed with clay (Fig. 10.1(b)). Precautions to be observed in maintaining the inside slopes and in designing drainage ditches, filters, and pumping sumps are described in Chapter 11.

Protection against wave action and erosion by flowing water can be obtained by blanketing the slopes with material such as large gravel, broken rock, canvas tarpaulins, or geotextiles, but the economics of providing and placing blanketing material rather than adopting a type of cofferdam not affected by erosion must be considered. Erosion of the top and inside face by overtopping is sometimes a cause of failure. The alternative to blanketing the inside slopes is to provide sluices through the embankments of sufficient capacity to allow rapid flooding of the cofferdam if a sudden rise in outside level threatens to overtop it.

An earthfill cofferdam consisting of an embankment formed by pumped sand with a clay core was used for the Haringvliet Sluice forming part of The Netherlands' Delta Plan (Packshaw[10.2]). Protection against wave action and tidal scour below low water was given by a fascine mattress, and protection against wave action above high water by a layer of asphaltic concrete, as shown in Fig. 10.2.

If earthfill cofferdams are to be built across a flowing waterway, there may be difficulties in the 'closure' of an earth embankment as the velocity of flow through the gradually narrowing gap may be sufficient to carry away the filling as it is being placed. These conditions can also occur if embankments are constructed in tidal waters, when the water rushing in and out of the 'tidal compartment' will cause erosion through the gap between the ends of a partly completed embankment. It may be necessary to drive sheet piles into the ends of the bank at the closure section and to complete the closure by sheet piling only, or to adopt such expedients as sinking barges or caissons in the gap. It is important to keep a careful watch on the state of earthfill cofferdams during the whole period they are in operation. Any cracks should be made good and seepages of water dealt with by stone-filled or piped drains.

10.1.2 Rockfill cofferdams

Rockfill embankments for cofferdams are similar in construction to earthfill dams, but because of the inherent stability of the material they can be formed with steeper slopes than earthfill dams. In fact, slopes equal to the natural angle of repose of tipped rock (about 1 in $1\frac{1}{2}$) can be used and even steeper slopes of up to 1 in $1\frac{1}{4}$ may be adopted if the rock on the face is handpacked to the required profile. Rockfill dams have the disadvantage of

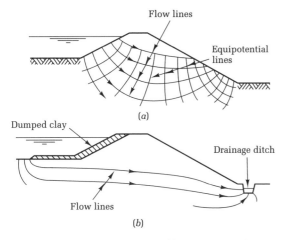

Fig. 10.1   (a) Seepage through permeable embankment. (b) Controlling seepage by clay blamket and toe drainage.

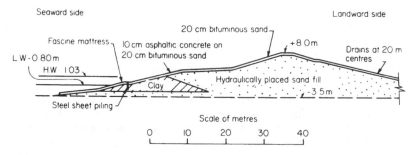

Fig. 10.2 Earthfill cofferdam for Haringvliet Sluice, The Netherlands.

not being impervious. In the case of low height dams sufficient watertightness can be achieved by dumping pf ash or loamy soil on to the outer face. Seepage will carry this material into the interstices of the rock and a fair degree of watertightness will gradually be attained (Fig. 10.3(*a*)). A more positive form of cut-off is required for high heads of water; this can be achieved by a clay or concrete core wall (Fig. 10.3(*b*) and (*c*)) or by sheet piling. The clay or concrete wall type of cut-off is usually restricted to cofferdams built in the dry. For construction under water a sheet pile wall can be built out from the bank using guide walings supported by a piled trestle. Rockfill can then be end-tipped out from the bank on both sides of the sheet piling (Fig. 10.3(*d*)).

For rockfill cofferdams constructed in the dry the provision of a sloping watertight core or cut-off has certain advantages. For example, the rockfill of a dam with a sloping clay core as shown in Fig. 10.3(*b*) can be built up in weather unfavourable to placing and rolling clay, leaving the clay to be brought up in better weather periods. With this type of construction it is necessary to have a reasonably high stability in the clay by using, say, a well-compacted and fairly dry lean clay such as boulder clay in order to avoid a slip developing through the clay layer. It is also necessary to protect the outside face of a clay core against wave action by blanketing with rockfill.

10.1.3 Timber sheet pile cofferdams

Timber sheet piling was extensively used for cofferdam work in the nineteenth century, but it has been gradually replaced by steel piling. However, where timber is readily available it can still be used with advantage and economy in low head dams.

Fowler[10.3] gives details of eight types of timber sheet piles, as shown in Figs 10.4(*a*)–(*g*). He gives the following rule of thumb for the spacing of walings: 75 mm planks require walings every 1.8 m for less than 1.5 m head of water, or every 0.9 m for 6.4 m head; 100 mm planks require walings every 2 m for 2.7 m head and every 1.5 m for 5.4 m head; 225 mm planks require walings every 2.7 m for 6 m head. Timber sheet piling of the Wakefield pattern (Fig. 10.4(*g*)) is used extensively in Mexico City and in Canada where timber is readily

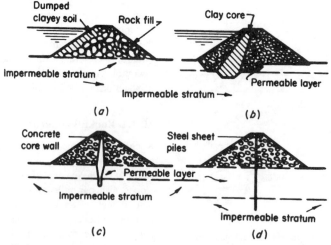

Fig. 10.3 Types of rockfill cofferdam.

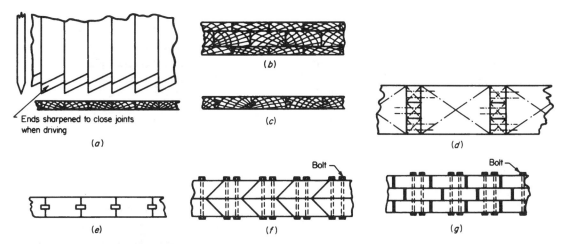

Fig. 10.4 Types of timber sheet piling (after Fowler[10.3]), (*a*) Plain butt-jointed sheeting. (*b*) Lapped butt-joint. (*c*) Tongue-and-groove joint. (*d*) Joint formed by nailing or spiking strips to sheet piles. (*e*) Keyed joint with keys inserted and driven after driving piles. (*f*) Birdsmouth joint formed by bolting together double-bevelled planks. (*g*) Wakefield sheet piling.

available. Timber sheet piling was also used extensively for the Stockholm Underground Railway extensions constructed in 1953–57 where the depth of water was less than 3.6 m and where there was only a small thickness of overburden to rock. The cofferdam was constructed by driving a row of H-section king piles on each side of the excavation. The piles were supported on their outer face by raking steel H-section struts. Timber walings were fixed at two levels between flanges of the king piles, and 75 mm tongue-and-groove timber sheet piles were driven in front of these walings through the thin overburden of sand and gravel to bedrock. The thrust transferred from

the walings to the king piles was carried by steel trusses spanning across the excavation. These trusses also served as a working platform for plant and materials and to support the sheet steel cover over the whole of the cut-and-cover excavation. This allowed work to continue in severe winter conditions. Soil was tipped in front of the sheet piles to increase the watertightness. The arrangement of the timber piled cofferdam and supporting truss is shown in Fig. 10.5.

Where timber sheet pile dams are required to retain high heads of water it is the usual practice to use a double wall and to fill the space between the two with puddled clay.

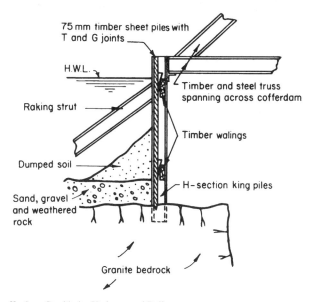

Fig. 10.5 Timber sheet pile cofferdam, Stockholm Underground Railway.

This type of construction was commonplace before the advent of steel sheet piling and the older engineering journals contain many examples of this type of work.

The main disadvantage of timber sheet pile dams is their limited depth of cut-off since the piles cannot be driven deeply into granular soils or stiff clays without risk of splitting or 'brooming'. They are most useful for low heads of water and for cofferdams founded on an irregular bedrock surface.

10.1.4 Timber cribwork dams

Cribwork dams are usually constructed in the form of preassembled timber cribs which are lifted by crane or floated into position and sunk on to the river bed. Stone filling is then placed in the crib to give stability, and steel sheet piling or other forms of sheeting are driven on the outer face to give watertightness. They are essentially free-standing dams used for wide excavations and are suitable for irregular and rocky river bottoms where the cribs can be carefully 'tailored' to suit the river-bed profile. It is essential that they are placed on a foundation of adequate bearing capacity to resist the heavy weight of contained rockfill and the bearing pressures caused by the water and earth pressure. This form of construction also facilitates closure in fast-flowing water, where the cribs can first be placed and filled with rock and the flow allowed to pass between them. The gaps between cribs can then be filled with sheet piling or horizontal timber stoplogs.

White and Prentis[10.1] give a detailed description of the Ohio River type of articulated timber cofferdam which was designed for use with the river of that name. The dam consists of two rows of timber sheeting with earthfill between and earth banking on both sides to give it stability (Fig. 10.6(a)). Because of its width this type of dam is stable against seepage when constructed on a permeable bottom as well as on a rock foundation. The distinctive feature of the Ohio dam is its articulated construction in the form of a continuous linked framework preassembled on a barge and 'paid-out' as the barge is towed along the line of the dam (Fig. 10.6(b)). The framework is tailored to suit the profile of the river bed as determined by close-spaced soundings. A gang working from a second barge follows up by setting the sheeting planks which are nailed to the walings above the water line, and a dredger follows behind pumping fill inside and outside the cofferdam. Drainage holes are provided in the sheeting to draw down the water level in the fill as the cofferdam is pumped dry, so avoiding excessive hydrostatic pressure between the timber sheeted walls. The cofferdam is removed by a reverse process to its installation. A dredger pumps out the filling, starting at one end. A gang takes away the

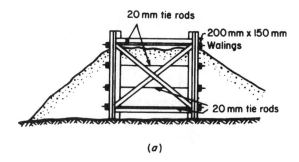

(a)

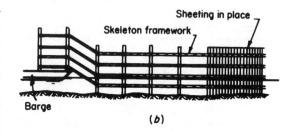

(b)

Fig. 10.6 Constructing the Ohio-type cofferdam (after White and Prentis[10.1]).

sheeting, then the framework is lifted by floating crane and dismantled section by section. The Ohio-type cofferdam is quick to install and requires no falsework, but it is difficult to construct in a fast-flowing current. It is also sensitive to erosion, and heavy slope protection is required at critical points such as closure sections and corners.

10.1.5 Steel sheet pile cofferdams

Steel sheet piling is widely used for cofferdams, including the deepest types, because of its structural strength, the watertightness given by its interlocking sections, and its ability to be driven to deep penetration in most types of ground. The deep penetration gives resistance to inward movement at the bottom of the excavation and sufficient cut-off to prevent piping in permeable ground. The principal types of sheet pile in general use around the world are shown in Fig. 10.7. The *Larssen* or trough-type sections (Fig. 10.7(a)) are widely used for cofferdam construction because of the positive character of their 'wrap-around' clutch which is beneficial in minimizing declutching when driving long piles in difficult ground. They have the disadvantage that the interlocks are located on the neutral axis of the combined section of a pair or continuous wall of piles. BS Code of Practice 8004 states that the piles may be assumed to develop substantially the strength of the undivided section only when they are driven fully into the ground with the exception of the following conditions for which the section modulus of a single pile (i.e. not a pair of piles acting as an undivided section) should be used:

(a) Where the pile passes through very soft clay or other weak material;

(b) Where the pile is prevented by rock from penetrating to the required cut-off depth;

(c) Where the pile is used as a cantilever or if it is cantilevered to a substantial height above the level of the highest waling;

(d) If backfilling is placed against one side of the piling after it has been driven.

The properties of single Larssen sheet pile sections are shown in Table 10.1(*c*).

Items (a), (b), and (c) can be overcome by welding or pressing together the interlocks of the inner and outer piles of each pair so that they develop the necessary shear resistance. In any event, the interlock resistance may not be a critical factor in design because in many cases a section modulus much higher than that required to resist soil or hydrostatic pressure is needed to ensure that the piles can be driven to the required penetration depth without damage or loss of integrity of the completed wall.

The *Frodingham* (or Hoesch) section is shown in Fig. 10.7(*b*). This has the advantage that the interlocks are at the outside face of the wall, thus permitting the development of the full bending and shear resistance of the single pile. However, the clutch is not as positive as the Larssen section and is liable to part if the piles twist when driven deeply into stiff or hard clays. Nevertheless the Frodingham section is used quite widely for cofferdam work.

The *Peine pile* (Fig. 10.7(*c*)) consists of broad-flanged H-sections with the tips of the flanges deformed. When linked together by separate I-section locking bars the piles

form a wall of considerable stiffness. The interlock is not as positive as the Larssen pile, but the close spacing of the locking bars on both sides of the wall provides a good resistance to twisting.

Straight web, or flat web, piles (Fig. 10.7(*d*)) are clutched together to resist tensile forces to form circular cofferdams of the types shown in Fig. 10.8(*g*) and described in Section 10.5.

The individual piles described above can be combined in various ways to increase the section modulus of the wall. Thus the Peine pile can be linked by the locking bars to Larssen sections (Fig. 10.7(*e*)), and pairs of Frodingham sections can be welded to universal beams (Fig. 10.7(*f*)). In a process developed by Dawson Construction Plant Ltd the flanges of standard I-sections are crimped and linked together by separate locking bars to form a wall similar to that shown in Fig. 10.7(*c*). When pitching and driving the combined sections of the type shown in Fig. 10.7(*f*), it is essential to restrain the beam sections from twisting by means of cross straps welded to the guide waling at the rear flange section. The dimensions and properties of the types rolled by British Steel are given in Table 10.1(*a*–*c*); reference should be made to manufacturers' handbooks for information on sheet piles from other countries.

For low heads of water and shallow excavations, the cofferdam can consist of a single wall which is self-supporting by the cantilever action of the piling (Fig. 10.8(*a*)). Higher heads can be withstood if a bank of earth is left on the inside face (Fig. 10.8(*b*)). If an earthfill bank is built up against the inside face, instability may occur if the outside water level falls, when an unbalanced pressure will develop on the inside. Single-wall construction can also be used where it is possible to strut across the excavation. For low heads a single top frame can be used

Table 10.1 Sections and properties of various makes of steel sheet piling. (*a*) Larssen steel sheet piling (manufactured by British Steel).

Section	b mm (nom)	h mm (nom)	d mm	t mm (nom)	f Flat of pan mm	Sectional area cm²/m of wall	Mass kg/linear metre	kg/m² of wall	Combined moment of inertia cm⁴/m	Section modulus cm³/m
6W	525	212	7.8	6.4	331	108	44.7	85.1	6459	610
LX8	600	310	8.2	8.0	250	116	54.6	91.0	12861	830
LX12	600	310	9.7	8.2	386	136	63.9	106.4	18723	1208
LX16	600	380	10.5	9.0	365	157	74.1	123.5	31175	1641
LX20	600	430	12.5	9.0	330	177	83.2	138.6	43478	2022
LX25	600	450	15.6	9.2	330	200	94.0	156.7	56824	2525
LX32	600	450	21.5	9.8	328	242	113.9	189.8	72028	3201
20W	525	400	11.3	9.2	333	188	77.3	147.2	40180	2009
25W	525	454	12.1	10.5	317	213	87.9	167.4	56727	2499
GSP3	400	250	13.0	8.6	271	191	60.0	150.0	16759	1340
4A	400	381	15.7	9.6	219	236	74.0	185.1	44916	2360
6	420	440	22.0	14.0	248	370	122.0	290.5	92452	4200
6	420	440	25.4	14.0	251	397	131.0	311.8	102861	4675
6	420	440	28.6	14.0	251	421	138.7	330.2	111450	5066

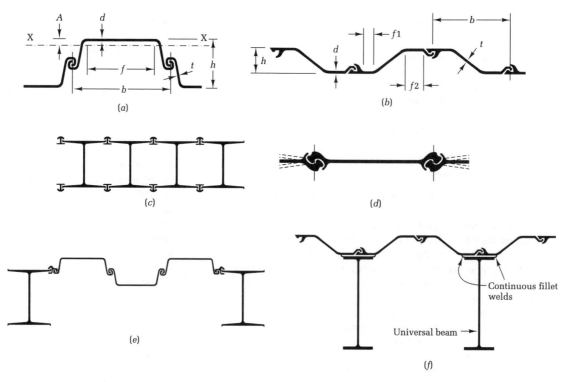

Fig. 10.7 Types of sheet pile sections. (*a*) Larssen. (*b*) Frodingham. (*c*) Peine piles. (*d*) Straight web. (*e*) Larssen sections clutched to Peine piles. (*f*) Frodingham sections welded to universal beams.

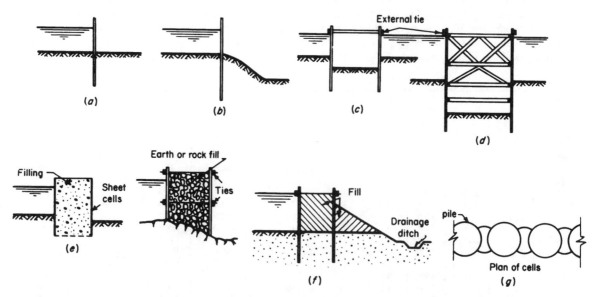

Fig. 10.8 Types of sheet pile cofferdam. (*a*) Single wall. (*b*) Single wall with earth bank. (*c*) With single top frame. (*d*) With multiple frames. (*e*) Double wall. (*f*) Double wall on permeable soil. (*g*) Cellular.

Table 10.1 (cont.)   (*b*) Frodingham steel sheet piling (manufactured by British Steel).

Section	b mm (nom)	h mm (nom)	d mm	t mm (nom)	f1 mm (nom)	f2 mm (nom)	Sectional area cm²/m of wall	Mass kg/linear metre	Mass kg/m² of wall	Combined moment of inertia cm⁴/m	Section modulus cm³/m
1 BXN	476	143	12.7	12.7	78	123	166.5	62.1	130.4	4919	688
FX13	675	300	9.5	9.5	127.7	154.5	137.9	73.1	108.3	19693	1313
FX18	675	380	9.5	9.5	135.4	160.3	147.6	78.2	115.9	34201	1800
FX26	675	430	13.2	12.2	127.1	146.6	194.9	103.3	153.0	55983	2603
FX36	675	460.	18.0	14.0	149.0	158.3	244.8	129.7	192.2	82915	3605
4N*	483	330	14.0	10.4	77	127	218.0	82.4	170.8	39831	2414
5*	425	311	17.0	11.9	89	118	302.0	100.8	236.9	49262	3168

* To be withdrawn during 1995.

(*c*) Properties of single Larssen sheet pile sections.

Section	Cross-sectional area (cm²)	Dimension A (Fig. 10.7(a)) (mm)	Moment of inertia about axis X–X (cm)	Section modulus about axis X–X (cm³)
6W	57	45	1247	155
9W	65	50	2031	204
12W	77	62	3240	288
16W	87	66	4495	347
20W	98	79	6737	467
25W	112	93	9605	607
32W	132	82	10580	626
GSB3	77	50	2275	228
4A	94	76	5650	412
6 (122 kg)	155	90	12437	770
6 (131 kg)	167	85	12882	777
6 (138.7 kg)	177	82	13185	780

(Fig. 10.8(*c*)), but for high heads or deep land cofferdams it is necessary to provide multiple frames (Fig. 10.8(*d*)). Where the excavation is too wide for cross-bracing an earth- or rock-filled double-walled cofferdam is adopted (Fig. 10.8(*e*)). Where such cofferdams are built on permeable ground it is usual to place earthfill on the inside, thus lengthening the seepage path and preventing piping (Fig. 10.8(*f*)). The interior earth bank also gives added stability where there is a risk of tilting due to overturning moments on soils of moderate bearing capacity. Self-supporting cofferdams to withstand high heads of water are sometimes constructed in cellular form, the cells being filled with rock or earth (Fig. 10.8(*g*)).

10.2 Design of single-wall sheet pile cofferdams

10.2.1 Design conditions

Cofferdams must be designed to withstand the following conditions:

(1) Hydrostatic pressure of water outside cofferdam.
(2) Hydrostatic pressure of water *inside* cofferdam if the water level outside falls at a faster rate than can be attained by pumping down or the discharge from valves or sluice gates.
(3) Earth pressure outside cofferdam.

They must have overall stability against the following conditions:

(1) Base failure from bottom heave
(2) Failure from inward yielding
(3) Failure from piping or blows
(4) Failure during removal

The distribution of hydrostatic pressure on a cofferdam is triangular, and because hydrostatic pressure generally provides the major proportion of the external pressure on a cofferdam and is generally three or four times (or more in certain types of ground) the earth pressure at any given depth, it is the normal practice to calculate earth pressure on cofferdams in the same way as retaining walls, i.e. a triangular pressure distribution calculated by the methods given in CP 2: *Earth-retaining Structures** and in sheet

* Revised code BS 8002: *Earth-retaining Structures* was published in 1994. See also Section 5.7.

pile manufacturers' handbooks. It must be remembered that the density of a soil below the water table is taken as its submerged density which is approximately half its fully saturated density, where seepage into the cofferdam takes place beneath the toe of the sheeting the effect of the pore pressures on the hydrostatic pressure distribution behind and in front of the cofferdam wall should be taken into account as shown in Fig. 5.25(c).

Much detailed and practical information on the design and construction of sheet pile structures is given in the recommendations of the German Committee for Waterfront Structures.[10.4]

10.2.2 Spacing of frames

For economy of materials and fabrication costs of bracing frames it is desirable to space the horizontal frames so that they each carry, as nearly as possible, an equal load from the hydrostatic pressure transmitted to them by the sheet piling. The exception is the top frame which is provided at or just above top water level as a guide frame and prevents inward movement when pumping down the cofferdam before the second frame is fixed. The required spacing of

the frames for equal loading can be expressed in terms of multiples of the spacing between the first and second frames. Thus, if the top waling is set at water level and assuming the sheet piling to be simply supported over the walings, then:

Distance from top to 2nd waling $= h$
Distance from top to 3rd waling $= 1.60h$
Distance from top to 4th waling $= 2.03h$
Distance from top to 5th waling $= 2.38h$
Distance from top to 6th waling $= 2.69h$
Distance from top to 7th waling $= 2.97h$
Distance from top to 8th waling $= 3.22h$
Distance from top to 9th waling $= 3.45h$
Distance from top to 10th waling $= 3.67h$

Where the strength of the sheet piling is the governing factor, the distance h is determined by the moment of resistance of the piling. Allowable working stresses for sheet piling are usually selected within a range of values depending on the grade of steel. The lower figures in each range are adopted for permanent structures and deep cofferdams for which consequences of failure due, say, to separation of pile clutches in hard driving conditions

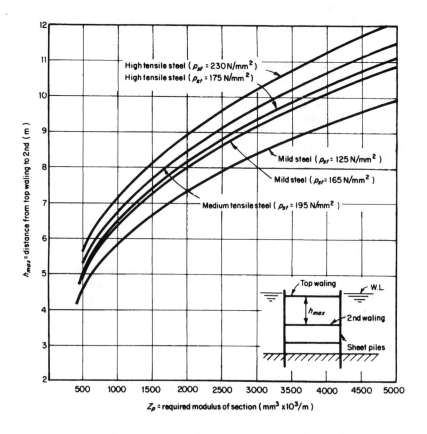

Fig. 10.9

would be serious. The higher figures in the range are used for temporary works including cofferdams in which failure or yielding would not have serious consequences. The range of values corresponding to each grade of steel are shown below. Also shown is the formula for calculating h_{max} assuming the sheet piling to be simply supported at the walings, and values of Z_p, the combined modulus of section of the sheet piling in cubic centimetres per metre of wall.

Grade of steel	Allowable tensile stress (N/mm^2)	Distance from top 2nd waling h_{max} (m)
Mild steel (BS 4360 Grade 43)	125–165	$\sqrt[3]{0.1937 \times Z_p}$ $\sqrt[3]{0.2557 \times Z_p}$
Medium tensile steel	165–195	$\sqrt[3]{0.2557 \times Z_p}$ $\sqrt[3]{0.3022 \times Z_p}$
High-yield steel (BS 4360 Grade 50 B and C)	175–230	$\sqrt[3]{0.2712 \times Z_p}$ $\sqrt[3]{0.3565 \times Z_p}$

Values of h_{max} are plotted against Z_p for the various allowable tensile stresses (Fig. 10.9).

A severe condition of hydrostatic pressure may occur in deep cofferdams when the water level within the cofferdam is lowered to allow the second bracing frame to be fixed in the dry. At this stage there is an excess pressure between the inner water level and the point of fixity of the sheet piling in the ground equal to the difference in head between the inner and outer levels. Thus in a deep cofferdam the bending moments due to this excess pressure over the long span between the top waling and the point of fixity in the ground may be considerable, and may well be the most severe condition for the flexural stress in the piling and the loading on the top bracing frames. If the stresses in the sheet piling or the loading on the top frame are too severe for this pumped-down condition, then heavier section piling can be used or the distance between the top and second frame can be reduced. It is also possible to get over the difficulty by using divers to fix the second frame under water. In very deep cofferdams it is advisable to check the conditions for the top and second frames when the cofferdam is pumped down to fix the third frame.

This problem may not arise in tidal waters when the second frame can be fixed during neap tides, or while the water level outside the cofferdam is at low water of spring tides. Sluices or valves of adequate size must be provided in the walls of the cofferdam to let the water flow out as the tide is falling, otherwise an internal pressure sufficient to burst the cofferdam may develop. It is always advisable to provide an external waling as a safeguard against outward movement of the piles due to unexpected head differences. The provision of sluices or valves to unwater a cofferdam on the falling tide also saves a considerable amount of unnecessary pumping and enables the cofferdam to be flooded rapidly if it shows signs of distress or bottom failure.

Driving in long lengths or to deep penetrations into difficult ground will require heavy sections of sheet piling or alternatively the adoption of medium or high-tensile steel. Thus, if the full flexural strength of the heavy or high tensile steel piling is to be developed, the spacing between walings will be large, with a correspondingly high loading in the frames. The loading may be too great for the economical use of timber frames and it will be necessary to use steel or reinforced concrete.

10.2.3 Design of frames

Walings and struts are constructed from timber or steel or a combination of these materials (Fig. 10.10). Walings are designed as beams simply supported between struts unless they are continuous over at least three spans or unless the joints are designed to develop the necessary shear and bending strength. In circular cofferdams on land the walings can be designed as ribs carrying radial thrust. This arrangement, with its freedom from cross-bracing, can give an economical design of cofferdam. Packshaw[10.2] states that the maximum diameter of cofferdams for which the frames are designed as circular ribs without cross-bracing is 45–60 m. The loads on the walings from trough-section sheet piling should be treated as a series of point loads in direct contact with the waling. Where walings act as struts to other walings (i.e. at corners) they should be designed to carry the combined stresses of bending and direct compression. Struts should be designed as round-ended members between walings or cross-members. If the length between cross-members is taken the connections should be adequately braced both in the horizontal and vertical planes. The structural design requirements of timber, steel, or concrete bracing members is given in Chapter 9. Bracing frames designed as prefabricated units must be accurately constructed with tolerances to allow easy fitting inside the sheet piling. They should be made 25–50 mm smaller than the plan dimensions inside the sheet piling to allow for inward deflexion of the sheet piling during pumping down.

When considering the bending stresses on the sheet piling below the bottom frame, the point of fixity can be taken in good ground as one-tenth of the height of the piling, but in weak ground such as soft mud the point of fixity should be taken as 6 m below ground level or the depth to a firm stratum, whichever is the least. Below the bottom frame the pressure on the sheet piling is carried by the passive resistance of the soil below excavation level.

Fig. 10.10   Single-wall steel sheet pile cofferdam with steel walings and timber knee strutting.

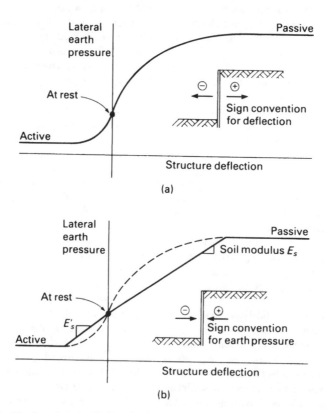

Fig. 10.11   Non-linear relationships for soil retained behind flexible structure: (*a*) Probable relationship; (*b*) Simplified relationship (after Haliburton[10.5]).

The method of calculating this passive resistance is given in CP2: *Earth-retaining Structures**. Heavy and continuous pumping from a cofferdam founded on permeable ground will reduce the passive resistance of the soil below excavation level because the upward seepage towards the pumping sump will cause the soil to be in a partly 'quick' or live condition. Piping or incipient 'blows' will completely eliminate all passive resistance and the active earth pressure and hydrostatic pressure on the piling will be carried solely by cantilever action from the bottom frame. Consideration should be given to the possibility of scour occurring during pitching the sheet piles, which will lower the soil level within the cofferdam, thus reducing the passive resistance.

* See footnote on p. 421.

10.2.4 Computer-aided design for cofferdams in braced sheet pile and diaphragm wall construction

Apart from the computerization of routine manual design techniques the application of computer-based methods to this topic has been somewhat limited. There has been some use of the beam-column on non-linear horizontal Winkler springs method: this is identical in principle to the $p-y$ approach for the isolated pile described in Section 7.17.2, but the form of horizontal spring characteristic is different and representation of struts or anchors has to be introduced. One approach to the technique has been described by Haliburton[7.50, 10.5] who adopted Terzaghi's[10.6] concept of non-linear curves of soil lateral pressure and displacement (Fig. 10.11). One problem which arises in defining the curves is the value of effective width or depth to be used for a flexible earth-retaining structure in the evaluation of E_s and E'_s. Haliburton[7.50]

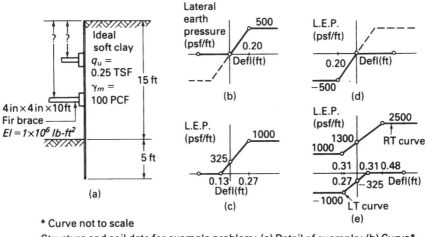

* Curve not to scale

Structure and soil data for example problem: (a) Detail of example; (b) Curve* at surface; (c) Curve* at H_c = 5ft; (d) Curve* at 15 ft depth for LT. Soil Mass; (e) Curves * at 20 ft depth

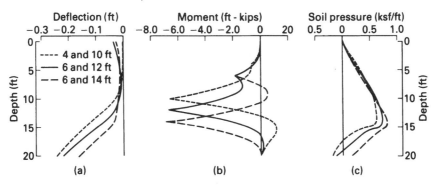

Fig. 10.12   Analysis of strutted sheet pile wall using non-linear springs to represent soil (example problem by Haliburton[10.5]).

suggests that the total depth of the final embedment (after completion of excavation) be taken as the effective depth for the soil on the excavation side of the wall; while for the retained soil, in the case of a strutted excavation, the effective depth should be taken as the largest vertical distance between struts or the distance between the top of the wall and the topmost strut, whichever is the greater. Figure 10.12 shows results obtained by Haliburton for the analysis of a strutted excavation.

An application of the Haliburton approach was undertaken by Clark and Prebaharan (see ref. 10.13) for cofferdams associated with the construction of part of the Singapore MRT underground railway. The cofferdam walls were of steel sheet piling which was driven through soft marine clays generally overlain by 8–12 m of hydraulically placed sand fill. Excavation within the cofferdams was to depths of the order of 15–18 m. Figure 10.13(a) shows predicted and measured values of strut loads for one cofferdam while Fig. 10.13(b) gives a comparison of predicted and measured sheet pile deflections for four of the cofferdams. The predictions appear to have been very accurate. However, the authors give no details of the precise derivation of the non-linear curves of lateral soil pressure and displacement – predicted values of sheet pile deflections in particular could be sensitive to assumptions in the derivation. In another study Wood and Perrin[10.7] reported predictions of displacements, bending moments, and strut forces for diaphragm walls enclosing an 11 m deep excavation, predominantly in London Clay. The site was extensively instrumented and preliminary measurements led to the following interim conclusions:

(a) Observed horizontal displacements were of the same magnitude as predicted, but agreement between measured and predicted deformation profiles was less good.
(b) Predicted bending moments were in excess of values derived from diaphragm wall strain gauge readings.
(c) For the upper two levels of struts there was good agreement between predicted and measured values of compressive force, but for the third and lowest level measured values were approximately half those predicted.

Application of the boundary element method to the case of a strutted diaphragm wall has been reported by Wood[10.8]. Both linear and non-linear analyses were made, the latter allowing for limiting values of soil pressure, and the results compared with results from Winkler spring linear and non-linear analyses. Wood concluded that for the particular case considered the Winkler spring approach gave a more realistic prediction of the behaviour of the wall.

The finite element method has been used to model the

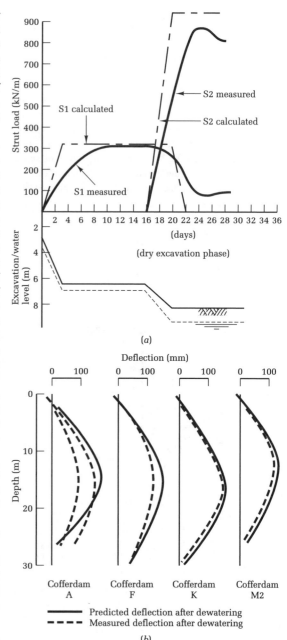

(a)

(b)

Fig. 10.13 Comparison of predicted and measured values of strut loads and sheet pile deflections in cofferdams for Singapore MRT underground railway (after Clark and Prebaharan[10.13]) (a) Strut loads – cofferdam K. (b) Sheet pile deflection.

behaviour of strutted excavations for elastic soil properties (see Burland and Hancock[9.32]) and also for non-linear and inelastic soil properties (see Simpson et al.[9.34]). The method has the potential to model with a high degree of

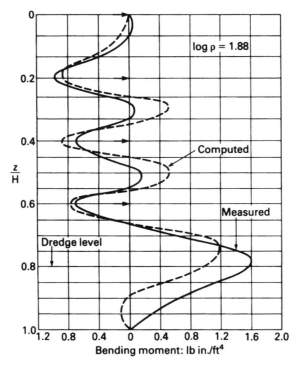

Fig. 10.14 Comparison of computed and measured bending moments in model of strutted excavation in loose sands (after Smith and Boorman[10.9]).

realism the complicated stress paths associated with the sequence of excavation stages, the insertion of struts or anchors, and the unloading/reloading response in the soil mass. Fig. 10.14 shows the results obtained by Smith and Boorman[10.9] from the finite element analysis of a model of a strutted wall in loose sand and gives a comparison with the model test results obtained by Rowe and Briggs.[10.10]

10.2.5 Piping in single-wall cofferdams

Stability against base failure by heaving and against inward yielding should be checked by the procedure described in Section 9.7. The stability against 'piping' or 'boiling' in the base resulting from heavy upward seepage of ground water into the cofferdam must be checked in cases where the sheet piles are not driven into an impermeable stratum. It is important either to prevent piping or to reduce it to an insignificant amount since severe piping can lead to loss of ground outside the cofferdam and even to undermining and collapse of the sheet pile wall. As already mentioned, it can cause loss of passive resistance of the ground to the inward thrust of the sheet piling. Piping occurs when the exit gradient, or the ratio of the head to the length of seepage path, at the point of outflow into the cofferdam reaches unity; at this stage the velocity of the upward-flowing water within the cofferdam is sufficient to lift and displace the soil through which it is

flowing. Since, by Darcy's law,

$$\text{Velocity} = \frac{\text{Constant} \times \text{Head}}{\text{Length of seepage path}},$$

it follows that the tendency to piping can be eliminated by lengthening the seepage path, for example by driving the sheet piling to a sufficient depth or if width is available by providing a bank of permeable soil within the cofferdam. Lengthening the seepage path by driving the piles to a deeper penetration may be ineffective in a dense granular soil, particularly if large gravel or cobbles are present, causing the sheet piles to split at the clutches. This will cause a large reduction in the length of the seepage path and a severe risk of piping. The other method of preventing piping is to reduce the head of the water causing seepage; this can be done by pumping from wellpoints or deep wells placed at or below the level of the bottom of the sheet piles. This procedure is described in Chapter 11. Minimum values for the depth of cut-off sheet piling are as follows:

Width of cofferdam	Depth of cut-off, D
$2H$ or more	$0.4H$
H	$0.5H$
$0.5H$	$0.7H$

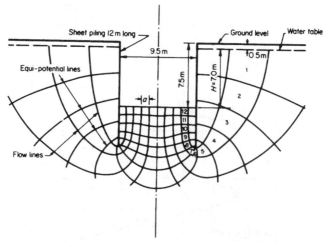

Fig. 10.15   Flow net for cofferdam.   Exit gradient $= \dfrac{H}{N_p \times a} = \dfrac{7}{12 \times 9.5/10} = 0.6.$

Factor of safety against piping $= 1/0.6 = 1.7.$

Piping is most likely to take place in loose fine sands which are permeable enough to allow seepage through them, but whose grain size is small enough to be disturbed by the seepage forces. Piping is unlikely to occur in gravels since the draw-down in water level outside the cofferdam as a result of their high permeability usually lowers the head, so reducing the hydraulic gradient. Piping does not take place in silty or sandy clays due to their low permeability. Doubtful cases can be analysed by drawing flow nets similar to that shown in Fig. 10.15. The use of finite difference and finite element methods of analysing flow into excavations is described in Chapter 11.

If piping takes place unexpectedly, for example by an unforeseen increase in external water head due to abnormally high tides or flood levels, or due to split clutches in sheet piling, emergency action must be taken to prevent its spreading. A small localized inflow can soon erode a large channel into the cofferdam, causing undermining and collapse of the wall. If boiling is seen to occur in an isolated patch of more permeable soil it can sometimes be controlled by building a wall of clay bags around the seepage, thus reducing the effective head. However, this procedure often results in a boil breaking out in another place. The most effective procedure is to flood the cofferdam and then take measures to reduce the hydraulic gradient as already described.

10.3 Construction of single-wall sheet pile cofferdams

10.3.1 Pitching and driving piles

To ensure verticality in pitching piles and to prevent them

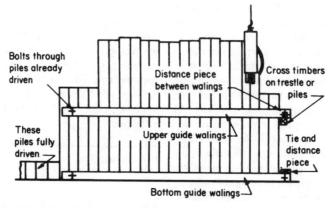

Fig. 10.16   Guide walings for driving sheet piles.

from moving off the vertical when driving past obstructions, it is most important that sheet piles should be driven through guides. In the case of land cofferdams the guides are provided simply by a lower pair of guide walings at ground level and an upper pair at the highest level permitted by the position of the hammer on completion of the first stage of driving. These guide walings are supported at one end by bolts through the partly driven piling and at the other by cross members (Fig. 10.16). Alternatively, a timber or steel trestle can be built and moved as a unit by crane. The trestle should have sufficient stability against overturning from wind forces on a panel of sheet piles pitched as shown in Fig. 10.16.

Where sheet piles are driven over water the guide walings are supported by timber or steel piled stagings constructed by floating plant. It is unsatisfactory to drive sheet piling by floating plant without fixed guides for the piles. The guide walings and staging can consist of pairs of guides at low water level and at top waling level (Fig. 10.17(a)) or the cofferdam walings can be assembled in a prefabricated unit which is supported from the river bed (Fig. 10.17(b)). Where short piles are driven by frame the guide walings need only be at the top waling level (Fig. 10.17(c)).

Guide walings should consist of substantial timbers, say 300 mm × 300 mm or H-sections placed with the web horizontal. The walings are fixed at a distance 5 mm wider apart than the overall depth of the piles, and this dimension is maintained by distance pieces. Three methods of pitching and driving sheet piles are shown in Fig. 10.18. For the *single* (or successive) method shown in Fig. 10.18(a) a single or pair of piles is clutched to the top of the last pile driven and then driven to the full design penetration. This method is suitable only for short piles or

piles driven wholly into loose sands, because it provides little restraint to the tendency for piles to twist as a result of unbalanced pressure from the soil packed into the trough of the pile and for the development of a secure interlock with the pile already driven. Also there is no restraint to the tendency of piles to creep and lean forward in the direction of driving.

In *echelon* driving (Fig. 10.18(b)) a panel of five to seven piles is driven to part penetration and the next succeeding panel is pitched. The hammer is then returned to the first panel which is again driven to part penetration followed by the second panel, and then a third panel is pitched. The hammer is again returned to the first panel which can be driven to its final penetration, and the work continues in two, three, or more stages of driving each successive panel. The German Waterfront Committee[10.4] states that echelon driving 'especially for long and deeply embedded piles should be required by the specification and always insisted on', the depth of stagger between adjacent piles should not be more than 5 m 'at extreme depths'. For three-stage driving the recommended stages are 0.4, 0.35, and 0.25 respectively times the full penetration depth.

The *panel* method (Fig. 10.18(c)) is described in BS 8004 and in the Institution of Structural Engineers' CP2: *Earth Retaining Structures*. At the commencement of driving a pair of piles is set up in the guide frame, carefully plumbed in two directions and partly driven. An adjoining panel of 6–12 piles is then set up and clutched with the first partly driven pair. The hammer is next placed on the last pile of the panel which is partly driven followed by the remaining piles working back towards the initial pair. The second panel is then assembled and clutched to the last pile of the first panel followed by part driving all

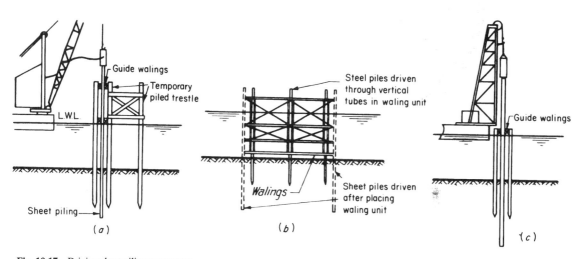

Fig. 10.17 Driving sheet piling over water.

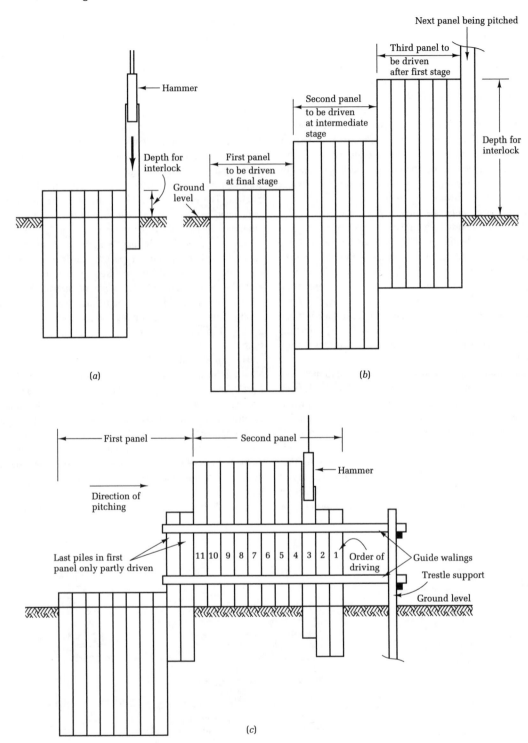

Fig. 10.18   Methods of pitching and driving sheet piles. (*a*) Single (successive). (*b*) Echelon. (*c*) Panel.

piles in this panel, again working in the reverse direction to the advance of the wall. Driving is continued in one or more stages of part driving until the first panel and then the succeeding panels are driven to the required penetration. The panel method is slower than echelon driving, but it provides the best security against the piles becoming declutched in difficult ground conditions.

Using either echelon or panel driving an important feature is the use of a restraining device on the leading pile to prevent deviations from line at the early stages of penetration. The Frodingham or other Z-section piles are driven with the male clutch leading, and the flat of the pile held against the face of the guide waling by a spacer block hooked over the flange of the guide (Fig. 10.19(a)). The leading clutch of a Larssen pile is similarly restrained by a cleat slotted over the two flanges of the waling (Fig. 10.19(b)). It may be necessary to provide a second spacer block in the trough of the second pile. If the piles have crept out of verticality there will be difficulty in making the closure. A slope of more than 1 in 300 will give difficulty in closing, depending on the length of the piles,

their flexibility, and the ground conditions. Corrections in verticality can be made by driving with an eccentric blow or by straining the piles by block and tackle. If these measures cannot give enough correction it will be necessary to use a tapered pile to make the closure. It is therefore convenient to make the first pile to be pitched a corner pile, when a special section riveted or welded tapered corner pile can be provided as detailed in the handbooks of pile manufacturers.

In small cofferdams it is advantageous to pitch and interlock all the piles before driving any of them. This is usually impossible in large cofferdams because of the length of guide walings required and the likely instability of a large area of unsecured piling exposed to flowing water in a river or tidal stream.

Piles are usually driven in pairs, except in hard driving conditions. However, they are usually pitched as single piles.

Sheet piles may be driven by drop-hammer, single-acting hammer, hydraulic (Fig. 8.8) or diesel hammer operating in a pile frame or with false leaders, or by

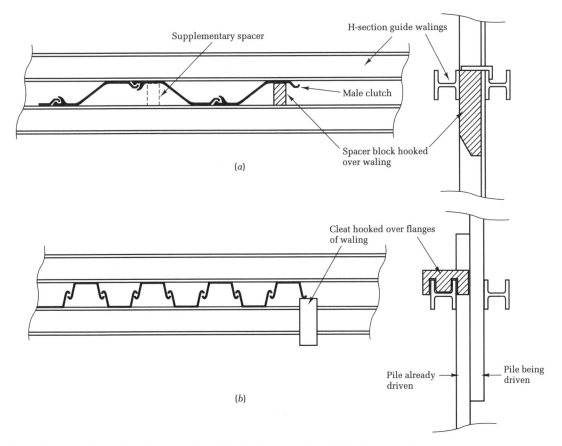

Fig. 10.19   Spacer blocks for guiding sheet piles. (*a*) For Frodingham sections. (*b*) For Larssen sections.

double-acting hammer with or without leaders. Double-acting, hydraulic or diesel hammers are generally preferred for speed and efficiency, and this type functions well in most types of ground, especially in sandy and gravelly soils.

Double-acting hammers are preferred for use with floating plant, and Type 'B' hammers have the advantage of being able to work under water to depths of up to 15 m. Hydraulic or drop-hammers are preferable in some circumstances, especially when driving in heavy clays.

Explosives can be used to loosen dense or hard soils and rocks to facilitate the penetration of sheet piles. The Rosenstock shock blasting process,[10.11] operated in the UK by the British Steel Corporation, employs low-energy explosive charges detonated simultaneously in close-spaced boreholes located along the line of the sheet piles. Driving of the piles into the loosened ground follows as quickly as possible before it has time to reconsolidate.

Vibratory hammers (Section 8.4) can be used for sheet pile driving in sandy or gravelly soils. Their relatively low noise level is advantageous for pile driving in built-up areas. The so-called 'silent' pile-driving equipment operates on the principle of pushing sheet piles into the soil by hydraulic rams against a reaction provided by the skin friction developed on piles already in place. Initially this reaction is obtained by the penetration of the panel of piles under the 10 t mass of the equipment. As the piles in the panel are alternately pushed down the skin friction resistance is progressively increased. The maximum thrust of each of the eight rams is 2250 kN. The equipment operates without ground vibrations and is virtually silent. The 'Hush' piling system employs a drop-hammer within a sound-absorbent box, which is also effective in achieving relatively quiet driving.

The noise from sheet pile driving is an important consideration in the selection of the type of hammer. As yet there is no legislation in Britain for specific noise levels which must not be exceeded in areas accessible to the general public. Local authorities adopt their own standards and maximum day and night noise levels of 70 and 60 dBA, respectively, are frequently stipulated for urban areas. Control of noise is also necessary to protect the health of site operatives. The Department of Employment recommends that no person should be exposed to a noise level of more than 90 dBA for 8 hours a day in a 5-day week. It is recognized that pile-driving noise exceeds 90 dBA, but the operations are not continuous throughout the working day and the observed noise level can be converted to an equivalent sound level which takes account of the duration of noise emission.

Some observed noise levels from pile-driving hammers are shown in Table 10.2. It will be seen that a distance of more than 1000 m is required from the noisiest group of hammers before the sound is attenuated to a permitted level of 70 dBA.

It is important to protect the heads of sheet piles during driving. Where double-acting hammers are used for driving piles singly or in pairs, a double anvil block is supplied with the hammer. For driving with single-acting or drop hammers, a single or double cast-steel driving cap should be used in conjunction with a hardwood or plastic dolly. Driving caps are sometimes used with the large double-acting hammers for driving heavy sections of Larssen piling.

Long lengths of sheet piling are conveniently driven in two stages. In the first stage the piles are pitched by long jib derrick or crawler crane and driven to part penetration by a light or medium weight hammer. In the second stage

Table 10.2 Noise from pile-driving operations

Type of hammer	Type of pile	Observed noise level		Approx. distances (m) for noise levels of	
		dBA	Distance from source (m)	70 dBA	90 dBA
'Silent'	Steel sheet	70	1.5	1.5	0.15
Vibratory (med. frequency)	H-section	90	1	14	1
BSP Impulse	Steel sheet	90	3	45	3
Drop (in 'Hush' tower)	Steel sheet	70	18	18	1.5
Drop (internal)	Cased	90	18	170	18
Drop (5 t)	Precast concrete	98–107	7	200–400	20–50
Diesel (light with shroud)	Steel sheet	95	18	274	27
Vibratory (high frequency)	Steel sheet	113	2	287	29
Diesel (light)	Steel sheet	97	18	345	34
Double-acting (air)	Steel sheet	90	110	1050	110
Single-acting (air)	Precast concrete	90	140	1350	140
Diesel (medium)	Steel sheet	116	7	1400	150
Diesel (medium)	Steel tubes	121	7	2400	280

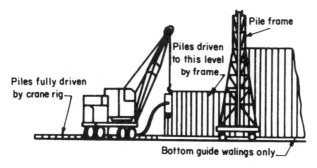

Fig. 10.20   Typical two-stage pile driving.

a heavy hammer suspended from a shorter jib crane is used to complete the driving. Alternatively, the two-stage driving can be accomplished with two pile frames; the first a tall light frame and the second a short sturdy one. Using a pile frame for the first-stage driving facilitates placing the hammer on the pile, which is not an easy job for a man working high on a ladder lashed to a swaying pile when working without a frame. A typical method of two-stage driving is illustrated in Fig. 10.20.

Where long runs of sheet piling have to be driven, the two-stage method, with simultaneous driving of the two stages, gives rapid progress. Another method is to use specially designed travelling rigs incorporating multiple hammers.

10.3.2 Driving with limited headroom

Difficult problems arise when piles have to be driven where there is limited headroom in which to pitch the piles and place the hammer. One method is to drive the piles in two lengths with dovetailed joints. Alternate piles are driven 1 or 2 m deeper than the adjoining ones; the upper row of piles can then be interlocked with the upstanding portions (Fig. 10.21) and the two lengths driven down together. Where watertightness is required the butt joint must be welded up. If it is desired to utilize the full flexural strength of the piling it may be necessary to add fishplates. The dovetailing method can also be used where the sheet piles are too long to be handled and pitched by the available crane.

10.3.3 Example of single-wall cofferdam design

The Thames Barrier cofferdams   The construction of the foundations for the abutments and piers for the Thames Barrier in 1977–80 required excavations in 11 cofferdams ranging in depth from a top level at +5.2 m to an excavation level at −13 to −24 m OD. Eight cofferdams were constructed from Larssen 6 sheet piles with bracing frames at one to four levels, but the two deepest

and one other deep excavation were supported by interlocked Peine piles (Fig. 10.7(c)). The change from the Larssen to the Peine sections was made after experiencing various problems with hard driving in chalk and dense sands which resulted in numerous split clutches.[10.12] It was also considered that the double skin of the piles would provide a more watertight structure and if declutching did occur the spaces between the two skins could be readily sealed with concrete. The need for only a single level bracing frame greatly facilitated the excavation operations and the subsequent foundation construction.

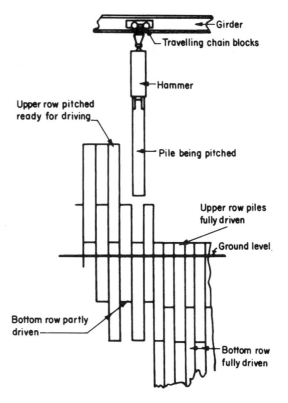

Fig. 10.21   Driving sheet piles in limited headroom.

As a first operation a contiguous bored pile wall was installed along the centre line of the cofferdam walls. The pile boreholes were filled with a low-strength mix of pf ash–cement. The Peine piles in lengths up to 38.4 m were then pitched by crane through piled trestle guides and driven into the 'soft' contiguous piles by Delmag D62 diesel hammers operating in the leaders of two Menck MK60 piling rigs. The piles were driven in pairs with 10 per cent stitch welds along the clutches. The piles were provided with toe plates to reduce skin friction and every eighth pile was provided with an inclinometer tube to check verticality. The closing section of two of the cofferdams was achieved by welding Larssen clutches to each pile on either side of the gap and then forming a 'blister' of Larssen piles to form the closure. The interior of the blister was sealed by cement and chemical grouting.

The stiffness of the sheet pile sections made it possible to excavate under water to depths of 10–12 m below river bed before fixing the bracing frame at −1.0 m (Fig. 10.22). During this period it was necessary to balance the interior and exterior (tidal) water level by pumping. Excavation was undertaken by a combination of grabbing,

reverse-circulation drilling, and airlifting. The variable gap between the waling and the sheet piles was filled by grout bags consisting of plastic fabric in a wire mesh container placed by diver and filled with a sand–cement–plasticizer mix injected into the bag from the surface. A 25 mm hole was formed in each bag to receive a 60 g explosive charge which was placed and detonated at the stage of removing the walings. The arrangement of the grout bags is shown in Fig. 10.23. The rocker bearers between the bags and the sheet piles were provided to accommodate the deflected shape of the piles.

Chalk and sand adhering to the piles was removed by divers using water jets and scrapers and the final excavation to formation level was performed by a combination of augering and airlifting. The underwater base slab was placed through 300 mm tremie pipes on a 7.5 m grid. The concrete contained 50 per cent cement replacement in the form of pf ash and a retarding admixture. The weight of the slab was insufficient to counterbalance the water pressure in the underlying strata, and pressure relief pipes discharging above the top of the slab were provided. In some cases the tubular piles supporting the piling trestle were used to form the pressure relief pipes. A section through the cofferdam after completing the base slab is shown in Fig. 10.22. The Thames Barrier was constructed by a joint venture of Costain, Tarmac, and Hollandsche Beton Maatschappij to the designs of Rendel Palmer and Tritton, consulting engineers to the then Greater London Council.

Marina Bay crossing and station, Singapore MRT

The 1.1 km cut-and-cover section of the Singapore MRT railway with a 300 m over-water crossing and a station on land reclaimed from the sea is a good example of the complexities of cofferdam construction in poor ground condition and the use of mathematical modelling techniques to analyse the effects of the construction sequence on strut loadings at each stage of the excavation.[9.7, 10.13] Southwards of the bay crossing the soil conditions consisted of about 10–12 m of sandy reclamation fill and 8–15 m of soft marine clay overlying firm to stiff clay. Ground-water level was about 2 m below ground level.

The full length of the section was divided into 16 cofferdams separated by sheet pile bulkheads. Six stages in the construction of the Marina Bay station cofferdams are shown in Fig. 10.24. Stage 1 comprised driving sheet piles consisting of Frodingham I BXN sections welded to 610 × 305 mm universal beams similar to the arrangement shown in Fig. 10.7(f). The sheet piles were 30 m long and were driven in lengths of 16 and 14 m to a depth of 27.5 m below ground level. The initial excavation was taken out to a depth of 2 m and the top level bracing frame consisting of twin broad-flange beam section walings and struts

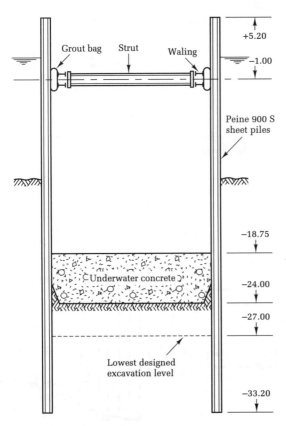

Fig. 10.22 Section through pier 4 cofferdam, Thames Barrier (after Grice and Hepplewhite[10.12]).

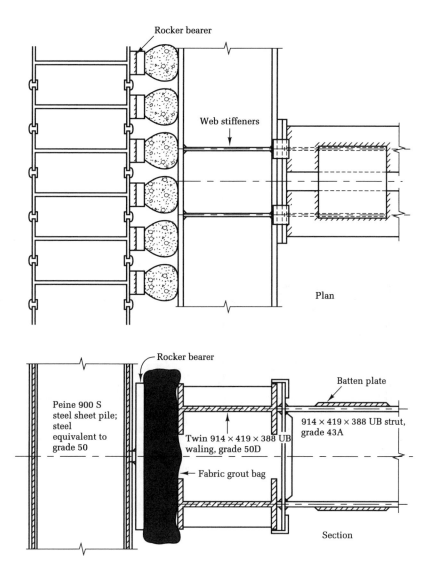

Fig. 10.23 The use of grout bags to fill the gap between piles and walings, Thames Barrier cofferdams (after Grice and Hepplewhite[10.12]).

was installed at ground level. The excavation was then deepened to 7–11 m and the second level bracing frame was installed at 6 m depth. The cofferdam was then flooded to within 0.5 m of the tops of the sheet piles and the excavation was taken out to the full depth of 18 m by grabbing, jetting, and airlifting (stage 2). At this stage bored piles were installed to support the station structure. The upper parts of the piles were permanently lined with steel casings vibrated into place. The exterior of the casings was provided with shear connectors to provide a restraint to uplift of the base slab.

The slab which was placed under water by tremie pipe (stage 3) was only 1.5 m thick and of insufficient weight to

overcome hydrostatic uplift. In addition to restraint by the piles pressure relief pipes were installed through the slab and a compressible void former made up from five layers of corrugated steel sheets (Fig. 10.25) was placed on a 300 mm sand bed before placing the tremie concrete. The purpose of the void former was to permit swelling of the stiff clay by about 50 mm, thus reducing further the uplift pressure on the slab. The cofferdam was dewatered at stage 4 and the lower part of the station box was constructed. At stage 5 the tops of the already constructed walls were strutted by the permanent intermediate floor slab which allowed the second level bracing frame to be removed. This was followed by completing the walls and

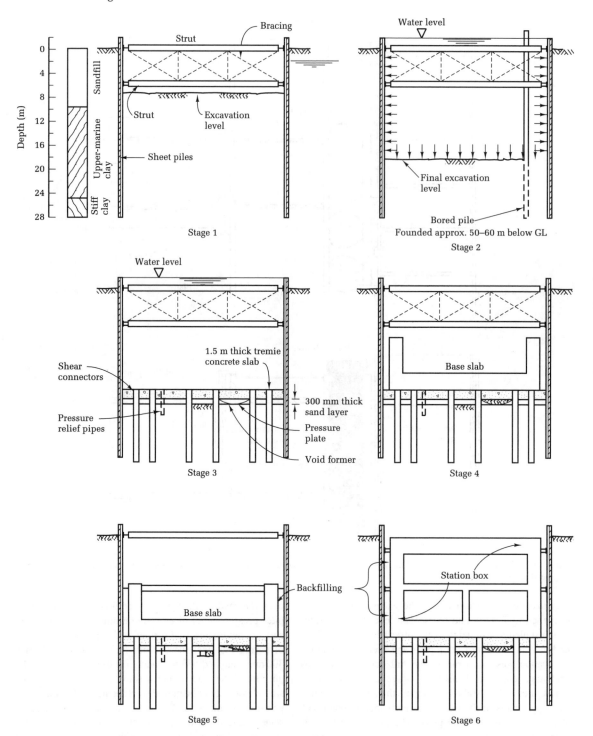

Fig. 10.24　Stages in the construction of Marina Bay Station, Singapore MRT (after Clark and Prebaharan[10.13]).

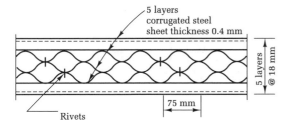

5 layers
corrugated steel
sheet thickness 0.4 mm

5 layers @ 18 mm

75 mm

Rivets

Fig. 10.25 Compressible void former beneath base slab, Marina Bay Station, Singapore MRT (after Clark and Prebaharan[10.13])

top slab (stage 6). The high modulus combined section sheet piles were needed to span the 8 m depth of wall between the second bracing frame and the tremie slab.

The tunnels on either side of the station were constructed in the same sequence except that it was possible to construct the complete tunnel box below the level of the second bracing frame.

10.4 Double-wall sheet pile cofferdams

10.4.1 Design principles

Double-wall sheet pile cofferdams are essentially free-standing walls used for wide excavations where cross-bracing would be uneconomically long. Cofferdams of this type are not used for land excavations since in these conditions support to a single wall can be given by raking shores, or by components of the permanent structure, as described in Section 9.5.7. Double-wall cofferdams consist of two lines of steel sheet piling (or for low heights an outer line of steel and an inner line of timber sheet piling) driven into the ground and tied together at the top by walings and tie bolts. For high heads it may be necessary to install sets of walings and tie bolts at lower levels. Dams on impermeable soil need not have an inner banking but, on permeable soils and soils of moderate bearing capacity, an inner banking is a desirable addition (Fig. 10.8(*f*)). The space between the piling is filled with sand, gravel, crushed rock, or broken bricks. The depth of penetration of the piles depends on the pressure exerted by the filling and on the need to prevent horizontal sliding. Each line of sheet piling is considered to be anchored at the top or intermediate levels and restrained at the bottom by the passive resistance of the soil. Where cofferdams are constructed on permeable soils, the depth of penetration of the piles is also governed by considerations of piping. White and Prentis[10.1] have made an extensive study of double-wall cofferdams on such ground and they take the view that lengthening the seepage path in a horizontal direction is more effective and is more economical than lengthening it by deeper penetration of the sheet piling. They regard inner earth banking as an essential feature in

the stability of double-wall cofferdams against collapse following on piping. Another important feature is the wide drainage ditch (Fig. 10.8(*f*)) carefully laid to falls with check weirs as necessary to keep the ditch full of water and the velocity of flow low enough to avoid erosion.

White and Prentis state that the inner row of sheet piles contributes nothing to the watertightness of the dam and that its main function is to prevent collapse in the event of overtopping causing erosion on the inside face. Their double-wall cofferdams for the work on the Mississippi River were designed to be overtopped during the flood season and they make the point that it is important to provide adequate sluices through the walls to allow the interior to be flooded as the water level outside rises, so reducing the risk of erosion or piping under high external heads. The need for rapid lowering of the internal water level during the period of a falling river level is also important, as the reverse head may cause collapse of a cofferdam which is not designed for this condition. Generally, the detailed and practical observations of White and Prentis on these double-wall cofferdams are of absorbing interest and should be studied carefully by engineers concerned with major cofferdam construction.

10.4.2 Filling materials

Where double-wall dams are constructed on unstable soils it is necessary to excavate between the sheet pile rows and remove the soft material before placing the filling. This procedure is facilitated by providing diaphragms between the rows at a spacing of four to eight times the width. Individual cells between diaphragms can then be pumped down and excavated. Clay and other materials which must consolidate before they offer much resistance to external pressure should not be used as filling. Watertightness is given by the outer row of sheet piles and not by the filling material. Weep holes should be provided at various levels in the inner wall to drain down the filling, thus avoiding internal hydrostatic pressure. Rubber washers can be provided around the tie bolts between the external walings at the face of the sheet piling for added watertightness.

The sheet piling of double-wall cofferdams founded on rock must be restrained from outward movement by banks of earth or rockfill or by bagged concrete on both inner and outer faces. The pressure exerted by the filling can be reduced by using concrete in the lower part. The risk of sliding on a sloping rock bottom can be prevented by a bank of rockfill on the inner face or by dowelling a base slab of concrete to the rock by means of short steel rods or old rails fixed into holes drilled in the rock.

The double-wall cofferdam shown in Fig. 10.26 was constructed for the Pitlochry Dam and Power Station, Scotland, in difficult conditions of bedrock covered by

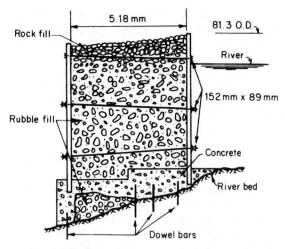

Fig. 10.26   Double-wall sheet pile cofferdam for Pitlochry Dam.

gravel and large boulders. Rock level in the bed of the River Tummel fell steeply towards the excavation. The cofferdam was constructed at a low stage in the river and diaphragms permitted individual cells to be pumped out and excavated to rock level. Sealing concrete was rammed under the piles and dowel bars were used to key the dam to the rock. Further concrete was placed beneath the inner row of piles. Records of the flow in the river showed the depth to vary between 0.3 and 3.8 m. Flows higher than 2.4 m were infrequent and the cofferdam was designed to be overtopped at this level.

10.5 Cellular sheet pile cofferdams

Cellular cofferdams for work in rivers and the sea consist of complete circles of interlocking piling connected by short arcs (Fig. 10.27(a)) or a series of arcs in a double wall connected by straight diaphragms (Fig. 10.27(b)) or modified circles (Fig. 10.27(c)).

They have the advantage that little falsework is required in their construction; the only requirements are top and bottom templates for pitching the piles. They can be constructed working out from land, each successive cell, after filling, forming the working platform for driving the next (Fig. 10.28). Trough-shaped piling is unsuitable and special straight web sections must be used. Sand, gravel, crushed stone, or broken bricks are suitable materials for filling. Clay is unsuitable for the reasons already stated for double-wall cofferdams. Cellular cofferdams can be used on an irregular rock bed when the sheet pile lengths can be tailored to suit the rock profile. They are also suitable for founding on stiff clays or sandy or gravelly soils. In the latter two cases it is usually necessary to provide an inner earth banking to give the necessary length of seepage path to avoid failure by piping.

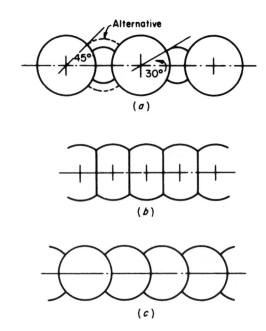

Fig. 10.27   Layout of cellular sheet pile cofferdams. (a) Circular cells. (b) Diaphragm cells. (c) Modified circular cells.

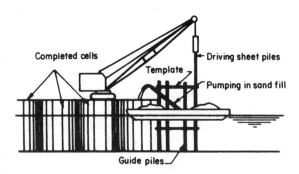

Fig. 10.28   Constructing cellular cofferdams.

Cellular cofferdams suffer from an important hazard in that a major failure at any interlock may result in complete failure of a cell with serious consequences. Thus it is inadvisable to use cellular construction on ground containing boulders or other obstructions which might cause splitting of the piles or parting of the interlocks. Great care must also be taken in the accurate pitching and driving of the piles. Packshaw[10.2] states that the ratio of the average width (i.e. the area of a cell divided by the distance between centres of cells) to the height should be 0.85 or 0.9 if the filling is sand, or 0.8–0.85 if the filling is rock.

The design of a cellular cofferdam requires consideration of the following factors:

(a) Resistance to overturning (Fig. 10.29(a));

(b) Resistance to sliding where there is little or no penetration of the sheet piles (Fig. 10.29(b));

(c) Resistance to tilting provided by vertical or horizontal shear resistance of fill material (Fig. 10.29(c));

(d) Resistance to tilting provided by shear resistance on a curved rupture surface at or near the base of the cell (Fig. 10.29(d));

(e) Resistance to failure by interlock separation (Fig. 10.29(e));

(f) Resistance to shear failure provided by the soil beneath the base (Fig. 10.29(f));

(g) Resistance to tilting due to escape of soil from beneath the base (Fig. 10.29(g));

(h) Resistance to general shear slide of the structure and soil retained by and beneath the structure (Fig. 10.29(h));

The failure modes in Fig. 10.29(c–g) have been analysed by Dismuke.[10.14] The failure modes in Fig. 10.29(a, b, and h) can be analysed by conventional methods.

10.6 Concrete-walled cofferdams

10.6.1 Bored cast-in-place piling

A cofferdam formed by a row of bored and cast-in-place piles sunk in close contact with one another is a useful construction expedient for situations where headroom limitations prevent the driving of steel sheet piling or where it is necessary to avoid vibrations from pile driving. Bored pile cofferdams can also be used in ground containing boulders which would split steel sheet piles or cause them to come out of interlock. However, bored piles sunk in bouldery ground are in themselves a costly form of construction.

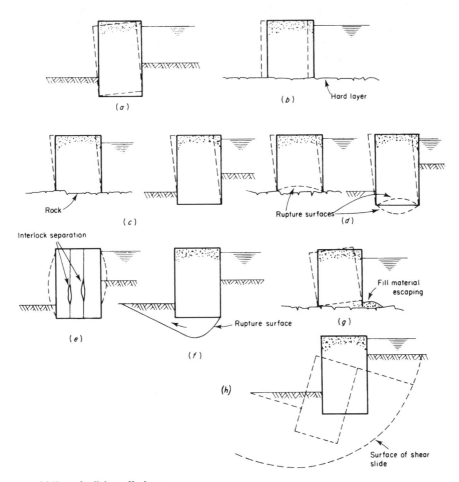

Fig. 10.29 Modes of failure of cellular cofferdams.

Fig. 10.30   A contiguous bored pile retaining wall supported by ground anchors and reinforced cast-in-place walings.

The size of the piles depends on the pressure exerted by the retained water and soil and the spacing of the walings. Diameters of 300 mm to 1 m are commonly used and the piles are reinforced against bending stresses. Fig. 10.30 shows a deep excavation supported by contiguous bored piles. Where this form of support is used in cofferdams to exclude a heavy flow of ground water, the 'secant' system of interlocking bored piles must be used in permeable soils, as described in Section 5.4.5.

10.6.2 Continuous diaphragm walls

The use of grabbing and drilling methods to excavate trenches supported by bentonite mud for the construction of basement walls has been described in Section 5.4.4. Similar techniques can be used for the construction of concrete-walled cofferdams. The ground within the cofferdams is subsequently excavated and support given to the concrete walls by strutting or tied-back anchorages.

Another method of forming a watertight diaphragm is the ETF process. This employs a heavy steel H-section which is driven with a guide device to the required penetration. Clay–cement grout is then injected at the bottom through a pipe fixed to the web of the H-section. The section is withdrawn while injection is continued; thus the grout fills the void left on withdrawal to form a section of the continuous diaphragm (Fig. 10.31). In a

process developed by GKN–Keller a vibrating unit with wide fins is used to form the slot in the ground. If the soil is suitable, a diaphragm wall can be made by stabilizing the soil *in situ* with cement using a churn drill or a rotary drill fitted with a fish-tailed drilling bit. This method can be used for small excavations in sands or gravels.

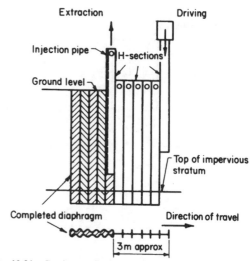

Fig. 10.31   Continuous diaphragm constructed by the ETF process.

10.6.3 Precast concrete blockwork

Precast concrete blocks can be used to form gravity dams. This method is limited to special circumstances, such as repetition work in short lengths of cofferdam where it is undesirable to use bracing or where for one reason or another the double-wall steel sheet pile cofferdam cannot be adopted. As blockwork dams are heavy and sensitive to fairly small differential settlements which might cause the blockwork to open at the joints, they cannot be used on ground of low bearing capacity.

It is normally necessary to construct the blocks to fine tolerances with joggled or dovetailed joints to give water-tightness and stability against sliding. They must also be set on an accurately screeded tremie concrete bed, and successive tiers must be carefully placed to ensure close joints. This involves much work by divers and the slinging and setting of heavy units. Therefore blockwork dams are generally a costly expedient only used in special circumstances.

10.6.4 Cofferdam construction by jet grouting

The jet grouting process consists of forming a cylindrical hole in the soil by jetting followed by injecting a sand–cement mix to fill the void. It can be used to produce a wall of overlapping grout columns to enclose an excavation. The process is described in Section 11.3.7. It has the advantage of being capable of forming a plug across the base of an excavation in addition to the cofferdam walls. This avoids the need for underwater concreting for a tremie concrete slab, or a ground-water lowering system to enable the base slab to be placed in dry conditions. The jet grouting process can also be used to seal openings in a sheet pile wall caused by declutching, particularly at corners of cofferdams.

Jet grouting was used to form cofferdams for constructing buoyant foundations to support the viaduct piers of a new expressway at Monte Barro in northern Italy.[10.15] A 450 m length of the viaduct crosses a valley where 8–10 m of peat were found to be overlying a great depth of sand and clay. Ground water was at 1 m below the ground surface. A circle of 1.2 m diameter × 25 m deep jet grouted columns was installed around the site of the foundation, followed by forming a 5 m thick jet grouted base slab. Then a second ring of columns was formed outside the first to depths between −13 and −25 m. The excavation within the grouted ring was taken out in 2 m stages and lined with 300 mm of reinforced concrete at each stage. After reaching the bottom of the foundation, a PVC watertight liner was fixed to the concrete lining and a second 300 mm wall was constructed followed by placing a 2.5 m thick concrete base slab and a capping beam. The completed foundation is shown in Fig. 10.32.

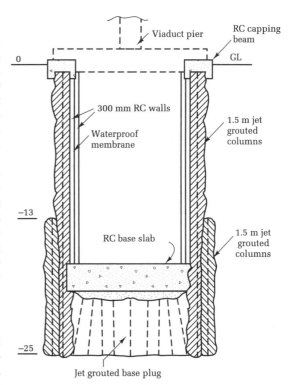

Fig. 10.32   Jet grouted cofferdam for viaduct pier, Monte Barro, Italy.

10.7 Movable cofferdams

Where there is a considerable amount of repetition work in underwater foundations, such as in the piers of multi-span river bridges or in long jetties, it is economical to design the cofferdam to be moved as a single unit from one foundation to another. These consist typically of a ring of steel plate segments or an assembly of sheet piles which are secured to the top of a caisson forming the permanent foundation of the bridge pier. After sinking the caisson to founding level the bridge pier is then constructed within the cofferdam which is then disassembled or floated off as a single unit to be used for the next pier of the bridge. An example of the movable cofferdam used for construction of the piers of the Sjaelland Färo–Falster Bridge in Denmark is shown in Fig. 6.26.

10.8 Underwater foundation construction

10.8.1 Working conditions

The following are some notes on methods of concreting under water and on underwater construction generally. Typical circumstances for the adoption of underwater construction are in relatively small scale or non-repetitive foundation work in deep water where the cost of coffer-dams or other temporary work might greatly exceed that of the permanent foundation structures; or on sites where stability conditions for cofferdams or caissons are such that security against 'blows' or piping can only be obtained at a high cost. Other underwater work includes fixing bracing to piles driven from floating plant or stagings, cutting off sheet piling or bearing piles when removing temporary structures or in site clearance, and underwater drilling and shot firing for rock removal.

One of the ways to overcome the problems of under-water working is to construct as much as possible of the permanent work above water level and then to sink it into place. The technique for performing such operations for over-water bridge foundations were described in Chapter 6. An example of the method for constructing a pumping station in deep water is described by Calkin and Mundy.[10.16] The pumping station is located in the land-locked Plover Cove Reservoir in Hong Kong where water levels could vary between the operating levels of +14 and −3.85 m PD. The reservoir bed after removing soft silt was at −12 m. As an alternative to constructing a very deep cofferdam the contractor, the Gammon–Keir Joint Venture, proposed to construct the pumphouse above

water on a piled staging while progressively lowering it on to piles which had been cut off about 1 m above bed level. The 24 1.5 m piles were taken down to sound rock by drilling methods through permanent casing and 13 of the pile casings remained to provide supports to a construction platform at +12 m which was the prevailing reservoir level at the time. Previously the jacket structure (Fig. 10.33) had been assembled around the foundation piles. The four sets of jacks at the corners of the jacket had a total lifting capacity of 8500 t. Each of the 16 centre hole jacks were threaded with suspension cables consisting of 37 18 mm Dyform strands.

The stages of lowering the structure are shown in Fig. 10.34. The 2.7 m thick base slab was constructed on the platform and holes were formed in the slab to match the positions of the piles (stage 1). The slab was then lifted off the staging, which was dismantled, and partly lowered into the water to reduce its effective weight (stage 2). The walls were then commenced and were constructed in 3.5 m lifts during which time the structure was lowered in stages to conform to construction progress and changes in the reservoir level (stage 3). Finally the base slab was lowered on to collapsible cylindrical forms surrounding the pile heads. These formed a seal with the base slab, allowing an underwater connection to be made between the slab and the piles (stage 4).

Underwater construction demands a special technique and the permanent work should be designed with this in view. Simplicity is the essential feature, for example mass concrete instead of reinforced concrete and simple out-lines to avoid complex underwater formwork. Although underwater work can proceed with little hindrance in sheltered waters, such work in the open sea is entirely at the mercy of the weather. The scheme should allow for the occurrence of the worst predicted storms during the construction period. It is folly merely to trust to luck that such storms will not occur if their effect would be to wreck the construction plant and partly completed work. The safe method is to design the work so that construction vessels can be quickly disengaged and taken to shelter, leaving the partly constructed foundations in a stable condition. Adequate fendering should be provided to the permanent and temporary structures and to the floating construction plant. It is impossible to make a hard and fast programme for work undertaken in open waters. Even if storms do not occur, the following are limiting factors:

(a) Transfer of men from launches or small ships to fixed stagings is difficult in wave heights of more than 1 m and virtually impossible in wave heights of more than 1.5 m unless special devices are used.

(b) Lifting heavy material by crane on a fixed staging from the deck of small ships or handling materials by

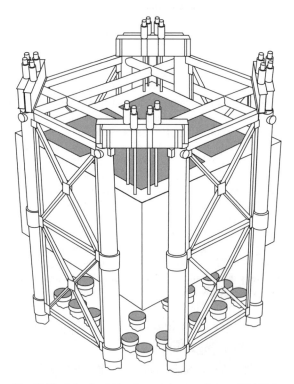

Fig. 10.33 Lowering jacket for pumphouse structure of Toi Mei Tuk B Pumping Station, Hong Kong (after Calkin and Mundy[10.16])

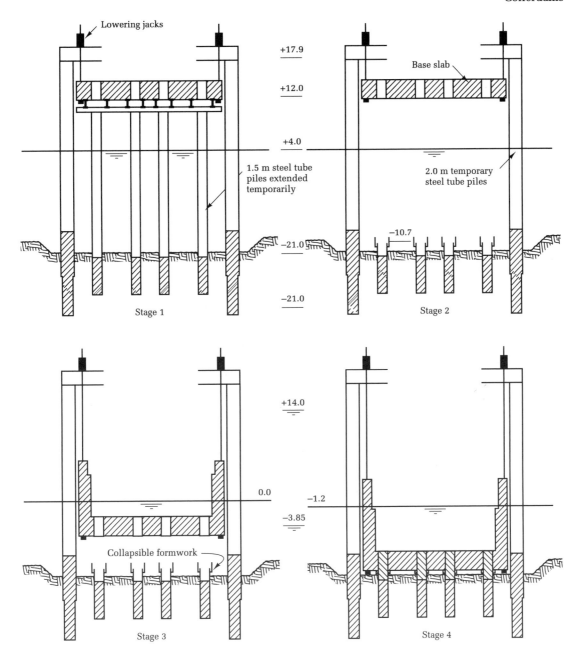

Fig. 10.34 Stages in the construction of Toi Mei Tuk B Pumping Station, Hong Kong (after Calkin and Mundy[10.16])

a pontoon-mounted crane is difficult in wave heights of more than 1 m and dangerous in waves more than 1.5 m high. 'Snatching' of the crane rope can result in a temporary load of the crane of several times the dead weight of the load being lifted.

(c) It is difficult to bring divers wearing conventional diving suits out of the water on to a launch or fixed ladder in wave heights of more than 1 m, although they can work in rather more severe conditions in self-contained equipment.

(d) Diving operations are difficult in currents of more than 2 knots.

(e) Divers cannot undertake long spells of work in water deeper than about 15 m without long decompression periods during their ascent or in decompression chambers on the site.

The above factors are, of course, applicable to ordinary construction work. Very large crane barges, including semi-submersible work platforms, underwater vehicles, and deep-water diving equipment have been developed for construction work in the open seas and in deep water for the North Sea and other offshore oil fields. These facilities as described in detail by Gerwick[6.14] are capable of operating in sea conditions of greater severity than those listed above.

10.8.2 Excavation

Excavation in limited areas on foundation work is normally carried out by grabbing and airlifting with the cranes operating from pontoons or stagings. Large hydraulically operated backacter excavators have been adapted for mounting on barges and working in water depths of 15–20 m. Small quantities of excavation can, however, be dealt with by divers using airlift equipment of the types illustrated in Section 6.4.6.

Where it is necessary to drill and blast rock under water, the work is carried out wholly by divers if the area to be excavated is small. Divers can drill with hand-held jackhammers in depths of water up to 15 m without modification to standard tools, but it may be desirable to carry a large-diameter exhaust pipe up to the surface to prevent exhaust bubbles from obscuring the diver's vision and to overcome back pressure on the tool. In deep water it is advantageous to increase the pressure of the air supply to ensure efficient operation of the tools. Underwater drilling with hand-held jackhammers is not difficult in hard rocks, but in weak rocks, such as weathered sandstone, marl, or weak limestone, the tools are liable to clog and jam and the diver has not the sensitivity of touch nor the ability to ease up a heavy drill to free the drill steel. Excavation by divers in weak rock produces such turbidity in the water that visibility is nil and the divers rely entirely on feel. Soft and slurried material should be cleaned off by air/water jet and raised to the surface by an ejector. For drilling and blasting in large areas the most efficient method is to drill from pontoons or barges with multiple drills mounted on them, or from transportable stagings lifted and moved from point to point by floating crane. The final stage of trimming the rock excavation for the anchorage pier of the South Bisan–Seto Bridge (Fig. 6.57) was undertaken by a horizontally rotating milling cutter operated from a jack-up barge.

10.8.3 Underwater concreting

Concrete deposited under water is never as sound as that placed in the dry. There is always a tendency for cement to be washed away or for the fine aggregate to become segregated from the coarse. Therefore, underwater concrete should normally be regarded as a sealing layer to enable a cofferdam, caisson, or hollow box foundation to be pumped dry for subsequent placing of the main structural concrete. However, where the circumstances are such that the whole of the concrete work must be carried out under water, every care must be taken to ensure a sound job. Extra cement over and above that required for compressive strength alone should be added to the mix, but too much cement results in excessive laitance forming on the surface of the layers. It is also important to place underwater concrete on a clean firm surface. After dredging, the surface should be levelled by dragging a heavy steel beam across it. If the surface is on rock, any high spots shown by the drag should be dressed down by a diver. If the surface is on clay, mud, or loose sand, a blanket of crushed graded rock should be placed and levelled by drag. Where the surface has to be levelled accurately to receive a box caisson or blockwork, it should be screeded by divers working with a strike-off template across a pair of carefully levelled steel screed rails. A blinding layer of concrete or sand–cement grout over the crushed stone layer is advisable where blockwork or reinforced concrete has to be placed.

Formwork   For small thicknesses of concrete where the outline need only be rough, such as sea-bed protection to piles or piers, the formwork can consist of concrete in bags (Fig. 10.35(a)). Greater heights of lift, or successive lifts, can be achieved with formwork provided by precast concrete blockwork (Fig. 10.35(b)). Steel formwork is required for high lifts and for all underwater structures where regularity of outline is required. The forms should be designed to be assembled in units above water, preferably with the reinforcement fixed within them before the whole assembly is lowered to the bottom. Gaps between the bottom of the forms and the crushed rock bed or blinding layer can be filled with bagged concrete or heavy plastic sheeting. The forms should not be designed to be shored up externally but as self-supporting units with the thrust on the panels carried by cross ties above or below the finished surface (Fig. 10.35(c)). Any underwater connections made between formwork units should be as simple as possible with pins or toggle connections rather than bolts.

Placing   Underwater concreting is carried out by the following methods:

(a)   Bags
(b)   Bottom-dumping skips

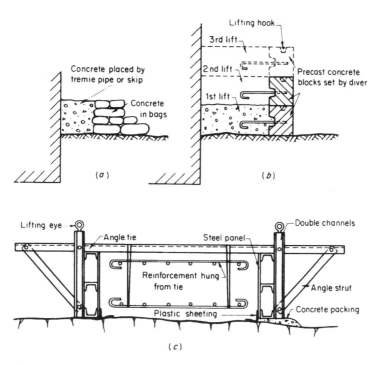

Fig. 10.35 Types of underwater formwork. (*a*) Concrete bagwork. (*b*) Precast concrete blocks. (*c*) Braced steel panels.

(c) Tremie pipe
(d) Grouting pre-placed aggregate (prepacked concrete)

Concrete placed by divers emptying it from *bags* is limited to small quantities since the work is slow and laborious. The bags are made of canvas or polythene about 0.5 m in diameter and 1.2 m long. The bottom is closed by a length of chain, or rope, with a slip knot which is released by the diver. Where the concrete is built up in *bagwork* the bags should only be partly filled (two-thirds to three-quarters). The diver builds the bags up in bonded courses with the mouths of the sacks away from the face. The bags are rammed flat while laying them. Steel spikes may be driven through successive courses to hold them together.

Bottom-dumping skips are designed to discharge their contents after being lowered on to the bottom. The door-release mechanism can be actuated by the skip coming to rest on the bottom, or the skip can be provided with an outside latch operated by diver or by a trip wire to the surface. The skips should be provided with covers to stop cement being washed away as they are lowered through the water. The concrete must not be dropped through the water, and the mechanism should be checked to ensure that the door has opened before the skip is raised through the water.

A *tremie pipe* is made up from a number of lengths of 200–300 mm diameter steel pipes with quick-release joints. A receiving hopper is fixed to the upper end. The tremie pipe is slung from a crane or gantry and lowered until the lower end is almost touching the bottom. A stopper of expanded polystyrene, cut in a roughly spherical shape, or an expendable steel plug, is placed in the top of the pipe and concrete is dumped into the hopper. This forces down the stopper until it is expelled from the bottom of the pipe and the concrete flows out. Another method of preventing segregation in the first charge of concrete is to introduce a bubble of compressed air into the tremie pipe just below the concrete. A plug of vermiculite granules can also be used as a stopper. The granules become dispersed in the concrete. As it is essential to keep the bottom of the pipe buried in concrete the hopper must be kept well charged and, if there is any sign of the pipe emptying, it must be lowered quickly to close the bottom. If there is any tendency for the concrete to jam in the pipe, the tremie must be 'dollied' up and down to release it. For concreting in large areas a number of tremie pipes are set at 2.5–7.5 m centres, the object being to limit the flow of concrete in a horizontal direction so preventing the build-up of excessive laitance. Concrete is placed in the hoppers one by one in turn, moving to the next one as soon as the level of the concrete has been raised by 0.5–1 m. The tremie pipe should be emptied after each turn and the stopper used for the refilling. If the area is not large enough to warrant a number of pipes, it should be compartmented by steel or precast concrete slab panels to ensure that the concrete does not have to flow laterally by more than about 3 m. Concrete pumps, either discharging into the hoppers or directly into the tremie pipes, are useful equipment for dealing with large volumes of concrete under water.

The layout of the tremie pipes used for placing concrete in the Thames Barrier cofferdams[10.12] is shown in Fig. 10.36. The largest underwater pours had a volume of 6600 m³ which was completed in a 3-day period. The rate of placing was limited to a rise of not more than 2 m in 24 hours.

The Dutch Hydrovalve system can be used for fairly shallow pours. The tremie pipe consists of a collapsed rubber pipe. A dry concrete mix with a slump of about 80 mm is fed into the hopper in batches which expand the pipe in discrete amounts and are allowed to discharge on to the surface of the concrete already placed. The cohesive character of the mix prevents the cement from being washed out. The pipe is allowed to rest on the concrete and is moved horizontally to keep a fairly level surface on the pours.

Concrete for placing by tremie should have adequate workability (125–200 mm slump) to allow it to flow readily. Bouvier[10.17] has made a detailed study of mix design and operational techniques for tremie concrete.

Prepacked concrete has its advantages in underwater construction, since the volume of materials to be passed through the mixer is only one-third of the total volume of concrete placed. This can give worthwhile savings in the size of the mixing and batching plant which, in turn, saves in the size of pontoons or barges or temporary stagings. These savings are important in work over water, where the cost of purchase or hire of floating plant represents a high proportion of the cost of the work.

Grout pipes in the form of perforated steel or PVC tubes or cylinders made up from expanded metal are placed at intervals over the area to be concreted. Then coarse aggregate of 12.5 mm or preferably 25 mm minimum size is dumped inside the formwork by skip or chute. The grout injection pipes are lowered down the perforated pipes until their ends are nearly touching the bottom. Alternatively, the injection pipes can be used without the perforated outer casings. Grout is then pumped in and displaces the water as the level rises. The injection pipes are slowly raised until the whole mass has been grouted. Soundings are taken in intermediate perforated pipes to check that an even flow is being maintained. The grout consists of a mixture of cement and sand and sometimes pulverized fuel ash (fly-ash). Workability aids are often added to improve flow characteristics. The grout is mixed in an ordinary grout pan and transported by cementation pump, or in a colloidal mixer with an attached displacement pump.

The method must be used with caution in flowing water, or in water containing silt or organic matter, since there is a risk of the grout being washed away from the aggregate

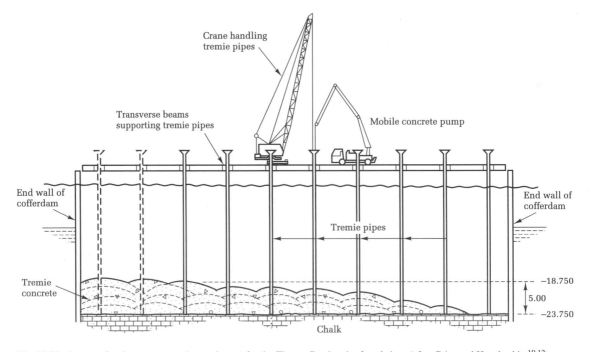

Fig. 10.36 Layout of underwater concreting equipment for the Thames Barrier pier foundations (after Grice and Hepplewhite[10.12])

or of the latter becoming coated with silt or other material which would prevent a proper bond with the grout. Silty coatings on aggregates are liable to occur when underwater excavation has been undertaken in weak rocks such as chalk or marl. Although the silt tends to settle slowly and eventually the water will clear, subsequent placing of the aggregate would stir up the silt layer on the bottom and cause it to be deposited on the lower layers of the aggregate. Drilling inspection holes into the prepacked concrete of a pier of the Arthur Kill Bridge in New York[10.18] showed that the 2300 m³ pier base consisted of layers or pockets of loose aggregate surrounded by hard concrete. The mass of the material could not be relied upon to carry the loading from the bridge structure and it was necessary to drill large-diameter holes through it to enable the pier base to be underpinned by piles.

The 6 m thick floor of a graving dock in Scaramanga, in Greece,[10.19] was concreted by 'prepacked' methods. In spite of injection points on a 3 m grid and great care in the selection of the 80–100 mm 'single-size' aggregate and in the design of the cement–sand grout with added plasticizer, extensive areas of leakage were experienced when the dock was first pumped out.

In tropical waters there is a risk of marine growth around the aggregate occurring in a relatively short period and therefore the grouting should be done within a few days of placing the aggregate.

Because of troubles with excess of laitance or of the likelihood of patches of aggregate remaining ungrouted, the author would not recommend the prepacked method for structures constructed entirely underwater where external water pressure must be resisted or for any structure where high-strength concrete is required. The process is best suited to mass concrete work, such as the hearting of piers or piles, where the material is placed within a structural shell. In all cases careful selection of materials and close supervision of the work are required, ensuring a uniform rise in grout level over the whole area being concreted.

In some circumstances it is advantageous to place mass filling as a pumped sanded grout without any coarse aggregate.

10.9 Examples on Chapter 10

Example 10.1   The cofferdam shown in Fig. 10.37 is required for an excavation to a design depth of 7.5 m. The sequence of construction will be

(1) Drive sheet piles to form the cofferdam and excavate to 1.5 m.
(2) Fix upper level bracing frame above water level.
(3) Excavate to design depth under water and fix lower level bracing frame at 7 m depth (Fig. 10.37(b)).
(4) Pump-out cofferdam, trim excavation and place base slab.
(5) Complete permanent work within cofferdam.

A fairly stiff sheet pile section will be required to prevent damage during driving of 12 to 15 m long piles. Try a Larssen LX16 section. Table 10.1(a) shows a section modulus Z_p of 1641 cm³/m. From Fig. 10.9 the maximum depth from the upper to the lower waling is about 7.5 m for Grade 43 steel (f_{st} = 175 N/mm² for temporary work).

The required depth of embedment of the sheet piles into the stiff clay will be determined by limit equilibrium methods using the strength factor method of CIRIA report 104 described in Chapter 5. An excavation depth of 8 m will be adopted to provide for over-excavation below the design depth.

From the standard penetration test N-values in Fig. 10.37(c) take a characteristic N-value of 8 blows/0.3 m. Fig. 2.13 gives ϕ' = 29°. The safety factor in the sand is taken as 1.35 to allow for disturbance and the safety factor in the clay as 1.2. Drained soil conditions will be assumed in the clay with an effective angle of shearing resistance of 24°. From Tables 5.4 and 5.9 the coefficients of active earth pressure and passive earth resistance are:

Table 10.3   Pressures and resistances at second stage of excavation

| Level | Active pressure (kN/m²) | | Passive resistance (kN/m²) | |
	unfactored	factored	unfactored	factored
a	$0.31 \times 10 = 3.1$	$0.40/0.31 \times 3.1 = 4.0$		
b	$0.31 \times 18 \times 1.5 = 8.4$	$0.40/0.31 \times 8.4 = 10.8$		
c	$a + b = 11.5$	$0.40/0.31 \times 11.5 = 14.8$		
d	$0.31 \times 8 \times 7.5 = 18.6$	$0.40/0.31 \times 18.6 = 24.0$		
e	$0.375/0.31(11.5 + 186) = 36.7$	$0.43/0.40 \times (14.8 + 24.0) = 41.7$		
f	$0.375 \times 9 \times d = 3.4d$	$0.43/0.375 \times 3.4d = 3.9d$		
g			$4.49 \times 8 \times 1 = 35.9$	$2.9/4.49 \times 35.9 = 23.2$
h			$3.26 \times 8 \times 1 = 26.1$	$2.6/3.26 \times 26.1 = 20.8$
i			$3.26 \times 9 \times d = 29.3d$	$2.6/3.26 \times 29.3d = 23.4d$

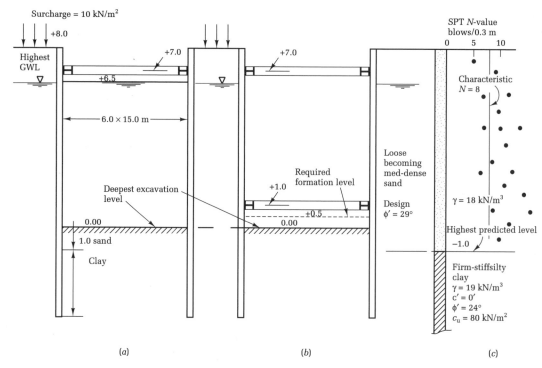

Fig. 10.37 Sections through cofferdam and soil conditions. (*a*) At second stage excavation. (*b*) At final stage. (*c*) Soil conditions.

Unfactored ϕ' values:
  Sand ($\phi' = 29°$) $K_a = 0.31$ for $\delta = 19°$ $K_p = 4.49$ for $\delta = 14.5°$
  Clay ($\phi = 24°$) $K_a = 0.375$ for $\delta = 16°$ $K_p = 3.26$ for $\delta = 12°$
Factored ϕ' values:

Sand ($\phi'_F = \tan^{-1}\left(\dfrac{\tan 29°}{1.35}\right) = 22.3°$)

$K_{aF} = 0.40$ for $\delta = 15°$ $K_{pF} = 2.9$ for $\delta = 11°$

Clay ($\phi'_F = \tan^{-1}\left(\dfrac{\tan 24°}{1.2}\right) = 20.4°$)

$K_{aF} = 0.43$ for $\delta = 14°$ $K_{pF} = 2.6$ for $\delta = 10°$

The active and passive pressures are tabulated in Table 10.3.

Taking moments about the upper waling frame and using the pressures given by the factored soil strengths:

Area	Force (kN/m)	Lever arm about upper waling (m)	Moment about upper waling (kN−m/m)
I	$4 \times 1.5 = 6.0$	$7.0 - 7.25 = -0.25$	-1.5
II	$\frac{1}{2} \times 10.8 \times 1.5 = 8.1$	$7.0 - 7.0 = 0$	0
III	$14.8 \times 7.5 = 111.0$	$0.5 + \dfrac{7.5}{2} = 4.25$	472
IV	$\frac{1}{2} \times 24.0 \times 7.5 = 90.0$	$0.5 + \dfrac{2 \times 7.5}{3} = 5.5$	495
V	$41.7 \times d = 41.7d$	$0.5 + 7.5 + \dfrac{d}{2} = 8 + \dfrac{d}{2}$	$333.6d + 20.85d^2$
VI	$\frac{1}{2} \times 3.9 \times d^2 = 1.85d^2$	$0.5 + 7.5 + \dfrac{2d}{3} = 8 + 0.67d$	$14.8d^2 + 1.24d^3$
	Total = $\overline{215.1 + 41.7d}$ $+1.85d^2$		Total = $\overline{965.5 + 333.6d}$ $+35.65d^2 + 1.24d^3$
VII	$\frac{1}{2} \times 23.2 \times 1 = 11.6$	$0.5 + 6.5 + 0.67 = 7.67$	89
VIII	$20.8 \times d = 20.8d$	$0.5 + 7.5 + \frac{1}{2}d = 8 + 0.5d$	$166.4d + 10.4d^2$
IX	$\frac{1}{2} \times 29.3d \times d = 14.65d^2$	$0.5 + 7.5 + \frac{2}{3}d = 8 + 0.67d$	$117.2d^2 + 9.8d^3$
	Total = $11.6 + 20.8d + 14.65d^3$		$\overline{89 + 166.4d + 127.6d^2 + 9.8d^3}$

For equilibrium between active pressure and passive resistance:

$$965.5 + 333.6d + 35.65d^2 + 1.24d^3 = 89 + 166.4d + 127.6d^2 + 9.8d^3$$

Equilibrium is given by a depth d of 3.5 m.

Checking for factor of safety on moments (Method 3) using unfactored soil strengths:

Area	Force (RN/m)	Lever arm about T (m)	Moment about T (kN−m/m)
I	$3.1 \times 1.5 = 4.7$	−0.25	−1
II	$\frac{1}{2} \times 8.4 \times 1.5 = 6.3$	0	0
III	$11.5 \times 7.5 = 86.3$	4.25	367
IV	$\frac{1}{2} \times 18.6 \times 7.5 = 69.8$	5.5	384
V	$36.4 \times 3.5 = 127.4$	9.75	1242
VI	$\frac{1}{2} \times 3.4 \times 3.5^2 = 20.8$	10.34	215
	Total = 315.3		
			Total overturning moment = 2207
VII	$\frac{1}{2} \times 35.9 \times 1 = 18.0$	7.67	138
VIII	$26.1 \times 3.5 = 91.3$	9.75	890
IX	$\frac{1}{2} \times 29.3 \times 3.5^2 = 179.5$	10.34	1856
	Total = 288.8		
			Total restoring moment = 2884

Therefore the factor of safety on moments $= \dfrac{2884}{2207} = \underline{1.31}$

Table 5.1 shows this to be satisfactory for temporary work.

The overturning and resisting moments, using unfactored soil strengths, are in equilibrium for a depth d of approximately 2.6 m (toe level −3.6 m). Thus:

Area	Force (kN/m)	Lever arm about T (m)	Moment about T (kN−m/m)
I–IV	167.1	(4.49)	750
V	$36.4 \times 2.6 = 94.6$	9.3	880
VI	$\frac{1}{2} \times 3.4 \times 2.6^2 = 11.5$	9.7	112
	Total = 273.2		Total = 1742
VII	$\frac{1}{2} \times 35.9 \times 1 = 18.0$	7.67	138
VIII	$26.1 \times 2.6 = 67.9$	9.3	631
IX	$\frac{1}{2} \times 29.3 \times 2.6^2 = 99.0$	9.7	960
	Total = 184.9		Total = 1729

Equating the overturning and resisting forces:

Load on upper waling $= 273.2 - 184.9 = 88.3$ kN/m

The shear forces and bending moments corresponding to the pressures and resistances in Fig. 10.38(a) are shown in Figs 10.38(b) and (c) respectively from which the maximum bending moment is 228 kN−m/m.

The allowable stress for temporary work using Grade 43 steel is 175 N/mm².

Required section modulus $= \dfrac{228 \times 10^6}{175}$

$= 1303 \times 10^3 \text{ mm}^3/\text{m}$

compared with the modulus provided of 1641 mm³/m for the Larssen LX16 pile. Checking for the individual pile:

Bending moment on single pile
$= 228 \times 0.60 = 137$ kN−m

Table 10.1(b) gives moment of inertia about axis of 5615×10^4 mm³ and centroid of pile is $190 - 72 = 118$ mm from centre line of wall

Moment of inertia about centre line of wall
$= 5615 \times 10^4 + 94 \times 10^2 \times 118^2$
$= 18704 \times 10^4 \text{ mm}^4$.

Modulus of section provided
$= \dfrac{18704 \times 10^4}{190} = 984 \times 10^3 \text{ mm}^3$

Modulus of section required
$= \dfrac{137 \times 10^6}{175} = 783 \times 10^3 \text{ mm}^3$.

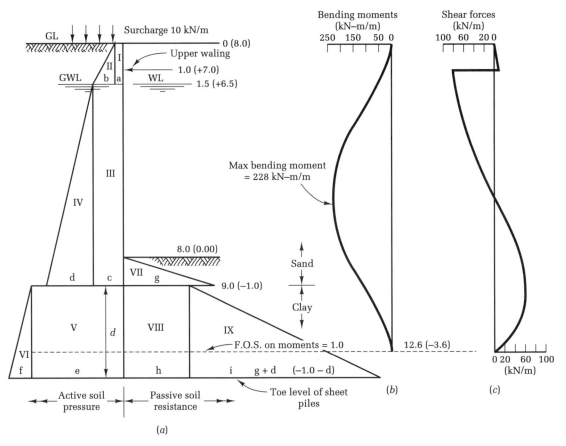

Bending moments (kN–m/m)
250 150 50 0

Shear forces (kN/m)
100 60 20 0

Surcharge 10 kN/m

GL

0 (8.0)

Upper waling

1.0 (+7.0)

GWL

WL

1.5 (+6.5)

Max bending moment = 228 kN–m/m

8.0 (0.00)

Sand

d   c   VII   g

9.0 (−1.0)

Clay

V   VIII   IX

F.O.S. on moments = 1.0

12.6 (−3.6)

VI

0 20 60 100
(kN/m)

f   e   h   i   g + d   (−1.0 − d)

Toe level of sheet piles

(b)   (c)

Active soil pressure   Passive soil resistance

(a)

Fig. 10.38 Conditions at second stage excavation. (a) Soil pressures and resistances. (b) Shear forces. (c) Bending moments.

The conditions for the second and final stages of excavation taken down to +0.5 m plus 0.5 m for over-excavation, and with the cofferdam pumped down to zero datum with the lower waling set at +1.0 m are shown in Fig. 10.39(a).

Pore pressures are calculated for conditions of steady state seepage around the toe of the sheet piles using Fig. 5.25(d). For a final excavation level at zero datum (8 mbgl)

$$\text{Pore pressure at } 12.6\,\text{m} = \left[\frac{2(4.5 + 6.5) \times 4.5}{2 \times 4.5 + 6.5} \right] \times 10$$

$$= 64\,\text{kN/m}^2.$$

Rate of increase in pore pressure behind wall

$$= \frac{64}{11.0} = 5.82\,\text{kN/m}^2/\text{m}.$$

Rate of decrease in pore pressure in front of wall

$$= \frac{65}{4.5} = 14.4\,\text{kN/m}^2/\text{m}.$$

Calculating combined soil pressures and pore pressure on wall:

From equation (5)

Active pressure

at +6.5 m = 0.31(18 × 1.5 + 10) − 0
= 11.5 kN/m²

at +1.0 m = 0.31[(18 × 7 + 10) − 5.82 × 5.5]
+ 5.82 × 5.5 = 64 kN/m²

at −1.0 m (in sand) = 0.31[(18 × 9 + 10) − 5.82
× 7.5] + 5.82 × 7.5
= 83 kN/m²

at −1.0 m (in clay) = 0.375[(18 × 9 + 10) − 5.82
× 7.5] + 5.82 × 7.5
= 92 kN/m²

at −4.5 m (in clay) = 0.375[(18 × 9 + 19 × 3.5 +
10) − 5.82 × 11.5] + 5.82 × 11.5
= 131 kN/m².

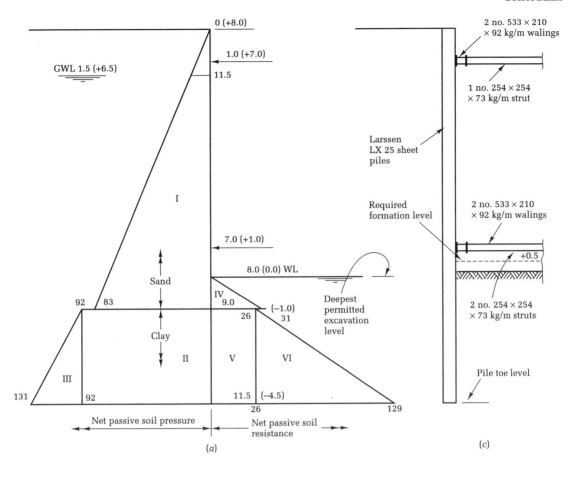

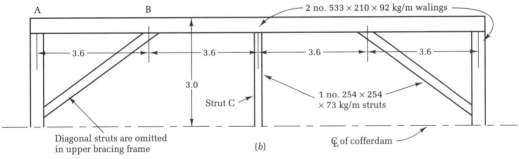

Fig. 10.39 Conditions at second stage excavation. (*a*) Soil plus pore-water pressures and resistances. (*b*) Arrangement of lower bracing frame. (*c*) Bracing and final stage excavation.

In front of wall:

Passive resistance in sand
$$\text{at } -1.0 = 4.49(18 \times 1 - 14.4 \times 1) + 14.4 \times 1$$
$$= 31 \, \text{kN/m}^2$$

Passive resistance in clay
$$\text{at } -1.0 = 3.26(18 \times 1 - 14.4 \times 1) + 14.4 \times 1$$
$$= 26 \, \text{kN/m}^2$$

Passive resistance in clay
$$\text{at } -4.5 = 3.26(18 \times 1 + 19 \times 3.5 - 14.4 \times 4.5)$$
$$+ 14.4 \times 4.5 = 129 \, \text{kN/m}^2.$$

The above combined soil pressures and resistance and pore pressures are shown in Fig. 10.39(*a*).

Taking moments about and below lower waling at +1.0:

Overturning moment

$$= 64 \times 2 \times 1 + \tfrac{1}{2} \times 19 \times 2 \times \tfrac{2}{3}.2 + 92$$

$$\times 3.5 \,(2 + \frac{3.5}{2}) + \tfrac{1}{2} \times 39 \times 3.5$$

$$\times (2 + \tfrac{2}{3}.3.5)$$

$$= 1656 \, \text{kN} - \text{m/m}$$

Resisting moment

$$= \tfrac{1}{2} \times 31 \times 1 \times (\tfrac{2}{3} \times 1 + 1) + 26 \times 3.5(2 + \frac{3.5}{2})$$

$$+ \tfrac{1}{2} \times 103 \times 3.5(2 + \tfrac{2}{3}.3.5)$$

$$= 1148 \, \text{kN} - \text{m/m}$$

Bending moment to be carried by sheet piles = 1656 − 1148 = 508 kN−m/m
Required modulus of section

$$= \frac{508 \times 10^6}{175}$$

$$= 2903 \times 10^3 \, \text{mm}^3/\text{m}.$$

This is excessive for the Larssen LX16 section. Use sheet piles in BS 4360 Grade 50A steel, required Z_p

$$= \frac{508 \times 10^6}{225} = 2258 \times 10^3 \, \text{mm}^3/\text{m}.$$

Provide Larssen LX25 section
$(Z_p = 2525 \times 10^3 \, \text{mm}^3/\text{m})$

Checking for individual pile bending moment
= 508 × 0.6 = 305 kN−m.

Centroid of pile is 225 − 82
= 143 mm from centre line of wall.

Moment of Inertia of pile about centre line of wall
$$= 9607 \times 10^4 + 120 \times 10^2 \times 143^2$$
$$= 34146 \times 10^4 \, \text{mm}$$

Modulus of section provided

$$= \frac{34146 \times 10^4}{112.5} = 3035 \times 10^3 \, \text{mm}^3$$

Modulus of section required

$$= \frac{305 \times 10^6}{225}$$

$$= 1355 \times 10^3 \, \text{mm}^3.$$

Load on lower waling is calculated by taking moments about upper waling (Table 10.4)

Load on lower waling

$$= \frac{(1770 + 3139 + 704) - (119 + 887 + 1861)}{6}$$

$$= 458 \, \text{kN/m}$$

Equating horizontal forces,
load on upper waling = (15.5 + 91.0 + 180.2 + 458)
$$- (332 + 322 + 68.2)$$
$$= 22.5 \, \text{kN/m}.$$

Therefore the maximum load on the upper waling is 88 kN/m for the first stage of excavation.

The above high values of the bending moment in the lower part of the sheet piles and the high forces on the lower walings after pumping-out the cofferdam are the result of assumption of 'long-term' drained conditions in the clay stratum with steady seepage into the excavation below the toes of the sheet piles. If it is accepted that 'short-term' undrained conditions are maintained in the period between pumping-out the cofferdam and completing the lower part of the permanent works then it is evident by inspection that the resisting moments on the sheet piles below the lower waling are potentially higher than the disturbing moments. The decision to assume either drained or undrained conditions in the clay is a matter of judgement by the engineer possibly aided by instrumentation to observe changes in forces carried by the lower walings and struts.

The general arrangement of the walings and struts forming the upper and lower bracing frames is shown in Fig. 10.39(b).

Designing the frames:

For lower waling AB fixed at B and freely supported at A:

Table 10.4

Area in Fig. 10.39(a)	Force (kN/m)	Lever arm about upper waling (m)	Moment (kN−m/m)
I	$\tfrac{1}{2} \times 83 \times 8 = 332.0$	$\tfrac{2}{3} \times 8 = 5.33$	1770
II	$92 \times 3.5 = 322.0$	$8 + 3.5/2 = 9.75$	3139
III	$\tfrac{1}{2} \times 39 \times 3.5 = 68.2$	$8 + \tfrac{2}{3} \times 3.5 = 10.33$	704
IV	$\tfrac{1}{2} \times 31 \times 1 = 15.5$	$7 + \tfrac{2}{3} \times 1 = 7.67$	119
V	$26 \times 3.5 = 91.0$	$8 \times 3.5/2 = 9.75$	887
VI	$\tfrac{1}{2} \times 103 \times 3.5 = 180.2$	$8 + \tfrac{2}{3} 3.5 = 10.33$	1861

Maximum bending moment

$$= \frac{458 \times 3.6^2}{8} = 742 \, \text{kN-m}.$$

For 165 N/mm² allowable stress, Z_p required

$$= \frac{742 \times 10^6}{165} = 4498 \times 10^3 \, \text{mm}^3.$$

Use 2no. 533 × 210 × 92 kg/m UB Sections
($Z_p = 2 \times 2076 \times 10^3 = 4152 \times 10^3 \, \text{mm}^3$)
which is sufficiently close for temporary work.

Strut C carries load of 458 × 3.6 = 1649 kN.
British Steel handbook shows that two 254 × 254 × 73 kg/m UC sections carry a safe load of 1948 kN over a length of 5 m. Take upper waling as strutted at centre point only and assume to be simply supported at end and centre:

$$\text{Bending moment} = \frac{88 \times 7.2^2}{8} = 570 \, \text{kN-m}.$$

Modulus of section required $= \dfrac{570 \times 10^6}{165} = 3454$
$\times 10^3 \, \text{mm}^3$.

Use 2no 533 × 210 × 92 kg/m UB Sections
($Z_p = 4152 \times 10^3 \, \text{mm}^3/\text{m}$)

Load on centre strut in upper frame = 88 × 7.2
$\qquad\qquad\qquad\qquad\qquad\qquad = 634 \, \text{kN}.$

Use single 254 × 254 × 73 kg/m UC section
(allowable load = 974 kN)

Web stiffeners will be needed in lower walings where struts are abutting.

Example 10.2   The 21 m × 9 m cofferdam for the piers of a river bridge shown in Fig. 10.40(a) is sited in 15 m of water at highest estimated flood level. The river bed consists of 5.5 m of soft silt overlying water-bearing dense

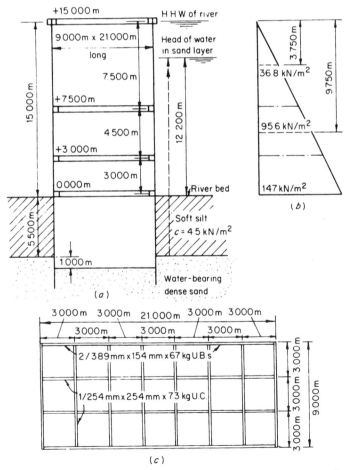

Fig. 10.40   (a) Cross-section. (*Note*: Vertical bracing members and other details not shown.) (b) Pressure diagram. (c) Layout of walings and struts.

sand. Borings show that the maximum head of water in the sand layer is 12.2 m above river-bed level. It is necessary to found the pier in the dense sand stratum, taking the foundation level 1.0 m into the sand.

In the conditions shown in Fig. 10.40(a) it will clearly be impossible to unwater the cofferdam to place the base-concrete in the dry. In the first place, pumping down the level to fix the first bracing frame will give excessive bending moments in the piling unless an uneconomical close spacing of frames is adopted, and in the second place, the sheet piles would have to be driven to at least 15 m below excavation level in order to avoid piping after pumping down. This would be impossible in the ground conditions. There fore the cofferdam must be designed for the base slab to be concreted under water. The sheet piling can be driven around a prefabricated bracing frame hung on to bearing piles driven through the soft silt into the dense sand stratum.

A heavy section of high tensile steel piling is required in view of the length to be handled and to economize in the underwater work in fixing the bracing frames.

It will also be desirable to work to the lower range of stress for high-tensile steel in view of the serious consequences of failure.

If Frodingham No. 5 is adopted, from Table 10.1(b)

$$Z_p = 3168\,cm^3/m\ run:$$

Maximum height between top and second waling from Section 10.2.2 is

$$h_{max} = \sqrt[3]{(0.2712 \times 3168)} = 9.5\,m.$$

Depth from top waling to third frame
$$= 1.6 \times 9.5 = 15.2\,m.$$

Depth from top waling to fourth frame
$$= 2.03 \times 9.5 = 19.3\,m.$$

The fourth frame will have to be raised to 15 m level to bring it above the river bed and no further frames can be fixed until after the base slab has been concreted and the cofferdam pumped down.

From pressure diagram (Fig. 10.40(b)). Average pressure on second frame is

$$\frac{3.75 \times 9.81 + 9.75 \times 9.81}{2} = 66.2\,kN/m^2.$$

Therefore total load on second waling is

$$66.2 \times 6.0 = 397.2\,kN/m\ run\ of\ waling.$$

For struts at 3 m centres, it will be satisfactory to assume a uniformly distributed load. Bending moment for simply supported spans is

$$\frac{397.2 \times 3^2}{8} = 447\,kN\,m.$$

Taking two-thirds of this value for continuous spans, bending moment is

$$\tfrac{2}{3} \times 447 = 298\,kN\,m.$$

For 165 N/mm² safe working stress in mild steel bracing, required modulus of section $= 1.81 \times 10^3\,cm^3$.

Structural steel handbooks show that two 389 mm × 154 mm × 67 kg universal beams can be used (net $Z = 2 \times 1.095 \times 10^3 = 2.19 \times 10^3\,cm^3$).

Checking shear stress, maximum shear on walings is

$$\frac{397.2 \times 3}{2} = 596\,kN,$$

shear stress on webs of walings is

$$\frac{596 \times 10^3}{2 \times 9.7 \times 389} = 78.9\,N/mm^2,$$

which is within safe limits.

If two rows of longitudinal struts are provided, effective length of cross struts will be about 2.7 m. Maximum load on struts is 595 × 2 = 1190 kN. It will be convenient for detailing the steelwork not to have the struts deeper than the walings. A universal column 254 mm × 254 mm × 73 kg in mild steel will support a safe concentric load of 1240 kN with an effective length of 2.7 m.

It will be necessary to provide web stiffeners to the walings at the junctions with the struts and diagonal bracing in a vertical plane between the four frames to give the necessary stiffness against distortion. Diagonal bracing in a horizontal plane is undesirable in this case, since it will obstruct the 3 m square bays through which underwater grabbing and concreting must take place. The frames should be designed to be assembled and lifted from a barge on to the river bed by floating crane. The number of units into which the whole assembly is divided will be governed by the capacity of the crane. The general arrangement of the completed bracing frames is given in Fig. 10.40(c).

It will also be advisable to check the overall stability of the cofferdam against overturning from the forces of winds, waves, and currents.

We will now consider the problems concerning the tremie concrete slab. Its thickness is given by the weight required to hold down 18.7 m artesian head of water in the sand stratum after the cofferdam has been pumped down. Maximum uplift pressure on underside of slab is

$$9.81 \times 18.7 = 183\,kN/m^2.$$

Required thickness of concrete, assuming density of 2.40 Mg/m³, is

$$\frac{183}{2.40 \times 9.81} = 7.80\,\text{m}.$$

This will come above the fourth frame which will have to be either left buried in the concrete or raised above tremie concrete level. Checking stresses in sheet piling between the fourth frame raised to 1.8 m above silt level and bottom of excavation before placing tremie concrete.

Hydrostatic pressures are balanced (assuming water levels inside and outside cofferdam are kept the same).

Earth pressure at the fourth frame is zero.

Earth pressure at excavation level (by Bell's equation) is given by

$$\gamma z - 2c = 8.8 \times 6.5 - 2 \times 4.5 = 48.2\,\text{kN/m}^2.$$

Total pressure on sheet piling is

$$\tfrac{1}{2} \times 48.2 \times 6.5 = 157\,\text{kN/m run of piling}.$$

Approx. maximum bending moment is

$$0.128 \times 157 \times 8.3 = 167\,\text{kN m}.$$

Required modulus of section is

$$\frac{167 \times 10^6}{175} = 0.9 \times 10^3\,\text{cm}^3/\text{m run}.$$

The modulus of section of Frodingham piling is $3168\,\text{cm}^3/\text{m}$ run, which is adequate for this condition.

References

10.1   White, P. and Prentis, E. A., *Cofferdams*, Columbia University Press, New York, 1950.

10.2   Packshaw, S., Cofferdams, *Proceedings of the Institution of Civil Engineers*, **21**, 367–398 (1962).

10.3   Fowler, C. E., *Engineering and Building Foundations*, John Wiley, New York, 1920, Vol. 1, pp. 72–76.

10.4   *Recommendations of the Committee for Waterfront Structures*, EAU 1985, 5th edn, Ernst & Sohn, Berlin, 1986, English translation available from publishers.

10.5   Haliburton, T. A., Numerical analysis of flexible retaining structures, *Proceedings of the American Society of Civil Engineers*, **SM6** (1968).

10.6   Terzaghi, K., Evaluation of coefficients of subgrade reaction, *Géotechnique*, **5**(4), 297–326 (1955).

10.7   Wood, L. A. and Perrin, A. J., Observations of a strutted diaphragm wall in London Clay: a preliminary assessment, *Géotechnique*, **34**(4), 563–579 (1984).

10.8   Wood, L. A., LAWPILE – a program for the analysis of laterally loaded pile groups, *Proceedings of the 1st Conference on Engineering Software*, **1**(4), Southampton University, England, 1979.

10.9   Smith, I. M. and Boorman, R., The analysis of flexible bulkheads in sand, *Proceedings of the Institution of Civil Engineers*, **57**(2) (1974).

10.10   Rowe, P. W. and Briggs, A., Measurements on model strutted sheet pile excavations, *Proceedings of the 5th International Conference on Soil Mechanics*, Paris, **2** (1961).

10.11   Blasting technique aids steel pile driving, *Ground Engineering*, **17**(7), 25–27 (1984).

10.12   Grice, J. R. and Hepplewhite, E. A., Design and construction of the Thames Barrier Cofferdams, *Proceedings of the Institution of Civil Engineers*, Part 1, **74**, 191–223 (1983).

10.13   Clark, P. J. and Prebaharan, N., Marina Bay Station, Singapore, excavations in soft clay, *Proceedings of the 5th International Seminar on Case Histories in Soft Clay*, Singapore (1987), pp. 95–107.

10.14   Dismuke, T. D., Cellular structures and braced excavations, in *Foundation Engineering Handbook* (Winterkorn, H. F. and Yang, N. C., eds), Van Nostrand-Reinhold, New York, 1975, pp. 445–480.

10.15   Steadman, R., Floating vote, *New Civil Engineer*, Thomas Telford, London, 14 Jan. 1993, pp. 16–17.

10.16   Calkin, D. W. and Mundy, J. K., Temporary works for pumping stations at Plover Cove Reservoir, Hong Kong, *Proceedings of the Institution of Civil Engineers*, **82**(1), 1121–1144 (1987).

10.17   Bouvier, J., Etude et perfectionnement d'une technique de béton immergé, *Annales de l'Institut Technique du Bâtiment et des Travaux Publiques*, No. 146, 151–180 (1960).

10.18   Thornley, J. H., Building a foundation through a foundation, *Engineering News Record*, **161**(9), 40–46 (1958).

10.19   Martin, G. P. and Irvine, K. D., Graving dock at Scaramanga, Greece, *Proceedings of the Institution of Civil Engineers*, Part 1, **52**, 269–290 (1972).

11 Geotechnical Processes

11.1 Ground improvement by geotechnical processes

Various processes are available whereby the character-istics of the ground can be improved either to facilitate construction operations or to permit the adoption of increased allowable bearing pressures or to reduced settle-ments under a given foundation loading.

The control of ground water in excavations either by pumping, freezing, or grout injections is an example of how construction operations can be facilitated by one or more of a number of geotechnical processes. Others include various compaction techniques for ground improvement.

11.2 Ground water in excavations

11.2.1 The seepage of water into excavations

Ground water is usually regarded as one of the most difficult problems in excavation work. Heavy and con-tinuous pumping from excavations is a costly item and the continual flow from the surrounding ground may cause settlement of adjacent structures. Heavy inflow is liable to cause erosion or collapse of the sides of open excavations. In certain circumstances there can be instability of the base due to upward seepage towards the pumping sump, or instability can occur if the bottom of an excavation in clay is underlain by a pervious layer containing water under artesian pressure. However, from a knowledge of the soil and ground-water conditions and an understanding of the laws of hydraulic flow, it is possible to adopt methods of ground-water control which will ensure a safe and economical construction scheme in any conditions. It is important to obtain all the necessary information before commencing work, and this aspect should not be neglec-ted at the site investigation stage (see Section 1.6). All too often it happens that after an excavation is commenced

ground water is met in larger quantities than anticipated, more pumps are brought to the site, and with a great struggle the excavation is taken deeper until the inflow is so heavy that the sides start collapsing with imminent danger to adjoining roads and buildings, or 'boiling' of the bottom is so widespread that a satisfactory base for foundation concreting cannot be obtained. At this stage the contractor gives up the struggle and calls for outside help in installing a wellpointing or bored well system of ground-water lowering or resorts to underwater construc-tion. The ground-water lowering systems may do the work efficiently, but the overall cost of abortive pumping, extra excavation of collapsed material, making good the dam-age, and standing time of plant and labour will have been much higher than if the ground-water lowering system had been used in the first instance. There have also been cases where, due to lack of knowledge of the capabilities of modern ground-water lowering systems, caissons have been used for foundations in water-bearing soils where conventional construction with the aid of such systems would have been perfectly feasible and much less costly.

The flow line of ground water under a fairly low head into an open excavation in a permeable soil is shown in Fig. 11.1(a). The water surface is depressed towards the pumping sump and, because of the low head and flat slope, the seepage lines do not emerge on the slope and the conditions are perfectly stable. If, however, the head is increased or the slopes are steepened, the water flows from the face, and if the velocity is high enough it will cause movement of soil particles and erosion down the face leading to undermining and collapse of the upper slopes (Fig. 11.1(b), see also Fig. 9.7). The remedy is to flatten the slopes and to blanket the face with a graded gravelly filter material which allows the water to pass through but traps the soil particles (Fig. 11.1(c)). The design of graded filters is referred to in Section 11.3.4. This form of instability by erosion is most liable to occur in fine or

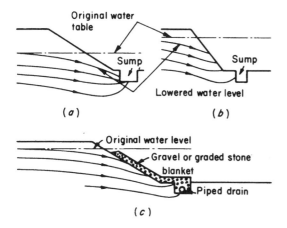

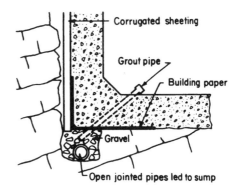

Fig. 11.3 Concreting adjacent to water-bearing rock formation.

Fig. 11.1 Seepage into open excavations. (*a*) Stable conditions. (*b*) Unstable conditions. (*c*) Increasing stability of slope by blanketing.

uniformly graded sands. There is much less risk of trouble with well-graded sandy gravels since these materials act as filters in themselves, and the draw-down given by their higher permeability prevents the emergence of the flow lines on the excavated face.

In the case of close-timbered or sheet piled excavations, the flow lines run vertically downwards behind the sheet piling and then upwards into the excavation (Fig. 11.2). This condition of upward seepage is particularly liable to cause instability, referred to as 'piping' or 'boiling', when the velocity of the upward-flowing water is high enough to bring the soil particles into suspension (Section 10.2.5).

The foregoing cases are mainly applicable to ground-water flow in permeable soils such as sands and gravels, or similar materials containing fairly low proportions of silt and clay. Little or no trouble occurs with excavation in clays. Where ground water is present it will usually seep from fissures and it can be dealt with by pumping from sumps. The velocity of flow is usually so low that there is no risk of erosion. Silts, on the other hand, are highly troublesome. They are sufficiently permeable to allow water to flow through them, but their permeability is low enough to make any system of ground-water lowering by

wellpoints or bored wells a slow and costly procedure. Ground water in rocks usually seeps from the face in the form of springs or weeps from fissures or more permeable layers. There is no risk of instability except where heavy flows take place through a weak, shattered rock. Generally, water in rock excavations can be dealt with by pumping from open sumps, and the only trouble is given when foundation concrete is designed to be placed against the rock face. If the springs or weeps are strong the water will wash out the cement and fines from the unset concrete and flow out through the surface of the concrete layer. The usual procedure is to construct the pumping sump at the lowest point in the excavation and continue to pump from it until the concrete has hardened, while at the same time allowing seepage to take place from the face towards the sump through a layer of 'no fines' concrete, or behind corrugated sheeting, or bituminous or plastic sheeting fixed to a wire mesh frame (Fig. 11.3). After completion of the concrete work the space behind the sheeting is grouted through pipes left for this purpose. Occasional weeps can often be dealt with by plastering with quick-setting cement mixtures, or by placing dry cement on the face before placing the concrete.

11.2.2 Calculation of rates of flow of seepage into excavations

In large excavations it is necessary to estimate the quantity of water which has to be pumped to draw down the water below formation level. This quantity must be known so that the required number and capacity of pumps can be provided.

In the case of trench excavations which are long relative to their width, the calculation of the quantity of flow can in most cases be treated as a condition of gravity flow to a partially penetrating slot (Fig. 11.4(*a*)). For this flow condition the ground water in the pervious layer is not

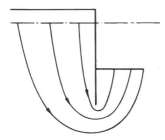

Fig. 11.2 Seepage into sheeted excavation.

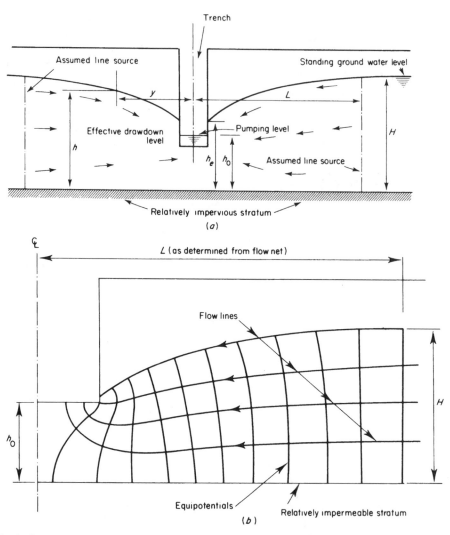

Fig. 11.4 Gravity flow to trench excavation (partially penetrating slot). (a) Draw-down for gravity flow from line sources remote from trench. (b) Flow net for conditions in (a).

confined under artesian head by an overlying impervious stratum, and the trench does not extend completely through the pervious water-bearing layer. Also the source of ground water is remote from the excavation; that is, there is no large body of water such as a river or the sea in close proximity to the trench. For the conditions in Fig. 11.4, flow to the trench from both sides of the excavation is given by the equation

$$q = \left[0.73 + 0.27\frac{(H - h_o)}{H}\right]\frac{k}{L}(H^2 - h^2_o),\qquad(11.1)$$

where

q = quantity of flow per unit length of trench,
k = coefficient of permeability of the pervious layer,
H, h_o and L are dimensions as shown in Fig. 11.4.

The dimension L can be obtained by drawing a flow net as shown typically in Fig. 11.4(b) or approximately by substituting L for R_0 in equation (11.6) (see below). Where the draw-down $H - h$ has been measured in a trial excavation at a distance y from the slot, L can be obtained from the equation

$$H^2 - h^2 \simeq \frac{L - y}{L}(H^2 - h^2_e).\qquad(11.2)$$

Where the trench penetrates fully the water-bearing layer, as shown in Fig. 11.5, the flow to the trench from both sides of the excavation is given by the equation

$$q = \frac{k}{L}(H^2 - h^2_e).\qquad(11.3)$$

If there is a large body of water such as the sea or a river

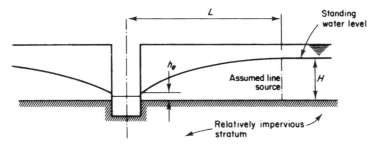

Fig. 11.5   Gravity flow to trench excavation (fully penetrating slot). Flow from line sources remote from trench.

close to one side of the excavation, the flow to the opposite side from a remote source of water will be small in comparison with that to the side close to the line source and it can be neglected. The flow from the nearby source, assuming this to be a line source of infinite length compared to the length of the trench, is equal to half the quantity as calculated by either equations (11.1) or (11.3), and the distance L is the distance from the trench to the nearby source.

Small excavations, say for a shaft, can be treated as a well. For the case of gravity flow from a circular source of seepage remote from any large body of water in contact with the pervious stratum, the flow to a fully or partially penetrating well can be calculated from the equation

$$Q = \frac{\pi k[(H - s)^2 - t^2]}{\log_e(R_0/r_w)}\left[1 + \frac{(0.3 + 10r_w)}{H} \sin \frac{1.8s}{H}\right],$$
(11.4)

where k is the coefficient of permeability and the other dimensions are as shown in Fig. 11.6.

Where a pumping test has been made in the excavation and the draw-down $H - h$ is measured at a radius r from

the well, the dimension R_0 for gravity flow can be calculated from the equation

$$H^2 - h^2 = \frac{(H^2 - h^2_w)}{\log_e(R_0/r_w)}\log_e\frac{R_0}{r}$$
(11.5)

Alternatively R_0 can be obtained by drawing a flow net or it can be obtained approximately from the empirical equation

$$R_0 = CH\sqrt{k}$$
(11.6)

where
$C = $ a factor equal to 3000 for radial flow to pumped wells and between 1500 and 2000 for line flow to trenches or to a line of well points,
$H = $ total draw-down in metres,
$k = $ coefficient of permeability in metres per second.

It is often necessary to determine the shape of the drawdown curve to a well, for example to assess the risk of settlement of existing structures near the excavation.

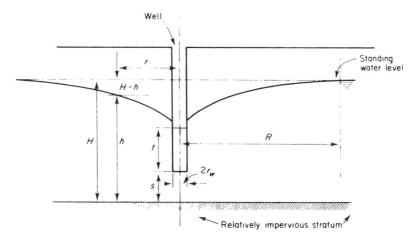

Fig. 11.6   Gravity flow to well from circular source remote from well.

The draw-down in uniform soils can be obtained by means of Fig. 11.7.

In the case of large rectangular or irregularly shaped excavations, the flow can be calculated by drawing a plan flow net of the type shown in Fig. 11.8, when for gravity flow to a fully penetrating excavation the flow per unit *thickness* of the pervious layer is given by the equation

$$q = k(H - h_e)\frac{N_f}{N_e} \tag{11.7}$$

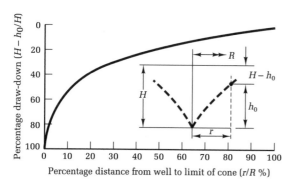

Fig. 11.7 Determination of shape of draw-down curve (after Somerville[11.1]).

or for a thickness D of pervious water-bearing soil, the total flow to the excavation is given by

$$Q = k(H - h_e)D\frac{N_f}{N_e} \tag{11.8}$$

where
  k = coefficient of permeability,
  N_f = number of flow lines,
  N_e = number of equipotential lines.

Equation (11.8) cannot be applied to partially penetrating excavations for which the flow is three-dimensional, since upward flow takes place at the base of the excavation as in the case of the partially penetrating slot (Fig. 11.4(b)). The flow as calculated from the plan flow net will not be greatly underestimated if the excavation penetrates the water-bearing layer by, say, 90 per cent, but for lesser penetrations the error is in inverse proportion to the square of the percentage.

It is important to note that the equations (11.1)–(11.8) are for the quantity of flow when steady-state conditions have been obtained. A higher pumping capacity will be required in large and deep excavations if the time required to achieve the necessary draw-down is not to be unduly protracted. The volume of water to be pumped out from

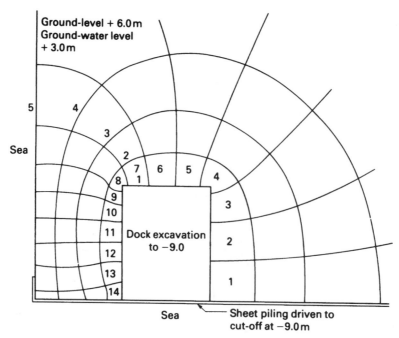

Fig. 11.8 Plan flow net for gravity flow to fully penetrating excavation for dry dock. In above example, $H = 12.0\,\text{m}$, $h_c = 0$, $k = 8 \times 10^{-4}\,\text{m/s}$, $D = 12.0\,\text{m}$, $N_f = 14$, $N_c = 5$. Therefore

$$Q = 8 \times 10^{-4}(12 - 0)12 \times 14/5$$
$$= 0.32\,\text{m}^3/\text{s}$$
$$= 19\,200\,\text{l/min}.$$

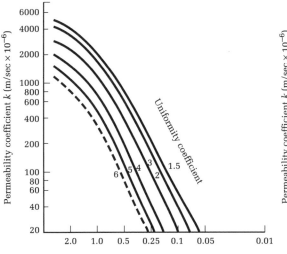

(a) Dense soils

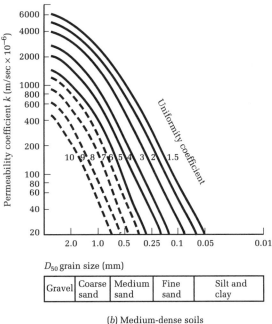

Gravel	Coarse sand	Medium sand	Fine sand	Silt and clay

(b) Medium-dense soils

Fig. 11.9 Determination of permeability coefficient from the grading curve (after Somerville[11.1]).

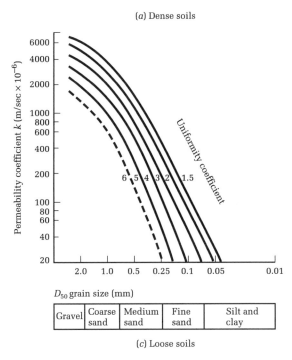

D_{50} grain size (mm)

Gravel	Coarse sand	Medium sand	Fine sand	Silt and clay

(c) Loose soils

standing water level to full draw-down conditions in the excavation should be calculated and divided by the time required by the construction programme. This will give the initial pumping capacity to achieve the planned draw-down.

It is necessary to know the coefficient of permeability of the ground. In loose uniform soils it may be calculated approximately by Hazen's formula:

Coefficient of permeability, $k = \dfrac{C_1}{10^4}(D_{10})^2$ m/s

(11.9)

where C_1 is a factor varying between 100 and 150, and D_{10} the effective grain size (mm). (*Effective grain size* is the sieve size through which 10 per cent of the material passes on the grading curve of the soil.)

More accurately the k-value can be obtained from Fig. 11.9 where it is related to the D_{50} size and the uniformity coefficient of the soil. The uniformity coefficient is the ratio of the sieve size through which 60 per cent of the material passes to the D_{10} size.

In variable or irregularly stratified soils it may be sufficient to make a rough estimate of k from the average grading of the deposits. On more important work field tests for permeability are justified. The most reliable information on the mass permeability of the soil is given by *in-situ* tests made in the site investigation boreholes using the methods listed in Section 1.6.3. A number of these tests made at different locations and various depths over a site can provide better information on the average permeability than a single full-scale pumping test, the results of which may not represent the average conditions on the site. A full-scale pumping test requires an array of observation standpipes around the pumping well, and is therefore rather costly to perform. However, the cost can be justified on important projects if only as a means of assessing the practical difficulties in installing the pumping wells for the dewatering installation. Where pumping tests are made in rock formations a large-diameter pumping well can be used as a means of facilitating geological examinations or plate-loading tests on the rock (see Section 1.5.3).

In the case of wellpointing schemes, it will normally be found that firms hiring the various wellpoint systems will, from their knowledge and experience, be able to give a fairly close estimate of the required number of wellpoints and pumping capacity, provided that they are supplied with borehole records and particle-size distribution curves of the soil strata. *CIRIA Report 113*, Control of ground water for temporary works[11.1] includes nomograms for determining the spacing of wellpoints for a given depth of ground-water lowering in soils of five different gradings. This publication provides a wealth of information on ground-water lowering and exclusion systems and is essential reading for any engineer concerned with designing and operating these systems.

Manual drawing of flow nets as described above is an accepted method of dealing with simple two-dimensional problems of ground-water seepage. However, as problem geometry becomes more complicated and for soils having anisotropic permeability (e.g. horizontal permeability not equal to vertical permeability) manual methods become unacceptable in terms of time taken and accuracy. Three-dimensional flow through a porous medium is governed by the differential equation

$$\frac{\partial}{\partial x}\left(k_x\frac{\partial \bar{u}}{\partial x}\right) + \frac{\partial}{\partial y}\left(k_y\frac{\partial \bar{u}}{\partial y}\right) + \frac{\partial}{\partial z}\left(k_z\frac{\partial \bar{u}}{\partial z}\right) = 0,$$ (11.10)

where $\bar{u}$ is the water head or potential and k_x, k_y, k_z the coefficients of permeability in the x, y and z-directions.

Many flow problems can be considered as two-dimensional, and in cases where permeability has the same value for all directions equation (11.10) reduces to Laplace's equation

$$\nabla^2 \bar{u} = 0 \quad \text{or} \quad \frac{\partial^2 \bar{u}}{\partial x^2} + \frac{\partial^2 \bar{u}}{\partial y^2} = 0.$$ (11.11)

For problems having orthogonal geometry and to which Laplace's equation applies, such as flow round an impervious barrier as shown in Fig. 11.10, the finite difference technique provides a ready and accurate means of solution. This involves replacing the continuous soil cross-section bounded by ABCD by a pattern of discrete points on an orthogonal grid within the cross-section. At each of the grid points the governing differential equation can be expressed approximately in terms of values of $\bar{u}$ at the particular grid point and at adjacent grid points. For example, considering the five grid points a to e shown in Fig. 11.10, the finite difference form of Laplace's equation at point C is

$$\nabla^2 \bar{u} = \frac{1}{h^2}(\bar{u}_a + \bar{u}_b - 4u_c + \bar{u}_d + \bar{u}_e)$$ (11.12)

where h is the grid point spacing.

Certain known values of $\bar{u}$ will exist for some grid points on the boundary, i.e. on lines AB, CD, DE, and GA $\bar{u}$ is equal to the hydrostatic potential, while along impervious boundaries BC, EF, and FG the conditions of zero flow across the boundary defines the values of $\bar{u}$ at imaginary grid points outside the boundary required to complete the equation (11.12). Consideration of this equation at each of the grid points where $\bar{u}$ is unknown results in a set of linear simultaneous equations. If the problem is to be solved

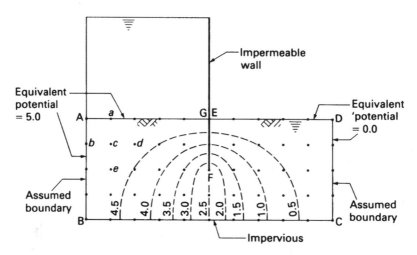

Fig. 11.10   Finite difference grid and results for seepage flow beneath an impermeable wall.

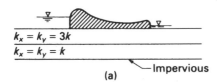
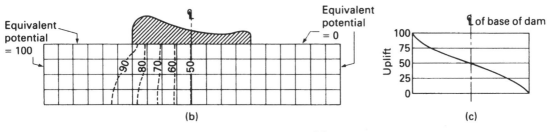

Fig. 11.11 Seepage flow through layered soil system under dam (after Desai[11.2a]). (a) Dam on layered system. (b) Finite element idealization results. (c) Uplift on base of dam.

with reasonable accuracy then a considerable number of grid points have to be considered and hence a considerable number of simultaneous equations need to be solved, for which a computerized method is a virtual necessity. As an alternative to solution of simultaneous equations by matrix inversion routines, Williams *et al.*[11.2] have described how the finite difference form of Laplace's equation may be solved iteratively by use of a spreadsheet program of the type readily available for use on desktop computers.

For more complicated cases (such as anisotropic permeability, non-orthogonal boundaries and multiple strata) the finite difference method becomes increasingly cumbersome even when computerized. In such instances the finite element method is probably the most powerful technique available. The porous medium within which the flow is occurring is modelled by subdivision into a number of zones or elements, each of which can have its own values of permeabilities in the principal directions. Boundary conditions in terms of known head or flow quantity across a boundary are defined and solution of the problem leads to values of head at the nodal points within the element array. Figure 11.11 illustrates a comparatively simple example of the finite element method applied to computation of seepage under a dam. Note the difference in permeability between the two strata under the dam; this presents no problem in a finite element idealization, but is awkward to handle in a finite difference approach.

The examples discussed so far have been for what is known as confined flow where all parts of the external boundary are defined by either a specified head or an impermeable boundary. A further class of problems involves flow having a free or phreatic surface, the position of which is not known at the outset. This feature is known as unconfined flow and examples are draw-down

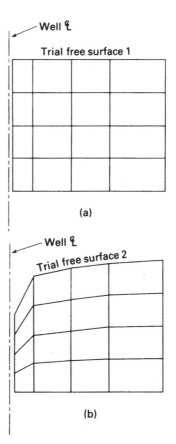

Fig. 11.12 Finite element analysis of free surface flow for flow of ground water to well (after Smith[2.48]). (a) Undeformed mesh. (b) Deformed mesh. (Reprinted by permission of John Wiley & Sons Ltd.)

curves in aquifers and in river banks or dam embankments. Here an assumption must be made for the position of the free surface and the solution proceeds iteratively. One approach involves specifying the trial-free surface as an impermeable boundary and then, when the first analysis has been made, adjusting the element mesh so that the top elevations equal the computed potentials on the first trial-free surface. Further iterations are made until mesh adjustments become negligible. The first two stages of the process are illustrated in Fig. 11.12. It will be appreciated that this technique can readily be applied to the investigation of problems such as the assessment of ground-water flow into excavations.

Detailed discussion of the use of finite element methods in seepage problems is given in references 2.47, 2.48, 11.2a and 11.3.

The reader should note that the Laplace equation is the governing differential equation for certain other engineering problems, notably that of heat flow. The majority of general-purpose finite element programs that are available include a capacity to analyse thermal problems. Hence, if the reader has access to such a general-purpose program it should be possible to use it for seepage problems with the program parameters for temperature, thermal flux, and thermal conductivity representing potential, seepage, and permeability respectively. However, in such instances trial problems with known solutions should always be run first to ensure that the alternative phenomenon is being represented correctly.

11.3 Methods of ground-water control

11.3.1 Effects of site and ground condition

The following methods of ground-water control and associated geotechnical processes can be used in excavation work:

(1) Pumping from open sumps
(2) Pumping from wellpoints
(3) Pumping from bored wells
(4) Pumping from horizontal wells
(5) Electro-osmosis

Elimination or reduction of ground-water flow by:

(6) Forming impervious barriers by grouting with cement, clay suspensions, or bitumen
(7) Chemical consolidation
(8) Compressed air
(9) Freezing

The choice of method depends to a great extent on site conditions. For example, pumping from open sumps can be used in most ground conditions provided that the site area is large enough to permit the excavation to be cut back to stable slopes, and that there are no important structures close to the excavation which might be damaged by settlement resulting from erosion due to water flowing towards the sump. Wellpointing or bored wells can be used in more restricted site conditions, and the special processes such as grouting, chemical consolidation, and freezing are used where it is necessary to safeguard existing structures, or in particular ground or rock conditions where pumping is impracticable.

The soil characteristics, and especially the particle-size distribution of the soil, also influence the choice of method. The range of soil types over which the various processes are applicable have been classified by Glossop and Skempton,[11.4] and are shown on the particle-size distribution curves in Figs 11.13 and 11.14. To use these curves, the particle-size distribution of the soil is obtained by sieving tests and the grading curve is plotted on one or other or both of the charts. For example, the coarse gravel (soil A in Fig. 11.13) may be unsuitable for wellpointing

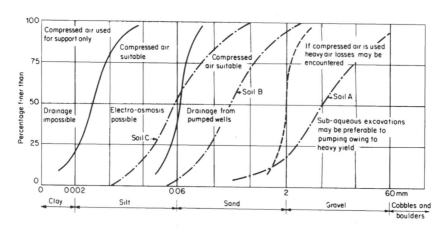

Fig. 11.13   Range of particle size for various ground-water lowering processes (after Glossop and Skempton[11.14]).

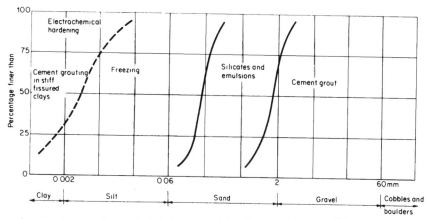

Fig. 11.14 Range of particle size for various geotechnical processes (after Glossop and Skempton[11.4]).

because of the heavy flow through the highly permeable ground. However, reference to the chart in Fig. 11.14 shows that the gravel is amenable to treatment by cement grouting to eliminate or greatly reduce the flow into the excavation. The coarse to fine sand (soil B) is suitable for wellpointing, but if there should be a risk of damage to adjacent structures due to lowering of the ground-water table, Fig. 11.14 shows that chemical consolidation can be used to solidify the soil and greatly reduce or prevent inflow. The sandy silt (soil C) is too fine for wellpointing, but we could use electro-osmosis, or the freezing process, or compressed air.

11.3.2 Pumping from open sumps

This is the most widely used of all methods of ground-water lowering. It can be applied to most soil and rock conditions, and the costs of installing and maintaining the plant are comparatively low. The method is essential where wellpointing or bored wells cannot be used because of boulders or other massive obstructions in the ground, and it is the only practical method for rock excavations. However, it has the disadvantage that the ground water flows *towards* the excavation; with a high head or steep slopes there is a risk of collapse of the sides. There is also the risk in open or timbered excavations of instability of the base due to upward seepage towards the pumping sump.

The essential feature of the method is a sump below the general level of the excavation at one or more corners or sides. To keep the floor of the excavation clear of standing water a small grip or ditch is cut around the bottom of the excavation, falling towards the sump. In large excavations which have to remain open over long periods, it is advantageous to pay special attention to the design of these drainage ditches. They should be sufficiently wide to

keep the velocity low enough to prevent erosion. Further safeguards against erosion can be given by check weirs (boards placed across the ditch), by stone or concrete paving, or by laying open-jointed pipes surrounded by graded stone or gravel filter material. Where the ground water is present in a permeable stratum overlying a clay, with the excavation taken down into the latter material, it is preferable to have the pumping sump at the base of the permeable stratum. This procedure reduces the pumping head, and avoids softening of the clay in the base of the excavation. The drain at the base of the permeable stratum above excavation level is known as a *garland drain*. Typical details of a garland drain for (*a*) open excavation over a timbered excavation, (*b*) a wholly timbered excavation, and (*c*) in rock excavation are shown in Fig. 11.15. The greatest depth to which the water table may be lowered by the open sump method is not much more than 8 m below the pump, depending on its type and mechanical efficiency. For greater depths of excavation it is necessary to reinstall the pump at a lower level, or to use a sinking pump or submersible deep well pump suspended by chains and progressively lowered down a timbered shaft or perforated tube. It is sometimes a useful procedure to sink the pumping sump for the full depth of the excavation by means of a shaft with spaces between the sheeting members to allow the water to flow into the shaft. Gravel filter material can be packed behind the sheeting members if excessive fine material is washed through. This method ensures dry working conditions for the subsequent bulk excavation, and it also provides an exploratory shaft for obtaining information on ground conditions to supplement that found from borings. Shallow sumps can be excavated below water level by back-acter excavator. Perforated precast concrete manhole rings can then be lowered through the water to form the sump lining. Ingress of sand or silt through the base of the sump

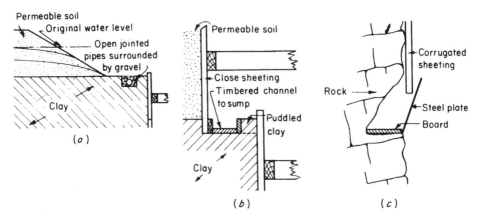

Fig. 11.15 Types of garland drain.

and through the perforations can be prevented by wrapping the units with porous nylon filter cloth before lowering them into the excavation.

Pumping plant Whenever flooding of an excavation can cause damage to partly constructed works, or where pumps have been installed below ground-water level, it is important to provide adequate standby pumping plant of a capacity at least 100 per cent of the steady pumping rate. It will often be found that the pumping capacity required to lower the ground water is greater than the capacity required to hold it down at a steady rate of pumping, depending on the time allowed in the construction programme. If, for example, double the required steady rate capacity is installed, the full installation can be used in the initial draw-down period, and then half the pumps can be shut down and used as the emergency standby and for regular maintenance. If the main pumps are electrically driven, it is useful to have the standby pumps driven by diesel engines in case of failure of the main electricity supply.

Types of pumps suitable for operating from open sumps are as follows:

(a) *Hand-lift diaphragm.* Output from 20 l/min for 30 mm suction, up to 250 l/min for 100 mm suction. Suitable for intermittent pumping in small quantities.

(b) *Motor-driven diaphragm.* Output from 350 l/min for 75 mm suction up to 600 l/min for 100 mm suction. Can deal with sand and silt in limited quantities.

(c) *Pneumatic sump pumps.* At 7 bar air pressure outputs range from 450 l/min against 15 m head to 900 l/min against 3 m head. Useful for intermittent pumping on sites where compressed air is available. Can deal with sand and silt in limited quantities.

(d) *Self-priming centrifugal.* Widely used for steady pumping of fairly clean water. Smallest units can be carried by one man. Outputs range from 750 l/min for 50 mm suction to 2000 l/min against a 15 m head for a 305 mm suction. Sand and silt in water cause excessive wear on impeller for long periods of pumping, therefore desirable to have efficient filter around sump or pump suction.

(e) *Rotary displacement (Monopump).* Can deal with considerable quantities of silt and sand. Output for 75 mm pump is 550 l/min against 6 m head.

(f) *Sinking pumps.* Suitable for working in deep shafts or other confined spaces where pumps must be progressively lowered with falling water table.

The usual type for present-day use is a self-contained pump and electrically driven motor unit capable of working below water and discharging through flexible armoured hose up to the ground surface. Outputs range from about 200 l/min against a 10 m head for 50 mm discharge hose, to 14 000 l/min against the same head for 250 mm discharge hose. The maximum head for this type of pump is about 70 m.

11.3.3 Pumping from wellpoints

The wellpoint system of ground-water lowering comprises the installation of a number of filter wells, usually about 50 mm diameter and 0.5–1 m long, around the excavation. These are connected by vertical riser pipes to a header main at ground level which is under vacuum from a pumping unit. The water flows by gravity to the filter well and is drawn by the vacuum up to the header main and discharged through the pump. The wellpointing system has the advantage that water is drawn *away* from the excavation, thus stabilizing the sides and permitting steep slopes. Indeed, slopes of 1 in $\frac{1}{2}$ (vertical to horizontal) are commonly used in the short term when wellpointing in fine sands, whereas with open-sump pumping where the water flows towards the excavation, the slopes must be cut back to 1 in 2 or 1 in 3 for stability. Thus, wellpointing can give a considerable saving in total excavation and permits working in fairly confined spaces. The installation is very rapid and the equipment is reasonably simple and cheap. There is the added advantage that the water is filtered as it is removed from the ground and carries little or no soil particles with it once steady discharge conditions have been attained. Thus the danger of subsidence of the surrounding ground is very much less than with open-sump pumping. One disadvantage of the system is its limited suction lift. A lowering of 5–6 m below pump level is generally regarded as a practical limit. Attempts at greater lowering result in excessive air being drawn into the system through joints in pipes, valves, etc. with consequent loss in pumping efficiency. For deeper excavations the wellpoints must be installed in two or more stages. Also, in ground consisting mainly of large gravel, stiff clay, or sand containing cobbles or boulders, it is impossible to jet down the wellpoints and disposable units have to be placed in boreholes or holes formed by a 'puncher' with consequent increase in installation costs. The use of a puncher can compact the soil around the wellpoint, thus reducing its capacity.

The filter wells or wellpoints usually consist of a 0.5–1 m long and 60–75 mm diameter gauze screen surrounding a central riser pipe. Where wellpoints are required to remain in the ground for a long period, e.g. for dewatering a drydock excavation, it may be economical to use disposable plastic wellpoints. These consist of a nylon mesh screen surrounding a flexible plastic riser pipe. Water drawn through the screen falls in the space between the gauze and the outside of the riser pipe to holes drilled in the bottom of this pipe and thence to the surface. The wellpoints are installed by jetting them into the ground, when the jetting water flows freely from the serrated nozzle. Various proprietary systems concentrate the jetting water through the nozzle rather than allowing it to be dissipated through the filter.

The capacity for a single wellpoint with a 50 mm riser is about 10 l/min. Their spacing around the excavation depends on the permeability of the soil and on the time available to effect the draw-down. In clean fine to coarse sands or sandy gravels a spacing of 1.0–1.5 m is satisfactory. In silty sands of fairly low permeability a 1.5–2.0 m spacing will be satisfactory. In permeable coarse gravels they may need to be as close as 0.3 m centres. As noted in Section 11.2.2, *CIRIA Report 113*[11.1] contains nomograms for determining wellpoint spacings for a range of depths of draw-down and soil gradings. The normal set of wellpointing equipment comprises 50–60 points to a single 150 or 200 mm pump with a separate 100 mm jetting pump. The wellpoint pump has an air/water separator and a vacuum pump as well as the normal centrifugal pump.

Installation   Water at a pressure of about 8 bar and at a rate of 1200 l/min is required for jetting down wellpoints. If a layer of clay overlies the water-bearing stratum it is often more convenient to bore through the clay by hand or power auger rather than to attempt jetting through it. When the wellpoint has been jetted down to the required level, the jetting water supply is cut down to a low velocity sufficient to keep the hole around the point open. Coarse sand is then fed around the annular space to form a supplementary filter around the point, and the water is then cut off. This process, known as 'sanding-in' the points, is an important safeguard against drawing fine material from the ground which might clog the system or cause subsidence. A wellpoint with a gauze fine enough to screen out the finest particles would have insufficient capacity. Sanding-in also increases the effective diameter of the wellpoint and hence its output. Therefore, sanding-in should never be neglected. There may be difficulty in sanding-in a wellpoint in highly permeable gravels when the jetting water is dissipated into the surrounding ground and does not reach the surface around the riser pipe. Normally, a wellpoint does not need sanding-in in these conditions but it sometimes happens that coarse gravels are overlying sand with the wellpoints terminating in the latter. In such cases the wellpoint must be inserted in a lined borehole, the lining tubes being withdrawn after the filter sand is placed. Wellpoints act most effectively in sands and in sandy gravels of moderate but not high permeability. The draw-down is slow in silty sands, but these soils can be effectively drained. There have been instances of silts and sandy silts being drained by the 'vacuum process', where the upper part of the riser pipe is surrounded by clay to maintain a high vacuum surrounding the wellpoint by means of which the soil is slowly dewatered.

Wellpoints are installed in the 'progressive' or 'ring' systems. The progressive system is used for trench work (Fig. 11.16). The header is laid out along the sides of the excavation, and pumping is continuously in progress in one length as further points are jetted ahead of the pumped-down section and pulled up from the completed and backfilled lengths. For narrow excavations it is often sufficient to have the header on one side only (Fig. 11.16(a)). For wide excavations or in soil containing bands of relatively impervious material, the header must be placed on both sides of the trench (Fig. 11.16(b)).

In the ring system (Fig. 11.17) the header main surrounds the excavation completely. This system is used for rectangular excavations such as for piers or base-

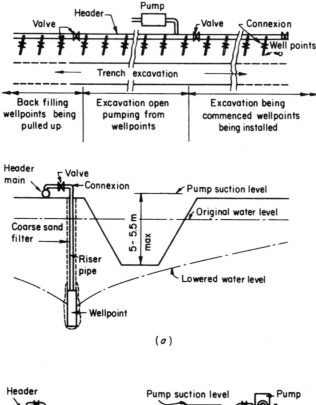

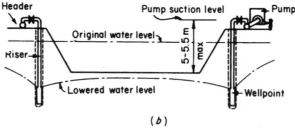

Fig. 11.16 Single-stage wellpoint installation by the progressive system. (*a*) Wellpoints on one side of trench. (*b*) Wellpoints on both sides of wide excavation.

ments. Continuous horizontal wellpoints consisting of 80–100 mm perforated plastic pipe wrapped with nylon mesh can be laid to depths of up to 6 m by a tractor-drawn machine which excavates a trench, lays the flexible filter tube at the base of the trench, and backfills the trench in a continuous operation. This technique is most suitable for dewatering long trenches, but the method can be used for large excavations such as dry docks. It has the advantage of a clear working space around the excavation, and the installation is not exposed to damage by construction operations. However, the depth of draw-down is limited to about 4 m. The filter pipes are liable to clogging if the soil contains layers of silt or clay and they cannot be readily lifted for cleaning. The costs of mobilizing and demobiliz-

ing the large and heavy trenching machine are high, and trenching to the required depth by a conventional back-acter may not be practicable in the water-bearing soils.

As an alternative to wellpoints with surface pumps, consideration can be given to the use of an eductor wellpoint system. This requires the provision of a double header main at ground level. Water pumped under pressure around one main flows down to the eductor at the bottom of each wellpoint where discharge from a nozzle creates a vacuum. Water is drawn towards the vacuum from the surrounding soil and flows up a riser pipe to the second header main in which water is collected for discharge by gravity to the outlet point. The method has the advantage that the draw-down of the ground water is

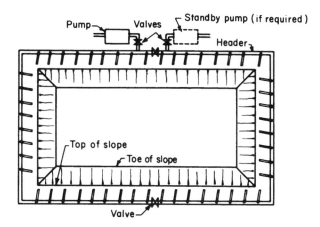

Fig. 11.17   Single-stage wellpoint installation by the ring system.

not limited in depth by the suction lift of a pump at ground level and it operates effectively in silty soils. Only clean water is pumped through the pressure main with reduced problems of pump maintenance compared with conventional wellpointing systems. However, because of the low efficiency of the system, pumping costs are about four times greater than that of conventional wellpointing[11.1] and the filter units are more expensive. Nevertheless, the eductor system can be economical for excavations deeper than 10 m when the ground-water lowering can be achieved by a single-stage installation instead of three or more stages required for conventional wellpointing.

Wellpointing of deep excavations   The limitation in the draw-down of water level to 5–6 m below pump level has already been mentioned; if deeper excavation below standing water level is required, a second or successive stages of wellpoints must be provided. A section through a three-stage installation is shown in Fig. 11.18. There is no limit to the depth of draw-down in this way, but the

overall width of excavation at ground level becomes very large. The upper stages can often be removed or reduced when the lower ones are working.

It is often possible to avoid two or more wellpoint stages by excavating down to water level before installing the pump and header. The procedure in its simplest form is shown in Fig. 11.19. A more complex case was provided for the excavation for the intake pumphouse for the Ferrybridge 'B' Power Station. At this site there were two aquifers, the upper one in a layer of ash and clinker fill being separated from deeper water-bearing sands and silts by a stratum of impervious soft clay. The head of water in the lower strata was 4.3 m above excavation level. The upper water table was cut off by light section sheet piling driven into the clay. Excavation was then taken down into the clay until a level was reached at which the weight of clay just counterbalanced the head of water in the underlying silty sands and sandy gravels. Wellpoints were installed at this level and the head in the water-bearing strata was lowered before the excavation was completed. A cross-section through the site is shown in Fig. 11.20.

Wellpoints used in sheet piled excavations are placed

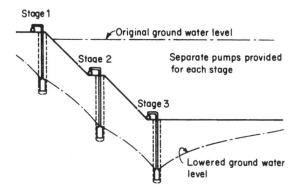

Fig. 11.18   Three-stage wellpoint installation.

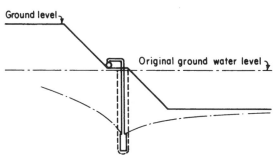

Fig. 11.19   Reduction in ground level before installation of wellpoint.

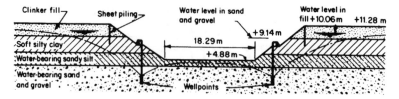

Fig. 11.20 Wellpoint installation for intake pumphouse, Ferrybridge 'B' Power Station.

close to the toes of the sheet piles. The risers being placed either inside or outside the sheet piles. This is done to ensure lowering the water level between the sheet pile rows. It is not necessary to lower the water table down to the tops of the wellpoints. Wellpoints are provided in conjunction with the sheet piles either to prevent 'boiling' of the bottom when the piles are of limited penetration, or to eliminate hydrostatic pressure on the back of a sheet pile cofferdam, thus allowing lighter bracing to be used. If wellpoints are provided for the latter purpose it is essential that standby pumping plant is provided, otherwise collapse of the cofferdam would quickly follow if pumping were to cease. In very permeable soil the water can rise to its original level in a matter of minutes. For this reason on completion of a dewatering scheme the wellpoints should always be shut off one by one to allow the water table to come gradually back to its normal level.

Very thin impervious layers of silt or clay are often met in water-bearing sandy soils. Layers as thin as 2 mm if continuous can be very troublesome in a dewatering scheme, causing breaks in the draw-down curve. These troubles can largely be overcome by jetting holes on the side of the wellpoints away from the excavation and filling them with coarse sand. These sand columns provide a path down which the water can seep to the wellpoints more readily than towards the sides of the excavation; thus weeping from the sides of the excavation is prevented. Difficulties can also be caused by layers of highly permeable material when the water will tend to bypass the wellpoints. This situation can be dealt with by jetting at close intervals in a row around the excavation to cause intermingling of the various layers.

Where the base of the excavation is in an impermeable stratum there is no point in installing the wellpoints below the interface between the pervious water-bearing layer and the impermeable stratum. It must be expected that some water will flow between the wellpoints over the top of the impermeable layer and emerge on the face of the slope. Erosion at this point can be checked by placing a row of sandbags along the line of the seepage with a garland drain (Fig. 11.15(a)) to collect the water. This arrangement was adopted for the ground-water lowering scheme used for constructing a large graving dock at Nigg in Scotland.[11.5] A relatively impervious marly sandstone outcropped

above the base of the excavation over the northern half of the dock, and a two-stage wellpoint installation was used in this part of the 15 m deep excavation. The upper stage was withdrawn leaving the lower stage in operation with a garland drain in the form of an open ditch on a berm at the base of the pervious sand layer (Fig. 11.21). The rock dipped below the southern half of the dock and in this area the ground water lowering was achieved by a row of deep bored wells on each side of the excavation. These can be seen in Fig. 11.22. A narrow strip of sand dunes separated the south side of the dock excavation from the sea. At this location the 20 m deep excavation for the entrance gate sill was dewatered by a three-stage wellpoint installation and because of the close proximity of the sea simultaneous pumping was necessary from all three stages. The dewatering installation remained in commission throughout the period of constructing three oil production platforms for the North Sea, and during this time the sandy slopes were protected against wind erosion by a thin blanket of bitumen–sand.

11.3.4 Pumping from bored wells

Pumping from wells can be undertaken by surface pumps with their suction pipes installed in bored wells. The depth of draw-down by this method is not much more than 8 m. The main uses of pumping from wells are when a great depth of water lowering is required or where an artesian head must be lowered in permeable strata at a considerable depth below excavation level. In such cases electrically powered submersible pumps are installed in deep boreholes with a rising main to the surface. Wells can be installed in a wider variation of ground conditions than is possible with wellpoints, since heavy boring plant is used to sink the wells enabling boulders, rock, or other difficult ground to be penetrated. The cost of installation of a deep well system is relatively high. Therefore, the process is generally restricted to jobs which have a long construction period such as dry docks or access shafts for long sub-aqueous tunnels, when the simplicity and trouble-free running of a properly designed deep well installation is advantageous.

The procedure in installing a bored well is first to sink a cased borehole having a diameter some 200–300 mm

Fig. 11.21 Wellpoint installation for the graving dock at Nigg, Scotland. (The upper stage has been removed, only the lower stage is working; note the garland drain at rockhead level.)

larger than the inner well casing. The diameter of the latter depends on the size of the submersible pumps. It should be noted that sinking a large-diameter well by cable percussion methods results in considerable loss of ground, resulting in subsidence around the well. It is preferable to use rotary reverse circulation drilling. Bentonite used for supporting unlined boreholes can cause clogging of the walls, but proprietary fluids can be used to break down the gel after completion of drilling. Nevertheless, large-diameter bored wells should not be installed close to

existing structures. After completion of the borehole the inner well casing is inserted. This is provided with a perforated screen over the length where dewatering of the soil is required and it terminates in a 3–5 m length of unperforated pipe to act as a sump to collect any fine material which may be drawn through the filter mesh. The perforated screen may consist of ordinary well casing with slots or holes burned through the wall and brass mesh spot-welded round the outside. A cheaper and quite effective well screen consists of slotted PVC tube sur-

Fig. 11.22 Ground-water lowering installations for the graving dock at Nigg, Scotland, (The two-stage wellpoint installation is in the background of the photograph, the bored wells can be seen in the foreground. A three-stage wellpoint system is operating across the entrance.)

rounded by a filter in the form of a nylon or terylene mesh sleeve. Slots are preferable to holes, since there is less risk of blockage from round stones. The effective screen area can be increased by welding rods longitudinally or spirally on to the casing to provide a clear space between the mesh and the casing. As an alternative to the mesh screen, purpose-made slotted casing with fine openings is marketed and is advantageous in many cases. For long periods of pumping the slotted casing can be constructed from bitumen-coated mild steel, stainless steel, or rigid PVC tubing.

After the well casing is installed, graded gravel filter material is placed between it and the outer borehole casing over the length to be dewatered. The outer casing is withdrawn in stages as the filter material is placed. The remaining space above the screen is backfilled with any available material. The water in the well is then 'surged' by a boring tool to promote flow back and forth through the filter, and at the same time any unwanted fines which

fall into the sump are cleaned out by baler before the submersible pump is installed. This is the last operation before putting the well into commission. The completed installation is shown in Fig. 11.23.

Design of filter The grading of the filter material is determined from the average grading of the soil to be dewatered. Sieve analyses are made on a number of samples, and an average grading curve is drawn or, in the case of a variably graded soil, the grading curves of the coarsest and finest layers are drawn. Then by Terzaghi's rules the filter material should be selected so that its grain size at the 15 per cent 'finer than' size (D.15) should be at least four times as large as the 15 per cent size of the coarsest layer of soil in contact with the filter, and not more than four times as large as the 85 per cent 'finer than' size (D.85) of the finest layer of soil in contact with the filter. The maximum size of the filter material should be at least twice that of the openings in the mesh screen (or

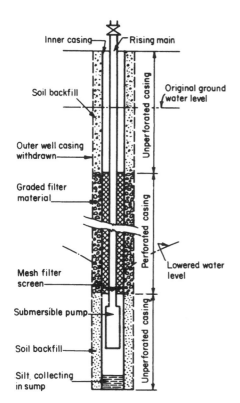

Fig. 11.23 Bored well installation.

perforated pipe if no mesh is provided). To avoid too great a loss of head through a filter it is frequently necessary to provide more than one layer of graded filter material with a minimum thickness of 150 mm for each layer. An

example of the design of a two-layer graded filter is shown in Fig. 11.24. This was used at Trafford Park, Manchester, for bored well dewatering of a glacial clay interbedded with thin seams of a water-bearing fine sand.

Great care is required in the placing of the filter and in avoiding damage to the perforated screen, since submersible pumps are very susceptible to clogging and breakdown if they have to pump dirty water.

Pumping plant Reference should be made to manufacturers' catalogues for the sizes and power requirements of submersible pumps for various duties. As examples, a pump in a 150 mm borehole can deliver 350 l/min against 30 m head, or a pump in a 350 mm borehole can raise 7500 l/min against 30 m head. *CIRIA Report 113*[11.1] gives pumping capacities and a range of operating heads for submersible pumps with outlet diameters from 38 to 150 mm. Bored wells can be spaced at much wider intervals than wellpoints, since the pumps can be installed at greater depths below excavation level, thus giving a wide area of draw-down for each individual well. The depth of the well depends on the depth to a lower impermeable stratum. Thus, in Fig. 11.25(a), the impermeable stratum is at a great depth, the pumps can be placed well below excavation level, and a wide spacing can be adopted. In Fig. 11.25(b), the impermeable stratum is at a shallow depth below excavation level and there is no point in placing the filter screen below the permeable stratum. This limits the depth of draw-down, hence a close spacing must be adopted. The shape of the draw-down curve depends on the permeability of the soil. For a coarse sandy gravel the draw-down curve is flat (Fig. 11.26(a)),

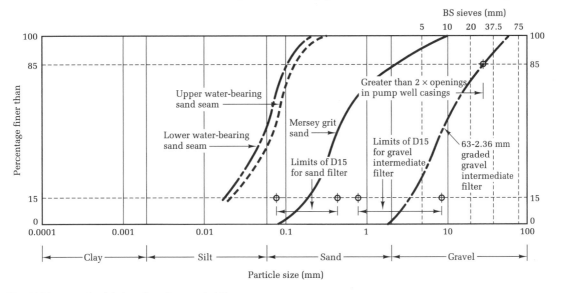

Fig. 11.24 Example of design of two-layer graded filter.

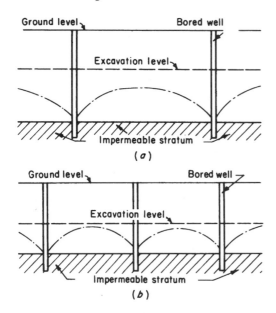

Fig. 11.25 Spacing of bored wells.

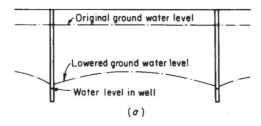

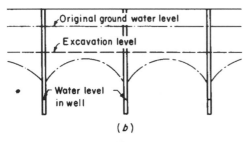

Fig. 11.26 Draw-down in permeable and semi-permeable soils. (a) Draw-down in sandy gravel. (b) Draw-down in silty sand.

and the free water surface is drawn well down towards the pumping level in the borehole, so a wide spacing is possible (though the pumping rate is correspondingly high). In the case of a soil of fairly low permeability, say a silty sand, the free water surface is not drawn down to the pumping level in the well, necessitating a closer spacing of wells, but with a lower output per well (Fig. 11.26(b)).

The spacing of the wells should not be so wide that

shutting an individual well down for repairs to the pump will cause the water level to rise above excavation level. In fact, the installation should be designed so that individual wells can be shut down as required for cleaning out the wells and maintenance of the pumps. There can be a tendency for incrustations and bacterial slimes to accumulate on pumps and well screens. The latter need to be cleaned by brushing and swabbing after lifting out the pumps. Where the electric power is supplied from the mains, standby diesel generators should be provided. If the installation has its own generating plant there should be ample standby capacity for the generators and prime movers. This standby capacity need not be 100 per cent of the maximum power requirements. As already mentioned in the case of pumping from open sumps, maximum power is only required during the time of lowering the water from its standing level to the fully drawn-down state. Thereafter the power required is considerably less and might be as little as 50 per cent of the maximum.

11.3.5 Pumping from horizontal wells

This process is applicable only in the special circumstances where well-pointing or bored well water lowering cannot be used. Typical of these conditions is where a deep excavation is to be sunk through heavily water-bearing ground and founded on or just within an impermeable stratum (Fig. 11.27). Because of the great depth or possible obstructions, sheet piling cannot be driven to give a cut-off in the impermeable layer. Bored wells cannot lower the ground water completely down to the rock stratum because of the shape of the draw-down curve. The procedure is then to sink a lined shaft by boring methods into the clay layer. The shaft is large enough to allow men to go down and install horizontal wells which are jacked or jetted out into the ground. These wells consist of an outer and inner casing, the space between them being

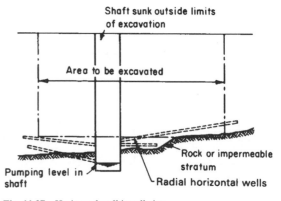

Fig. 11.27 Horizontal well installation.

filled with a gravel filter. They are provided with valves where they pass through the wall of the shaft, to allow them to be shut off until all are installed. The valves are then opened and pumping takes place from the shaft using a sinking pump or submersible pump.

This method of dewatering has also been used as a permanent drainage measure where leakage was occurring in a large and deep basement constructed in a rock excavation. A deep shaft was sunk near the basement and radial wells were drilled from it to intercept a layer of crushed stone which had, fortunately, been provided beneath the basement.

11.3.6 Ground dewatering by electro-osmosis

The various methods of ground dewatering described in the previous pages are used mainly in gravels and sands. Soils of finer particle size, i.e. silts and clays, are more troublesome to drain because capillary forces acting on the pore water prevent its flowing freely under gravity to a filter well or sump. The vacuum process of wellpointing in silts has already been mentioned, but if this process is ineffective and if, for some reason or another, sheet piling cannot be used then electro-osmosis is a possible expedient and may well prove to be less expensive than the last resort of freezing the ground. In the electro-osmosis system direct current is made to flow from anodes which are steel rods driven down into the soil, to filter wells forming cathodes. The positively charged particles of water flow through the pores in the soil and collect at the cathodes where they are pumped to the surface. Casagrande[11.6] has shown that the equation of flow is similar to Darcy's law, the rate of flow being dependent on the porosity of the soil and the electrical potential. In the few recorded instances of the process a current of 100 A has been used, the power requirements being some 0.5–1.3 kW/m³ of soil dewatered for large excavations and as much as 13 kW/m³ for small excavations.

A typical layout of an installation is shown in Fig. 11.28. The anodes are placed nearest to the excavation

causing the ground water to flow away from the slopes, which effectively stabilizes them and permits steep slopes even in soft water-bearing silts. The anodes corrode and require constant replacing, but the cathodes remain serviceable for long periods. The process was developed in Germany during the Second World War, and was used to stabilize a railway cutting at Salzgitter and in the construction of U-boat pens and other works at Trondhjem.[11.6] More recently, electro-osmosis was used to stabilize an embankment of sandy soil in the excavations for a dry dock at Singapore.[11.7] In this material the power consumption was 0.5 kW h/m³ of soil dewatered. The power was supplied at 25 A from portable welding generators.

Generally, electro-osmosis has been employed to remedy a difficult situation where other methods have failed rather than as a construction process in its own right. The main drawback is its high installation and initial running costs, but the power consumption, and hence running costs, decrease considerably after the ground is stabilized.

11.3.7 Reduction or elimination of ground-water flow by forming grouted barriers

In ground where the permeability is so high that wellpointing or bored wells need a very high pumping capacity, or in water-bearing rock formations where wellpointing cannot be used and bored wells are very costly, other expedients must be sought to control the ground water. One method is to inject fine suspensions or fluids into the pore spaces, fissures, or cavities in the soil or rock, so forming a barrier of reduced permeability. The type of injection material is governed by the particle size distribution of the soil or the fineness of fissuring in rock strata. Reference to Fig. 11.14 will give a guide to the suitability of various injection processes for soils. Another guide for the suitability of *suspension* grouts is given by the 'groutability ratio', Thus for a soil to accept a suspension grout the ratio of the D.15 size of the soil to the D.85 size of the grout particles must be greater than 11 : 25 for cement grouts and greater than 5 : 15 for clay grouts.

The most reliable index is the permeability coefficient of the soil. Practical lower limits of the coefficient for various grouting materials are:

Cement grout	5×10^{-4} m/sec
Clay–chemical grout	1×10^{-5} m/sec
Chemical grouts (non-particulate)	1×10^{-6} m/sec

Grouting is a fairly costly process. The aim should be to keep the volume of the basic material to a minimum. In this respect chemical grouts have certain advantages. Whereas they cost considerably more than cements or clays per tonne, they can be considerably diluted and still

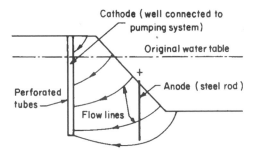

Fig. 11.28   Electro-osmosis installation.

work effectively in their role of reducing the permeability of the ground. Various additives can be employed to control the viscosity and gelling properties of suspensions and fluid chemical grouts, thus limiting their spread in the ground and keeping the thickness of the impermeable barrier to a minimum. Fluid grouts can be more effective than suspensions in reducing the permeability of the ground because all the pore spaces are filled, whereas suspensions may only fill the larger voids. A great deal of useful information on the principles and practices of grouting are included in the *Proceedings of the Symposium on Grouts and Drilling Muds in Engineering Practice*[11.8] and in the Code of Practice: *Foundations* BS 8004.

Care is needed in the adoption of injection processes, especially when working close to existing underground structures, since large quantities of materials injected into the ground under pressure must inevitably cause some displacement. The injected material tends to travel along the more permeable layers or along planes of weakness, often emerging at a considerable distance from the point of injection. Thus, a watch must be kept on sewers or cellars if any are existing in the vicinity of the work, and continuous records must be kept of surface levels.

Payment for grouting is normally made by the quantity of material injected. *CIRIA Report 113*[11.1] gives the following costs:

Material	Cost per m³
Clay	less than £20
Cement	less than £20
Cement/pf-ash	less than £20
Sodium silicate (Joosten process)	£15 (1976 price)
Sodium-aluminate	£40
Silicate-ester	£45

Prices on the basis of cost per cubic metre of ground stabilized are given as follows:

Material	Cost per m³ of stabilized soil
Cement/bentonite	£20
Bentonite gel	£22
Resins	£200–£300
Chemicals	Up to £300–£350

Three principal methods are used for injecting grout into the soil. The simplest is the *open-hole* method in which the grout injection pipe or lance is driven into the ground with its lower end closed by an expandable plug and the upper part sealed at the ground surface by a ring of caulking material. Grout is injected through the pipe to force out the plug. The method is best suited to very coarsely graded soils or rocks with wide fissures where the grout will enter the ground rather than by flowing back to the surface around the lance.

Stage grouting is performed by lowering the lance into a pre-drilled borehole and injecting the grout in stages. This can be done by top-down methods when the drill hole is taken down at each stage through the previously grouted zone after the latter has set. In the bottom-up method the hole is drilled to the full depth and a packer is used around the lance to seal off the top of the length to be grouted.

Sleeve grouting is best suited to grouting in soils. It employs the *tube-à-manchette* which allows grouting or regrouting in stages of varying length and at various intervals of time and using a range of different materials. The *tube-à-manchette* is shown in Fig. 11.29. A cased borehole is first drilled to the required depth, and the sleeve pipe is lowered into the hole and surrounded by a semi-plastic grout. The casing is then withdrawn. The sleeve pipe has openings at 300 mm spacing closed by an external close-fitting rubber sleeve. The perforated injection pipe is lowered into the sleeve pipe with a double packer sealing the top and bottom of the length to be grouted. When the grouting material is pumped down the injection tube the pressure pushes off the sleeve and grout flows through the openings and ruptures the plastic grout and then permeates the surrounding ground. Normally, grouting commences at the bottom of the hole and proceeds upwards in stages, but by using the double packer

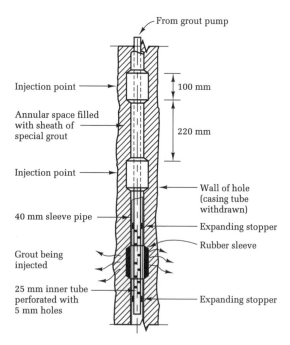

Fig. 11.29 *Tube-à-manchette* for sleeve grouting.

selected zones can be regrouted, varying the fluidity or viscosity of the mixes to suit the permeability of the ground.

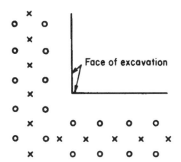

Fig. 11.30 Layout plan of grout injection holes to form 'grout curtain' around excavation. ○, Primary injection holes; ×, secondary injection holes.

Cement grouting   Cement is suitable for injection where strengthening is required in addition to a reduction in the permeability of soil or rock strata. It is necessary for soil to have a very coarse grading (see Fig. 11.14) to permit effective grouting with cement. The process is largely ineffective in sands, except for its consolidating or compacting effect when injected at close intervals. In coarse materials or rocks the excavation is surrounded by a 'grout curtain' consisting of two rows of primary injection holes at 2.5–5 m centres in both directions, with secondary holes (and possibly tertiary holes) between them (Fig. 11.30).

Cement grouting was used for the foundation excavations for extensions to the North Point Generating Station, Hong Kong. This site was on ground which had been reclaimed by tipping granite–rubble sea walls and filling up the ground behind them with sand and clay. A series of such reclamations had been made on the generating station

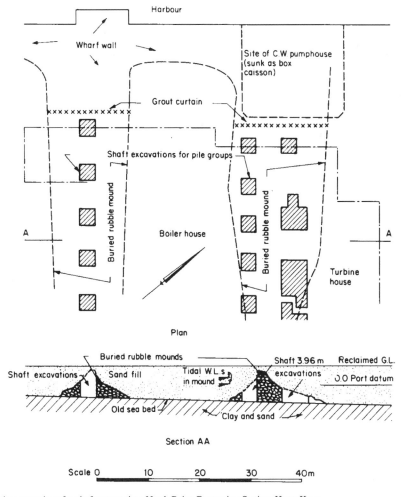

Fig. 11.31 Grouting operations for shaft excavation, North Point Generating Station, Hong Kong.

site, which was crossed by three buried rubble walls. These walls intersected at right angles a wharf wall of granite masonry built on a rubble mound (Fig. 11.31). It was necessary to sink a number of shafts through two of the buried walls to enable piles to be driven for the foundations of the generating station buildings and plant. It was clearly impossible to drive piles through the rubble, which included 5 t boulders. It was equally impossible to drive sheet piles around the shaft excavations to exclude the ground water. Accordingly, it was decided to excavate in timbered shafts. The site investigation showed that ground water in the rubble mounds had a connection to the sea in the harbour through the wharf wall. Therefore, in order to avoid pumping large quantities of water from the shafts, a grout curtain was formed across the two buried rubble walls close to their intersection with the wharf wall. A single row of 9 m deep holes was drilled at 3.7 m centres to form each curtain. The first holes drilled took about 30 t of cement, but the quantities for succeeding holes became less and, in all, over 150 t of cement were used for the two grout curtains. To economize in cement a sanded grout was used in the proportions of three parts cement to one part of sand. The effectiveness of the grout curtain can be judged from the fact that only two 150 mm pumps were needed to dewater the shafts nearest the harbour, with a single 150 mm pump for the shafts further away. A total of 22 shafts was excavated to a maximum depth of 8.5 m.

Jet grouting with cement Jet grouting provides a means of forming a series of columns of cement or cement-stabilized soil in the ground to produce an impervious barrier or load-bearing elements. To form a column a borehole is drilled to the required depth by water flush. Then a valve is operated to direct an air-shrouded high-pressure water jet in a horizontal direction while simultaneously rotating the drill pipe. The jets erode the soil, forming a cylindrical cavity, and the soil is flushed to the surface through the annulus around the drill pipe. Cement grout is then jetted horizontally at a pressure of 20 MN/m^2 either from the water jet nozzle or from a separate nozzle. The double tube equipment (Fig. 11.32(a)) employs a grout or water jet and an air jet while the triple tube equipment (Fig. 11.32(b)) has separate air, water and grout jets. In suitable conditions the grout jet can be used to cut the soil, but this method results in more contamination of the grout with soil than the triple tube system. Pairs of jet outlets are provided to cut and grout the column in diametrically opposite directions.

The soil can either be wholly removed and replaced by cement to produce a strong column with a diameter in the range of 0.8–2 m, or there can be partial mixing of the soil with cement to form a stabilized soil column. Typically clayey silts can be stabilized to give an unconfined

compression strength of 2–15 N/mm^2 and sands or sandy gravels produce strengths in the range of 5–25 N/mm^2. The columns are installed at centre to centre spacings of 0.75 times the cavity diameter to form single, double or triple overlaping rows, depending on the particular requirement. Applications of the jet grouting process include:

(a) Forming an impervious barrier around an excavation;
(b) Forming a cofferdam around and beneath an excavation (see Fig. 10.32);
(c) Sealing gaps around a contiguous bored pile wall (see Section 5.4.5);
(d) Sealing gaps between declutched sheet piles;
(e) Forming load-bearing elements (as individual columns);
(f) Forming struts at one or more levels to support a sheeted excavation;
(g) Forming an arched support over a tunnel excavation;
(h) Strengthening soft clays in advance of excavating at a tunnel face;
(j) As underpinning supports (Fig. 12.23).

The process can be costly relative to conventional cement grouting because of the large quantities of cement used. However, the process provides a more positive barrier to ground-water flow, particularly in layered pervious and semi-pervious soils. It also has the advantage of providing load-bearing elements in the ground. However, careful control is necessary to ensure that the required

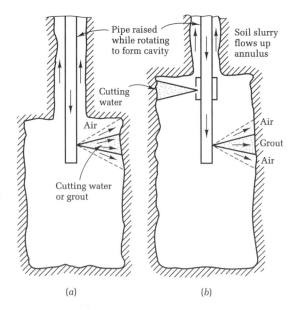

Fig. 11.32 Forming a jet grouted column. (a) By double-tube equipment. (b) By triple-tube equipment.

objectives are achieved. The jetted cavity can be irregular in shape where easily erodible soil layers alternate with strongly cohesive layers, producing fins of grout which reduce the load-bearing capacity of the column. Borehole callipers can be used to check cavity dimensions. The principal drawback is the heave of the ground surface which is caused by pressure developed in the cavity when the upflow of jetting fluid and soil becomes throttled or completely blocked in the annulus around the drill pipe above the cavity. Other problems concerned with the use of the process in an urban environment are the space necessary for the bulky equipment and the disposal of large quantities of water and soil slurry ejected at the surface. Another possible problem is the heat generated in the ground by the hydration of the cement.

A notable example of the use of the jet grouting process to stabilize soft clays in advance of tunnelling for the Singapore MRT is described by Berry *et al.*[11.9] The work comprised the installation of 4700 columns to form a 1.5 m grouted annulus around a 1 km length of the twin 6 m diameter tunnels near Raffles Place in the city centre. Surface heave gave considerable problems. An average surface heave of 70 mm was recorded over the eastbound tunnel driven at 10–15 m below street level and an average of 140 mm above the westbound tunnel at depths between 15 and 25 m. The maximum heave was 550 mm.

Trials showed that heave could be reduced by precutting the zone to be treated by high-pressure water jetting, then re-entering the borehole and cutting the remaining clay in the cavity by high-pressure grout. Heave was also reduced by using triple-tube instead of double-tube equipment and by eliminating as far as possible the installation of holes at angles greater than 10° from the vertical. Heat

generated in the ground produced temperatures up to 45 °C in the tunnels, and cement slurry reacted with ammonium salts in the soil to produce free ammonia as the tunnels were being driven. In spite of these drawbacks the use of the process greatly facilitated progress in tunnel driving compared with adjacent sections where it could not be used because of the congestion of underground utilities.

Clay grouting   Injections of bitumen emulsion or slurries of clay or bentonite, sometimes with added chemicals to aid dispersion and suspension of these materials, have been used in ground where the grading is too fine for cement grouting, and in gravels where reduction in permeability is required without the need for any strengthening of the ground. Grouting with a slurry of chemically treated bentonite clay has been used extensively for creating impermeable cut-offs in alluvial strata beneath dam foundations, and to create impermeable barriers around excavations in water-bearing alluvial strata. The principle of the method is to use bentonite clay in combination with Portland cement, soluble silicates, and other agents in differing proportions to produce a grout, the characteristics of which can be varied to suit the permeability of the ground into which it is injected. The larger voids are first filled with clay–cement grout followed by clay–chemical grouting to fill the spaces in the finer materials.

The construction of the Auber Station on the Paris Metro is a good example of the use of several grouting techniques to enable excavations to be performed in very difficult ground and site conditions. An adit was first driven at the level of the crown of the station tunnel. Clay cement grout was injected from radial drill-holes to treat the ground surrounding a further adit which was then

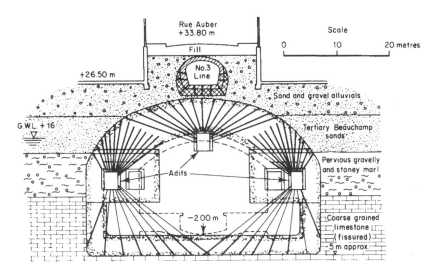

Fig. 11.33   Grout injection pattern around excavations for Auber Metro Station, Paris.

driven at springing level on each side of the station tunnel (Fig. 11.33). Second-stage injections of clay cement were made above the level of the crown adit to form an 'umbrella' of treated ground above the tunnel, and from above and below the side adits to complete the curtain of grout around the excavation. Simultaneously with the first stage grouting an ethyl-acetate grout was injected beneath the existing tunnel of No. 3 Line of the Metro to form a strengthening gel. The work has been described by Glossop[11.10] and Janin and le Sciellour.[11.11].

Jefferis[11.12] has described methods of producing cut-off walls by trenching with a backacter and pumping in a bentonite slurry to prevent the sides of the trench from collapsing. A trench width of 600 mm is suitable. Guide walls may be needed to prevent water from bleeding away, and the top of the slurry should be covered as soon as possible to prevent drying and the formation of shrinkage cracks. The following mix proportions will provide a strength in the range of 100–300 kN/m^2.

Cement	100–350 kg
Bentonite	200–600 kg
Water	1000 kg

Up to 60 or 75 per cent of the cement can be replaced by granulated blast furnace slag or up to 30 per cent by pf ash. The wall is constructed in panels and the backacter is used to cut back into the previously placed material when forming the joint between panels. Generally the aim should be to produce a barrier with a permeability of 10^{-8}–10^{-9} m/sec.

Chemical consolidation   The chemical injection process or chemical consolidation is applicable to sandy gravels, and sands of all but the finest gradings (see Fig. 11.14). The chemical most commonly used is sodium silicate which in conjunction with other chemicals forms a fairly hard and insoluble 'silica-gel'. In the 'two-shot' process, pipes are driven into the ground about 0.5 m apart and calcium chloride is injected down one and sodium silicate down the other as the pipes are slowly withdrawn in stages. Alternatively, one chemical can be injected as the pipe is driven down followed by the other chemical as it is withdrawn.

The 'two-shot' method has been largely superseded by the 'one-shot' technique in which all chemicals are mixed together immediately before injecting them. The grout is formulated so that the gel formation is delayed for a sufficient time to allow for complete penetration of the ground. Many other chemical processes based on the 'one-shot' principle have been developed with the aim of obtaining a very low viscosity at the time of injection with only a slow increase in viscosity until gelation occurs, thus ensuring maximum penetration. The chemicals include acrylic polymers, resins and lignins. Chemical consolidation has applications in underpinning work and an example is given in Section 12.3.7.

Complete cut-off of water by chemical consolidation or clay injection can only be obtained by repeated injections at close spacing over a considerable width of treated ground. The cost of such work is rarely justifiable, and in most practical cases a partial cut-off, say 80–95 per cent, is all that is required. It must be remembered that permeable water-bearing soils contain some 30–40 per cent void spaces, which must be filled with expensive chemicals. Thus, considering the volume of ground to be treated in a grout curtain some 2–2.5 m thick, the process is necessarily an expensive one as given by the prices at the beginning of this section.

11.3.8 Excavation under compressed air

The use of compressed air for excluding water from foundation excavations in caissons and shafts has been described in Chapter 6.

11.3.9 The freezing process

Because of its high cost, freezing of the ground to prevent inflow of water into excavations is usually regarded as a last resort when all other expedients have failed or are impracticable for one reason or another. The high cost of the freezing method is due to the necessity of sinking a large number of boreholes at close spacing around the excavation. The boreholes must be drilled to a high order of accuracy in verticality to avoid the risk of a gap in the enclosure of frozen ground, and the refrigeration plant is costly to install and maintain. The system also has the drawback that it takes several months to drill holes, install the plant, and freeze the ground; also, freezing certain types of ground causes severe heaving. There are difficulties in operating compressed-air tools in the low temperatures prevailing in the excavation, and there are also difficulties in concreting the permanent work. Nevertheless, in some situations freezing is the only practicable method of dealing with ground water, such as in very deep shaft excavations where the pressure of water is too high to allow men to work in compressed air and where fissures in rock are too fine for injection. The freezing process also has the considerable advantage of certainty in its effectiveness of excluding water from an excavation, whereas ground-water lowering or grout injection methods may not be fully effective because of variability in the ground conditions.

Basically, the system involves sinking a ring or rectangle of boreholes at 1–1.5 m centres around the excavation. The boreholes are lined with 100–150 mm steel or

plastic tubing with closed bottoms, and inner tubing of
38–75 mm diameter open at the bottom is then inserted.
The tops of the inner tubes are connected to a ring main
carrying chilled brine from the refrigeration plant. The
brine used to freeze the ground is pumped down the inner
tubes and rises up the annular space between them and the
outer casing; it then returns to the refrigeration plant via
the return ring main. Usually it takes some 6 weeks to 4
months to freeze the ground. A borehole lined with
perforated pipes is drilled near the centre of the treated
area to act as a tell-tale. As the ice wall forms and closes
up it compresses the ground water within the wall. When
the water rises up the tell-tale pipe and overflows at
ground level, the ice wall has closed and excavation can
commence.

It is most important when using the freezing process to
ensure that the ice wall is continuous before commencing
excavation. Difficulties which were encountered when
excavating in an incomplete ice wall have been described
by Ellis and McConnell.[11.13] A number of examples of
its use in civil engineering works have been given by
Harris.[11.14]

A paper by Collins and Deacon[11.15] gives a detailed
account of the installation methods, calculations for ice
wall thickness and refrigerant consumption, and also
information on costs. This paper describes the use of the
freezing process for sinking deep shafts through water-
bearing chalk for the Ely Ouse–Essex water scheme.

Considerable savings in time of freezing can be
achieved by using liquid nitrogen fed directly from insu-
lated containers into pipes driven into the ground, as
described by Harris.[11.14] Liquid nitrogen is expensive, but
the installation costs may be considerably lower than those
of the brine method, and its use may therefore be more
economical than brine installation where a short-term
expedient is required to overcome a localized patch of bad
ground. Liquid nitrogen can freeze the ground about five
times faster than chilled brine, allowing the ice wall to be
formed in days (or even hours).

11.4 The settlement of ground adjacent to excavations caused by ground-water lowering

11.4.1 Causes of settlement

The problems of settlement of the ground surface adjacent
to excavations due to piping of soil beneath sheet piling,
erosion from sloping sides of excavations, and infiltration
of fines into unscreened pumping wells has been mentioned in the preceding pages. However, there is one cause
of ground settlement which is liable to occur in some types
of soil no matter how carefully the ground-water lowering
is executed. This type of settlement is due to an increase in

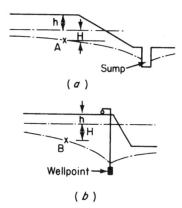

Fig. 11.34

density of the soil as a result of a general lowering of the
water table. Thus, at a point A near an excavation where
the ground water is being lowered by pumping from a
sump (Fig. 11.34(a)) or B adjacent to a wellpoint system
(Fig. 11.34(b)), the effective overburden pressure before
lowering the ground-water table is given by

$$p_e = \gamma_{sub}H + \gamma_{sat}h,$$

and after ground-water lowering p_e becomes

$$p'_e = \gamma_{sat}(H + h),$$

which is an increase in pressure of

$$\gamma_{sat}H - \gamma_{sub}H = (\gamma_{sub} + \gamma_\omega)H - \gamma_{sub}H$$
$$= \gamma_\omega H.$$

In other words, the effective pressure at A or B is
increased by an amount equivalent to the head of water
which existed above these levels before dewatering. If
compressible clay or peat layers exist above the water-
bearing layer the increase in their effective weight causes
them to consolidate, with accompanying settlement of the
ground surface. Similarly, the increase in effective pressure on compressible strata below the lowered water table
will also cause consolidation of these strata with corresponding settlement at ground level.

Although the effects are severe in soft clays and peats,
appreciable settlement can occur in sands, especially if
they are loose and if the lowered water table is allowed to
fluctuate. Little or no trouble need be feared with dense
sands and gravels, provided that the ground-water lowering system has efficient filters to prevent loss of fines
from the soil. Consideration must be given to the possibility of the settlement of piled foundations if they are
bearing on compressible materials, when the lowering of
the water table will cause negative skin friction or drag-

down on the piles. If the increase in load on the piles is small, and if they are toed into stiff or hard clay, the resulting settlement may well be negligible. However, if the increase in load is sufficient to exceed their carrying capacity, heavy settlement is inevitable.

11.4.2 Recharging wells

Precautions against such effects can be taken by a system of 'recharging wells', i.e. by pumping water into the ground near the excavation to keep up the water table. This procedure was adopted for the basement excavation for the Tower Latino Americana in Mexico City.[11.16] Lowering of the deep water table in the city has caused heavy settlements of the lacustrine volcanic clays, and it was feared that a steep draw-down in the shallow water-bearing strata for the foundation excavations would cause serious settlement of neighbouring streets and buildings. The excavation, which was taken down almost to the top of the first water-bearing sand layer (about 2.5 m below ground level), was surrounded with tongued and grooved timber sheet piling (Wakefield piling) and four 35 m deep pump wells were installed. The water level was then lowered to just below the final excavation level of 12 m and the water was discharged into an 'absorption' ditch at ground level to maintain the existing ground-water conditions in the shallow deposits. It was also discharged into injection or recharging wells to maintain the existing head in the lower pervious layers (Fig. 11.35). These wells consisted of 75 mm pipes perforated at 12, 16, 21, and 28 m where they passed through the pervious layers and were surrounded by a 90 mm thick sand filter.

The recharging process prevented settlement of adjacent structures, and, at the same time, the increase in effective pressure on the clay layer below basement level due to the reduction in the water table compensated for the decrease in overburden pressure. Thus, no swelling of the basement excavation occurred. The recharging process is standard practice in Mexico City.

The construction of the 22 m deep basement for the Hong Kong and Shanghai Banking Corporation headquarters in Hong Kong provides an example of the use of the eductor system for both ground-water lowering and recharge.[4.9] The area surrounding the 47-storey building was occupied by busy city streets and buildings. Previous experience of the construction of the nearby Chater Road Underground Railway Station had shown that large settlements could be caused by deep excavations and that significant ground movements could be caused by excavation under bentonite for diaphragm walls, even though the slurry level was kept well above the minimum relative to the external water level to maintain stability. Ground movements were mainly related to compression and consolidation of a deep stratum of completely decomposed granite which was overlain by about 5 m of sandy reclamation fill and on the seaward side of the site by up to 4 m of sandy marine deposits. Ground-water level was about 2–3 m below street level.

The close proximity of the streets and underground services and obstructions formed by the basement retaining walls and vaults of the former bank building made it impracticable to raise the guide walls for the diaphragm wall in order to increase the excess head of slurry. This could be achieved only by lowering the ground-water

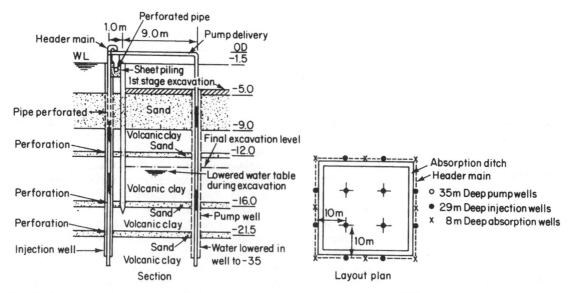

Fig. 11.35 Pumping and recharge system for basement excavation, Mexico City (after Zeevaert[11.16]).

level on both sides of the wall which carried a risk of causing settlements to the surrounding buildings and streets. Accordingly, close control of the relative levels between external ground-water levels and the slurry level was necessary. Control to maintain an excess head of 3.5 m was achieved by installing 150 mm diameter wells at 6 m centres around the external periphery of the basement and 15 similar wells within the basement area. Each well was provided with an eductor which served as a means of either lowering or recharging the ground water. As each diaphragm wall panel was being excavated ground-water levels close to the wall were recorded at 6-hourly intervals, and the slurry level was measured at hourly intervals to maintain the level at a constant depth of 0.3 m below the top of the guide walls with the external ground-water level lowered by about 2–3 m.

Each of the 51 diaphragm wall panels was toed-in to a depth of 0.3 m into moderately decomposed granite found at 25–35 m below street level. Numerical analyses using finite element methods had shown that any gap left between the base of the wall and the rock could cause a substantial lowering of the external ground-water level at the stage of dewatering the basement for the bulk excavation. A gap 1 m wide was calculated to cause a reduction in external water level within the decomposed granite of as much as 10 m. Consequently, each wall panel was provided with four vertical ducts. Grout holes were drilled through the ducts to a depth of 0.5 m into rock and injected with cement to refusal at 10 bar pressure. Alternate holes were redrilled 5 m into rock and grouted to refusal at 15 bar.

Maximum surface settlements close to the exterior of the diaphragm wall were generally in the range of 20–35 mm after completion of basement construction. These were within 60–70 per cent of the settlements predicted by finite element analyses using values of soil stiffness in the decomposed granite obtained by back analysis of movements resulting from sinking a caisson near the site.

11.5 Ground water under artesian head beneath excavations

11.5.1 Problems with artesian head

The problem of ground water under artesian head beneath an impervious layer has been mentioned. This caused partial failure of a cofferdam for a pumping station at Cowes Generating Station, Isle of Wight.[11.17] When excavation was almost down to the final level of −6.63 m OD (Fig. 11.36) the floor rose 100–150 mm in one corner and a crack opened up across this corner with water seeping up through it. Counterweighting of a 3.5 m ×

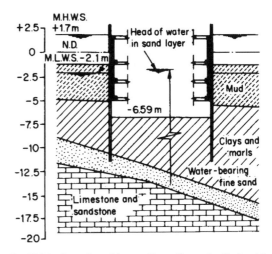

Fig. 11.36 Ground conditions at Cowes Generating Station (after Coates and Slade[11.17]).

3.5 m area was carried out with steel plates and 6 t of kentledge, but over the next 2 days the conditions deteriorated with further cracking, and depressions occurred in the ground around the outside of the cofferdam. It was then decided to flood the excavation to prevent further deterioration. Subsequent borings showed a 1.2–1.8 m thick layer of fine sand beneath the clay at excavation level. This sand layer was charged with fresh water under a 10 m head. The artesian head had lifted the clay, opening up fissures through which water flowed, thus softening the clay and allowing inward yielding of the sheet piling.

The foundations were completed by casting the foundation slab under water and anchoring it to a lower stratum of bedrock by prestressing wires located inside steel box piles.

Similar problems occurred in the intake pumphouse of Ferrybridge 'B' Power Station, but these were foreseen and the artesian head in the underlying permeable layer was lowered by wellpointing. However, unforeseen trouble did occur in sheet piled excavations for the circulating water discharge culverts on the same site. They were excavated to a lesser depth than the intake pumphouse, and it was thought that the weight of the 1.5–1.8 m layer of soft clay overlying the permeable layer, together with the adhesion of the clay to the sheet piles, was more than sufficient to counterbalance the 4.6 m head of water in the underlying water-bearing strata (Fig. 11.37). This proved to be the case, but trouble was experienced at one location where sheet piles forming a bulkhead across the trench were withdrawn to allow the base slab of the culvert to be placed. Water boiled strongly up the holes left by the sheet piles in the clay. The difficulty was overcome by excavating ditches on each side of the trench bottom and these led the water to a pumping sump.

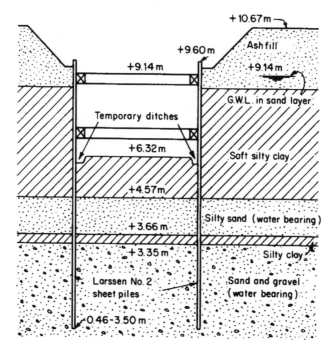

Fig. 11.37   Ground water conditions at Ferrybridge 'B' Power Station.

11.5.2 The use of relief wells

Where an excavation is located wholly within a clay formation underlain by a permeable stratum containing ground water at an artesian or sub-artesian pressure, uplift of the base of the excavation similar to the examples in the previous section can be prevented by installing bleeder wells through the clay in advance of excavation. A bleeder well system is feasible only if the water-bearing stratum is of low to medium permeability. Heavy discharge from a high permeability layer such as coarse sandy gravel would be likely to cause wash-out of fine material from the water-bearing layer and erosion of the filter material used for backfilling the wells. There would also be difficulties in constructing the permanent works.

A typical installation is shown in Fig. 11.38. The spacing and diameter of the wells depend on the permeability and water head in the water-bearing layer. The

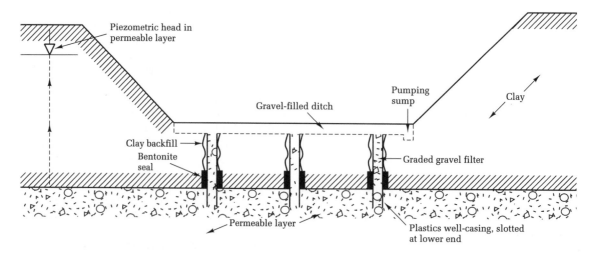

Fig. 11.38   Bleeder well installation. (Note: plastics well-casing and bentonite seal not needed for low-velocity discharge.)

grading of the filter material is determined from the D_{15} and D_{85} sizes of this layer (Fig. 11.24). Erosion of soil around the bleeder well by the upflowing water can be prevented by a permanent plastics well-casing slotted over its length within the permeable layer. Water should be prevented from piping around the annulus by means of a bentonite seal. After completing the excavation the discharge from the wells should be collected by a gravel-filled ditch and led to a pumping sump or discharged through flexible pipes laid across the base of the excavation or tremie concrete slab to the sump.

11.6 The use of geotechnical processes for ground improvement

11.6.1 Applications

Various geotechnical processes can be used to strengthen weak and compressible soils, thereby permitting an increase in foundation bearing pressures, or reducing settlements. Because of the cost of these types of ground treatment they are not normally used, since it is usually much cheaper to increase the size of foundations or to use bearing piles. However, there are circumstances where it may be expedient to adopt such methods, for example where the area available for foundations is limited by the presence of existing structures, or where piling cannot be adopted because of risks of settlement due to vibrations or loss of ground.

Methods of compacting or consolidating the ground to increase allowable bearing pressure or to reduce settlements include:

(a) Consolidation by preloading;
(b) Shallow compaction;
(c) Deep compaction;
(d) Dynamic consolidation;
(e) Stone columns;
(f) Stabilization by grout injections;
(g) Electro-osmosis and electrochemical hardening;
(h) Lime columns.

Theoretical and practical aspects of all the above processes are discussed in the state-of-the-art papers in the 8th European Conference on Soil Mechanics.[11.18] Methods (c), (d), and (e) are reviewed critically in a publication of the Institution of Civil Engineers.[11.19]

11.6.2 Consolidation by preloading

Total and differential settlements of foundations on soft compressible clays and loose granular soils or fill materials can be greatly reduced by preloading the area of the structure. The preload is applied by means of a mound of soil or rubble imposing a bearing pressure on the ground equal to or higher than that of the permanent structure. The preloading material is kept in place until level measurements show that the time–settlement curve has flattened or that the settlement has decreased to a very slow rate. An example of preloading a loose colliery waste fill has been given in Section 3.6.

When preloading soft clays or clay fills the rate of settlement of the mound may be rather slow, requiring the load to be in place for many months. In such cases consideration should be given to accelerating the rate of consolidation of the soil by the introduction of vertical drains. These can take the form of drilled holes filled with sand, 'sand-wicks', or preformed channels made from a plastic or cardboard duct covered with paper and driven into the ground by a special machine-operated mandrel. Vertical drains of these types can be installed to depths up to 20–30 m. Water expelled from the soil during the process of consolidation travels upwards through the drains to a blanket of pervious material placed beneath the preload fill. Drainage in a downward direction also takes place if the drains can be taken down to a pervious soil layer beneath the soft clay stratum.

The principle of the vertical drainage system is to obtain a reduction in the length of the seepage path for water squeezed from the soil. Thus when the drains are spaced at, say, 2–5 m centres the length of the seepage path is reduced from either the full depth or half the depth of the compressible stratum to half the spacing of the vertical drains, with a proportionate reduction in the time for consolidation of the soil.

A vertical drainage system is costly and it should not be adopted without making a detailed investigation of the permeability characteristics of the soil *in situ*. Rowe[1.1] has shown that many natural soil formations contain fissures or laminations of sand and silt which in themselves act as natural drainage channels at a close spacing, making it unnecessary to install vertical drains. Preloading with or without the assistance of vertical drains is not fully effective in peats or highly organic soils for which a high proportion of the total settlement is due to secondary compression (creep). The rate of secondary compression is not dependent on the drainage characteristics of the deposit.

Where large areas are to be preloaded it can be advantageous to construct two test embankments, one with and one without artificial drainage, and to compare the amounts and rates of settlement. It is sometimes found that while the rate of settlement can be accelerated by vertical drains the total settlement is much higher, therefore little advantage is gained. Preloading can be performed in underwater locations, but the process is much less effective than on land because of the lower effective weight of the preload material and its tendency to spread when dumped through water. A layer of geotextile fabric is needed below the fill where it is placed on a very soft clay or silt deposit.

11.6.3 Shallow compaction

Loose or disturbed granular soils at the base of excavations for strip or pad foundations can be compacted by rolling or ramming. Vibratory rollers or plate compactors work efficiently in granular soils, but the depth of compaction with ordinary equipment is unlikely to exceed about 300 mm. In thick deposits of loose granular soil or fill above ground-water level, it can be cheaper to remove these deposits and replace them in compacted layers than to use the specialist processes described in the following section.

11.6.4 Deep compaction

Heavy vibrators can be inserted into loose granular soils and then withdrawn, leaving a column of compacted soil in the ground. By inserting the vibrators at close spacing a state of uniform compaction can be obtained thereby greatly reducing the differential settlements which would otherwise occur with a natural soil deposit of widely varying density. In Britain the method has been used extensively to compact fill material consisting of brick rubble, broken concrete, timber, piles, paving materials, soil, and other miscellaneous materials resulting from site clearance of old houses. Usually, little attention is paid to uniform compaction of such materials when clearing and levelling a site scheduled for urban redevelopment. However, by means of close-spaced insertions of a heavy high-frequency vibrating unit along the line of the proposed foundations, the materials are induced to attain a closer state of packing and voids formed by arching of the fill are broken down. Additional granular material is fed into the depression formed around the vibrator. By means of repeated insertions and withdrawals of the unit the granular material is compacted into any remaining voids, thus forming a strip of dense granular fill upon which the new foundations can be constructed without risk of appreciable differential settlement.

Two principal methods are used for deep compaction. In the vibroflotation process a heavy vibratory unit (Fig. 11.39) is jetted down into the soil (Fig. 11.40(*a*)). On reaching the desired depth, a rotating vibrating machine within the unit is set in motion and the direction of jetting is reversed to carry the soil particles downwards. The vibrator compacts the soil around the unit and sand is shovelled in to fill the cone-shaped depression which appears at the surface (Fig. 11.40(*b*)). The unit is withdrawn in 0.25 m stages, vibration being applied at each stage and ground surface made up as required. The unit is

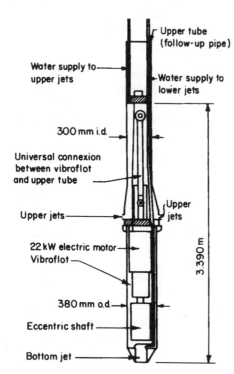

Fig. 11.39   Diagram of vibroflot.

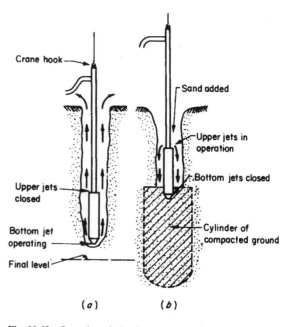

Fig. 11.40   Operation of vibroflot. (*a*) Jetting down vibroflot. (*b*) Withdrawal of vibroflot.

put down again about 2 or 3 m away and the process repeated until the whole area to be treated is covered by overlapping cylinders of compacted soil. The units can be jetted to a depth of about 10 m.

The vibroflot is most effective in clean sands or gravels or granular fills, but it can operate in silty or clayey sands containing up to 25 per cent of silt or 5 per cent of clay.

In the vibro-replacement process the vibratory unit consists of a large vibrating tube which relies on dead weight combined with high-frequency vibration to obtain the desired penetration. No water is used, otherwise the procedure is generally similar to vibroflotation. In another variation the stone is fed from the bottom of the vibrator and a pull-down unit is used to compact the column. The Swedish Vibro-Wing process consists of a vertical mandrel with horizontal blades of various lengths set in different directions around its length. A heavy vibrating unit is mounted on top of the mandrel which drives the mandrel down to the required depth, and it is then slowly withdrawn by crane while vibrating continues. The process is suitable for sandy soils.

The economics of dynamic compaction in granular soils require careful consideration. Settlements of foundations with moderate loadings on loose to medium-dense soils are not large, and a high proportion of the settlements are 'built-out' at the construction stage such that the reduction in settlements at the end of construction achieved by the process may be so small as not to be worth the cost involved. Deep vibratory compaction of natural granular soil deposits is best applied to the foundations of structures sensitive to small differential settlement when the compaction can be used to produce uniform density in a variable soil deposit, thereby reducing the amount of differential rather than total settlement.

The process was used under water beneath the piers of the 3 km Oosterschelde Storm Surge Barrier.[11.20] The 50 m × 25 m piers weighed up to 18 000 t in air and were sunk in water depths up to 30 m. They were located in an area of deep deposits of loose to medium-dense sands, where densification was necessary to reduce deformation of the barrier structure caused by differential settlement under loading from gate operations and storm surge forces. Vibratory compaction was required to a depth of 15 m below dredged level. This was undertaken by four 'needles' suspended from a pontoon covering an area of 25 m × 6 m. The pontoon was moved in stages across an area 80 m wide, after which the sea bed was levelled by dredger and covered by an upper and lower mattress each consisting of a three-layer sand and gravel filter, and then by a final concrete block mattress. The hollow pier units were sunk on to the sea bed followed by injecting cement grout between the base slab and the block mattress, and then ballasting the interior of the pier with sand.

Control of deep compaction in granular soils is best achieved by measuring the *in-situ* density before and after treatment. The measurements are made by means of a grid pattern of static cone penetration tests supplemented by the cheaper dynamic cone tests at a closer grid spacing (Section 1.4.5).

11.6.5 Dynamic consolidation

Deep deposits of loose granular and silty soils can be consolidated by dropping heavy weights from a considerable height on to the ground surface. Weights of up to 200 t falling from a height of 40 m have been used. The weights are first dropped at a wide spacing to compact the deeper soil layers, then the spacing is decreased to compact the surface layers. It is desired to avoid the formation of a thick surface crust at too early a stage in the operations. The impact of the weights form deep imprints in the ground surface and water squeezed from the underlying soil collects in these holes. Drainage is necessary to remove the accumulated water and fill must be imported to level the ground surface.

It is claimed that the method is effective in soft clayey soils when the impact of the falling weight causes rupture and cracking of the soil mass. The cracks then form drainage channels for the expulsion of water in subsequent passes of the falling weight. The process requires careful evaluation and control in clayey and silty soils, particularly in soft clays which are sensitive to disturbance, when destruction of the soil fabric will cause loss of shear strength. Pore pressures are built up during the tamping process and they are slow to dissipate due to blockage of natural drainage paths. If the tamping causes only heave of the ground surface and loss of strength of the previously undisturbed soil, then little useful work will have been done.

The method has been used successfully to break down soil or weak rock which has arched over voids in a cavernous rock formation. An example of this has been given by Holt.[11.18]

Explosives have been used to compact soils at depth.[11.21] The techniques used in the former USSR to compact deep deposits of loess by means of explosives have been described in Section 3.2.

11.6.6 Strengthening the ground by stone columns

The vibratory units described in Section 11.6.4 can be used to introduce strengthening materials into loose sands, soft silts, or clays and fill materials to enable them to carry foundation loads without excessive settlements. The vibratory units are used to form columns of clean graded stone or blast furnace slag at a close spacing over the foundation area. In sands and silty sands the vibration and lateral displacement of the soil causes densification of the mass. The efficacy of the treatment can be expressed in terms of a factor n which is the ratio of the settlement of a given area A subjected to treatment to the settlement of the same area without treatment. Priebe[11.22] has established curves relating n to the ratio of A to the total cross-sectional area of the stone column as A_c. The curves are shown in Fig. 11.41 for granular soils in a range of angles of shearing resistance. They are based on the assumption that the stone columns are formed in a regular triangular pattern.

It is claimed that the lateral displacement increases the shear strength of soft clays, but in soils sensitive to disturbance the net effect is more likely to be a loss of strength accompanied by ground heave if the columns are installed by tamping or by a poker unit without water jetting to displace the soil. In a soft compressible clay the stone columns have a much greater stiffness than the surrounding soil, hence they will attract load applied to the ground surface and initially they will try to act as piles. The columns will then bulge, causing them to settle, and some of the load will be transferred to the surrounding clay. The relative proportions of load carried by the columns and the clay will depend on the extent of the bulging. If the ratio of the length of the column to its diameter is too large the column will fail by bulging. Hughes and Withers[11.23] showed that this critical state occurred at a vertical stress in the columns of 25 times the undrained shear strength of a soft clay when the length of the column was four times the diameter. A settlement of 58 per cent of the diameter occurred before bulging failure took place. At this length the column failed simultaneously by bulging and in end-bearing resistance on the soft clay. Hughes and Withers have established the design chart (Fig. 11.42) relating the *allowable* load on a stone column to the undrained shear strength of the surrounding clay. They point out that the bearing capacity of the

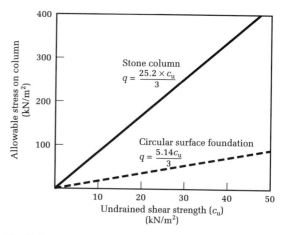

Fig. 11.42   Chart for determining allowable stress on stone column (after Hughes and Withers[11.23]).

individual column does not increase for length to diameter ratios greater than 6.

Balaam and Poulos[11.24] have shown that finite element methods can be used to predict the settlement behaviour of single or groups of columns based on the modulus ratio of the column and the soil and the column spacing. The Balaam and Poulos curves are shown in Fig. 11.43 for a range of column length to diameter ratios of 5–20. They are compared with the Priebe curves (Fig. 11.43) and the empirical method of Greenwood.[11.25] The Greenwood curves assume that the stone columns are founded on firm clay, sand, or harder ground.

Bulging failure can be prevented by adding cement to the stone as the column is formed in the ground. However, this adds considerably to the stiffness of the column and a corresponding reduction in the proportion of load carried by the soil. If no bulging occurs all the load is carried by the columns when they act only as piles with a risk of crushing near the head where the stress on the column is a maximum, although the stress can be reduced by enlarging the shaft diameter at the top of the column.

Stone columns are used for the foundations of structures located on loose fill materials which can be advantageous where the vibrations break down assemblies of particles cemented by mineral compounds. However, there can be risks of serious settlement if the fill contains soluble compounds where the pervious columns act as vertical drains for surface water which leaches out the soluble material giving lateral support to the columns. There is also a risk of contamination of ground water at deeper levels as a result of seepage through the column.

The stone column process is best applied to large loaded areas such as the approach embankments to bridges where the earthworks are placed over soft clays and settlements due to bulging of the columns are not detrimental.

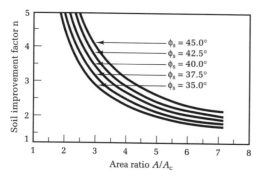

Fig. 11.41   Chart for determining spacings of stone columns for densification of sands (after Priebe[11.22]).

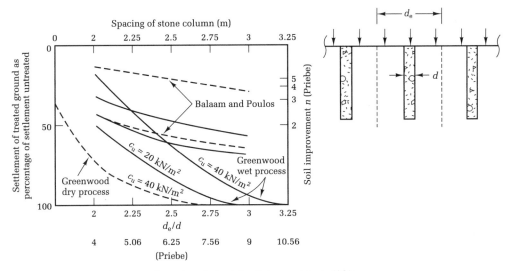

Fig. 11.43   Settlement diagram for stone columns in soft clay (after Balaam and Poulos[11.24])

Large quantities of stones are required to form columns in very soft soils. In such conditions the process is only marginally cheaper than conventional piling, bearing in mind the higher working load which can be safely carried on a pile compared with a stone column.

11.6.7 Stabilizing the ground by grout injections

Using the methods described in Section 11.3.7, various forms of grout can be introduced into granular soils or cavernous rock formations to increase their strength and reduce their compressibility. However, the cost of the grout materials and injection processes are high, making the methods uneconomical compared with deep compaction. Their use is generally restricted to site conditions where deep vibratory or dynamic compaction cannot be used because of the presence of existing structures, for example in underpinning work (see Section 12.3.7).

These techniques of injecting grouts into granular soils or rocks are referred to as *permeation* grouting. In the case of silts and clays the grouts cannot be induced to flow into the void spaces between the soil particles, but if high pressures are used flow takes place into natural fissures, or new fissures are opened where planes of weakness exist. The injection pressures expand the fissures by a process known as *hydrofracture* or *claquage* and because the vertical overburden stress is usually lower than the *in-situ* horizontal stress the soil mass is lifted. Using close-spaced points and high pressures the claquage grouting technique can be used in clays to raise structures which have undergone settlement, or to carry out the injections simultaneously with excavations or tunnelling to correct the ground movements as they occur. Used in this way the

techniques are referred to as *compensation grouting*. Where no natural fissures exist, or cannot be induced to form, a limited expansion or compression of the soil mass can be achieved by the bulb of grout which forms around the injection tube. This process is known as *compaction grouting*.

The techniques of compensation and compaction grouting are used mainly in underpinning work and will be described in greater detail in Section 12.3.7.

11.6.8 Electro-osmosis and electrochemical hardening

The electro-osmosis process described in Section 11.3.6 can be used to increase the shear strength and reduce the compressibility of soft clayey and silty soils beneath foundations. By introducing an electrolyte such as calcium chloride at the anode, the base exchange reaction between the iron anode and the surrounding soil is increased resulting in the formation of ferric hydroxides which bind the soil particles together.[11.26] However, because of the costs of electric power and wastage of electrodes, electro-osmosis with or without electrochemical hardening can be considered only for special situations where the alternative of piling cannot be adopted.

11.6.9 Stabilization by lime columns

Introducing lime in solution or powder form into a soft clay improves its bearing characteristics and reduces its swelling potential. The lime acts on the clay by a combination of flocculation and base exchange to produce a more friable material with a reduced plasticity index, a higher shear strength, and a higher permeability. Some cementation of the agglomerated clay particles also occurs.

Lime added in powder form can be slaked or unslaked. The latter is more effective because the heat of hydration has a drying effect, thereby reducing the pore-water content of the clay. The stabilization process is slow, but it can be accelerated by the addition of powdered gypsum. The addition of gypsum increases the strengthening effect particularly in organic clays. Eggestad[11.27] reports that lime and lime–gypsum mixtures are added to the soil in the proportions of 3–10 per cent by dry weight of soil. The lime to gypsum ratio is 75 : 25, where long-term stability is required, or 50 : 50 for short-term temporary stabilization.

Broms and Anttikoski[11.28] have reviewed the methods of introducing lime into the soil. These include:

(a) Pumping slaked lime slurry into boreholes;
(b) Filling boreholes with powdered unslaked lime;
(c) Extruding unslaked lime powder from tubes punched into the ground;
(d) Mixing unslaked lime powder into the soil by rotary drill where the material is added through the drill pipe as it is being withdrawn and mixed with the soil by a wing bit.

Generally the lime/soil columns are 0.3–0.5 m in diameter spaced at 1–2 m centres. Where the action of the lime causes a marked increase in permeability of a clay the columns function as vertical drains.

References

11.1   Somerville, S. H., Control of ground water for temporary works, *Construction Industry Research and Information Association, Report 113* (1986).

11.2   Williams, B. P., Smyrell, A. G., and Lewis, P. J., Flownet diagrams – the use of finite differences and a spreadsheet to determine potential heads, *Ground Engineering*, **25**(5), 32–38 (1993).

11.2(a)   Desai, C. S., Finite element procedures for seepage analysis using an isoparametric element, *Proceedings of the Conference on Applications of the Finite Element Method in Geotechnical Engineering*, US Army Waterways Experiment Station, Vicksburg, Mississippi, 1972.

11.3   Naylor, D. J. and Pande, G. N., *Finite Elements in Geotechnical Engineering*, Pineridge Press, Swansea, 1981.

11.4   Glossop, R. and Skempton, A. W., Particle size in silts and sands, *Journal of the Institution of Civil Engineers*, **25**, 81–105 (1945).

11.5   Worth, R. E. J. and Trenter, N. A., Graving dock at Nigg Bay for offshore structures, *Proceedings of the Institution of Civil Engineers*, **58**, 361–376 (1975).

11.6   Casagrande, L., The application of electro-osmosis to practical problems in foundations and earthworks. *Building Research Establishment Technical Paper No. 30* (1947).

11.7   Chappell, B. A. and Burton, P. L., Electro-osmosis applied to unstable embankments, *Proceedings of the American Society of Civil Engineers*, **101**(GT-8), 733–739 (1975).

11.8   *Proceedings of the Symposium on Grouts and Drilling Muds in Engineering Practice*, London, Butterworths, 1963.

11.9   Berry, G. L., Shirlaw, J. N., Hayata, K., and Tan, S. H., A review of grouting techniques utilized for bored tunnelling with emphasis on the jet grouting method, *Proceedings of the Singapore Mass Rapid Transport Conference*, Institution of Engineers, Singapore, 1987, pp. 207–214.

11.10   Glossop, R., The rise of geotechnology and its influence on engineering practice, *Géotechnique*, **18**, 107–150 (1968).

11.11   Janin, J. J. and le Sciellour, G. F., Chemical grouting for Paris rapid-transit tunnels, *Proceedings of the American Society of Civil Engineers*, **96**(CO-1), 61–74 (1970).

11.12   Jefferis, S., Cut-off walls, practice and problems, *Notes taken at meeting of the British Geotechnical Society*, 10 April 1991.

11.13   Ellis, D. R. and McConnell, J., The use of the freezing process in the construction of a pumping ststaion and storm-water overflow at Fleetwood, Lancashire, *Proceedings of the Institution of Civil Engineers*, **12**, 175–186 (1959).

11.14   Harris, J. S., Ground freezing, in *Underpinning* (ed. S. Thorburn and J. F. Hutchison), Surrey University Press, 1985, pp. 222–241.

11.15   Collins, S. P. and Deacon, W. G., Shaft sinking by ground freezing: Ely Ouse–Essex scheme, *Proceedings of the Institution of Civil Engineers*, Suppl. Paper No. 7506S (1972).

11.16   Zeevaert, L., Foundation design and behaviour of Tower Latino Americana in Mexico City, *Géotechnique*, **7**, 115–133 (1957).

11.17   Coates, R. H. and Slade, L. R., Construction of circulating water pumphouse at Cowes Generating Station, Isle of Wight, *Proceedings of the Institution of Civil Engineers* **9**, 217–232 (1958).

11.18   Various authors in the *Proceedings of the 8th European Conference on Soil Mechanics*, Helsinki, 1983.

11.19   Various authors and discussion, *Ground Treatment by Deep Compaction*, Institution of Civil Engineers, London, 1976.

11.20   van Driel, P. J., Oosterschelde Storm Surge Barrier, *The Netherlands Commemorative Volume*, 11th International Soil Mechanics Conference, San Francisco, 1985, pp. 231–248.

11.21   Prugh, D. J., Densification of soils by explosive vibration, *Proceedings of the American Society of Civil Engineers*, **89**(C-O1), 79–100 (1963).

11.22   Priebe, H., Abschätzung des Setzungverhaltens eines durch Stopfverdichtung Verbesserten Baugrundes, *Die Bautechnik*, Berlin, **53**, 160–162 (1976).

11.23   Hughes, J. M. O. and Withers, N. J., Reinforcing of soft cohesive soils with stone columns, *Ground Engineering*, **7**(3), 42–49 (1974).

11.24   Balaam, N. P. and Poulos, H. G., The behaviour of foundations supported by clay stabilized by stone columns, in *Proceedings of the 8th European Conference on Soil Mechanics*, Helsinki, **3**, 199–204 (1983).

11.25   Greenwood, D. A., Mechanical improvement of soils below ground surface, *Proceedings of the Conference on Ground Engineering*, Institution of Civil Engineers, London, 1970, pp. 11–22.

11.26   Farmer, I. W., Electro-osmosis and electrochemical stabilization, in *Methods of Treatment of Unstable Ground* (F. G. Bell, ed.), Newnes-Butterworth, London, 1975, pp. 26–36.

11.27   Eggestad, A., Improvement of cohesive soils, in *Proceedings of the 8th European Conference on Soil Mechanics*, Helsinki, **3**, 1–17 (1983).

11.28   Broms, B. B., and Anttikoski, J., Soil stabilization, in *Proceedings of the 8th European Conference on Soil Mechanics*, Helsinki, **3**, 141–153 (1983).

12 Shoring and Underpinning

12.1 Requirements for shoring and underpinning

Shoring of a structure is required:

(a) To support a structure which is sinking or tilting due to ground subsidence or instability of the super-structure;

(b) As a safeguard against possible settlement of a structure when excavating close to and below its foundation level;

(c) To support a structure while making alterations to its foundations or main supporting members.

Underpinning of a structure is required for the reasons given in (a) and (b) above and, in addition,

(d) To enable the foundations to be deepened for structural reasons, for example to construct a basement beneath a building;

(e) To increase the width of a foundation to permit heavier loads to be carried, for example when increasing the storey height of a building;

(f) To enable a building to be moved bodily to a new site.

Shoring and underpinning are highly skilled operations and should be undertaken only by experienced firms. No one underpinning job is like another and each one must be given individual consideration for the most economical and safest scheme to be worked out. For this reason the author does not intend to describe the methods in detail, but merely to state the general principles which are followed. Much interesting and detailed information is given in Prentis and White's book *Underpinning*[12.1] which is a classic work on the subject. Also, *Underpinning* (ed. Thorburn and Hutchison)[12.2] treats the subject comprehensively with many case histories of modern practice.

12.2 Methods of shoring

Shoring by external props is only required in connection with underpinning work for structures which are out of plumb, or for structures which are sensitive to the effects of small settlements. It is generally unnecessary to provide extensive shoring to steel or reinforced concrete-framed buildings while excavating close to their foundations since the structural framework effectively ties the building together. It is, of course, necessary to support individual columns while underpinning their foundations. Buildings with load-bearing walls, if in sound condition, can generally be secured against harmful movement while underpinning by means of horizontal ties at the various floor and roof levels supplemented as necessary by internal bracings. These methods avoid obstructing the ground around the building where the underpinning operations or excavations are taking place.

Seymour[12.3] has pointed out the risks of demolishing a building of framed construction where the party wall between this and the abutting building predates both of them. He stated that it was and still is the practice to leave a party wall in this state to serve as a fire-stop between adjoining buildings and to preserve the common ownership of the wall. In the three cases shown in Fig. 12.1 demolition of building B will involve a risk of outward collapse of the party wall unless it is supported by external shoring. As well as being unbonded to either existing building the party wall may consist of several unbonded skins of brickwork or two outer leaves with rubble infill. Brickwork forming old chimney breasts and flues may be in a weak decayed condition.

Raking shores (Fig. 12.2) are generally used where external support is necessary. The angle of the shores is generally 60–75°. If the feet of raking shores obstruct construction operations then *flying shores* (Fig. 12.3) can be used provided that there is a conveniently placed wall

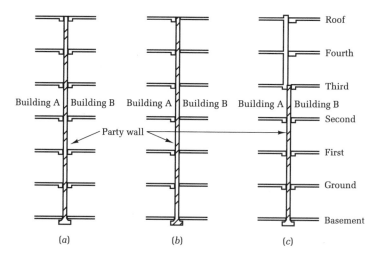

Fig. 12.1 Typical sections of old party walls (after Seymour[12.3]).

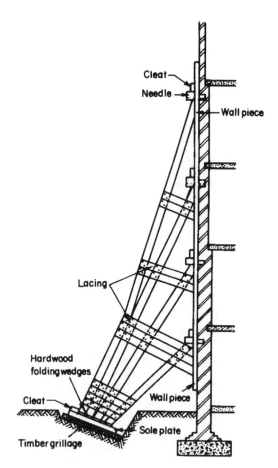

Fig. 12.2 Arrangement of raking shores to five-storey building.

or other structure to strut against. For practical reasons the length of timber flying shores is generally limited to about 10 m, but shores of latticed tubular construction can span up to 25 m. The levels of raking shores and flying shores are so arranged so that they bear on the walls at the floor levels of the buildings. This ensures that the thrust is transmitted as far as possible to the whole structure and avoids putting a bending load on to a wall. If there is no floor at a suitable level at the end of a flying shore a stiff vertical member should be provided to distribute the load (Fig. 12.4). Because flying shores do not bear on the ground they cannot carry the weight of a wall; they merely provide a restraint against bulging or tilting. Support to the vertical load of a wall, where it is required in conjunction with flying shores or horizontal ties, can be given by means of *dead shores* (Fig. 12.5). Dead shores are vertical struts bearing on the ground at the required distance away from the wall to be clear of underpinning operations and surmounted by a horizontal beam or *needle* spanning between a pair of shores.

Raking or dead shores should be designed to carry the whole weight of the walls and any loads transmitted to them from the floors and roof of the building. The loading should be lightened as far as possible by removing machinery installations or stored materials. The needles are inserted through holes cut in the wall. A gap is left between the needles and the underside of the wall into which fine dry concrete or fairly dry mortar is rammed to ensure a good bearing over the full width of wall. Pad stones may be necessary to avoid concentrations of loading on the brickwork when supporting heavy walls. Where raking shores are used, wall pieces are placed over the needles and the upper ends of the shores are restrained

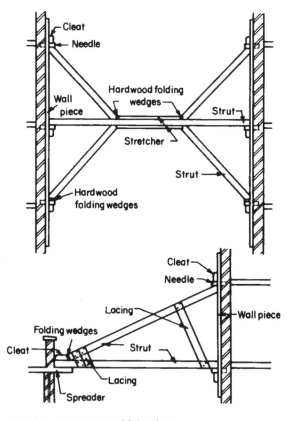

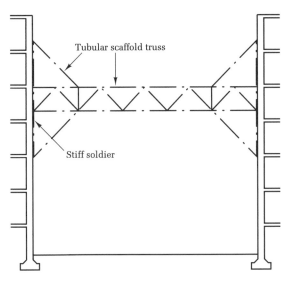

Fig. 12.3   Arrangements of flying shores.

Fig. 12.4   Tubular steel lattice flying shore for multi-storey building (after Seymour[12.3]).

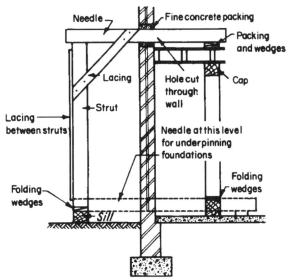

Fig. 12.5   Arrangement of dead shores.

from kicking up by means of *cleats* nailed to the wall pieces.

Raking and dead shores must be securely braced together and provided with a firm bearing on a *sill* or *sole plate*. If the ground has a low bearing capacity, a mat or grillage should be provided in timber, concrete, or steel construction. Hardwood folding wedges should be inserted at the feet of shores to take up any yielding of the ground and elastic shortening of the strutting materials. A careful watch should be kept on the shores, and the wedges should be tightened or loosened as necessary when the timber shrinks in dry weather or swells in wet weather.

Columns of framed structures can be shored up individually by needle beams. Where steel beams are supported, cleats are bolted or welded to the opposite flanges to provide a bearing to the needles (Fig. 12.6(*a*)). Alternatively, the beams at first floor level can be shored up and the column left hanging from them (Fig. 12.6(*b*)). Reinforced concrete or brick columns can be supported by girts tightly bolted around them and prevented from slipping by chases cut in the faces of the columns (Fig. 12.7(*a*)). Alternatively, chases or sockets can be cut in the sides of the column bases to permit the insertion of jacks (Fig. 12.7(*b*)). It is preferable to place jacks between the needle beams and the cleats or girts. The pressure from the jacks is increased until upward movement of the column is just detected. This prestresses the supports to the needles and their bearing on the ground before the foundation of the column is cut away or undermined. After the needle system has been given time to settle down, the spaces

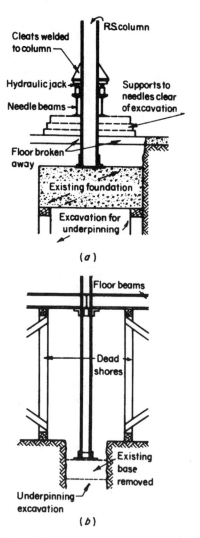

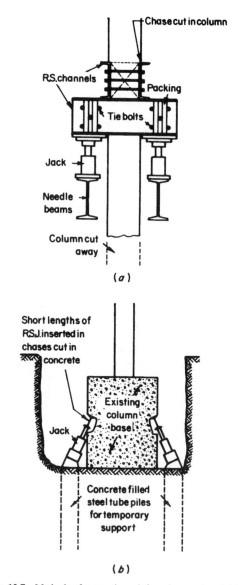

Fig. 12.6 Methods of supporting steel columns. (*a*) Support by needles. (*b*) Support by shoring.

Fig. 12.7 Methods of supporting reinforced concrete or brick columns.

between the needles and cleats can be packed and tightly wedged and the jacks removed.

Before any shoring work is commenced the building should be carefully surveyed. Records should be made of the levels of floors and the inclination of walls and sills, noting, marking, and photographing any cracks. Tell-tales and datum points should be placed where necessary for observing the movement of cracks and settlements. The records of levels, photographs, and notes should be agreed between the building owners, engineers, and contractors. The observations and measurements should be continued throughout the period of shoring and any subsequent excavation or underpinning, and until such time thereafter as all detectable movements have ceased.

12.3 Methods of underpinning

12.3.1 Preliminary investigations

Before undertaking any scheme of underpinning whether in connection with adjacent construction operations, or to prevent further settlement as a result of ground subsidence or overloading of foundations, it is important to carry out a careful soil investigation by means of boring or trial pits with laboratory tests on soil samples to determine allowable bearing pressures for the new foundations.

If underpinning is necessary to arrest settlement it is

essential that the underpinned foundations should be taken down to relatively unyielding ground below the zone of subsidence. For example, if the bearing pressures of existing foundations are such that excessive settlement is occurring due to consolidation of a compressible clay soil it is quite useless simply to widen the foundations by shallow underpinning. This will merely transmit the pressures to the same compressible soil at a lower level and the cycle of settlement will start all over again. The underpinning must be taken down to a deeper and relatively incompressible stratum, if necessary by piers or piles.

It is essential to recognize the true cause of settlement. Shallow underpinning may be quite satisfactory if the settlement is due to shrinkage of a clay soil due to dry weather; for example if settlement has occurred due to soil shrinkage in London Clay it is sufficient to underpin the foundations to a depth of 1.5–2.0 mm. If, however, the soil shrinkage is due to the drying action of the roots of trees or hedges the underpinning will have to be taken well below the root system (see Section 3.1.2).

Underpinning is ineffective if the settlement is due to some deep-seated cause such as coal mining. In such cases all that can be done is to wait for the movement to cease, then restore the structure to level (if practicable) before repairing it. Alternatively, a jacking system can be installed to maintain the foundations at a constant level (see Section 3.5.5).

12.3.2 Underpinning to safeguard against the effects of adjacent excavations

It has been noted in Section 9.8 that settlement and inward yielding of the ground surface adjacent to deep excavations is generally at a maximum close to the face of the excavation, and decreases with increasing distance from the face. The tables in Appendix B and Table 9.3 can be used as guides to enable a decision to be made whether or not to underpin the shallow foundations of a building near a deep excavation. It is not always necessary to underpin a building if it lies within the zone of potential settlement as indicated by Figs 9.42 and 9.43. Judgement is required to assess the consequences of the estimated settlement and inward yielding. Small settlements may not damage a structure, or the cost of repairing damage and compensating the owner may be less than the cost of underpinning. Much will depend on the use of the building, its structural form and its age and general condition (see Section 2.6.2). It is also important to remember that the observations of inward yielding shown in the tables were made on properly strutted or anchored supports to the excavations. Much greater movements can occur if the support system is poorly designed or constructed. Underpinning to a greater distance from the face than indicated by the figures

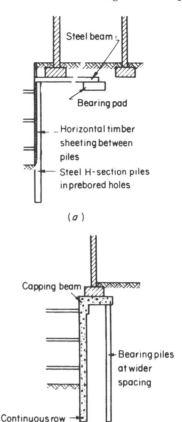

Fig. 12.8 Methods of underpinning by piling adjacent to deep excavatioons.

may be necessary if a ground-water lowering system is in operation unless other safeguards are taken (see Section 11.4).

If the structure to be underpinned is close to the excavation it is often convenient to combine the underpinning with the supports to the excavation. For example, the supports can consist of steel universal beam soldiers inserted in prebored holes with horizontal sheeting members. The load of the building can then be transferred to the tops of the piles (Fig. 12.8(a)). Alternatively, a system of close-spaced bored piles can be used (Fig. 12.8(b)).

In all cases where underpinning is provided close to excavations it is important to design the underpinning members to carry any lateral loads transmitted to them from the retained earth or ground water.

12.3.3 Underpinning by continuous strip foundations

In its simplest form, underpinning consists in excavating

rectangular pits or 'legs' at intervals beneath the existing strip foundation. The pits are then filled with concrete or brickwork up to the underside of the existing foundation, after which the intervening legs are excavated and the concrete or brickwork constructed within them to bond on to the work already in place, so forming a continuous strip of underpinning at the required depth. The maximum length of wall which can be left unsupported above each leg is usually taken as 1.2–1.5 m for brick walls of normal construction. The unsupported lengths should be equally distributed over the length of the wall and in no circumstances should the sum of the unsupported lengths exceed one-quarter of the total length of the structure. Jordan[12.4] recommends that the legs should be dealt with in groups of six in the sequence 1, 2, 3, 4, 5, 6 (Fig. 12.9), the legs of the same number being excavated simultaneously.

If the wall is heavily loaded or shows signs of structural weakness, the unsupported length at any given point should not exceed one-fifth to one-sixth of its total length. Seymour[12.3] states that district surveyors in London will normally allow between one-sixth and one-eighth of a length of a wall to be underpinned in 1 m sections at any one time.

The concrete should be placed as quickly as possible after completion of excavation in each leg. If there is likely to be any delay in commencing concreting, the last 100–150 mm of soil should be left and removed only when the concreting is ready to start. Alternatively, a layer of sealing concrete can be placed at the bottom.

Concrete underpinning should be carried up to within 50–100 mm of the underside of the existing foundation, then left to set and shrink before fine dry concrete or mortar is rammed in to make full contact with the old and new work. Chases or pockets formed from expanded polystyrene should be left in the vertical stop ends of the legs to bond in with the concrete of the adjacent legs. Horizontal chases can conveniently be formed by the walings supporting the poling boards lining the excavation. Additional bonding can also be provided in the form of short lengths of reinforcing steel.

Brickwork can be used to fill the spacing between the top of the underpinning concrete and the underside of the existing foundation. The bricks should be bedded in cement mortar and well rammed into position. Under-

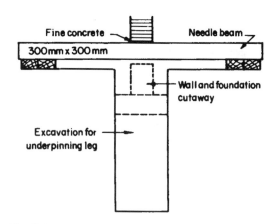

Fig. 12.10   Supporting wall by needling.

pinning in wide foundations should be undertaken in steps working from back to front.

Excavation for intermediate legs will throw additional load on to legs already in place. Therefore, to avoid lateral flow of a silty or clayey soil from beneath heavily loaded first series legs, the excavation for the second series legs should not be taken as deep as that for the first series. It should also be noted that by excavating from one side only of the wall to be underpinned, as shown in Fig. 12.9, followed by placing concrete over the whole base area of the pit, the existing wall will then impose its load eccentrically to the central axis of the new foundation. The effect of this eccentricity should be considered in relation to the allowable bearing pressures (see Section 4.1.3).

Sometimes as a final stage in underpinning, pressure grouting of any remaining voids between the old and new work is carried out. This gives some measure of prestressing the ground beneath the new foundation. Pressure grouting is not normally used for ordinary walls, but it may be advantageous for wide foundations or irregular shapes where it is difficult to ensure thorough packing of the underpinning concrete or brickwork beneath the structure. Generally, dry packing gives a sounder job than grouting.

If it is desired to excavate beneath brick walls to lengths greater than 1.2–1.5 mm or if the load distribution is not uniform, it may be necessary to give direct support to the walls by needle beams (Fig. 12.10).

12.3.4 Underpinning with piers or bases

If a building has a framed structure with the wall loading transferred to the columns, it is convenient to underpin the columns in individual pits beneath each column foundation. The columns or column bases are shored up or supported by needles as described in Section 12.2, while the pit is dug to the required level and backfilled with the

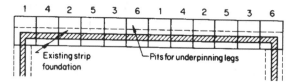

Fig. 12.9   Sequence of excavating underpinning legs for continuous strip foundation.

underpinning material. It may be convenient to break out the bases of steel columns and construct entirely new ones.

Pier foundations may be a desirable method of underpinning walls where deep excavations are required in difficult ground, for example water-bearing or bouldery soils. This method should only be considered where there is a stratum of good bearing capacity at a reasonable depth, because heavy loads will have to be carried by the piers. It is necessary to provide a beam beneath the wall to transfer the load to the piers. The beam can be constructed in a number of ways; these include the following:

(a) Supporting the wall by needles for the full length between piers and inserting steel or precast concrete beams (Fig. 12.11(a)).
(b) Inserting steel beams into chases cut into both sides of the wall foundation (Fig. 12.11(b)).
(c) Inserting precast concrete blocks or stools into pits excavated beneath the walls in the manner described for underpinning wall foundations by continuous strips. The blocks are provided with longitudinal holes which are lined up to form continuous ducts along the length of the beam. Mild or high-tensile steel rods or wires are threaded through the ducts.

Either a normal reinforced concrete beam can then be formed by concreting around the blocks or stools, or the high-tensile steel rods or wires can be tensioned and grouted up to form a prestressed concrete beam spanning between the piers (Fig. 12.11(c)).

(d) Providing tie rods or prestressing cables at the sides of or in chases cut in the sides of the wall foundation. These are tensioned to produce a prestressed beam from the wall structure (Fig. 12.11(d)). This method would not be contemplated if the existing wall and foundation were in poor condition.

After completion of the beam by one or another of the above methods, the needles are carefully removed to transfer the weight of the wall to the piers. If the existing structure is sensitive to settlement it may be desirable to provide jacks at the ends of the beams. These enable the wall to be lifted clear of the needles while the piers and the ground beneath them take up their immediate settlement.

The Pynford system[12.5] of underpinning with beams comprises as a first step cutting pockets in the existing wall to take precast concrete stools. These stools are used to underpin load-bearing walls carrying up to 60 mkN/m run. The stools are about 75 mmm deeper than the depth of

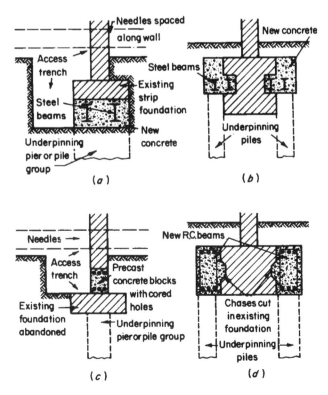

Fig. 12.11 Methods of underpinning wall foundations using beams spanning between piers or piles.

Fig. 12.12   Underpinning with Pynford stools.

the beam to allow for concreting and pinning up. The intermediate brickwork is then cut away leaving the building supported entirely by the stools, but bearing on its old foundation. Reinforcing steel is then threaded through the spaces provided in the stool and the beam is concreted (Fig. 12.12). This stiffens the foundations, enabling excavations to be made beneath the beams for individual piers or bases taken down to the desired level for underpinning. Finally the load is transferred to the piers or bases to complete the underpinning.

Piers or bases are constructed in concrete or brickwork in timbered shafts. Alternatively, cylindrical shafts can be sunk with cast iron or precast concrete segmental lining (see Section 4.6.3) backfilled with concrete. In difficult ground compressed-air caissons have been used for underpinning, but such methods would only be considered if underpinning by piles were impractical due to the site conditions.

12.3.5 Underpinning by piles

Underpinning by piles is similar in principle to the method of piers described above. It is a convenient method to use if the bearing stratum is too deep for economical excavation in shafts or if the ground conditions are difficult for hand excavation.

Where walls are underpinned by piles, the piles can be placed at close centres when they are relatively short. If, however, the piles must be taken down to a considerable depth to reach a satisfactory bearing stratum it will be more economical to space them widely with a corresponding increase in working load. Where the piles are closely spaced, the wall and its foundation, if in good condition, can be relied upon to span between the piles. For widely spaced piles a beam will have to be provided by one of the methods described in Section 8.16.2.

Underpinning piles to walls are normally provided in pairs, one on each side of the wall (Fig. 12.13(a)). Piles underpinning columns are placed in groups around or at the sides of the column (Fig. 12.13(b)). If, due to access conditions, it is impractical to install piles within a building it is necessary to provide them in pairs on the outside with a cantilevered capping beam (Fig. 12.13(c)). This arrangement puts a heavy compression load and bending moment on the inner (fulcrum) pile and tension on the outer pile.

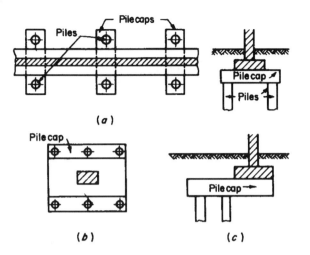

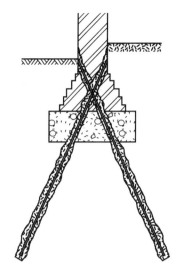

Fig. 12.13 Underpinning with piles. (*a*) Strip foundations. (*b*) Column base. (*c*) Cantilevered pile cap.

Fig. 12.14 Underpinning old wall with micropiles.

Bored piles (as described in Section 8.14) are generally used since they cause little or no noise and vibration and the piling rig can be operated in conditions of low headroom. In water-bearing sandy soils a careful boring technique is required to prevent sand being drawn into the borehole from the surrounding ground. It is not desirable to pump from the pile shaft before concreting, as this may cause settlement due to 'blowing' at the bottom of the pile borehole caused by external water pressure. Uncased augered holes are suitable for underpinning in stiff clays, for example where settlement has occurred due to deep-seated soil shrinkage resulting from the drying action of tree roots.

If it is necessary to transfer load to the pile at a very early age it may be desirable to leave the casing in the borehole or to use universal beams or precast concrete columns inserted in prebored holes. Steel tube or universal beam piles are useful where bending moments or lateral loads are carried, for example where underpinning piles are used as retaining walls for basements or where they are required to carry cantilevered loads at an early age.

Minipiles or micropiles (Section 8.14.8) are a useful means of underpinning, particularly where the existing footing walls are of massive construction. In these conditions the holes for the underpinning piles can be drilled through the existing foundations and the reinforcing bars and infilling concrete is bonded to them (Fig. 12.14). The system has the advantage that the work can be carried out from the ground surface and deep excavations to form pile capping beams are not needed. One proprietary system of micropiling for underpinning work is the 'Pali Radice' (root pile) system described with many case histories by Lizzi[12.6]. The equipment for installing the piles is suitable for working in confined spaces and the low-frequency vibrations of the drilling machine need not be damaging. The rotary drilling process results in much less loss of ground than conventional drilling by cable percussion methods.

Jacked piles Piles can be installed directly beneath foundations if jacked types are used, when the weight of the building provides the reaction for the jacks. The Franki Miga system employs a proprietary form of jacked pile which consists of a number of 305 mm × 305 mm × 762 mm long precast concrete units with a 50 mm diameter hole running down the centre (Fig. 12.15). The piles are installed by first excavating a pit beneath the foundation. The bottom pile section, which has a pointed end, is placed in the pit and a hydraulic jack, with steel packing plates and short steel beam sections to spread the load, is used to force the pile into the ground until it is nearly flush with the ground at the bottom of the pit. The jack is removed, the next section is added, and the process repeated until the desired pile-carrying capacity is reached. Adjacent units are bonded together by grouting short lengths of steel tube into the central hole at each stage. When the pressure gauge on the jack indicates that the required pile-carrying capacity has been reached (i.e. working load plus a safety factor) short lengths of steel beam or rail are driven hard between the head of the pile and the existing foundation. The jack is then removed and the head of the pile and packings are solidly concreted.

In the USA it is the usual practice to jack down pipe piles in short sections in a similar manner to the Franki Miga pile, except that a pair of jacks is used as shown in

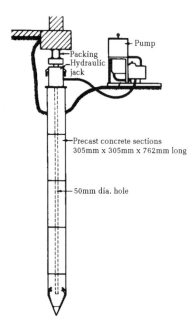

Fig. 12.15 Underpinning with the Franki Miga pile.

Fig. 12.16. The pipe piles are usually installed with open ends and the soil removed from time to time to facilitate the entry of the pipe sections. If this is not done a plug of soil tends to consolidate at the bottom of the pile, which greatly increases the required jacking force. On reaching the required level as indicated by the jacking force, the pipes are finally cleaned out and filled with concrete. The space between the top of the pipe and the existing foundation is filled with a short steel column and packing plates after which the jacks are removed.

Where the soil below or close to the existing founda-

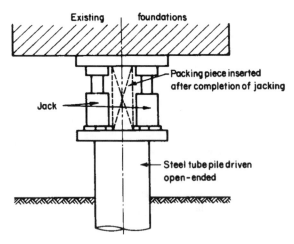

Fig. 12.16 Underpinning with steel tube piles.

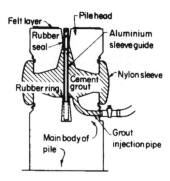

Fig. 12.17 Underpinning pile with adjustable head.

tions is soft it may be possible to jack down the pipe with a closed end to reach a firm stratum. If the end is closed by a shoe with a diameter larger than the pipe the jacking force is reduced. On reaching founding level grout is introduced under pressure through a hole in the bottom of the pipe wall. This fills the annulus, thus increasing the effective diameter of the underpinning member and increasing the skin friction resistance.

When installing jacked piles it is the normal practice to work to a safety factor of 1.5, i.e. the jacking force is equal to the calculated working load plus 50 per cent. The final jacking load is maintained for a period of at least 12 hours before the packing is inserted. The adoption of a higher safety factor may lead to excessive loads on the existing structure, although additional weights may be added to increase the jacking reaction.

It is important to insert the packing between the pile and the structure *before* releasing the load on the jacks. In this way elastic rebound of the pile is prevented and the settlement minimized. A method of forcing up the head of a pile into contact with a structure above it is shown in Fig. 12.17. This shows the device as used for piles supporting tunnel sections beneath the River Maas at Rotterdam. The head of the pile is connected to the lower part only by a peripheral nylon sleeve. Cement grout is injected under pressure into the space between the two components, thus lifting the pile head and forcing it into contact with the superstructure.

12.3.6 'Pretest' methods of underpinning

Mention has been made in the preceding pages of the use of jacks in underpinning work. By jacking between the existing structure and the new underpinning, the underlying ground is preloaded before the load of the structure is finally transferred to the underpinning. This method has been given the name 'pretest' by its originators, Messrs Spencer White and Prentis of New York, although several proprietary underpinning systems involve the use of jacks. The application of the pretest method to pile underpinning has already been described. The 'legs' are brought up to the required level and, after hardening, jacks are installed and a series of beams are placed on top of the jacks. A layer of concrete is then deposited on top of the beams. Before this sets the beams are jacked up, thus bringing the concrete into close contact with the underside of the existing foundations. The concrete is then allowed to harden, after which the loads on the jacks are increased to the desired pretest value. Brickwork or steel pinning is then built up on both sides of the jacks to enable them to be released and removed. The spaces occupied by the jacks are filled with further pinning or, if continuing movements of the underlying strata are expected, they can be left open to serve as jacking pockets. The jacks are reinstalled in the pockets from time to time, as required to correct further settlement.

The time required to install jacks and underpinning beams can be shortened, if required, by the use of proprietary brands of rapid-hardening cement for the concrete work.

The required jacking loads should be carefully calculated, due allowance being made for the number of legs which are open at any given time. It may be necessary to adjust the jacking loads as the work proceeds, and extensometers or strain gauges have been used as an accurate means of controlling the jacking operations.

There are various patented systems of jacking which involve interconnections of the jacks in conjunction with a centralized pumping plant and a hydraulic accumulator. The Pynford Pedatifid system[12.7], referred to in Section 3.5.5 in connection with mining subsidence, is controlled by water-level gauges at various jacking points around the building. These gauges actuate electrical relays which control the motor-operated jacks.

Hydraulic jacks used in ordinary underpinning work should preferably have screwed rams fitted with collars. The collars are kept screwed down against the cylinders so that in the event of failure in the hydraulic pressure the ram will not fall. The Freyssinet 'flat jack', a hollow light-gauge metal canister, has useful applications to underpinning work. Its shallow depth enables it to be inserted into small spaces. The jack is designed to be expendable. Air introduced under pressure expands the thin metal box, which jacks up the structure to the desired amount. Cement grout is then introduced and the injector-pressure pipe is sealed off.

Freyssinet flat jacks were used for major underpinning work to the central tower of York Minster, as described by Dowrick and Beckmann[12.8]. This tower is mainly supported on four main piers each carrying a vertical load of about 38 500 kN. These piers were supported by the remains of the walls of an earlier cathedral, which were in turn bearing on strip foundations of Norman age underlain by a stiff to hard sandy silty clay and in places by even deeper foundations of Roman origin. The old masonry foundations imposed rather high and eccentric loading on the soil, and the underpinning scheme was essentially one which spread the load over a wider foundation area. This was achieved by constructing a large reinforced concrete spread foundation about 14.5 m square under each of the four main piers. This foundation surrounded the old masonry footing walls, and four layers of 32 mm stainless steel prestressing rods passing underneath the pier unified the cracked masonry and surrounding concrete (Fig. 12.18). After the whole foundation had been prestressed, flat jacks were introduced between this combined foundation and the new concrete underpinning below and the jacks were inflated to achieve uniform bearing pressure on the underside of the combined foundation.

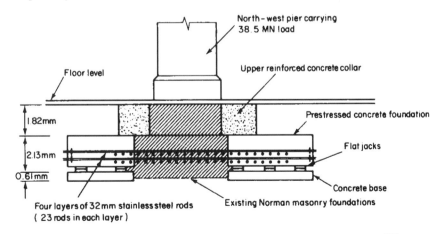

Fig. 12.18 Underpinning the foundations of the central tower of York Minster (after Dowrick and Beckmann[12.8]).

12.3.7 Underpinning by injections

Injections of the ground with cement or chemicals to fill voids or to permeate and strengthen the ground are sometimes used as a means of underpinning. The work may be done wholly by injections or the ground treatment may be used as a means of temporary strengthening while excavating for normal underpinning legs or shafts. Cement grouting is a useful expedient to fill voids in the ground beneath foundations which have been caused by erosion or by vibration effects in loose granular soils. It is also a useful means of strengthening old rubble masonry foundation walls before excavating beneath them for normal underpinning operations. Cement grouting can be used to consolidate open-textured gravel soils beneath machinery foundations in cases where settlement is occurring due to vibration effects.

Chemicals can be used for injection into coarse sands or sandy gravels to produce a wall or block of consolidated ground beneath the foundations to the desired level for underpinning. In favourable ground conditions this is a useful method of underpinning in connection with deep excavations close to existing structures. The injections are made from ground level, thus avoiding the necessity of

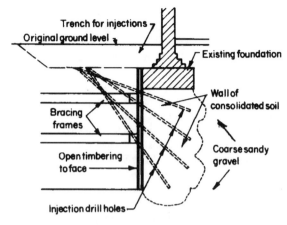

Fig. 12.19   Underpinning by chemical injections.

shoring or needling, and the wall of consolidated ground acts as a retaining wall when excavating close to the existing foundations (Fig. 12.19). The consolidated ground is supported where necessary by bracing frames, but the need for sheet piling or close timbering is often eliminated. An example of this is shown in Fig. 12.20. The

(a)

(b)

Fig. 12.20 Use of chemical injections to facilitate underpinning of an old building at Kirkcaldy, Fifeshire. (*a*) General view of underpinning work showing wall supported by tubular steel raking shores. (*b*) Excavating in fine sand stabilized by chemical injections.

range of soil types in which chemical consolidation can be used is given in Fig. 11.14 and it is also described in detail by Littlejohn[12.9] who gives examples of the application of grouting to underpinning operations.

It was noted in Section 11.6.7 that cement or chemical grouts cannot be induced to permeate clays or clayey silts, but that the technique of hydrofracture can be used to cause uplift of a mass of clay or silt, thus providing a means of raising a structure. A notable example of this compensation grouting technique is its use in combination with other techniques to control ground movements beneath three structures at Waterloo Station during tunnelling for an escalator shaft.[12.10] The inclined shaft passed beneath the Victory Arch entrance to the station, and the twin tunnels of the Waterloo and City lines (Fig. 12.21(*a*)). The Victory Arch is founded on Thames Gravel where chemical grouting was used to underpin the structure by injections through vertical and inclined holes at 1.5 m spacings. This was followed by compaction grouting of the gravels using a cement paste to form bulbs around the close-spaced injection points drilled from the

cellars below the arch. Regrouting of the same points was made possible by the stiff paste which was cracked by high pressure water after completing each injection.

The Waterloo–City line tunnels were also founded on a thin layer of Thames Gravel underlain by London Clay. Hydrofracture was induced at the gravel–clay interface by cement injections through *tubes-à-manchette* set at 1.5 m centres below the two outer and the central walls of the brick tunnels. The grouting tubes were 'primed' by injecting grout before tunnelling commenced so that the treatment made to correct ground movements during tunnelling would have immediate effect. The key to the success of these techniques in preventing extensive settlement of the structures was the extensive instrumentation to monitor the ground movements. Electrolevels were installed at 3 m centres along three 50 m long beams fixed to the tunnel walls. Water gauges were set across the tunnels to measure differential settlements between the walls. Other instrumentation included 34 vertical extensometers through the tunnel invert along the centre of the rail tracks, and an array of electrolevels set in boreholes at 1 m above

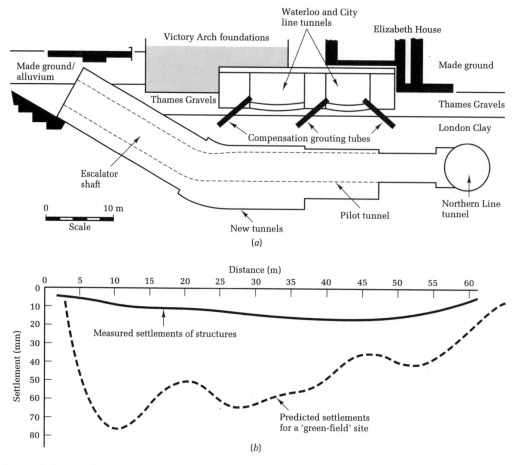

Fig. 12.21   Section through escalator tunnel at Waterloo Station. (*a*) Showing location of grouting tubes. (*b*) Measured and predicted settlements.

the crown of the tunnel to detect early ground movements. Monitoring was undertaken throughout the three stages of driving a 4 m pilot tunnel, enlarging it to the full size of 7.5–8.25 m diameter, and the opening of cross passages. Although the tunnelling continued throughout the 24 hours the compensation grouting to correct the resulting settlements could be undertaken only at period of night-time possession. Levels taken at a point on the Victory Arch are shown in Fig. 12.22 indicating an end of construction settlement of 10 mm. The measured settlements along the length of the new tunnels are compared with the predicted ground movements for an open site without treatment in Fig. 12.21(*b*).

The use of jet grouting for underpinning was mentioned in Section 11.3.7. Jet grouting was used in conjunction with micropiles to underpin old arched cellars before excavating for a 16.1 m deep underground car park in Milan.[12.11] Direct support was given to the cellar foundations by three angled micropiles and consolidation grout-

ing of the fluvial sands and gravels. Then upper and lower stages of 800 mm angled jet grout columns were formed as the excavation was taken down (Fig. 12.23). The columns were reinforced by 73 mm steel tubes pushed in before the grout had set. The final stage was forming the outer basement wall by 20 m deep vertical micropiles consisting of 139 mm reinforcement tubes in 200 mm diameter drilled holes. Because of the flexibility of the wall it was necessary to install ground anchors at five levels as the excavation was taken down.

It is evident from the description of the jet grouting process in Section 11.3.7 that it must be undertaken with great care when used as an underpinning technique, to avoid the damaging effects of ground heave.

12.3.8 The use of the freezing process in underpinning work

The applications of the freezing process to excavation

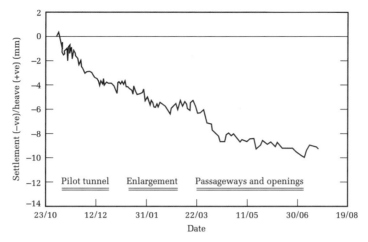

Fig. 12.22   Settlement–time curve for measuring point on Victory Arch, Waterloo.

work has been described in Section 11.3.9, and examples of its application to underpinning have been described by Harris.[11.14] Freezing has been used to solidify ground which was consolidating beneath the foundation loading, thus arresting further settlement and enabling permanent underpinning to be carried out. Dumont-Villares[12.12] has described the underpinning measures taken when a 26-storey building at São Paolo, Brazil, settled and tilted in

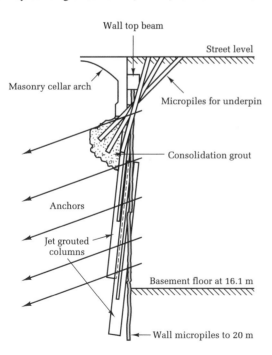

Fig. 12.23   Underpinning cellar by micropiles and jet grouting, Milan.

1941. Although the building was founded on piles, later investigation showed that the piles were underlain by a wedge of soft silty clay. The wedge was thickest at one corner of the building where the total settlement was nearly 0.3 m with a differential settlement (tilting) almost as great. The freezing was restricted to an area around the corner of the building where maximum settlement occurred, with the object of forming a solid block of ground around the piles. This arrested the settlement in a period of about 3 months from the commencement of freezing.

A number of pits were then sunk through the frozen ground to a stratum of sand below the soft clay. Pier foundations were constructed in these pits and the building was restored to level by jacking from the tops of the piers. The existing piles had to be cut free where necessary as the building was raised. Selected piles were restored to use by inserting precast concrete sections (Franki–Miga–Hume piles) between the old piles and the underside of the building. The precast sections were jacked down to the required pretest load. No further settlement was recorded over a 10-year period following the completion of the underpinning work.

12.4 Moving buildings

Underpinning methods are normally used in connection with moving buildings (without demolition) from one site to another. The weight of the building is transferred to a system of beams placed on both sides of the walls and wheeled carriages or rollers are installed beneath the beams running on rail tracks laid in the desired direction. Changes in direction of movement along the rails can be achieved by jacking up the carriages in turn and rotating them in another direction to run on a new set of tracks.

These operations are usually undertaken by specialist firms who adopt a careful technique based on a thorough survey of the building, accurate calculation of loads, a detailed soil investigation along the line of the tracks and at the new site of the building, and controlled jacking in conjunction with strain-measuring and equalizing devices.

When buildings with load-bearing walls are moved, holes are cut in the walls and needles are inserted at the appropriate spacings over which the brickwork can safely span. Longitudinal beams are then laid on each side of the wall and jacked up against the needles, shims and wedges being added as necessary to ensure even distribution of the load. Where necessary individual pairs of jacks can be provided between each needle and the beams. The carriages and rollers are then installed beneath the longitudinal beams, or a second set of girders can be provided beneath the wall beams in order to reduce the number of carriages and hence the number of tracks.

Fig. 12.24(*a*) shows an old building supported at first floor level by steel beams bearing on a chassis fabricated from 300 mm × 300 mm timbers. The building was moved by Pynford Ltd on this chassis to a new location further along the street.[12.13] On reaching this location four pairs of steel 'A'-frame jacks were set up beneath the steel beams (Fig. 12.24(*b*)). Operation of the jacks raised the building by 1.2 m and moved it 0.75 m laterally on to the new foundations.

The system of support used by Christiani and Nielsen for moving houses near Paris has been described by Olsen[12.14] and is illustrated in Fig. 12.25. Three pairs of steel beams were placed alongside the longitudinal walls and short needles were fixed to their lower flanges at two intermediate points on these walls. Two pairs of main cross girders were then laid over the longitudinal beams, one pair to each end wall. Wheeled carriages were placed at four points, i.e. at the ends of the outer cross girders as shown in Fig. 12.25. The pairs of longitudinal girders were 'preflexed' by small hydraulic jacks placed at six points above the short needle beams on the longitudinal

(*a*)   (*b*)

Fig. 12.24 Moving an old building at Hereford. (*a*) Building supported on a temporary timber chassis before moving to new location. (*b*) Building being jacked into its final position.

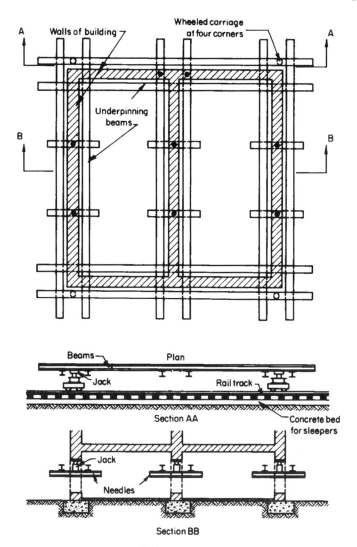

Fig. 12.25 Moving a house on four carriages (after Olsen[12.14]).

walls and at four points on the end walls. When the total load on these small jacks equalled the total load of the building, the jacks were locked and the load was transferred through the cross girders to the four main jacks and through them to the carriages. Two of the main jacks were coupled together to enable them to act as one unit. Thus the system was on a statically determinate three-point support. If one of the two jacks sank due to unevenness or settlement of the track the pressure in the cylinder fell, causing oil to flow from the cylinder to the other jack. Thus the first jack was raised again while the other fell by a corresponding amount to equalize the movement of the pair of supports.

The thrust required to move structures along the tracks

is usually provided by long-stroke hydraulic jacks. The cylinders are fixed to the rails and the rams operate against the carriages. Jacking is preferable to winching, since the stretch and snatch of the cables of the latter method are liable to set up a dangerous jerky motion.

Prentis and White[12.1] state that the propulsive force to start the movement need only be 1–2 per cent of the weight of the structure, using steel rollers on a well-laid track bearing on a good foundation. The force to keep the building moving may only be about two-thirds of the force required to start the movement.

Framed buildings are moved by needling and propping individual columns, and installing roller carriages either under each column or under a combined support to a row

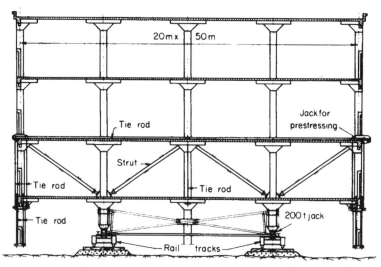

Fig. 12.26   Moving a four-storey warehouse (after Olsen[12.14]).

of columns. Olsen[12.14] has described the installation of temporary struts and ties in the lower storey of a four-storey warehouse to enable the loading of 55 columns to be transferred to 22 main jacking points in two rows, as shown in Fig. 12.26. The 22 jacks were provided with equalizing connections forming three independent groups, so giving the desirable three-point support previously mentioned. The warehouse, which measured 100 m × 20 m, was split into two sections. Each section of 2500 t mass was moved half a mile – including the crossing of a road and a railway. The propulsion was by six 50 t jacks.

The 'air-cushion' principle as employed in hovercraft can be used for moving structures, as described by Pryke.[12.13] Flexible skirts are fitted around the periphery of the structure and air is blown under pressure beneath the base to raise the structure which can then be rotated and towed to a new location. This method is best suited to structures having a level base such as petroleum storage tanks. Complications can arise due to loss of air from beneath the skirts or through the substructure if attempts are made to raise irregularly shaped or permeable foundations. Care is also needed to obtain a fairly smooth non-erodible track on which to move the structure on its cushion of air.

References

12.1   Prentis, E. A. and White, L., *Underpinning*, 2nd edn., Columbia University Press, New York, 1950.

12.2   Thorburn, S. and Hutchison, J. F. (eds), *Underpinning*, Surrey University Press, 1985, 296 pp.

12.3   Seymour, A. J., The influence of party walls on site redevelopment in Central London, *The Structural Engineer*, **71**(17), 299–304 (1993).

12.4   Jordan, S. K., Foundations for basement buildings adjoining existing property, *Journal of the Institution of Civil Engineers*, **15**, 119–140 (1940).

12.5   Pryke, J. F. S., The Pynford underpinning method, in *Underpinning*, Surrey University Press, 1985, pp. 162–221.

12.6   Lizzi, F., Pali radice (root piles) and reticulated pali radice, micropiling, in *Underpinning*, Surrey University Press, 1985, pp. 84–161.

12.7   Pryke, J. F. S., Eliminating the effects of subsidence, *Colliery Engineering*, **31**(37), 501–507 (1954).

12.8   Dowrick, D. J. and Beckmann, P., York Minster structural restoration, *Proceedings of the Institution of Civil Engineers*, Suppl. paper no. 74155, pp. 93–156 (1971).

12.9   Littlejohn, G. S., Underpinning by chemical grouting, in *Underpinning*, Surrey University Press, 1985, pp. 242–275.

12.10   Wheeler, P., Waterloo compensation, *Ground Engineering*, **26**(7), 14–16 (1993).

12.11   Anon., Wall change, *Ground Engineering*, **25**(8), 14–15 (1992).

12.12   Dumont-Villares, A., The underpinning of the 26–storey 'Companhia Paulista de Seguros' Building, São Paolo, Brazil, *Géotechnique*, **6**, 1–14 (1956).

12.13   Pryke, J. F. S., Moving structures, *The Consulting Engineer*, **31**(9), 85–89 (1967).

12.14   Olsen, K. A., The re-siting of structures, in *Proceedings of the Institution of Structural Engineers' 50th Anniversary Conference*, 1958, pp. 364–371.

13 Protection of Foundation Structures against Attack by Soils and Ground Water

13.1 Causes of attack

Foundations are subject to attack by destructive compounds in the soil or ground water, by living organisms, and by mechanical abrasion or erosion. Thus, timber piles in jetties are attacked by organisms in the soil and water causing decay of the timber; they suffer from the depredations of termites and the insect-like marine 'borers'; they are abraded by ships, ice, or other floating objects and they may suffer severe damage by the movement of shingle if they are sited on beaches exposed to wave action. Concrete in foundations may have to withstand attack by sulphates in the ground or in chemical wastes. Steel piles can be subject to corrosion in certain conditions. The severity of attack on foundations depends on the concentration of the aggressive compounds, the level of and fluctuations in the ground-water table and the climatic conditions. Immunity against deterioration can be given to a varying degree by protective measures. Some of these may be very costly and the engineer may have to seek a compromise between complete protection over the working life of the structure and partial protection at a lower cost, but with the added expense of periodical repairs and renewals.

13.2 Soil and ground-water investigations

13.2.1 Methods of investigation

In considering protective measures the first step is to undertake a detailed investigation of the soil and ground-water conditions. The latter are of particular importance in evaluating the risk of attack on timber, concrete, or steel structures. Generally the required data can be obtained in conjunction with the general investigations for foundation design using the methods described in Chapter 1. Samples of the ground water can be taken for chemical analysis, and portions of the disturbed and undisturbed soil samples can be set aside for this purpose.

If necessary stand-pipes should be left in boreholes for long-term measurements of the ground-water table. It is important to establish the highest level to which the ground water can rise, either from the stand-pipe recordings or the results of local inquiry. It is essential to make a sufficiently large number of tests to determine the average and range of the sulphate content, and in particular the variations in sulphate content with depth. Money can be wasted if expensive precautions are taken in foundation concrete based on the results of only two or three sulphate content tests which may give values unrepresentative of the average conditions. Also there is no need to extend the protection below the sulphate-bearing zones, unless sub-soil drainage causes aggressive ground water to be drawn down to lower levels. The possibility should be borne in mind of marked changes in the ground-water level occurring at some time in the future, say as a result of drainage or irrigation schemes.

In considering sulphate attack on concrete foundations it is usually sufficient to determine the sulphate content and pH value of the soil and ground water. If the sulphate content is found to be more than 0.24 per cent it is necessary to determine the concentrations of the relevant cations.[13.1] A full chemical analysis is required where the soil or ground water is contaminated with chemical wastes in order to identify compounds potentially aggressive to concrete.

13.2.2 Sulphates in soils

In Great Britain sulphates occurring naturally in soils are generally confined to the Keuper Marl, the Lias, Oxford,

Kimmeridge, Weald, Gault, and London Clays. The superficial deposits or drift are generally free of sulphates except where they are in close proximity to sulphate-bearing soils. Some peats have a high sulphate content. Sulphates occur in Europe, the USA, and in the semi-arid gypsiferous soils of the Middle East.

Because of wide variations in the sulphate content of soils, the best indication of possible aggressive conditions is given by analysis of the ground water. The highest concentrations of sulphates in the ground water will occur towards the end of a long dry spell. If the sampling is done during heavy winter rains the concentration may be unrepresentative of the most severe conditions. In hilly ground where there is a flow of ground water down the slope across sulphate-bearing soils, the highest concentration of sulphates will be on the downhill side. Flow of ground water from sulphate-bearing soils will result in concentration of sulphates in soils which are not naturally sulphate-bearing.

13.2.3 Investigations for corrosion

For investigating the possibility of corrosion of buried steel structures the most rapid and economical method is to carry out an electrical resistivity survey of the site by the method described in BS 5930. The readings of the apparatus used give the conductivity of the soil which is a measure of the concentration of soluble salts in the soil and which in turn is a measure of its corrosive action. The presence in the soil of sulphate-reducing bacteria causes serious corrosion which commonly occurs in tidal mud flats contaminated by sewage or in ground containing organic refuse. In these conditions a bateriological analysis is required, for which purpose the soil must be sampled using the techniques described in Section 1.4.3, but with the added precaution of using sterilized unoiled tubes. The ends of the tubes must be waxed immediately on withdrawal from the ground.

Sea water should be sampled in tidal waters, estuaries, or the open sea at various stages of the tide, both at neap and spring tide periods, to determine the worst conditions of salinity and bacteria content. Again sterilized bottles should be used if bacteriological examination is thought necessary. A full chemical analysis together with determinations of pH value is required on the water samples.

13.3 Protection of timber piles

13.3.1 Timber decay in land structures

Nowadays the only timber in foundations is in piles or in bracing and fenders to marine structures. Biological decay of buried timber does not occur if the timber is kept wholly wet, but the decay may be severe if the timber is kept in a partly wet and partly dry (or moist) state, for example in the zone of a fluctuating water table for buried timber piles, or in the half-tide zone for piles in the sea or rivers. Timber which is properly air-seasoned will, if kept dry, be immune to biological decay.

Potter[13.2] has classified various grades of durability in terms of their approximate life when in contact with the ground (see Table 13.1). Using this table Potter has classified the durability of the timbers suggested as suitable for timber piles in Section 8.8 as follows:

Douglas fir	Moderately durable
Pitch pine	Durable
Douglas fir (larch)	Moderately durable
Western red cedar	Durable
European oak	Durable
Greenheart	Very durable
Jarrah	Very durable
Opepe	Very durable
Teak	Very durable

Table 13.1   Durability classification of the heartwood of untreated timbers

Grade of durability	Approximate life in ground contact (years)
Very durable	>25
Durable	15–25
Moderately durable	10–15
Non-durable	5–10
Perishable	<10

To ensure freedom from decay and infestation of timber, precautions must be taken from the time that the timber is felled. Dry rot fungus is present in forests due to the accumulation of dead wood, and it may also occur in heaps of sawdust and scrap wood in timber yards. Therefore, felled timber should be cleared from the forests as quickly as possible and stacked in the timber yards on firm well-drained elevated ground from which all vegetable soil has been removed. The ground beneath and around the stacks should be kept clear of weeds.

Timber should be stacked clear of the ground and placed so that air can circulate freely around all baulks and planks. It should season in the stacks until the moisture content of the outer layers is less than 30 per cent, when it is ready for treatment by creosote or other preservative.

Alternatively, if the timber is to be used in permanently waterlogged ground it can be stored wholly immersed in water.

13.3.2 Preservation by creosote

Impregnation by creosote under pressure is the most effective method of preserving timber in foundations against decay and it gives a degree of protection against termites and some marine borers. Creosote is preferable for foundation work to other preservatives such as water-soluble and solvent types. Softwoods can be completely penetrated by pressure treatment, but hardwoods such as Douglas fir can only be properly impregnated by incising them and subjecting them to long sustained pressure. Even with this treatment the heartwood cannot be fully penetrated. In piling work the absorption and penetration of the creosote is greatly improved by the use of round piles in which the outer sapwood is retained. Richardson[13.3] states that the heartwood is absorptive to creosote and with Scots pine or Baltic redwood an outer band of creosoted wood of up to 75 mm depth is readily obtained which will protect the piles for a very long time. He points out the disadvantage of using squared timber for piles (a common British practice) which results in most of the sapwood being cut away, thus exposing the heartwood which in practice rarely receives adequate creosote treatment. Because complete impregnation of hardwoods cannot be obtained, all bolt holes and incisions made by cant hooks, dogs or slings should subsequently be re-creosoted. Creosote should be poured by funnel down bolt holes, or better still they should be given pressure treatment by specially designed equipment. Particular attention should be given to the end-grain.

13.3.3 Protection by concrete

Because of the likelihood of severe decay of timber piles in the zone of a fluctuating water table, the creosote treatment is not likely to be effective in this zone over the life of the structure. If the water table is fairly shallow the piles can be cut off at the lowest water level and the pile cap taken down to this level. If the lowest water level is too deep for this to be economical it will be advisable to use composite piles, the portion permanently submerged being in timber and the upper portion in concrete similar to that illustrated in Fig. 8.37.

13.3.4 Action of marine borers

The burrowing of molluscan and crustacean organisms which inhabit saline or brackish waters is responsible for the most severe damage of piles and fendering in marine and river work. An illustrated account of the species of these organisms with photographs of their depredations and an account of protective methods is given by the Building Research Establishment.[13.4] They list the principal species as follows:

Molluscan borers	Crustacean borers
Teredo ('shipworm')	*Limnoria* ('gribble' or
Bankia	'sea-louse')
Xylophaga dorsalis	*Chelura*
Martesia (in tropical waters only)	*Sphaeroma*

The molluscan borers enter through minute holes when young and grow to a size of up to 25 mm in diameter and 1 m or so long, destroying the wood as they grow. An example of *Teredo* attack is shown in Fig. 13.1. The crustaceans are mainly surface workers and form a net-

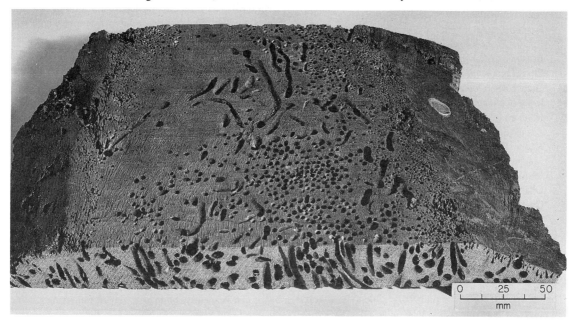

Fig. 13.1 *Teredo* attack on timber. (Crown copyright reproduced with the permission of BRE, Princes Risborough Laboratory.)

work of branching and interlacing holes. The borers are found throughout the oceans of the world. Salinity, temperature, current action, depth of water, pollution, pH value, dissolved oxygen, and sulphuretted hydrogen, all affect the presence or absence of borers. Usually saline or brackish water with salinities of more than 15 parts per 1000 (sea water normally has a salinity of 30–35 parts per 1000) is essential to their survival, but *Sphaeroma* are found in almost fresh tropical waters in South America, South Africa, India, Ceylon, New Zealand, and Australia.

The destructive action of marine borers can be extremely rapid. Timber jetties have been destroyed after only a few months of exposure, therefore some protection is necessary in all cases where timber piles are driven in salt or brackish water. Such protection should only be omitted if there is strong evidence that borers do not exist in a particular locality. The investigation should cover an examination of all timber structures and driftwood and advice should be sought from a marine biologist.

13.3.5 Protection against marine borers

The best protection is to use a timber which is known to be resistant to borers.

The Building Research Establishment[13.4] lists the following timbers as having heartwood resistant to borer attack. Those marked with an asterisk are believed to be the best for marine work.

> Afrormosia (*Pericopsis elata*)
> African padauk (*Pterocarpus soyauxii*)
> Andaman padauk (*Pterocarpus dalbergioides*)
> *Basralocus (Angelique) (*Dicorynia guianensis*)
> *Belian (*Eusideroxylyn zwageri*)
> *Brush box (*Tristania conferta*)
> *Ekki (*Lophira alata*)
> *Greenheart (*Ocotea rodiaei*)
> Iroko (*Chlorophora excelsa*)
> Ironbark (*Eucalyptus siderophloia*)
> Jarrah (*Eucalyptus marginata*)
> Kapur (*Dryobalanops* spp.)
> *Manbarklak (*Eschweilera longipes*)
> Muninga (*Pterocarpus angolensis*)
> *Okan (*Cyclicadiscus gabunensis*)
> *Opepe (*Nauclea diderrichii*)
> *Pyinkado (*Xylia xylocarpa*)
> *Red louro (*Ocotea rubra*)
> *Southern blue gum (*Eucalyptus globulus*)
> Teak (*Tectona grandis*)
> *Turpentine (*Syncarpia laurifolia*)

The sapwood of the above timbers is liable to be attacked by borers. If it is impossible to remove all sapwood the timber should be treated with creosote as a precautionary

measure. Greenheart fenders in Milford Haven were attacked in the sapwood by *Teredo*, causing about 10 mm of damage in 5 years.

The *21st Report of the Sea Action Committee* of the Institution of Civil Engineers[13.5] stated that no species of timber is absolutely free from borer attack, but certain species are highly resistant and in many conditions of exposure they can be considered to have practical immunity. The report listed the more resistant species as greenheart, pynkadou, turpentine, totara, and jarrah. These timbers would be expected to have a life of many years in British waters. In commenting on the suitability of various types of preservative, the report concludes that ordinary coal tar creosote is the most satisfactory and states that in British waters any timber which is *efficiently* impregnated with creosote should be practically immune to borer attack.

The efficacy of preservative treatments depends on their complete continuity over the outer layer of the timber. If the treated shell is perforated by lifting hooks or by saw cuts or bolt holes, the borers can make an entry. Even though such cuts or abrasions are treated before the piles are finally driven, there is still the possibility of damage by floating objects or by the erosive action of shingle.

Some hardwoods cannot be impregnated with creosote, and in these cases protection can be given by jacketing the piles with concrete. Chellis[13.6] describes several methods of jacketing including placing concrete inside movable steel forms or precast concrete shells. The use of gunite (sand–cement mortar sprayed on to the timber) is also mentioned. The essential practice in any form of concrete jacketing is to provide a dense fairly rich mix which will not craze or crack.

Chellis states that metal sleeving is not used to any great extent because of the impossibility of finding a corrosion-resistant metal at an economical price and difficulties at joints and bracing connections.

13.4 Protection of steel piling against corrosion

13.4.1 Wastage due to corrosion

Corrosion of iron or steel in an electrolyte (for example, soil or water) is an electrochemical phenomenon with different areas of the metal surface or structure acting as an anode and a cathode in cells. Pitting will occur at the anodic areas, and rust will be the final product in the cathodic areas. Both air and water are essential to the occurrence of corrosion. The general wastage of bare structural steel in dry unpolluted climates is practically nil, but it can be nearly 1 mm per year in humid and saline conditions on tropical surf beaches. Pollution is an important factor in atmospheric corrosion, the rate of which is greatly accelerated by the presence of dust, acids (derived from smoke), and sodium chloride (from sea spray). The rate of corrosion in soils is a function of the electrical conductivity of the soil, but other factors, including the presence in the soil of sulphate-reducing bacteria, can greatly accelerate corrosion.

The relative aggression of water depends on the salinity, pollution if any, temperature, and oxygen content.

Morley and Bruce[13.7] quote the following rates of corrosion of bare steel in different environments:

Undisturbed soil, negligible
Fresh water, 0.05 mm/year
Normal sea water, 0.08 mm/year
Sea water splash zone, 0.1–0.25 mm/year
Industrial atmosphere in Britain, 0.1–0.2 mm/year

Corrosion rates in disturbed soil can be in the range of 0.01–0.08 mm/year due to the oxygen present. A protective coating can be applied to piles driven into the soil, but this gives only partial protection since the coating is liable to be stripped off by stones or other obstructions in the soil. However, if the higher range of 0.08 mm wastage per year is taken a steel pile with a web thickness of 20.6 mm will have a life of more than 60 years before 50 per cent of its thickness is lost (although there is likely to be some deeper local pitting). Since the stresses on the cross-sectional area of the steel are generally low there is unlikely to be failure of the pile due to over-stressing. Longer life can be given by increasing the grade of steel (e.g. grade 50 instead of grade 43) to allow for the higher stresses caused by corrosion losses. Romanoff[13.8] examined many steel piles which had been in the ground for periods of up to 40 years. The examination showed that where the piles had been driven into *undisturbed* soil, i.e. where no oxygen was present, the corrosion of the steel was so light that it had no significant effect on the life of the piling. This indicates that where piles are driven wholly in undisturbed soil no form of protection need be given in the portion below the ground surface. However, as already mentioned, there are certain factors, such as the presence of sulphate-reducing bacteria, which thrive in the absence of oxygen and which can greatly accelerate corrosion. In these cases additional protection, for example jacketing H-section piles, filling hollow piles with concrete, or applying cathodic protection, must be considered.

For piles projecting above the soil line, a good paint treatment will give adequate protection of the exposed portion, provided that the paintwork can be renewed from time to time. However, maintenance of paintwork is impossible below the water line in jetty structures, and the paintwork in this part of a pile is especially liable to damage by the action of waves, barnacles, and floating objects. Little[13.9] pointed out that unprotected steel piles in jetty structures can have a life of 17 years, but the best paint coatings last for only 7 years. Therefore additional protection, usually in the form of cathodic protection, is necessary below the 'splash zone' where a long life is required from the piles.

13.4.2 Protection by paint treatment

Protective measures in the form of paint coatings in the 'atmospheric zone' above the splash zone in marine structures should first consist of applying a sand- or grit-blasting treatment to obtain a white or near-white metal condition. For the paintwork Hedborg[13.10] recommended a zinc silicate priming coat on the clean metal surface to a dry film thickness of 50–75 μm, followed by top coats of vinyl or epoxy paints. Epoxy coal tar paints are usually employed for the latter and they are applied in three or more coats to obtain an overall film thickness (including the primer) of 175 μm. For less important structures the finishing coats can consist of hot coal tar or bituminous enamel paints heavily applied by brush or swab to give a finished coating thickness of 1.5–2.5 mm. The primer should be compatible with the type of paint used for the finishing coats.

Morley[13.11] recommended the following protective measures for marine structures:

Atmospheric and splash zone: Coal tar epoxide paint to 250 μm film thickness or preferably isocyanate cured coal tar epoxide paint to 400 μm film thickness to give estimated 10-year life.

Intertidal zone: Bare steel to nominal or increased thickness to allow for corrosion loss (because of uncertainty of driving depths it may be necessary to extend the paint treatment from the splash zone into the intertidal zone).

Continuously immersed zone: Bare steel or cathodic protection.

Underground zone: No protection necessary.

Paint coatings cannot give a long life in the splash zone and cathodic protection is ineffective in this region. Therefore, if a long life is required either the thickness of metal should be increased to allow for wastage by corrosion or high-tensile steel can be provided at mild steel working stresses. Alternatively, cover plates should be provided. The cover plates can either be in steel to the same specification as the piles, or a corrosion-resistant metal such as monel metal can be used.

13.4.3 Cathodic protection

The principle of cathodic protection is the utilization of

the characteristic electrochemical potential possessed by all metals. The behaviour of any two metals in the presence of an electrolyte (that is, in a cell) is governed by their relative electropotentials. For example, where zinc and iron are the electrodes in a cell, zinc, being higher in the electromotive series, will act as the anode. In this way a complete structure can be protected by connecting it electrically to anodes at suitable intervals. The current escapes to the soil or solution by way of the anode, making the whole structure cathodic and so preventing the escape of metallic ions from the structure to the soil or solution. In time this action causes polarization of the surface of the structure which prevents rusting and thus provides an additional benefit.

Cathodic protection can be applied either by using sacrificial anodes or by a power-supplied system. Sacrificial anodes consist of large masses of metal which corrode away in the course of their protective action and have to be replaced from time to time. They must be higher in the electromotive series than the structure being protected. Magnesium anodes are generally used for the protection of steel structures. Sacrificial anode systems are often preferred for marine structures since they do not require cables which can be damaged by vessels or by objects dropped from the structure. However, it is necessary to renew the anodes using divers. This can be difficult in deep-water structures.

In the power-supplied (or current-impressed) systems the anodes take the form of large pieces of scrap-iron or lumps of carbon. The d.c. current required for flow from the anode to the cathode is supplied from the mains through a transformer/rectifier or directly from a d.c. generator.

The rate of wastage of anodes or the power requirements for power-supplied systems can be minimized by keeping the area of exposed steel as small as possible. The cathodic protection does not supplant proper paint or other surface treatment of the metal. The more this surface coating can be kept intact the less will be the wastage of anodes and, in the case of power-supplied systems, the requirements for power. Thus thorough treatment by paint or bituminous enamel should be given to the piles as described above. It is particularly important to ensure thorough surface protection between high and low water mark in marine structures and in the 'splash zone' above, since cathodic protection is ineffective above water level and the unprotected parts, if uncoated, may act anodically to the parts below water with accelerated corrosion.

13.5 Protection of concrete structures

The principal cause of deterioration of concrete in foundations is attack by sulphates present in the soil, in the ground water, or in sea water. Other agencies causing deterioration include chemical wastes, organic acids, frost, sea action, certain deleterious aggregates, and corrosion of reinforcement.

13.5.1 Sulphate attack

Sulphates in solution react with Portland cement to form insoluble calcium sulphate and calcium sulpho-aluminate. Crystallization of the new compounds is accompanied by an increase in molecular volume which causes expansion and disintegration of the concrete at the surface. This disintegration exposes fresh areas to attack and if there is a flow of ground water bringing fresh sulphates to the affected area, the rate of disintegration can be very rapid. Easily soluble sulphates such as those of magnesium, sodium, and ammonium are more aggressive than calcium sulphate (gypsum). Ammonium sulphate does not occur naturally in the soil, but it may be present in agricultural land where it has been used as a fertilizer. Insoluble sulphates do not attack concrete.

The sulphate content of sea water in the deep oceans expressed as SO_4 is about 2.6g/l which is within the range where precautions are recommended for concrete exposed to sulphate-bearing ground waters. However, sea water is not regarded as being markedly aggressive to concrete because sodium chloride has an inhibiting or retarding action on the expansive reaction of the sulphates. It is usually satisfactory to adopt a good-quality rich mix using normal Portland cement, although, in the case of reinforced concrete structures in sea water it is advisable to adopt the added precaution of using sulphate-resisting cement. This will prevent cracking of the concrete followed by corrosion of the reinforcement.

In the soil or in ground water, the higher the concentration of sulphates the more severe is the attack. The disintegration is particularly severe if the foundation structure is subjected to one-sided water pressure, for example a basement or culvert with a head of ground water on the outside. Other factors increasing the severity of attack are porosity of the concrete, the presence of cracks, sharp corners, and disruption of the surface by barnacles and similar growths.

Attack does not take place if there is no ground water, and for the disintegration to continue there must be replenishment of the sulphates. If the ground water is absolutely static the attack does not penetrate beyond the outer skin of concrete. Thus there is little risk of serious attack on structures buried in clay soils provided that there is no flow of ground water such as might occur along a loosely backfilled foundation trench. There is risk of attack in clayey soils in certain climatic conditions, for example when hot, dry conditions cause an upward flow of water by capillarity from sulphate-bearing waters below foundation level. Similar conditions can occur when water is drawn up to the ground floor of a building due to the drying action of domestic heating or furnaces.

13.5.2 Protection against sulphate attack

Recommendations on precautions to be taken against sulphate attack on buried structures are given in *Building Research Station Digest 363*.[13.1] This is a considerable improvement on the previous *Digest 250* because it takes account of the environmental conditions and the workability requirements for different types of structure. The same five classes of concentration of sulphates have been revised and the magnesium concentration has been added as an additional factor. A feature of the new classification system is the recommendation to modify the class for the environmental conditions at the particular site and for the particular structure under consideration. Thus if the site is dry or the soil has a low permeability where the ground water is static, a reduction in the class is recommended for classes 2–4, but where a concrete structure is subjected to one-sided water pressure without an impermeable membrane the class is advanced. The class is reduced for massive concrete construction or properly cured precast concrete units, but advanced for thin cast-*in-situ* concrete.

Minimum cement contents and maximum water–cement ratios are given for each class with different values for different types of cement, depending on whether ordinary Portland cement or a sulphate-resistant cement is used. Types of sulphate-resisting cement include Portland cement in combination with various proportions of blast furnace slag, pf ash, and pozzolanic pf ash. The cement contents and maximum water–cement ratios given in *Digest 363* provide mixes with adequate workabilities for massive cast-*in-situ* concrete in strip and pad foundations, pile caps and basement retaining walls, and for precast concrete in shaft linings and piles. However, cement contents somewhat higher than those recommended by the *Digest* for cast-*in-situ* concrete in piles may be needed where the concrete is required to be placed under water through a tremie pipe. The recommendation to reduce the classification by one class where the ground-water table is static could be questioned in the case of cast-*in-situ* concrete in bored piles where there is a *shallow* ground-water table. This is because the quality of concrete placed under water through a tremie pipe tends to deteriorate towards the head of the pile as a result of bleeding of excess water washing out the cement. Weak concrete used as a blinding layer beneath reinforced concrete in foundations is vulnerable to sulphate attack. The resulting expansion of the blinding concrete could cause an uplift force

tending to lift shallow structures such as pile caps and slender ground beams.

In American practice, type I cement to ASTM C 150–71 is similar in chemical and physical properties to British ordinary Portland cement. The American type II cement has a moderate sulphate resistance, and type V has a high sulphate resistance equivalent to that of British sulphate-resisting cement to BS 4027. High-alumina and supersulphate cements give practically complete immunity against attack from the severest conditions normally encountered in the ground or in sea water. Supersulphate cement is, however, attacked by ammonium sulphate. It is not manufactured in Britain at the present time and the imported brands are expensive. High-alumina cement also suffers from the disadvantage that the compressive strength of concrete made with it can fall drastically, due to a phenomenon known as 'conversion'.[13.12] The fall can occur at any time in the life of the structure and is accompanied by a marked decrease in the resistance of the concrete to sulphate attack. This disadvantage can be overcome to some extent if over-rich mixes are avoided and precautions are taken to prevent the exposure of concrete to heat. Steam curing must not be used and stocks of piles made with high-alumina cement should be shaded from the sun in the casting yard and on site. Piles should be driven when they are between the ages of 3 and 7 days. It should be noted that high-alumina cement concrete, even when it has undergone 'conversion', possesses a residual strength which may be adequate for the working stresses in the structural member. However, collapses of a few high-alumina concrete structures in Britain in 1973–74 led to withdrawal of Code of Practice approval of this cement for structural concrete which includes the foundations. Accordingly, the use of high-alumina cement for protection against sulphate attack has not been shown in *Digest 363*. The best form of protection for high sulphate concentrations in ordinary foundation work is to use well-compacted dense impermeable concrete made with sulphate-resisting cement or in severe conditions to use a protective membrane.

Where the sulphate content of the ground water exceeds 30 parts per 100 000, but is less than 300 parts per 100 000, sulphate-resisting cement should be used in brickwork mortar in the proportions of one part of cement to three parts of sand. If the workability of the mortar needs to be increased a plasticizer can be used with this mix or lime can be added in the proportions of $1 : \frac{1}{4} : 3$ cement–lime–sand. Special advice must be sought for the precautions necessary when the sulphate content of the ground water exceeds 300 parts per 100 000.

Asphalt tanking gives complete protection against sulphate attack, and if basements are tanked in the manner described in Chapter 5, no other form of protection is

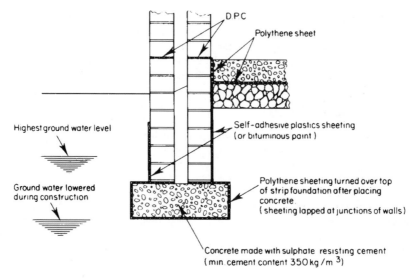

D.P.C

Polythene sheet

Highest ground water level

Ground water lowered during construction

Self-adhesive plastics sheeting (or bituminous paint)

Polythene sheeting turned over top of strip foundation after placing concrete. (sheeting lapped at junctions of walls)

Concrete made with sulphate resisting cement (min. cement content 350 kg/m^3)

Fig. 13.2   Protection of strip foundation against attack by acidic ground water

needed in sulphate-bearing ground water. Concrete in strip or pad foundations can be protected by wrapping them with bituminous felt or plastic sheeting. The self-adhesive types of plastic sheeting are very suitable for this purpose, and an example of protection using this type of sheeting in conjunction with an ordinary polythene membrane is shown in Fig. 13.2. Bituminous emulsion does not provide a satisfactory coating.

In conditions where ground water can be drawn up to ground floor slabs by heat in buildings, adequate protection can be given by bitumen damp-course material or polythene sheeting laid below the ground slab.

Ordinary concrete foundations above the ground-water table in Kuwait (where the soils are heavily gypsiferous) have been given satisfactory protection by a heavy coat of hot bitumen on the exposed surfaces.

Concrete in the shafts of driven or bored and cast-in-place piles can be protected by an outer membrane formed from a heavy duty plastics sheeting. This is laced on to the bottom of the reinforcing cage. However, it is difficult to prevent the fastenings from tearing and the sheeting from rising in a concertina fashion as the temporary outer casing is withdrawn. Rigid PVC tubing or galvanized corrugated cylindrical steel sheeting give better though more costly protection. Precast concrete piles can be protected by epoxy-coal tar paints. The above forms of protection either for cast-in-place or precast concrete piles will reduce the skin frictional resistance on the pile shaft.

13.5.3 Sea action on concrete

In an extensive review of the literature and inspection of

structures which have been immersed in the sea for 70 years, Browne and Domone[13.13] found no disintegration in permanently immersed reinforced concrete structures, even though severe damage had occurred in the 'splash zone'. The main cause of deterioration of concrete structures exposed to sea water is spalling of the surface in this zone due to corrosion of the reinforcement as a result of poor design or faulty construction. Other causes are attack by sulphates in the sea water, disruption by marine growths, frost action, and erosion by wave action. Deterioration from almost all these causes can be prevented by providing 50 mm or more cover to the reinforcement and by using a rich mix (not leaner than $1 : 1\frac{1}{2} : 3$) well compacted to give a dense impermeable concrete. Sea water is not recommended for mixing in reinforced concrete structures because of the risk of corrosion of the reinforcement. It is, however, satisfactory for mass concrete.

Erosion at beach level due to shingle moving under wave action may be severe. Protection can be given by dumping mass concrete or rubble around vulnerable points.

13.5.4 Organic acids

The naturally occurring organic acids are mainly lignic or humic acids found in peaty soils and waters. These acids form insoluble calcium salts by reaction with free lime of normal Portland cements, so that the risk of attack is slight with dense concrete which is relatively impermeable. In certain marsh peats, oxidation of pyrites or marcasite can produce free sulphuric acid which is highly aggressive to concrete. The presence of free sulphuric acid is indicated

Table 13.2   Precautions against attack by organic acids in peaty or marshy soils on concrete in building foundations

pH value	Precautions for foundations above ground-water table in any soil and below ground-water level in impervious clay	Precautions for foundations in contact with flowing ground water in permeable soil
6	None necessary	None necessary
6–4.5	Use ordinary Portland cement at ≮370 kg/m^3, max W/C ratio 0.5	Use ordinary Portland cement at ≮380 kg/m^3 or sulphate-resisting cement at ≮350 kg/m^3, max. W/C ratio 0.5
4.5	Use ordinary Portland cement at ≮400 kg/m^3, or sulphate-resisting cement at ≮390 kg/m^3, max. W/C ratio 0.5	Use supersulphate cement or ordinary Portland cement at ≮400 kg/m^3, or sulphate-resisting cement at ≮390 kg/m^3, max. W/C ratio 0.5; *in addition* protect by external sheathing

Notes
(a) The cement contents recommended in the table are suitable for medium-workability concrete (50–75 mm slump). Lower cement contents can be used if placing conditions are satisfactory for concrete mixes of lower workability, provided that the concrete can be compacted to produce a dense impermeable mass.
(b) The precautions apply only to concrete exposed to organic acids in peaty and marshy soils. Special advice should be sought where mineral acids or acid industrial wastes are present.

by pH values lower than 4.3 and a high sulphate content. The pH value of the ground water provides a rather crude measure of the potential aggressiveness to concrete of naturally occurring organic acids. In an extensive review of precautionary methods adopted in various countries, the Construction Industry Research and Information Association[13.14] noted that in European countries reliance is generally placed on obtaining a dense impermeable concrete as a means of resisting attack by organic acids rather than the use of special cements. Recommendations for precautions against acid attack are given in *Digest 363* in relation to the pH value of the ground water. Where the acids are organic and derived from natural sources the reader may find Table 13.2 more convenient to use. This table has been based on practice in Germany and The Netherlands where extensive deposits of marshy soil are present. Ordinary Portland or rapid-hardening Portland cement combined with ground-granulated slag or pf ash provides better resistance to acid attack than concrete made only with ordinary Portland cement.

13.5.5 Chemical and industrial wastes

Aggressive chemicals present in the ground in chemical works or in waste dumps give the most difficult problems in protecting foundation structures. These problems usually require the help of specialist advice. The chief difficulty is in identifying the full range of deleterious compounds from a limited number of samples, and the concentration of the chemicals can vary widely from point to point. Another complication is given by some of the chemical processes, for example the foundations may incorporate underground tanks containing hot acids or alkalis. High-alumina and supersulphate cements can withstand attack by acids having pH values as low as 3.5,

but high-alumina cement cannot be used where alkalis are in strong concentrations. The chemical resistance of concrete has been reviewed by Gutt and Harrison[13.15] and Eglinton[13.16]. On a site where fill material was contaminated with high concentrations of acid waste the author recommended piled foundations consisting of a precast concrete shell, the hollow interior having a PVC pipe insert filled with concrete. The outer shell of these end-bearing piles was regarded as sacrificial over the shaft length in contaminated ground.

Industrial wastes such as slag, burnt colliery shale, and fly ash used as filling below floors can cause dangerous concentrations of sulphates in ground floor slabs as a result of the drying action previously described. Cases have occurred where colliery shale fill containing sulphates has caused the expansion of ground floor concrete resulting in outward tilting of foundation walls. This trouble can be overcome by providing bituminous or plastic sheeting below the floors and carried through to the damp-proof course in the walls.

Severe damage was caused to a large number of houses in the north of England as a result of expansion of naturally occurring pyritic shales.[13.17] The expansion was a chemical–bacterial action accelerated by conditions of warmth and moisture beneath the floors. The adoption of a polythene membrane beneath the floors did not prevent the expansion of the shale.

Hawkins and Pinches[1.3] have described the combination of chemical and microbiological action in causing heave in thinly bedded mudstones beneath the Llandough Hospital near Cardiff. Relative heave up to 70 mm was measured in 1982 beneath the floors of part of the pre-war building which was severely damaged, and required shoring followed by extensive remedial works. The heave of the mudstone was found to be caused by the growth of gypsum in the form of crystalline selenite within the laminations of the moderately weathered rock. At the stage of excavation for the original construction relaxation of stress in the rock would have caused some parting of the laminations and exposed them to access by weathering agents in the form of moisture and bacteria, causing oxidation of pyrites to ferrous sulphate and the formation of sulphuric acid. Over the years the heat transmitted through the floors of the building and from heating ducts below the floor would have promoted the bacterial action, causing the expansive crystalline growths in the rock mass. Remedial action included underpinning the internal walls by piles, the use of collapsible void formers between the ground floors and the underlying rock, and the provision of an air gap around heating ducts.

Some observed values of sulphate content and pH of industrial wastes are as follows:

Material	Sulphate (as SO_3) (%)		pH
	Range	Average	
Clinker	0.07–0.9	0.3	
PF ash	0.03–1.5	0.5	
Brick hardcore	0.1–0.3	—	
Railway ballast	0.2		
Colliery waste	0.2–4.4	2.3	3.5–7.0
Black ash (Le Blanc process)	4.4		>11
Gasworks waste	1.5–5.0		4
Dyeworks waste	0–4.0		2.9–4.3

13.5.6 Frost action

Concrete placed 300 mm or more below ground level in Great Britain is unlikely to suffer any deterioration from frost. However, for severe conditions of exposure careful attention must be paid to the design of the mix and the placing of the concrete. The disruptive effect of frost is due to expansion of the water while it is freezing in the pores in the concrete. The effect is most severe if the freezing is rapid when there is no time for water to be extruded from the surface or into the interior of the mass.

For severe exposure conditions Murdock and Brook[13.18] recommend using the lowest possible water–cement ratio compatible with having sufficient workability for compaction. They state that a water–cement ratio of less than 0.5 can be regarded as safe against frost damage, between 0.5 and 0.6 there is a risk in isolated cases and above 0.6 the risk becomes progressively greater. Air entraining agents are reputed to increase the frost resistance of concrete.

13.5.7 Corrosion of reinforcement

Plain concrete is not affected chemically by chlorides present in the aggregates and mixing water or drawn into the mass by exposure to silt-laden air or saline ground water. However, damage to reinforced concrete can occur due to corrosion of the embedded reinforcing steel. The expansion of the corrosion products on the surface of the steel causes cracking of the concrete and in severe cases to spalling of the concrete cover, thus giving an increasing risk of further corrosion. In normal circumstances where the aggregates and mixing water are free or are within specified maximum concentrations of chlorides, the concrete itself provides a protective environment due to reactions with substances in the cement with the highly alkaline pore water. This provides a passivating film on the metal surface.[13.17] However, if chlorides enter the concrete through cracks or honeycomb patches the alkalinity of the pore water is reduced, allowing the breakdown of the protective film. Concrete made with sulphate-resisting Portland cement is more prone to attack than ordinary Portland cement concrete.

Reinforced concrete in foundations is less liable to attack by chlorides than that in above-ground structures because it is not exposed to salt-laden air or to cycles of wetting and drying causing surface cracking. However, there is a risk in bridge foundations if concentrations of de-icing salts seep into the foundations from the superstructure. Provided that the chloride contents of the materials used for mixing the concrete are within specified limits, that admixtures containing chlorides are not used, and that the cover to the steel is appropriate to the exposure conditions (see BS 8110) there should be little risk of corrosion damage. However, it is essential that the concrete should be dense and well compacted.

References

13.1 Sulphate and acid resistance of concrete in the ground, *Building Research Establishment Digest 363* (1991).
13.2 Potter, F. H., Timber design, *Civil Engineers' Reference Book*, 4th edn. (L. S. Blake ed.), Butterworths, 1989, Part 16, pp. 5–6.
13.3 Richardson, N. A., Wood preservation, *Proceedings of the Institution of Civil Engineers, Part 1*, **2**, 649–678 (1953).
13.4 Marine borers and methods of preserving timber against their attack, *Building Research Establishment Technical Note No. 59* (1972).
13.5 Deterioration of timber, metal, and concrete exposed to the action of sea water, *21st Report of the Sea Action Committee*, Institution of Civil Engineers, London, 1965.
13.6 Chellis, R. D., Finding and fighting marine borers, *Engineering News Record*, pp. 344–347, 422–424, 493–496, and 555–558 (4 March, 18 March, 1 April, and 15 April 1948).
13.7 Morley, J. and Bruce, D. W., Survey of steel piling performance in marine environments, final report, *Commission of the European Communities, Document EUR 8492 EN* (1983).
13.8 Romanoff, M., *Corrosion of Steel Pilings in Soils*, National

Bureau of Standards Monograph 58, US Department of Commerce, Washington, DC, 1962.

13.9   Little, D. H., Discussion in *Proceedings of the Conference on the Behaviour of Piles*, Institution of Civil Engineers, London, 1970, p. 191.

13.10   Hedborg, C. E., Corrosion in the offshore environment, in *Proceedings of the Offshore Technology Conference*, Houston, 1974, paper no. OTC 1949, pp. 155–168.

13.11   Morley, J., The corrosion and protection of steel piling, *British Steel Corporation*, Report No. T/CS/1115/1/79/C, Nov. 1979.

13.12   High-alumina cement concrete in buildings, *Building Research Establishment Current Paper No. 34/75* (1975).

13.13   Browne, R. D. and Domone, P. L. J., The long term performance of concrete in the maritime environment, in *Proceedings of the Conference on Offshore Structures*, Institution of Civil Engineers, London, 1974, pp. 31–41.

13.14   Eglinton, M. S., Review of concrete behaviour in acidic soils and ground waters. *Construction Industry Research and Information Association Technical Note No. 69* (1976).

13.15   Gutt, W. H. and Harrison, W. H., Chemical resistance of concrete, *Concrete*, **11**(5), 35–37 (1977).

13.16   Eglinton, M. S., *Concrete and its Chemical Behaviour*, Thomas Telford, London (1987).

13.17   Nixon, P. J. Floor heave in buildings due to the use of pyritic shales as fill material, *Chemistry and Industry*, pp. 160–164 (1978).

13.18   Murdock, L. J. and Brook, K. M., *Concrete Materials and Practice*, 5th edn., London, Edward Arnold, 1979.

Appendix A Properties of Materials

Material	Density when drained above ground water level, γ (Mg/m^3)	Density when submerged below ground water level, γ_{sub} (Mg/m^3)
Gravel	1.60–2.00	0.90–1.25
Hoggin (gravel–sand–clay)	2.00–2.25	1.00–1.35
Coarse to medium sands	1.70–2.10	0.90–1.25
Fine and silty sands	1.75–2.15	0.90–1.25
Loose fine sands	Down to 1.50	Down to 0.90
Stiff boulder clay	2.00–2.30	1.00–1.35
Stiff clay	1.80–2.15	0.90–1.20
Firm clay	1.75–2.10	0.80–1.10
Soft clay	1.60–1.90	0.65–0.95
Peat	1.05–1.40	0.05–0.40
Granite	2.50†	—
Sandstone	2.20†	—
Basalts and dolerites	1.75–2.25	1.10–1.60
Shale	2.15–2.30	1.20–1.35
Stiff to hard marl	1.90–2.30	1.00–1.35
Limestone	2.70†	—
Chalk	0.95–2.00	0.30–1.00
Broken brick	1.10–1.75	0.65–0.95
Ashes	0.65–1.00	0.30–0.50
Pulverized fuel ash	1.20–1.50	0.65–0.80

† Measured in the solid (i.e. not crushed or broken).

Appendix B Ground Movements Around Excavations

Table B.1 Inward yielding of excavation support systems

No.	Soil conditions	Location	Depth of excavation (m)	Type of support	Maximum movement Amount (mm)	Maximum movement Movt depth (%)	Ref
1	Soft	Chicago	11.4	SPS	58	0.51	B1
2	to	Oslo, Vaterland	9	SPS	23	0.25	B2
3	firm	San Francisco	15	SPS	21	0.14	B3
4	NC	South Africa	22.9	DWA	38	0.16	B4
5	clays	South Africa	14.7	DWA	76	0.52	B4
6		South Africa	14.7	DWA	19	0.13	B4
7		Singapore, Newton	15.0	DWS	88	0.59	B5
8		Singapore, Bugis	18.0	DWS	140	0.78	B6
9		Oslo, Olav Kyrres	12.4	SPS	31	0.25	B7
10		Oslo, Kongengate	19.1	DWS	15	0.05	B8
11		Oslo, Studentelunden	15.0	DWS	40	0.27	B9
12		Oslo, Jerbanetorget	10	DWS	20	0.20	B9
12a		York	8.6	SPS	42	0.49	—
13	Stiff	London, Bloomsbury	16	DWS	12	0.07	B3
14	to	London, Westminster	12	DWS	20	0.17	B10
15	hard	London, Vauxhall	14.5	DWA	22	0.15	B3
16	OC clays	London, Neasden	8	DWA	50	0.62	B11
17		London, City	25	DWS	32	0.13	B12
18		Singapore, Tieng Bahru	15.3	SShS	25	0.16	B5
19		Genoa	26.3	DWA	60	0.23	B3
20		Humber Bridge	30	DWS	22	0.07	B3
21		London, St Pancras	20	DWS	40	0.13	B13
22		Bell Common	10	BPS	30	0.30	B14
23	Sands	Buffalo	6.4	TA	10	0.16	—
24	and	Buffalo	11.2	TA	53	0.47	—
25	gravels,	Zurich	20	DWS	36	0.18	B15
26	sandy	New York, WTC	17.7	DWA	66	0.37	B3
27	decom-	New York, WTC	19.2	DWA	146	0.76	B3
28	posed	London, Westminster	19	BPS	35	0.18	B16
29	rocks	London, Guildhall	9.7	DWA	10	0.10	B3
30		London, Victoria Street	8	DWS	3	0.04	B3
31		Paris, Beauborg	16.5	SShA	20	0.12	B3
32		Hong Kong Central	19	DWS	43	0.23	B17
33		Hong Kong Central	19	DWS	30	0.17	B17
34		London, Harewood Av.	8.5	SShS	8	0.09	B16
35		Hatfield	9.3	SPA	12	0.13	B18

Key: SPS Sheet pile strutted; DWA Diaphragm wall, anchored; DWS Diaphragm wall, strutted; SShA Soldiers and shotcrete anchored; SShS Soldiers and shotcrete strutted; BPS Bored pile strutted; BPA Bored pile anchored; TA Timber, anchored.

Table B2 Settlement of ground surface close to excavations

No.	Soil conditions	Location	Depth of excavation (m)	Type of support	Settlement		
					Amount (mm)	Settlement depth(%)	Distance depth
2	Soft	Oslo, Vaterland	9.5	SPS	23	0.2	0.7
7	to	Singapore, Newton	15.0	DWS	250	1.7	0.3
8	firm	Singapore, Bugis	18.0	DWS	270	1.5	0.5
9	NC	Oslo, Olav Kyrres	12.4	SPS	70	0.6	0.2
10	clays	Oslo, Kongengate	19.1	DWS	40	0.2	0.5
11		Oslo, Studentelunden	15.0	DWS	70	0.5	0.3
12a		York	8.6	SPS	30	0.35	0.12
14	Stiff to	London, Westminster	12	DWS	19	0.2	0.8
16	hard	London, Neasden	8	DWA	24	0.3	—
17	OC	London, City	25	DWS	20	0.1	—
22	clays	Bell Common	10	BPS	65	0.6	0.3
28	Sands	London, Westminster	19	BPS	15–25	0.1	0.2
32	and	Hong Kong Central	19	DWS	25	0.1	0.6
33	gravels	Hong Kong Central	19	DWS	32	0.2	—

Key: SPS Sheet pile strutted; DWA Diaphragm wall, anchored; DWS Diaphragm wall, strutted; SShA Soldiers, shotcrete anchored; SShS Soldiers and shotcrete strutted; BPS Bored pile strutted; BPA Bored pile anchored; TA Timber, anchored.

References to Tables B.1 and B.2

B.1 Peck, R. B., Earth pressure measurements in open cuts, Chicago Subway, *Transactions of the American Society of Civil Engineers*, **108**, 1008–1036 (1943).

B.2 Measurements at a strutted excavation, *NGI reports 1–9*, Norwegian Geotechnical Institute, Oslo, 1962.

B.3 Various authors, *Proceedings of the Conference on Diaphragm Walls and Anchorages*, Institution of Civil Engineers, London, 1974.

B.4 Jennings, J. E., Discussion on deep excavations and tunnelling in soft ground, *Proceedings of the 7th International Soil Mechanics Conference*, Mexico, **3**, 331–335 (1969).

B.5 Various authors, *Proceedings of the Singapore Mass Rapid Transport Conference*, Institution of Engineers, Singapore, 1987.

B.6 Hulme, T. W., Potter, L. A. C., and Shirlaw, J. N., Singapore Mass Rapid Transit System: Construction, *Proceedings of the Institution of Civil Engineers*, London, **86**(1), 709–770 (1989).

B.7 Karlsrud, K. and Myrvoll, F., Performance of a strutted excavation in quick-clay, *Norwegian Geotechnical Institute, Bulletin No. 111*, 9–15 (1976).

B.8 Di Biagio, E. and Roti, J. A., Earth pressure measurements on a braced slurry-trench wall in soft clay, *Norwegian Geotechnical Institute, Bulletin No. 91*, 1972.

B.9 Karlsrud, K., Performance and design of slurry walls in soft clay, *Norwegian Geotechnical Institute, Bulletin No. 149*, 1–9 (1983).

B.10 Burland, J. B. and Hancock, R. J. R., Underground car park at the House of Commons, London, *The Structural Engineer*, **55**, 87–100 (1977).

B.11 Sills, G. C., Burland, J. B. and Czechowski, M. K., Behaviour of an anchored diaphragm wall in clay, *Proceedings of 9th International Soil Mechanics Conference*, Tokyo, **2**, 147–154 (1977).

B.12 Marchand, S. P., A deep basement in Aldersgate Street, London, Part 2, Construction, *Proceedings of the Institution of Civil Engineers*, London, **97**, 67–76 (1993).

B.13 Simpson, B., Retaining structures: displacement and design, *Géotechnique*, **42**(4), 541–578 (1992).

B.14 Symons, I. F. and Tedd, P., Behaviour of a propped embedded retaining wall at Bell Common Tunnel in the longer term, *Géotechnique*, **39**(4), 701–710 (1989).

B.15 Huder, J., Deep braced excavation with high ground water level, *Proceedings of the 7th International Soil Mechanics Conference*, Mexico, **2**, 443–448 (1969).

B.16 Mair, R. J., Developments in geotechnical engineering research: applications to tunnels and deep excavations, *Proceedings of the Institution of Civil Engineers*, London, **93**, 27–41 (1993).

B.17 Humpheson, C., Fitzpatrick, A. J. and Anderson, J. M. D., The basements and substructure for the new headquarters of the Hong Kong and Shanghai Banking Corporation, Hong Kong, *Proceedings of the Institution of Civil Engineers*, London, **80**, 851–883 (1986).

B.18 Symons, I. F. and Carder, D. R., Field measurements on embedded retaining walls, *Géotechnique*, **42**(1), 117–126 (1992).

Appendix C Conversion Tables

Imperial–metric

Imperial	Metric (or metric force)	SI
1 in	25.4 mm	
1 ft	304.8 mm	
1 yd	914.4 mm	
1 in^2	645.16 mm^2	
1 ft^2	92 903 mm^2	
1 yd^2	0.8361 m^2	
1 in^3	16 387 mm^3	
1 ft^3	0.028 317 m^3	
1 yd^3	0.7645 m^3	
1 ft^3 of fresh water	28.32 l	
1 Imperial gallon	4.546 l	
1 lb	0.4536 kg	4.448 22 N(= 1 lbf)
1 ton (2240 lb)	1016.05 kg	9.964 02 kN(= 1 tonf)
1 lbf/in^2	0.0703 kgf/cm^2	6.894 76 kN/m^2
1 lbf/ft^2	4.882 kgf/m^2	47.8803 N/m^2
1 lb/ft^3	16.02 kg/m^3	
1 ft^3/s	0.0283 m^3/s	
1 Imperial gall/min	4.546 l/min	
1 in unit of moment of inertia	41.6198 cm units	
1 in unit of modulus of section	16.3860 cm units	

Metric–imperial

1 mm	0.039 37 in
1 m	39.37 in = 3.281 ft
	= 1.0936 yd
1 mm^2	0.001 55 in^2

SI	Metric (or metric force)	Imperial
	$1\,m^2$	$10.764\,ft^2 = 1.196\,yd^2$
	$1\,mm^3$	$0.0610 \times 10^{-3}\,in^3$
	$1\,m^3$	$35.315\,ft^3 = 1.308\,yd^3$
	$1\,m^3$ of fresh water	220.4 Imp. gall
		$= 2204\,lb$
	$1\,l$	0.220 Imp gal
		$= 61.026\,in^3$
$1\,kgf = 9.806\,65\,N$	$1\,kg$	$2.2046\,lb$
	$1\,t$	$1000\,kg = 0.9843$
		ton (long)
	$1\,kgf/cm^2$	$14.22\,lbf/in^2$
	$1\,kgf/m^3$	$0.062\,45\,lb/ft^3$
	$1\,m^3/s$	$35.315\,ft^3/s$
		$= 132\,27$ Imp gal/min
	$1\,l/s$	0.22 Imp gal/s
	1 cm unit of moment	
	of inertia	$0.024\,03$ in units
	1 cm unit of modulus	
	of section	$= 0.061\,03$ in units
$1\,N$	$0.10197\,kgf$	$0.224\,81\,lbf$
$1\,bar$	$10197\,kgf/m^2$	$14.5038\,lbf/in^2$
$1\,kN/m^2$	$101.97\,kgf/m^2$	$20.8854\,lbf/ft^2$
$1\,kN$	$101.64\,kgf$	$0.10036\,tonf$

USA equivalents of the most frequently used British Standard sieves

British Standard sieves		American Standard sieves	
Sieve size	Sieve opening (μm)	Sieve opening (μm)	Sieve number
6.3 mm	6300	6350	No. 3 ($\frac{1}{4}$ in.)
5 mm	5000	4760	No. 4
3.35 mm	3350	3360	No. 6
2.36 mm	2360	2380	No. 8
1.18 mm	1180	1190	No. 16
600 μm	600	590	No. 30
425 μm	425	420	No. 40
300 μm	300	297	No. 50
212 μm	212	210	No. 70
150 μm	150	149	No. 100
75 μm	75	74	No. 200

British Standard sieve openings are as given in BS 410.

Metric–USA

$$1\,t = 1.102 \text{ US ton}$$
$$1 \text{ litre} = 0.2642 \text{ US liquid gallon}$$

USA–metric

$$1 \text{ US ton} = 907.18\,kg$$
$$1 \text{ US liquid gallon} = 3.785 \text{ litres}$$

Author Index

Subject Index